Petroleum Geochemistry and Geology

Petroleum Geochemistry and Geology

Second Edition

John M. Hunt

W. H. Freeman and Company
New York

To Phyllis

Photo credits:
Cover: South Marsh Island, Louisiana, Mobil Oil Corporation
Part 1: Thermal cracking unit, Exxon Corporation
Part 2: Offshore drilling
Part 3: Belridge Field, California, Mobil Oil Corporation
Part 4: Onshore drilling, Egypt.

Library of Congress Cataloging-in-Publication Data

Hunt, John Meacham.
Petroleum geochemistry and geology / John M. Hunt.—2nd ed.
p. cm.
Includes bibliographical references (p.) and indexes.
ISBN: 0-7167-2441-3 (hard cover)
1. Petroleum—Geology. 2. Natural gas—Geology 3. Petroleum—Prospecting. 4. Natural gas—Prospecting. 5. Geochemical prospecting. I. Title.
TN870.5.H86 1995
553.2'8—dc20 95-32291
CIP

Printed in the United States of America

First printing 1995, VB

Contents

Color Plate Legends

Plate 1A
Maturation map of the Late Jurassic Kimmeridge Shale and overlying oil and gas fields in the North Sea. The brown represents mature areas with Kimmeridge Shale temperatures above 93°C (200°F) and depths greater than 3,048 m (10,000 ft). Yellow represents immature areas with lower temperatures. Oil fields are green, and gas fields are red. (See p. 5 for discussion.) [Demaison 1984]

Plate 1B
Maturation map on top of the New Albany shale, Illinois Basin, with associated oil fields. Brown represents areas with vitrinite reflectance values of $R_o > 0.6\%$; yellow is $< 0.6\%$. Dark brown is postmature $R_o > 1.3\%$. (See p. 6.) [Demaison 1984]

Plate 1C
Maturation map on top of the Permian coal measures of the Cooper Basin, Australia, with associated gas fields. Brown represents areas mature for gas (R_o = 0.9 to 2%). Yellow is areas immature for gas, and dark brown is postmature. (See p. 6.) [Demaison 1984]

Plate 1D
Maturation map on the base of the Mississippian Bakken Shale source rock in the Williston Basin of the United States and Canada. Brown is mature for oil; yellow is immature. (See p. 6.) [Demaison 1984]

Plate 2A
Structure of an asphaltene molecule based on various analyses. White represents paraffins, naphthenes (cycloparaffins) are red, and aromatic structures are yellow. A porphyrin structure is in the center at the bottom. Nitrogen atoms are blue, oxygen are green, and sulfur are red. (See p. 49.). [Pelet et al. 1986]

Plate 2B
Generation of oil (green), wet gas (light red), and dry gas (dark red) from the La Luna Formation at the end of the Middle Miocene. The immature nongenerating area is yellow. The postmature for gas area is dark brown. (See p. 183.) [Talukdar et al. 1986]

Plate 2C
Generation of hydrocarbons from the La Luna Formation at the end of the Miocene. Same color code as in Plate 2B. (See p. 183.) [Talukdar et al. 1986]

Plate 2D
Generation of hydrocarbons from the La Luna Formation at the present time. Same color code as in Plate 2B. (See p. 183.) [Talukdar et al. 1986]

Plate 3A
Chemical model of an immature type I kerogen (Green River Shale). Paraffins are white, naphthenes are red, and aromatics are yellow. (See p. 209.) [Behar and Vandenbroucke 1987]

Plate 3B
Chemical model of a type II kerogen (Toarcian Shale) at the beginning of catagenesis. Color codes are the same for Plates 3A, 3B, and 3C. (See p. 209.) [Behar and Vandenbroucke 1987]

Plate 3C
Chemical model for a type I kerogen (Green River Shale) at the end of catagenesis. (See p. 209.) [Behar and Vandenbroucke 1987]

Plate 3D
Fluorescent oil in microfractures of the Querecual Formation, Eastern Venezuelan Basin. (See p. 268.) [Courtesy of S. C. Talukdar]

Plate 4A
Vertical migration of fluorescent oil in Wolfcamp Formation, Permian Basin, Lea County, New Mexico. (See p. 268.) [Courtesy of M. Malek-Aslani]

Plate 4B
Fluorescent bitumen (C) migrating through fractures crosscutting the mineral matrix (M) and kerogen (K). (See p. 268.) [Talukdar et al. 1986]

Plate 4C
Light fluorescent oil migrating from bitumen into fractures in the matrix of the Jurassic Smackover source rock. (See p. 269.) [Courtesy of M. Malek-Aslani]

Plate 4D
Bitumen from a deeper mature part of the Woodford Shale source rock of Oklahoma moving along microfractures (arrows) into an immature part of the Woodford Shale. (See p. 273.) [Courtesy of R. K. Olson]

Plate 5A
Bitumen filling a polygonal fracture network and migrating laterally along a stylolite in the Woodford Shale source rock of Oklahoma. Core sample is from 3,056 ft (932 m), Marshall County, Oklahoma. Truncated white grain is probably recrystallized radiolarian. (See p. 273.) [Comer and Hinch 1987]

Plate 5B
North-to-south cross section, Southern Norwegian Central Graben, showing approximate location of overpressured compartment seals and gas chimneys over major fields. (See p. 302.) [Courtesy of R. C. Leonard]

Plate 5C
Comparison of TOC with factors affecting it in modern sediments of the Black Sea. (See p. 326.) [Huc 1988]

Plate 5D
Chromaticity diagram with maturation tracks of alginite and coal/source rock extracts. (See p. 362.) [Courtesy of T. van Gijzel]

Plate 6A
Senonian spore and cuticle in transmitted white light at 25×. (See p. 362.) [Courtesy of H. M. Heck]

Plate 6B
Same Senonian spore and cuticle slide in reflected blue light. (See p. 362.) [Courtesy of H. M. Heck]

Plate 6C
Oil-prone amorphous kerogen from the Kimmeridge Shale of North Sea at 40× in transmitted white light. The sample width is 160 microns. (See p. 362.) [Courtesy of C. Thompson-Rizer]

Plate 6D
Same kerogen sample in reflected blue light. Sample width is 160 microns. (See p. 363.) [Courtesy of C. Thompson-Rizer]

Plate 7A
Whole-core fluorescence of the Antrim Shale of the Michigan Basin. (See p. 363.) [Courtesy of H. Hinch]

Plate 7B
Thermal alteration index scale with Munsell color standards. (See p. 364.) [Courtesy of D. L. Pearson]

Plate 7C
Photomicrography of pellet mount of Indonesian coal showing liptinite (black) vesicles imbedded in inertinite (white) under reflected white light. Vitrinite is the gray mass at the top and the side. Sample height is 160 microns. (See p. 400.) [Courtesy of C. Thompson-Rizer]

Plate 7D
Same Indonesian coal sample showing fluorescence of liptinite vesicles in inertinite and liptinite fragments in vitrinite under reflected blue light. (See p. 400.) [Courtesy of C. Thompson-Rizer]

Plate 8A
Pyrobitumen coating grains of chlorite in a deep Norphlet gas reservoir in the Mobile area, offshore Mississippi. (See p. 440.) [Courtesy of Chevron, U.S.A., Inc.]

Plate 8B
Star plots of adjacent oil accumulations. The dissimilar plots on the left indicate two separate pools, whereas the similar plots on the right indicate a single continuous accumulation. (See p. 540.) [Courtesy of DGSI, The Woodlands, Texas]

Plate 8C
Seismic line in Southern Norwegian Central Graben showing formation tops and the base of the Upper Jurassic in six pseudo-wells, with a control well on far left. (See p. 605.) [Courtesy of R. C. Leonard]

Plate 8D
Photomicrograph showing extreme silicification in the top seal of the first overpressured compartment at Venture field, Nova Scotia, Canada. Depth, 4,436 m (14,550 ft). Primary porosity was completely lost by massive welding of the quartz framework (q) via quartz dissolution along grain contacts (suturing—see arrows) and syntaxial quartz overgrowths (qo). **Note:** (1) Barely visible "remnant" of authigenic chlorite (ch) whose precipitation preceded porosity loss through compaction and (2) isolated porosity (p) in epoxy blue. Crossed nicols. Magnification × 100. Venture B-43 well. (See p. 304.) [Courtesy of V. H. Noguera Urrea]

Foreword

Everything has already been written to describe John Hunt's extraordinary career. In summation, he is the grandfather of world petroleum geochemistry and dean emeritus of the profession. He could have earned that title only because of his pioneer research in the 1950s and 1960s. After all, he authored the first successful oil-to-source rock geochemical correlation demonstration (Hunt and Jamieson 1956) at a time when gas chromatography and mass spectrometry did not exist. It was all done with simple bulk distillation and glass column chromatography equipment that today would be seen as deserving of a place in the Smithsonian Museum. Later, in 1962, Hunt brought the critical proof that the abundant light volatile hydrocarbons present in crude oils could not originate from recent sediments and had to be derived from thermally transformed, deeply buried, ancient sediments. The rest is the history of progress, by pioneers such as himself at Exxon, Philippi at Shell, Tissot and his IFP team in France, Vassoevich in Russia, Eglinton in England, Seifert at Chevron, and so forth. This progress was made possible by the development of better technologies (gas chromatography, mass spectrometry, quantitative pyrolysis) from the mid-1960s to the late 1970s. Since his pioneer days, John Hunt has always been in the forefront as a researcher and educator, through well over 100 publications and numerous participations in professional meetings, schools, and international forums.

All the fundamental knowledge on petroleum formation mechanisms was well established by the mid-1970s, but it took another ten years for geochemistry to begin to create a significant impact on exploration decision making. This is relatively short if one reflects that discoveries such as antisepsia by Lister and the discovery of germs as a cause of disease by Pasteur took over twenty years to be unreservedly accepted by the medical profession. Any scientific development in its beginnings is, rightfully so, put on trial until proven workable by practical experience.

Any activity critical to the acceptance of new scientific methods is end-user education. Few have done as much as John Hunt to propagate geochemical knowledge among petroleum geologists. The first edition of his textbook *Petroleum Geochemistry and Geology* was perfectly tailored to reach a geological audience, whose forte is not necessarily organic chemistry. Completeness of facts, openness to quote other people's work, economy of words, extreme clarity of style, and a focus on the important are trademarks of John Hunt's writings. He is doubtlessly among the very few great communicators in this field. It was always a privilege to have him agree to lecture at Chevron, as part of our international geochemistry curriculum for explorationists. Invariably, the professional audiences were highly responsive and felt enriched by his contact. We are fortunate today that he has totally updated and completed his original textbook into the second edition that you are now holding in your hands. I consider it must reading for petroleum explorationists and graduate students in petroleum geology. Furthermore, with current interest on a better understanding of oil and gas geochemistry in reservoirs, the "Petroleum in the Reservoir" and "Crude Oil Correlations" chapters alone justify purchasing the book as a basic reference. For instance, the subject of biomarkers, which has become so important for petroleum system identification and even reservoir continuity studies, is very clearly explained with a geological readership in mind. Finally, the chapter "Abnormal Pressures," a subject often bypassed in geochemistry textbooks, receives a comprehensive treatment that will prove most helpful to explorationists in some basins.

Thanks to John Hunt for providing us once again with a clear, complete, and useful geochemical compendium in his revised edition of *Petroleum Geochemistry and Geology*. It should help petroleum geologists and geochemists at all levels of experience, from graduate student to expert level. For my part, I have learned a lot of valuable new information from the numerous topical updates disseminated through this very user-friendly volume.

Gerard J. Demaison

Preface

The objective of this book is to explain the basic principles of petroleum geochemistry and to show how they can be integrated with geology and geophysics to reduce the risk in petroleum exploration. The book was written so as to be readily understood by students and by practicing geologists, geophysicists, and petroleum engineers who have had the basic college courses in geology and chemistry. Like the first edition, it can be used as both a text and a reference book.

This is not just a revision of the first edition; it is a whole new book. Only Chapter 3 on petroleum refining and a few short sections in other chapters are similar to the first edition. Most of the text describes developments of the past decade that are useful in exploring for and producing petroleum. The book answers such questions as: how to rank geophysical prospects by the probability that they will contain oil, gas, or water; how to model the generation and expulsion of oil, condensate, and gas from the thermal history of a basin; what geochemical measurements need to be made on outcrops and cuttings from wildcat wells to guide further exploration; how to read and evaluate the validity of geochemical service reports; how to use biomarker fingerprints to determine the source rocks of oils, oil shows, asphalts, and seeps and to correlate oils with seeps and shows; how to use hydrocarbon ratios to determine if individual reservoirs are connected or separated from each other; how to use carbon isotopes to define gas sources, gas maturity, reservoir continuity, and the effectiveness of fault seals in gas reservoirs.

A major problem in writing this book was in selecting examples of key concepts from the large number of good papers published in the last few years. In 1978 the new journal *Organic Geochemistry* consisted of 60 pages. In 1994 it covered 1,256 pages. At the same time there was a considerable increase in petroleum geochemistry papers in the *AAPG Bulletin* and other geoscience journals. Summarizing all this became a monumental task.

Fortunately, I was ably assisted by my wife, Phyllis Laking Hunt, who handled all the administrative aspects of getting the book out. Also helping were many friends who submitted papers of their latest work. In addition, the downsizing by several oil companies resulted in the release of more proprietary data than was available for the first edition. Finally, my students in 33 countries provided valuable feedback.

I am particularly grateful to the many reviewers who offered critical comments that greatly improved the book. Three Petroleum Exploration Consultants; Thane H. McCulloh from Dallas, Texas; Gerard J. Demaison from Capitola, California; and Albert V. Carozzi of the University of Illinois reviewed the entire manuscript. Kenneth E. Peters of Mobil Oil who coauthored *The Biomarker Guide* provided detailed comments on Chapters 4, 10, 14, and 15. Wallace G. Dow, President of DGSI reviewed Chapters 10, 11, and 14. Other geochemists who submitted critiques of selected chapters or sections include Earl Baker of Florida Atlantic University, Kenneth F. Steele of the University of Arkansas, Michael D. Lewan of the U.S. Geological Survey, Joseph A. Curiale of Unocal Corporation, and Jean K. Whelan, Timothy I. Eglinton, and Lorraine B. Eglinton, of the Woods Hole Oceanographic Institution. Ray Leonard, Vice President of the Amoco Eurasia Petroleum Company, generously provided data and illustrations from his evaluation of prospects on the Southern Norwegian Shelf (Chapter 16).

This book represents the culmination of a fascinating half-century in which I followed petroleum geochemistry from its birth in the late 1940s to its acceptance in exploration decision making in the 1980s and now its use in reservoir development in the 1990s. It is a science that has come of age.

John M. Hunt
Scientist Emeritus
Woods Hole Oceanographic Institution

Abbreviations Used in the Text*

°API	degrees API gravity (American Petroleum Institute)
aro	aromatic hydrocarbons
ASTM	American Society for Testing and Materials
bbls	barrels
bp	boiling point
BSR	bottom simulating reflector at the base of a gas hydrate
C_4–C_7	butanes, pentanes, hexanes, and heptanes
C_{15+}	hydrocarbons containing 15 or more carbon atoms
^{12}C	stable isotope of carbon with an atomic mass of 12
^{13}C	stable isotope of carbon with an atomic mass of 13
CAI	conodont alteration index
COST	Coastal Offshore Stratigraphic Test
CPI	carbon preference index
CSIA	compound-specific isotope analysis
d.a.f.	dry ash free
DSDP	Deep Sea Drilling Project
DST	drill-stem test
E_a	activation energy
Eh	redox potential—a measure of the oxidizing or reducing intensity of the environment
EOM	extractable organic matter
FID	flame ionization detector
g	gram
Ga	10^9 years ago
GAE	generation-accumulation efficiency
GC	gas chromatography

*See Table 10-5 (p. 360) for abbreviations used in pyrolysis.

GCMS	gas chromatography–mass spectrometry
GOR	gas–oil ratio
Gt	10^9 metric tons
H	hydrogen
HC	hydrocarbon
H/C	atomic hydrogen to carbon ratio
HP	hydrous pyrolysis
HPLC	high performance liquid chromatography
HTGC	high temperature gas chromatography
IPOD	International Program of Ocean Drilling
JOIDES	Joint Oceanographic Institutions for Deep Earth Sampling
kcal	1,000 calories
kHz	kilohertz (1,000 cycles/sec frequency)
kJ	kilojoule (10^3 joules). The joule is the SI unit of energy. One calorie = 4.18 joules.
kPa	kilopascal
L	liter
LC	liquid chromatography
LNG	liquefied natural gas (methane, ethane)
LPG	liquefied petroleum gas (propane, butanes, pentanes)
ip	isoprenoid
ls	limestone
LOM	level of organic maturation
Ma	10^6 years ago
MA	monoaromatic
Mcfd	thousand cubic feet per day
md	millidarcy
mg	milligram
µg	microgram (10^{-6} g)
µm	micrometer (micron) 10^{-6} m
mi	mile
ml	milliliter
mm	millimeter
MMcf	million cubic feet
MPa	megapascal (10^6 pascals)
MPI	methylphenanthrene index
MS	mass spectrometry

m.y.	million years
m/z	the mass to charge ratio of an ion in mass spectrometry
ng	nanogram (10^{-9} g)
nm	nanometer (10^{-9} m)
NMR	nuclear magnetic resonance
NSO	nitrogen, sulfur, and oxygen-containing compounds
O/C	atomic oxygen to carbon ratio
OCS	Outer Continental Shelf
OEP	odd–even predominance
OM	organic matter
Pa	pascal (SI pressure unit: 1 psi = 6.9 kPa, 1 psi/ft = 22.5 kPa/m)
PAH	polycyclic aromatic hydrocarbon
PAL	present atmospheric level
Pd	displacement pressure
P/D	pressure/depth profile
PDB	Peedee belemnite (carbon isotope standard)
pH	the negative logarithm of the hydrogen ion concentration: a measure of the acidity or alkalinity of a solution (acids, less than 7; bases, more than 7)
PMP	porphyrin maturity parameter
PNA	polynuclear aromatics
ppb	parts per billion
ppm	parts per million
ppt	parts per thousand (‰)
Pr/Ph	pristane to phytane ratio
psi	pounds per square inch
PVT	pressure–volume–temperature
PYGC	pyrolysis–gas chromatography
QCF	quadrillion (10^{15}) cubic feet
RFT	repeat formation tester
R_m	statistical mean reflectivity
ROF	reservoir oil fingerprinting
R_o	reflectance in oil immersion
sat	saturated hydrocarbons (alkanes)
SANS	small angle neutron scattering
SCF	standard cubic feet of gas (measured at STP)
SCI	spore color index
sh	shale

SI	the international system of units
SMOW	standard mean ocean water
SPI	source potential index
ss	sandstone
STP	standard temperature and pressure, 60°F (15.6°C) and 760 torr (133.3 Pa)
SWC	side wall core
TA	triaromatic
TAI	thermal alteration index
TCF	trillion cubic feet
TD	total depth
TDS	total dissolved solids
TOC	total organic carbon
TOM	total organic matter
TSR	thermochemical sulfate reduction
TTI	the time–temperature index developed by Lopatin
VI	viscosity index

PART ONE

INTRODUCTION

Chapter

1

The Development of Petroleum Geochemistry and Geology

Petroleum geochemistry is the application of chemical principles to the study of the origin, migration, accumulation, and alteration of petroleum (oil and gas) and the use of this knowledge in exploring for and recovering petroleum. Although the existence of petroleum has been known since ancient times, only in this century have we had the technology to obtain the enormous quantities of this fossil fuel now needed to meet the energy demands of the world's expanding economy. The world's requirement for oil alone is now about 70 million barrels (9.3 million metric tons) per day, and it is estimated that 80 million barrels (10.7 million metric tons) per day will be required in the year 2000 (Thompson et al. 1986). Finding and producing such a vast amount of oil requires the ingenuity and hard work of many people—petroleum exploration geologists, geophysicists, geochemists, and drilling and production engineers.

There is no clear evidence of when geological or geochemical principles were first applied to the search for oil. Since the beginning of history, the discoveries of natural seeps of oil and gas have been recorded, and hand-dug wells were common on the sites of such seeps. Petroleum was also a frequent by-product of early drilling for saltwater. In 600 B.C. Confucius mentioned wells a few hundred meters deep (Owen 1975, p. 2). In some saltworks, gas was burned to evaporate the brine. Chinese drilling tools reached depths of 1,000 meters by 1132. By the end of the eighteenth century, the Yenangyaung oil field in Burma had more than 500 wells and produced about 40,000 tons of oil annually. The spectacular surface flows of oil and gas at Baku resulted in the early development of the petroleum industry in Azerbaijan, and in 1870, the annual production at Baku reached about 28,000 tons. Colonel Edwin L. Drake is credited with starting the American oil industry on its remarkable career by drilling near the Titusville, Pennsylvania, seep in 1859. By 1871, 700,000 tons of oil—91% of the world's production—was coming from the Pennsylvania area opened by the Drake well. Annual production at Baku had steadily increased to nearly 4 million tons in 1890, almost equal to the production in Pennsylvania and New York that year (Owen 1975, p. 101).

These early explorations were carried out by wildcatters with little or no geological knowledge. Eventually, however, geological principles evolved and were used, the first and foremost being the *anticlinal theory*. Stated simply, because oil is lighter than water, it seeks the highest part of an underground structural fold. Thus, it is more favorable to drill an anticline for oil than a syncline.

This basic principle of looking for petroleum high on a structure is still the first criterion of exploration in rank wildcat areas. But in most parts of the world, surface geological mapping has long since been replaced or supplemented by the three-dimensional geophysical mapping of subsurface structures.

As oil-producing areas began to be discovered throughout the United States and Canada in the early 1900s, geologists soon realized that oil could be found in a variety of geological conditions not explained by the anticlinal theory. For a while, the seemingly erratic nature of oil accumulations made it difficult for geologists to convince drillers of the importance of using geological principles to locate well sites. Some of the early, disillusioned drillers felt that the best way to drill a "dry hole" was to employ a geologist. During these difficult times, the unpredictable nature of petroleum was clearly expressed by a Pennsylvania judge in an early court decision when he said, "Oil is a fugacious mineral." What he meant was that oil could move from its point of origin, thereby making it difficult to define legally its geographic boundaries.

The discovery of oil beneath large anticlinal structures in Kansas, Oklahoma, and California about the time of World War I brought with it a resurgence of geological structural prospecting, with geologists firmly in control of exploration decisions. The successful application of the reflection seismograph to subsurface structural mapping in the 1920s further strengthened faith in the anticlinal theory. Then, in the middle 1930s, the great East Texas pool was found. The discovery of this stratigraphically trapped oil and gas reservoir made drillers and geologists alike realize that finding oil requires a knowledge of all available principles of earth science. No longer could exploration geologists simply look for folded structures and anticlinal culminations; they had to understand sedimentation, stratigraphy, paleontology, geochemistry, mineralogy, petrology, geomorphology, and historical geology. The day of the wildcatter with a nose for oil was waning, as fields were now being found more efficiently by intensive detailed studies using all the scientific data available.

Then, as oil became increasingly more difficult to find, it became clear that geologists needed to understand the geochemistry of petroleum. What is the composition of petroleum? How does it originate, and how does it migrate in the subsurface? How does it change with depth, temperature, and pressure? How can we use such knowledge to help us find commercial accumulations?

For more than 100 years, the source of petroleum has been known as being the organic matter deposited with the sediments of a basin. Karl G. Bischof, professor of chemistry at the University of Bonn, Germany, first stated that petroleum originated from the slow decomposition of organic matter. T. Sterry Hunt, who has been called the world's first authority on petroleum geology (Owen 1975, p. 54), elaborated on this theory by defining the lower forms of marine life as the probable sources of petroleum (Hunt 1863, p. 527). Later, Russia's father of geochemistry, V. I. Vernadskii, who was the inspiration for the development of Russia's vast mineral resources, reemphasized the organic origin of oil: "The general features of oil genesis are clear. We should consider oils as sedimentary minerals genetically related to organic matter. Organisms are undoubtedly the source material of oils–oils cannot contain any significant amounts of juvenile (primordial) hydrocarbons" (Vernadskii 1934, pp. 152–153). There is now a vast amount of geochemical data showing that essentially all petroleum hydrocarbons—oil and gas—originate from the organic matter deposited in sedimentary rocks.

The concept of bituminous shales as source beds for the Pennsylvania oil accumulations was proposed as early as 1860 by state geologist John Newberry. Later, Newberry and others suggested that in Ohio and Kentucky the oil should be sought where sandstones are in contact with the Ohio (Devonian) black shale. But the thermal history of the source rock was not a factor recognized until David White gave his classic paper on the carbon-ratio theory to the Washington Academy of Sciences (White 1915). He showed that in the eastern United States there is a geographic relationship between the occurrence of oil and gas fields and the maturity of coal beds in the same area. Oil fields are confined to areas of low-maturity coal (coal with less than 60% fixed, or nonvolatile, carbon), whereas gas fields occur at higher maturity (fixed carbon 60 to 70%), and no oil or gas fields are found where the fixed carbon of coal is over 70% (anthracite).

This early use of coal as a time–temperature indicator of the maturity of associated source rocks has been replaced by several more definitive indicators within the rock itself. One of these is the reflectance of vitrinite, the main component of coal, which also is found disseminated in about 80% of all sedimentary rocks. Other indicators are the color of microfossils (spores, pollen), which change from yellow to brown to black with increasing temperature (depth). The hydrogen-to-carbon ratio of the organic matter and the spatial arrangement of the atoms in fossil molecules (stereoisomers) also change with the temperature. These maturation indicators are used to define source rocks as immature, mature, or postmature with respect to their ability to generate and expel oil. (*Immature* means little or no generation; *mature* refers to the principal generation; and *postmature* means beyond generation.)

Although the bulk of the world's oil is in anticlinal traps, the majority of anticlinal structures that have been drilled in the last 100 years have no recoverable hydrocarbons. In addition, entire areas of some basins contain no recoverable hydrocarbons, even though they do contain good reservoir rocks and seals. Why is this so? Is it due to a lack of organic material, insufficient heat to generate oil, or an absence of suitable migration pathways to the reservoirs?

To answer these questions, we should first note that exploration risk is the probability of spending exploration funds without economic success. Reducing this risk depends not only on finding a trap but also on determining how high the probability is that oil has migrated from a mature source rock into that trap and has not escaped or been destroyed. In 1984, Demaison summarized this problem by writing that successful exploration depends on the simultaneous occurrence of three independent factors: (1) the existence of a trap (structure, reservoir, seal); (2) the accumulation of a petroleum charge (source, maturation, migration to the trap, timing); and (3) the preservation of the entrapped petroleum (thermal history, meteoric water invasion). The probability of success in finding petroleum is the product of the probabilities of all three of these factors. Therefore, if any one is zero, a "dry hole" will be the result, no matter how favorable are the other two factors.

Petroleum Generative Depressions

The concept of an organic origin of oil has led to the mapping of source rock organic facies. Jones (1987) defined an organic facies as a mappable subdivision of a stratigraphic unit distinguished from the other adjacent subdivisions by the character of its organic matter (OM). Different organic facies generate and migrate different amounts and types of oil and gas (Demaison 1984). Organic facies maps can be used to estimate the amounts and types of petroleum formed from mature source rocks.

Historically recognized zones of great wildcat successes, the so-called fairways, are closely related to what are called *petroleum generative depressions* or *hydrocarbon kitchens*. These are areas where organic-rich source beds are buried at high enough temperatures to generate and migrate substantial quantities of petroleum. High-potential petroleum fairways, such as the Central and Viking Grabens of the North Sea, owe their huge petroleum accumulations to the presence of the Late Jurassic Kimmeridge Shale buried 3,048 m (10,000 ft) deep at temperatures above 93°C (200°F) in a generative depression. By mapping the organic parameters of the Kimmeridge, it was possible to determine the probability that a prospective trap contained oil, gas, or no petroleum. When this technique first was mentioned in a North Sea news report (North Sea Letter, 1981), it was met with skepticism. But time has shown that the concept does reduce exploration risk.

A *generative basin* may have one or more petroleum-generating depressions. The depressions are recognized by making overlays, or organic facies maps, and maturation maps of each source rock interval in a basin. Four examples of maturation mapping are shown in Plate 1. The term *success* on these maps does not mean economic success; rather, it means that it was possible to produce a flow of oil or gas from a subsurface accumulation.

Plate 1A shows the success ratios for wildcat wells drilled between the 55th and 62nd parallels in the North Sea Basin, plus the known oil and gas fields at the time of this study. The immature area of the Kimmeridge Shale source rock where temperatures have not risen above 93°C (200°F) and no significant oil has been generated are in yellow. Mature areas of generation above 93°C are tan. Within the mature areas, one in three wildcats discovered producible hydrocarbons, whereas only one in thirty made a discovery outside this area. Furthermore, fields with the largest reserves are usually near the center of the generative depression where the Kimmeridge is the thickest and most mature. Some of the discoveries in the immature area, like the Beatrice Field, were of oil that originated in a different and more deeply buried source rock than did the Kimmeridge (Bailey et al. 1990; Peters et al. 1989). The close proximity of the largest petroleum reserves to the Kimmeridge depocenter is due to vertical migration from the overpressured source rock to the lower-pressured Cretaceous chalk above (Hunt 1990).

The use of vitrinite reflectance as an approximate maturation indicator is shown in Plate 1B for the Illinois Basin. The 0.6%R_o (reflectance in oil immersion) represents the approximate beginning of the oil generation "window." This map shows that 90% of the oil reserves in the Illinois Basin are within about 30 miles updip from 0.6 and 0.7%R_o values for the New Albany Shale source rock. Most of the large accumulations, shown by the large circles, are immediately updip from the depocenter of the New Albany generative depression.

Plate 1C is a maturity map for gas, in contrast to the other illustrations, which are for oil. The source rock is the Permian coal of the Cooper Basin in Australia, which is a gas-generating organic facies. The gas generation window is shown in tan at a reflectance between 0.9 and 2%R_o. The immature gas area (which is mature for oil), is in yellow, and the postmature area for gas is in brown. The map shows the gas window because the Permian coals generate and expel primarily gas, with very little oil. All the giant gas fields sourced by these coals and coaly shales occur in or near the tan area of maximum gas generation. Moomba, the largest gas field, lies immediately updip of the main gas "kitchen." One of every two wildcat wells in the gas mature area yielded methane gas, compared with one in twenty-three in the immature area and none in the postmature zone. Drill-stem tests in the postmature zone yielded some gas high in CO_2. This example demonstrates that a basin may contain primarily either oil or gas, depending on whether the major source beds generate oil or gas.

Figures A, B, and C in Plate 1 all represent either vertical migration or short-distance lateral migration from source to reservoir. Plate 1D shows the long-distance migration updip on the northeastern flank of the Williston Basin. The mature kitchen of Bakken source rock is shown in tan, and the immature area is in yellow. About half the oil has moved vertically upward to reservoirs on the Nesson anticline, and half has migrated updip about 160 km (100 miles) under the Charles salt and the Jurassic red shale to oil reservoirs in Canada. This example demonstrates that understanding the migration pathways is as important as defining the source rock. In all the examples of Plate 1, the location of the largest oil and gas fields can be correlated with geochemically identified oil or gas generative depressions.

Today, prospect appraisal in sparsely drilled basins involves modeling the entire process of hydrocarbon generation, expulsion, migration, trapping, and preservation. Figure 1-1 is an example of a cross section through a hydrocar-

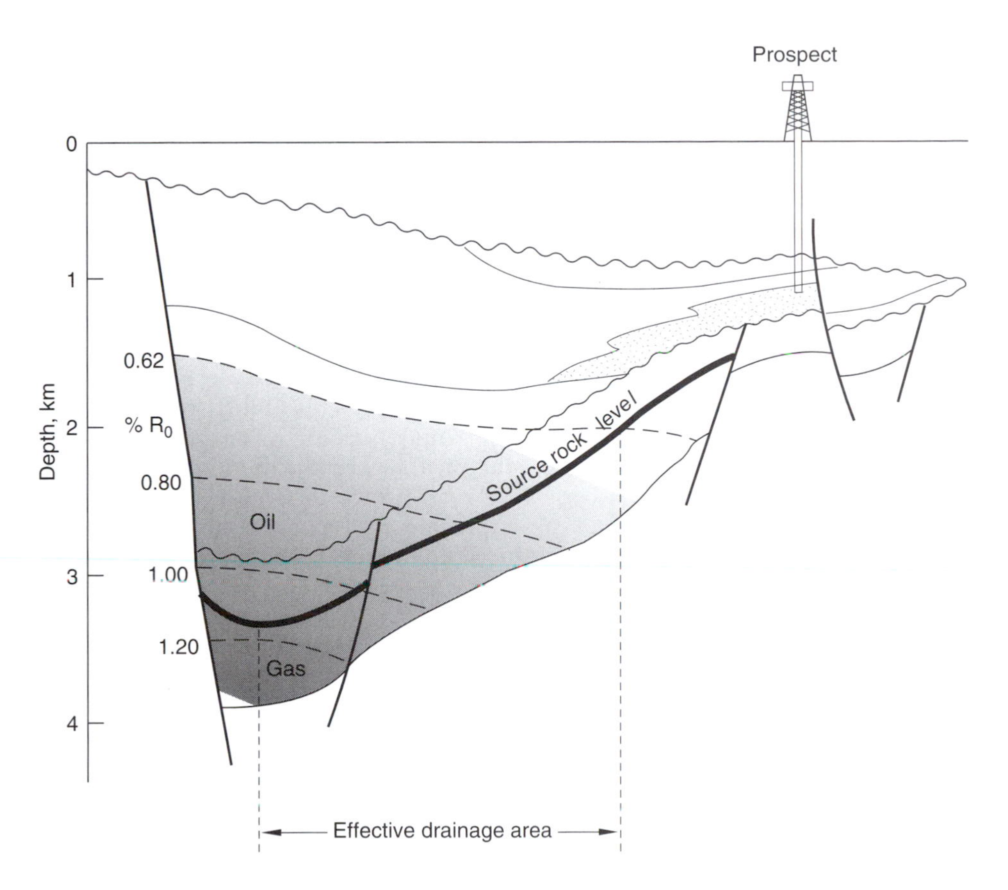

Figure 1-1

Cross section through the drainage area from a Jurassic source rock to a Cretaceous sandstone on a prospective structure. The effective drainage area starts at a vitrinite reflectance level, R_o of 0.62%, where expulsion is believed to begin. [Sluijk and Nederlof 1984]

bon drainage area to a prospective structure (Sluijk and Nederlof 1984). The dashed isomaturity lines shown by vitrinite reflectance (%R_o) values follow the geothermal gradient rather than the stratigraphic position of the source rock. In this example, the expulsion of oil starts at a vitrinite reflectance level of R_o = 0.62% and gradually changes to the expulsion of condensate and gas for that portion of the source rock buried to reflectance values beyond R_o = 1.0%. By constructing drainage maps and burial history graphs for the source rock and using such data as the hydrocarbon yield as a function of maturity and the time of trap formation, it is possible to estimate the volume of hydrocarbon delivered to the trap.

This technique of appraisal, although it has some uncertainties, has proved to be less risky than comparisons made on trap size only. For example, Murris (1984) reported that before drilling, Shell International ranked 165 prospects in a seriatum based on Sluijk and Nederlof's modeling system (1984), which we just discussed. The drilling resulted in forty-five discoveries and 120 dry holes. After drilling, the results were compared with what would have happened if the prospects had been drilled based on trap size only, with the largest trap first. It was found that ranking the trap size had a forecasting efficiency of 18%, compared with 0% for random drilling. However, the ranking based on geochemical modeling had a forecasting efficiency of 63%.

The first twenty wildcats discovered about 3.4 billion barrels of oil in place. If they had drilled according to the ranking of the trap size, they would have discovered only 800 million barrels of oil. This ability to reduce risk early in an exploration program has been a major factor in convincing oil companies to use both geological and geochemical model studies in prospect appraisal.

Geochemistry is not a cure-all, however. Dry holes will still outnumber discoveries in wildcat programs no matter what techniques are used. But geochemistry is helping reduce the risk of drilling dry holes and will continue to do so as its applications are increased and extended. As Demaison has often said, "Most of the world's oil is in geophysically mappable anticlines, but most oil producing anticlines are in geochemically mappable generation–migration fairways."

SUMMARY

1. Petroleum geochemistry is the application of chemical principles to the study of the origin, migration, accumulation, and alteration of petroleum and the use of this knowledge in the exploration and recovery of oil and gas.

2. More than 100 years of investigations and research have demonstrated that the bulk of the world's petroleum originated from the decomposition of the organic matter deposited in sedimentary basins.

3. The field observations of geologists in the late nineteenth century led to the idea that oil originated in bituminous shales and migrated into sandstones.

4. The carbon-ratio theory was the first geochemical concept that related oil and gas accumulations to metamorphism. Oil fields gave way to gas where the fixed carbon contents of coal exceeded 60%, and gas fields could not be found where the values exceeded 70%. Source rocks are now defined as immature, mature, or postmature for oil and gas generation, based on a host of maturation indicators, of which vitrinite reflectance is the most widely used.

5. Successful exploration depends on the simultaneous occurrence of three independent factors: (1) the existence of a trap (structure, reservoir, seal), (2) the

accumulation of a petroleum charge (source, maturation, migration to the trap, timing), and (3) the preservation of the entrapped petroleum (thermal history, meteoric water invasion). The probability of success in finding petroleum is the product of the probabilities of all three of these factors.

6. Organic facies are mappable subdivisions of stratigraphic units, which are distinguished from the adjacent subdivisions by the character of their organic matter. Different organic facies generate and expel different amounts and types of oil and gas.

7. Petroleum generative depressions are areas where organic-rich source beds are buried at high enough temperatures to generate and expel substantial quantities of petroleum.

8. Prospect appraisal requires modeling the entire process of hydrocarbon generation, expulsion, migration, trapping, and preservation.

SUPPLEMENTARY READING

Owen, E. W. 1975. Trek of the oil finders. In *A history of exploration for petroleum.* Amer. Assoc. Petrol. Geol. Memoir 6. Tulsa: American Association of Petroleum Geologists, pp. 1–4.

Demaison, G., and R. J. Murris (eds.). 1984. *Petroleum geochemistry and basin evaluation.* Amer. Assoc. Petrol. Geol. Memoir 35. Tulsa: American Association of Petroleum Geologists, 426 p.

Chapter

2

Carbon and the Origin of Life

Carbon (from *carbo,* meaning "charcoal") is in the fourth group of the periodic table of elements, which means that it has four electrons in its outermost electron shell. Carbon is unusual in that it forms strong carbon–carbon bonds, which remain strong when the carbon groups combine with other elements. The most stable elements, or combinations of elements, are those that contain eight electrons (an octet) in the outer shell. Carbon assumes this configuration by forming covalent bonds, that is, by sharing electrons with itself and other elements. For example, carbon is readily reduced with hydrogen, or oxidized with oxygen, to form the two most common carbon compounds in the earth's crust—methane and carbon dioxide:

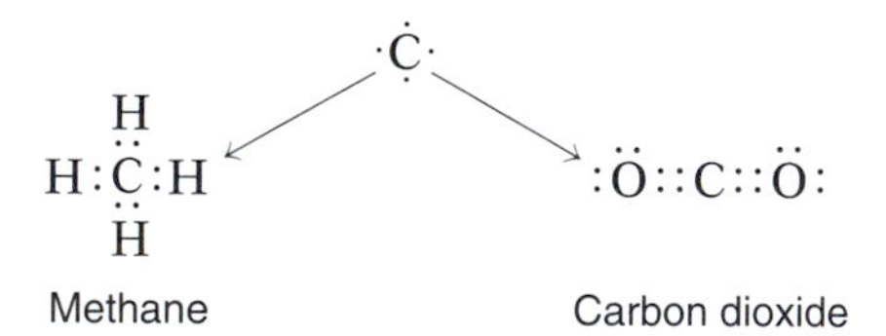

The carbon atoms form an octet of electrons around them by sharing one electron from each of four hydrogen atoms or by sharing two electrons from each of two oxygen atoms.

The uniqueness of carbon, which enables it to be the basic element of all life, lies in its ability to combine with itself to form long carbon chains, rings, and complex, bridged structures. Only one other element, silicon, also with a valence bond of 4, can do this. Silicon chains can be made in the laboratory, but they do not exist in nature for the following reasons: (1) The Si–Si bond energy of 53 kcal/mole is much weaker than the C–C bond energy of 83 kcal/mole; (2) the outermost electron shell of silicon is readily attacked by water, oxygen, or ammonia, so silicon chains are unstable in the presence of these compounds; and (3) silicon is unable to form double bonds with oxygen to yield an SiO_2 monomer in the same manner that carbon forms CO_2 gas. Silicon oxides exist only as high-molecular-weight solid polymers. These crystalline solids do not circulate through the hydrosphere and biosphere, as does CO_2.

Carbon has been the basic structure of all life as we know it since the beginning of life on earth. Consequently, the chemistry of carbon is often referred to as *organic* chemistry, whereas the chemistry of all other elements is called *inorganic* chemistry. The 100-plus elements other than carbon combine with one another to form about 70,000 inorganic compounds, whereas carbon combines with itself and the other elements to form about 4 million organic compounds. Carbon is in the food we eat, the air we breathe, the clothes we wear, the houses we build, the fuel to heat those houses and power our cars, trains, airplanes—carbon is the most ubiquitous element on earth; it is everywhere in the earth's crust.

The Primitive Earth

Since oil is organic in origin, as will be discussed later, it is important to understand the origin and development of life on earth. Speculations about the prevalence of Precambrian oil really depend on the extent to which organic matter was formed and preserved in Precambrian times.

The earth is believed to be as old as the oldest known meteorites and terrestrial lead, about 4.6 Ga (10^9 years ago) (Patterson 1956). At first, the earth was probably composed of about 90% iron, oxygen, silicon, and magnesium and 10% all other natural elements. One model proposes that the earth heated up during this first billion years because of the impact energy of falling planetesimals, the compression of the earth due to gravity, and the disintegration of radioactive elements. The rise in temperature caused the iron to melt and sink to the center while the lighter material floated to the surface. In effect, this converted the earth from a relatively homogeneous body to a heterogeneous layered body with a dense iron core, a mantle of original body, and surface crust of light material, as shown in Figure 2-1. In commenting on this model, Press and Siever stated: "Differentiation is perhaps the most significant event in the history of the earth. It led to the formation of a crust and eventually the continents. Differentiation probably initiated the escape of gases from the interior, which eventually led to the formation of the atmosphere and the oceans" (1986, p. 12). Although the details of this model are speculative, there is general agreement that during this first billion years the earth underwent cataclysmic changes that eliminated the original crust. No earth rocks have

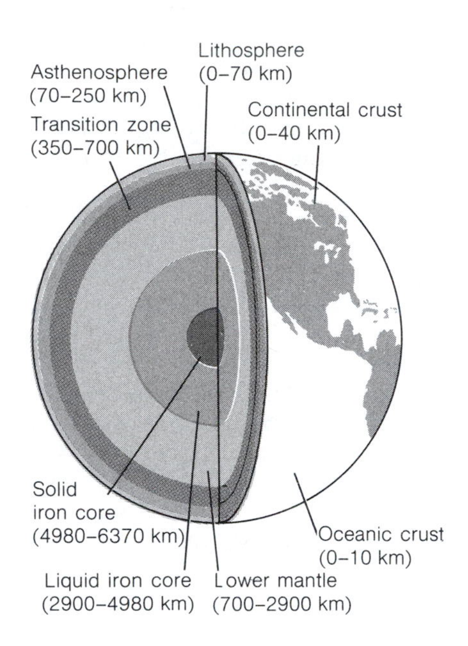

Figure 2-1

The earth is density zoned, with a dense iron core followed by a zoned residual mantle with a superficial crust of light rock on top. [Press and Siever 1986]

been found older than about 3.8 Ga, whereas moon rocks range in age from 3.1 to 4.6 Ga.

As the earth heated up and lighter materials came to the surface, volcanism contributed enormous quantities of water vapor, carbon dioxide, nitrogen, hydrogen sulfide, and hydrogen to the atmosphere. The hydrogen sulfide formed iron sulfides and also was destroyed by photochemical reactions, so its partial pressure would have been low (Holland 1984, p. 107). The hydrogen gradually diffused into outer space, and the water vapor condensed, leaving nitrogen and carbon dioxide as the major atmospheric components. Ammonia and methane could not have been present except in trace amounts because they also would have been destroyed by photochemical reactions. By about 3 Ga there was enough carbon dioxide in the atmosphere to cause extensive chemical weathering because of the higher acidity of surface waters. This weathering caused the dissolution of enough silica to form large chert and quartz deposits in the Precambrian. Holland (1984, p. 332) estimates that between 2 and 3 Ga, the CO_2 pressure was probably between $10^{-3.1}$ and $10^{-1.9}$ atmospheres.

In an atmosphere devoid of oxygen, life was limited to unicellular organisms that could live under reducing conditions, such as the sulfate-reducing bacteria that are found today in anoxic, stagnant waters. Life as we know it now did not develop until oxygen became an important atmospheric component.

The geochemical evidence supports the idea that a nonoxidizing atmosphere existed until about 2 Ga. Extensive banded iron formations, which consist of alternate layers of iron and silica, are unique to the Precambrian, with more than 95% deposited before 2 Ga (Holland 1984, p. 378). These formations, which occur on all continents, imply a nonoxidizing atmosphere under which large quantities of iron could be transported in a soluble ferrous state to sites of deposition. Examples are the Hamersley Group of western Australia (2.5 Ga), the Soudan Formation of North America (earlier than 2.5 Ga), the Dharwar Formation of India (about 2.5 Ga), the Krivoi Rog of Russia (about 2.1 Ga), and deposits of the Huronian period of North America (1.7 to 2.5 Ga).

Other evidence for a nonoxidizing atmosphere is the presence of uraniferous conglomerates and detrital uraninite before about 2 Ga. Holland et al. (1986) estimate the partial pressure of oxygen to have been about 3×10^{-5} atmospheres near 2.5 Ga. The present-day partial pressure of oxygen is 0.2 atmospheres, and that of carbon dioxide is 0.0003 atmospheres.

Red beds—sediments whose grains are coated with ferric oxide—first appeared about 2.6 Ga. Their appearance marks the beginning of a significant increase of oxygen in the atmosphere. Photosynthesis by primitive organisms produced this oxygen, as discussed later.

Primitive Life

The earliest evidence of life is the stromatolites found in the 3.5 Ga Warrawoona Group of northwestern Australia (Walter 1983). However, about 3.8 Ga there was an increase in the ratio of light (^{12}C) to heavy (^{13}C) carbon isotopes in the organic matter of sediments, compared with the ratio in primordial carbon. This suggests that life may have begun about that time (Schidlowski 1988). The stable carbon isotopes, which are described more fully in Chapter 3, are shifted toward the lighter ^{12}C in all photosynthesis processes.

These early organisms are called *prokaryotes* because their genetic material is disarranged in the cell nucleus and they are asexual. The earliest prokaryotes were *anaerobic photoautotrophs.* A photoautotroph is an organism that uses light as an energy source and CO_2 as a major source of cellular carbon (CH_2O), as follows:

$$CO_2 + 2H_2S \xrightarrow{\text{Light}} [CH_2O] + 2S + H_2O$$

Bacterial photosynthesis

Such organisms need only a source of reducing power such as H_2S in a suitable aqueous environment in order to proliferate.

The second most important event after the origin of prokaryotes was the development of a "chlorophyll-like" reaction center in a prokaryote with a redox potential capable of splitting water in the presence of light. Such a prokaryote uses water as a reducing agent while forming oxygen as a by-product, as follows:

$$CO_2 + H_2O \xrightarrow[]{\text{Light + Chlorophyll}} [CH_2O] + O_2$$

Green plant photosynthesis

Chapman and Schopf (1983, p. 318) suggest that a primitive precursor of the purple nonsulfur bacteria may have been the first oxygen producer. Its evolution would have given rise to the cyanobacteria (blue-green algae) which require molecular oxygen to produce some of their metabolic products, such as sterols, fatty acids, and some carotenoids. It is unlikely that the cyanobacteria could have developed without an initial small buildup of environmental oxygen.

Nevertheless, the blue-green algae are believed to be the major cause of the accumulation of oxygen in our atmosphere. As these cyanobacteria spread across the earth's oceans, the oxygen they emitted was taken up by various oxygen sinks, such as ferrous iron and sulfides. Eventually, free oxygen began to build up in the atmosphere, as shown in Figure 2-2, although the oceans remained anoxic through the Early Proterozoic (Holland 1984, p. 514). These conditions led to the third most significant biological event, the origin of *eukaryote* organisms, whose chromosomes and nucleated cells are comparable to those of all higher life. The early eukaryotes (1.4 Ga) were asexual, incapable of genetic variability. Around 0.8 to 1 Ga, the sexual eukaryotes appeared. According to Schopf (1983), they were the "evolutionary trigger" that caused an explosive increase in both the diversity and the evolutionary development of life. Within a few hundred million years, megascopic green, red, and brown algae were populating the world's oceans.

These multicellular organisms allowed the widespread development of fauna and flora in the Early Cambrian. Cambrian strata contain at least 1,200 different kinds of life, including brachiopods, gastropods, calcareous sponges, algae, worms, and complex trilobites up to 5 kilograms (11 lb) in weight. Life was still limited to the lakes, rivers, and oceans. Not until the Late Silurian were the land surfaces invaded by plants.

Holland (1984, p. 511) pointed out that an imbalance of only 5% between oxygen generation and oxygen consumption can cause a 50% increase or decrease in the atmosphere's oxygen levels in about 40 million years. It is probable that the spread of higher land plants in the Devonian led to a significant increase in atmospheric O_2.

As oxygen entered the deep parts of the ocean, another process developed in proximity to the hydrothermal vents. *Aerobic chemoautotrophs,* which are capable of synthesizing organic matter in the absence of light, became active as follows (Jannasch and Wirsen 1979):

$$CO_2 + O_2 + 4H_2S \longrightarrow [CH_2O] + 4S + 3H_2O$$

Bacterial chemosynthesis

This process is responsible for the rich faunal communities clustered in complete darkness on the seafloor, at spreading centers around submarine thermal springs. In terms of evolution, aerobic chemoautotrophs came after oxygen was

Age Ga	Eon	Era	O_2 % of PAL	Precambrian History
0.5		Paleozoic	>30 >10	Ocean oxic $>1 m\ell O_2/\ell H_2O$
	Proterozoic	Sinian		Oldest invertebrates Oceans suboxic $<1 m\ell O_2/\ell H_2O$
1.0		Riphean	4	Oldest fossils of higher algae
				Origin of eukaryotes
1.5			2	
				Widespread blue-green algae (cyanobacteria)
2.0			~1	Oceans anoxic $<0.1 m\ell O_2/\ell H_2O$
		Huronian		
2.5	Archean		0.003	Oldest redbeds
		Randian		Origin of sulfate reducing bacteria and cyanobacterium-like fossils Abundant stromatolites
3.0		Swazian		
3.5				Oldest stromatolites
		Isuan		Origin of bacterial photosynthesis Oldest sedimentary rocks Atmosphere of N_2 and CO_2
4.0	Priscoan	Hadean		
4.5				Origin of Earth

Figure 2-2

The evolution of life and oxygen in the atmosphere. Ga = 10^9 years ago, PAL = present atmospheric level. Definitions of oxic, suboxic, and anoxic environments are based on biofacies associations. [Rhodes and Morse 1971] Data from Cloud 1983, Holland 1984, Schidlowski 1986, Schopf 1983, Vidal 1984.

available from green plant photosynthesis. Anaerobic chemoautotrophs, however, such as methanogens that generate methane from CO_2, have probably existed since the beginning of the Archean. The fossil record is sparse, but carbon isotope data indicate that these bacteria were active about 2.8 Ga.

The oxidation of the hydrosphere has never been complete, even through the Phanerozoic. The richest petroleum source rocks have resulted from periodic widespread anoxic conditions, which are discussed in more detail in subsequent chapters.

Petroleum Potential of Precambrian Rocks

Petroleum accumulations and shows have been found in Proterozoic rocks in the United States, Russia, China, Canada, Australia, Venezuela, Oman, and Morocco (Becker and Patton 1968; Gorin et al. 1982; Jackson et al. 1986; P'an 1982). Many of these hydrocarbons originated in younger rocks and migrated into Precambrian reservoirs. But some—such as the Markovo condensate and oil of eastern Siberia, the Nonesuch oil of Michigan, some of the Oman oils, and the Precambrian oil shows of the McArthur Basin, Australia—appear to be of Proterozoic origin, Also, Lopatin (1980) mentions oil shales of Proterozoic age in Russia. Vassoevich et al. (1971) give many other examples of Precambrian hydrocarbons, including a map containing nine potential Precambrian oil and gas basins in Eurasia, Africa, and Australia.

The Markovo field is one of several in the Lena–Tunguska petroleum province of eastern Siberia that has produced gas and condensate from Proterozoic and Lower Cambrian strata. Proved plus probable reserves total 622 billion cubic feet of gas and 16 million barrels of condensate (Meyerhoff 1980). Proved reserves in Russia are those reserves that are certain to be recovered with existing facilities, whereas probable reserves are those that may be recovered with as yet uninstalled improved recovery systems (Khalimov 1980). The Markovo (Riphean age, 1 Ga) and Parfenovo (Sinian age, 0.8 Ga) producing horizons are overlain by shales, dolomites, and evaporites (halite, anhydrite) of Precambrian age, which provide a near-perfect seal. Some Parfenovo reservoirs are completely surrounded by Proterozoic shales, thus eliminating the possibility of a younger source. Proved plus probable reserves for all Proterozoic and Lower Cambrian fields in the province are small, however. The total production of all eastern Siberian and Russian Far East petroleum fields was less than 0.5% of the total Russian production in 1986, and most of this was from post-Cambrian sediments (*Oil and Gas Journal,* March 9, 1987).

Russia also produces oil from Precambrian Sinian age formations in the northern Volga–Urals region. Chemical analyses (Balashova et al. 1983) indicate that the oils are genetically different from all younger oils in the area, which suggests that these also may be from a Precambrian source.

There are two groups of South Oman oils. One is generated by the Precambrian Huff Formation and the other by Paleozoic sediments (Klomp 1986). The 1.0 Ga Nonesuch shale (United States) is so organic rich that it seeps oil from the White Pine copper mine in Michigan. It is typical of many younger paraffinic oils in composition except that it has no steranes or triterpanes (Hoering 1976). The oldest oil shows reported as of 1986 are from unmetamorphosed Precambrian sediments of Australia. They are about 1.4 Ga in age (Jackson et al. 1986).

Kelly and Nishioka (1985) found oil trapped as primary fluid inclusions in Precambrian calcite crystals. The calcite occurred in copper-iron sulfide—bearing veins that crosscut the Nonesuch shale at the White Pine Mine. The calcite age is about 1.05 Ga, which also sets a minimum age for the oil.

These examples demonstrate that oil and gas accumulations will continue to be found in unaltered or only slightly metamorphosed Precambrian sediments worldwide, so such strata should not be ignored as potential petroleum sources. The amounts remaining to be discovered, however, will not be large. From a geochemical standpoint, several factors increase the risk of dry holes. First, the microbes of the Proterozoic era appear to have deposited less carbon in sediments worldwide than they did during equivalent times in the Phanerozoic. No contribution came from land, and aquatic contributions were highly variable. Total organic carbon (TOC) analyses by Hayes et al. (1983) demonstrate this variability with values for the Hamersley Group shales (2.5 Ga) ranging from 0.01 to 6.5 wt%. The Hayes mean value for nearly 300 Archean and Proterozoic sediments was around 0.3 wt%, but Hayes pointed out that these samples may be biased toward carbon-rich rocks, as they were not selected to represent all major rock types. Ronov's (1982) data, which include all rock types, indicate that TOC deposition more than doubled from the Proterozoic to the Phanerozoic. Ronov reported 0.26 wt% for TOC in Upper Proterozoic rocks. These values compare with 0.56 wt% for the Phanerozoic, including 0.81% for the Cenozoic. Of course, these represent the means for all selected rocks in an era rather than those that may have been source rocks, but they do suggest a trend. Most good petroleum source rocks have TOC values from 2 to 3% even after they have passed through the catagenesis, oil-generating stage. Less than 10% of Hayes's Precambrian samples were in this range, and these may be biased toward high values.

The availability of hydrogen. which is the key to petroleum generation, is the second geochemical factor raising risk. The ratio of hydrogen to carbon (H/C) for plankton is around 1.6. During increasing burial, the plankton is partially converted to petroleum, and the H/C ratio of the residual organic matter (kerogen) steadily decreases. This is because the oil and gas with ratios of 1.8 and 4, respectively, require more hydrogen. When the H/C of kerogen drops below 0.3, the amount of available hydrogen is so low that no oil and only trivial amounts of gas are generated.

Precambrian kerogens in general are low in hydrogen. Most have been extensively dehydrogenated, with many having H/C ratios of less than 0.2 (Hayes et al. 1983). Structurally, such a kerogen with the formula $H_{30}C_{150}$ would contain 61 fused aromatic rings, like a precursor of graphite. For detailed analyses, the Hayes group separated the kerogens from 62 Precambrian rocks, ranging in age from 0.8 to 3.8 Ga. Only 6 had H/C ratios above 0.4, and only 1 of these 6 had more than 2 wt% TOC. A combination of low TOC and low H/C kerogen ratio is not conducive to generating oil. Earlier studies by

McKirdy et al. (1980) and later studies by Hoering and Navale (1987) confirmed this extensive dehydrogenation in ancient kerogens. There is just not enough hydrogen left in many of these very old rocks to form appreciable quantities of oil and gas.

Finally, there is a tendency for oil accumulations to be lost over geologic time. Lopatin (1980) cites evidence that there may have been large petroleum accumulations formed during the Proterozoic and later destroyed. He mentions black slates with thick lenses and layers of graphitic carbon material in the Proterozoic of southwest Greenland, the Krivoy Rog Series of the Ukrainian Shield and the Upper Huronian Series of the Canadian Shield. These could be the residues of rich, oil-generating kerogens. Also, there are thick Proterozoic oil shales in the Onega Basin of Russia, now dehydrogenated but still containing over 2 wt% organic carbon. Lopatin estimates that they could have yielded over 50 billion barrels of oil. The problem is that in the subsequent 1 billion years, most of these oil fields would have been destroyed by tectonic activity, resulting in the reservoirs' leakage. Only a few residual traces of once major accumulations would be left in the metamorphosed rocks.

In summary, analyses of Precambrian sediments indicate that they do not have the source rock quality of Phanerozoic sediments, in either the quantity of kerogen or its hydrogen content. Oil and gas will continue to be found, particularly in unaltered Precambrian sediments, but the quantities will not be large unless the source rocks are organic-rich, their kerogens are not extensively dehydrogenated, and the reservoir rocks are unusually well preserved.

Inventory of Carbon in Sedimentary Rocks

Carbon is cycled through the biosphere by means of photosynthesis and oxidation:

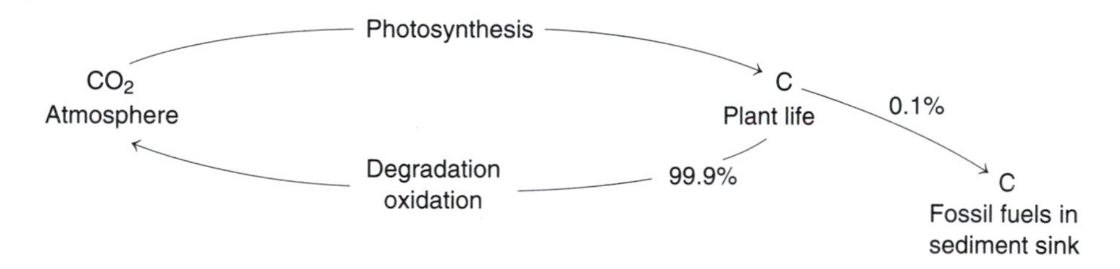

Plants (phytoplankton) utilize CO_2 to form the carbon of their cells, and animals (zooplankton) eat the plants and give off the excess carbon as CO_2. Dead organisms are microbially or chemically oxidized to CO_2. Out of this cycle, about 0.1% of the carbon is withdrawn and buried with the sediments (Ryther 1970). Since the beginning of life, about 1 of 11,000 parts of this 0.1% carbon has become a commercial petroleum accumulation that has survived to the present.

Carbon exists in sedimentary rocks in two forms: as reduced carbon, principally organic matter formed by biological processes over geologic time, and as oxidized carbon, principally carbonate. Table 2-1 shows the distribution of these forms of carbon in the earth's sedimentary rocks (Hunt 1972, 1977). The quantity in sedimentary rocks was determined by using the mass of rocks in the earth's crust, as calculated by Ronov and Yaroshevsky (1969) and by using organic carbon data from the general literature. Asphalt data from the literature have also been used. The asphalt in nonreservoir rocks is defined as largely the nonhydrocarbon fraction soluble in lipid solvents such as toluene. The figures in Table 2-1 must be considered approximate, since there are inherent errors in all such calculations.

The 64×10^{21} g of carbonate carbon compares with 71×10^{21} g from Ronov's (1982) data, and the 12×10^{21} g of insoluble organic carbon compares with Ronov's 9×10^{21} g. Perrodon (1983) estimated that the world's produced, proven, and yet to be discovered oil totaled 290 Gt. To this should be added unrecoverable oil of 680 Gt, for a total of 970 Gt (0.97×10^{18} g). Adding natural gas (136×10^{12} m^3, or 0.2×10^{18} g) gives a total of 1.17×10^{18} g, similar to the 1.1 published in Hunt (1979). The heavy oil and asphalt value of 0.6×10^{18} g includes 0.3×10^{18} g from the giant reserves and resources of Canada and Venezuela.

The large differences between the quantities of kerogen in sediments, dispersed (nonreservoired) petroleum and reservoired petroleum in Table 2-1 indicate that the generation, migration, and accumulation of oil is an inefficient process. According to Table 2-1, the yield of bitumen (petroleum plus asphalt)

TABLE 2-1 Carbon in 10^{18} g in Sedimentary Rocks

	Organic carbon (reduced)	*Carbonate carbon (oxidized)*
All sedimentary rocks	Insoluble organic carbon	
Shales	8,900	9,300
Carbonates	1,800	51,100
Sandstones	1,300	3,900
Coal beds thicker than 4.6 meters (15 feet)	15	
Nonreservoir rocks	Soluble organic carbon	
Asphalt	275	
Petroleum	265	
Reservoir rocks		
Heavy oil and asphalt	0.6	
Petroleum	1.1	
Total	~12,600	~64,000

from kerogen appears to be about 5% overall, and the ratio of accumulated to dispersed petroleum is about 1/300. The process is more efficient than these data suggest, however, because most of the reservoired bitumen is from a few stratigraphic intervals of good petroleum source rocks instead of from all the sediments. Thus, most of the North Sea oil is from the Kimmeridge Shale. The Williston Basin, United States, oil is from the Bakken Shale, and the heavy oils of California are from the Monterey Shale. Generation efficiencies of 10 to 40% are more typical of good petroleum source rocks. Also, expulsion, accumulation, and trapping efficiencies can vary tremendously, as will be explained later. Furthermore, the distribution of good source rocks through geologic time has not been uniform.

Figure 2-3 shows an interesting comparison between the deposition of carbonate carbon and that of total organic carbon (TOC) in sedimentary rocks of

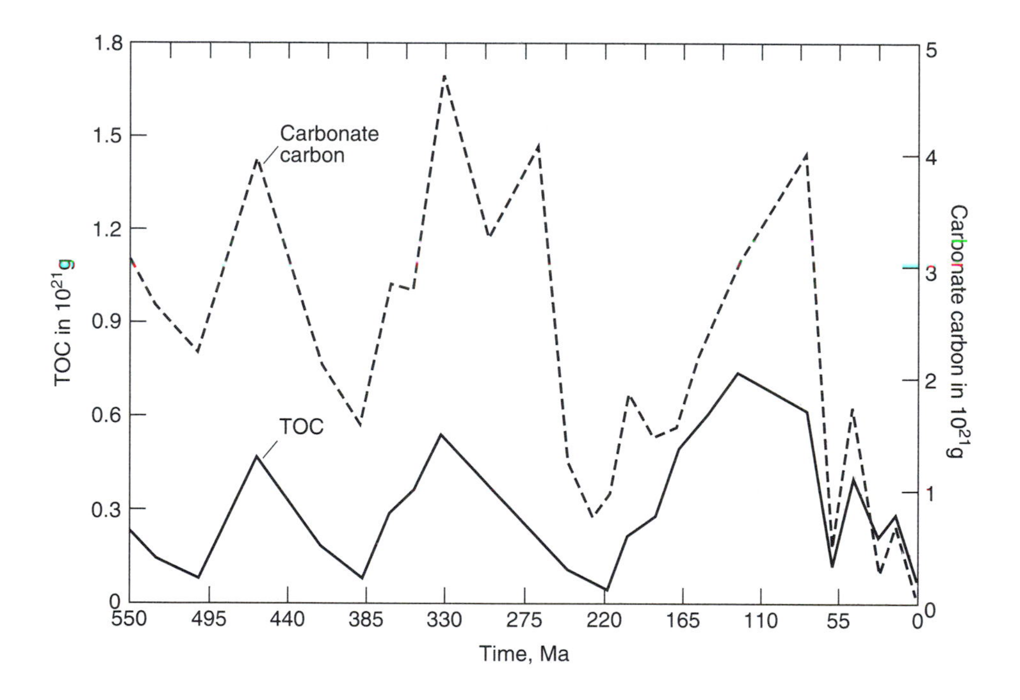

Figure 2-3

Change in the mass of total organic carbon (TOC) during the Phanerozoic compared with the mass of C as CO_2 in carbonate and noncarbonate rocks in sediments of the continents. [Ronov 1982]

the continents (Ronov 1982). The highs in this graph roughly correspond to periods of widespread deposition of good petroleum source rocks or coal beds in the case of the Carboniferous and Permian (Bois et al. 1982). They also correlate with periods of increased volcanism. Ronov believes that volcanism's periodic transport of additional deep-seated CO_2 to the surface also periodically increased the deposition of both organic and carbonate carbon in the sediments. There are other interpretations, as will be seen later in Chapters 5 and 16. Nevertheless, it is still true that more than three-fourths of the world's conventional and heavy oils were generated from Jurassic or Cretaceous source rocks, with the remainder mostly from the Tertiary and Devonian–Carboniferous periods of the Paleozoic (Bois et al. 1982). These also were the periods of highest TOC deposition, according to Figure 2-3. This does not mean that other periods produced no hydrocarbons. In fact, Bois et al. (1982) show that petroleum source rocks can be found to some extent in all Phanerozoic time periods except possibly the Triassic. The nonuniformity does show, however, that not all time periods were equivalent in depositing such rocks, just as not all rocks are equivalent in generating oil. An important objective of geochemistry is to define the areas and stratigraphic units incorporating the best petroleum source rocks in a sedimentary basin.

SUMMARY

1. The three most important events in the early evolution of the biosphere were (1) the origin of prokaryote organisms capable of forming cellular material from $CO_2 + H_2S$, (2) the origin of green plant photosynthesis capable of forming cellular material from $CO_2 + H_2O$, and (3) the origin of eukaryotes capable of evolving into all forms of higher life.

2. Unaltered or slightly metamorphosed Proterozoic rocks are prospective for petroleum accumulations originating in both Proterozoic and younger source rocks.

3. The risk of dry holes based on Proterozoic source rocks is greater than for the Phanerozoic because (1) less organic carbon was deposited in the Proterozoic compared with equivalent times in the Phanerozoic, (2) Proterozoic kerogens have low H/C ratios indicating extensive dehydrogenation, and (3) there is a greater risk that early-formed oil accumulations have been destroyed by erosion, leakage of reservoirs over geologic time, and metamorphism.

4. More than 75% of the world's conventional and heavy oils were generated from Jurassic or Cretaceous source rocks. The remainder originated mostly from the Tertiary and the Devonian–Carboniferous of the Paleozoic. These also were the periods of highest organic carbon deposition on the continents.

SUPPLEMENTARY READING

Budyko, M. I., A. B. Ronov, and A. L. Yanshin. 1985. *History of the earth's atmosphere.* Berlin: Springer-Verlag, 139 p.

Cloud, P. 1983. The biosphere. *Scientific American* 29 (3), 176–189.

Holland, H. D. 1984. *The evolution of the atmosphere and oceans.* Princeton, NJ: Princeton University Press, 582 p.

Schopf, J. W. (ed). 1983. *Earth's earliest biosphere: Its origins and evolution.* Princeton, NJ: Princeton University Press, 543 p.

Chapter

3

Petroleum and Its Products

Petroleum is a form of *bitumen* composed principally of hydrocarbons and existing in the gaseous or liquid state in its natural reservoir.* The word *petroleum* originates from the Latin *petra* ("rock") and *oleum* ("oil"). In common usage, it has come to mean any hydrocarbon mixture that can be produced through a drill pipe. Thus some of the Duchesne oils produced in the Uinta Basin, Utah, come to the surface as liquids at their reservoir temperature of about 93°C (200°F) but soon cool to solids. The main forms of petroleum are *natural gas,* which does not condense at standard temperature and pressure (STP = 760 mm Hg or 101 kPa, 60°F or 15.6°C), *condensate,* which is gaseous in the ground but condenses at the surface, and *crude oil,* the liquid part of petroleum.

Petroleum is composed almost entirely of the elements hydrogen and carbon, which is in the ratio of about 1.85 hydrogen atoms to 1 carbon atom in crude oil. The minor elements sulfur, nitrogen, and oxygen constitute less than 3% of most petroleum. Traces of heavy metals such as vanadium and nickel also are present. Table 3-1 compares the elemental composition of gas, oil, asphalt, coal, and the dispersed organic matter (kerogen) in sedimentary rocks. In Table 3-1, going across from gas to kerogen, there is a marked decrease in hydrogen and a corresponding increase in sulfur, nitrogen, and oxygen relative to carbon. The origin of petroleum from kerogen depends on many factors, but *the quantity of petroleum generated is determined mainly by*

*Definitions of the various bitumens and their fractions are given in the glossary at the end of the book.

TABLE 3-1 Elemental Composition of Fossil Fuels and Kerogen (wt%)

	Gas	*Oil*	*Asphalt*	*Coal*	*Kerogen*
Carbon	76	84.5	84	83	79
Hydrogen	24	13	10	5	6
Sulfur	0	1.5	3	1	5
Nitrogen	0	0.5	1	1	2
Oxygen	0	0.5	2	10	8
	100	100	100	100	100

the hydrogen content of the kerogen. A high-hydrogen kerogen (7 to 10% H) generates far more oil and gas than does a low-hydrogen kerogen (3 to 4% H).

Since hydrogen is a much lighter element than the other elements in Table 3-1, oils with a higher hydrogen content have lower specific gravities. Thus a Pennsylvania crude with a hydrogen content of 14.2% has a specific gravity of 0.862 (33°API), compared with a Coalinga, California, crude with 11.7% hydrogen and a specific gravity of 0.951 (17°API). The elemental analysis in Table 3-1 is about the average for oils worldwide. Some oils have much higher contents of nitrogen, sulfur, and oxygen (NSO) than are shown here. This gives the oil a higher specific gravity, since these elements are heavier than carbon or hydrogen.

The elements carbon and hydrogen are combined as hydrocarbons that vary in both size and type of molecule in crude oil. Differences in the physical and chemical properties of petroleum are due to the variations in the distribution of the different sizes and types of hydrocarbons and the percentage of NSO compounds.

Molecular Size Variation

The smallest molecule in petroleum is methane, with a molecular weight of 16. The largest molecules are the asphaltenes, with molecular weights in the tens of thousands. Between these two extremes are thousands of compounds having simple to very complex structures. Hydrocarbons form *homologous series,* that is, families of molecules whose members have similar properties and differ in size by the CH_2 group. The formula for the paraffin series is C_nH_{2n+2}, where n is any number from 1 to about 100. As the molecular size increases, the individual members change from gases to liquids to solids. In the paraffin series, n equals 1 to 4 for gases, 5 to 16 for liquids, and above 16 for solids for the straight-chain paraffins.

Petroleum is separated into its various molecular sizes by means of distillation. A typical refining tower yields products from the smallest to the largest size of molecule, as follows: gas, gasoline, kerosine, light gas oil (diesel fuel), heavy gas oil, lubricating oil, and residuum. These molecular size groups are described in more detail in the section on the uses of petroleum.

Molecular Type Variation

Hydrocarbon molecules occur in different structural forms with the following names: *Alkanes* are open-chain molecules with single bonds between the carbon atoms; *cycloalkanes* are alkane rings; *alkenes* contain one or more double bonds between the carbon atoms; and *arenes* are hydrocarbons with one or more benzene rings. Most petroleum geologists and engineers are more familiar with the terms *paraffins* for alkanes, and *naphthenes* or *cycloparaffins* for cycloalkanes, *olefins* for alkenes, and *aromatics* for arenes. Consequently, these terms will be used in this text.

In discussing molecular structures, both conventional and shorthand skeletal formulas are used, as shown in Figures 3-1 and 3-2. Figure 3-1 gives these

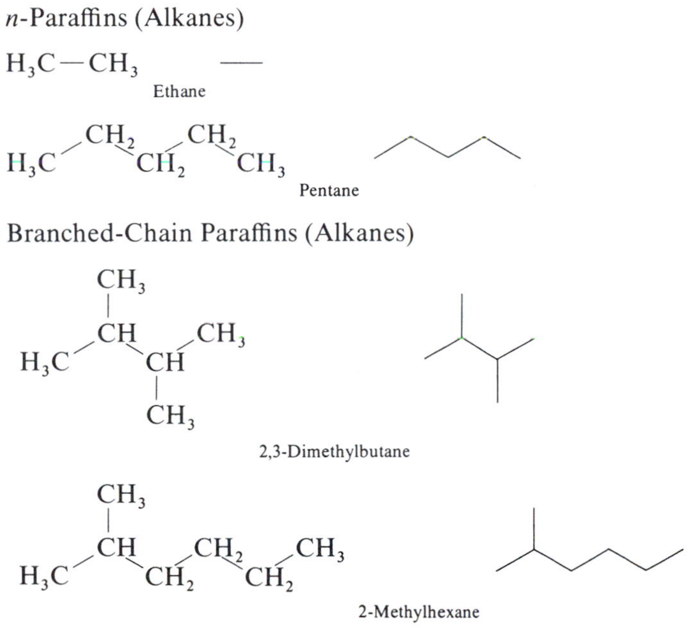

n-Paraffins (Alkanes)

$H_3C{-}CH_3$

Ethane

$H_3C{-}CH_2{-}CH_2{-}CH_2{-}CH_3$

Pentane

Branched-Chain Paraffins (Alkanes)

$H_3C{-}CH(CH_3){-}CH(CH_3){-}CH_3$

2,3-Dimethylbutane

$H_3C{-}CH(CH_3){-}CH_2{-}CH_2{-}CH_2{-}CH_3$

2-Methylhexane

Figure 3-1

Hydrocarbon formulas for normal and branched-chain alkanes. The conventional formula is on the left, and the skeletal formula is on the right.

Olefin (Alkene)

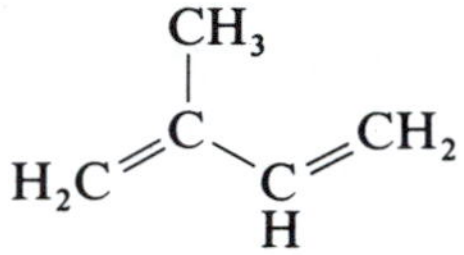

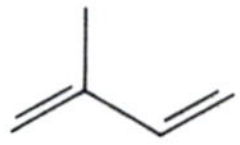

Isoprene

Naphthene (Cycloalkane)

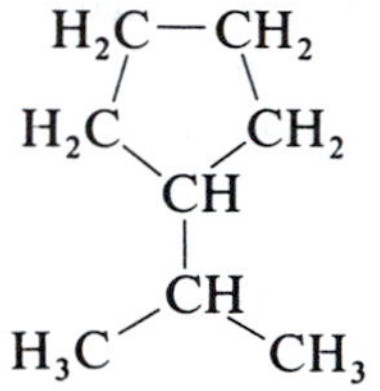

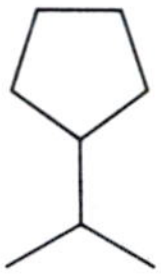

Isopropylcyclopentane

Aromatics (Arenes)

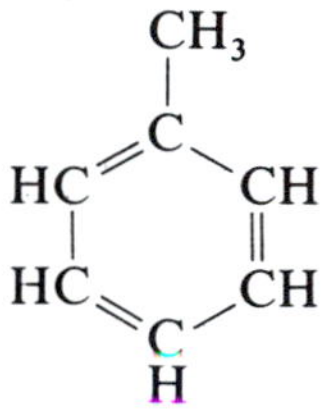

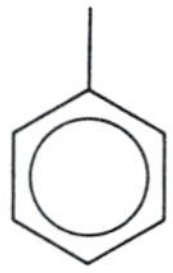

Toluene

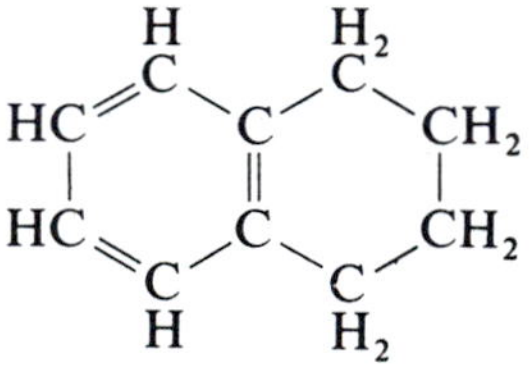

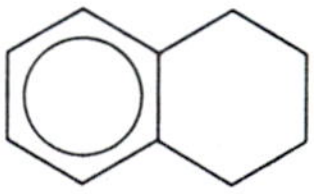

Tetralin

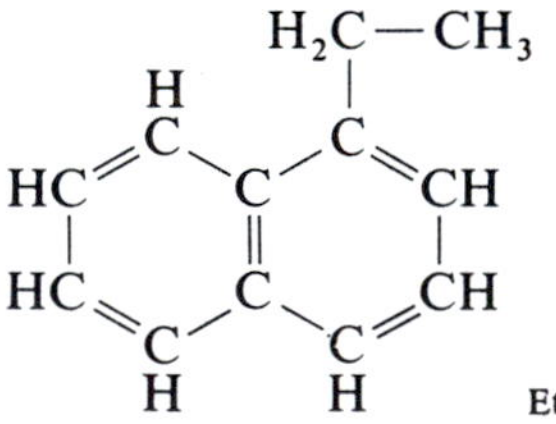

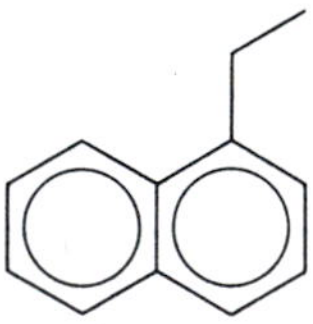

Ethylnaphthalene

Figure 3-2

Formulas for an olefin, a naphthene, and three aromatic hydrocarbons.

structures for normal and branched-chain paraffins. Figure 3-2 depicts an olefin, a naphthene (cycloparaffin), and aromatic hydrocarbons. In the skeletal formulas shown on the right, a carbon atom with enough hydrogen atoms to give a total of four bonds is implied at each corner or end of the structures.

Paraffins (C_nH_{2n+2})

Next to naphthenes, the paraffin-type hydrocarbons are the second most common constituents of crude oil. Paraffins dominate the gasoline fraction of crude oil, and they are the principal hydrocarbons in the oldest, most deeply buried reservoirs. The terms *saturated* and *aliphatic* hydrocarbons are also used for this group. The straight-chain paraffins shown in the first two examples of Figure 3-1 are called *normal paraffins,* or *n*-paraffins. The normal paraffins form a homologous series, as described earlier. The *n*-paraffins on the left side of Figure 3-3 form a homologous series, since each hydrocarbon differs from the succeeding member by one carbon and two hydrogen atoms. The members of the series are called *homologs*. Since a normal paraffin is defined as a straight chain, there are a limited number of them in crude oil—usually fewer than eighty (n = 1 to 80).* This makes them the most easily identified compounds in petroleum. All other molecular types contain hundreds of different molecules, so identification is much more difficult.

The word *paraffin* is derived from the Latin *parum affinis,* which means "of slight affinity." The normal paraffins are relatively inert with strong acids, bases, and oxidizing agents. Sulfuric acid, for example, is used to remove impurities from normal paraffins so that they may be used for medicines and as coatings for food containers. Plants began to synthesize paraffin waxes early in geologic history as coatings for seeds, spores, leaves, and other cells for protection during storage. As sediments undergo diagensis, much of the original organic matter is altered or destroyed, but the paraffin coatings last unless the rock is subjected to high-temperature metamorphism. Plants growing in dry desert areas form particularly hard waxes as coatings in order to minimize the loss of water. Consequently, many ancient sediments deposited in desert areas contain a preponderance of the paraffin-type hydrocarbon waxes.

In addition to straight chains, the paraffins can form branched-chain paraffins, as shown in Figure 3-3. Whereas only about eighty structures of straight-chain normal paraffins exist in petroleum, it is theoretically possible to have more than a million branched-chain structures, as shown in Table 3-2. This table lists the number of possible isomers representing different kinds of branching, all containing the same number of carbon atoms and corresponding to the formula C_nH_{2n+2}. Isomers are different compounds with the same molecular formula.

*A few very waxy crudes, such as from the Altamont and Bluebell fields of the Uinta Basin, Utah, contain traces of paraffin chains with up to 200 carbon atoms.

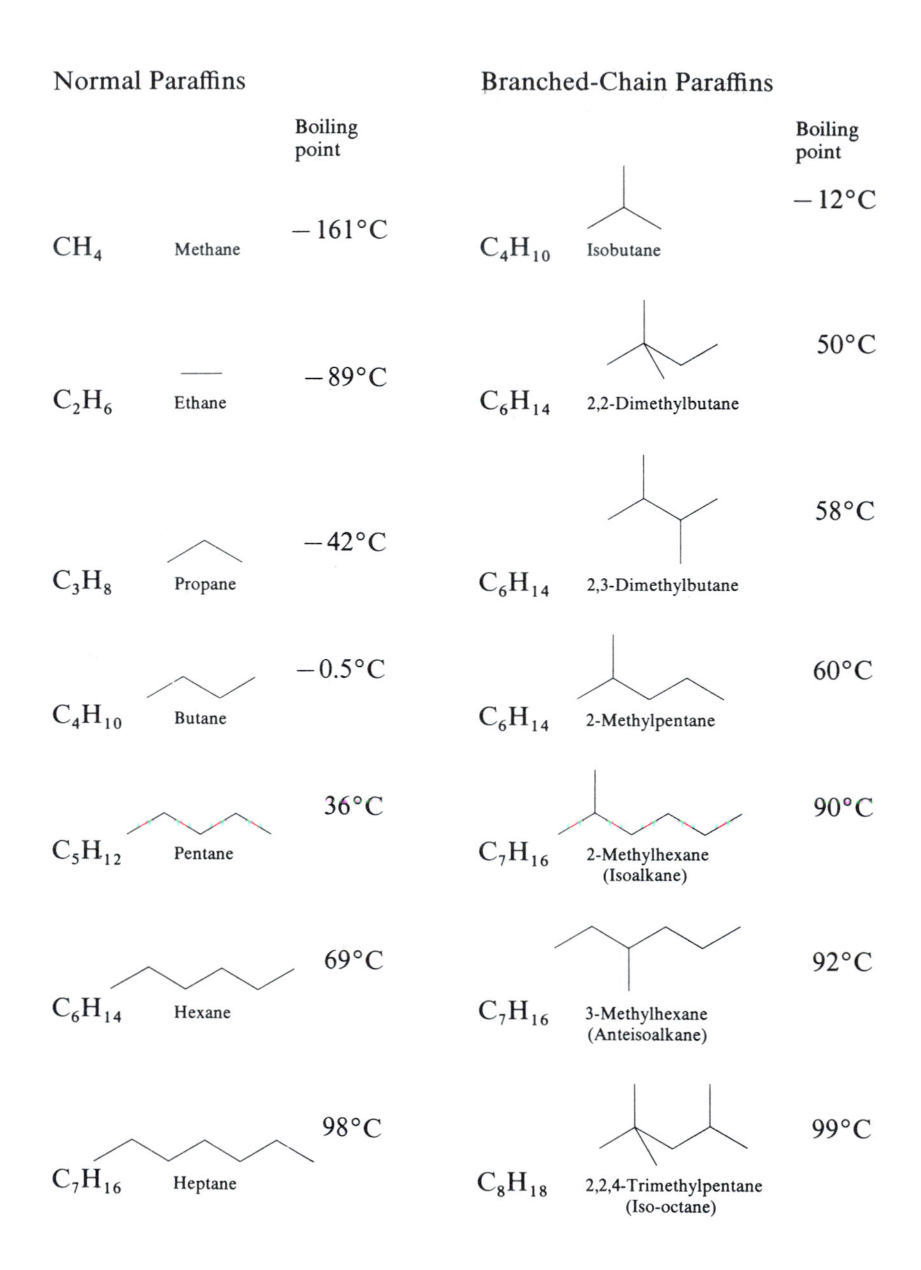

Figure 3-3

Skeletal formulas for some normal paraffin and branched-chain paraffin hydrocarbons in petroleum. Iso-octane, at the bottom of the figure, is the standard for motor fuel octane ratings.

TABLE 3-2 Possible Number of Paraffin Isomers for Each Size of Molecule

Size	*Isomers*	*Size*	*Isomers*
C_1, C_2, C_3	1 each	C_{10}	75
C_4	2	C_{11}	159
C_5	3	C_{12}	355
C_6	5	C_{13}	802
C_7	9	C_{15}	4,347
C_8	18	C_{18}	60,523
C_9	35	C_{25}	36,797,588

Fortunately, since crude oil is derived from a finite number of structures in living things, it is not as complex as the theoretical number of isomers seems to indicate. However, since a number of these isomers can form through the cracking and rearrangement of organic structures over geologic time and since there are an equally large number of isomers possible with naphthenes and aromatics, it obvious that the composition of petroleum is very complex.

The boiling point of a normal paraffin is slightly higher than that of any isoparaffin with the same molecular formula. Thus, normal heptane boils at a higher temperature than do its two isomers, shown in Figure 3-3.

Those paraffins most commonly synthesized by plants are the normal alkanes and the 2- or 3-methyl isomers. The 2-methylalkanes are sometimes referred to as the *isoalkanes* and the 3-methyl as the *anteisoalkanes.* Examples of these are on the right side of Figure 3-3.

In paraffin hydrocarbons, the covalent bonds of the carbon atom are normally at an angle of 109.5°. This is the angle of C–H bonds in methane.

Naphthenes or Cycloparaffins (C_nH_{2n})

The cycloparaffins that are formed by joining the carbon atoms in a ring are the most common molecular structures in petroleum. Naphthene rings (Figure 3-4) generally contain five or six carbon atoms, because in these ring sizes the carbon–carbon bond angles approach 109.5°. The 5-membered cyclopentane ring has bond angles of 108°. The carbon atoms lie in a plane, and the ring is not strained. The 6-membered cyclohexane ring, however, would form valence angles of 120° if it were in a plane. In order to eliminate this strain, the cyclohexane ring is actually a puckered configuration not in a plane. It is theoretically possible to form rings with more than six carbon atoms, by warping the ring further. A few cycloheptanes (C_7H_{14}) have been identified in petroleum, but no rings smaller than C_5 or larger than C_7 have been found. A few rings outside this range are formed by living things, but they have not been identified in petroleum.

The average crude oil contains about 50% naphthenes, with the quantities increasing in the heavier fractions and decreasing in the lighter fractions. In the heavier fractions, the naphthenes tend to fuse into polycyclic rings, that is, a group of rings in which two or more carbon atoms are shared among the rings. In Figure 3-4, decalin is an example. The most common naphthenes are methylcyclopentane and methylcyclohexane, which together represent 2% or more of the average crude.

The naphthenes and paraffins are also referred to as *saturated hydrocarbons* because all available carbon bonds are saturated with hydrogen. If hydrogen is removed from a paraffin, it will form one, two, or three double bonds, depending on whether two, four, or six hydrogen atoms are removed. Removing hydrogen from naphthenes forms either cyclo-olefins or aromatics. During World War II, methylcyclohexane was concentrated in refinery runs and stripped of half its hydrogen to form toluene, the starting material for trinitrotoluene (TNT).

Aromatic Hydrocarbons (C_nH_{2n-6})

The term *aromatic hydrocarbon* originated when some early, pleasant-smelling compounds such as cymene were isolated from natural fragrant oils. However, most hydrocarbons have very little odor in the pure state. The strong odor of petroleum is due to the nonhydrocarbons. All aromatic hydrocarbons contain at least one benzene ring. This is a flat 6-carbon ring (top of Figure 3-4), in which the fourth bond of each carbon atom is shared throughout the ring. For simplicity, the ring is shown with an inner circle, which indicates that the fourth bond's unpaired electrons are delocalized over all carbon atoms in the ring. The aromatics are unsaturated hydrocarbons that react to add hydrogen or other elements to the ring. The aromatics rarely amount to more than 15% of a total crude oil. They tend to be concentrated in the heavy fractions of petroleum, such as gas oil, lubricating oil, and residuum, in which the quantity often exceeds 50%. Toluene (Figure 3-4) and metaxylene are the most common aromatic hydrocarbons in petroleum. Aromatics have the highest octane ratings of the hydrocarbon types, so they are valuable in gasoline blends. However, they are undesirable in the lubricating-oil range because they have the highest change in viscosity with temperature of all the hydrocarbons.

The heavy gas oil, lubricating oil, and residuum of petroleum contain increasing amounts of polycyclic (condensed-ring) hydrocarbons (Figure 3-4). McKay and Latham (1973) found that the 335–530°C boiling range of Recluse, Wyoming, crude contains four to eight condensed-ring systems, and the higher boiling ranges appear to have even larger systems of condensed rings. It is probable that the number of rings in polycyclic molecules of residuum increases continuously up to the size of asphaltene particles (Plate 2A).

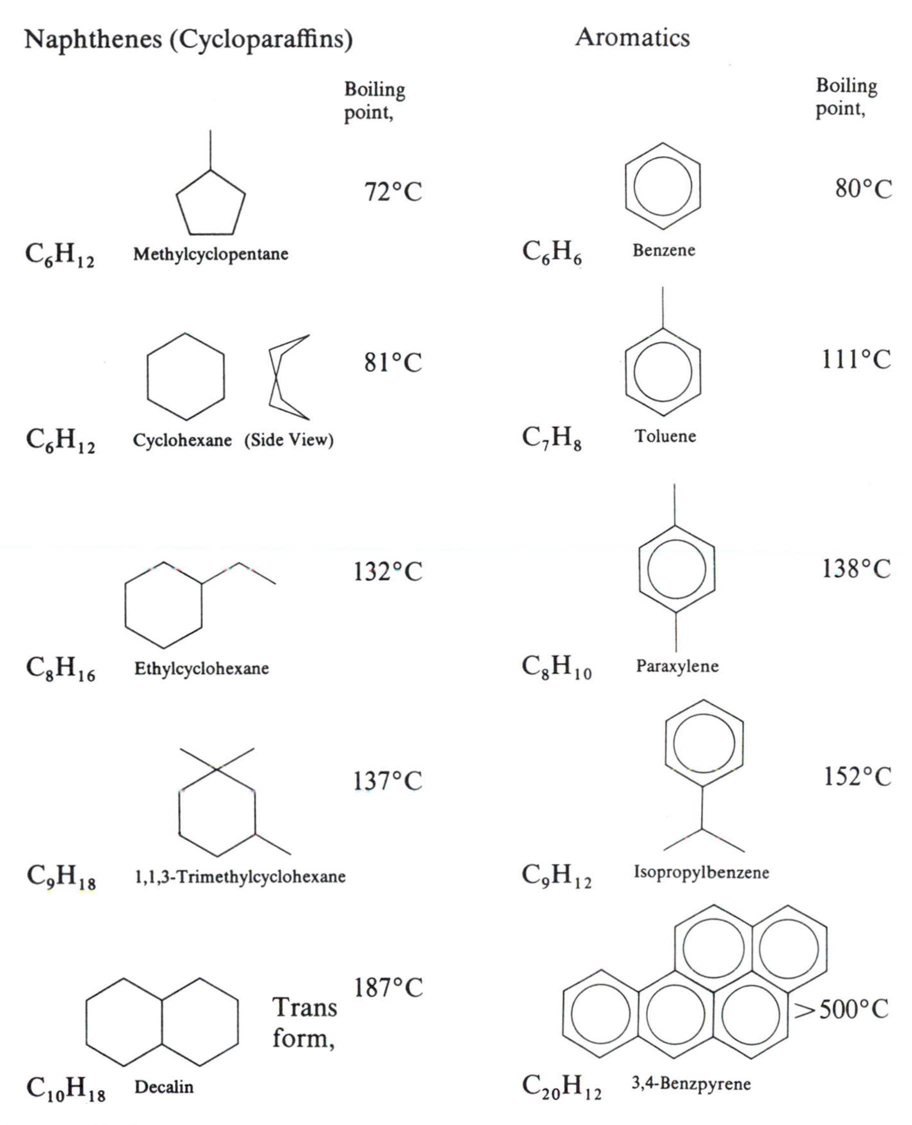

Figure 3-4

Skeletal formulas for cycloparaffin (naphthene) and aromatic hydrocarbons in petroleum. All C_6 naphthene rings are warped. Cyclohexane is shaped like a chair. Aromatic rings are flat.

Polycyclic aromatic hydrocarbons in natural products are being studied more intensively because some of them, such as 3,4-benzpyrene, 3,4benzphenanthrene, and 1,2,3,4-dibenzphenanthrene, are potent carcinogens. Besides being found in coal tars and petroleum, they are a common constituent of the burning of most organic material. The benzpyrene in coal smoke is about 300 mg/kg, in petroleum 0.5 to 2 mg/kg, in polluted urban air 100 μg/1000 m^3, in smoke from 100 cigarettes 10 to 15 μg, and in smoked meats 2 to 10 μg/kg. Nearly all smoked or burned foods, such as charcoal-broiled meats, contain carcinogens such as benzpyrene.

Olefin Hydrocarbons (C_nH_{2n-2})

Olefin hydrocarbons contain double bonds between two or more carbon atoms, as shown in Figure 3-5. This causes them to be very reactive, compared with the other hydrocarbon types. The unsaturated state of the olefin is much more unstable than that of the aromatics. If hydrogen or other elements are not available to react with the unsaturation, some olefins will react with themselves to form high-molecular-weight polymers.

Many hydrocarbons formed by plants and animals are olefins. Ethylene (C_2H_4) is the major gas formed by the ripening of fruits and vegetables. Apples, pears, tomatoes, and corn all yield ethylene upon ripening. In fact, ethylene is now used to control the ripening of bananas when they are ready to be marketed.

Fish oils and vegetable oils are high in olefins, which are believed to be useful in controlling the deposition of fatty material on human arteries. Peanut oil, olive oil, fish liver oil, and wheat germ oil all contain some olefinic hydrocarbons. Squalene, a natural component of human tissues that is an intermediate in the biosynthesis of cholesterol, is an olefin. Vitamin A and many pigments, such as the orange of carrots and the red of tomatoes, are olefins.

Olefins are uncommon in crude oil because they are readily reduced to paraffins with hydrogen or to thiols with hydrogen sulfide in the sediments. The two reactions are as follows:

$$\underset{\text{Propylene}}{H_2C{=}CH{-}CH_3} \xrightarrow{H_2} \underset{\text{Propane}}{H_3C{-}CH_2{-}CH_3}$$

$$\underset{\text{Propylene}}{H_2C{=}CH{-}CH_3} \xrightarrow{H_2S} \underset{\text{2-Propylthiol}}{H_3C{-}\overset{\overset{\large SH}{|}}{C}{-}CH_3}$$

Isoprene is a diolefin whose basic structure is one of the most important in nature (Figure 3-5). The isoprene structure appears to have been formed by the first photosynthetic organisms about 3.6 Ga. It is the basic building block for many hydrocarbon structures in living things, including the terpenes, rubber, most pigments, vitamin A, and the sterols. It is the precursor of the essential oils of flowers, fruits, seeds, and leaves. Of all the biological structures that have formed the hydrocarbons of recent sediments, isoprene is undoubtedly the most important.

Although olefins are uncommon in petroleum, they are formed in refinery processes, in which they are major starting materials for petrochemicals.

Nitrogen, Sulfur, and Oxygen Compounds (Asphaltics)

The fifth molecular type is the nonhydrocarbon, that is, compounds containing atoms of nitrogen, sulfur, or oxygen in the molecule. Although these elements are present in small amounts, they disproportionately increase the nonhydrocarbon fraction of a crude oil by being incorporated in the molecules. For example, if an asphalt were composed of a single compound having the formula $C_{30}H_{60}S$, it would contain by weight 80% carbon, 13% hydrogen, and 7% sulfur. Yet there would not be a single hydrocarbon in this asphalt. Most crude oil residua contain a high percentage of nonhydrocarbon compounds.

Small amounts of nonhydrocarbons are scattered through the entire boiling range of crude oil, and a few of these are listed in Figure 3-5. Sulfur compounds include thiols, sulfides, thiophenes, and benzothiophenes. Nitrogen compounds include pyrroles, indoles, pyridines, quinolines, and carbazoles. Oxygen compounds are mainly chain or ring acids, as shown in Figure 3-5, where R equals a straight or branched paraffin chain. Carboxylic (chain or ring) acids and phenols represent 3.5% of the Midway Sunset, California, crude oil. Seifert and Teeter (1970) identified 40 classes of carboxylic acids, including some 200 compounds, in this oil. Rall et al. (1972) identified 13 classes of sulfur compounds, including 176 individual structures in four crude oils.

Molecules in the high boiling ranges of petroleum frequently contain more than one of the molecular types just described. To avoid misunderstanding, a molecule is called *aromatic* if it contains at least one aromatic ring. It is called *naphthenic* if it contains at least one cycloparaffin ring, and *paraffinic* if it does not contain either an aromatic or a cycloparaffin ring. For example, a combination of an aromatic ring with the other two types would be called an *alkylaromatic* and *cycloalkylaromatic*.

The Composition and Uses of Petroleum

Distillation is the principal method for separating crude oil into useful products. When promoters were trying to raise money to drill the Drake well in 1859, they submitted a sample of the oil from the Titusville seep to Professor Benjamin Silliman of Yale so that he could determine its value. Silliman placed the oil in a distillation flask and boiled off eight fractions, each of which he described in detail. His results showed that the seep would make an illuminating oil that would be superior to most of the oils then available. This ensured the financing of the Drake well.

Olefins

C_2H_4 Ethylene

C_4H_8 Butylene

C_5H_8 Isoprene

C_7H_{12} Methylcyclohexene

$C_{30}H_{50}$ Squalene

Nitrogen, Sulfur, and Oxygen Compounds

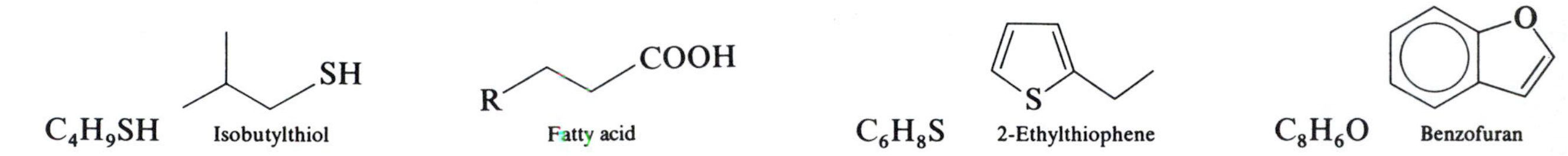

C_4H_9SH Isobutylthiol

Fatty acid

C_6H_8S 2-Ethylthiophene

C_8H_6O Benzofuran

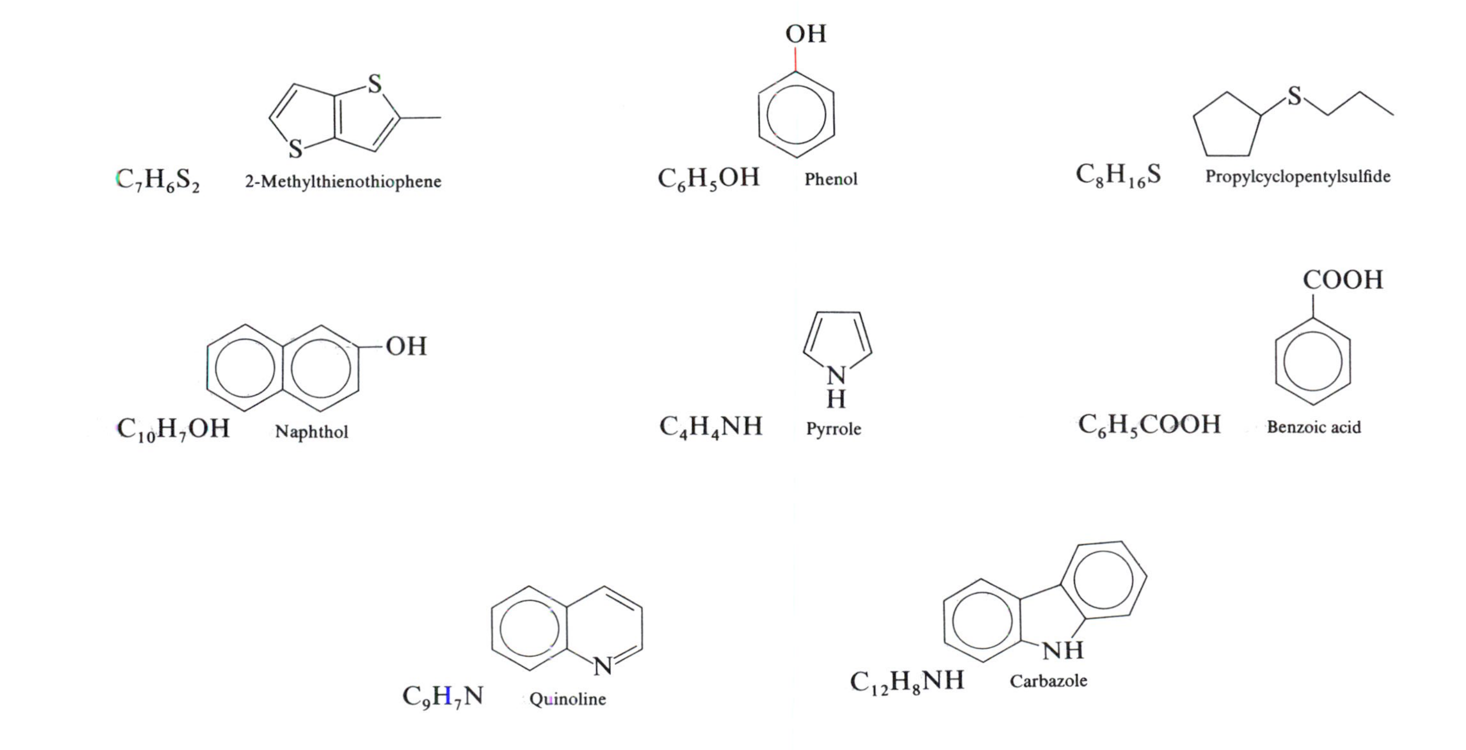

Figure 3-5

Skeletal formulas for olefin hydrocarbons and for nitrogen, sulfur, and oxygen compounds. Squalene, the precursor of sterols, is formed from six isoprene units. Heterocyclics are compounds with N, S, or O in the ring and are common in the residuum of petroleum.

Today a modern refinery distills thousands of barrels of oil a day through continuously operating distillation towers that are based on the same principle as Silliman's distillation flask. A refinery tower is equivalent to a series of individual distillation flasks, in which the distillate from the first flask is condensed in the second flask and redistilled to produce a distillate for the third flask. Instead of flasks, the tower has condensation plates, as shown in Figure 3-6. The vapor distilled from one of the chambers rises to the chamber above and passes through the condensed liquid of that chamber, as depicted in the inset of Figure 3-6. Each overlying chamber in the tower condenses successively lighter and smaller molecules, until only the light gasoline escapes from the top. At the bottom of the tower are those molecules that are so large and heavy that they cannot penetrate as gases through the first plate and so end up in the residuum. Refining towers may have different internal designs for condensing the vapors, but the efficiency of all of them is measured in terms of the number of plates, each bubble plate being the equivalent of the original distillation flask.

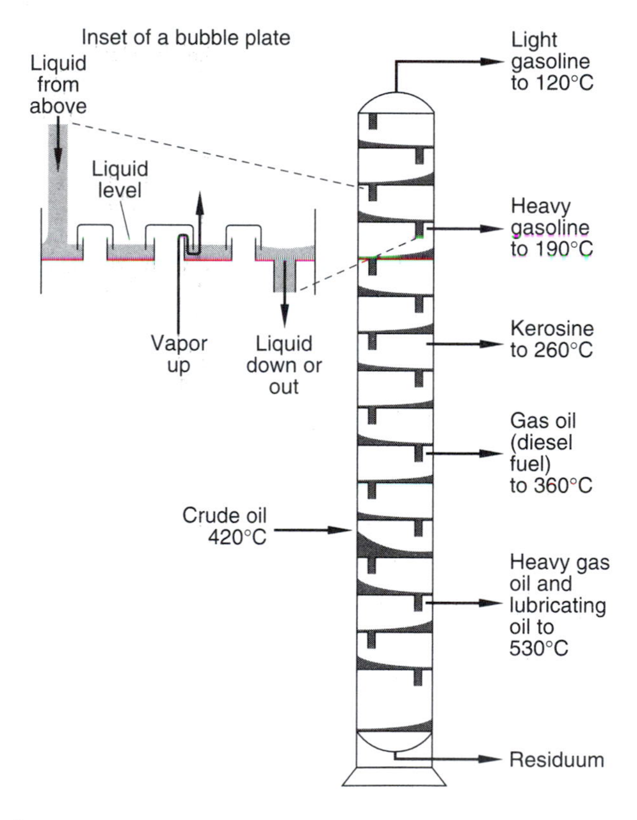

Figure 3-6

A distillation tower.

The refining tower in Figure 3-6 is run continuously by taking products out at various levels in the tower while continuously introducing fresh crude oil. The boiling ranges for the various crude oil fractions are for a typical Gulf Coast refinery. Refineries in other areas show some variation in products and boiling ranges.

The composition of a typical 35°API gravity oil is given in Table 3-3. The molecular types can vary considerably from the figures listed. The average oil tends to have more paraffins in the gasoline fraction and more aromatics and asphaltics in the residuum. However, in a highly parafffinic oil, the waxes predominate over the asphaltic compounds in the residuum. The density, °API gravity, of an oil varies with both the size and the types of molecules. Since the element carbon is heavier than hydrogen, the density of hydrocarbons generally increases with a decreasing ratio of hydrogen to carbon atoms. Thus, in Table 3-4, normal hexane, cyclohexane, and benzene have increasing density (decreasing °API) as the H/C ratio decreases. A highly paraffinic oil is lighter than an aromatic or asphaltic oil with the same molecular size distribution. However, different molecular sizes have a larger effect on gravity than do different molecular types. An oil with 50% gasoline is always lighter than one with 50% lubricating oil and residuum, irrespective of molecular type distribution. The greater density of the large molecules outweighs any type difference.

TABLE 3-3 Composition of a 35°API Gravity Crude Oil

Molecular size	*Volume percent*
Gasoline (C_5 to C_{10})	27
Kerosine (C_{11} to C_{13})	13
Diesel fuel (C_{14} to C_{18})	12
Heavy gas oil (C_{19} to C_{25})	10
Lubricating oil (C_{26} to C_{40})	20
Residuum ($> C_{40}$)	18
Total	100

Molecular type	*Weight percent*
Paraffins	25
Naphthenes	50
Aromatics	17
Asphaltics	8
Total	100

TABLE 3-4 Change in Hydrocarbon Gravity with Molecular Type

Hydrocarbon	*Molecular type*	*Formula*	*H/C atomic ratio*	*Gravity*	
				Density, d_4^{20}	°API
n-Hexane	Paraffin	C_6H_{14}	2.3	0.6594	82
Cyclohexane	Naphthene	C_6H_{12}	2.0	0.7786	50
Benzene	Aromatic	C_6H_6	1.0	0.8790	29

Figure 3-7 shows the distribution of various hydrocarbon types with boiling range (molecular size) in a naphthenic crude oil. The light-gasoline fraction of the oil is dominated by the normal, iso-, and cycloparaffins because there are only two aromatics, benzene and toluene, that boil below 130°C (266°F). Moving from gasoline into the heavier fractions of oil, there is a marked increase in the aromatic content of kerosine, after which the aromatic content increases slowly until reaching the heavy lubricating-oil range. Aromatics and NSO compounds represent about 75% of the residuum.

Even though normal paraffins and isoparaffins decrease in the heavier fractions, there are both straight and branched carbon chains attached to the naphthene and the aromatic rings. Consequently, cracking these heavier fractions of oil releases large quantities of paraffins.

The resins, waxes, and asphaltenes are not shown in Figure 3-7 because they are not distillation products. Rather, they are extracted from the heavy lubricating oil and residuum by means of solvents. The resin fraction is defined as the propane-insoluble, pentane-soluble fraction, and the asphaltenes are pentane insoluble, benzene soluble.

Rossini (1960) found that about 60% of the Ponca City crude oil was composed of only 295 compounds. Although hundreds of compounds are possible in the range up to C_{15}, only certain hydrocarbons are abundant. Bestougeff (1967) proposed designating these more common hydrocarbons as "predominant constituents," several of which are listed in Table 3-5. It is interesting that the 2- and 3-methylalkanes and pristane are the dominant isoparaffins. These are the same types of structures synthesized by living organisms. Among the biologically produced alkanes, the 2- and 3-methyl isomers usually dominate over all other structures except the normal alkanes.

The cycloparaffin rings with a single methyl group attached are more common than the unsubstituted rings. It is not unusual to find twenty times as much methylcyclopentane as cyclopentane in crude oil. The methyl-substituted aromatics (toluene and ethylbenzene) are also in higher concentration than the unsubstituted benzene ring.

The major products of petroleum are described next.

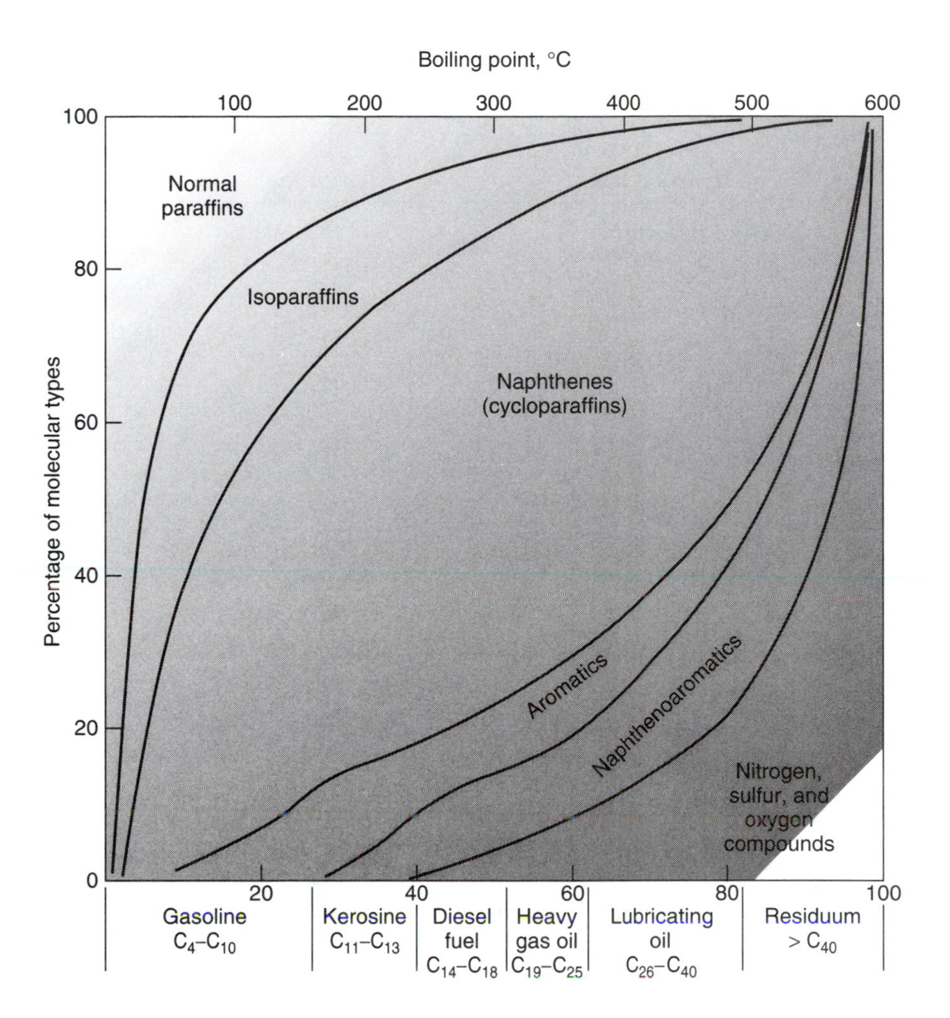

Figure 3-7

Chemical composition of a crude oil.

Gases

Gas at the wellhead usually consists of methane (CH_4) with decreasing amounts of the heavier hydrocarbons, sometimes including traces as high as nonane (C_9H_{20}). The principal nonhydrocarbon gases are nitrogen, carbon dioxide, and hydrogen sulfide. Small amounts of helium are also found in some gases. Dry gas is predominantly methane and ethane, whereas wet gas may

TABLE 3-5 Predominant Constituents of Petroleum (1 to 3% range, maximum concentration)

Hydrocarbon	*Formula*	*Maximum wt% in crude oil*
Normal paraffins		
Pentane	C_5H_{12}	3.2
Hexane	C_6H_{14}	2.6
Heptane	C_7H_{16}	2.5
Octane	C_8H_{18}	2.0
Nonane	C_9H_{20}	1.8
Decane	$C_{10}H_{22}$	1.8
Branched-chain paraffins		
2-Methylpentane	C_6H_{14}	1.2
3-Methylpentane	C_6H_{14}	0.9
2-Methylhexane	C_7H_{16}	1.1
3-Methylhexane	C_7H_{16}	0.9
2-Methylheptane	C_8H_{18}	1.0
Pristane (isoprenoid)	$C_{19}H_{40}$	1.1
Cycloparaffins (naphthenes)		
Methylcyclopentane	C_6H_{12}	2.4
Cyclohexane	C_6H_{12}	1.4
Methylcyclohexane	C_7H_{14}	2.8
1,2-Dimethylcyclopentane	C_7H_{14}	1.2
1,3-Dimethylcyclopentane	C_7H_{14}	1.0
1,3-Dimethylcyclohexane	C_8H_{16}	0.9
Aromatics		
Benzene	C_6H_6	1.0
Toluene	C_7H_8	1.8
Ethylbenzene	C_8H_{10}	1.6
m-Xylene	C_8H_{10}	1.0

Source: Data primarily from Bestougeff 1967.

contain 50% or more of propane and butanes. If the gas cap on an oil accumulation has a high content of wet gas, the oil will contain more gasoline than a field with a dry-gas cap will. Normal butane usually predominates over isobutane in the older, more deeply buried gases. If the carbon dioxide content of a gas is high, it may be used for the secondary recovery of petroleum, providing that the CO_2 can be delivered economically to nearby oil fields.

Restrictions on the emission of sulfur gases to the atmosphere have hastened the construction of plants to recover H_2S from natural and refinery gases. Hydrogen sulfide is one of the most poisonous gases known. A 0.1% concentration in air is fatal in less than 30 minutes. Drillers have died from sniffing H_2S at the wellhead. At the refinery, it is converted to sulfur as follows:

$$2\ H_2S + 3\ O_2 \rightarrow 2\ SO_2 + 2\ H_2O$$

$$2\ H_2S + SO_2 \rightarrow 3\ S + 2\ H_2O$$

The sulfur is used to manufacture sulfuric acid and other sulfur products.

Gases are taking an increasing share of the energy market from oil, and this may continue because gas is the cleanest fossil fuel. Also, many countries now require that wellhead gas be saved rather than flared, so processing plants are being built in the areas of major oil fields. Most gas used to be transported to the consumer by pipeline, but increasing amounts are now being carried as liquids in tankers. Liquefied natural gas (LNG) is primarily methane with a boiling point of –161°C (–258°F). Liquefied petroleum gas (LPG) is largely propane and butane. It can be liquefied under pressure at room temperature. The bottled-gas tanks used on farms and to run some automobiles are at pressures of about 200 psi (pounds per square inch), or 585 kg/cm^2 (kilograms per square centimeter). Because of the costs of maintaining the low temperature, LNG is much more expensive to process and transport than LPG is. Consequently, LNG shipped by tanker will continue to be used primarily for peak gas demands in urban areas until the costs become more competitive.

Ethylene, propylene, and butylene (Figure 3-5) are olefins, not present in natural gas but formed in the refinery by cracking the gas oils and heavier hydrocarbons to make gasoline. High-temperature (700–900°C or 1,292–1,652°F), low-pressure (5 psi, or 34.5 kPa) vapor-phase cracking favors the production of olefins.

Gasoline

Gasoline is composed of hydrocarbons mainly ranging from C_5 to C_{10}. From the time of the Drake well until the advent of the automobile, gasoline had no value and was usually discarded. Automobiles required an enormous increase in gasoline production. Since crude oils contain only 10 to 40% gasoline, the cracking process, which involved breaking large molecules into gasoline-sized ones at high temperatures, was developed. Also, molecules smaller than the gasoline range were polymerized to the larger gasoline size. Combined cracking and polymerization would yield as much as 70% gasoline from a barrel of crude. Later, more esoteric processes were developed, such the cyclization of paraffins to form naphthenes and the dehydrogenation of naphthenes to form aromatics. These processes reform the molecules. Today a refinery operation can be shifted to produce almost any molecular type and size range of hydrocarbons from a single crude feedstock, although each operation naturally adds to the cost of the product. Examples of these processes are shown in Figure 3-8.

Reforming hydrocarbon molecules is particularly important to producing high-octane gasolines without the use of such additives as tetraethyl lead, which is banned in most markets. High-octane gasolines prevent engine knock. When pure hydrocarbons were first tested in gasoline engines, it was discovered that iso-octane (2,2,4-trimethylpentane) caused the least knock, so it was given a rating of 100. Normal heptane, which caused the most knock, was rated 0. The knock characteristics of all hydrocarbons and fuels were then

Cracking

$$\underset{\text{Gas oil}}{C_{30}H_{60}} \rightarrow \underbrace{CH_4 + C_2H_4 + C_2H_6 + C_3H_6}_{\text{Gases}} + \underbrace{C_7H_8 + C_7H_{14} + C_8H_{18}}_{\text{Gasoline}}$$

Polymerization

$$\underset{\text{Propylene + Butylene}}{C_3H_6 + C_4H_8} \rightarrow \underset{\text{Heptene}}{C_7H_{14}} \xrightarrow{H_2} \underset{\text{Heptane}}{C_7H_{16}}$$

Alkylation

$$\underset{\text{Propylene + Butane}}{C_3H_6 + C_4H_{10}} \rightarrow \underset{\text{Heptane}}{C_7H_{16}}$$

Reforming: Dehydroisomerization

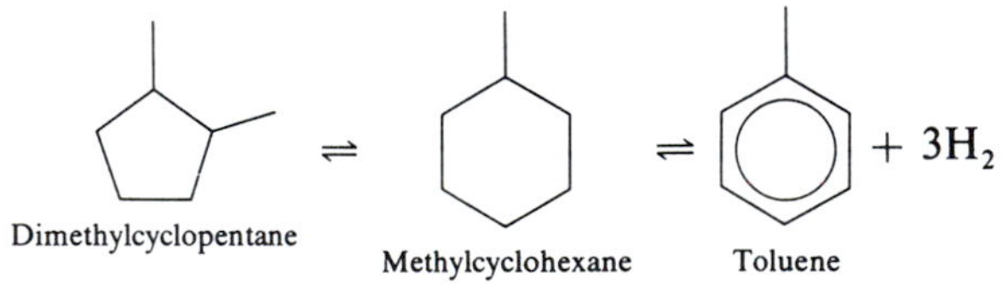

Dimethylcyclopentane $\rightleftharpoons$ Methylcyclohexane $\rightleftharpoons$ Toluene $+ 3H_2$

Reforming: Dehydrocyclization

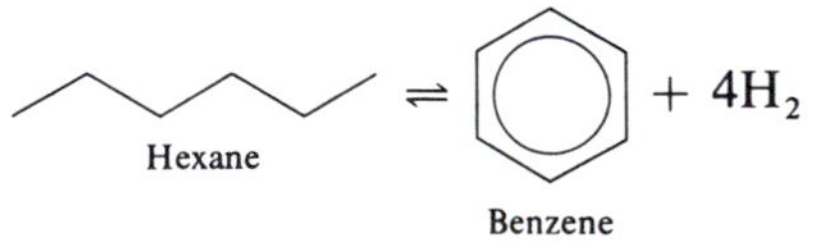

Hexane $\rightleftharpoons$ Benzene $+ 4H_2$

Figure 3-8

Refinery processes for rebuilding hydrocarbons.

rated relative to the *n*-heptane–iso-octane scale. It was later learned that those structural groups that retard oxidation had the least knock (Livingston 1951). Thus, ring compounds (aromatics and naphthenes) and highly branched paraffins would not oxidize until the temperatures were high enough for complete combustion. Long-chain paraffins would start oxidizing at a lower temperature, and later combustion would cause knock.

Today, high-octane fuels are made without chemical additives, by converting the long-chain hydrocarbons into highly branched chains and ring compounds by means of reforming processes.

Kerosine

Kerosine replaced whale oil in the lamps of the world during the late nineteenth century. It, in turn, was replaced by the gas lamp and electric light. In order to prevent kerosine from being diluted with the more dangerous gasoline, the *flash point* test was developed. The flash point is the temperature to which an oil can be heated before its vapors will flash when a flame is passed over the oil. Flash points are still determined for crude oils because they indicate the temperature below which an oil can be handled without danger of fire.

The first high-temperature cracking process was developed to increase the yield of kerosine by cracking the heavier fractions of crude oil. As the use of gasoline increased and that of kerosine diminished, the latter was cracked to make gasoline. The increased use of both jet and diesel fuels have reversed this trend, so that kerosine and light gas oils are now in heavy demand.

The kerosine fraction of crude oil (C_{11} to C_{13}) is the first to show an appreciable increase in the cyclic hydrocarbons that dominate the heavier fractions of crudes. The aromatics in kerosine range from 10 to 40%, considerably higher than for the total crude. Sachanen (1945) reported kerosine fractions of some California, Mexico, and Borneo crudes as ranging from 25 to 40% aromatics. Condensed bicyclic naphthenes and aromatics, such as tetralin and naphthalenes, are common in this range. Naphthenic acids, phenols, and thiophenes are among the nonhydrocarbons of kerosine.

Gas Oil

Light gas oils (C_{14} to C_{18}) are used in both jet fuels and diesel fuels. A diesel engine is a compression ignition engine because hot compressed air ignites the fuel. Fuel requirements for high thermodynamic efficiencies are exactly the opposite of the gasoline engine. The long-chain paraffin hydrocarbons that knock badly in spark ignition are the best fuels for the diesel engine. Cetane, normal hexadecane ($C_{16}H_{34}$), is the standard for the diesel engine, just as iso-octane is for the gasoline engine. Branched and cyclic hydrocarbons have low cetane numbers and high octane numbers, whereas long-chain hydrocarbons have high cetane numbers and low octane numbers. Consequently, a paraffin-base Pennsylvania crude oil yields an excellent diesel fuel but a poor gasoline; an aromatic-base California crude yields an excellent gasoline but a poor diesel fuel.

The composition of the gas oil fraction is given in terms of the grouping of molecular types. Robinson (1971) published a detailed mass spectral analysis of the gas oil–lube fraction of West Texas crude and was able to identify nineteen molecular compound types: four saturated, twelve aromatic hydrocarbon, and three nonhydrocarbon types. He found a marked decrease in paraffins (saturates) and an increase in aromatics in going from the diesel oil through the lube-oil range. There was also an increase in the hydrocarbons with three or more condensed rings, in both the paraffins and the aromatics, in going to the higher boiling fractions.

Gas oil also contains the sulfur compounds depicted in the following figure:

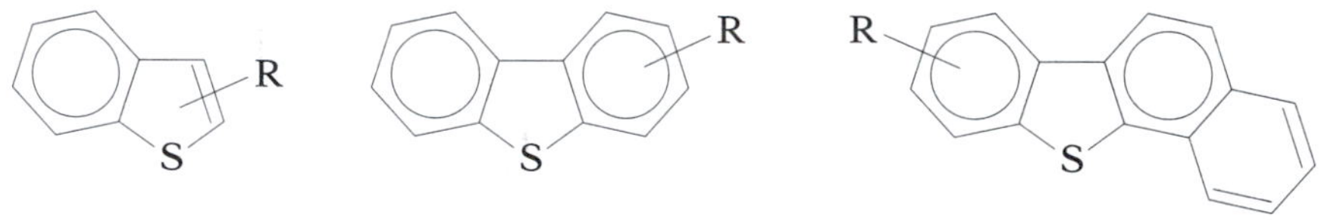

R S R S R S

Benzothiophenes Dibenzothiophenes Naphthobenzothiophenes

The letter R represents any substituent group, such as a paraffin chain.

Poirer and Smiley (1984) found that the sulfur compounds in the gasoline fraction of a Lloydminister (Canadian) crude oil were mainly thiophenes and aliphatic sulfides. The kerosene–diesel oil fraction contained mostly benzothiophenes and dibenzothiophenes.

Lubricating Oils and Waxes

Lubricating oil normally ranges from about C_{26} to C_{40}, but it can go as low as C_{20} and as high as C_{50}, depending on the distillation process. This range contains the normal paraffin waxes (C_{22} to C_{40}) and some asphaltics (NSO compounds). Highly paraffinic crude oils frequently have a high wax content in this range and a correspondingly high *pour point*. The pour point is determined by heating an oil in a tube to 46°C (115°F) to dissolve all the wax and then gradually cooling it in a bath that is held about 11°C (20°F) below the estimated pour point. The temperature at which the oil will not flow when the tube is horizontal is the pour point. Crude oils have pour points ranging from about –57 to 43°C (–70 to 110°F). The pour point is raised by straight-chain hydrocarbons and lowered by branched-chain hydrocarbons, cyclic compounds, and asphaltic substances. The pour points of lubricating oils are lowered by removing the waxes by means of solvents such liquid propane or by ketones such as methylethylketone.

Not all the waxes extracted from lubricating oil are normal paraffins. For example, Levy et al. (1961) cite a wax containing 39% *n*-paraffins, 32% isoparaffins, 27% naphthenes, and 1% aromatics. The isoparaffins and cyclic hydrocarbons tend to lower the melting point. Levy identified sixty-seven compounds in a commercial paraffin wax with a melting point of 53°C (128°F). This included about 79% *n*-paraffins, 10% 2- and 3-methylalkanes, 1.3% other branched alkanes, 9% naphthenes, and 0.2% aromatics. It is interesting that the 2- and 3-methylalkanes are the dominant branched hydrocarbons. They are also the principal branched paraffins synthesized by living things.

An important property of good lubricating oils is the change in viscosity with temperature, or *viscosity index* (VI). This index is a series of numbers ranging from 0 to 100. A VI of 100 indicates that an oil does not tend to become viscous at low temperatures or become thin at high temperatures. Paraffin-base lubricating oils containing long-chain hydrocarbons have a VI of nearly 100, whereas naphthene-base oils composed of rings have VIs around 40, with the more naphthenic aromatic oils going as low as 0. Before oil additives were widely used, the best lubricating oils came from the paraffin-base Pennsylvania crudes, and the worst lube oils came from the more aromatic California crudes. For years, Pennsylvania crude oils sold at a premium because they had a higher VI and better lubricating qualities than did other available oils. The introduction of additives that raised the VI of both Mid-Continent and California oils made them more competitive.

The color and odor of crude oil are largely caused by the nitrogen, sulfur, and oxygen (NSO) compounds concentrated in the lubricating-oil and residuum fractions. Most of the pure hydrocarbons in petroleum are colorless and odorless, but traces of the NSO compounds can impart a host of colors and odors.

Hydrodesulfurization is a process for removing the sulfur and nitrogen from crude oil. The organic sulfur compounds are decomposed to hydrogen sulfide and a hydrocarbon as follows:

$$C_4H_4S \text{ (thiophene: HC—CH, HC=CH ring closed by S)} + 4\,H_2 \rightarrow C_4H_{10} + H_2S$$

Strong acids and bases also are used to clean up oil fractions. A lubricating-oil fraction can be made completely colorless and odorless by treating it with fuming sulfuric acid until no further reaction occurs. This is the way that white oil or Nujol, a laxative, is prepared.

Residuum: Resins, Asphaltenes, and Waxes

The most complex and least understood fraction of petroleum is the residuum. The principal constituents are some of the very heavy oils, resins, asphaltenes, and high-molecular-weight waxes. The wax fraction in most residua is about half that in the lube-oil fraction. Treatment of the residuum with liquid propane at temperatures not exceeding 21°C (70°F) precipitates the resins and asphaltenes. This fraction is then treated with normal pentane, which dissolves the resins and precipitates the asphaltenes, as shown:

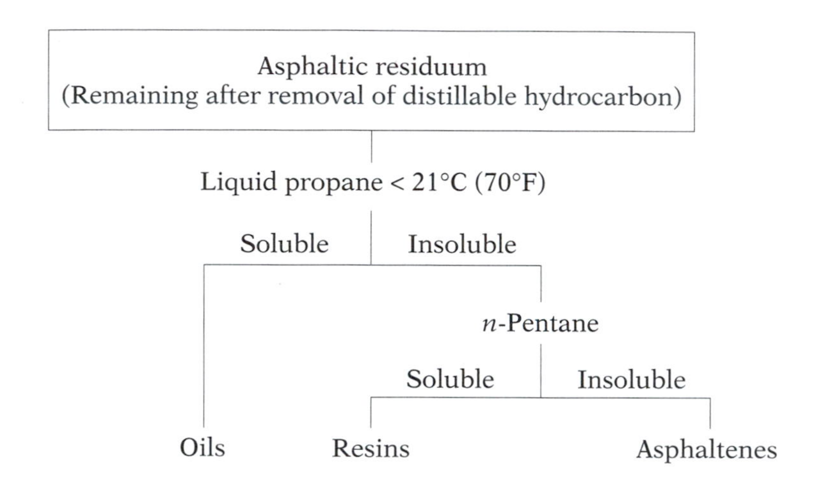

The asphaltenes are dark brown to black, amorphous solids. The resins may be light to dark colored, thick, viscous substances to amorphous solids. The resins and the asphaltenes contain about half the total nitrogen and sulfur in crude oil. Most of this is in the form of heteromolecules, which are condensed to both aromatic and naphthene rings. In going from oils to resins to asphaltenes, there are increases in molecular weight, in aromaticity, and in nitrogen, oxygen, and sulfur compounds. Heavy (low API gravity) crude oils invariably contain more nitrogen and sulfur, as shown in Figures 3-9 and 3-10. These are the approximate ranges for nitrogen and sulfur for a variety of total crudes. Oxygen follows the same trend, with residua frequently containing more than 5% oxygen.

A simple, rapid way to compare the paraffinicity or aromaticity of any fossil fuel or its fraction, such as resins and asphaltenes, is by determining the ratio of hydrogen to carbon atoms. As shown earlier in Table 3-4, the paraffin, *n*-hexane, has 14 hydrogen atoms to 6 carbon atoms, whereas the aromatic benzene has 6 to 6. The H/C ratios are 2.3 and 1, respectively. As a hydrocarbon becomes more compact, with more condensed aromatic rings having less hydrogen, the H/C ratio continues to drop. Figure 3-11 shows several hydrocarbon structures with their H/C ratios. Condensed saturated rings, such as the cycloparaffins decalin and cholestane, have lower ratios compared with a single ring. The lowest ratios occur with condensed aromatic rings such as anthracene, pyrene, and coronene. For comparison, a typical oil, an asphalt, and a coal have H/C values of 1.85, 1.5, and 0.6, respectively.

Resins have H/C ratios ranging from about 1.3 to 1.6, whereas most asphaltenes range from about 1.0 to 1.3. This would be for typical asphaltenes

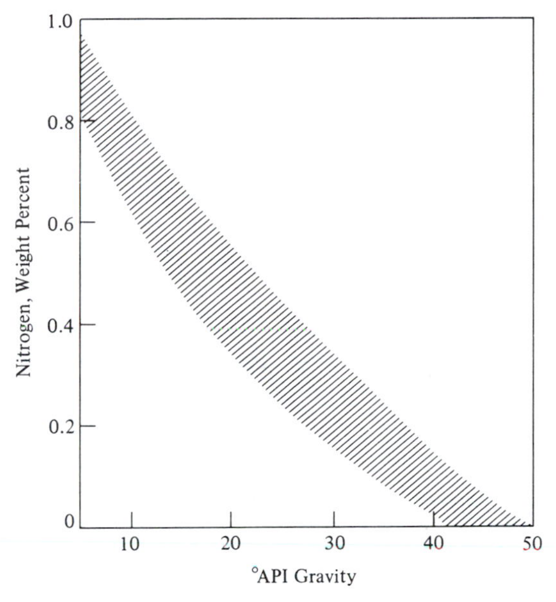

Figure 3-9

Variation in nitrogen content with °API gravity for crude oils. [Nelson 1974]

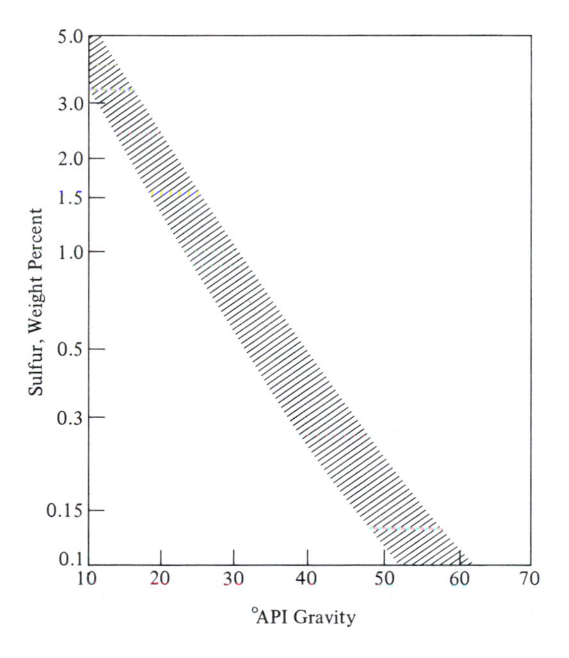

Figure 3-10

Variation in sulfur content with °API gravity for crude oils. [Nelson 1972]

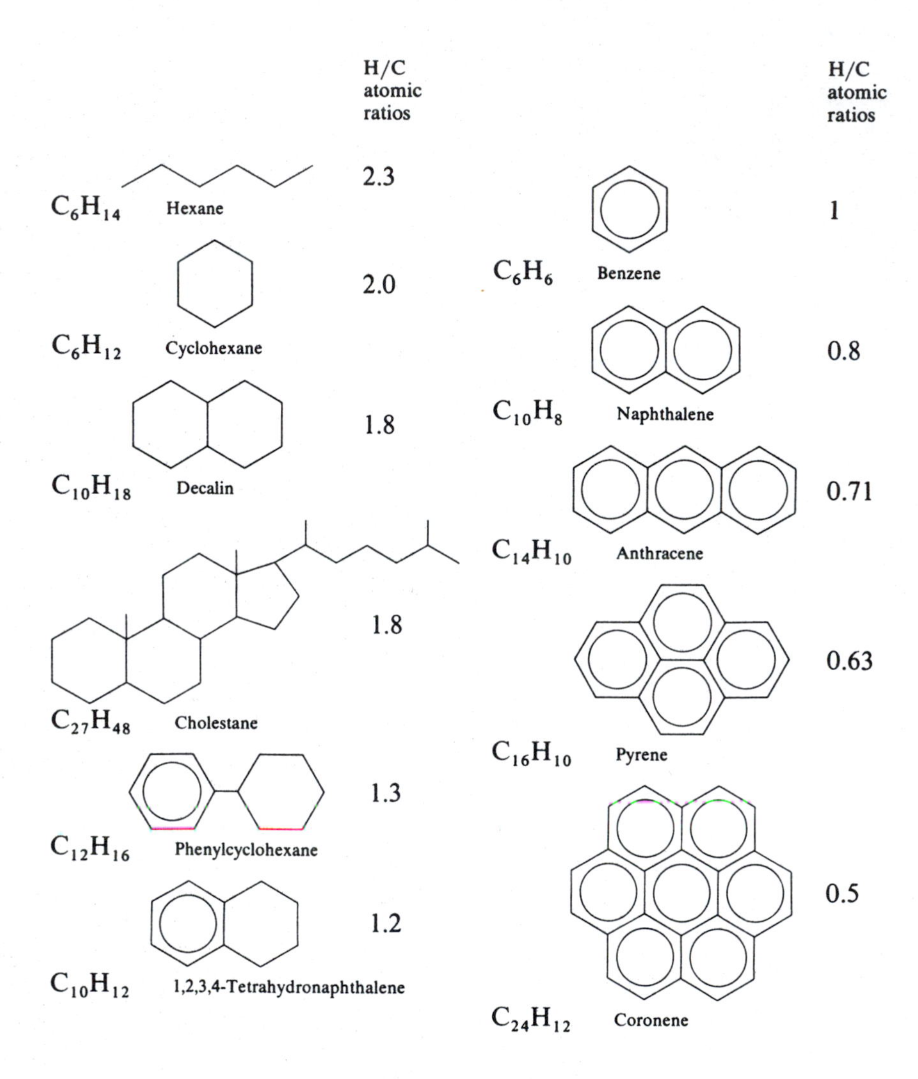

Figure 3-11

Hydrogen-to-carbon (H/C) atomic ratios shown as numbers to the right of the structures. The ratio decreases with increasing cyclization and aromaticity.

with 40 to 50% aromatic carbon atoms. Oils that originate from hydrogen-rich kerogens have higher H/C ratios in both resins and asphaltenes than do oils that originate from hydrogen-poor kerogens (Behar and Vandenbroucke 1986). H/C ratios decrease in resins and asphaltenes during an oil's maturation, since they are a major source of the hydrogen required to form light oils and condensates (Pelet et al. 1986).

The physical nature of resins and asphaltenes has been deduced from a combination of nuclear magnetic resonance (NMR) studies, X-ray diffraction, and molecular-weight distributions. Asphaltic components such as asphaltenes exist in petroleum as colloidal particles dispersed in an oily medium. As the oily medium is removed by distillation, the particles become more concentrated, to form an asphalt. At standard temperatures, asphalts are highly viscous, appearing to be solids. At high temperatures, they behave like Newtonian liquids. The viscous behavior depends on the quantity and size of asphalt particles. Reerink and Lijzenga (1973) showed that the molecular-weight distribution of asphaltene particles in Kuwaiti bitumen is very wide, ranging from 2,000 to 200,000, depending on the method of preparation. Air blowing of bitumens widened the molecular-weight distribution of the asphaltenes.

The probable structures of resins and asphaltenes have been worked out through combined NMR and X-ray studies (Bersohn 1962; Pelet et al. 1986; Winniford and Wetmore et al. 1966; Yen 1974). An asphaltene molecule in petroleum consists of groups of condensed aromatic and naphthenic rings connected by paraffin chains, which may contain sulfur or oxygen bridges. Plate 2A shows a proposed model for an asphaltene molecule with an H/C ratio of 1.33 and a molecular weight of 7,819 based on a statistical analysis of the available data (Pelet et al. 1986). Combinations of these molecules form particles. In going from light oils to dark oils to resins to asphaltenes, the asphaltene particles increase in size, with a corresponding increase in molecular weight. The range in observed molecular weights is caused by the variation in the dispersion and clustering of asphaltene particles and micelles in an asphalt.

The condensed aromatic structures of asphaltenes also contain free-radical sites. These sites contain highly reactive, unpaired electrons. Some of these sites in asphaltenes are capable of complexing metals, and Erdman (1962) demonstrated that the Boscan asphaltene is capable of complexing an additional 1,200 ppm (parts per million) of vanadium, even though it already contains 4,000 ppm. Both vanadium and nickel appear to be complexed with large molecules in the lubricating oil–residuum range. The presence of these metals even in trace amounts has adverse effects on the refinement and use of these fractions. Both vanadium and nickel tend to poison cracking catalysts. Gas turbines corrode badly with high-vanadium fuels. Fortunately, most of the vanadium and nickel in the high lubricating oil–residuum fraction is in the asphaltenes. Consequently, deasphalting is particularly important with high-vanadium crudes.

The main use of asphalt residuum is as a blend for making furnace oils and for road construction. Miscellaneous uses of asphalt include applications as binders, fillers, water-insulating materials, and adhesives in construction and manufacturing. Asphalt contents as high as 50% have been reported for some Middle East and South American crudes, although most crudes have less than 15% asphaltic constituents. The world production of straight-run asphalt is less than 3% by weight of the total petroleum production.

Gravity of Crude Oil

The specific gravity of a substance is the ratio of its mass to the mass of an equal volume of water at a specified temperature. The specific gravity of oil fractions is generally determined at 15.6°C (60°F).

Many years ago the American Petroleum Institute (API) developed a linear gravity scale to determine gravity by reading a simple hydrometer floating in oil. This API gravity was calculated to give water the value of 10°API for a specific gravity of 1 (see the formula in the Appendix). Table 3-6 compares the gravities and viscosities of petroleum fractions and water at 15.6°C (60°F). The heavier and more viscous fractions of petroleum have lower API gravities. Natural heavy oils and asphalts with API gravities of 8 to 12 are very much like the heavy ends of a refined crude oil in their physical and chemical properties. Such viscous heavy crudes must be mixed with lighter oils in order to transport them by pipeline. About 50% of the molecules in a 35°API oil are larger than C_{20} (bp = 343°C at STP). In a 25°API oil, about 75% of the molecules are larger than C_{20}.

Water has about the same viscosity as gasoline, but its gravity is comparable to that of residuum. This is because the oxygen atoms in water make it much heavier than any hydrocarbon molecules. Likewise, if nitrogen, sulfur, or oxygen atoms are introduced into a petroleum fraction, they will increase the specific gravity, thereby decreasing the API gravity. For example, Figure 3-12 shows the gravities of a series of hydrocarbons compared with NSO compounds (W. L. Orr, 1988, personal communication). Hexadecane, n-$C_{16}H_{34}$, with 16 carbon atoms, has an API gravity of 52°, whereas thiophene, with 4 carbon atoms and 1 sulfur atom, has an API gravity of about 2°. Pyridine, with 5 carbon atoms and 1 nitrogen atom, has an API gravity around 10°. Clearly, even a small percentage of NSO compounds can have a noticeable effect on the gravity of an oil fraction. NSO compounds are concentrated in the resin and asphaltene fractions of crude oil, so that when these components increase, the API gravity decreases (specific gravity increases).

TABLE 3-6 Properties of Petroleum Products and Water

	API gravity	*Specific gravity*	*Viscosity (millipoise)*
Gasoline	60	0.74	6
Kerosine	50	0.78	20
Diesel fuel	45	0.79	100
Lubricating oil	30	0.85	500
Residuum	10	1	$>10^5$
Water	10	1	10

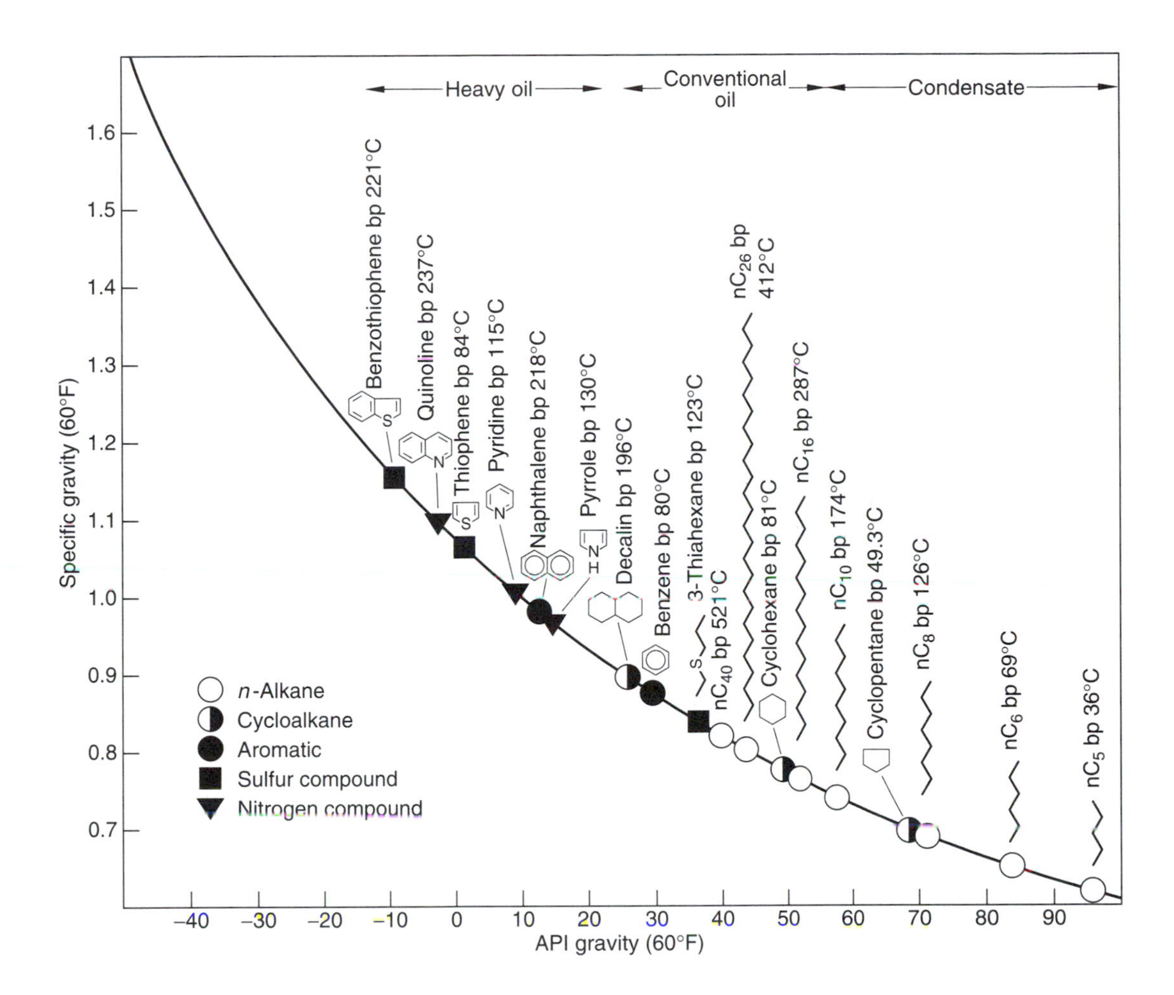

Figure 3-12

Cross plot relating API and specific gravities of individual hydrocarbons and NSO compounds in oils and condensates. [Courtesy of Wilson L. Orr] See Appendix I for equation relating API gravity to specific gravity.

It was mentioned earlier when discussing Table 3-4 that paraffins are the lightest hydrocarbons and aromatics are the heaviest, due to the difference in hydrogen content relative to carbon. This also is shown in Figure 3-12, in which tetracontane, with 40 carbon and 82 hydrogen atoms, is lighter than benzene, with 6 carbon and 6 hydrogen atoms. Solid paraffin waxes separated from lubricating oils are frequently lighter than kerosene, which usually has 20 to 40% aromatic hydrocarbons. Both the NSO concentration and the hydrocarbon type distribution cause a shift in the gravity of a distillation fraction, but gasoline is always lighter than diesel fuel because they both are mainly hydrocarbon mixtures and the gasoline molecules have more hydrogen.

The terms *light* and *heavy* also are used in marketing crude oils. The price of light, sweet crude is quoted in the newspapers. Currently, light is greater than 31.1°API gravity; medium is 22.3 to 31.1°API; heavy is 10 to 22.3°API; and extra heavy is less than 10°API gravity (Martinez 1984).

Condensates

Reservoir gases under very high pressure are able to dissolve large amounts of liquid hydrocarbons. When such gases are produced, the liquid condenses and is collected in a separator. The condensate generally has an API gravity higher than 55°. Most condensates are composed of saturated hydrocarbons in the light gasoline range (butanes, pentanes, and hexanes) which causes the high API gravities (Nelson 1978). This gives them inferior octane numbers for gasoline. Since the condensates require reforming, they sell for less than 40°API crude oil.

Six typical condensates from the U.S. Gulf Coast and western states contained 62 to 80% gasoline and were classified as paraffinic (McKinney et al. 1966). Connan and Cassou (1980) published data on twenty condensates from France, Australia, Tunisia, Switzerland, Nigeria, New Zealand, Colombia, and Indonesia. The saturates in fifteen of these ranged from 73 to 95%, of which over 30% were *n*-alkanes. About 80% of most condensates boil below 204°C (400°F), equivalent to C_{11}, although components above C_{40} have been identified. Condensates are found in nearly all oil-producing areas.

A condensate is distinguished from a light oil in that the former usually, but not always, has a gas–oil ratio greater than 5,000 cubic feet of gas per barrel of oil (Levorsen 1967, p. 462). In addition, by definition a condensate must exist in the gaseous state in its reservoir.

Katz (1983) reviewed the phase behavior of hydrocarbon mixtures in oil, gas, and condensate production. This is important to prevent retrograde condensation, that is, the formation of separate oil and gas phases due to pressure reduction in the reservoir. From 80 to 90% of the hydrocarbons in a reservoir can be produced in the gaseous state by primary production. Less than 30% can be produced in the liquid state. Retrograde condensation also occurs in pipelines and in gas storage reservoirs if pressures drop below the critical point.

Condensates with a high percentage of aromatic or naphthenic hydrocarbons have been found in basins of the Gulf Coast, Canada, Russia, and Israel (Gavrilov and Dragunskaya 1963; Hitchon and Gawlak 1972; Nissenbaum et al. 1985; Snowdon and Powell 1982). The Alberta, Canada, condensates of the Rundle and Wabamun formations averaged about 21% aromatics, with a high of 45%. The Israeli oil contained about 47% naphthenes and 4% aromatics. Such condensates would be superior to the typical paraffin type as gasoline feedstocks. These aromatic condensates are believed to form by *evaporative fractionation* during vertical migration, particularly in deltas. The origin of condensates is discussed in later chapters.

Stable Isotopes

Carbon

Nearly all the mass of an atom is in the dense nucleus as protons with a positive electrical charge of +1 and as neutrons that are electrically neutral. The protons are balanced by an equal number of electrons with charges of −1, orbiting around the nucleus. The number of protons is unique to each element and is called the *atomic number*. The sum of the masses of protons and neutrons is the *atomic weight*. Protons and neutrons have the same mass.

Atoms whose nuclei contain the same number of protons but a different number of neutrons are called *isotopes*. All carbon atoms have 6 protons, but there are 3 carbon isotopes containing 6, 7, and 8 neutrons, giving them atomic masses, respectively, of 12, 13, and 14. The distribution of these three isotopes in the biosphere is shown in Table 3-7. Carbon-12 and -13 are the original forms of carbon in the earth. Carbon-14 is formed from the bombardment of atmospheric nitrogen with neutrons produced by cosmic radiation, and it enters the biosphere as $^{14}CO_2$.

The isotopes ^{12}C and ^{13}C are stable, but ^{14}C decays to ^{14}N. One neutron (n) in a ^{14}C atom spontaneously decays, giving off an electron (β particle) and leaving a new proton (p). This produces ^{14}N, since nitrogen atoms have 7 protons in their nuclei. This reaction is ^{14}C (6 p + 8 n) — β → ^{14}N (7 p + 7 n). A mass of ^{14}C atoms disintegrates at a fixed rate, so that half the mass is changed from carbon to nitrogen in 5,570 years (the half-life of ^{14}C). Consequently, the age of a carbon-containing substance can be determined by measuring its output of β particles. This technique is good for only about five half-lives (30,000 years) because β-particle emission becomes too low in older materials to distinguish it from "background noise." The isotope ^{14}C can be used to distinguish marsh gas methane from methane seeping from a petroleum accumulation, since the

TABLE 3-7 The Distribution of ^{12}C, ^{13}C, and ^{14}C in the Biosphere

Symbol	*Protons*	*Neutrons*	*Atomic mass*	*Weight percent of carbon in biosphere*
^{12}C	6	6	12	98.89
^{13}C	6	7	13	1.11
^{14}C	6	8	14	1×10^{-11}

latter is too old to contain any ^{14}C. It also has been used to date hydrocarbons found in very recent sediments.

The isotope ^{13}C is distributed through sediments of all geological ages, in contrast with ^{14}C, which is limited to very young sediments. Although it cannot be used for dating, ^{13}C can solve many geochemical problems because its difference in mass relative to ^{12}C results in fractionation by both biological and physical processes. The ratio ^{13}C to ^{12}C is determined on an isotope-ratio-mass spectrometer, using the following equation to calculate the ratio difference (δ) in parts per thousand (parts per mil, or ‰), relative to a standard.

$$\delta^{13}C = \left[\frac{(^{13}C/^{12}C)\text{ sample}}{(^{13}C/^{12}C)\text{ standard}} - 1 \right] \times 1000$$

The standard that has been used most widely in the literature over the years is a belemnite from the Peedee Formation in South Carolina. Table 3-8 shows a composition of the actual ^{13}C content as a percentage of total carbon for three materials. Modern isotope-mass spectrometry can determine the ^{13}C content with a precision of better than 1 part in 10,000. As seen in Table 3-8, it is unwieldy to compare these numbers in the fourth decimal place, but it is easy to understand them in terms of the reference standard in the last column. The limestone contains 5‰ more ^{13}C than PDB does (Peedee belemnite), whereas the lipids contain 25‰ less ^{13}C than PDB does.

The common $\delta^{13}C$ ranges of some carbon reservoirs on earth relative to PDB are given in Figure 3-13. The bicarbonate of seawater is about the same as PDB. The CO_2 in the atmosphere has 7‰ less ^{13}C than PDB does. The U.S.

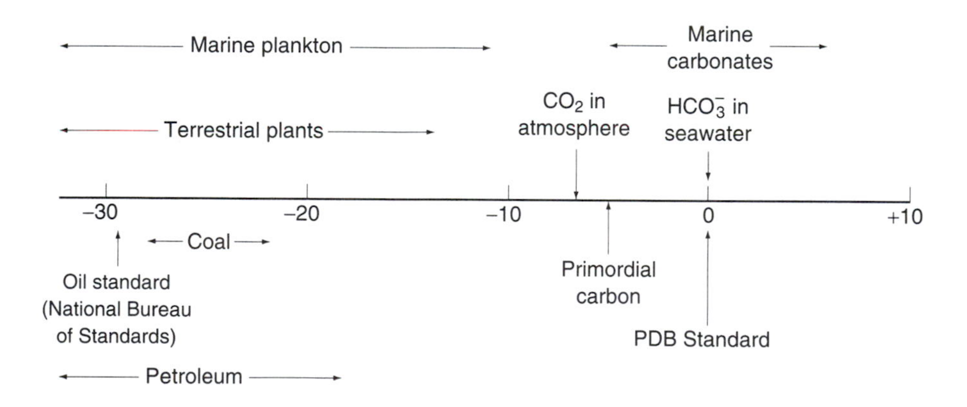

Figure 3-13

The range in the carbon-13 content of carbon reservoirs (in parts per thousand relative to the Peedee belemnite standard).

National Bureau of Standards has an oil standard with a $\delta^{13}C$ of –29.8‰ on the PDB scale (Schoell et al. 1983). Fossil organic matter falls in the range covered by marine plankton and terrestrial plants.

The value for primordial carbon was calculated by using the data on the mass of carbon in the earth's crust, from Hunt (1972) and the $\delta^{13}C$ data from Hoefs (1969). Since the quantity of ^{13}C relative to ^{12}C in the earth's crust has probably not changed over geologic time, it is possible to estimate primordial $\delta^{13}C$ values from the mean of various carbon reservoirs, assuming that there is no exchange of carbon between these reservoirs and the mantle. The mean $\delta^{13}C$ from this estimate is –5‰. This value is close to the mean for diamonds, carbonatites (Degens 1969), volcanic CO_2, and gas inclusions in igneous rocks (Galimov 1968).

Hydrogen, Sulfur, and Nitrogen

Stable isotopes of these elements also are used in petroleum geochemistry. The heavy isotopes are deuterium (^2H or D) and ^{34}S and ^{15}N whose isotopic abundances are 0.015%, 4.22%, and 0.37%, respectively. This compares with 1.11% for ^{13}C, as shown in Table 3-8. The standard for D is mean ocean water (SMOW). For ^{34}S it is FeS (troilite) from the Canyon Diablo meteorite (CD), and for ^{15}N it is atmospheric nitrogen. In each case the isotope content of a sample is reported as the ratio of the heavy isotope to the light isotope, compared with a standard, as given in the previous equation for $\delta^{13}C$.

Petroleum components are greatly depleted in deuterium relative to SMOW. Crude oil values for δD range from –60‰ to –180‰, and gases go to –300‰ (Schoell 1984). Plots of δD versus $\delta^{13}C$ have proved useful in characterizing both crude oils and gases (Schoell 1983).

In crude oils, $\delta^{34}S$ ranges from about –7.5‰ to +25‰ relative to the CD standard (Orr 1986; Vredenburgh and Cheney 1971). Orr (1977) used $\delta^{34}S$ to evaluate mechanisms of formation of H_2S and its relationship to associated sulfates. The $\delta^{15}N$ of petroleum varies from about –8‰ to + 18‰ (Macko 1981; Stahl 1977).

The use of stable isotope ratios in solving geochemical problems is explained in subsequent chapters.

TABLE 3-8 Variation in Carbon-13 in Natural Materials

	Percent of carbon-13	*$\delta^{13}C$‰ relative to PDB*
Peedee belemnite (PDB)	1.1112	0
A typical limestone	1.1162	+5
Plankton lipids	1.0862	–25

SUMMARY

1. Petroleum has an average composition of 85% carbon, 13% hydrogen, and 2% of sulfur, nitrogen, and oxygen.

2. Petroleum molecules differ in size and type. Distillation separates petroleum into molecular groups of different sizes: gas C_1 to C_4, gasoline C_5 to C_{10}, kerosine C_{11} to C_{13}, light gas oil (diesel fuel) C_{14} to C_{18}, heavy gas oil C_{19} to C_{25}, lubricating oil C_{26} to C_{40}, and residuum $> C_{40}$. High API gravity crudes have a high gasoline and a low residuum content, whereas low API gravity oils are low in gasoline and high in residuum.

3. The different types of hydrocarbon molecules in crude oil are paraffins (alkanes with a single bond between carbon atoms), naphthenes or cycloparaffins (cycloalkanes with carbon rings), olefins (alkenes with one or more double bonds between carbon atoms), and aromatics (arenes with one or more benzene rings). Olefins with one or more double bonds are found in refinery cracking products but rarely in crude oil. Lubricating oil and residuum contain polycyclic ring hydrocarbons with multiple aromatic or mixed naphthene–aromatic rings fused together, two rings sharing one side in common.

4. Nonhydrocarbons in petroleum are composed mainly of carbon and hydrogen but also contain one or more of the elements nitrogen, sulfur, and oxygen (NSO). Small amounts of NSO compounds occur throughout petroleum, with the largest quantities in lubricating oil and residuum. Heterocyclic compounds containing N, S, or O in a ring are common in residuum.

5. Less than 3% of the output of oil refineries is used to produce more than two-thirds of the organic chemicals used in the United States for 7,000 end-use materials.

6. Petroleum waxes are extracted from the lubricating-oil fraction and residuum of high-wax crudes. Wax is predominantly *n*-paraffins and branched paraffins in the range from C_{20} to about C_{60}.

7. Petroleum asphalts are either straight-run residues from distilling crude oils or "blown" asphalts produced by air oxidation of crude residues. Asphalts contain heavy oils, resins, asphaltenes, and high-molecular-weight waxes. Asphaltenes are agglomerations of molecules with condensed aromatic and naphthenic rings connnected by paraffin chains. They have molecular weights in the thousands.

8. A crude oil becomes heavier and API gravity decreases as the percentage of aromatic and naphthenic hydrocarbons increases relative to paraffins and as the percentage of NSO compounds increases.

9. The terms *light* and *heavy* also are used in marketing crude oils. Currently, light is greater than 31.1°API gravity; medium is 22.3 to 31.1°API; heavy is 10 to 22.3°API; and extra heavy is less than 10°API gravity.

10. Condensates generally have API gravities above 55 and gas–oil ratios greater than 5,000 cubic feet of gas per barrel of oil. Condensates with some exceptions are composed of saturated hydrocarbons in the light gasoline range.

11. The stable isotopes most commonly used in petroleum research and applications are ^{13}C, D, ^{34}S, and ^{15}N. The $\delta^{13}C_{PDB}$ ranges for crude oil are approximately –18‰ to –32‰, and the values for gas extend to about –90‰.

SUPPLEMENTARY READING

Hobson, G. D. (ed.). 1984. *Modern petroleum technology.* 5th ed., vol. 1. New York: Wiley.

Neumann, H.-J., B. Paczynskaya-Lahme, and D. Severin. 1981. *Composition and properties of petroleum.* Chichester: Halstead Press. New York: Wiley, 137 p.

Petrov, A. A. 1984. *Petroleum hydrocarbons.* Berlin: Springer-Verlag, 255 p.

PART TWO

ORIGIN AND MIGRATION

Chapter

4

How Oil Forms: Natural Hydrocarbons

The *organic theory* of petroleum origin is based on the accumulation of hydrocarbons from living things plus the formation of hydrocarbons by the action of heat on biologically formed organic matter. The *inorganic hypothesis* assumes that oil forms from the reduction of primordial carbon or its oxidized form at elevated temperatures deep in the earth. The overwhelming geochemical and geological evidence from both sediment and petroleum studies of the past few decades clearly shows that most petroleum originated from organic matter buried with the sediments in sedimentary basins. This chapter documents the evidence for an organic origin. Although a few hydrocarbons in the crust may be derived from inorganic sources, the quantities are negligible compared with those from organic sources. The problems with the inorganic hypotheses are discussed in more detail by Hunt (1979, pp. 69–73), North (1985, pp. 37–40), and Bromley and Larter (1986).

The origin of petroleum follows two pathways from living material, as shown in Figure 4-1. Around 10 to 20% of the petroleum is formed directly from the hydrocarbons synthesized by living organisms or from their molecules, which are readily converted to hydrocarbons. This is the pathway on the left of Figure 4-1. Most of these early-formed hydrocarbon molecules contain more than 15 carbon atoms, and they include easily recognized biological structures. The second pathway on the right involves the conversion of the lipids (fats), proteins, and carbohydrates of living material into the organic matter (kerogen) of sedimentary rocks. When this kerogen is buried deeper at higher temperatures, it cracks to form a bitumen that breaks down further to form petroleum. Some hydrocarbons also form directly from

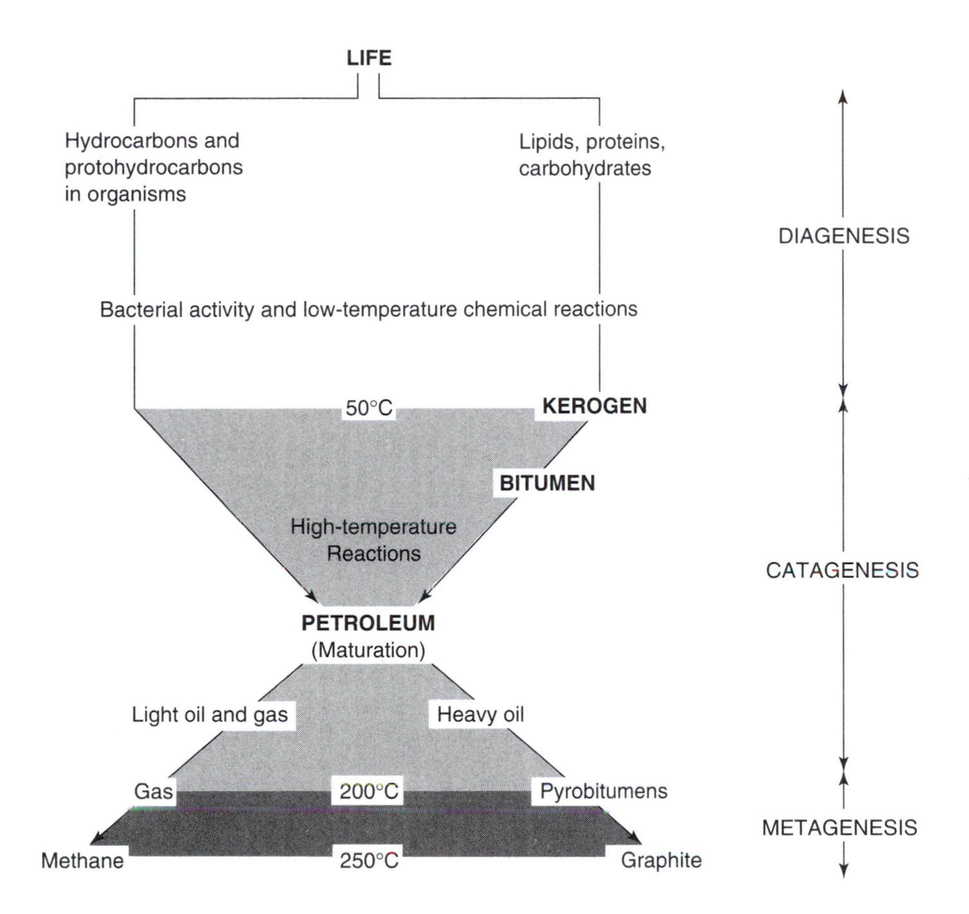

Figure 4-1

The origin and maturation of petroleum.

kerogen. If the petroleum is buried deeper at higher temperatures, it changes along two pathways, one leading to increasingly smaller hydrogen-rich molecules and the other leading to larger, hydrogen-deficient molecules. The end products are methane and graphite, as shown.

The first pathway from life simply represents an accumulation of the free hydrocarbons from dead organisms, plus hydrocarbons formed by bacterial activity and low-temperature chemical reactions in Recent unconsolidated sediments. These are the hydrocarbons that can be extracted from Recent sediments with organic solvents. In the second pathway, there are no free hydrocarbons until the kerogen is heated to a high enough temperature to crack it and release the hydrocarbons. From 80 to 90% of petroleum is formed along this pathway.

Organic *diagenesis* is the biological, physical, and chemical alteration of the organic debris before a pronounced effect of temperature. It covers the temperature range up to approximately 50°C (122°F) and is represented in Figure 4-1 by the unshaded area. The stage in which increasing temperatures cause kerogen to thermally decompose to bitumen and bitumen to oil, condensate, and gas is called *catagenesis.* The catagenesis range is from about 50°C (122°F) to 200°C (392°F). It is the lightly shaded area between these temperatures in Figure 4-1. The zone of higher temperatures, from 200 to 250°C (392 to 482°F) in which small amounts of methane continue to be formed and the remaining organic matter is converted to graphitic residues is called *metagenesis.* The end of metagenesis is approximately the beginning of *metamorphism,* in which kaolinite is converted to muscovite and the greenschist facies begins to appear. The process resulting in the formation of oil is described in detail in Chapters 4 and 5.

The Composition of Living Organisms

All living things are formed from a few, simple, molecular building blocks that have changed relatively little through geologic time. The structures formed by these basic molecules contain widely varying amounts of carbon and hydrogen relative to oxygen, nitrogen, and sulfur. Consequently, although any organic matter may contribute to the formation of oil, there are valid reasons for believing that certain compounds are the principal precursors of petroleum and that others make up the mass of residual organic matter in sedimentary rocks. The major building blocks of life are carbohydrates, proteins, lipids (fats), and lignin. The carbohydrates include sugar, starches, and cellulose, all of which are important to the maintenance of life in both plants and animals. They have the general formula $C(H_2O)_n$, with n equal to or greater than 4.

Proteins are polymers of amino acids, all of which contain an amino (NH_2) and an acid (COOH) group in their structure. They constitute more than 50% of the dry weight of animals and account for most of the nitrogen compounds in living organisms.

Lipids are biological substances insoluble in water but soluble in fat solvents such as ether, chloroform, and benzene. The term *lipid* is derived from the Greek word meaning "fat." The most common lipids are the animal fats and vegetable oils. Fats are formed by the combination of fatty acids and glycerol. Palmitic, $C_{15}H_{31}COOH$, and oleic, $C_{17}H_{33}COOH$, acids are the most common fatty acids of the animal and vegetable fats. When an organism is under stress, the percentage of fat in its body tends to increase, since it represents a source of energy if the food supply is cut off. Fats also are used for body insulation and for controlling buoyancy in marine animals.

The waxes of plants and animals are formed by the combination of high-molecular-weight alcohols and fatty acids. Waxes also include some of the uncombined higher alcohols, acids, and saturated hydrocarbons. Beeswax, for example, is formed from paraffin alcohols containing from 24 to 34 carbon atoms.

The word *sterol* is derived from the Greek *steros,* meaning "solid," and *-ol,* meaning "alcohol." Cholesterol is the best known of the solid alcohols. Sterols are widespread in nature, and their reduction products, the "biological marker" steranes, are now widely used in crude oil correlation.

The resins that bleed from the cut and damaged surfaces of tree trunks are among the most highly resistant to chemical and biological attack of all plant products. Trees also contain resins in their heartwood and on their leaf surfaces. Many of these resins are composed of unsaturated polycyclic acids that polymerize on exposure to air, forming a hard, tough layer over the wounded surface. Because of their resistance to decay over geologic time, resins (amber) containing perfectly preserved insects have been found in ancient sediments. Thomas (1969) published a detailed survey of resins from the genus *Agathis,* a group of conifers found in land areas of the South Pacific.

Lipids also include the essential oils of plants and the plant and animal pigments, many of which contain a variety of natural hydrocarbons. These are discussed in more detail in the section on biological markers.

The average chemical composition of the major constitutents of living organisms is given in Table 4-1 and compared with petroleum. It is evident that the lipids can be converted to oil by the loss of a small amount of oxygen, whereas considerable oxygen would need to be removed in order to form hydrocarbons from carbohydrates or lignin. Both oxygen and nitrogen would need to be removed from proteins. The ratio of carbon atoms to the N, S, and O atoms is approximately 1/1 in carbohydrates, 3/1 in proteins, and 10/1 in lipids. Diagenetic degradation reactions acting on equal amounts of all four of these substances in a reducing environment would produce more hydrocarbons from the lipids than from the other materials.

The quantity of lipids and other major constitutents of typical fauna and flora are shown in Table 4-2. Plants contain predominantly carbohydrates, with the higher forms containing lignin for strength. Lignins are found only in vascular land plants, not in any marine organisms. Lignin is a major precursor of

TABLE 4-1 The Average Chemical Composition of Natural Substances

	Elemental composition in weight percent				
	C	*H*	*S*	*N*	*O*
Carbohydrates	44	6			50
Lignin	63	5	0.1	0.3	31.6
Proteins	53	7	1	17	22
Lipids	76	12			12
Petroleum	85	13	1	0.5	0.5

TABLE 4-2 The Composition of Living Matter

Substance	*Weight percent of major constituents*			
Plants	Proteins	Carbohydrates	Lipids	Lignin
Spruce wood	1	66	4	29
Oak leaves	6	52	5	37
Scots-pine needles	8	47	28	17
Phytoplankton	23	66	11	0
Diatoms	29	63	8	0
Lycopodium spores	8	42	50	0
Animals				
Zooplankton (mixed)	60	22	18	0
Copepods	65	25	10	0
Oysters	55	33	12	0
Higher invertebrates	70	20	10	0

Note: Dry, ash-free basis. There is great variability for different species of each organism. For example, Blumer et al. (1964) reported *Calanus* copepods to contain 27 to 57% lipids. Lipid variability is partly due to the nutrient availability and health of the organisms. If the food supply is limited or there is crowding during growth, the organism will increase its lipids. For example, *Chlorella* grown in a favorable environment contains 20% lipids, in an unfavorable environment, 60% lipids.

humic coals. Animals are made up mainly of protein. Marine organisms such as corals and sponges also contain mostly protein in their $CaCO_3$ matrix.

The lipid content of all forms of living matter is more than enough to account for the origin of oil. As mentioned in Chapter 2, less than 1% of the organic matter deposited in sediments over geologic time is required to form all the known petroleum. Lipids are more resistant to degradation in a reducing environment than are proteins and carbohydrates. Hydrocarbons are the most stable part of the lipids. Consequently, there is generally more than 1% of hydrocarbons or protohydrocarbons in the organic matter of petroleum source rocks.

The Biological Markers

The organic compounds in sediments, rocks, and crude oils whose carbon structures, or skeletons, can be traced back to a living organism are called *biological markers* or *biomarkers*. They are micro-microfossils generally less than 30 nm in diameter and are highly variable in their stereochemistry, that is, the spatial arrangement of the atoms and groups in their molecules. Because of this variability, fossil biomarkers frequently can be linked directly to the specific group of plants, animals, or bacteria from which they originated.

Not all of the hydrocarbons following the pathway on the left of Figure 4-1 qualify as biomarkers. In addition to having a characteristic structure, a biomarker must occur in high enough concentrations in the original organism that it will be at easily detectable levels in the petroleum ultimately formed. Also, the biomarker structure must be sufficiently stable to survive the diagenetic and catagenetic processes that eventually create petroleum. For example, although proteins and many carbohydrates are formed at high concentrations, they cannot be used as biomarkers because they are largely destroyed during diagenesis. Even many biomarkers do not survive long beyond the oil-generating stage. They are most useful in following the geohistory of an oil from its origin to maturity. The rather complex molecules such as the 4- and 5-ring steranes and hopanes provide the most geohistory information. Consequently, they are widely used in crude oil maturation and correlation studies. Other important biomarkers are the *n*-paraffins, the isoprenoids such as pristane, phytane, and the carotanes and porphyrins.

Several hundred biomarkers have been identified in oils, sediments, and rocks. Nearly all of them originated along two biosynthetic pathways that have existed since the Proterozoic. The first involves the enzyme-controlled condensation of the 2-carbon acetic acid structures (CH_3CO_2H) to form long carbon chains whose lengths are in multiples of two, for example, C_{12}, C_{14}, and C_{16}. The second biosynthetic pathway involves the polymerization of isoprene, a 5-carbon building block with the structure shown in Figure 4-9. It undergoes repetitive condensations via a compound called isopentenyl pyrophosphate to form the highly branched and cyclic isoprenoids in multiples of five carbon atoms, for example, C_{10}, C_{15}, C_{20}, C_{25}, C_{30}, C_{35}, and C_{40}. Compounds composed of these five carbon units are called *terpenoids, isoprenoids,* or *isopentenoids.* The alkane–alkene nomenclature described in Chapter 3 applies to all hydrocarbon structures, including terpenoids. Thus pristane (Figure 4-8) has no double bonds (saturated), whereas pristene may have several double bonds (unsaturated).

Polyterpenoids with up to 1,000 isoprene units are represented in nature by rubbers, such as guttapercha and caoutchouc. The latex of rubber trees and the fossil rubber balls occasionally found in lignites are examples of these natural polyterpenoids.

In Chapter 3 (Table 3-2), it was mentioned that crude oil is formed from a finite number of structures in living things. Therefore it is not nearly as complex as it would be if it were abiogenic in origin. Only two structures, the 2- and 5-carbon skeletons just described, are the progenitors of a large percentage of the hydrocarbons found in petroleum.

Odd–Even Periodicity in Straight-Chain Hydrocarbons

An intriguing aspect of the biosynthesis of hydrocarbons is the ability of organisms to regulate the length of the carbon chain and the unsaturated (double-bond) carbon content. Long-chain saturated hydrocarbons are solids (waxes). Short-chain hydrocarbons are liquids. Olefin (unsaturated) hydrocarbons are liquids. Thus, the alkane $C_{21}H_{44}$ is a solid (melting point = 41°C, or 106°F), but the olefin $C_{21}H_{42}$ is a liquid (MP = 3°C, or 37°F).

Marine organisms need liquid fats (lipids) for food storage, insulation, and buoyancy in water. Land plants need solid waxes as external lipids to prevent water from leaving the organism (transpiration) and to minimize mechanical damage and inhibit fungal and insect attack on external surfaces. Consequently, marine organisms synthesize mostly liquid paraffins and olefins up to about C_{31}. Land plants synthesize hydrocarbon waxes to C_{37}. Waxes also contain esters, fatty acids, and alcohols. The waxes in crude oil come almost entirely from land plants.

The lipids of both marine and land plants contain from 1 to about 80% paraffin hydrocarbons. One of the most hydrocarbon-rich algae is the single-celled *Botryococcus braunii,* which Maxwell et al. (1968) reported can contain 76% of its dry weight as hydrocarbons. The percentage of hydrocarbons varies with the availability of nutrients during growth. Boghead coals, such as the Scottish torbanite and coorongite, consist almost entirely of colonies and remains of *B. braunii.* At the other extreme is the Brazilian palm leaf (carnauba) wax, with only 1% hydrocarbon. Animal lipids contain 1 to 10% hydrocarbons, with insect lipids having up to 75% hydrocarbons.

Odd-Numbered Normal Paraffin Chains

A significant discovery regarding biosynthesized normal paraffins was made by Chibnall and his associates in 1934. He found that plants synthesize almost exclusively paraffins with an odd number of carbon atoms in the chain. He identified odd-numbered paraffins from C_{25} to C_{37} in a variety of plants. Accompanying the paraffins were alcohols and acids with exclusively even-numbered carbon chains from C_{24} to C_{36}. Even-numbered paraffin chains and odd-numbered acids and alcohols were present, but in very small amounts (Waldron et al. 1961). Table 4-3 shows the data of Chibnall et al. (1934), Waldron et al. (1961), and others on paraffins in land plant lipids. The most dominant hydrocarbons are C_{27}, C_{29}, and C_{31}. About 90% of the paraffin fraction of apple skin, brussel sprouts, broccoli, and cabbage is a C_{29} alkane. Tree leaves and grass waxes have about equal amounts of C_{29} and C_{31} alkanes. The odd-carbon chain length strongly predominates over the entire range from C_{21} to C_{37}. The dominant hydrocarbon of the cactus leaf, which needs a hard wax of particularly high molecular weight to conserve water, is at C_{35}.

Low-molecular-weight, odd-carbon alkanes with C_7, C_9, and C_{11} chain lengths are found in the oil of certain pine trees but are not common in other plants. They also are found in Ordovician crude oils, as explained later.

Compared with land plants, marine plants synthesize smaller odd-carbon chain lengths of C_{15}, C_{17}, C_{19}, and C_{21}, as shown in Table 4-4. Furthermore, the marine plants generally contain a higher percentage of odd-numbered, straight-chain olefins having from one to six double bonds in the C_{15} to C_{21} range. The olefins, which are shown in boldface in Table 4-4, are composed

TABLE 4-3 Straight-Chain Hydrocarbons in Land Plants

Source	*Carbon atoms in chain*																
	21	*22*	*23*	*24*	*25*	*26*	*27*	*28*	*29*	*30*	*31*	*32*	*33*	*34*	*35*	*36*	*37*
Apple skin							12	2	86								
Grape skin	1	1	6	4	17	4	20	3	22	2	15	1	2				
Rose petal	3		3		3	1	18	2	21	1	33	1	11				
Palm tree leaf (carnauba)	1	1	5	2	6	1	15	3	25	1	29		6				
Sempervivoideae tree leaf			1	2	4	5	12	7	22	6	31	1	4				
Sugarcane wax			1	1	7	5	56	3	13	2	4		2				
String bean wax						1	3	2	20	3	60	2	9				
Runner bean leaf							2	2	42	2	45	2	5				
Brussels sprout									88	2	10						
Turkish tobacco leaf					2	1	12	2	11	7	34	10	19				
Cactus leaf											2	1	15	6	65		11
Sunflower seed oil	1	2	3	3	3	3	12	3	37	3	25	1					
Barley				2	7	3	16	3	28	3	27	3	5	3			
Clover			6				7	1	23	7	38	3	8				
Cocksfoot grass							1	1	35	4	49		10				
Rye grass							7	1	40	5	39		8				
Oats				2	18	2	22	2	23	2	22	1	5	1			

Note: The values shown are the percentage of total straight-chain hydrocarbons. Values less than 1 are not shown. Up to 4% of green leaves and up to 15% of dried leaves is wax, of which 1 to 80% is hydrocarbon.
Source: Data from Vandenburg and Wilder 1970 and Waldron et al. 1961 and a compilation by Clark 1966.

TABLE 4-4 Straight-Chain Hydrocarbons in Marine Plants

Plant species	*Carbon atoms in chain*																			
	14	*15*	*16*	*17*	*18*	*19*	*20*	*21*	*22*	*23*	*24*	*25*	*26*	*27*	*28*	*29*	*30*	*31*	*32*	*33*
Rhizosolenia setigera				10				86 **4**												
Eutrepiella sp.		4		19				**77**												
Ascophyllum nodosum		98																		
Corallina officinalis		14		76												1		1		
Sargassum	2	55	2	7	2	2	1	1	1	1	1	2	2	3	5	6	4	3		
Spongomorpha arcta						**17**		**82**												
Ectocarpus fasciculatus		7		1				**91**												
Pilayella littoralis								**98**												
Scytosiphon lomentaria		38		11 **20**		1 **23**					1									
Chorda tomentosa		31				**5**		**64**												
Laminaris digitata		64		2		**17**		**14**												
Ascophyllum nodosum		56 **4**		1 **29**		**2**	2	**6**												
Fucus vesiculosis		65		**2**				**16**												

Porphyra leucosticta		**15**		17		**68**												
Tribonema aequale	14	32 **54**																
Rhodymenia palmata		1		99														
Polysiphonia urceolata				96		**3**												
Coelastrum microsporum				100														
Scenedesmus quadricauda				26		7								**43**				
Tetrahedron sp.		1		30				40			3	20		6				
Anacystis nidulans		23	8	44 **20**	2													
Anacystis montana				12				**9**		**8**		**15**	**4**	**38**				
Brackish and marine coastal plants (grasses)																		
Ruppia										8	1	22	2	51	2	13		1
Diplanthera				16	4	12	2	3	1	16	1	20	1	13	1	8	1	
Syringodium				6	1	4		20	2	44	1	7		3	1	5	1	5
Halophila						14	4	23	1	20	2	36						
Thalassia				9	1	11	2	34	4	26	3	6	1	3				

Note: The values shown represent the percentage of total straight-chain hydrocarbons. Values less than 1 are not shown. Saturated paraffins are in roman type; olefins are in bold. Hydrocarbon concentrations in dry algae ranged from about 10 to 500 ppm, with the average around 200 ppm. *Tribonema aequale* is the only freshwater alga on this list. *Ruppia* grows in both fresh and saline waters.

Source: Data from Attaway et al. 1970; Blumer et al. 1971; Clark and Blumer 1967; Gelpi et al. 1970; Youngblood et al. 1971.

TABLE 4-5 Hydrocarbons in Bacteria and Higher Organisms

Source	*Carbon atoms in straight-chain hydrocarbons*										
	15	*16*	*17*	*18*	*19*	*20*	*21*	*22*	*23*	*24*	*25*
Nonphotosynthetic bacteria											
E. coli		2	6	28	12	10	6	6	8	7	6
P. shermanii	2	3	13	4	4	4	4	4	3	2	1
Clostridium acidurici	1	14	50	5	5	3	2	2	1	1	1
Desulfovibrio essex 6			2	3	16	34	26	9	3	1	1
Desulfovibrio Hildenborough			2	3	10	17	11	5	3	5	8
Photosynthetic bacteria											
Vibrio marinus	4	2	24	2							
	3	**56**									
Rhodopseudomonas spheroides	1	3	43	19	19						
Chlorofrium sulfurbacteria	2	1	50	1	1	1	2	3	4	7	11
Rhodospirillum rubrum			4		1						
Rhodomicrobium nanniclii								1	1	1	1
Higher organisms											
Zooplankton lipids[a]	11		7		**112**	**79**					
Beeswax									4	1	8
Cow manure										1	2

[a]The values are the percentage of total hydrocarbons, except for zooplankton, which is in parts per million (ppm). The percentage or ppm of olefins is shown in bold. Hydrocarbons in bacteria range from 100 to 400 ppm dry weight of cells.
Source: Data from Blumer 1967; Han and Calvin 1969; and Oro et al. 1967 and a compilation by Clark 1966.

mainly of C_{19} and C_{21} chain lengths. Blumer et al. (1970) discovered that several species of planktonic algae contain only one hydrocarbon, an olefin with six double bonds called 3,6,9,12,15,18-*heneicosahexaene* ($C_{21}H_{32}$).

Grasses growing in marine and brackish coastal waters show an odd-numbered hydrocarbon predominance intermediate between the marine algae and

										Isoprenoid hydrocarbons		
26	*27*	*28*	*29*	*30*	*31*	*32*	*33*	*34*	*35*	*Pristane*	*Phytane*	*Squalene and others*
3	3		1									
										47	1	
										2	1	
13	11	5		1						1		
										10	1	
13	2											
												95
1												**93**
										10[4]		**140**
1	30	1	17	1	19	2	16					
2	8	4	18	5	27	5	22	2	4			

the land plants. The dominant chain lengths appear to be C_{21}, C_{23}, and C_{25}, The *Ruppia,* which grows equally well in fresh and marine waters, peaks at C_{27}.

Although plants contribute most of the organic matter to sediments, it is of interest to compare hydrocarbon distributions in bacteria and higher organisms (Table 4-5). In general, the hydrocarbons of bacterial lipids show no odd-carbon predominance, although in several species the C_{17} chain length is dominant. More than half the hydrocarbon in the marine bacteria *Vibro marinus* is a C_{17} olefin.

The hydrocarbon fractions of both the marine and land plants previously discussed contain only traces of the isoprenoid hydrocarbons. In contrast, some bacteria contain mainly pristane, squalene, and related isoprenoid hydrocarbons and few straight-chain hydrocarbons. This same difference is noted in zooplankton lipids, which contain more than fifty times as much pristane and squalene as they do straight-chain hydrocarbons.

Bacteria partially modify the organic matter in sediments, but they do not seriously alter the odd-chain-length preference of land-derived hydrocarbons. Knoche and Ourisson (1967) compared the stability of this odd-carbon preference in a fresh horsetail plant (*Equisetum brongniarti*) growing today with one fossilized in the Triassic (2×10^8 years old). In both the fresh and fossil horsetail plants, the dominant hydrocarbons were C_{23}, C_{25}, C_{27}, and C_{29}. Furthermore, the relative proportions of these hydrocarbons in the living and fossil plants were almost identical.

Figure 4-2 shows the relative percentages of *n*-paraffins of different chain lengths in plant and insect waxes (Chibnall et al. 1934) and in Recent sediments of the Catalina Basin (Bray and Evans 1961). Note the strong predominance of

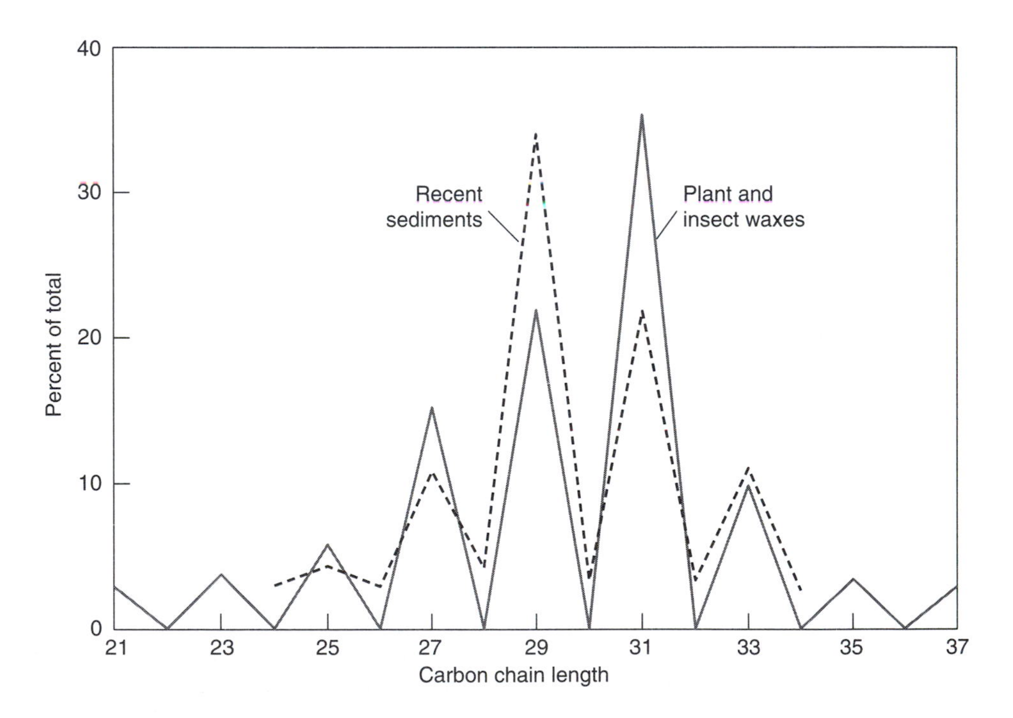

Figure 4-2

Relative percentages of *n*-paraffins of different chain lengths in recent sediments and in plant and insect waxes.

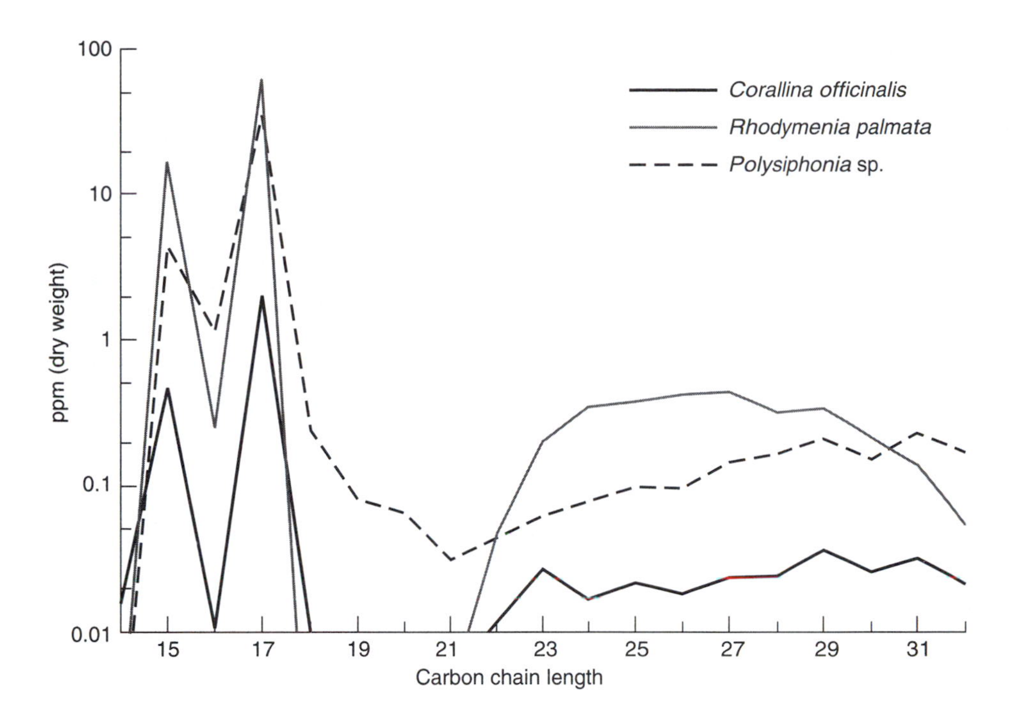

Figure 4-3

Dominant *n*-paraffin chain lengths of three species of red algae.

the odd-carbon chain length in the C_{27}-through-C_{33} range in both organisms and sediments. The even-numbered chain lengths were in too low a concentration to be detected by Chibnall's chromatographic method, but they did appear in Waldron's later mass spectral analyses (Table 4-3).

In contrast, the dominant *n*-paraffin chain lengths for three species of red algae are C_{15} and C_{17} (Figure 4-3). In the C_{27}-through-C_{31} range, in which terrestrial organisms show an odd predominance, there is no significant difference in the odd to even chain lengths of the red algae. This difference between land plants and aquatic organisms is summarized in Figure 4-4. Waxes in the high-molecular-weight range, C_{27} through C_{31}, end up in coal and petroleum that is formed from land-derived organic matter. The C_{15}-through-C_{19} hydrocarbons are found in petroleum whose source is largely aquatic organisms.

There are a few exceptions to the differences between land and aquatic organisms indicated in Figure 4-4. Some aquatic organisms synthesize unsaturated liquid straight-chain hydrocarbons in the C_{27}-to-C_{31} range. During burial and diagenesis, these alkenes can be reduced to the waxes, which are no different from those from land plants. They are more restricted in quantity, however,

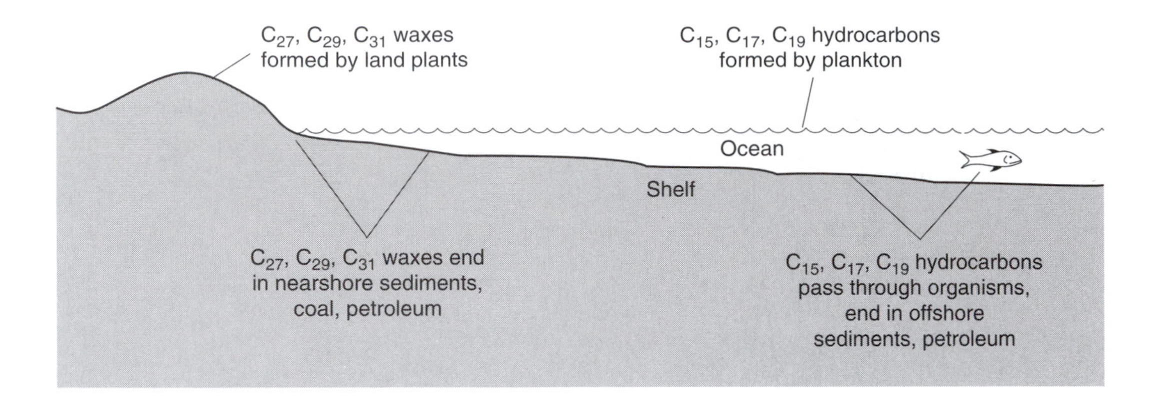

Figure 4-4

The contributions of odd-chain-length *n*-alkane molecules to sediments by land and marine organisms. [Hunt 1968]

than the vast amount of waxes contributed by land plants. Gelpi et al. (1970) identified a C_{27} mono-olefin in about a third of the hydrocarbons of green and blue-green algae but found no C_{29}-to-C_{33} straight-chain hydrocarbons. The most common aquatic alga to generate straight-chain C_{27}-to-C_{33} olefins is the previously mentioned *B. braunii,* the major constituent of many boghead coals. It exists as two physiological races, one of which generates odd-numbered unbranched alkadienes and trienes in the C_{25}-to-C_{33} range, with the other race generating mainly triterpenes and botryococcenes (Metzger et al. 1985a). Botryococcane from the alkene is shown in Figure 4-9. Botryococcane plus *n*-C_{27}-to-*n*-C_{33} hydrocarbon waxes were identified in oil seeps presumably originating from lacustrine sediments in the Otway Basin of Australia (McKirdy et al. 1986). Since *B. braunii* grows primarily in lacustrine and coastal environments, its contributions of C_{27}-to-C_{33} hydrocarbons would augment those of higher plants, as shown in Figure 4-4.

Although the odd-chain-length paraffin waxes are concentrated in nearshore deposits, they can be carried far out to sea. Spores and pollen, which have a high wax content (Table 4-2), can be transported by trade winds thousands of miles from shore. For example, Gagosian and Peltzer (1986) found a strong odd carbon chain length preference in the *n*-alkanes collected at the top of a 20-meter tower located on Enewetak Atoll in the Pacific Ocean about 5,000 km southeast of Asia.

An odd-carbon predominance in the C_9-through-C_{19} chain-length range appears to originate from a primitive prokaryote organism unique to the early Paleozoic, specifically the Ordovician. Reed et al. (1986) noted that *n*-paraffins of more than thirty Ordovician oils from various basins had an odd predominance restricted to the C_9-to-C_{19} range. The oils also were unique in having almost no pristane or phytane, suggesting that the source was a primitive bacterium that did not use chlorophyll to synthesize cellular carbon. Organic matter isolated from the presumed Ordovician source rocks contained *Gloecapsamorpha prisca,* which is the dominant algal material in kukersite, an oil shale found in Estonia. *G. prisca* is widespread in the Ordovician, occurring mainly in organic-rich lamina, indicating localized anoxic conditions. It was probably an anaerobic photoautotroph that could proliferate only in an H_2S environment. It has not been identified in sediments other than Ordovician in age. The C_9-to-C_{19} odd-carbon predominance has not been reported in post-Ordovician sediments, but a similar range may exist in the pre-Ordovician. A chromatogram published by Klomp (1986) indicates that there is a slight odd *n*-alkane predominance in the C_{13}-to-C_{23} range of a Precambrian Oman crude oil.

In their study of the hydrocarbons in Recent sediments and ancient source rocks, Bray and Evans (1965) recognized a decrease in the odd/even ratio of C_{25}-to-C_{33} *n*-paraffin chain lengths in going from Recent sediments to ancient sediments to crude oil. They calculated the carbon preference index (CPI), specifically for this C_{25}-to-C_{33} carbon range (Table 4-6). They found that the *n*-paraffins in the Gulf Coast muds had five times as much C_{25}-to-C_{33} odd-carbon molecules as even-carbon molecules. The odd/even ratio of *n*-paraffins for ancient shales was between 1 and 3, whereas in oils it was 1. They deduced that a source rock is thermally mature when it has generated enough hydrocarbons to reduce the odd/even ratio of its C_{25}-to-C_{33} *n*-paraffins to the ratio in crude oil, which is about 0.9 to 1.3. Consequently, the carbon preference index (CPI) can be used for evaluating the maturity of source rocks.

There are some problems in using it, however. One is the variability of the organic source material, because most aquatic organisms synthesize odd-carbon chains only in the low-molecular-weight range, not in the C_{25}-to-C_{33}

TABLE 4-6 Ratio of Odd-to-Even Carbon Chain Lengths of C_{24}-to-C_{33} *n*-Paraffins

Continental plants	*CPI*[a]	*Marine organisms*	*CPI*
Barley	7	Sponges	1.2
Maize	5	Coral	1.1
Tree leaves	4	Plankton	1.1
Nearshore sediments		*Deep-sea sediments*	
Basins off southern California	2.5–5.1	Cariaco Trench	1.0
Offshore Texas, Louisiana	2.6–5.5		

[a] $$\text{CPI} = \frac{\%C_{25} - C_{33}\text{ odd} + \%C_{23} - C_{31}\text{ odd}}{2(\%C_{24} - C_{32}\text{ even})}$$

Source: Data from Bray and Evans 1961; Koons et al. 1965.

range. Consequently, their CPIs are very close to 1 (Table 4-6). Sediments composed only of marine source material have a CPI of 1 at the surface and at all depths. In contrast, the CPI of continental plants ranges up to about 20, and samples with any appreciable contribution from land have CPI values considerably greater than 1. In practice, examples like the Cariaco Trench sample in Table 4-6 are rare. They require that essentially no land plant material be transported to the sediment.

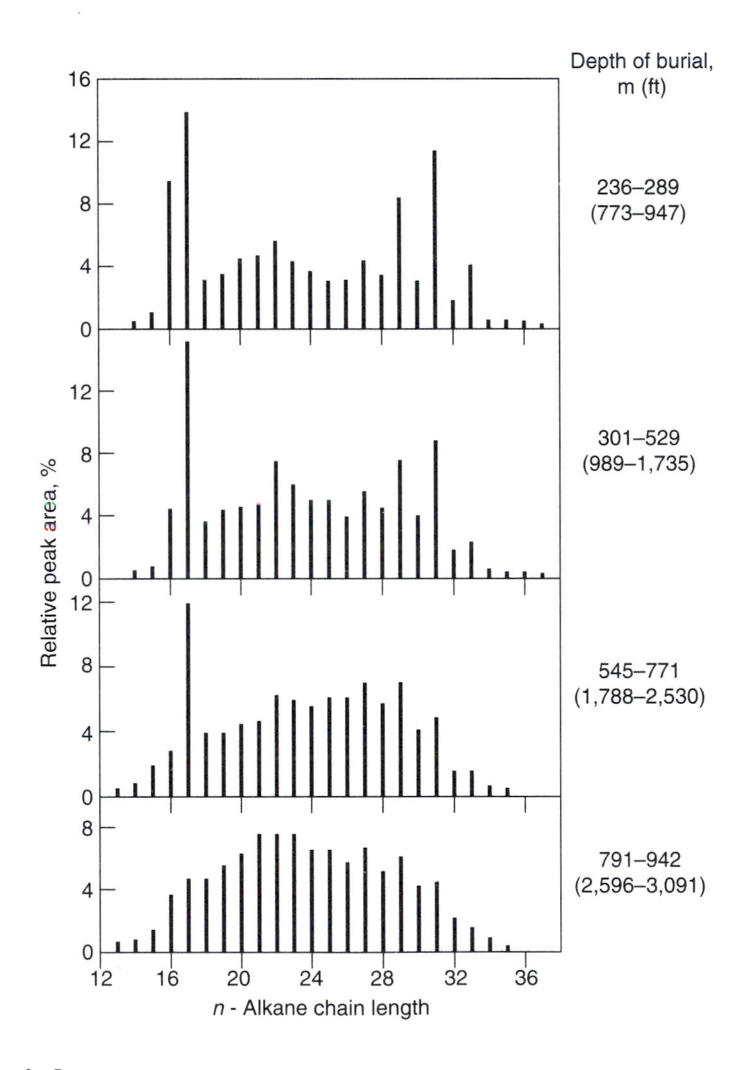

Figure 4-5

The relative concentration of *n*-alkanes in the C_{14}-to-C_{36} range of paraffins extracted from Green River oil shales of Colorado. [Anders and Robinson 1973]

The n-paraffins in many shales have an odd-carbon predominance in both the high- and low-molecular-weight ranges. For example, Figure 4-5 shows the quantity of n-alkanes extracted from four samples of the Green River oil shale of Colorado. In the most shallow sample, the C_{17} and C_{31} alkanes are dominant. The C_{17} is probably coming from lacustrine plankton or from bacteria. Many bacteria in Table 4-5 have a high content of the C_{17} alkane. The C_{27}, C_{29}, and C_{31} alkanes are from land plants.

Figure 4-5 also illustrates that the odd-carbon predominance gradually disappears with increasing depth of burial until it is barely noticeable at 914 m (3,000 ft). This is because of the thermal generation of the complete suite of alkanes from the oil-shale kerogen. The depths shown are present day rather than the maximum depths of burial.

The odd–even predominance of the n-paraffin hydrocarbons in the shales of the Green River and Wasatch Formations of the Uinta Basin is sufficiently strong to appear in the waxy oils from that basin (Figure 4-6). The CPI of 1.6 for some Uinta Basin oils is among the highest recorded for oils. Other oils in

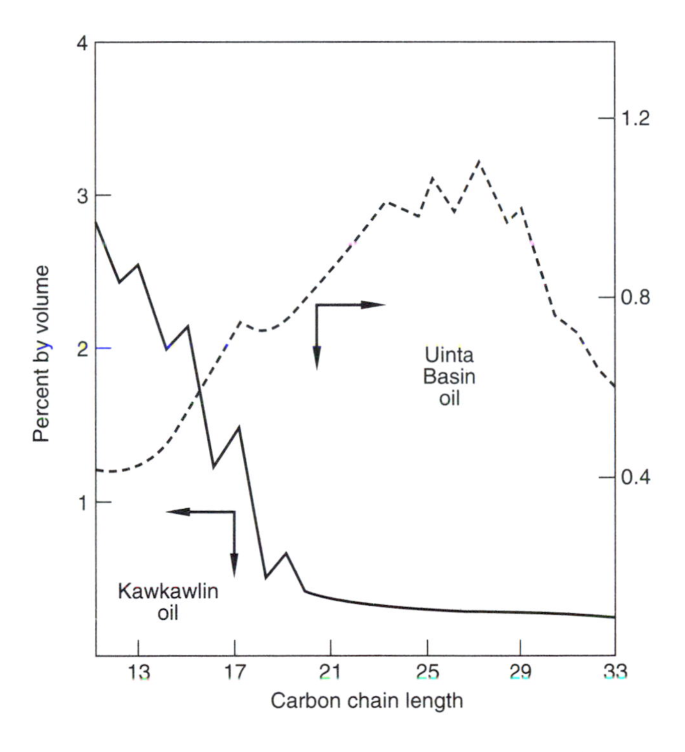

Figure 4-6

Odd–even carbon predominance in the low range for the marine Kawkawlin oil of Michigan and in the high range for the nonmarine Uinta Basin oil of Utah. [Martin et al. 1963]

this basin, such as Red Wash, have CPI values within the normal 0.9-to-1.3 range. The Kawkawlin oil in Figure 4-6 shows a predominance in the marine odd-carbon range from C_{13} to C_{19}, but this is not the range in which the CPI is calculated, as shown in Table 4-6.

A very strong odd predominance has been observed in the C_{15}-to-C_{19} range of Ordovician crude oils in the U.S. Midcontinent region (Longman and Palmer 1987; Reed et al. 1986). The ratio of C_{15}, C_{17}, C_{19}, to the C_{14}, C_{16}, C_{18} *n*-alkanes in some of these oils ranges up to 1.8. This is due to the presence of the previously mentioned *G. Prisca*. But the CPI of the oils in the C_{25}-to-C_{33} range is still close to 1.

Even-Numbered Normal Paraffin Chains

A predominance of even-chain-length *n*-alkanes in the C_{20}-to-C_{30} range is known to occur specifically in anoxic carbonate or evaporite sediments. Tissot and Welte (1984, p. 106) reported examples of even-chain-length predominances in hydrocarbons extracted from Paleocene and Eocene carbonates of south Tunisia, Upper Jurassic carbonates of the Aquitaine Basin in France, and the Zechstein evaporitic sequences of northwest Germany. Connan et al. (1986) found a strong even-carbon number predominance in *n*-alkanes from carbonate–anhydrite sediments of Guatemala. The Green River Lake of the Uinta Basin, Utah, became highly saline in late Eocene time. Siliceous dolomites were deposited with complex sodium and potassium carbonates (nahcolite). The organic matter of these rocks generated the pyrobitumen wurtzilite, which also has an even *n*-alkane chain-length predominance (Douglas and Grantham 1974).

Dembicki et al. (1975) reported an even-carbon-numbered predominance in the C_{20}-to-C_{30} *n*-alkanes in a Recent and fossil algal mat, both of which were deposited in a highly saline anoxic carbonate environment. Shen Guoying et al. (1980) found a high even-carbon number predominance in the C_{20}-to-C_{32} range of source rock extracts and crude oils of the Shahejie Formation of northern China. The source rocks were deposited in a highly saline environment containing interbedded evaporites, gypsum, and argillaceous limestone. Grabowski (1984) reported an even-carbon number predominance in extracts of the Cretaceous Austin Chalk of south central Texas.

The formation of the C_{20}-to-C_{32} even-versus-odd periodicity appears to be related to reducing conditions during diagenesis. As mentioned earlier, living organisms create mainly even-numbered carbon chain lengths. These chains always have a functional group at the end, such as an acid or alcohol. In oxic or suboxic environments, the acid or alcohol can be decarboxylated (loss of CO_2). This results in the even-carbon-numbered acid or alcohol being reduced to an odd-carbon-numbered hydrocarbon. In a strongly reducing environment, the oxygen of the acid or alcohol is removed as H_2O with no loss of a carbon atom. This results in the formation of an even-chain-length hydrocarbon. Such environments frequently are associated with highly siliceous phosphatic limestones and dolomites, like the lower Monterey Formation in California.

In some parts of the Monterey Formation, the organic matter generates heavy oils with an even-carbon-number predominance. Figure 4-7 shows the *n*-paraffin distribution for three oil samples of different API gravity from the Santa Maria Valley field (Petersen and Hickey 1987). The strong even predominance in the 8.9°API gravity oil becomes much less apparent in the 15.2°API oil. The probable reason for this is that the heavier oil is generated at an early stage of the kerogen breakdown to bitumen where the biogenic contributions dominate. As the bitumen is cracked further to higher levels of maturity, the lighter oils with less *n*-alkane periodicity are formed.

Although even-chain-length *n*-alkanes in the C_{20} to C_{32} range are found mostly in anoxic environments, this is not the case with the C_{12}-to-C_{24} range. The short even-chain-length *n*-alkanes have been found in a variety of environments: oxic, anoxic, lacustrine, and marine (Grimalt and Albaiges 1987). They are believed to be metabolic products of bacteria and other microorganisms.

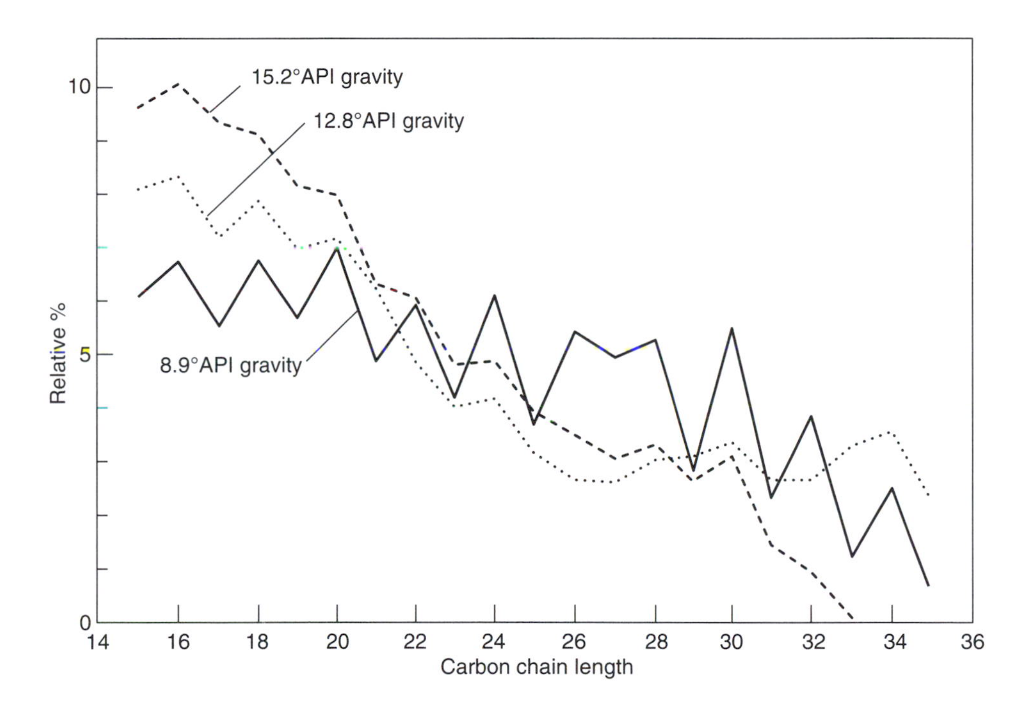

Figure 4-7

Normal paraffin distributions in three oils from Santa Maria Basin, California. [Petersen and Hickey 1987]

For example, Table 4-5 shows that the dominant *n*-alkanes in *E. coli* and in *Desulfovibrio* are C_{18} and C_{20}, respectively. Simoneit et al. (1980) found that the lipids formed by aerobes in the soil by a natural gas seep in Chile show an even predominance in the C_{18}-through-C_{24} range. Volkman et al. (1980) reported a slight even-chain-length predominance in the C_{16}-through-C_{24} *n*-alkanes of diatoms cultured under normal laboratory conditions. He attributed this to bacteria in the culture. Nishimura and Baker (1986) reported a strong even predominance in the C_{16}-through-C_{24} *n*-alkane range of organic lean nearshore clastic sediments rich in diatoms. They suggest that the *n*-alkanes were a direct metabolic product of bacteria associated with the diatoms. This concept could explain part of the even-numbered periodicity in the Monterey crude oils (Figure 4-7), since the Monterey Formation is highly diatomaceous throughout. Bacteria also are believed to form C_{20}-to-C_{35} *n*-alkanes with no odd or even predominance in many environments (Volkman et al. 1983, and references therein).

Chlorophyll

The most common pigment in plants is chlorophyll, the green coloring matter involved in the photosynthesis process, as explained in Chapter 2. Chlorophyll (Figure 4-8) is formed by joining four pyrrole rings to form the basic porphyrin structure and then by adding various carbon substituents to the ring, including the long phytol chain shown. This is an isoprenoid chain containing four 5-carbon isopentenyl skeletons for a total of twenty carbon atoms. The chlorophyll molecule contains a magnesium atom chelated to the nitrogens in the center of the ring. Chlorophyll is in the plant chloroplasts and is usually accompanied by one or more of the yellow pigments in the carotenoid group. Chlorophyll is hydrolyzed in soils and in the digestive system of animals to yield the long-chain alcohol phytol $C_{20}H_{43}OH$. Phytol can be further reduced to form the hydrocarbon phytane (C_{20}) or oxidized and decarboxylated to form the hydrocarbon pristane (C_{19}), as shown in Figure 4-8. Both of these compounds occur in petroleum and are widespread in sediments. They are important biological markers.

Hemin is the red coloring matter of animal blood. It contains the same parent porphyrin ring structure as chlorophyll does, but with iron in place of magnesium and different carbon groups on the rings in place of phytol and other substitutents of chlorophyll.

During diagenesis, chlorophyll loses its magnesium to become pheophytin a. It then follows one or another of several diagenetic pathways to form a wide spectrum of alkyl petroporphyrins in sediments and crude oils. In many petroporphyrins, nickel, copper, or the vanadyl group (V = O) fills the position in the molecule formerly occupied by magnesium. This causes a high concentration of these elements, particularly vanadium and nickel, to occur in kerogen, bitumen, and crude oils (Louda and Baker 1986). The petroporphyrins are particularly useful in correlating biodegraded oils with their source rocks or other oils, since neither the ratios of different porphyrins nor their metal contents are signifi-

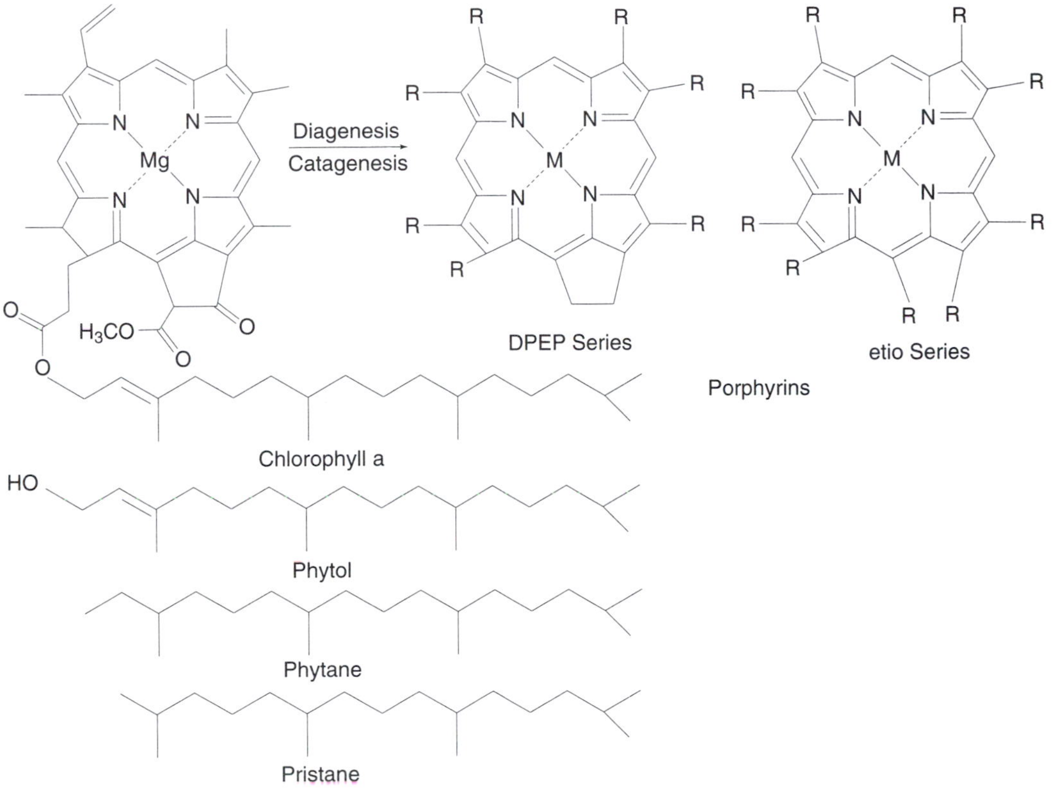

Figure 4-8

Chlorophyll and five of its derivatives: phytol, phytane, pristane, DPEP porphryins, and etio porphyrins. R = H or alkyl, M = metals such as vanadium, nickel, iron, copper, and manganese. Porphyrins containing a metal also are known as *metalloporphyrins, geoporphrins*, and *petroporphyrins*. A detailed description of the processes forming these compounds can be found in Baker and Louda 1986.

cantly altered by microbial degradation. Petroporphyrin ratios also have been used as maturity indicators. Their applications are discussed in later chapters.

The C_5 to C_{45} Acyclic Isoprenoids

The acyclic isoprenoid hydrocarbons pristane, C_{19}, and phytane, C_{20}, are widely used in crude oil and source rock correlation studies (Figure 4-8). The smaller isoprenoid molecules norpristane (C_{18}) and farnesane (C_{15}) are present in lower concentrations than are pristane and phytane, so they are rarely used in correlation. Abbreviations are sometimes used for the isoprenoids, such as Pr and Ph or ipC_{19} and ipC_{20}. Thus, farnesane would be ipC_{15}.

Pristane and Phytane: C_{19}, C_{20}

Since pristane represents a product of decarboxylation, the ratio Pr/Ph tends to be high in more oxidizing environments such as peat swamps and low in strongly reducing environments (Powell and McKirdy 1973). Organic-rich anoxic carbonate sequences generally form oils with ratios less than 2, whereas the more organic-lean lacustrine, fluvial, and deltaic sediments generate oils and condensates with ratios greater than 3. Connan and Cassou (1980) were able to separate high-wax crude oils and condensates derived from humic kerogen from crude oils formed from algal kerogen deposited in carbonate sequences by plotting their pristane/n-C_{17} ratios against phytane/n-C_{18}.

Data from other investigators have generally supported this concept. Rohrback (1983) reported Pr/Ph ratios to be less than 1 for 52 crude oils derived from marine limestones and marls. Illich and Grizzle (1983) found Pr/Ph ratios to be less than 1 for about 40 oils generated from Paleozoic carbonate source rocks of the Michigan basin. Oils generated by Mesozoic carbonates in the South Florida basin (Palacas et al. 1984), eastern Denver basin (Rice 1984), and south central Texas (Grabowski 1984) all had ratios less than 2. The organic-rich calcareous Kimmeridge source rocks of the North Sea generated oils with Pr/Ph ratios around 1 (Williams and Douglas 1980).

In contrast, six waxy oils and condensates in the Cooper–Eromanga Basin had Pr/Ph ratios from 4 to 6 (Kantsler et al. 1984). The source rocks in this basin contain almost entirely land-derived humic organic matter dominated by vitrinite. Puttmann et al. (1986) noted that an algal-rich boghead coal from the Ruhr had a Pr/Ph ratio of 2.2, whereas most of the associated humic, vitrinitic coals of the same age and rank had ratios around 9.5.

There are exceptions to this concept due to extraneous sources of pristane or phytane altering the ratio. The extractable organic matter of laminated calcareous shales of the Cretaceous Greenhorn formation in Colorado deposited under reducing conditions had about the same Pr/Ph ratios (1 to 3.8) as did moderate to highly macroburrowed sediments deposited under more oxidizing conditions (Pratt 1984). Such anomalies may be due to pristane's having sources other than the degradation of phytol. Illich (1983) examined the interrelationships of the C_{16}-through-C_{20} isoprenoids in more than 188 crude oils from eight different basins. The oils were generated by a variety of sedimentary environments and represented a wide range of maturities. He concluded that some of the pristane and smaller isoprenoids, C_{16} and C_{18}, were derived from a source other than phytol, probably higher-molecular-weight structures in the kerogen. This conclusion is supported by Goosens et al. (1984) who found α-tocopherol to be an alternative source of pristane in sediments. Additional inputs of phytane, independent of phytol, may result in the degradation of C_{40} isoprenoidal ethers identified in organisms such as methanogenic bacteria (Albaiges et al. 1985; Chappe et al. 1982).

Phytane concentrations were twenty to fifty times greater than pristane in near-surface sediments of Ace Lake, Antarctica, where methanogens were the dominant bacteria (Volkman et al. 1986). In addition, some pristane or phytane may be preferentially removed from sediments by the reaction of isoprenoids with H_2S to form thio-isoprenoids (Brassell et al. 1986b).

Another problem is that the Pr/Ph ratio increases during diagenesis as increasing amounts of pristane are formed by the decarboxylation of phytanic acid. The ratio reaches a maximum at the beginning of catagenesis and then decreases as cracking products from kerogen begin to dominate (Boudou 1984).

Clearly, the Pr/Ph ratio is affected by factors other than the degradation of phytol under different environmental conditions. Consequently, the ratio should be used with caution in interpreting oil source-bed environments.

Other Acyclic Isoprenoids and Anteisoprenoids

A whole range of acyclic isoprenoid hydrocarbons have been identified in petroleum with carbon numbers extending from C_5 up to C_{45} (Albaiges 1980). Four types of acyclic isoprenoids have been identified, depending on how the isoprene units are linked together. The isoprene skeleton is described as having a head at the branched end and a tail at the straight-chain end, as shown in Figure 4-9. Regular isoprenoids such as pristane, phytane, and farnesane represent the head-to-tail linking of isoprene skeletons. The four linkages are head to tail, tail to tail, head to head, and irregular (Figure 4-9). Squalane, whose olefin squalene is found in most marine organisms, represents a tail-to-tail linkage. It is a nonspecific biomarker; that is, it is distributed almost ubiquitously in all forms of life. The head-to-head isoprenoid in Figure 4-9 was discovered by Moldowan and Seifert (1979) in a California Miocene crude oil. This type of linkage has been identified in bacterial cell-wall lipids, and it has been recovered from the chemical degradation of kerogen (Chappe et al. 1980). The irregular branched isoprenoid on the left side of Figure 4-9 has been found in the green alga *Enteromorpha prolifera*. This C_{20} hydrocarbon and the corresponding C_{25} isoprenoid are widespread in freshwater and marine sediments (Rowland et al. 1985). The C_{20} isoprenoid also is found in the Rozel Point oil of Utah, whereas the C_{30} isoprenoid has been identified in the Maoming oil shale of China (Brassell et al. 1986a). These hydrocarbons are discussed further in the section on sesterterpenoids.

Botrycoccane is a C_{34} hydrocarbon derived from *B. braunii* (Figure 4-9). High concentrations of botryococcane have been found in Sumatra crude oils (Moldowan and Seifert 1980) and in the bitumen seeps of the Otway Basin of Australia (McKirdy et al. 1986). The C_{31}-to-C_{33} hydrocarbons of similar structure were found in the lacustrian Maoming oil shale of China (Brassell et al. 1986a). This shale contains abundant algal remains of both dinoflagellates and *B. braunii.* Metzger et al. (1985b) isolated nine botryococcenes in the C_{30}-to-C_{37} range from natural blooms and laboratory cultures of *B. braunii.* This suggests that additional hydrocarbons with these structures will be found in petroleum generated in lacustrine or brackish-water sediments.

Regular *isoprenoids* with head-to-tail structures have their methyl groups attached to the even carbon numbers on the chain. Thus, farnesane (Figure 4-9) has methyl groups on the second, sixth, and tenth carbon atoms in the chain. Regular *anteisoprenoids* have the same structures, but the methyl groups are attached at the odd-carbon positions. For example, the C_{21} hydrocarbon discovered by Albaiges (1980) in Spanish oils has a 17-carbon chain with

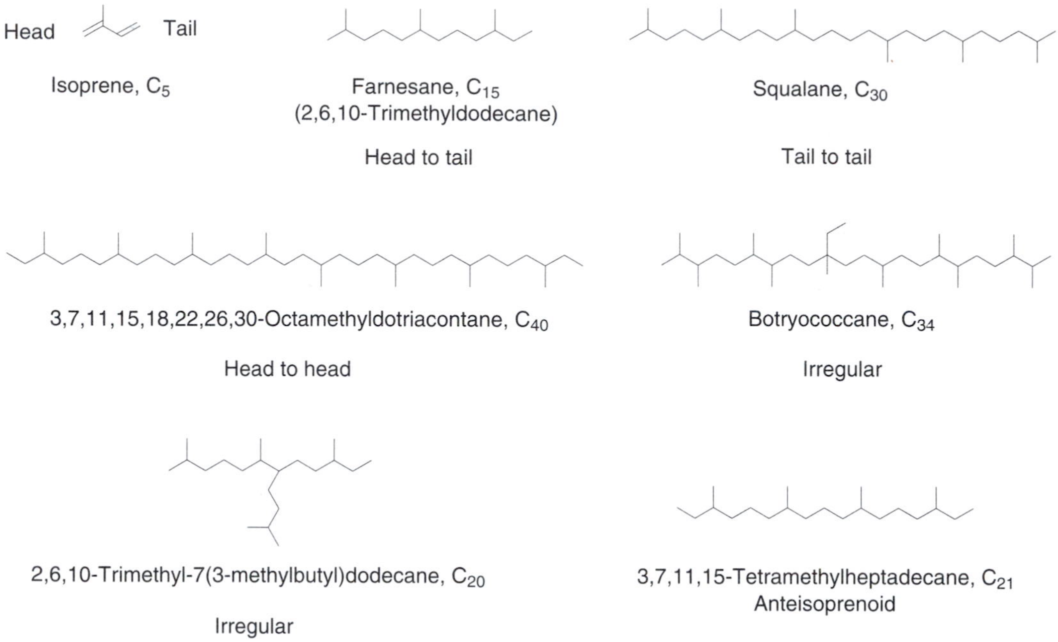

Head Tail

Isoprene, C_5

Farnesane, C_{15}
(2,6,10-Trimethyldodecane)

Head to tail

Squalane, C_{30}

Tail to tail

3,7,11,15,18,22,26,30-Octamethyldotriacontane, C_{40}

Head to head

Botryococcane, C_{34}

Irregular

2,6,10-Trimethyl-7(3-methylbutyl)dodecane, C_{20}

Irregular

3,7,11,15-Tetramethylheptadecane, C_{21}
Anteisoprenoid

Figure 4-9

Biological markers: acyclic isoprenoid and anteisoprenoid hydrocarbon skeletons.

methyl groups at the 3-, 7-, 11-, and 15-carbon positions (Figure 4-9). Likewise, a single CH_3 group attached to a carbon chain will be called an *iso* or *anteisoalkane* if the group is attached to the second or third carbon atom, respectively.

Terpenoids: The Cyclic Isoprenoids

The most common terpenoid structures include the C_{10} (mono-), C_{15} (sesqui-), C_{20} (di-), C_{30} (tri-), and C_{40} (including carotenoids) tetraterpenoids. They are initially formed in living organisms by the previously mentioned polymerization of the 5-carbon isopentenyl pyrophosphate. Consequently, all terpenoids in living organisms start with a carbon structure divisible by the 5-carbon isoprene unit. This is the "isoprene rule" developed to assist in structure analysis. After burial, however, the molecular sizes vary because of the addition or loss of carbon atoms during biological and low-temperature chemical processes.

Literally hundreds of terpenoids have been identified in natural products. Some of those more important to petroleum geochemistry are examined next. Simoneit (1986) provides a more complete discussion of naturally occurring cyclic terpenoids, including 146 chemical structures and 370 references. A book by Philp (1985) gives the structures of 373 biomarkers found in oils, natural products, and sediments, along with their mass spectra. Of this group the steroids and the pentacyclic triterpenoids, including hopanes, are the most widely used in petroleum geochemistry.

Monoterpenes (C_{10}) are not currently used in geochemistry, so our discussion will start with the C_{15} terpenoids.

Sesquiterpenoids, C_{15}

About a thousand sesquiterpenoids have been identified, most of which come from terrestrial plants. Cadalene is an aromatized sesquiterpenoid that is widespread in the essential oils of many higher plants (Figure 4-10). Cadalene and its precursor hydrocarbons occur in deltaic sediments in Spain (Albaiges et al. 1984). Several crude oils from the Eocene of south Texas (Bendoritis 1974) and a subbituminous coal (Baset et al. 1980) also contain cadalene. Drimane (Figure 4-10) was identified by Alexander et al. (1983) in sixteen crude oils from worldwide sources, including Cambro-Ordovician samples. This suggested that it was not a higher plant source but more likely a product of microbial degradation of higher terpenes. In contrast, eudesmane appears to occur only in oils that have a major contribution of terrestrial source material. Simoneit (1986) listed other sesquiterpenoids that have been identified in fossil resins, sediments, and petroleum. The sesquiterpenoids are not widely used in crude oil correlation, possibly because of their relatively simple structures. Philp (1985, pp. 21–22), however, found significant variations in the mass spectra of sesquiterpenoids in crude oils from Australia, New Zealand, and Alaska.

Diterpenoids, C_{20}

Resins are widespread in land plants. They are the progenitors of the resinite maceral of coal. The main constituents of the resins are diterpene acids, which on reduction during diagenesis yield bi-, tri-, and tetracyclic hydrocarbons, both naphthenic and aromatic. They are unambiguous indicators of terrestrial source material contributing to source rocks, oils, and coals. The most common are tricyclic diterpanes with structures related to abietane and pimarane (Figure 4-10). Kaurane is the progenitor of the tetracyclic series. Barnes and Barnes (1983) followed the diagenesis of diterpenes in oxic and anoxic sediments of Lake Powell in western Canada. The anoxic sediments contained the entire series of aromatized hydrocarbons leading to retene (Figure 4-10) derived from abietic acid. Barrick and Hedges (1981) identified ten tricyclic diterpenoids of the general abietane and pimarane structural types in Puget Sound on the west coast of the United States. Livsey et al. (1984) used

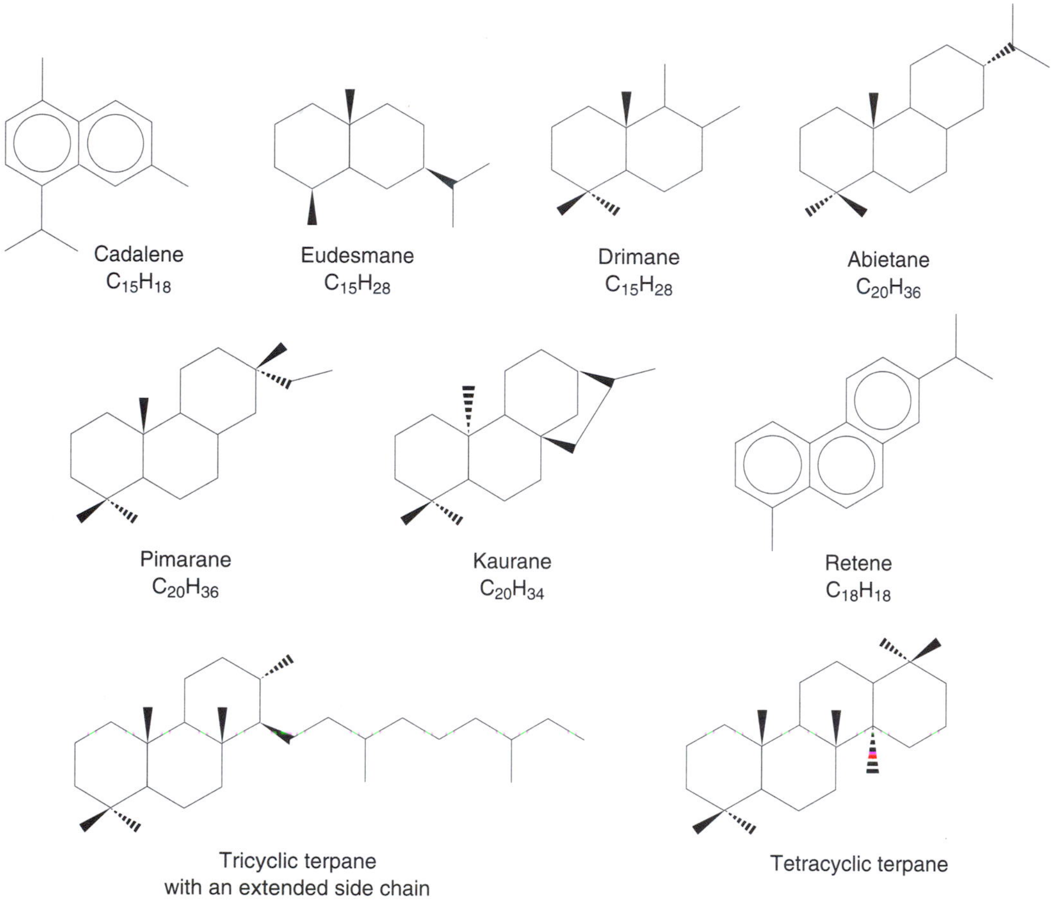

Cadalene
$C_{15}H_{18}$

Eudesmane
$C_{15}H_{28}$

Drimane
$C_{15}H_{28}$

Abietane
$C_{20}H_{36}$

Pimarane
$C_{20}H_{36}$

Kaurane
$C_{20}H_{34}$

Retene
$C_{18}H_{18}$

Tricyclic terpane
with an extended side chain

Tetracyclic terpane

Figure 4-10
Biological markers: terpenoid structures.

diterpanes of the pimarane, abietane, and kaurane types to identify a stratigraphic horizon containing coaly sediments in a well offshore from Labrador.

Diterpenoids occur in peat, bituminous coal, and crude oils, although the quantities in the latter are often small. Snowden and Powell (1982) identified diterpanes in crude oils and rocks extracts of the Beaufort–Mackenzie Basin in Canada. The brown coals and crude oils of the Gippsland Basin of Australia have similar tricyclic diterpanes (Philp et al. 1983). The diterpanes identified in a Gippsland oil by Noble et al. (1986) included bicyclic labdane, tricyclic pimarane, and tetracyclic kaurane type structures, among others. All these compounds are derived from conifers, although labdane may have multiple sources. Labdane occurs in the Athabasca heavy oil associated with hopanes and other biomarkers having a microbial origin (Dimmler et al. 1984).

Alexander et al. (1987) classified the diterpanes most commonly found in crude oils into six families based on structural similarities: the labdane, abietane, pimarane, beyerane, kaurane, and phyllocladane families. The precursor diterpenoids of all these hydrocarbons occur in gymnosperms (mainly conifers), and all but phyllocladane are found in angiosperms (flowering plants). Representatives of all six diterpenoid families occur in Australian crude oils (Alexander et al. 1987).

Sesterterpenoids, C_{25}

The sesterterpenoids, which are carbon skeletons with five isoprene units, occur as tetra- and pentacyclic compounds in sponges (Moore 1979).

Acyclic sesterterpenoids with the irregular C_{20} structure shown in Figure 4-9 have been identified as important components of the hydrocarbons of a variety of aquatic sediments from around the world (Robinson and Rowland 1986). As mentioned earlier, the parent structure has been detected as a C_{25} diene in the green alga, *E. prolifera.* The presence of the corresponding C_{20} and C_{30} branched alkanes in the presumably hypersaline Great Salt Lake, Utah, oil and in Chinese oil shale, previously mentioned, indicates that this unusual biomarker may be more useful in future crude oil and source rock correlation studies.

Tricyclic Terpanes, C_{19} to C_{45}

The tricyclic terpanes are widely distributed in crude oils and source rocks of marine or lacustrine origin. They are considered to be diagenetic products of prokaryote membranes (Ourisson et al. 1982). Their presence in the crude oils of South Oman, which are thought to have been generated by the late Precambrian Huff Formation, indicates that primitive bacteria are one of the sources (Grandville 1982; Grantham 1986). They also are important constituents of organic-rich black shales from the Ordovician of the United States (Fowler and Douglas 1984). They are usually not found in oils predominantly derived from terrestrial source material, such as the Australian oils from the Upper Cretaceous–Tertiary Gippsland Basin (Philp and Gilbert 1986).

A homologous series of tricyclic terpanes extending from C_{19} to C_{30} was identified by Aquino Neto et al. (1983) in about forty crude oils and sediment extracts ranging in age from Jurassic to Tertiary from a variety of countries. All the terpanes had the ring system shown in Figure 4-10 with an isoprenoid side chain of increasing length. Later, Moldowan et al. (1983) identified tricyclic homologs extending out to C_{45} in several crude oils and source rocks.

The mass fragmentograms of tricyclic terpanes having an m/z (mass/charge) ratio of 191 frequently show a maximum peak at C_{23}, with C_{24} or C_{21} being the second most dominant peak. Crude oils of South Oman, the Williston Basin, the Western Canada Basin, the South Florida Basin, and the Middle Magdalena Valley of Colombia all have this characteristic (Grantham 1986; Leenheer 1984; Palacas et al. 1984; Zumberge 1983, 1984). Palacas et al. used the C_{23} tricyclic terpane pattern to correlate the Cretaceous Sunniland oils with their source rocks, and Leenheer did the same with the Mississippian oils and Bakken source rocks of the Western Canada Basin.

The concentration of tricyclic terpanes in crude oils appears to increase with increasing maturity, probably because of the breaking off of tricyclic terpane moieties in the asphaltenes and kerogens (Kruge 1986). Ekweozor and Strausz (1983) obtained an entire series of homologous tricyclic terpanes in the C_{19}-to-C_{26} range by pyrolysis of the asphaltenes, resins, and heavy oils extracted from the Athabasca oil sands of Alberta. The highest yields were from the asphaltenes. The C_{23} and C_{24} tricyclic terpanes were dominant in all products.

Tricyclic terpanes are less subject to microbial alteration than many other biomarkers are, so they tend to stand out in biodegraded oils (Behar and Albrecht 1984; Connan et al. 1980). Demethylated tricyclic terpanes have been observed in some slightly biodegraded and nonbiodegraded crude oils, but it is not clear whether they originated in the source rocks or are products of microbial alteration in the reservoir (Howell et al. 1984). Because of their resistance to biodegradation, tricyclic terpanes allow the correlation of intensely biodegraded oils (Seifert and Moldowan 1979).

Increasing maturation also causes a change in the stereoisomers at the 13- and 14-carbon positions on the C ring. Aquino Neto et al. (1986) suggested that the tricyclic terpane stereoisomers may be useful in assessing maturity along with the more widely used steranes and hopanes, since the former are more resistant to biodegradation.

Tetracyclic Terpanes, C_{24} to C_{27}

The m/z 191 mass fragmatograms of tricyclic terpanes commonly have tetracyclic terpanes in the C_{24}-to-C_{27} range. Ekweozor et al. (1981) recognized these compounds in several crude oils from the Niger Delta. They concluded that these were degradation products of pentacyclic triterpenoid precursors in the petroleum. The hypothesis was that thermal cleavage of terminal rings in pentacyclic hopanes would yield the tetracyclic terpanes that could further degrade to tricyclic compounds. The presence of a C_{25} tricyclic alkane structurally related to the tetracyclics in the oils supported this concept. The tetracyclic series, which has the structure of 17,21-secohopanes, commonly extends from C_{24} to C_{27}, although there is tentative evidence for extension to C_{35} (Aquino Neto et al. 1983). The previously mentioned study of tricyclic terpanes by these authors in about forty crude oils and sediments also included the tetracyclics. Both the tetracyclic and tricyclic terpanes are fairly resistant to biodegradation, and both increase with maturity relative to the pentacyclic hopanes. Thermal alteration experiments by the same authors indicated that both tetracyclic and tricyclic terpanes could be formed from various crude oil fractions, such as resins and asphaltenes.

Palacas et al. (1984) observed that some carbonate rocks in the South Florida Basin contained organic matter characterized by a relatively high concentration of C_{24} tetracyclic terpanes. These rocks were not the main source of the Sunniland crude oils in this basin, but they did correlate with a noncommercial oil in a more shallow reservoir. Palacas et al. used the relative amounts of the C_{23} tricyclic and C_{24} tetracyclic terpane peaks to delineate different source facies in the basin.

Philp (1985a) found the C_{24} tetracyclic terpanes (Figure 4-10) to be very abundant in the majority of Australian oils and source rock extracts, whereas he frequently did not find tricyclic terpanes. The tetracyclics are thought to come from the degradation of C_{30} pentacyclic triterpenoids, although an independent biosynthetic route to the tetracyclic terpanes may exist in bacteria.

Pentacyclic Triterpenoids: Hopanes, C_{27} to C_{40}

The pentacyclic compounds are classified into two groups, *hopanoids* and *nonhopanoids*. The hopanoids are the most widespread biomarkers in the biosphere and geosphere. They occur in the membranes of prokaryotes, where they provide rigidity and strength in the same way that the sterols do for the eukaryotes. Hopanoids, such as the C_{35} bacteriohopanetetrol (Figure 4-11), have about the same molecular dimensions as the sterols (1.9 nm long and 0.77 nm wide), so they can fit into the lipid bilayers of membranes and perform similar strengthening functions (Ourisson et al. 1979; Rohmer et al. 1979). Actually, since the prokaryotes evolved first, the hopanoids were performing this function long before the sterols were.

Several hundred hopanoids have been isolated from organisms and sediments. They are used more than any other biomarkers for fingerprinting crude oils and source rocks. They are widely distributed among bacteria, cyanobacteria (blue-green algae), and other primitive organisms with prokaryote cells. Hopanoids also occur in ferns, lichens, and a few higher plants (Ourisson et al. 1982). Some hopanoids occur in almost every sample containing organic matter, regardless of its age or origin. The C_{31}-to-C_{35} extended (side-chain) hopanoids, however, occur only in microorganisms, so a fossil hopanoid with a 1- to 5-carbon chain is indicative of a microbial origin (Ourisson et al. 1984). Hopanoids are not present in most methanogens or other archaebacteria. The structural components of methanogen membranes include acyclic isoprenoids such as the C_{40} compound with the head-to-head linkage shown in Figure 4-9 (Brassell et al. 1981).

Naturally occurring hopanoids have the thermodynamically less stable $17\beta(H),21\beta(H)$ stereochemistry as the bacteriohopanetetrol in Figure 4-11. As the organic matter is buried and undergoes diagenesis and maturation, saturated hopanes with the more stable $17\alpha(H),21\beta(H)$ stereochemistry are formed. This is the configuration found in oils and mature source rocks. The extended

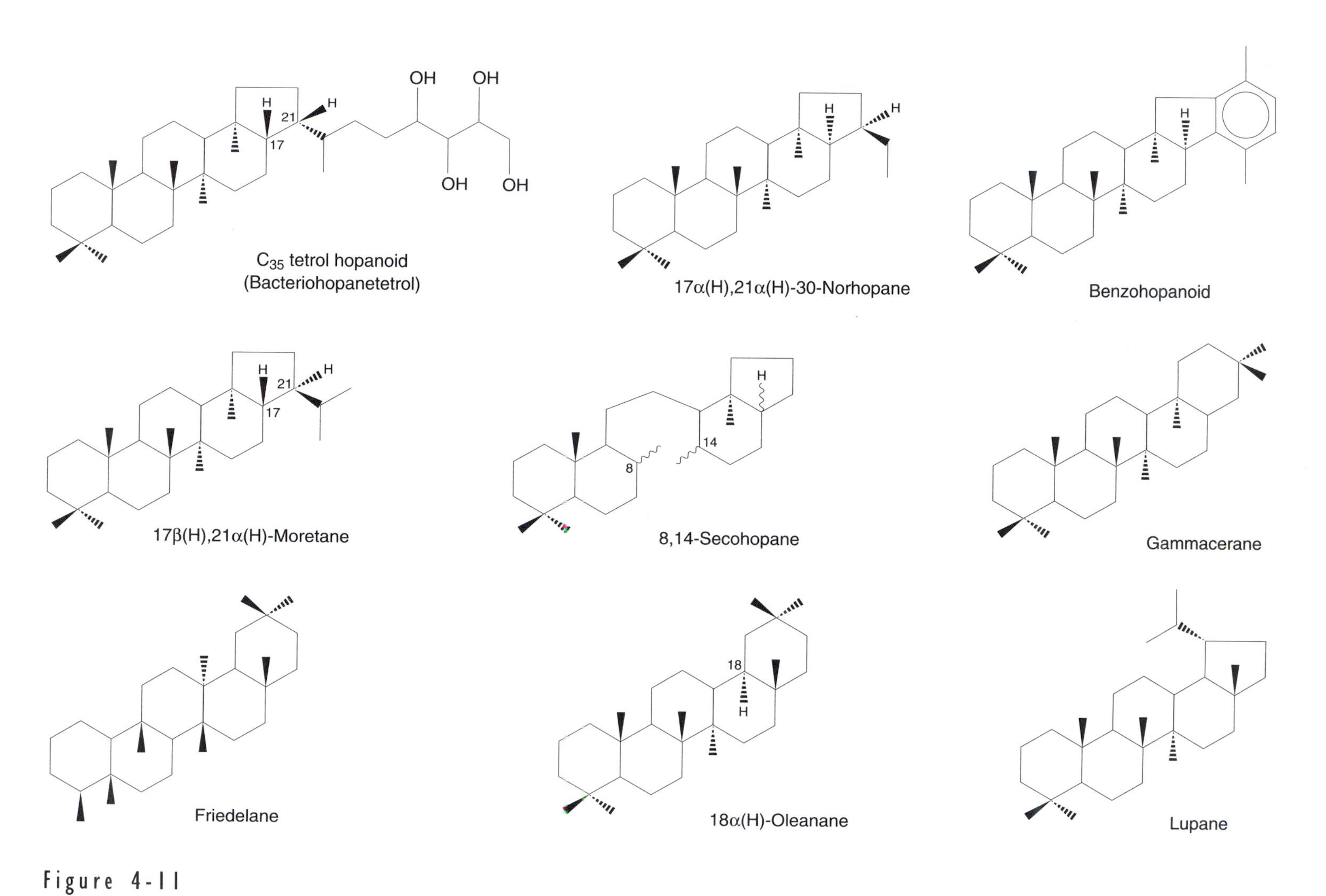

Figure 4-11
Biological markers: pentacyclic triterpenoid structures.

Box
4-1

Stereochemistry of Biomarkers

Isomers are molecules with the same number and kinds of atoms but with different arrangements of the atoms in space. For example, *n*-butane and isobutane are structural isomers in which atoms are connected differently (straight and branched; see Figure 3-3). The atoms of stereoisomers are connected in the same way, but their spatial three-dimensional arrangements are different. For example, when a molecule like cholestane is drawn in three dimensions as follows, the hydrogen at the 3 position points up above the plane of the molecule and that at the 5 position points down below the plane (Peters and Moldowan 1993, p. 30):

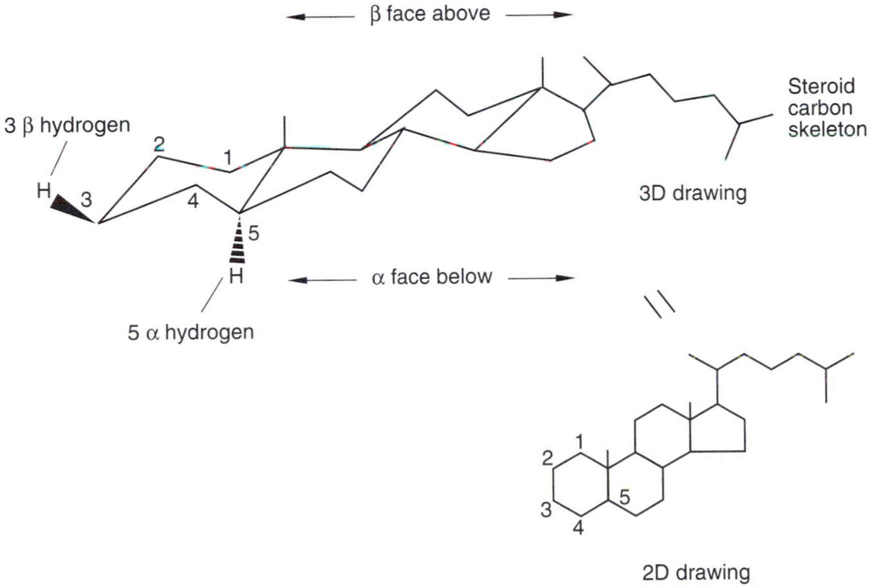

β face above
3 β hydrogen
H 3 2 1 4 5 H
5 α hydrogen
α face below
Steroid carbon skeleton
3D drawing
1 2 3 4 5
2D drawing

A hydrogen atom (H) or methyl group (CH_3) attached to the ring system is in the β position if it points up above the rings and in the α position if it points down below the rings. The α(H) and β(H) stereoisomers are identical except for the position of the substituent. This difference often shows up as separate peaks on a mass fragmentogram because of a difference in retention time. In publications, the following symbols are used to indicate whether a C–H or a C–C bond is pointing up or down.

A C–H bond in the beta (up) position is shown by H▼ in the figure above or by ● in the oleanane structure below. A beta C–C bond is shown as ▼ without the H. The oleanane has 5 beta C–C bonds. A C–H bond in

Box 4-1

Stereochemistry of Biomarkers *(continued)*

the alpha (down) position is the H in the figure above and the ○ in the oleanane below. Oleanane has 3 alpha C–H bonds. A C–C bond in the alpha position is shown as or as . There are 3 alpha C–C bonds on the oleanane. When the bond configuration is not specified, it is shown as a wiggly line 〰 .

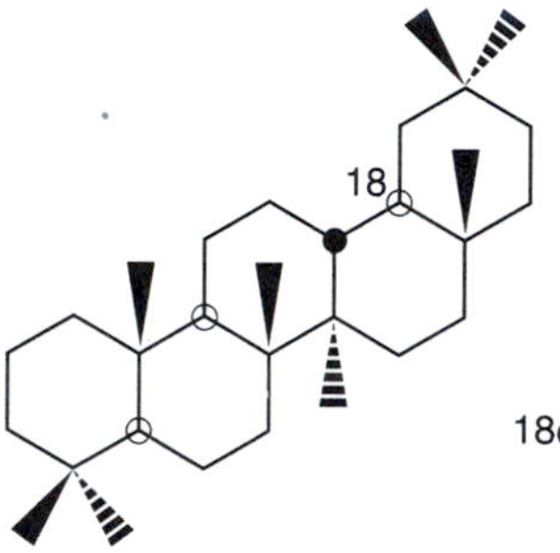

18α (H) oleanane

In addition to the three-dimensional difference in substituents on the rings, there is a difference in the way that the side chain is hooked to the ring system at the 20-carbon position for steranes and the 22-position for hopanes. The three-dimensional (3D) side-chain structure is called *sinister* (S, left-handed) or *rectus* (R, right-handed), depending on the spatial orientation of the groups. The carbon at the 20-position is at the center of a tetrahedron with three bonds showing and the fourth bond (to hydrogen) behind the page.

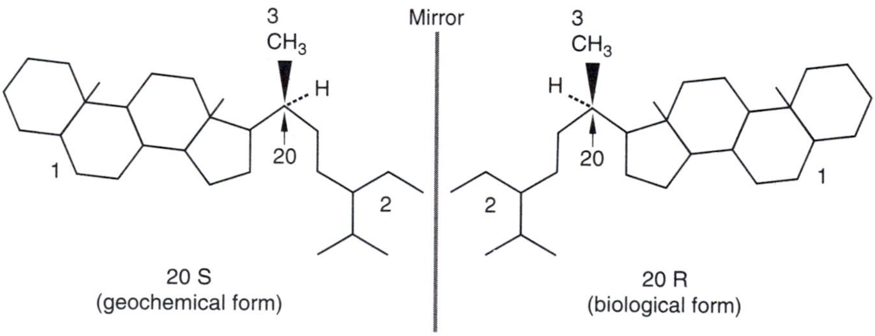

Mirror

20 S
(geochemical form)

20 R
(biological form)

If you place a pencil perpendicular to the page at the 20 position of the molecule on the left, you will see that as you look counterclockwise around the pencil, the three visible groups (1, 2, and 3) attached to carbon number 20 get smaller. This is the S isomer. If you do the same with the molecule on the right, the groups (1, 2, and 3) get smaller going around clockwise. This is the R isomer. These two isomers are mirror images of each other and are not superimposable, so they are called enantiomers. The carbon at the 20 position that has four different substituents attached to it is called a chiral center. If two attachments are the same, it is not a chiral center.

The main isomerisation shifts from α to β in steranes occur with the hydrogens at the 5-, 14-, and 17-carbon positions, as shown in the following illustration. The mass fragmentogram (mass/charge, m/z 218) is of sterane stereoisomers in the Cheng oil from China (Shi et al. 1982). The peaks of stereoisomers for the C_{27}, C_{28}, and C_{29} molecules are prominent. Within each of these molecular sizes are four large peaks representing

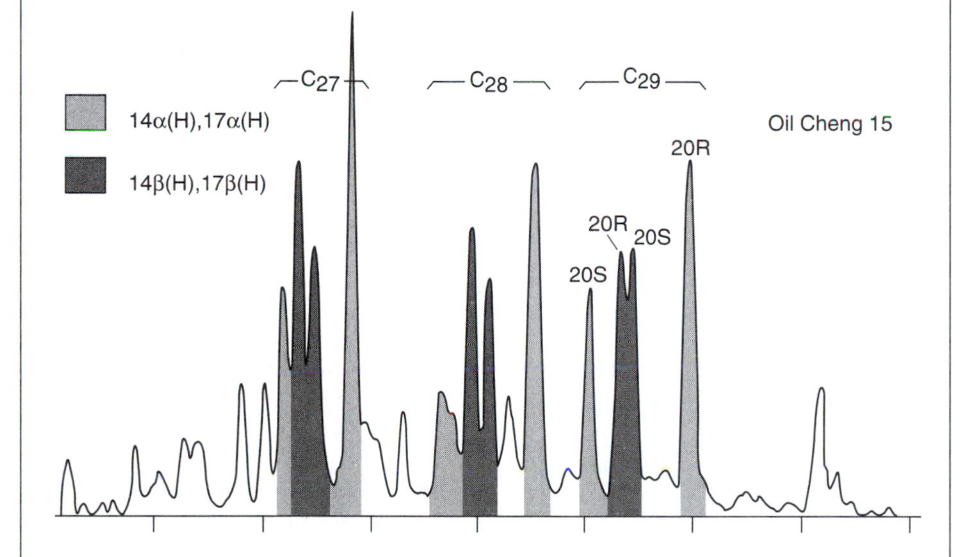

the 14α(H),17α(H),20R and S and the 14β(H),17β(H),20R and S molecules. The same four stereoisomers are present for the C_{27} and C_{28} steranes. The original biological stereoisomers are 14α(H),17α(H),20R. They are partially converted to the 14β(H),17β(H),20S isomers with increasing thermal stress.

The complexity of these molecules with many peaks for each biomarker class is what makes them so valuable in applications of petroleum geochemistry. Differences in the compound class and the number of carbon atoms are very useful in correlating crude oils with one another and with source rock extracts. Differences in the three-dimensional configurations of the side chains and the rings can be used to determine maturity differences between oils and rock extracts.

hopanoids ranging from C_{31} to C_{35} also occur in the 17β(H) configuration in immature sediments with only the 22R epimer. This is converted to a mixture of 22S and R with diagenesis (Peters and Moldowan 1991). Norhopane is formed by the loss of one carbon at the 30 position on the side chain (Figure 4-11) and trisnorhopane from the loss of three carbons, such as at the 22, 29, and 30 positions (see Box 4-2 for the nomenclature of biomarkers, pp. 100–101). Homohopane has one additional carbon on the hopane side chain, and trishomohopane has three additional carbons. Moretanes are not present in living organisms but form from hopanes at higher levels of maturity. They differ from hopanes in having a 17β(H),21α(H) stereochemistry (Figure 4-11).

These compounds occur in most rocks and crude oils in varying quantities, depending on their source, and with different stereoisomers, depending on the extent of maturation. Black shales such as the Cretaceous shales of the Angola Basin and oil shales like the Messel of Germany, the Green River of Utah, and the Cretaceous of Jordan (Kimble et al. 1974; Michaelis et al. 1986; Wehner and Hufnagel 1986) contain a variety of these compounds. Hopanoids were the major hydrocarbon assemblages extracted from the highly bituminous sapropelic Serpiano oil shale of Switzerland (McEvoy and Giger 1986). A total of fifty-six hopanoids were identified, including hopanes, hopenes, benzohopanes, monoaromatic secohopanoids, and high concentrations of methylhopanoids. The high abundance of hopanes indicated a major bacterial input to this sediment. Pyrolysis of the asphaltenes and kerogen of the sediment yielded hopane and methylhopane mixtures similar to those extracted from the shales.

An extended series of C_{31}-to-C_{35} hopenes and hopanes with a 17α(H),21β(H) configuration were identified by ten Haven et al. (1985) in the hypersaline sediments of a Messinian sedimentary basin of Italy. The authors concluded that the hopanes were formed from the reduction of the hopenes rather than the isomerization of the unstable 17β(H),21β(H) stereoisomer. This was based on the distribution of hopenes and hopanes combined with their similar isomerization at the 22-carbon position.

Hopanes, which have a benzene ring condensed to their E ring, are called *benzohopanoids* (Figure 4-11). These compounds have been associated mainly with carbonate–evaporite depositional environments of restricted circulation (Belayouni and Trichet 1984; Hussler et al. 1984). They also have been found in organic-rich shales such as the previously mentioned Messel of Germany, Serpiano of Switzerland, and Cretaceous shales of the Angola Basin. The benzohopanoids are formed during early diagenesis, presumably from low-temperature catalyzed dehydration and cyclization reactions. These compounds occur in many carbonate source rocks and their associated crude oils in France, Spain, Italy, Iraq, Libya, and Central America. Trace amounts also were present in deltaic deposits of Nigeria and New Zealand (Hussler et al. 1984, and references therein). These authors also identified monoaromatic D-ring, 8,14-secohopanes in the aforementioned carbonate rocks and associated crude oils. These are hopanes in which the C ring is cut open and the D ring aromatized. They are believed to be formed from hopane precursors by thermocatalytic cleavage of the weak 8,14-carbon bond at a late stage of maturation.

Secohopanes with the E ring broken at the 17,21 bond and others with the C ring broken at the 8,14 bond (Figure 4-11) have been identified in several crude oils (Philp 1985a). Schmitter et al. (1982) found a series of C_{27} to C_{31} 8,14-secohopanes in a Nigerian crude oil. The C_{28} homolog was missing. This distribution is similar to the regular hopanes, which range from C_{27} up but do not include the C_{28} member in many crude oils and source rocks. There are exceptions in which the C_{28} hopane is prominent, such as the Monterey Shale extract and crude oils from the northern Volga, Russia, and the North Sea (Philp 1985a, and references therein).

A C_{31} to C_{35} homologous series of saturated hexacyclic hydrocarbons occurs in crude oils from Paleozoic carbonate reservoirs of eastern Montana (Rinaldi 1985). The hexacyclic hydrocarbons have a structure similar to the benzohopanoid in Figure 4-11, but with a cyclohexane ring in place of the benzene ring. They are thought to form during early diagenesis from cyclization of the side chain of extended hopanoids.

Connan and Dessort (1987) identified four hexacyclic hopanoid alkanes (hexahydrobenzohopanes having 32 to 35 carbon atoms in crude oils and rock extracts from anoxic evaporitic rocks). The rocks consisted of marls, massive anhydrites, and bituminous, argillaceous carbonates. The hexacyclic hopanoids also occur in sulfur-rich heavy oils, both unaltered and biodegraded from the Ales, Camargue, and Aquitaine Basins of southern France. Various biomarker ratios indicated that both the oils and rock extracts originated from a kerogen with a high content of bacterial by-products. The presence of these hexacyclic hopanoids in severely biodegraded crude oils suggests that they may be useful paleoenvironmental indicators of carbonate–anhydrite settings when other biomarker indicators are destroyed or extensively altered.

Hopanes with two, three, or four aromatic rings have been isolated from immature rocks such as the Messel oil shale (Greiner et al. 1976). These compounds are not found in living organisms but are believed to develop from the aromatization of triterpenoid precursors of the hopanes.

Pentacyclic Triterpenoids: Nonhopanoids

Most of the pentacyclic triterpenoids without the characteristic hopane structure originate from higher plants. Some of the nonhopanoids used in geochemistry are gammacerane, friedelane, oleananes, and lupanes (Figure 4-11).

Gammacerane is highly resistant to biodegradation, so it tends to be present in a variety of oils and source rocks. Seifert et al. (1984) found gammacerane in several heavily biodegraded oil seeps from western Greece. They used a "gammacerane index," which is the ratio of the concentration of gammacerane to that of the $17\alpha,21\beta$(H)-hopane times 100, to distinguish the seeps and relate them to possible source rocks. Gammacerane also is present in several biodegraded oils from the Shengli oil field of China and in shales thought to be the source rocks (Shi et al. 1982). Gammacerane is frequently found in hypersaline carbonate–evaporite deposits. Sediments deposited by the highly saline lakes that existed in much of eastern China during the Paleogene have high contents of gammacerane in their extracts (Fu et al. 1986, and references therein). Sediments of the previously mentioned Messinian evaporitic basin of Italy also contain gammacerane (ten Haven et al. 1985).

The saline, siliceous carbonates of the Green River Formation in the Uinta basin of Utah contain gammacerane. High gammacerane concentrations were found in eighteen crude oils originating from various hypersaline lacustrine basins of China, such as the Jianghan Basin (Philp and Fan 1987). One source of gammacerane is believed to be tetrahymanol. It has the structure of gammacerane with an OH at the 3-carbon position. Tetrahymanol occurs in the protozoan *Tetrahymana* (Henderson et al. 1969) and in phototrophic bacteria. It is the only pentacyclic triterpenoid identified in the animal kingdom, although tetracyclic triterpenoids of the lanosterol type also occur in aquatic animals. Gammacerane will continue to be used for crude oil correlation, but its value as a source indicator depends on further clarification of its various possible sources.

Pentacyclic triterpenoids with the lupane and oleanane structures (Figure 4-11) are derived from both angiosperms and higher plants of Upper Cretaceous age and younger (Henderson et al. 1969). Both are found in soft brown coal, such as the lignites of the Lower Rhenish Bay of Germany (Hagemann and Hollerbach 1980).

Relatively high concentrations of C_{28}-bisnorlupanes were found in extracts of rocks from west Greenland and the Gulf of Suez by Rüllkotter et al. (1982). The organic matter of the analyzed rocks was largely of terrestrial origin, indicating that the bisnorlupanes were products of an early diagenetic alteration of higher plants. A relationship between bisnorlupane and oleanane concentrations in petroleums from the Mackenzie Delta, Canada, supports their terrestrial origin (Brooks 1986).

Among the more widely used nonhopanoids in petroleum geochemistry are the oleananes. The 18αH-oleanane was used for crude oil—source rock correlations in the Niger Delta (Ekweozor et al. 1979). Marine shales of the Tertiary Akata Formation, which contain terrestrial higher plant material, are believed to be the source of most of the Niger Delta oils.

The presence of oleananes, lupanes, and hopanes in an oil from a basin offshore from Malaysia was used to define the oil as being of Tertiary age (Richardson and Miller 1983). No Cretaceous rocks were in this area, and since the precursors of these hydrocarbons are angiosperms, the only possible source is Tertiary rocks.

Many Australian oils originate from organic materials with a high terrestrial input. Analyses of twelve Australian oils showed 18α(H)-oleanane to be present in most of those of Upper Cretaceous–Tertiary age from the Gippsland, Sydney, Surat–Bowen, and Cooper–Eromango Basins, whereas the oils of Permian, Jurassic, and Triassic age in those basins contained essentially no oleananes. Also, the Tertiary oils of the Mahakam and Niger Deltas and the Beaufort–Mackenzie (Canada) and New Zealand Basins all contained oleananes (Philp and Gilbert 1986, and references therein). All these oils had a high input of terrestrial source material. These various studies indicate that oleananes and lupanes may be considered reasonably reliable indicators of higher plant source material from an Upper Cretaceous–Tertiary sediment.

Pentacyclic triterpenoids containing three or four aromatic rings, usually the A, B, C, and D rings, were isolated from Recent sediments and low-rank coals (Philp 1985b, pp. 214–218, and references therein). They are early aromatized products of triterpenoid precursors, which so far have not been used much in geochemical studies.

Tetraterpenoids: Carotenoids, C_{40}

The light yellow to deep red colors of plants are caused by isoprenoid polyterpene pigments known as *carotenoids*. Examples are lycopene and β-carotene (Figure 4-12), which are the red and yellow pigments of tomatoes and carrots. Carotenoids are produced by all photosynthetic organisms, including higher plants, algae, bacteria, and fungi. They also are found in nonphotosynthetic bacteria.

Carotenoids include the hydrocarbons (carotanes), the xanthophylls (carotenoids with oxygen functions other than the carboxyl group), and the carotenoid acids. Many of these compounds form hydrocarbon derivatives when buried under reducing conditions. Carotane, the saturated form of carotene, was first identified in the Green River oil shale (Murphy et al. 1967), and lycopane, the saturated lycopene (Figure 4-12) was found in the Messel oil shales. Carotenoids occuring in a variety of both recent and ancient sediments were reviewed by Simoneit (1986). Because of their high molecular weight, the carotenoids are isolated by high-pressure liquid chromatography (HPLC) and characterized by spectrophotometry and chemical ionization mass spectrometry.

Carotenoids appear to be incorporated into kerogen early in diagenesis. Chemical degradation of lipid-free kerogens of modern lacustrine and marine sediments have yielded the breakdown products of β-carotene (Machihara and Ishiwatari 1987). Also, carotanes are found in relatively immature oil shales such as the Permian Irati shale of South America (Figure 4-14).

Carotane also was detected in six shales and an oil sample from the Shengli oil field of eastern China (Shi et al. 1982). Both α and β carotane, along with several carotenoid-derived alkanes occur in very high concentrations in more than 100 oil samples from the Kelamayi field of northwestern China (Jiang and Fowler 1986). They also are present in a black, calcareous, carboniferous shale that is believed to be the source of these oils. Jiang and Fowler's studies indicate that carotenoid-derived alkanes can survive over a considerable geological time and temperature range, providing that they are deposited in strongly reducing saline depositional environments to prevent degradation into smaller molecules. These authors noted that α-carotane is thermally more stable than β-carotane, so the ratio of these two increases with maturity. However, because β-carotane is more resistant to microbial attack than α-carotane is, the ratio decreases with increasing biodegradation.

Carotenoids with the end-member rings aromatized have been identified in both source rocks and crude oils. Ostroukhov et al. (1982) found a series

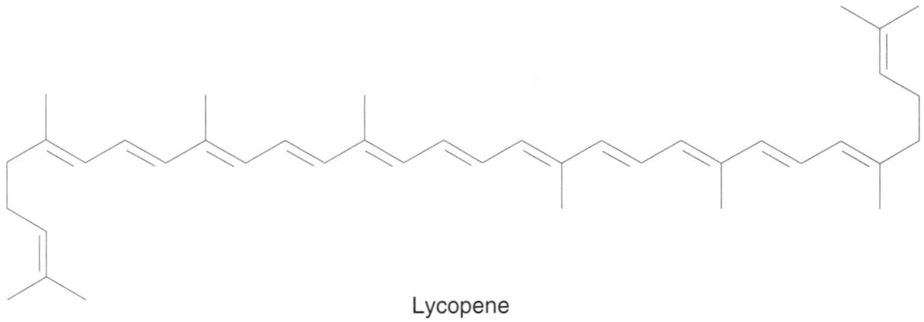

Lycopene

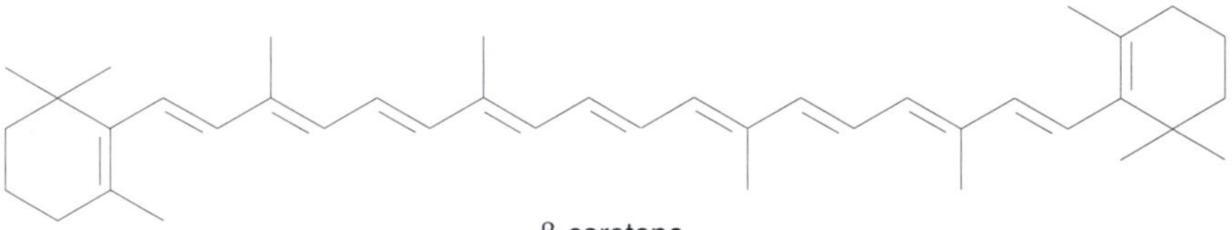

β-carotene

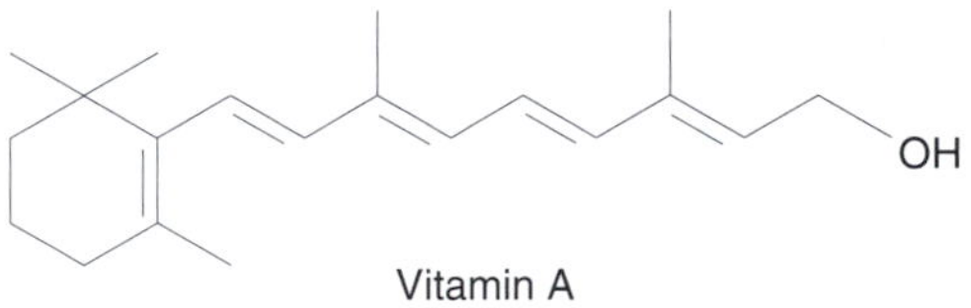

OH

Vitamin A

Figure 4-12

Carotenoids contain eight isoprene units. The beautiful colors of autumn leaves are due to carotenoids. Organisms synthesize two vitamin A molecules by cleavage of a ß-carotene molecule.

of monoaromatic isoprenoid hydrocarbons that appeared to be derived from β-carotene in the Bucharsky oil of Russia. Schaefle (1977) identified three aromatic carotenoid-derived hydrocarbons in the Toarcian shale of the Paris Basin.

Aromatic isoprenoids of carotenoid origin and their degradation products also were identified in oils from the Alberta Basin and the Canadian portion of the Michigan Basin by Summons and Powell (1987).

Their studies indicated that methyl substitution on the 2, 3, 4 and 2, 3, 6 positions on the aromatic rings of these compounds appears to be confined to photosynthetic green sulfur bacteria. This specific substitution was recognized in Russian crude oils and the oils from source rocks of the Silurian of the Michigan Basin and Middle Devonian of the Alberta Basin. Geological studies indicate that the source rocks of these oils were deposited under hypersaline conditions in sulfide-rich waters, which is consistent with the idea that sulfur bacteria were present.

Peters and Moldowan (1993, p. 164) consider carotenoids to be highly specific for hypersaline lacustrine deposition. They used β-carotane to assist in identifying a lacustrine Devonian source rock as the cosource for the Beatrice oil in the Inner Moray Firth, North Sea (Peters et al. 1989).

Steroids, C_{19} to C_{30}

Owing to the complexity of the tetracyclic steroids and the pentacyclic triterpenoids, they have become the most widely used biomarkers in evaluating the source, maturity, migration, biodegradation, and correlation of crude oils with source rocks. All these applications are examined in later chapters.

Steroids and their sterane derivatives are not terpenoids. They do not follow the previously mentioned "isoprene rule."

The reason is that the conversion of squalene (a terpene) to cholesterol in nature involves oxidation and decarboxylation steps that destroy part of the original isoprenoid structure of squalene. This origin of cholesterol was a major step in evolution because cholesterol is the metabolic precursor of many important steroids, including human hormones. The early sterols became a critical part of the extensive membrane structures of the eukaryotes. Consequently, sterols have been widespread in plants and animals since the Late Proterozoic. Cholesterol, the most important sterol (Figure 4-13), has eight asymmetric carbon centers, that is, eight carbon atoms, each surrounded by four different substituents. Each asymmetric carbon atom can form two stereoisomers that link the same substituents but differ in the substituents' spatial arrangement. This means that theoretically, cholesterol can exist in 2^8, or 256, different structural configurations. In nature, the organism synthesizes only one of these forms, namely, the flattest. This flatness enables cholesterol to fit in the cell membrane, where it acts as a stabilizer. But the flatness introduces strain into the molecular bonds that is not released until the organism dies and the membrane decays. During diagenesis and thermal maturation, cholesterol degradation products isomerize to stereoisomers that are less strained. These shifts in the configurations of the stereoisomers then provide a way of following maturation changes as the cholesterol derivatives undergo diagenesis and catagenesis.

Sterols have been identified in all major groups of living organisms. Although sterols are ubiquitous in the biosphere, they vary widely in size and structure. Also, microbial activity and low-temperature chemical reactions may form early alteration products that can be associated with specific sources or depositional environments. This makes them useful for fingerprinting crude oils and related bitumens in source and correlation studies.

The steranes most widely used in petroleum geochemistry contain from 27 to 29 carbon atoms. Marine sterols extending from C_{26} to C_{30} and sterane derivatives down to C_{19} have been identified in sediments and oils. More than 100

Box 4-2

Nomenclature of Biomarkers

The suffix *-oid* refers to all compounds with the same basic structure (alcohols, olefins, etc.), and *ane* refers only to saturated hydrocarbons. Thus, cholesterol, cholestene, and cholestane all are steroids, but only cholestane can be called a sterane. If a naphthene ring in a sterane changes to an aromatic ring, it is called a steroid because the ring is unsaturated.

Terpenoids are named by a semisystematic procedure that accounts for their origin as well as details of their functional groups and stereochemistry. The nomenclature system recognizes four variations in their molecular structure:

1. compound class
2. number of carbon atoms
3. three-dimensional (3D) side-chain configuration
4. 3D ring-system configuration

The carbon atoms in each molecule are numbered as shown for steranes (a steroid class) and hopanes (a triterpenoid class). The numbering enables the stereochemistry of specific atoms to be described in the name. The rings are labeled A through E.

C_{29} Sterane (stigmastane)

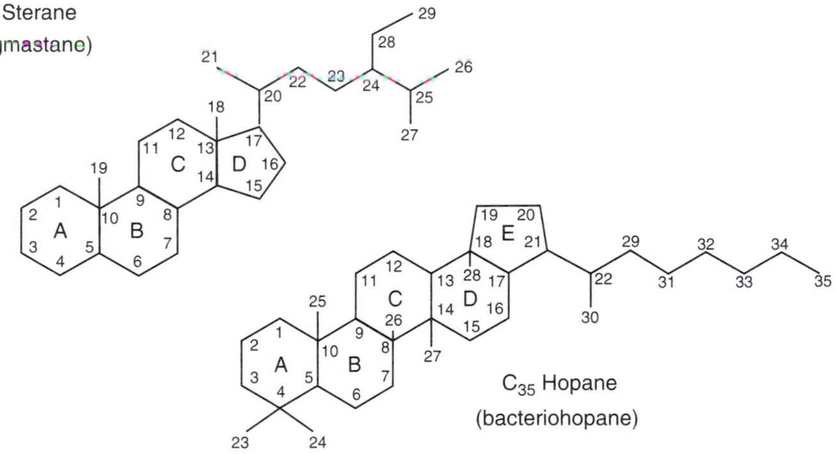

C_{35} Hopane (bacteriohopane)

When a compound is isolated from a natural source and its structure is unknown, it may be given a trivial name based on the family or genus or species of the biological material from which it was isolated. Thus, hopane is named after *Hopea,* a plant of Southeast Asia, and oleanane is from the plant *Oleus auropea.* Trivial names carry little structural information and do not indicate the major sources of biomarkers. For example, the dominant sources of hopanes are microorganisms rather than the rare plant it was named after. Almost all the names in Figures 4-10 and 4-11 are trivial names.

For example, $5\alpha(H),14\beta(H),17\beta(H),20S\text{-}C_{29}$ sterane is a typical name. The compound class is sterane, and the number of carbon atoms is 29. The side-chain stereochemistry is 20S, and that of the ring system is $5\alpha(H),14\beta(H),17\beta(H)$. The molecule would have the geological S form shown in Box 4-1.

Several prefixes can be attached to class names, depending on how the structures have changed from the original class. The names and their meanings are the following:

spiro: joining two rings by one carbon atom

seco: breaking rings

benzo: fusing a benzene ring

nor: subtracting a CH_2 or CH_3 group from the molecule

homo: adding an additional CH_2 or CH_3 group to the molecule

bis: two

tris: three

tetrakis: four

iso–CH_3: shifted on structure

neo–CH_3: shifted on hopanes from carbon number 18 to 17

de or *des:* replacement of a group or ring by hydrogen

ent: inversion of all asymmetric centers

Thus, as an example of de, des-A-androstane (a steroid) would be androstane without the A ring. And as an example of ent, if all alpha bonds in abietane (Figure 4-10) were changed to beta and all beta bonds to alpha, it would be called ent-abietane.

Hopane is a C_{30} triterpane, like the figure in this box but without the C_{31} to C_{35} side chain. Trisnorhopane is a C_{27} triterpane; that is, it is missing three methyl groups compared with hopane. The number immediately preceeding the prefix *nor* is the carbon number attached to the molecule that is eliminated. Thus, C_{29}-30-norhopane is hopane with the carbon at the 30 position missing. A hopane in which both the carbon atoms at the 28 and 30 positions are missing would be called C_{28}-28, 30 bisnorhopane.

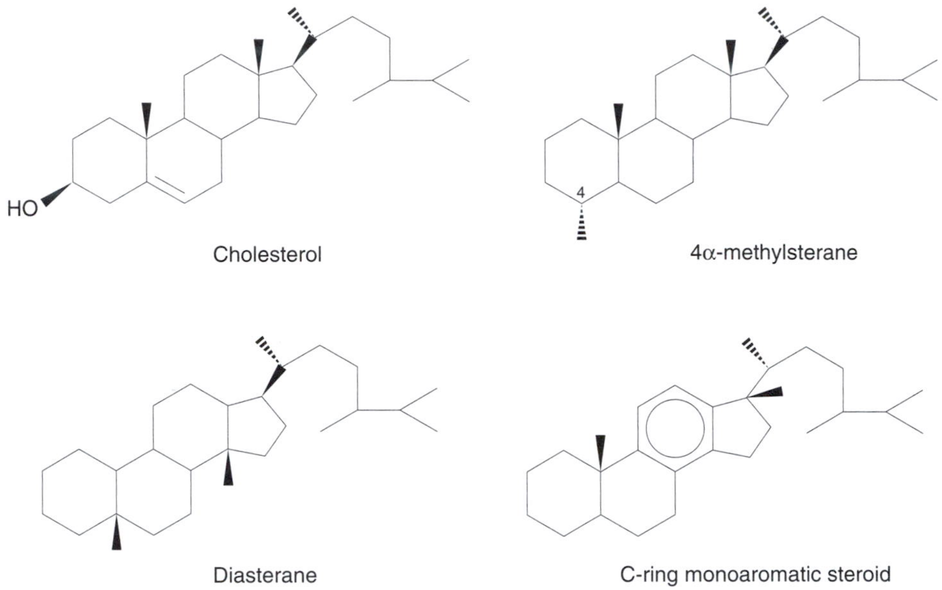

HO

Cholesterol

4

4α-methylsterane

Diasterane

C-ring monoaromatic steroid

Figure 4-13

Biological markers: steroid structures.

sterols have been isolated from marine and lacustrine organisms. About 40 of these in the C_{26}-to-C_{30} range were described by Scheuer (1973). Most of the variations in molecular structure occur in the side chain attached to the D ring.

Cholesterol, C_{27}, is the major sterol in thirty-five species of red algae, and fucosterol, C_{29}, was dominant in all brown algae examined (Patterson 1971). The sterols of green algae are more complex. Ergosterol, C_{28}, was predominant in five species of *Chlorella*. In five other species, chondrillasterol, C_{29}, was the major sterol, and two species contained poriferasterol, C_{29}, as the dominant sterol (Patterson 1971). Dinosterol, C_{30}, was found in some species of dinoflagellates.

The major sterol in higher plants is β-sitosterol, C_{29}. Stigmasterol, C_{29}, is abundant in soybeans and other terrestrial plants. Campesterol, C_{28}, and brassicasterol, C_{28}, are other sterols found in land plants. None of these four sterols is an important constituent of marine or lacustrine organisms.

Djerassi (1981) found that the bioalkylation (lengthening) of the sterol side chain at the 22- and 23-carbon positions is unique to marine sterols. Higher plants exclusively form sterols with side chains having alkyl substituents at C_{24}. Furthermore, no terrestrial sterols with quaternary carbons in the side chain are known, whereas this occurs frequently in C_{30} and C_{31} marine sterols (Djerassi 1981).

After the organism dies, sterols are converted to stanols, sterenes, and finally steranes, by means of microbial activity and low-temperature diagenetic reactions. Steranes do not exist in living organisms or at the sea bottom, but their concentration increases with deeper burial in the sediments (Mackenzie et al. 1982).

The reactions leading from sterols to steranes do not normally change the carbon number. Consequently, when Huang and Meinschein (1979) suggested that the C_{27} and C_{29} sterol concentrations in surface sediments could be used to distinguish terrestrial from marine plant sources, respectively, it followed that plots of the steranes should show the same relationship. Triangular plots of C_{27}, C_{28}, and C_{29} steranes in the source rocks and crude oils of many areas have shown a predominance of the C_{29}, in which the organic matter has a substantial terrestrial contribution (see Figure 15-20).

This source indicator does not work, however, when the marine contribution is dominated by an organism, such as brown algae, with C_{29} as its major sterol. Some marine environments such as the carbonate source rocks of the South Florida Basin (Palacas et al. 1984) have a C_{29} as the predominant sterane. This is also true of some Precambrian South Oman crude oils whose source rocks were deposited before the emergence of land plants (Grantham 1986). The C_{29} steranes also are dominant in the Precambrian Riphean crude oils of the Siberian Platform (Arefev 1980).

Moldowan et al. (1985) noticed a considerable overlap in the marine and nonmarine sterane distributions of about forty oils from worldwide sources when plotted on the C_{27}–C_{28}–C_{29} ternary diagram. Consequently, these diagrams should be used with caution as depositional source indicators. There was somewhat less overlap when plotting the monoaromatized steroids. The most reliable indicator in the forty oil samples was the 24-*n*-propyl-C_{30} steranes that were present in all oils from post-Devonian marine source rocks and absent in all oils derived solely from lacustrine or other nonmarine sediments. The precursor appears to be 24-propylidene cholesterol, which occurs in the marine algae *Chrysophyte*.

The 4α-methyl steroids found in both marine and lacustrine sediments are derived specifically from dinoflagellates. These are the only organisms in which these compounds have been recognized (Brassell and Eglinton 1986, and references therein). The 4-methyl steranes (Figure 4-13) formed from these steroids have been identified in lacustrine oil shales, such as the Messel of Germany (Kimble et al. 1974) and the Maoming of China (Brassel et al. 1986a) where they co-occur with botryococcanes. Both of these shales are organic-rich sediments deposited in shallow, swampy lakes. High 4-methylsterane-to-sterane ratios also have been observed in some bitumens of the Otway Basin of Australia (McKirdy et al. 1986) and in oils and shales from the Shengli field in China (Shi et al. 1982).

The diagenesis of sterols also leads to rearranged sterenes called *diasterenes* which then are gradually reduced to diasteranes (Figure 4-13). These reactions, which are described in detail by Mackenzie et al. (1982), are thought to be catalyzed by acidic sites on clays during diagenesis and early catagenesis. For example, the organic matter in the clay-rich post-Neocomian shales of the North Slope of Alaska have high diasterane contents (Seifert et al. 1980). Kruge (1986) observed that the quantity of diasteranes to steranes increased with depth in the Monterey Formation of the San Joaquin Basin. Hydrogenation of the diasterenes to diasteranes appeared to occur mainly during late diagenesis. An increase in the formation of diasteranes from sterenes also was observed by Brassell et al. (1984) in clastic sediments of the Falkland Plateau.

Diasterane/sterane ratios are commonly used to distinguish oils generated by carbonate source rocks versus those formed by clastic source rocks. Low ratios in oils indicate anoxic, clay-poor carbonate source rocks. High ratios are typical of oils derived from clastic source rocks with abundant clay minerals (Peters and Moldowan 1993). However, high ratios also occur in bitumens from organic-lean carbonate rocks.

Palacas et al. (1984) saw no correlation between the clay content and the abundance of diasteranes in the carbonate–evaporite low-TOC sediments of the South Florida Basin. Some samples with little or no clay contained large concentrations of diasteranes, whereas other carbonates with 20 to 40% clay minerals contained minor amounts of diasteranes. One problem is that both biodegradation and increasing thermal maturity tend to destroy steranes in preference to diasteranes, thereby increasing the ratio. Consequently, the ratios are most useful for comparing samples that have similar histories of thermal stress and bioalteration.

Some steranes form monoaromatic A rings and C rings during diagenesis. Both have been detected in immature sediments, but only the C-ring monoaromatics are prominent in crude oils and mature source rocks (Philp 1985a). Aromatization of the B ring has been observed only after the steroid rearranges to an anthrasteroid (linear fused ring) structure (Hussler and Albrecht 1983).

C-ring monoaromatic steroids occur in two groups, C_{27}–C_{29} and C_{20}–C_{21}, the former containing a longer side chain, as shown in Figure 4-13. Thermal maturation results in side-chain cleavage to give the smaller molecule. There are also C_{21} and C_{22} steroid natural products that can form short side-chain monoaromatic steroids. C-ring monoaromatic steroids were useful in correlating crude oils with source rocks in Prudhoe Bay, Alaska, the overthrust belt of Wyoming, and the Tarragona Basin of Spain (Seifert et al. 1983). The triaromatic steroids do not occur in Recent sediments. They are formed from monoaromatics with increasing depth of burial, as is explained in connection with maturation processes in Chapter 10.

Sulfur-Containing Biomarkers

Except for very waxy crudes and condensates, most crude oils contain some sulfur compounds. According to Chapter 3, the API gravity of crude oils varies inversely with sulfur content. High–API gravity oils are low in sulfur, and low–API gravity oils are high in sulfur. The reason for this is that most sulfur compounds are in the lubricating-oil and the asphaltene-bearing residuum fractions of the oil, whose gravities are in the 5-to-20°API range. Also, sulfur compounds are heavier than structurally equivalent hydrocarbons (Figure 3-12). High-sulfur oils can come from high-sulfur kerogens, such as some of the Monterey oils of California, or from the biodegradation of more conventional oils, such as the Athabasca heavy oils of Alberta. Biodegradation processes tend to selectively remove the nonsulfurous compounds.

A few organic sulfur compounds in crude oil may be formed in the living organisms. For example, sulfur bacteria contain 5 to 8% sulfur, and brown (*Phaeophyceae*) and red (*Phodophyceae*) algae contain 0.7 and 0.1% to more than 3% sulfur, respectively. However, the greater quantities of sulfur compounds and their structural diversity in crude oils compared with the biomass rule out organisms as an important direct source.

The idea that most of the sulfur compounds in crude oil are formed during early diagenesis by the interaction of inorganic H_2S and polysulfides with unsaturated organic structures and other unstable organic compounds was proposed by Ivlev et al. in 1973, based on thermodynamic considerations. Since then, detailed studies of sulfur compound structures and their sulfur isotope ratios have confirmed that this process is the major source of reduced sulfur in sediments and crude oils (Orr and Sinninghe Damste 1990). The H_2S is produced by microbial sulfate reduction under reducing conditions in all types of lithologies. In clastic depositional environments, detrital iron reacts with the H_2S, thereby minimizing the quantity of organic sulfur compounds formed. But in biogenic–pelagic sediments such as carbonates, diatomites, and bedded cherts, there is little detrital iron. Consequently, high-sulfur oils tend to be associated with carbonate sequences, and low-sulfur oils are usually associated with terrestrial clastics.

By 1978 more than 200 individual sulfur compounds were identified in crude oil, mostly in the low-molecular-weight range (Orr 1978). Then in the early 1980s, several homologous series of terpenoid sulfides were identified in Athabasca bitumen and a variety of crude oils. Payzant et al. (1986) summarized the isolation of bicyclic, tricyclic, and tetracyclic terpenoid sulfides plus some hexacyclic sulfides having the 17α(H),21β(H) hopane carbon skeleton from these oils. The sulfides were found to have the same structures as the hydrocarbons discussed in the previous sections, except that a sulfur atom had replaced a carbon atom in one of the naphthenic or aromatic rings of the molecules.

Since 1986 nearly all the well-known biological marker compounds, such as the long-chain alkanes, acyclic and cyclic isoprenoids, terpenoids, and steroids containing both naphthenic and aromatic rings, have been found to occur with one or more sulfur atoms in their structures. More than 1,500 sulfur compounds have been identified that can be related directly to biomarker precursors (Sinninghe Damste and de Leeuw 1990).

Sulfur-containing biomarkers are currently being investigated as possible paleoenvironmental indicators and for maturity and biodegradation evaluations and crude oil correlations. Several examples of these are discussed by Sinninghe Damste and de Leeuw (1990) and in the book *Geochemistry of Sulfur in Fossil Fuels* edited by Orr and White (1990).

Distribution of Biomarkers

Quantities in Crude Oil

The absolute concentrations of biological markers in crude oils vary widely depending on the depositional environment of the source rock plus the maturation and alteration of the crude oils in their source rocks and reservoirs. Biomarkers in general, however, are present in low concentrations compared with the major hydrocarbon groups such as *n*-paraffins. Table 4-7 gives the quantities of a few biomarkers reported for various oils, along with the average quantity of *n*-paraffins. An oil typically contains more than thirty times as much *n*-paraffins as steroids. Most individual biomarkers occur in the 10-to-1,000 ppm (0.001 to 0.1%) range in crude oils. Biomarkers as a group, excluding the *n*-paraffins and acyclic isoprenoids, generally represent less than 2% of a total crude oil.

As oils are buried deeper, higher temperatures are encountered, and the biomarkers are cracked to smaller molecules and gradually disappear. Hopane concentrations in a North Sea crude oil decreased from 0.178 to 0.003% when going through the vitrinite reflectance maturation range from R_0 = 0.66 to 0.93% (van Graas 1990). Steranes decreased from 0.141 to 0.019% through the same interval. This and previous studies show that biomarker applications in petroleum geochemistry are limited to maturation levels before and up to the end of the oil-generation window.

Biomarker Indicators

Table 4-8 summarizes some of the representative biological markers that have been used in petroleum geochemistry, showing either the source material or

TABLE 4-7 Biological Marker Concentrations in Weight Percent of Crude Oil

n-paraffins	16	Triaromatic steroids	0.1
Acyclic isoprenoids	6	Monoaromatic steroids	0.15
Pristane + phytane	0.5	C_{29} monoaromatic steroids	0.01
Hopanes	0.3	Benzohopanes	0.03
$17\alpha(H)C_{30}$-hopanes	0.02	Aromatic secohopanes	0.02
Steranes	0.3	C_{30} aromatic secohopanes	0.003
Diasteranes	0.04		

Note: Based on data for several oils from the literature plus personal communications from P. Albrecht and from J. M. Moldowan (1990).

TABLE 4-8 Biological Markers as Source and Paleoenvironmental Indicators

Biomarker	*C range*	*Indication*
n-alkanes		
CPI>5	C_9–C_{21}	Marine, lacustrine algal source, C_{15}, C_{17}, C_{19} dominant
	C_{25}–C_{37}	Terrestrial plant wax source, C_{27}, C_{29}, C_{31} dominant
CPI<1	C_{12}–C_{24}	Bacterial source: oxic, anoxic, marine, lacustrine
	C_{20}–C_{32}	Saline, anoxic environment: carbonates, evaporites
Acyclic isoprenoids		
Head to tail		
Pristane	C_{19}	Chlorophyll, α-tocopherol, oxic, suboxic environments
Phytane	C_{20}	Chlorophyll, phytanylethers of methanogens, anoxic, saline
Head to head	C_{25},C_{30},C_{40}	Archaebacteria, bacterial cell-wall lipids
Botrycoccane	C_{34}	Lacustrine, brackish
Sesquiterpenoids		
Cadalene, eudesmane	C_{15}	Terrestrial plants
Diterpenoids		
Abietane, pimarane, kaurane, retene	C_{19},C_{20}	Higher plant resins
Tricyclic terpanes	C_{19}–C_{45}	Diagenetic products of bacterial and algal cell-wall lipids
Tetracyclic terpanes	C_{24}–C_{27}	Degradation of pentacyclic triterpenoids
Hopanes	C_{27}–C_{40}	Bacteria
Norhopanes	C_{27}–C_{28}	Anoxic marine
2- and 3- methylhopanes	C_{28}–C_{36}	Carbonate rocks
Benzohopanoids	C_{32}–C_{35}	Carbonate environments
Hexahydrobenzohopenoids	C_{32}–C_{35}	Anoxic, carbonate-anhydrite
Gammacerane	C_{30}	Hypersaline environments
Oleananes, lupanes	C_{30}	Late Cretaceous and Tertiary flowering plants
Bicadinane	C_{30}	Gymnosperm tree resins
β-carotane	C_{40}	Arid, hypersaline
Steranes	C_{19}–C_{23} C_{26}–C_{30}	Eukaryote organisms, plants, and animals
24-*n*-propylsterane	C_{30}	Restricted to marine sediments
4-methylsteranes	C_{28}–C_{30}	Marine and lacustrine dinoflagellates
Dinosteranes	C_{30}	Marine, Triassic or younger

the environment of deposition that has generally been associated with the biomarkers listed. These are the compounds that are often found in the gas chromatograms and mass chromatograms of the organic extracts of immature rocks. For example, most of the major peaks for the immature Irati Shale extract shown in Figure 4-14 are biomarkers. This Permian shale extends for 1,700 km (1,056 miles) through Brazil. However, much of it has never been heated enough to generate significant thermal hydrocarbons and mask the immature biomarker signature. The extract shown is rich in terpanes, steranes, carotanes, and isoprenoid branched alkanes. As the organic matter of this shale is buried deeper and heated to higher temperatures, additional hydrocarbons form thermally from kerogen and gradually dilute the peaks shown in Figure 4-14.

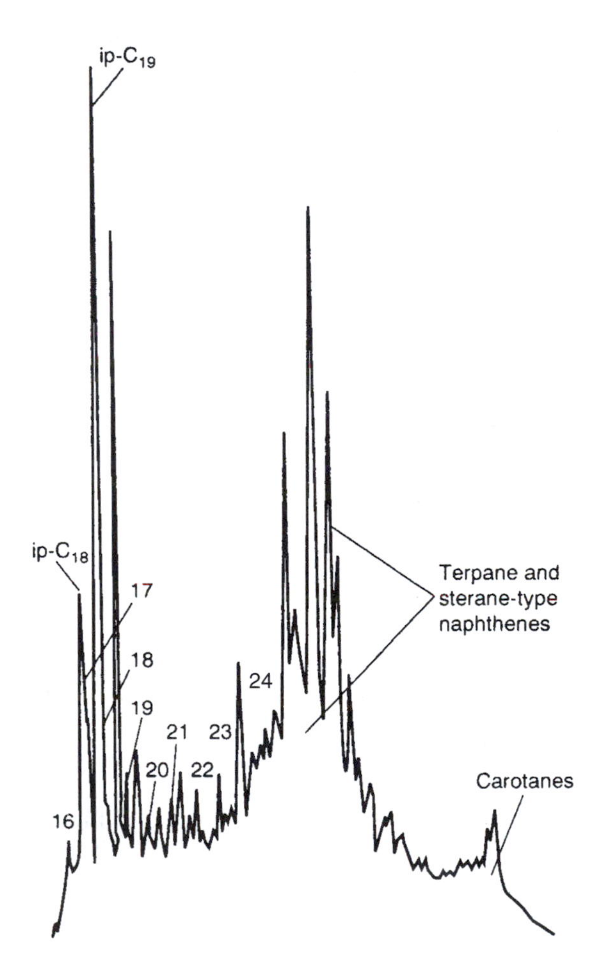

Figure 4-14
Gas chromatograph of the C_{15+} paraffin–naphthene hydrocarbon fraction of the Irati shale. Isoprenoid (ip-) hydrocarbons C_{18}, C_{19} (pristane), and C_{20} (phytane) are dominant, plus terpanes, steranes, and carotanes. The numbers are for the *n*-paraffins from C_{16} to C_{24}.

Most of the biomarkers listed in Table 4-8, starting with pristane, can be used in ratios for either oil–oil or oil–source rock correlation. As previously mentioned, however, ratios such as pristane to phytane must be used with caution because of multiple sources. Phytane generally forms from chlorophyll in anoxic environments, but it also comes from C_{40}-phytanylethers in the cell walls of methanogens. Bacterially derived hopanes can open a ring to form tetracyclic terpanes. But the tetracyclic structure also can come from the opening of a ring in the pentacyclic oleananes derived from flowering plants. This means that the source of some biomarkers cannot be unequivocally defined but that they still can be used in empirical correlations.

SUMMARY

1. Petroleum originates from a combination of the hydrocarbons formed directly by living organisms plus the hydrocarbons formed from the thermal alteration of the organic matter (OM) of sedimentary rocks. From an exploration standpoint, the quantity of hydrocarbons formed by inorganic processes is negligible.

2. The three major stages of petroleum formation are *diagenesis,* the biological, chemical, and physical alteration of OM before a pronounced effect of temperature; *catagenesis,* the thermal alteration of OM; and *metagenesis,* high-temperature thermal alteration. The approximate temperature ranges for these stages are diagenesis, up to 50°C (122°F); catagenesis, 50 to 200°C (122–392°F); and metagenesis, above 200°C (392°F).

3. Biological markers (biomarkers) are molecules in crude oils, rocks, and sediments whose carbon structures or skeletons can be traced back to living organisms. Biomarkers include the *n*-paraffins, porphyrins (chlorophyll), acyclic isoprenoids (pristane, phytane), terpenoids, and steroids. Although individual biomarkers are in concentrations of only 10 to 200 ppm in crude oil, they can be accurately measured, and their complexity and variety is so unusual that they are widely used in source rock correlation and maturation studies. They also are resistant to secondary processes, like biodegradation and high thermal maturity, and can thus be used to correlate heavily biodegraded or high-maturity oils with normal oils.

4. The two basic building blocks of biomarkers are (1) the 2-carbon acetic acid structure that combines to form long carbon chains and (2) the 5-carbon isoprene structure that combines to form all the isoprenoids and terpenoids as well as the precursor to the steroids. The isoprene rule states that organic compounds containing carbon atoms in multiples of five probably have isoprenoid structures.

5. Marine plants synthesize hydrocarbons with odd-carbon chain lengths in the C_{15}-to-C_{21} liquid range, whereas land plants synthesize odd-carbon chain lengths in the C_{25}-to-C_{35} solid range.

6. The carbon preference index (CPI) is a maturity indicator. It is high (five to ten) in immature sediments and is about one in mature shales and crude oils. It is less than one in anoxic carbonate rocks.

7. The ratio of the isoprenoids pristane to phytane is a source indicator. It is high in the coals and oils formed in suboxic environments and low in those formed in anoxic environments.

8. The biomarkers most commonly used in crude oil–source rock correlations and in maturation evaluations are the normal steranes, diasteranes, mono- and triaromatic steranes, and the tricyclic, tetracyclic, and pentacyclic terpanes.

9. The names of individual steranes, terpanes, and other biomarkers are based on four variations in their origin and chemistry: (1) compound class, (2) number of carbon atoms, (3) three-dimensional (3D) side-chain configuration, and (4) 3D ring system configuration.

SUPPLEMENTARY READING

Engel, M. H., and S. A. Macko (eds.). 1993. *Organic geochemistry.* New York: Plenum Press, 861 p.

Johns, R. B. (ed.). 1986. *Biological markers in the sedimentary record.* New York: Elsevier, 364 p.

Philp, R. D. (ed.). 1985. *Fossil fuel biomarkers—Applications and spectra.* New York: Elsevier, 294 p.

Waples, D. W., and T. Machihara. 1991. *Biomarkers for geologists: A practical guide to the application of steranes and triterpanes in petroleum geology.* AAPG Methods in Exploration Series 9. Tulsa: American Association of Petroleum Geologists, 91 p.

Whelan, J., and J. W. Farrington (eds.). 1992. *Organic matter: Productivity, accumulation, and preservation in recent and ancient sediments.* New York: Columbia University Press, 533 p.

Chapter

5

How Oil Forms: Generated Hydrocarbons

Chapter 4 described the hydrocarbons formed by living organisms, which is the pathway on the left in Figure 4-1. This chapter describes how hydrocarbons are generated from the organic matter (OM) deposited with the sediments. This is the pathway on the right in Figure 4-1. When organisms die, their organic matter undergoes a variety of reactions, some microbial, such as the formation of methane by anaerobes, and some purely physical or chemical, such as dehydration and oxidation. The combined attack of weathering and microbes converts much of the organic matter either to gasses that escape into the atmosphere or soluble products that are carried off by groundwater. In high-energy oxygenated enivornments, additional material is consumed by benthic filter feeders and burrowing organisms in nearsurface sediments. Environments that preserve unusually large amounts of organic matter in the sediments are stagnant lakes and silled basins, where the bottom waters are strongly reducing, as in the Black Sea today. In such areas, the organic content of the sediment frequently exceeds 15%. At the other extreme are the red clays of the oceanic abyssal plains, where slow rates of deposition, aerobic waters, and little contribution of organic matter results in sedimentary organic contents of less than 0.1%.

Neither of these extremes is typical of most petroleum-forming environments. Oil is formed from the organic matter deposited in sedimentary basins, where the water column is aerobic, but the bottom waters are periodically anaerobic and the sediments are nearly always anaerobic below the first few centimeters. The more resistant organic matter, including humic material, resins, waxes and lipids are preferentially preserved. The organic content of such sediments, which eventually become source rocks of petroleum, generally ranges between about 0.5 and 5%, with a mean around 1.5%.

Detailed studies of the organic structures in sediments and petroleum have made it possible to explain some of the mechanisms and conditions of oil formation. The overall process can be summarized briefly as follows: The simple molecular hydrocarbon spectrum of living organisms becomes the complex spectrum of petroleum through the diagenetic formation of a wider group of hydrocarbon derivatives from the original organic molecules and the addition of large quantities of hydrocarbons formed by thermal alteration of deeply buried organic matter. The largest quantity of petroleum hydrocarbons is formed from organic matter heated in the earth to temperatures between about 60 and 150°C (140 and 302°F).

The three main stages of OM alteration were defined in Chapter 4 as diagenesis, catagenesis, and metagenesis. Details of these three stages are described in the following sections.

Diagenesis of Organic Matter

Source and Quantity of Organic Matter

The quantity of petroleum generated is determined mainly by the quantity of hydrogen in the organic matter (OM) in the sediment. It follows that strongly reducing environments such as stagnant lakes and silled basins preserve and enhance the quantity and hydrogen content of OM, whereas oxidizing environments lower it. The OM deposited in sediments consists primarily of biopolymers from living things: carbohydrates, proteins, lipids, lignin, and subgroups such as chitin, waxes, resins, glycosides, pigments, fats, and essential oils. Some of this material is consumed by burrowing organisms; some may be complexed with the mineral matter; and some is attacked by microbes that use enzymes to degrade the biopolymers into the simple monomers from which they were originally formed. Some degraded biomonomers undergo no further reaction, but others condense to form complex high-molecular-weight geopolymers, which, along with undegraded biopolymers, become the precursors of kerogen. During diagenesis under reducing conditions, this complex mixture of geo- and biopolymers and monomers undergoes a whole series of low-temperature biological and chemical reactions that result in the formation of more hydrocarbon-like materials through the loss of oxygen, nitrogen, and sulfur. The stronger the reducing conditions are, the greater the hydrogen content of the resulting OM will be.

Some biologists (e.g., Rhoades and Morse 1971) defined marine and lacustrine environments as oxic if the waters contain more than 1 ml/l of dissolved oxygen, dysoxic or suboxic if the dissolved oxygen is in the range of 1 to 0.1 ml/l, and anoxic if it is less than 0.1 ml/l.* The organisms that live in these environments are called *aerobic, dysaerobic,* and *anaerobic,* respectively. Oxygen contents of 0.1 ml/l represent the approximate lower limit of benthic metazoa, whereas 1 ml/l represents the first appearance of a diverse benthic assemblage of calcareous fossils that need oxygen to survive. The dysaerobic organisms need less oxygen, and the anaerobic organisms, such as sulfate-reducing bacteria, are active only in the absence of oxygen.

There is no systematic correlation between primary biological activity and the total organic carbon content (TOC) of marine or lacustrine muds at the sediment–water interface (Demaison and Moore 1980). The reason is that the preservation of the OM is more important than the production of the OM. Worldwide studies have shown that on the average, only about 0.6% of the organic carbon produced in marine basins survives burial in the sediments (Hunt 1979, p. 105). Bralower and Thierstein (1984) used the TOC preservation factor to compare a large number of depositional environments. This is defined as the TOC accumulation rate in surface sediments expressed as a percentage of the primary production rate in overlying surface waters. The preservation factor for well-oxidized environments such as the Central Pacific was 0.01%, whereas strongly reducing environments like the Black Sea and Saanich Inlet had values of 10%. In general, nearshore sites had TOC preservation factors several orders of magnitude higher than hemipelagic and pelagic sites.

Phytoplankton (plants) and zooplankton (animals) constitute over 90% of the life in the oceans. The main producers of OM among the phytoplankton are unicellular diatoms with a siliceous skeleton, found primarily in the temperate and cold zones; peridineans with an algulose skeleton, found in warm waters; and coccolithophores, unicellular plants with a calcareous skeleton that are abundant in warm seas. The primary consumers of phytoplankton are small herbivorous zooplankton such as copepods. They serve as food for larger carnivorous zooplankton and for fish, the third step in the marine food chain. Added to this marine source is the often considerable contribution of land-derived OM. Wind-blown spores, pollen, other organic debris, and woody and recycled OM draining from continents by river, submarine discharges, and runoff can substantially alter the original marine matrix. OM that falls to the seafloor is reworked by benthic organisms and bacteria, which contribute their own products. For example, Lijmbach (1975) observed that the bacterial reworking of OM results in an amorphous biomass high in bacterial bodies that more readily forms oil.

In the oxic environment (Figure 5-1), dead OM falling through the water column is partly consumed by zooplanton and aerobic microorganisms. The more resistant particles, such as fecal pellets and woody particles, that reach

*Tyson and Pearson (1991, p. 7) recommended the following oxygen ranges for environments based on more recent observations: oxic, 8 to 2.0 ml O_2/l H_2O; dysoxic, 2 to 0.2 ml/l; suboxic, 0.2 to 0.0 ml/l; and anoxic, 0.0 ml/l (H_2S present). The corresponding biofacies are called *aerobic, dysaerobic, quasi-anaerobic,* and *anaerobic.*

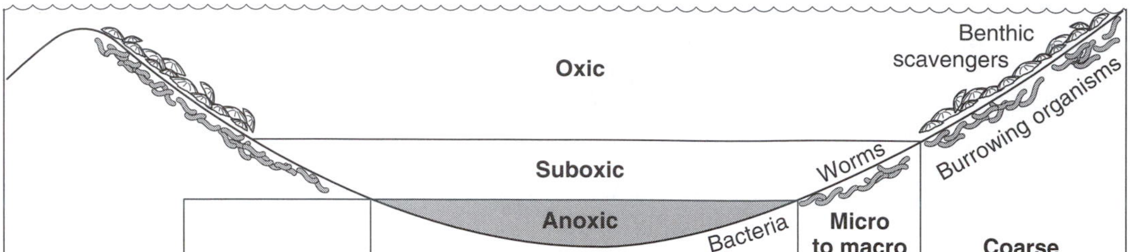

Oxic

Benthic scavengers

Suboxic

Worms

Burrowing organisms

Anoxic

Bacteria

Sediment	Laminated, no bioturbation	Micro to macro burrowed	Coarse bioturbation
% TOC	3 to 20	1 to 3	0.05 to 1
H/C of OM	1.6	1.2	0.8
OM type	I and II	II – III	III – IV

Figure 5-1

Marine and lacustrine benthic environments. Oxygen contents in ml/l H_2O are oxic greater than 1, suboxic 1 to 0.1, and anoxic less than 0.1 OM. Types I to IV are defined in Chapter 6.

the bottom are further scavenged by heavily calcified benthic organisms, such as brachiopods and molluscs. Soft-bodied worms and other burrowing organisms may extend several centimeters into the sediments, increasing the bioturbation and diffusion of oxygen below the sea bottom. Bioturbation has been observed at all water depths, including the deep-sea sediments of the abyssal plains where the waters are oxic (Demaison and Moore 1980). In such well-oxidized environments, the TOC of the bottom sediments rarely exceeds 1% and frequently is in the range of 0.05 to 0.5% (Hunt 1979, p. 105).

Even highly oxic sediments, however, become anoxic within a meter or so of depth if they are fine grained (Figure 5-1). The Eh is a measure of the oxidizing or reducing intensity of a chemical system. In sediments containing available oxygen, the Eh varies between 0 and +400 millivolts relative to the standard hydrogen electrode. In reducing sediments containing hydrogen sulfide, it varies from 0 to –400 mv. Emery (1960) noted that the surface sediments in some oxic basins offshore California had a positive Eh. At sediment depths greater than 2 m (6.6 ft) the Eh values generally became 0, and at the greater depths the values fell as low as –300 mv. The reason for this change is that aerobes living near the sediment–water interface consume all available oxygen as they decompose the OM. Further OM decomposition can take place only by anaerobic organisms, such as sulfate reducers, which generate H_2S in the process of oxidizing organic matter to CO_2, as follows:

$$2\ CH_2O + SO_4^{2-} \rightarrow H_2S + 2\ HCO_3^-$$

This reaction starts below the depth where oxygen is available, which can occur in the water column or in the sediments. Lindblom and Lupton (1961) observed that the switch from aerobes to anaerobes in a carbonate mud from Florida Bay occurred within the first meter (3.3 ft) of depth.

Bacterial activity decreases rapidly with depth in fine-grained sediments. Aerobes and anaerobes measure in the millions per cc at the surface, whereas only a few hundred exist at depths beyond 3 m (10 ft). Preliminary experiments on deep-sea drilling project sediments by Whelan et al. (1986) suggest that anaerobes such as sulfate reducers and methanogens, although few in number, can still be active at sediment depths of 167 m (548 ft), equivalent to an age of 10,000 years in some sediments.

Suboxic environments (Figure 5-1) exist in basins with a continual decrease in oxygen toward the basin floor, such as the Black Sea, Lake Maracaibo, and some of the borderland basins off Southern California. In going from oxic to suboxic environments, there is an exponential decrease in the species diversity of all benthic organisms (Rhoads and Morse 1971). Surviving members are very poorly calcified, and the quantity of burrowing organisms exceeds the deposit feeders. Finally, in the anoxic environment there are no organisms except anaerobic bacteria. The TOC values in the suboxic environment may range from 1 to 3 wt%, whereas values go up to 20% in the anoxic sediments. The high TOC of the latter is partly due to the fact that the oxic–anoxic boundary is up in the water column, which reduces exposure of the dead OM to oxygen during settling. The increase in TOC in the laminated anoxic sediments is generally accompanied by an increase in the H/C ratio of the OM (Figure 5-1).

Demaison and Moore (1980) defined sediments from the anoxic environments as strongly oil prone; those of the suboxic as moderately oil prone to gas prone, and of the oxic as gas prone to nonsource. These are generalized categories, since the shift between environments is not sharp, as suggested in Figure 5-1. However, it can change quickly in relatively short geological time periods.

Anoxic environments preserve more OM with the higher hydrogen content than others do. What geological conditions favor anoxic environments? The two most common factors are (1) restricted circulation and (2) a prominent oxygen-minimum layer.

Restricted Circulation

The patterns of water circulation in a basin determine whether or not vertical mixing is restricted. In all aquatic environments, organic matter is formed by the photosynthetic fixation of carbon in the photic zone, the interval penetrated by sunlight. This contributes to the high oxygen content in the near-surface waters, since oxygen is a by-product of green plant photosynthesis. Below the photic zone, however, there is a net consumption of oxygen due to the respiratory and biochemical decomposition of the organic matter formed above. *This net oxygen loss could cause all the world's waters to become anoxic below the photic zone if there were no circulation.* However, in most marine and lacustrine settings, the waters at depth are replenished by oxygen-bearing waters at such rates that oxygen consumption does not exceed oxygen renewal. For example, Figure 5-2 represents the type of basin in which there is a constant turnover of the bottom waters. In a warm climate, evaporation of the surface waters causes the formation of saltier, denser water that sinks to the bottom and flows out as a denser undercurrent into the ocean. Since this water comes from the photic zone, it is well oxygenated. As the deep water moves out, compensating seawater moves in at the surface. The Mediterranean and Red Seas are examples of this today. Processes of organic decomposition that consume oxygen can extend from below the photic zone, through the water column to the bottom. Consequently, any restriction in the supply of oxygen-bearing waters can cause a rapid shift from oxic to anoxic conditions. About 10,000 years ago the Black Sea was a freshwater lake, oxic to the sea bottom. Today it contains a 2,000-meter layer of hydrogen sulfide-rich water, the top of which is still rising. According to Jannasch, 1988, the top of the H_2S layer has risen from 125 m (410 ft) to about 90 m (295 ft) in the last thirteen years.

During the Wurm glacial stage of low sea level, the Black Sea was fresh because of an influx of water from rivers to the north and a sill to the south that

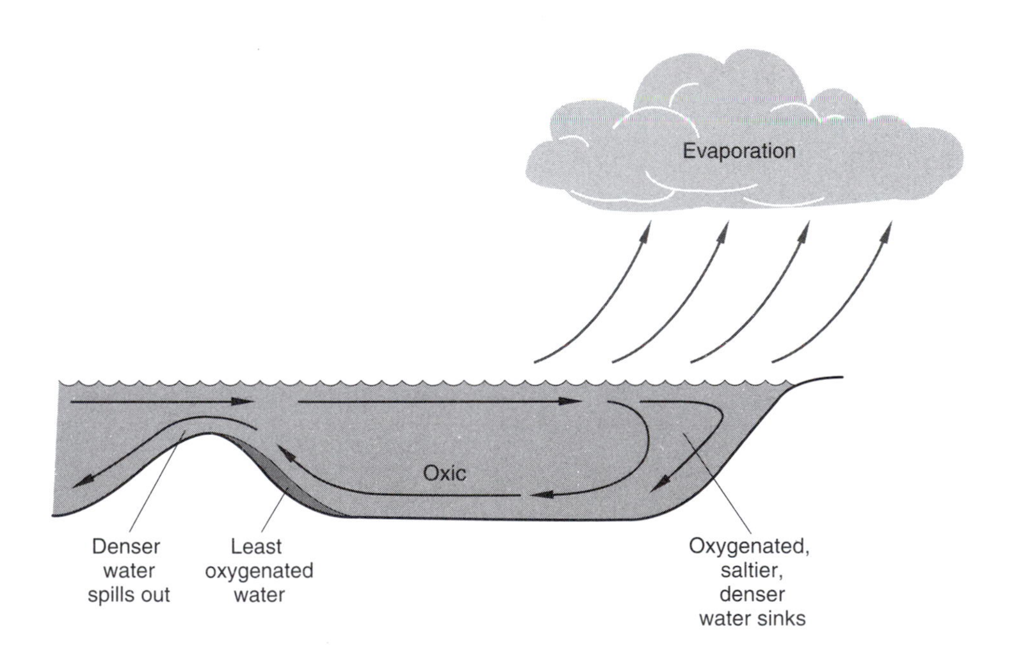

Figure 5-2

Oxic basin where the turnover of bottom water supplies oxygen to the bottom, as in the Mediterranean and Red Seas. [Demaison and Moore 1980]

prevented the invasion of seawater. As the glaciers melted and the global sea level rose, seawater flowed across the sill, forming a layer of bottom water in the basin that was too dense to mix with the overlying fresher river water. Fresh surface water flowed out to sea, and a permanent halocline developed (Figure 5-3) that prevented the influx of oxygenated waters. This caused the rate of oxygen consumption to exceed supply. Deuser (1974) estimated that below the photic zone, the supply of carbon from decaying OM was about 3,000 times greater than the available oxygen needed to decompose it. The TOC values range up to about 18% in the anoxic sediments and less than 2% in the oxic sediments of the Black Sea (Emery and Hunt 1974). A 1988 Black Sea cruise discovered a 30-meter suboxic zone between the oxic and anoxic water layers. It contained aerobic, chemoautotrophic bacteria that fix carbon dioxide in the absence of sunlight (Jannasch 1988).

Some basins that have a circulatory supply of oxygen-bearing waters may still become anoxic intermittently if the rate at which OM is introduced into the deeper waters is too high compared with the oxygen input. Examples of such oxygen-deficient waters today are found in Norwegian fjords, some basins off Southern California, the Cariaco Trench of Venezuela, and the Orca Basin in the Gulf of Mexico.

Lakes in cold northern regions are generally well oxygenated, as incoming cold river waters tend to sink to the bottom. Also, there is seasonal overturn that increases the oxygen supply. But in warm tropical climates, large anoxic lakes are common, such as Lakes Tanganyika and Kivu in the east African lake system. Lake Kivu contains more than 400 meters of anoxic H_2S waters, and

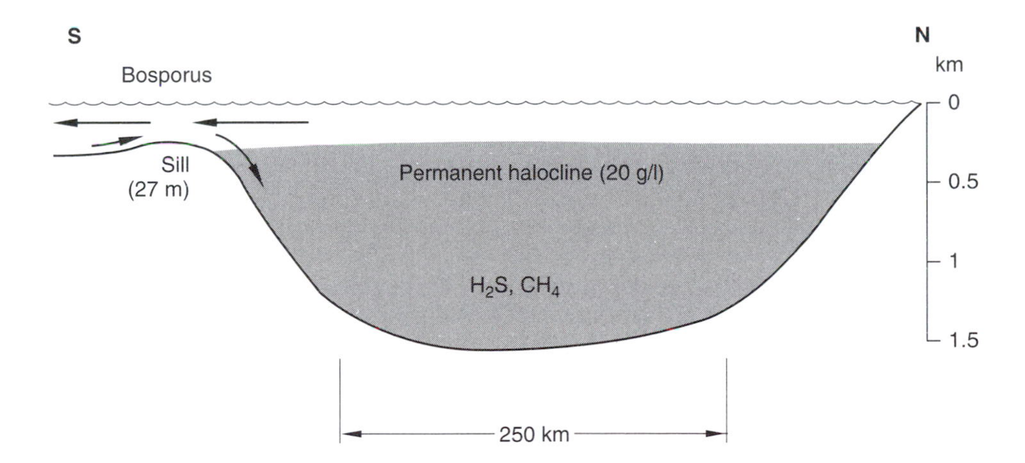

Figure 5-3

Anoxic basin (Black Sea) with stratified, stagnant H_2S bottom waters due to the lack of circulation, resulting in an insufficient input of oxygen. The Bosporus Strait of Turkey restricts circulation. [Demaison and Moore 1980]

Lake Tanganyika contains about 1,300 meters. Sediments in both lakes contain TOC values ranging as high as 10 to 15% (Degens et al. 1973). The Baltic Sea also contains H_2S waters in the lower part of the Gotland Deep. Romankevich (1984) found TOC values in this area above 3%. His book also includes the range of TOC values for sediments of the oceans and most inland seas of the world (pp. 105–160).

Known petroleum source rocks deposited in anoxic silled basins in the past include the Cretaceous Mowry Shale of the northwestern interior United States, the Upper Jurassic Bazhenov anoxic shales of the Western Siberian Basin, the Kimmeridge Shale of the North Sea, and the Toarcian Shales of the Paris Basin and Great Britain (Demaison and Moore 1980).

Changes in Sea Level

When the sea level rises or land subsides, the sea covers more of the continent's edges or even its interior than it did previously. This is defined as a *transgression.* A *regression* can result when the sea level drops because of the building out of nonmarine sediments, as in a delta. Simultaneous worldwide (*eustatic*) changes in sea level are caused primarily by tectonics, such as the variation in the speed of the seafloor's spreading. The extensive transgression of the seas on continental platforms during the Cretaceous period is thought to be related to the thermal expansion of the rapidly moving oceanic lithosphere. This would have reduced the general volume of ocean basins, thereby raising the sea level.

Emery and Aubrey (1991, p. 23) stated that the Cretaceous was probably the time of the fastest spreading of the seafloor, with the sea level reaching, between 110 and 85 M.Y. ago, a maximum of about 350 m above present ocean levels. This is the same stratigraphic interval (Aptian–Turonian) in which the source rocks of 29% of the world's conventional petroleum reserves were deposited (Klemme and Ulmishek 1991). The reason that transgressive cycles are important is because the greatest volume of the world's petroleum source rocks were deposited during transgressions.

Glacial ages are another important cause of eustatic changes, since the melting of large amounts of ice increases the volume of seawater, the depressed land area under the glacial mass rebounds, and the peripheral bulge sinks. However, Emery and Aubrey (1993) consider the ice ages to have had less effect on the sea level in the past than did movements of the seafloor.

The Oxygen Minimum Layer

For decades oceanographers have observed a worldwide phenomenon known as the *oxygen minimum layer.* Concentrations of oxygen in the seawater of many coastal areas decrease to a minimum at a depth of around 100 to 500 meters. The cause of this minimum is thought to be the great biological activity at these levels, resulting in large amounts of dead OM being recycled in the water column. This creates a proportionately large demand for oxygen to degrade the OM. Consequently, an oxygen minimum zone occurs in the water column, which causes anoxic conditions in underlying sediments, where it impinges on the coast.

For example, off the coast of Peru, the oxygen content of surface waters is around 5 ml/l. It drops to less than 2 ml/l between water depths of 30 and 1,000 meters and to less than 0.1 ml/l from about 75 to 500 meters. The TOC values in the sediments are as high as 10% where this oxygen minumum zone contacts the shelf. This organic-rich zone is about 75 km wide and extends along the shelf for over 1,000 km, as shown in Figure 5-4 (Demaison and Moore 1980). In contrast, TOC values are less than 3% where the zones of high-oxygen waters contact the shelf (Henrich and Farrington 1984). A similar situation exists in the upwelling zone along the southwest African Shelf, where the

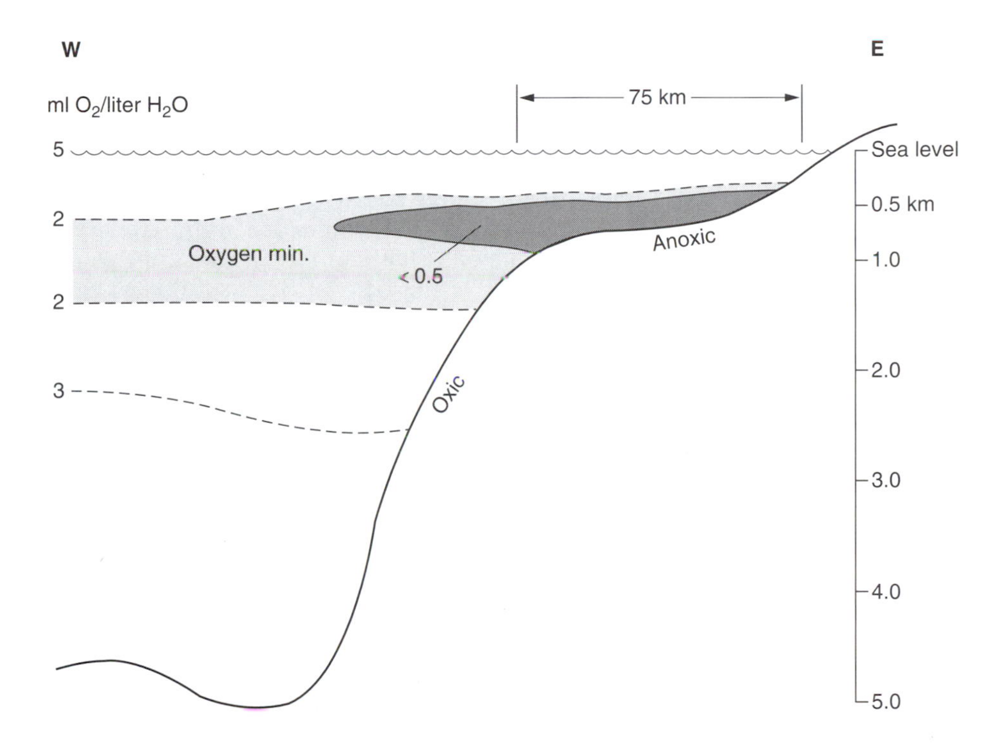

Figure 5-4

Anoxic sediments deposited along the shelf of Peru, owing to contact with the oxygen minimum layer containing <0.5 ml O_2/l H_2O. TOC values in anoxic sediments are 3 to 10%; in oxic, 0.5 to 3%. [Demaison and Moore 1980]

anoxic sediments cover an area approximately 150 × 700 kilometers. Accordingly, TOC values as high as 24 wt% have been recorded in the sediments of Walvis Bay where hydrogen sulfide periodically rises to the surface, causing mass mortality of marine life (Demaison and Moore 1980).

The oxygen minimum layer occurs worldwide wherever the consumption of oxygen from organic degradation exceeds the available supply. The Upper Continental Slope of the Indian Ocean and large regions of the eastern tropical Pacific Ocean have oxygen minimum layers today.

In the Gulf of California, the oxygen minimum zone of the Pacific Ocean impinges on the slope between about 300 and 1,300 meters. Within this interval are sediments of finely laminated diatomaceous earth containing over 5 wt% TOC. Above and below the oxygen minimum layer, the sediments are bioturbated and have TOCs generally below 2%, as can be seen in Figure 5-5.

Oxygen minimum zones with concentrations below 0.5 ml/l nearly always cause anoxic conditions where they contact sediments. But upwelling and other high productivity conditions do not always create an oxygen minimum. For example, the Atlantic Ocean off the Grand Banks of Newfoundland is one of the most productive areas of the world's oceans, yet there is no oxygen minimum zone because the water column is well oxygenated (over 6 ml/l). Likewise, in the upwelling area of the Northern Pacific offshore Japan, there are no low-oxygen waters (Demaison and Moore 1980).

The Permian Phosphoria Formation of North America is a good example of an organic-rich source rock formed by the effects of an oxygen minimum layer. The Phosphoria is the major source rock for much of the Paleozoic oil in the western interior United States. Other examples are in the Tertiary of California, such as the McClure and Antelope Shales of the San Joaquin Valley, the Middle Miocene Monterey Formation of the Coastal Basins, and the Puente Nodular Shale of the Los Angeles Basin. These sediments are typically laminated, phosphatic, bituminous, dark brown shales and siliceous, diatomaceous shales. Phosphatic material is abundant, both finely disseminated and in nodules. Also, the sediments are generally devoid of macrofossils (Demaison and Moore 1980). Other widespread phosphatic black shales formed by deep-water anoxia occur in the Pennsylvanian cyclothems of Kansas and Oklahoma (Wenger and Baker 1986). The TOC values are as high as 25% in the black shales, compared with less than 1% in the gray shales deposited under oxidizing conditions.

When the sea level is falling, the oxygen minumum zone tends to impinge on the continental slope rather than on the shelf, as illustrated in Figure 5-6. This means that a much smaller area of the seafloor becomes anoxic. With the rising sea level the oxygen minumum begins to spread across the shelf, greatly increasing the area of anoxicity.

Transgressive cycles produced far more petroleum source rocks in the world than did regressive cycles. The major transgression in the Cambrian (Figure 5-6) occurred before there was life on land and before marine life reached the high productivity levels of later periods. The first important transgression resulting in the deposition of widespread black shales was in the Ordovician, which has been the source of much oil in North America. The Late Devonian transgression produced major source rocks in the Urals–Volga Basin

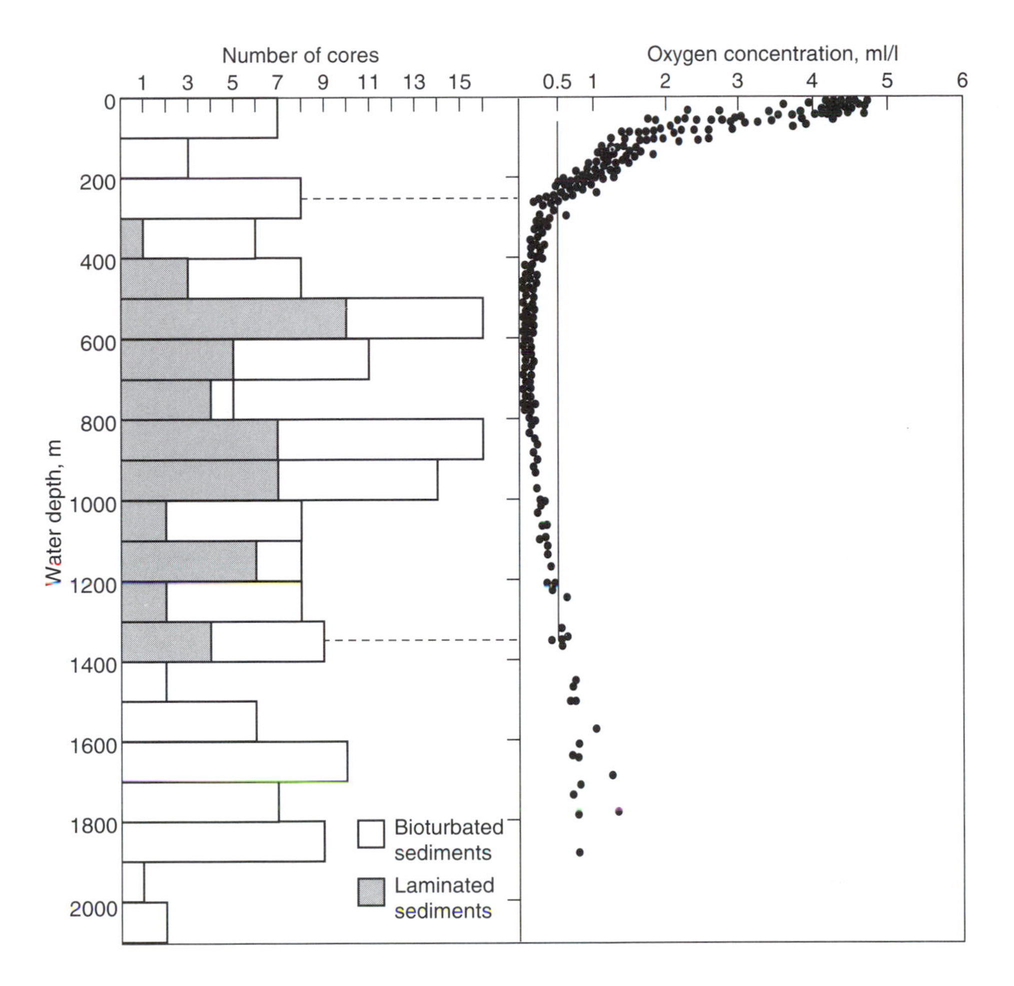

Figure 5-5

Increase in laminated sediments versus bioturbated sediments where Gulf of California bottom waters contain less than 0.5 ml O_2/l H_2O. [Demaison and Moore 1980]

of Russia and the Western Canada Basin, in addition to North Africa and parts of the United States. The largest transgression, according to Ronov (1994) and Vail et al. (1977), began in the Jurassic and continued until the end of the Cretaceous. The spreading of anoxic conditions during this period caused the deposition of the major source rocks of the North Sea, Middle East, western Siberia, eastern Venezuela, and much of middle America. Klemme and Ulmishek (1991) estimated that over half the world's conventional oil originated from Jurassic–Cretaceous source beds formed during this major

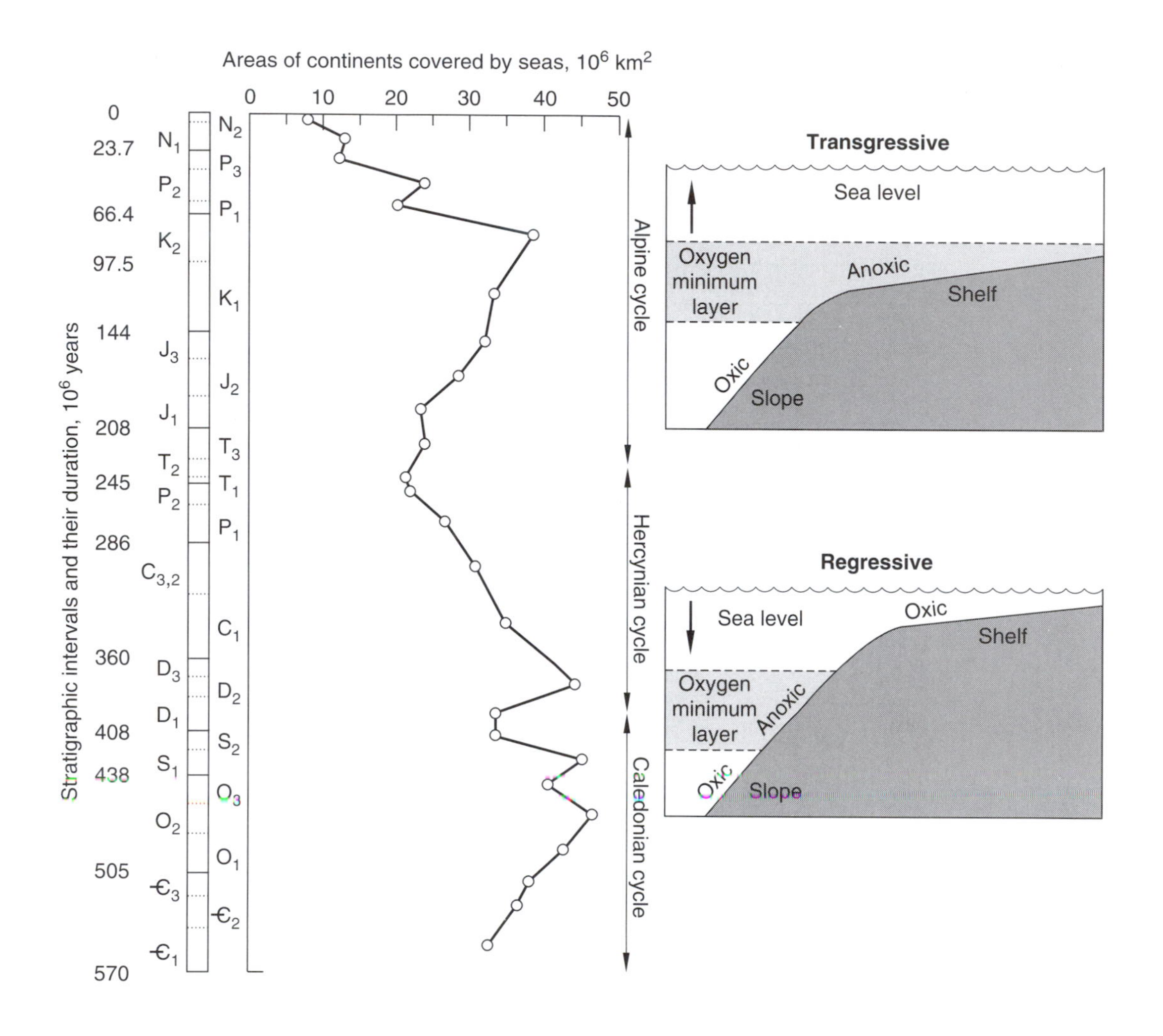

Figure 5-6

Transgressions and regressions of the seas on the continents during the Phanerozoic. Subscripts 1, 2, and 3 represent Early, Middle, and Late intervals, respectively, for the geologic periods listed. [Ronov 1994] Also shown are areas of slope and shelf covered by oxygen minimum zone during transgressive and regressive cycles.

transgression. Adding heavy oil raises this to over three-fourths of the world's oil (see Chapter 2).

The effects of transgressive and regressive cycles on the organic petrology and geochemistry of the Mancos shale of the San Juan Basin, New Mexico, was studied in detail by Pasley et al. (1991). They observed that the transgressive cycle sediments had higher TOCs and more hydrogen in their OM compared with the regressive cycle sediments. Figure 5-7 shows the hydrogen index plotted against TOC for five shale samples in each cycle plus five mudstones deposited at the beginning of the transgressive cycle. The hydrogen index (HI) is the milligrams of hydrocarbon (HC) per gram of TOC that is obtained on heating the shale sample to a high enough temperature to crack the kerogen to form oil (for more details, see Chapter 10). The HI values in the transgressive cycle of the Mancos Shale range from 433 to 623 mg HC/g TOC, whereas in the

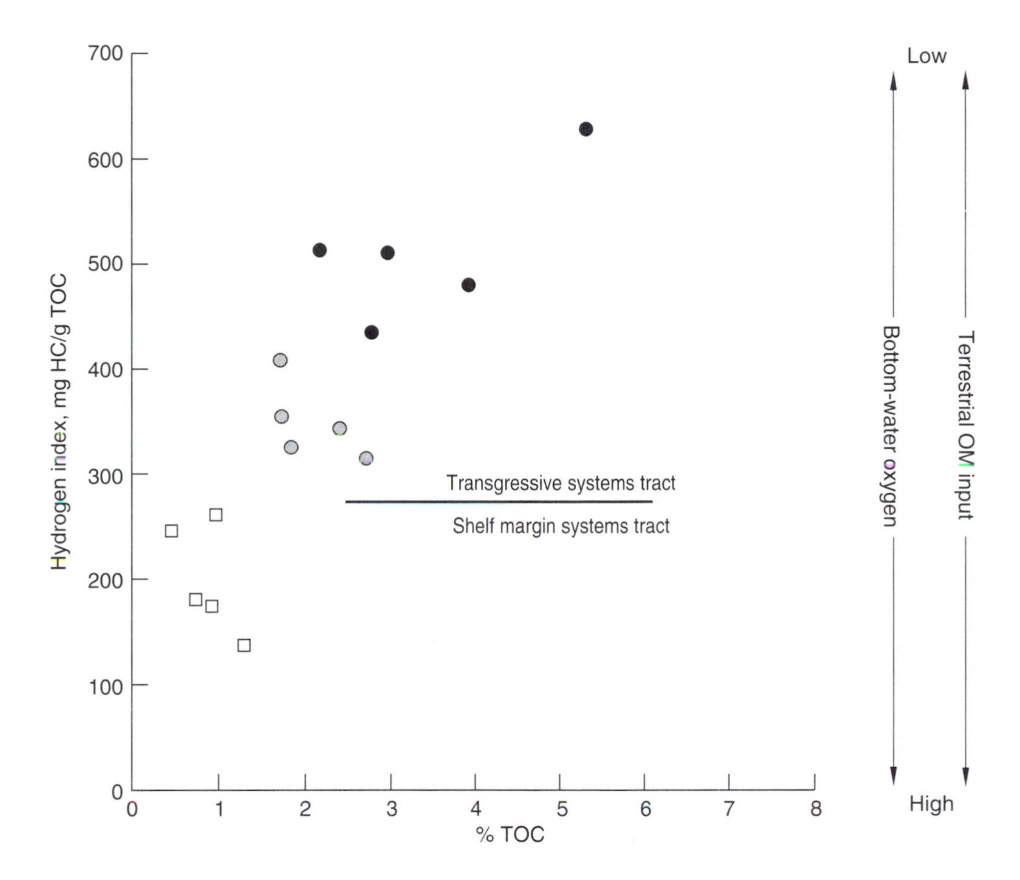

Figure 5-7

Comparison of TOC and hydrogen indices for transgressive and regressive cycles of the Mancos Shale of the San Juan Basin, New Mexico. Transgressive Mancos Shale ●; mudstone interbeds in transgressive Tocito Sandstone ●; and regressive Mancos Shale □. [Pasley et al. 1991]

regressive cycle they range from 135 to 263 mg HC/g TOC. The TOC values in the transgressive cycle range from 2.1 to 5.3 wt%, compared with 0.49 to 1.3 wt% for the regressive cycle.

The petrographic study by Pasley et al. (1991) of the Mancos Shale showed that samples in the transgressive cycle exhibited strong fluorescence and that those in the regressive cycle were weakly to nonfluorescent. The absence of burrowing in the transgressive cycle is indicative of anoxic conditions, whereas the regressive cycle shows burrowing with abundant *Zoophycos* and *Chondrites* characteristic of oxic or suboxic conditions. Macerals of terrestrial precursors such as phytoclasts, scleratoclasts, sporinite, and inertinite represented 67% of the total organic matter (TOM) in the regressive cycle but only 34% in the transgressive cycle.

Curiale et al. (1992) integrated isotopic, molecular, and paleontological data on rocks within the framework of a sequence stratigraphic model in the Western Interior Basin of New Mexico. They found that the best source rock intervals occur stratigraphically immediately above the condensed section, which is near the maximum rise in relative sea level. Molecular indicators such as desmethylsterane carbon number distributions and sterane/hopane ratios changed continuously as the transgression proceeded, whereas the hydrogen index and fluoramorphinite content showed a discontinuity near the transgressive–regressive boundary.

Chandra et al. (1993) observed that the more organic-rich shales in the Cauvery Basin of India were associated with the transgressive phases of five local sea level cycles. The best source rocks, however, were deposited during the previously mentioned global Aptian–Turonian rise in sea level caused by the spread of the seafloor. This underscores the fact that global transgressions are often superimposed on local transgressions during source rock deposition.

Coal-Forming Environments

Coal is a readily combustible rock containing more than 50% by weight, and more than 70% by volume, of organic material formed from the compaction or induration of variously altered plant remains. Coals are composed of *macerals,* which are microscopically recognizable constituents of coal that can be differentiated by their morphology. The dominant coal macerals in order of decreasing hydrogen contents are *liptinite, vitrinite,* and *inertinite* (see Glossary). *Humic* coals are formed under oxic conditions. Over 80% of the world's coals are humic, and vitrinite comprises more than 70% of their macerals. Humic OM is derived from plant cell and wall material, which is composed mostly of lignin and cellulose plus the aromatic tannins. Humification is accelerated by the presence of oxygen and heat, such as in tropical climates.

Less than 10% of all coals are *sapropelic.* This is a more specialized type of organic accumulation that occurs when the spores, pollen, cuticles, and resins of plant materials and the organic remains of algae and plankton accumulate in a swamp environment or the stagnant parts of lakes to form a watery ooze called *sapropel.* The term *gyttja* refers to any organic-rich sediment deposited in open waters, whereas *sapropels* are thought to form in waters low in or free of oxygen (suboxic to anoxic). Wind- or waterborne spores and the remains of surface algae are the major contributors to sapropel, which ultimately changes with depth of burial to form *cannel* and *boghead* coals. The approximate hydrogen to carbon atomic ratios of these major coal types are boghead 1.5, cannel 1.2, and humic 0.8.

The densest accumulation of OM occurs in the coastal swamps of high-vegetation areas. When large amounts of vegetation are deposited in shallow, stagnant, fresh-to-brackish water swamps, the pH falls to the range of 3.5 to 4. At this level, microbiological activity is so low that the rate of organic deposition exceeds microbiological decay and creates a peat bed, which is essentially pure OM. As long as a critical balance is maintained among drainage, sedimentation rate, and subsidence, the peat swamp may grow to considerable thickness. Muller (1964) estimated that it took 4,000 years to form 12 m (about 40 ft) of peat (equivalent to 1 m, or about 3.3 ft, of coal) in tropical Borneo. Peat deposits cover very large areas when the swamps are continuously formed behind a regressive shoreline.

Coal swamps originate in both regressive and transgressive cycles. Stach et al. (1982) cite the Miocene brown-coal swamps of central Germany and the Pennsylvanian coals in the eastern and central United States as examples of marine regressions. In contrast, many Carboniferous coal seams of the Ruhr and the present south coals of New Guinea are the result of forests drowned by marine transgressions.

Organic Matter, Kerogen, and Bitumen

The OM in Recent sediments is not the same as the kerogen in lithified rocks. About 40 to 60% of Recent OM is soluble in acids, bases, and organic solvents, compared with less than 20% of the kerogen in lithified rocks (Connan 1967; Huc et al. 1978). During diagenesis a whole series of low-temperature reactions occur, such as decarboxylation, deamination, polymerization, and reduction. These contribute the kerogen of ancient rocks (for more details, see Hunt 1979, pp. 112–119). In addition, some free bitumen is released, which consists largely of biological markers such as *n*-paraffins, isoprenoids, steranes, triterpanes, and porphyrins. The bitumen also contains nitrogen, sulfur, and oxygen (NSO) compounds. The ratio of hydrocarbons to these nonhydrocarbons in the bitumen of Recent sediments is in the range of 1/10 to 1/20. The extractable bitumen is usually only 5 to 15% of the TOC.

Light hydrocarbons are present at the parts-per-billion level in Recent sediments (Hunt 1984, and references therein). The paraffins in surface sediments are mainly straight chains. With increasing diagenesis, branched hydrocarbons become the dominant products from low-temperature carbonium ion or free-radical reactions. In most surface sediments the ratio of a branched pentane or hexane to its straight-chain homolog varies from 0.21 to 0.5. With increasing depth of burial, these ratios increase to a maximum between 2 and 5, as shown in Figure 5-8 for iso- and normal pentane (Hunt 1984). The change in the ratio shown is compared with the concentration of total C_6–C_7 hydrocarbons in fine-grained shales from a COST (coastal offshore stratigraphic test) well-drilled offshore South Padre Island, Texas, in the Gulf of Mexico. The peak in the ratio coincides with the threshold of intense oil generation, which is at

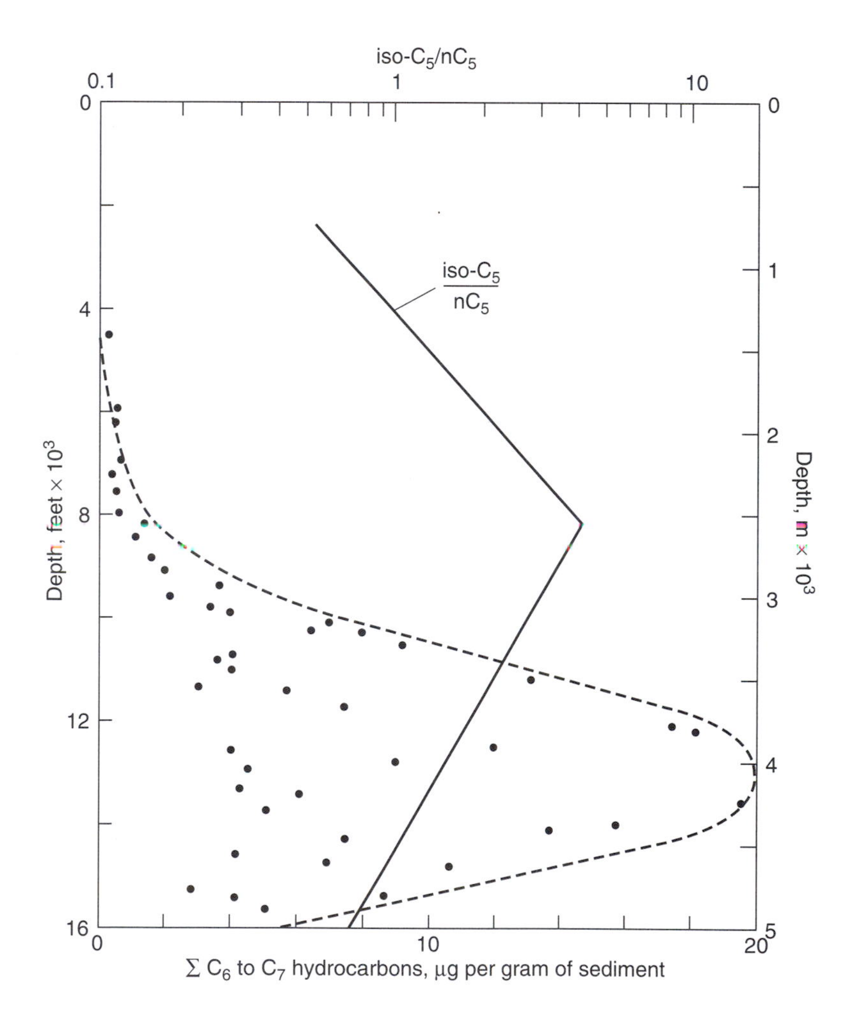

Figure 5-8

Distribution of C_6 and C_7 saturated hydrocarbons (dots) compared with the ratio of isopentane to *n*-pentane (solid line) in cuttings of the South Padre Island, Texas, COST No. 1 well. The dashed line encompasses the oil-generation window. [Hunt 1984]

about 2.5 km in Figure 5-8. The ratio decreases with depth below 2.5 km because high-temperature thermal cracking reactions form predominantly straight-chain hydrocarbons. A similar maximum is observed in the ratios of isobutane to *n*-butane and 2-methylpentane to *n*-hexane in Gulf Coast wells (Hunt 1984).

Biological markers undergo many reactions during diagenesis. For example, relatively high concentrations of C_{35} extended hopanes are formed from the reduction of bacteriohopanetetrol in highly reducing environments. Lower-numbered extended hopanes are formed in oxic environments. Tricyclic terpanes are demethylated in some environments. Sterenes formed from cholesterol are converted to both steranes and diasteranes in highly reducing environments, particularly in shaly clay-rich rocks. Monoaromatic steranes are formed from the aromatization of the A ring of the precursor sterols (Hussler et al. 1981). The many factors causing variations in individual biomarker concentrations in Recent sediments are still under investigation.

Catagenesis of Organic Matter

Organic matter is subjected to increasingly higher temperatures with greater depth of burial. Over time, these higher temperatures cause the thermal degradation of kerogen to yield petroleum-range hydrocarbons under reducing conditions. These hydrocarbon-forming reactions have been demonstrated many times in the laboratory and have been observed in the natural environment.

For example, Engler (1913) heated oleic acid and other organic materials at temperatures below 250°C and obtained paraffin, naphthene, and aromatic hydrocarbons in the entire petroleum range. Later, he described the generation of oil from OM as a two-step process involving bitumen as an intermediate. Subsequent studies by several investigators converted resins, acids, alcohols, ketones, and natural terpenes into petroleum hydrocarbons by heating at temperatures ranging from 50 to 200°C (Hunt 1979, pp. 120–121).

More recent studies on kerogen decomposition in the laboratory have demonstrated Engler's original description of a bitumen intermediate in oil generation to be valid for many, but not all, kerogens (Miknis et al. 1987).

$$\text{Kerogen} \rightarrow \text{bitumen} \rightarrow \text{oil} + \text{gas} + \text{residue}$$

In this equation the bitumen is defined as the nonvolatile organic matter soluble in organic solvents that is formed in the shale during initial heating, whereas kerogen is the insoluble organic matter. Further heating converts the bitumen to oil and gas. The residue is the nonvolatile portion of the kerogen and bitumen remaining after maximum heating in the laboratory or under natural conditions.

Kerogens are classified as types I, II, III, and IV, based on their C, H, and O contents (see Chapter 10 for details). Types I and II generate most of the world's oil, and type III generates primarily gas, condensate, and some waxy oil. Type IV generates only small amounts of methane and CO_2. Several studies cited by Miknis et al. (1987) illustrated that for lacustrine hydrogen-rich type I oil shales, the bitumen decomposition is the rate-controlling step for oil production. The formation of a bitumen intermediate seems to be typical of many type I and type II kerogens (Shen et al. 1984), but not necessarily of type III kerogens. Bitumen yield appears to be inversely proportional to the carbon aromaticity of the kerogen that is highest in type III (Miknis et al. 1987).

Strong evidence for a bitumen intermediate in oil generation comes from hydrous pyrolysis experiments that simulate oil generation and expulsion from source rocks. This technique involves heating the source rock isothermally in a closed system in the presence of water and measuring the extractable bitumen and expelled oil that is formed (M. D. Lewan 1985, 1993b). Figure 5-9 contains three examples of hydrous pyrolysis that show that bitumen is an intermediate between kerogen and oil.

In Figure 5-9 (A) the residual TOC is decreasing during bitumen formation but is relatively constant during oil generation. Similar results were observed in the other two experiments. This indicates that the loss of TOC is related mainly to bitumen formation rather than oil formation. In all three experiments, A, B, and C, the peak in bitumen occurs before the peak in expelled oil. Bitumen is declining while oil is increasing, indicating that oil is coming from the bitumen. The peak in bitumen formation for A, B, and C is about 290, 300, and 330°C, respectively. The peak in oil expulsion is around 340, 350, and 355°C, respectively. This would be expected, since the activation energies required to crack the kerogen increase in going from type II-S to type II to type I kerogens, as discussed in Chapter 6.

The Oil Window

The depth interval in which a petroleum source rock generates and expels most of its oil is called the *oil window*. Most oil windows are in the temperature range from 60°C (140°F) to 160°C (320°F).

Larskaya and Zhabrev (1964) were the first geochemists to demonstrate that the generation of hydrocarbons from the kerogen of shales increases exponentially with depth (increasing temperature). They found that the extractable bitumen content of shales in the western Ciscaspian region changed very little in the temperature range from 20 to 50°C, after which it increased markedly. In the Jurassic, Cretaceous, Paleocene, and Miocene deposits, the most intensive hydrocarbon generation started at 60°C, equivalent to present-day sediment depths of 1,200 to 1,500 m (3,940 to 4,920 ft). Hydrocarbon yields in the fine-grained shales increased by factors ranging from 3 to 7 in the deep sediments, compared with the shallow sediments. This extraction procedure involves pulverizing a source rock to a fine particle size and using hot organic solvents to remove the generated bitumen. After removing the solvents, the C_{15+} bitumen is separated into hydrocarbons and nonhydrocarbons (NSO compounds).

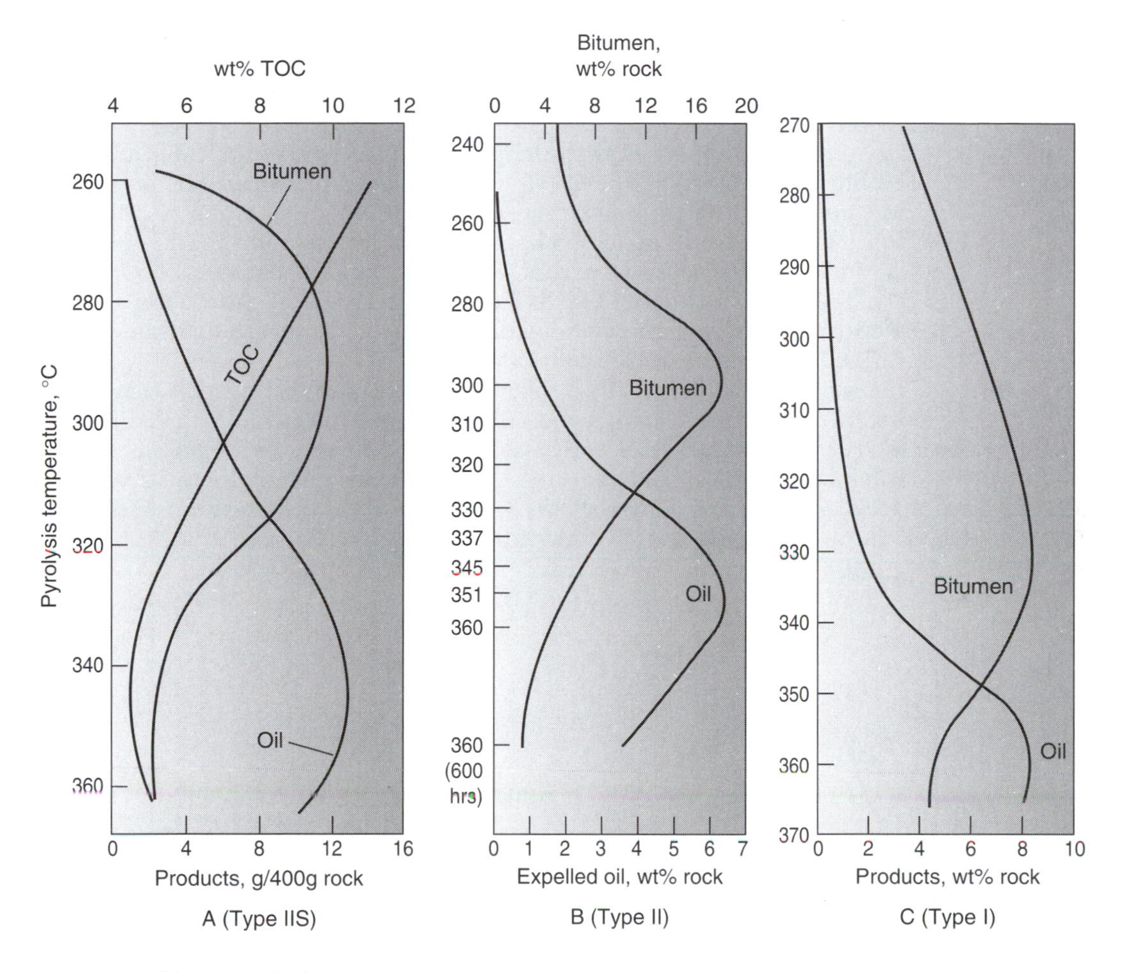

Figure 5-9

Quantity of bitumen and expelled oil generated at each pyrolysis temperature for 72 hours except for the 360°C sample in B, which was heated for 600 hours. A = Monterey Shale of California containing type II-S kerogen. The change in TOC is shown also. [Baskin and Peters 1992] B = Phosphoria Retort Shale of Montana containing type II kerogen. [Lewan et al. 1986] C = Eocene Green River Shale (Mahogany Ledge) of Utah containing type I kerogen. [Ruble 1995] Note that the temperature scales are different.

Some investigators show curves of hydrocarbons, but others show total bitumen yield. Gasoline-range hydrocarbons are determined simply by heating well cuttings to about 95°C and analyzing on a gas chromatograph the products released (see also Chapter 10).

Vassoevich et al. (1969) described this intensive generation zone as representing the "principal phase of oil formation." Connan (1974) called the depth at which a significant increase in hydrocarbons occurred the "threshold of intense oil generation."

The oil-generation window in the previously mentioned South Padre well is presented in Figure 5-8. The C_6–C_7 gasoline-range hydrocarbons occur only in trace amounts in Recent sediments. Consequently, the values (solid dots) are close to zero down to about 2.5 km (8,000 ft). At this point there is an exponential increase in the hydrocarbons formed, which peaks at about 4 km (13,000 ft), equivalent to 140°C (284°F). The interval from 3 to about 4.9 km (9,900 to 16,000 ft) is the oil window (dashed line). The threshold of intense oil generation is the beginning of the window.

The stratigraphic interval above 2.5 km in the South Padre well is referred to as *immature* with respect to oil generation, and the oil window is called *mature*. Below the oil window the rocks are *postmature* for oil but mature for gas.

The maturity differences are more visible in gas chromatograms (GCs) of rock extracts. Figure 5-10 shows the GC profile of an immature extract from the Kimmeridge Shale source rock of the North Sea (Thomas et al. 1985). The

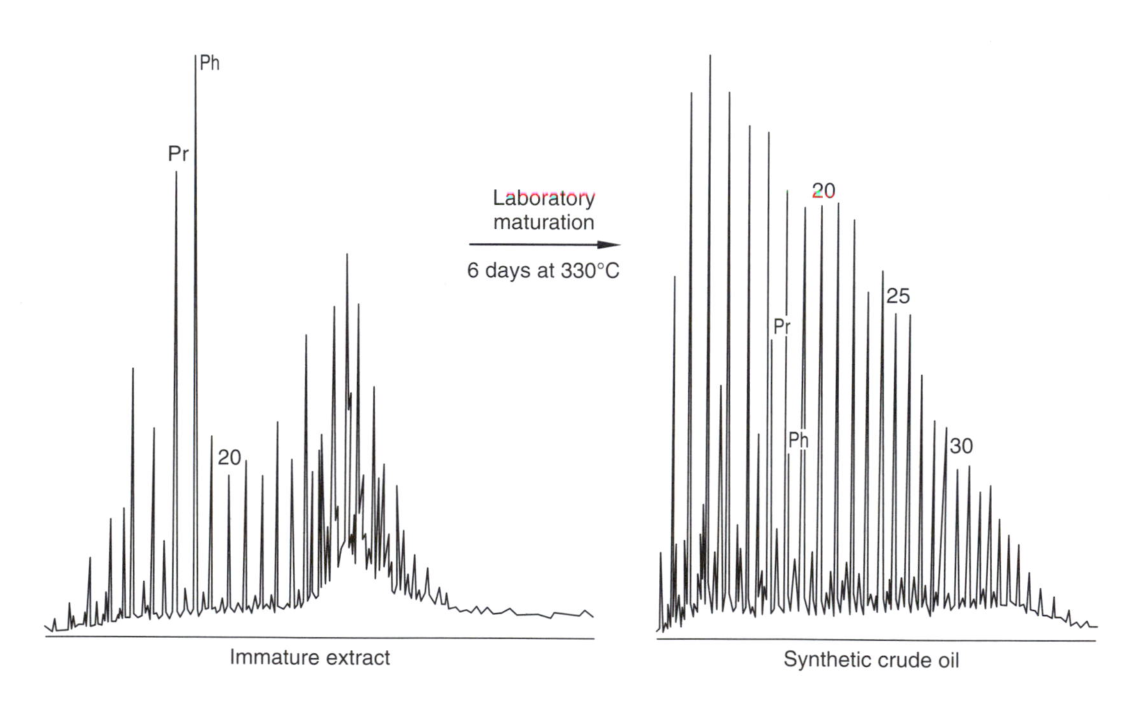

Figure 5-10

Gas chromatograms of the extract of an immature Kimmeridge Shale compared with the synthetic oil obtained from heating the shale in a sealed vessel for six days at 330°C. [Thomas et al. 1985]

isoprenoid hydrocarbons such as pristane and phytane stand out above the *n*-paraffin peaks, along with a high concentration of steranes and terpanes in the C_{30} range where the profile peaks on the right side of the left graph. The predominance of these biological markers indicates that this rock has not been heated high enough to generate oil. Maturation of the sample in the laboratory for six days at 330°C yields the GC profile of a typical mature North Sea crude oil, as shown on the right.

The oil window for the C_5–C_7 gasoline-range hydrocarbons in the Green River Formation of the Uinta Basin is in Figure 5-11 (Anders and Gerrild 1984).

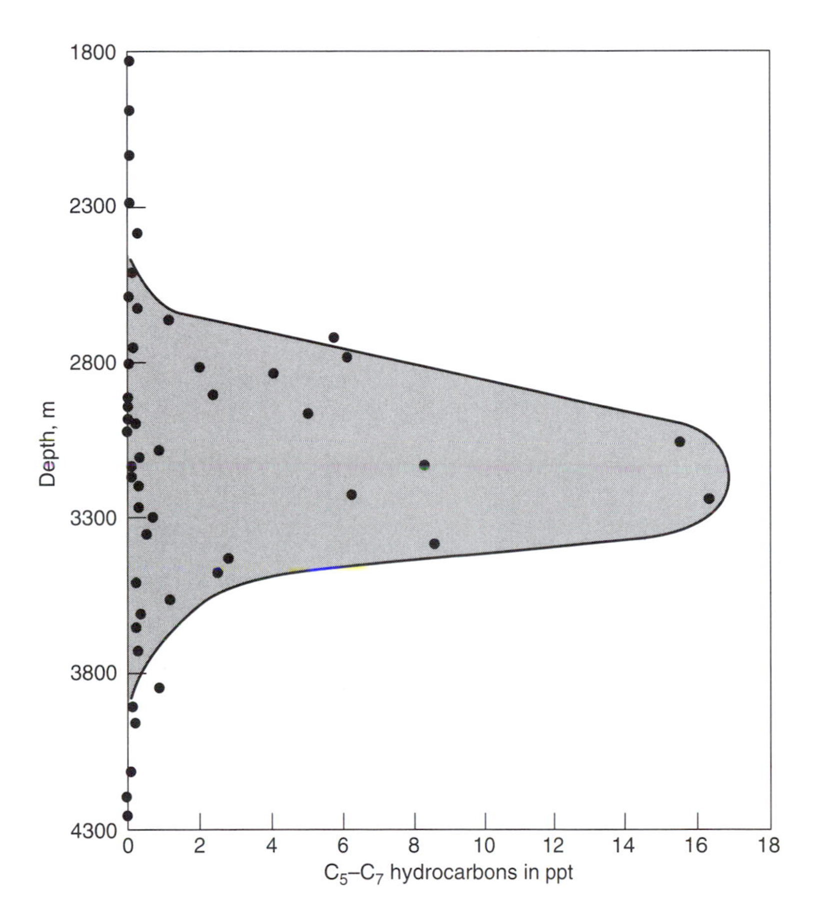

Figure 5-11

Gasoline-range hydrocarbon distributions in parts per thousand of the Eocene Green River shales of the Uinta Basin, Utah [Anders and Gerrild 1984]. Low concentrations in the oil window (solid line) are alluvial sediments that showed little change with depth. High yields are from lacustrine shales.

This resembles Figure 5-8 in that the shallower concentrations of these hydrocarbons are near zero, since only traces occur in living organisms and recent sediments. Starting around 2,600 m (8,530 ft) there is an exponential increase in the yield of hydrocarbons, followed by a decrease down to 3,700 m (12,140 ft) where the C_5–C_7 yields again approach zero. The yield peaks at a present-day depth of 3,000 m (9,840 ft), equivalent to 95°C (203°F). However, the basin has been uplifted about 1,800 m (5,900 ft; Sweeney et al. 1987) so the true peak-yield temperature is closer to 140°C (284°F). The Green River Formation contains mainly type I kerogen in the lacustrine Black Shale Facies, types II and III in the alluvial facies.

The oil window of an organic-rich type II kerogen is in Figure 5-12 (Claypool and Mancini 1989). This is the Jurassic Smackover Formation of the Mississippi Interior Salt Basin in the U.S. Gulf Coast. Data points represent the total extractable C_{15+} hydrocarbons relative to TOC for Cretaceous and Smackover source rocks with indigenous hydrocarbons. The threshold of

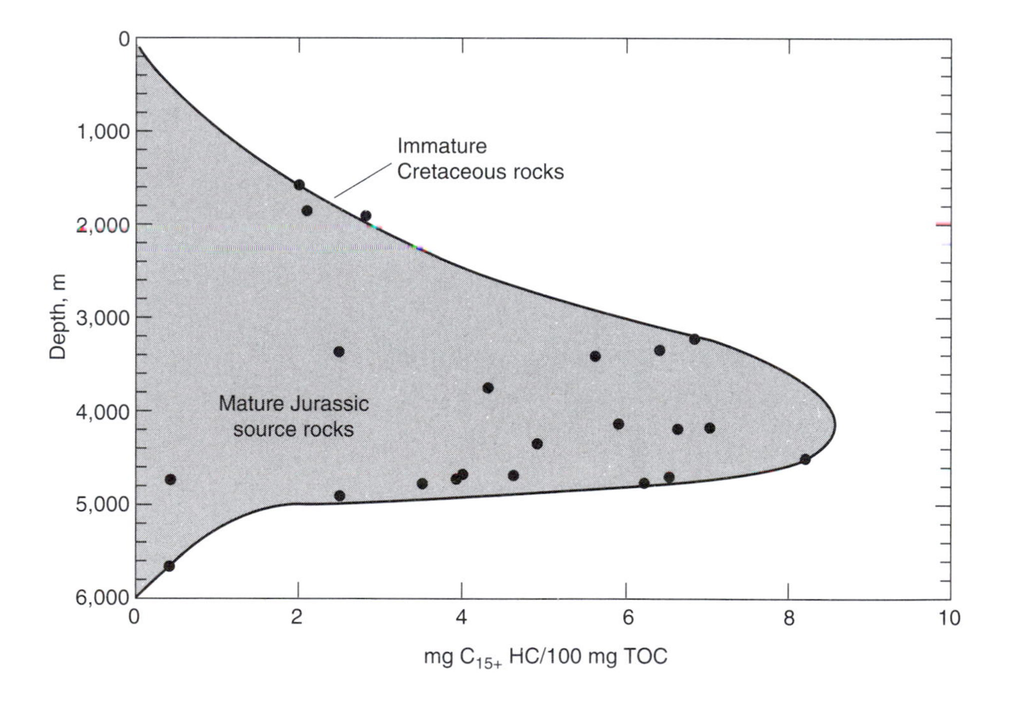

Figure 5-12

Plot of the C_{15+} extractable hydrocarbons in Jurassic Smackover and Cretaceous Tuscaloosa source rocks of the Mississippi Interior Salt Basin, Alabama, against burial depth. Samples containing only indigenous hydrocarbons are plotted. The solid lines in these figures represent the oil-generation window. [Claypool and Mancini 1989]

intense oil generation is around 3,048 m (10,000 ft), and the oil floor (base of the oil-generation window) is at about 5,180 m (17,000 ft).

The total saturated hydrocarbons and total *n*-alkanes in the C_{15+} extract of coals and coaly shales (type III kerogen) are shown in Figure 5-13 (Welte et al. 1984). This area has been uplifted about 1,800 m, so the maximum depth of the base of the oil window is more like 4,100 m (13,450 ft) rather than the 2,500 m (8,240 ft) shown in Figure 5-13. The peak yield temperature during maximum burial was around 150°C. Oil-generation windows often have been observed in many coals and coaly shales (for details, see Chapter 11). However, the oils generated are so strongly adsorbed on the coals that they are usually cracked to light hydrocarbons and gases before expulsion.

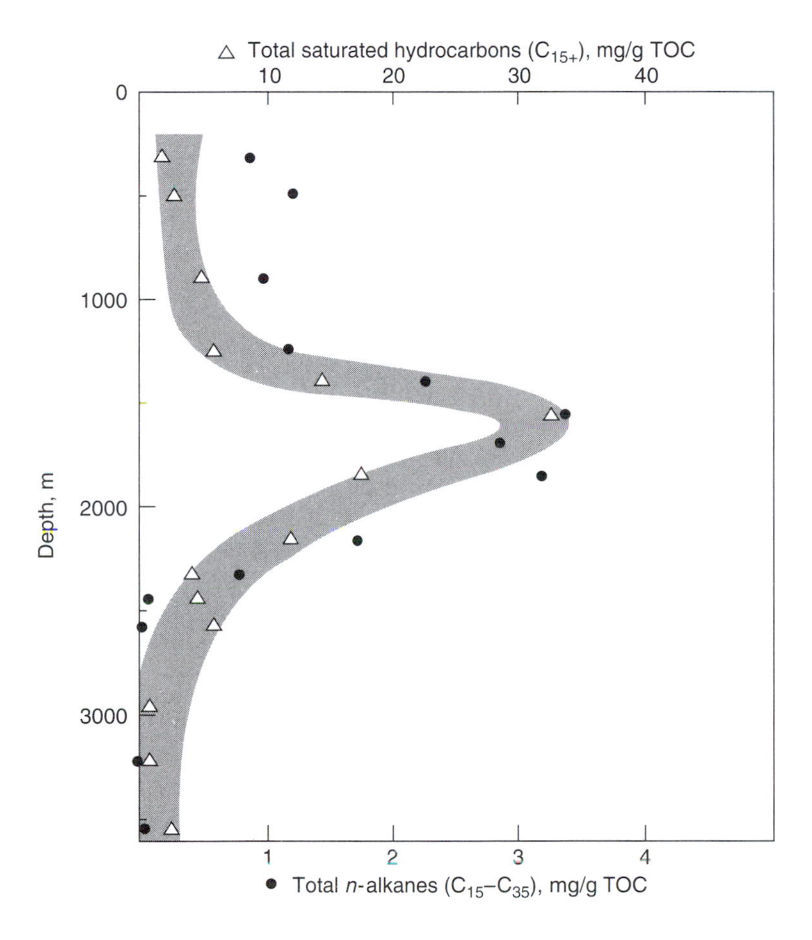

Figure 5-13

The yield of C_{15+} *n*-alkanes and total saturated hydrocarbons in mg/g TOC versus present-day depth in coaly shales of the Western Canada Deep Basin. [Welte et al. 1984]

In the Aquitaine Basin of France, the oil-yield curve increases at the present-day temperature of about 72°C (162°F), equivalent to a subsurface depth of 2,500 m (8,200 ft; Le Tran 1972). The yield peaks at around 100°C (212°F), as given in Figure 5-14. The beginning of intense gas generation (the gas window) occurs within the oil window, with the gas yield peaking around a present-day temperature of 150°C (302°F). This is one of the few examples of both the oil and gas windows being identified in the same section.

The same type of curve showing the depth range of oil generation can be determined on samples that have been buried deeply and then uplifted to the

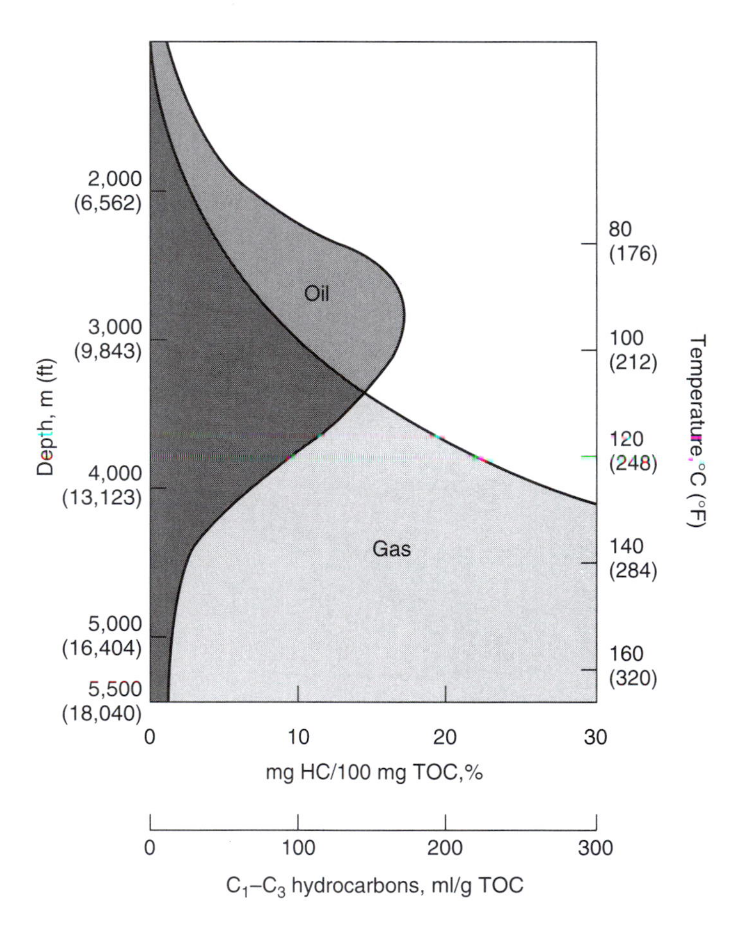

Figure 5-14

The yield of C_{15+} hydrocarbons in percentage of TOC with depth, compared with the yield of C_1–C_3 hydrocarbons (gas) in ml/g TOC for carbonate source rocks of the Aquitaine Basin, France. [Le Tran 1972]

surface. If weathering has not altered the organic matter, the imprint of increased hydrocarbon generation can still be observed in the rock samples. An example of this is the distribution of C_{15+} hydrocarbons in the Phosphoria Formation of Wyoming. A study by Claypool et al. (1978) was conducted largely on unweathered outcrop samples that at one time had been buried very deeply. Through a reconstruction of geological data, Claypool et al. were able to show the oil window and its peak zone of hydrocarbon yield (Figure 5-15). The peak occurred at an apparent depth of about 3.3 km (10,800 ft). The end of the oil generation window was at about 5 km (16,400 ft). The gas chomatograms indicated that the sterane and triterpane biomarkers gradually disappeared through this interval. Also of interest is the fact that the ratio of hydrogen to carbon in the kerogen decreased markedly through the oil-generation zone, from a value of 1.2 to 0.6. This occurs because the kerogen is providing the hydrogen to form the oil. The decrease in hydrogen then causes the color of the kerogen to change. High-hydrogen kerogen is light yellow in transmitted light, and low-hydrogen kerogen is black, as indicated in Figure 5-15.

Other examples of oil windows, including the Douala Basin of Cameroon, the Paris Basin, the west Pre-Caucasus of the former Soviet Union, and the Los Angeles Basin, are reviewed in Hunt 1979, pp. 133–140.

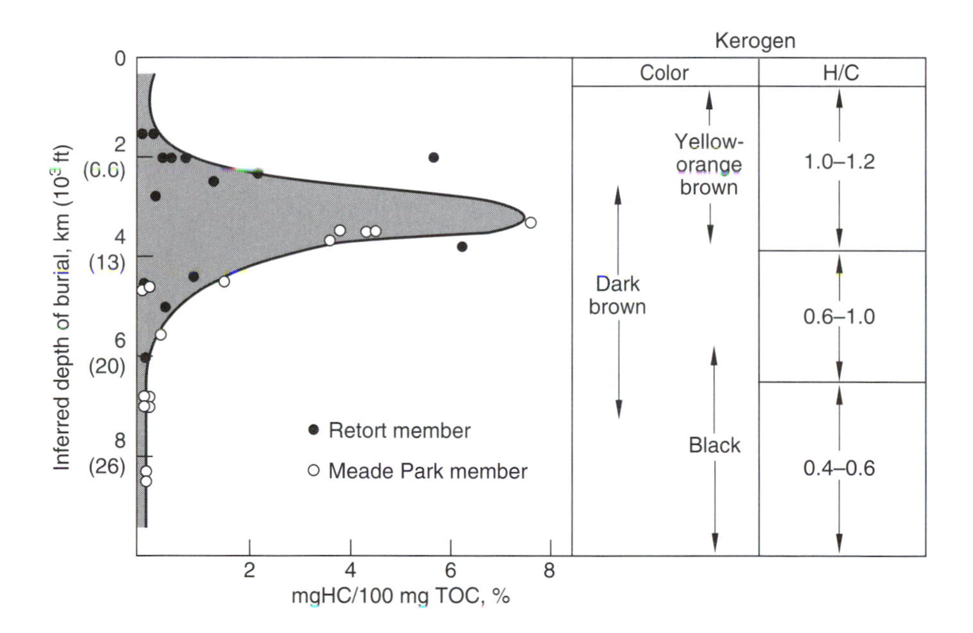

Figure 5-15

The C_{15+} hydrocarbon yield in percentage of TOC, compared with changes in kerogen composition for the Permian Phosphoria Formation, western United States. [Claypool et al. 1978]

The peak oil yield shown in Figures 5-11 through 5-15 is not necessarily the point of maximum oil generation. Three processes control the shape of the oil-yield curve: (A) the generation of oil, (B) the migration of oil out of the source rock, and (C) the conversion of oil to gas. The yield increases when A > B + C and decreases when A is < B + C. The peak is the point at which generation is equal to migration plus conversion of oil to gas.

In hydrous pyrolysis experiments, the bitumen window represents the extractable soluble OM before oil expulsion, and the oil window represents the expelled oil. The light hydrocarbon (C_5–C_7) windows in Figures 5-8 and 5-11 are generated hydrocarbons that have not been expelled. These are not the C_{15+} bitumens. However, Figures 5-12, 5-13, 5-14, and 5-15 are the total C_{15+} bitumen extracts, excluding any hydrocarbons that have migrated out of the rock before the analysis. Consequently, many published oil windows may be a partial combination of the separate bitumen and oil windows observed in hydrous pyrolysis (Figure 5-9).

An oil window based on the C_5 to C_8 or the C_{15+} hydrocarbon range is visible from analyses of most, but not all, source rocks. The reason is that in some source rocks the migration pathways are highly variable and the mix of kerogen types changes over short depth intervals. This is what causes the scatter in data points below the yield-curve envelope, as seen in Figures 5-8, 5-11, and 5-12. Hydrocarbons generated close to vertical microfractures, sandy partings, or other permeable pathways may move out of the source rock as soon as they are formed. Types III and IV kerogens, which are incapable of generating much oil, have low hydrocarbon yields, even in the oil window. Data points showing the highest yields in these curves represent good generating capabilities, but some of the oil may be unable to escape from the source rock until it is cracked to gas.

It is significant that the oil-window pattern is still observed in uplifted source rocks, such as in Figure 5-15. Uplift is expected to preserve the outlines of the oil window because there would be less tendency for the oil to crack to gas and the reduction in pressure differentials inhibit migration.

The oil or gas window observed in any sedimentary section is usually formed during the maximum depth of burial, so the present-day depth does not necessarily represent the depth interval of generation.

Oil windows are a vertical representation of the generation interval. The areal extent of a mature source rock that has expelled petroleum to one or more specific reservoirs is called the *generative depression, hydrocarbon kitchen,* or *cooking pot* or *pod* of active source rock, as discussed in Chapter 1. Thus, Thomas et al. (1985) described the Ekofisk and Troll kitchens as areas that supplied the hydrocarbons for these respective structures.

All these examples show that petroleum originated over a finite temperature range that can be observed in a natural environment. In exploring for oil and gas in wildcat areas, it is important to know whether or not the rocks have passed through this generation range and, if so, at what depths generation was initiated, peaked, and terminated. Such data alone cannot pinpoint the location of economic petroleum accumulations, since they are affected by migration and trapping. However, it does bracket the depth ranges in which the mature hydrocarbon source beds occur, and it does indicate the most likely subsurface zones in which to prospect for oil and gas. Maps of geochemical parameters, such as isopachs of mature source beds, significantly reduce the risk associated with exploration.

Plates 1A through 1D are maps based on vitrinite reflectance. Peters and Cassa (1994) show additional examples of mapping geochemical data. There is no point in drilling a hole into rocks where the only possible source is too immature to generate hydrocarbons or is so depleted in hydrogen that its generating capability is gone. The limited depth—temperature zone of hydrocarbon generation is a natural phenomenon resulting from the thermodynamic instability of the organic matrix in the rocks. Although we do not understand all aspects of the origin process, enough of it is clear to enable subsurface geochemical analyses to be effectively used in making exploration decisions.

Metagenesis and Metamorphism

Metagenesis is the last stage in the significant thermal alteration of organic matter. Here methane generation diminishes and graphitic structures begin to form. Metagenesis occurs in the temperature range of 200 to 250°C (392 to 482°F). At such temperatures, the atomic H/C ratio of the kerogen falls to less than 0.4, typical of the kerogen in a phyllite. The bottom of the 9,583 m (31,430 ft) well drilled in the Cambro-Ordovician Arbuckle Formation in the Anadarko Basin of Oklahoma contained kerogen with an H/C ratio of about 0.25. That is equivalent to a $C_{96}H_{24}$ hydrocarbon structure containing 37 fused aromatic rings. Such a structure could generate only trace amounts of methane before becoming pure graphite.

Inorganic geochemists term mineralogic changes that can be attributed to the action of heat and pressure at depth as *metamorphism*. The low-temperature end of the metamorphic scale is about 200 to 300°C (392 to 572°F). For example, kaolinite is converted to muscovite, and the greenschist facies begins to appear in this range. Consequently, the high-temperature end of organic thermal alteration overlaps with the low-temperature beginning of inorganic thermal alteration.

Distribution of Hydrocarbons in Fine-Grained Rocks

Figure 5-16 shows the relative quantities of hydrocarbons in fine-grained source rocks during the stages of diagenesis, catagenesis, and metagenesis. The areas under each curve represent the relative quantities of hydrocarbons, in terms of carbon, that are present at each stage. The hydrocarbons are divided into (1) the gases methane through propane; (2) gasoline, kerosene, and light gas oil, C_4 to C_{14}; and (3) gas oil through lubricating oil, C_{15} to C_{40}. About 76% of all hydrocarbon gases (C_1 to C_3) are formed in the catagenic stage, with the

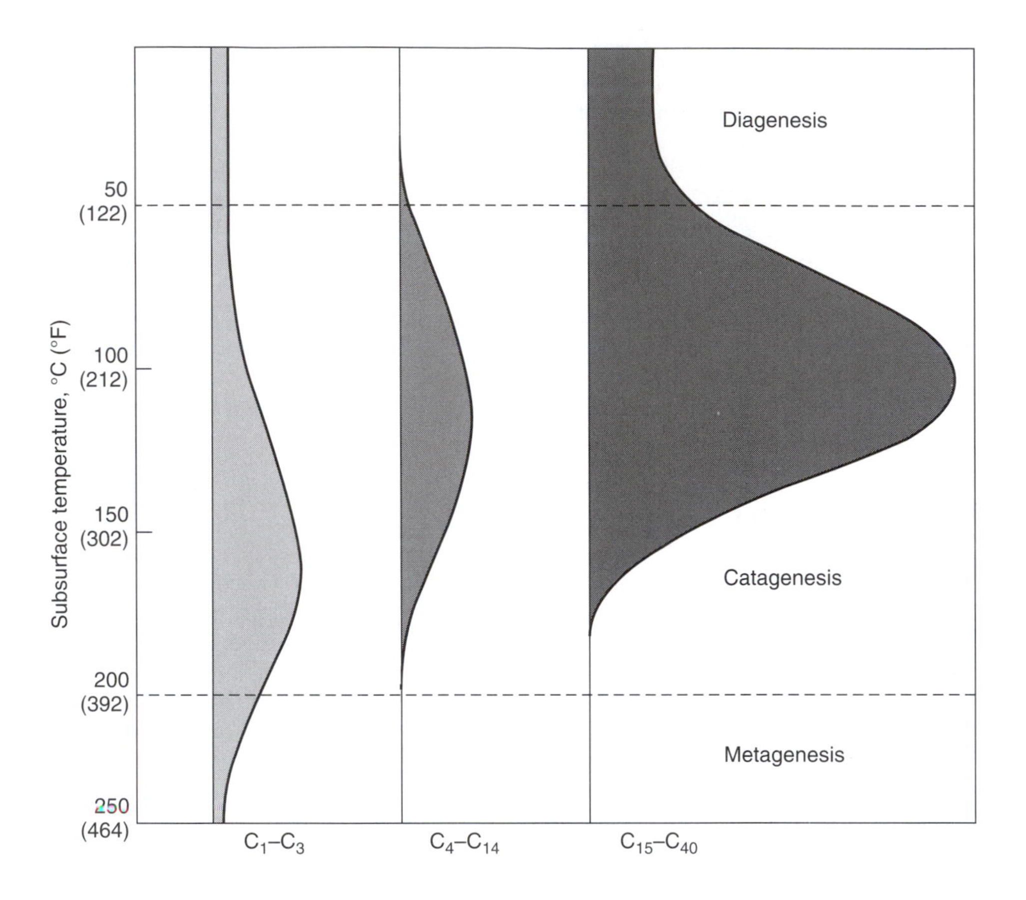

Figure 5-16

Relative quantities of hydrocarbons in fine-grained nonreservoir rocks. The areas under the curves are proportional to masses as carbon.

remaining 24% divided equally between diagenesis and metagenesis. The catagenic stage also forms about 97% of the C_4–C_{14} hydrocarbons and 86% of the C_{15+} hydrocarbons. Most of the remaining hydrocarbons in these two ranges are formed during diagenesis.

The diagenesis stage is characterized by biological markers in the C_{15}–C_{40} range as the dominant hydrocarbons. The gas is mostly biogenic methane, and only trivial amounts of C_2 to C_{14} hydrocarbons are present. Low-temperature diagenetic processes do not generate significant quantities of gasoline-range hydrocarbons even over long geological time periods, nor do they form the hydrocarbon distribution above C_{15} that is typical of crude oil.

In the catagenesis stage, all the hydrocarbons C_1 through C_{40} are formed in larger amounts than in other stages. Heavy-oil fractions are formed first, followed by the cracking of these fractions to yield light oil and gases. The generation of gases and the cracking of heavy hydrocarbons and bitumens create localized overpressures that force the hydrocarbons out of the source rocks, as discussed in Chapters 8 and 9. Finally, in the metagenesis stage, only methane is formed in significant amounts. Even the quantities of methane in rock cuttings decrease considerably as subsurface temperatures approach 250°C (464°F).

Early-generated oils tend to be heavy, asphaltic, and high in sulfur, like some of the Monterey oils of California, the Boscan oil of Venezuela, the Rozel Point oil of Utah, and several oils of Italy. High-sulfur kerogens generate and expel these oils earlier than they do more conventional oils because of lower-activation energies, as explained in Chapter 6. For example, the immature, heavy Emilio oil with over 4% sulfur was generated by the Triassic Emma Limestone (17% sulfur) of the Central Adriatic region (Mattavelli and Novelli 1990). The immature, heavy, high-sulfur Gela, Perla, and Prezioso oils of Sicily are from the high-sulfur kerogen of the Triassic Noto Formation. About 70% of the Italian oils are immature, nonbiodegraded, sulfur-rich, heavy oils, according to Matavelli and Novelli (1990). These oils still contain *n*-paraffins below C_{20}, indicating that the low API gravities (10 to 17) are due to early generation, not biodegradation.

The hydrocarbon composition of early- and late-generated oils from the same source rock also differs somewhat because of differences in the reactivity of different molecular structures in the kerogen. For example, hydrocarbons with tertiary carbon atoms such as ethylcyclopentane and 3-ethylpentane are formed earlier in the oil window than are hydrocarbons with quaternary carbon atoms such as 2,2-dimethylhexane (Hunt 1984). In the previously mentioned South Padre well off Texas, both the 2,2-dimethylbutane and pentane containing quaternary carbon atoms were formed about 900 m deeper than were the corresponding 2,3-homologs with tertiary carbon atoms. Also, 1,1-dimethylpentane with a quaternary carbon atom was formed 700 m deeper than were the 1,2-homolog and ethylcyclopentane, both of which contain tertiary carbon atoms.

SUMMARY

1. The most important factor determining the quantity of oil generated in a petroleum system is the hydrogen content of the kerogen.

2. Sediments are defined as *oxic, dysoxic, suboxic,* and *anoxic,* depending on the oxygen content of the overlying waters. Organisms that live in these environments are called *aerobic, dysaerobic, quasianaerobic,* and *anaerobic.*

Most petroleum is generated from source rocks deposited in anoxic-to-dysoxic environments because they contain more hydrogen-rich OM than do oxic sediments.

3. Anoxic environments are created from the lack of circulation below the photic zone in marine or lacustrine waters. They also form from the spreading of the oxygen minimum layer across land areas. Transgressive cycles have formed most of the important source rocks because they result in greater areas of anoxicity on the continents.

4. Hydrous pyrolysis experiments indicate that type I and II kerogens decompose to form bitumen and the bitumen decomposes to form oil.

5. The depth interval in which a petroleum source rock generates and expels most of its oil is called the *oil window.* Most oil windows are in the temperature range from 60°C (140°F) to 160°C (320°F).

6. The stratigraphic intervals above, within, and below the oil window are referred to as *immature, mature,* and *postmature,* respectively, for oil generation.

7. Postmature for oil is mature for the gas window. From one-half to two-thirds of thermogenic gas comes from the thermal cracking of previously formed oil in both source and reservoir rocks and in coal. Gas windows are in the 100-to-200°C (212-to-392°F) temperature range.

SUPPLEMENTARY READING

Degens, E. T., P. A. Meyers, and S. C. Brassell (eds.). 1986. *Biogeochemistry of black shales.* Vol. 60. Hamburg: Geologisch–Paläontologischen Institutes, 421 p.

Huc, A. Y. 1990. *Deposition of organic facies.* AAPG Studies in Geology 30. Tulsa: American Association of Petroleum Geologists, 234 p.

Katz, B. J., and L. M. Pratt (eds.). 1993. *Source rocks in a sequence stratigraphic framework.* AAPG Studies in Geology 37. Tulsa: American Association of Petroleum Geologists, 247 p.

Tyson, R. V., and T. H. Pearson (eds.). 1991. *Modern and ancient continental shelf anoxia.* Geological Society special publication 58. London: Geological Society, 470 p.

Chapter

6

Modeling Petroleum Generation

The oil window described in Chapter 5 is useful in evaluating the present-day depth and maturity of a specific petroleum source rock, especially when combined with other maturation indicators (see Chapter 10). However, it does not tell you when oil generation started, how long it lasted, or at what depth it actually occurred. Also, if good well samples are not available or if the analyses fail to show an oil window, you have no clear definition of the generation interval. The timing of petroleum generation is particularly important in relation to the formation of structures, stratigraphic traps, and faults that can act as migration pathways. Such information can be obtained by modeling the time–temperature history of the source rock, which is the topic of this chapter.

The carbonization of coal with increasing time and temperature is a phenomenon that was first described by Hilt in 1873. Hilt's law stated that the fixed (nonvolatile) carbon of coal increased with increasing depth and temperature. Later, White (1915) showed a relationship between the occurrence of oil and gas and the rank of coals in the eastern United States. Oil fields were found where the fixed carbon content of coals was less than 60% and gas fields were in the 60-to-65% range. Above 70% fixed carbon, there were no oil or gas accumulations. These early studies correlated the origin of oil with the time–temperature formation of coals of different rank. The first direct attempt at modeling petroleum generation was by J. K. A. Habicht (1964), who constructed a burial history curve of a Jurassic source rock and used Arrhenius equation kinetics to determine the time and depth of oil generation in the Gifhorn trough of northwest Germany. Habicht used an activation energy of 58 kcal/mol and an A factor of 5×10^{13} sec^{-1} in the Arrhenius equation. His burial profile indicated that generation started about 90 Ma (million years before the present) at 2,610 meters, and after further burial and uplift, the generation of oil was completed about 10 Ma.

Later Philippi (1965) documented the increase in the yield of hydrocarbons from source rocks in the Los Angeles and Ventura basins of California with increased time and temperature. Still later a series of source rock burial history curves were used to describe the origin of hydrocarbons in the North Sahara Hassi Messaoud area (Poulet and Roucache 1969). These early studies indicated that temperature alone cannot explain the different examples of oil generation. This gave kinetic theory involving both temperature and time an important role in evaluating the oil potential of sedimentary basins. The exposure time that a source rock experiences at varying temperatures with burial must be taken into account in a subsiding basin. Rapid burial with high geothermal gradients does not result in the same level of maturation as does slow burial with low geothermal gradients.

An example of these differences is shown in the burial history curves of Figure 6-1. These are similar to Habicht's curve in plotting time before the present (Ma) on the abscissa and source rock burial depth on the ordinate. The Cretaceous source rocks A and B are shown to have been at the surface 100 Ma. Subsequently, source rock A was buried slowly for the first 80 m.y. of its history to 50°C and rapidly during the last 20 m.y. to 150°C. Rock B was buried rapidly during the first 20 m.y. to 100°C and then slowly to 150°C during the last 80 m.y. Because A was cool (< 50°C) during most of its history, it is only about halfway into the oil window today. In contrast, B was heated at a high temperature for a very long time. Consequently, it is now in the gas window, postmature for oil. Both source rocks were deposited on the surface at the same time and both are at the same subsurface temperature today, but they have had entirely different burial histories.

The conclusions regarding the maturity of A and B were made using the Arrhenius equation, as Habicht did in 1964. The first mathematical model for oil generation using Arrhenius kinetic theory along a source rock burial history curve was published by Tissot in 1969. Applications of the method in various basins were demonstrated in subsequent publications, but the model did not find widespread use, as it was quite rigorous.

Meanwhile, Teichmüller's classic 1958 study of the Wealden Basin established the first relationship between the reflectance of the vitrinite maceral of coal and the occurrence of oil. Since then, vitrinite reflectance has become one of the most widely used maturation parameters for empirically defining the oil and gas windows. In the thermal alteration of vitrinite, the reflectance increases exponentially with a linear increase in temperature. This is why vitrinite reflectance generally plots as a straight line on a semilog plot (Dow 1977).

This empirical relationship between vitrinite reflectance and petroleum formation was used effectively by N. V. Lopatin (1971) to develop a simpler method than that of Tissot for using both time and temperature to calculate the thermal maturity of the organic matter in sediments. Lopatin superimposed on

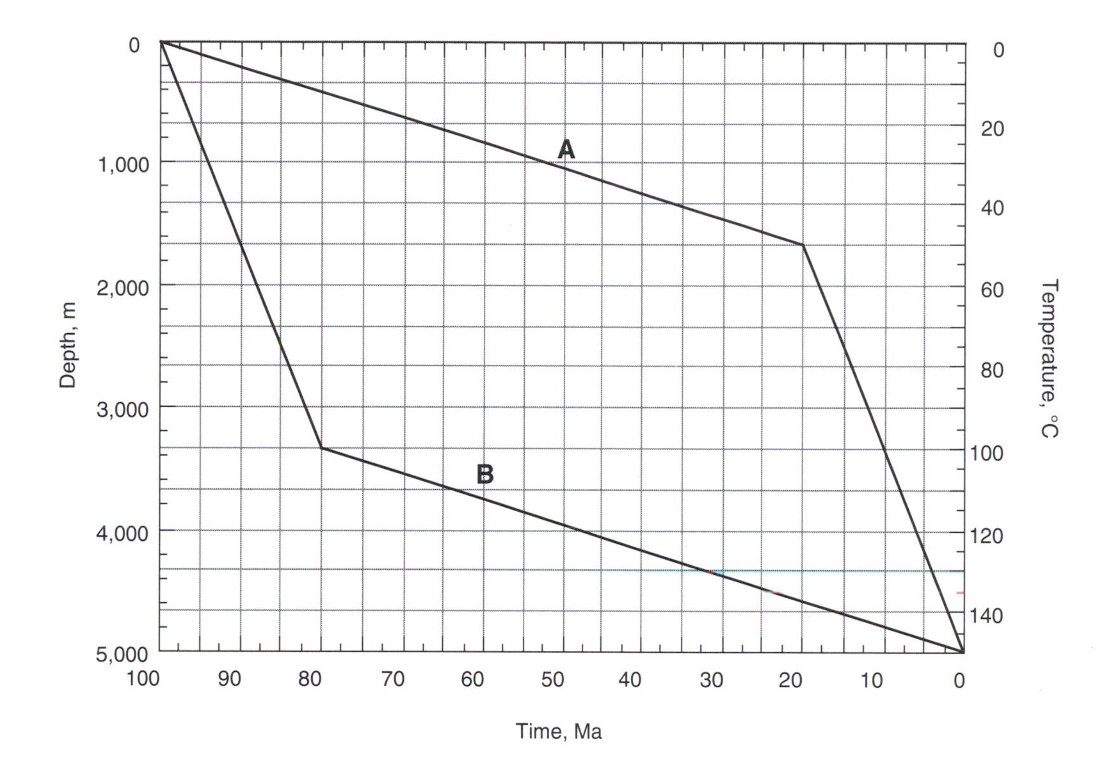

Figure 6-1

Burial history curves of hypothetical source rocks A and B overlain with a grid of isotherms at 10°C intervals.

a burial history diagram a grid of isotherms at 10°C intervals, as shown in Figure 6-1. He then determined the thermal exposure of the source rock in each time–temperature interval and summed them up to give the total exposure since the time the source rock was originally deposited. This is called the *time–temperature index* (*TTI*) of maturity (for more details, see Waples 1985, p. 123). The thermal exposure is calculated by multiplying each time interval by a temperature factor based on the old chemical rule that reaction rates double for each 10°C rise in temperature (Bergius 1913). This is why Lopatin used isotherms at 10°C intervals.

Lopatin's technique worked remarkably well, even though the reaction rates in oil generation increase by more than a factor of two with rising temperatures and they vary considerably among different kerogens exposed to different thermal conditions. Although Lopatin's initial calibration for the oil window was based on coals, Waples's revision (1985, pp. 135–136) enabled the method to be extended somewhat to other kerogen types. It became a widely used technique for defining the oil and gas windows in sedimentary basins.

Most techniques, however, begin to develop problems as their applications are increased. The Lopatin method tended to underestimate thermal maturity relative to the Arrhenius equation in basins with heating rates significantly higher than 1°C/m.y. and overestimate maturity with heating rates significantly lower than 1°C/m.y. (Wood 1988). This method also underestimated thermal maturity for fast-reacting kerogens, such as those in the Monterey Shale offshore California. Nevertheless, Lopatin's inovative time–temperature index based on 10°C isotherms was a logical approach to simplifying the rigorous maturity calculations required by the Tissot model. Consequently, an effort was made to combine the simplicity of the Lopatin approach with Arrhenius equation kinetics.

Determining Kinetic Parameters for Oil Generation

The exponential temperature dependence of the rate of kerogen decomposition can be expressed in theoretical terms by the Arrhenius equation, which may be written as

$$k = A \exp(-E/RT) \tag{6-1}$$

where k is the reaction rate constant (1/m.y.), A is the preexponential or frequency factor (1/m.y.), E is the activation energy (kJ/mol), R is the ideal gas constant, and T is the temperature in kelvins (°C + 273). The kinetic parameters E and A can be obtained experimentally by heating the source rock at various temperatures and measuring the yield of hydrocarbons. The most widely used techniques for this are an open, nonisothermal, dry-programmed temperature pyrolysis system such as Rock-Eval and a closed, isothermal, wet system such as hydrous pyrolysis (see Chapter 10 for more details).

There are advantages and disadvantages to both. The open system is rapid and uses relatively small sample sizes (< 250 mg). Due to the ease of operation it is possible to perform several programmed temperature runs at different heating rates in a relatively short time. The activation energy is determined by a computerized curve-fitting process, which uses a series of E values. Its supporters argue that a distribution of activation energies is obtained that is more realistic than a single value because kerogen breakdown represents a whole series of reactions (Burnham et al. 1987a). Its detractors argue that it does not monitor exclusively the rate-controlling step of bitumen to oil but, rather, all the reactions involving the formation of bitumen, oil, and gases (Lewan 1990). Also, as Wood (1988) stated, the initial products of kerogen decomposition do not constitute petroleum, and the late products are of less interest because they are usually decomposed to gas and constitute a small fraction of the products generated at peak conditions. Consequently the peak stage of generation is of most interest, and this is what a single set of kinetic parameters can model with some accuracy using the Arrhenius equation.

Another problem with dry, open pyrolysis is the decrease in product yield with decreasing heating rates (Evans and Campbell 1979). Lewan et al. (1995) extrapolated the data of these authors to lower heating rates and found that the product yield became zero before reaching geological heating rates. Finally, an open system pyrolysate does not generally resemble normal crude oil in that it has a disproportionately high percentage of polar compounds and unsaturates (Lewan et al. 1979). This may be because open pyrolysis is a dry system. There is some evidence (Lewan 1993a) that water terminates free radicals by means of hydrogen transfer during bitumen decomposition. Water also assists in the expulsion process.

Polymer chemists and others outside the field of geochemistry have criticized nonisothermal techniques as not providing sufficient information for reaction kinetics calculations. Wedlandt's (1985) analysis claims that Arrhenius parameters cannot be calculated correctly by means of nonisothermal curve-fitting methods. Lakshmanan et al. (1991) concluded that curve-fitting kinetic models from nonisothermal experiments cannot be extrapolated to geological conditions with any degree of certainty.

Hydrous pyrolysis is carried out isothermally in a closed system in the presence of excess water. It yields a liquid product similar to crude oil under conditions that are more comparable to natural burial than open-system pyrolysis is. The closed system and the presence of water imitate natural burial more closely by developing internal pressures within the system and by dissolving water in the bitumen. This enhances the hydrogen exchange between the water and bitumen phases (Lewan 1993a). A disadvantage of hydrous pyrolysis is that it is very time-consuming. It is not a practical method for looking at large numbers of samples. This constitutes a problem because of the variability in kinetic parameters in a source rock lithology sampled at several different locations (Lewan 1986). However, it is possible to develop a standard set of hydrous pyrolysis analyses that can be cross-correlated with a large number of open pyrolysis runs.

Qin et al. (1994) compared the thermal maturation of a brown coal from the Lower Tertiary of east China by using both open anhydrous and closed hydrous techniques. They used elemental analyses of the solid organic residues to follow the evolutionary pathways of the coal on a van Krevelen diagram (see Figure 11-4). The locus of the hydrous pyrolysis analyses closely followed the natural burial pathway of coal, but the anhydrous pyrolysis data deviated markedly from it. The results indicated that the anhydrous method was eliminating oxygen as water, whereas in the natural system it is eliminated mainly as CO_2. Other comparisons of the two methods on the same samples showed that including bitumen plus oil and gas generation in Rock-Eval kinetics results in an oil window covering a wider temperature range than does using only oil generation, as in hydrous pyrolysis kinetics (Burnham 1987a).

Variations in Kinetic Parameters with Kerogen Type

Burnham et al. (1987b) suggested that activation energies must be known to within about 13 kJ/mol to predict oil generation to within 10°C in a basin, assuming a constant heating rate of 3°C/m.y. This accuracy is required because of the large extrapolation from the laboratory to the geological environment. Clearly, with this limitation it is not possible to assume that a single set of kinetic parameters can describe all petroleum formation, as some writers have attempted to do (Quigley et al. 1987). Kerogens can be very different in composition, and the range in their kinetic values is considerable.

In 1974, Tissot et al. defined three major types of kerogen in sedimentary rocks based on their maturation track in a van Krevelen diagram (see Chapter 10 for details). This is a plot of the atomic H/C versus O/C used originally for coals and coal macerals by the coal scientist D. W. van Krevelen (1961). The Tissot group adopted the van Krevelen diagram for the dispersed kerogen of sedimentary rocks. The type I kerogen followed the maturation track of lacustrine oil shales and boghead coals. The track of type II was characteristic of most marine oil source rocks, and type III followed the original coalification track of van Krevelen for the gas-generating humic coals. These classifications resulted in a separate set of Arrhenius kinetics for each of these three major kerogen types (Tissot and Éspitalie 1975).

Later, Lewan (1985) showed that time–temperature relationships of oil generation also vary considerably for kerogens in the type II class. Therefore it is not possible to use the same Arrhenius kinetic parameters for all type II kerogens. This variability is attributed in part to the amount of organic sulfur incorporated into the kerogen during its early diagenetic development. In 1985, Orr reported that high-sulfur kerogens of the Monterey Formation in the Santa Maria Basin of California generated low-maturity oils at faster reaction rates (lower thermal exposures) than did typical type II kerogens. He called these kerogens *type II-S* and characterized them as having atomic S/C ratios > 0.04. Asphaltenes of the associated immature crude oils had S/C ratios generally > 0.035.

This early generation is explained by the fact that carbon–sulfur bonds are known to have lower bond energies than carbon–carbon bonds (Lovering and Laidler 1960). Consequently, kerogens with more sulfur in their molecular structures have lower activation energies. They decompose to low-maturity oils at lower time–temperature increments, as Orr observed.

This inverse relationship between organic sulfur content and activation energy (E) for type II kerogens is apparent in Table 6-1. This table contains Arrhenius kinetic parameters and sulfur data from the literature for representative samples of all three kerogen types. Type I and II kerogens are designated A through C or D, depending on their reaction rates, with A being the fastest and D the slowest (Hunt et al. 1991). The data shown for type II kerogens are from four source rocks representative of the range of compositions within the type II group. These are the Miocene Monterey Formation of California, the Permian Retort Shale Member of the Phosphoria Formation of Montana, the Cambrian

TABLE 6-1 Arrhenius Kinetic Parameters (*E* and *A*) for Generation of Expelled Hydrocarbons from Source Rocks with Varying Amounts of Organic Sulfur

Kerogens		*Standard samples*				
Type	*Reaction rate*	*Source*	*E (kJ/mol)*	*A (1/m.y.)*	S_{org} *(wt%)*	*Atomic S/C*
IA	Fast	Green R., CO[a]	194	1.01×10^{25}	2.2	0.011
IB	Medium	Green R., UT[b]	219	8.87×10^{26}	1.3	0.007
IC	Slow	Green R., WY[c]	269	7.48×10^{30}	1.4	0.007
IIA (II-S)	Fast	Monterey[d]	143.5	7.017×10^{20}	11	0.055
IIB (II-S)	Medium-fast	Phosphoria[e]	178.7	4.223×10^{23}	9	0.045
IIC	Medium	Alum[f]	201.3	1.546×10^{25}	7.4	0.036
IID	Slow	Woodford[g]	218.3	5.656×10^{26}	5.4	0.024
III	Medium	Tent Island, Canada[g]	230	3.98×10^{27}		

Sources: [a]Yang and Sohn 1984, [b]Sweeney et al. 1987, [c]Miknis and Turner 1988, [d]Lewan 1989, [e]Lewan 1985, [f]Lewan and Buchardt 1989, [g]Issler and Snowdon 1990.

Alum Shale of Sweden, and the Devonian–Woodford Shale of Oklahoma. Type IIA has the highest organic sulfur content (11%) and the lowest *E* value. It decomposes to oil, starting at temperatures as low as 50°C. Type IID has the lowest organic sulfur content (5.4%) and the highest *E* value. It starts to decompose at temperatures around 110°C. Types IIA and IIB are equivalent to the type II-S defined by Orr as having an atomic S/C > 0.04.

Some type II kerogens (marine and lacustrine) have organic sulfur contents as low as 2% (Lewan 1986), but no published kinetic data are yet available. However, at low-sulfur contents, other compositional variations may be more important than sulfur in affecting kinetics. For example, the type IB kerogen from the Green River Formation in the Uinta Basin of Utah in Table 6-1 has an *E* of 219 kJ/mole, an *A* of 8.87×10^{26}/m.y., and an organic sulfur content of 1.3 wt% (Sweeney et al. 1987). The *E* and *A* values are similar to the type IID Woodford Shale in Table 6-1, even though the sulfur content is much lower. This suggests that sulfur contents much below the type IID range do not materially alter the *E* values.

The designations A through D in Table 6-1 refer to reaction rates within each group, not between groups. As a group, type I kerogens react more slowly than type II kerogens. This is because type I kerogens usually have a higher ratio of aliphatic to aromatic carbons, which results in a higher level of activation energy (E) values, compared with type II kerogens. All the type I kerogens in Table 6-1 are from the Green River Formation; type IA is from the Anvil Points Mine in the Piceance Creek Basin of Colorado; type IB is from the Altamont field in the Uinta Basin of Utah; and type IC is from the Tipton Member oil shale in Wyoming. They range in E from 194 to 269 kJ/mol. This range is probably due to differences in composition resulting from the changing depositional environment. The Green River lakes fluctuated extensively in size and salinity from fresh to hypersaline (Smith 1983). Kerogens from nonmarine sections show greater variations in kinetics within a single formation than do those from mainly marine sequences.

The source rock of the type III kerogen in Table 6-1 is the Paleocene Tent Island Formation of the Beaufort–Mackenzie Basin in northern Canada (Issler and Snowdon 1990). It is typical of the type III source rocks in parts of this basin. Its kinetic parameters also are similar to those of the Miocene coals and coaly shales of the Mahakam Delta, Indonesia (Tissot et al. 1987). There are not enough published data on type III source rocks to determine whether this group shows the same kinetic variability as do the other two groups.

The importance of the range in kinetic parameters in Table 6-1 is better understood by comparing two examples of hypothetical subsiding basins with very different thermal histories. The first example in Figure 6-2 (a) and (b) represents source rocks subsiding at a rate of 100 m/m.y. in a pull-apart basin with a geothermal gradient of 45°C/km. Cumulative integration of the Arrhenius equation with the integral of the first-order rate expression over the subsidence history of this model basin results in the oil-generation curves shown. Figure 6-2 (a) is for kerogen types IIA, B, C, and D of Table 6-1. Note that oil generation is complete from kerogen type IIA in about 1,500 m of burial at temperatures below 100°C, whereas type IID requires 3,300 m of burial and temperatures above 150°C. The spread between the oil windows of types IIA and IID is about 1,800 m of burial equivalent to 81°C.

Figure 6-2 (b) shows oil-generation curves for types I and III kerogens in the same basin. The entire group of curves has shifted to greater depths corresponding to higher temperatures for oil generation than for the type IIs. Type III has a slower reactivity, requiring higher temperatures, than do all the type IIs in Figure 6-2 (a), and only the type IA and IB samples overlap the type II group. Also, there is a significant difference in reactivity within the type I group corresponding to as much as a 1,100 m (50°) difference in the oil windows of the fastest and slowest reacting kerogens.

Figure 6-2 (c) and (d) represent source rocks subsiding at a rate of 100 m/m.y. in a cratonic basin with a geothermal gradient of 25°C/km. Because of this lower gradient, the spread in depth between oil windows is greater than in the pull-apart basin. Kerogen type IID in Figure 6-2 (c) requires 2,800 m more of burial than does type IIA for oil generation.

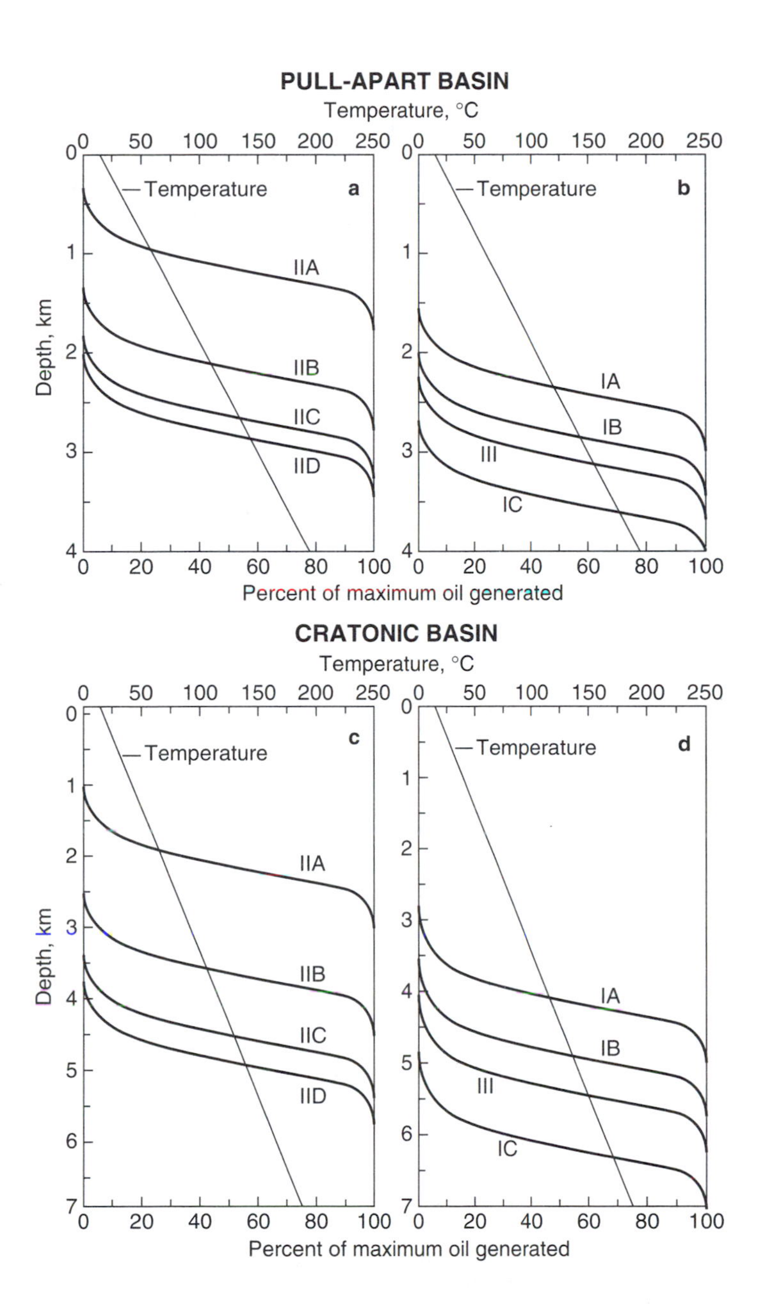

Figure 6-2

Oil-generation curves for kerogen types IA, IB, IC, IIA, IIB, IIC, IID, and III, in a pull-apart basin with a geothermal gradient of 45°C/km and in a cratonic basin with a gradient of 25°C/km. The burial rates in both basins are 100 m/m.y.

In Figure 6-2 (d) there is a 2,000 m spread within the type I kerogen group. The type I group worldwide includes many lacustrine oil shales, and Figure 6-2 (d) shows why only the deepest ones have gone beyond the bitumen stage. Time–temperature requirements have been too great for major oil generation and expulsion. For example, the Tipton Member of the Green River Formation (IC) would require over 6,000 m of burial at temperatures above 150°C to form oil in the cratonic basin of Figure 6-2. Large oil fields from lacustrine sediments such as the Daqing field in China are mainly formed from type II kerogens which break down at lower temperatures than do the type I. In summary, Figure 6-2 clearly emphasizes the importance of using different kinetic parameters for different kerogen types in defining the petroleum-generation intervals in sedimentary basins.

A Graphical Method for Modeling the Oil and Gas Windows

The integration of time and temperature with burial using the Arrhenius equation is a formidable task that is best suited to a computer. However, it is often desirable to make a hand-calculated determination of petroleum generation when a computer and the supporting software are not readily available. This section describes a simple but valid approach for determining the progress of oil generation using time–temperature index graphs based on the Arrhenius equation. In addition to presenting a calculation that is easy to understand, this approach also provides a working understanding of oil-generation kinetics, which minimizes the "black box" mentality fostered by some computer programs.

Graphical estimates of the time–temperature maturation of organic matter have been used since Karweil (1955) constructed the first time–temperature graph for coalification. Lopatin (1976) used a modification of his TTI method to construct a time–temperature graph for oil generation. Such graphs also can be constructed with Arrhenius kinetics. In 1988, Wood derived a time–temperature index based on the Arrhenius equation, which he called the TTI_{ARR} to differentiate it from Lopatin's TTI, which was designated TTI_{LOP}. He then compared the two methods for modeling temperature as a function of time in a large number of hypothetical burial histories involving different basin configurations and heating rates. Comparisons were made with fast-, medium-, and slow-reacting kerogens. As mentioned earlier, Wood concluded that the Lopatin method tended to underestimate thermal maturity relative to the Arrhenius equation for fast-reacting kerogens and that it did not adequately account for large differences in heating rates.

The derivation of TTI_{ARR} from the Arrhenius equation is explained in Wood (1988) and Hunt et al. (1991) (see Box 6-1). Wood states that his derivation is an approximate analytical solution of the Arrhenius equation integral and has a solution error of less than 1% at temperatures less than 300°C and E values greater than 50 kJ/mol. Wood's expression assumes a linear heating rate within each 10°C interval.

Box 6-1

Construction of Graphs

As mentioned earlier, a time–temperature index (TTI) based on the Arrhenius equation was derived by Wood (1988). Wood expressed this index as

$$TTI_{ARR} = \frac{A(t_{n+1} - t_n)}{T_{n+1} - T_n} \left\{ \left[\frac{RT_{n+1}^2}{E + 2RT_{n+1}} \exp\left(\frac{-E}{RT_{n+1}} \right) \right] - \left[\frac{RT_n^2}{E + 2RT_n} \exp\left(\frac{-E}{RT_n} \right) \right] \right\} \times 100 \quad (6\text{-}2)$$

where t_n and t_{n+1} are, respectively, the time (m.y.), and T_n and T_{n+1} are, respectively, the absolute temperature (°C + 273) at the start and end of a 10°C interval. R, E, and A are the same as in the Arrhenius equation (6-1). This expression assumes a linear heating rate within each 10°C interval. Multiplication by 100 is simply for the convenience of eliminating values with significant decimals.

Some burial-history curves have long time periods at a constant temperature with negligible subsidence or uplift. To account accurately for those time periods at a constant temperature on a given burial-history curve, a TTI_{ARR} value can be calculated.

$$TTI_{ARR} = [(t_{n+1} - t_n) A \exp(-E/RT)] \times 100 \quad (6\text{-}3)$$

Or, as an approximation, a constant-temperature line may be drawn on the graph between two diagonal lines at a distance of 45% from the line on the left and 55% from the line on the right. For example, in Figure 6-3 the 80°C constant-temperature dashed line between 70 and 80°C and 80 and 90°C crosses the 8-m.y. line at a TTI_{ARR} of 300.

As soon as the kinetic parameters *E* and *A* are known for a source rock, graphs may be constructed for determining TTI_{ARR} along the burial history curve of the rock in 10°C increments (Hunt et al. 1991). The graphs in Figures 6-3 through 6-11 were constructed for this purpose, with each based on the kinetic parameters listed in Table 6-1. The TTI_{ARR} values for each 10°C interval that a source rock experiences are determined graphically on the basis of the amount of time a source rock resides in each 10°C interval. As an example using the solid diagonal lines in Figure 6-3, a source rock subjected to a

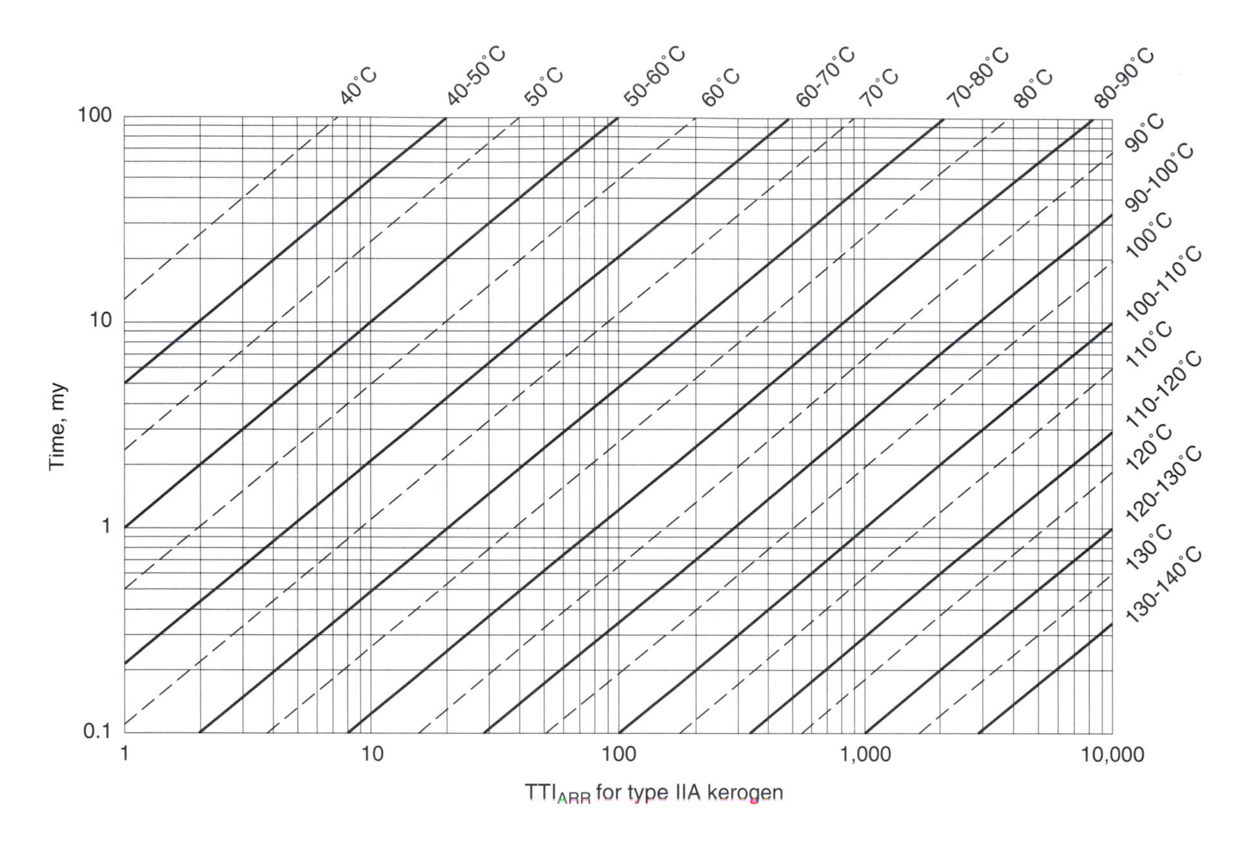

Figure 6-3

Graph showing the relationship among time, temperature, and TTI_{ARR} for kinetics of type IIA kerogen as represented by the Monterey Formation of California. The solid slanted lines represent the changing temperatures, and the dashed lines are the constant temperatures in Figures 6-3 through 6-11. [Hunt et al. 1991]

temperature range of 70 to 80°C for 1 m.y. would have a TTI_{ARR} of 20 for that 10°C temperature interval. Burial history curves sometimes include long time periods at a constant temperature with negligible subsidence or uplift. In order to account for this more accurately, the TTI_{ARR} at constant temperature may be read directly from the graphs using the dashed diagonal lines between the solid lines.

In addition to the graphs for oil generation, Figure 6-11 was made for the cracking of oil to gas. This is the main source of gas in noncoaly sediments. Gas

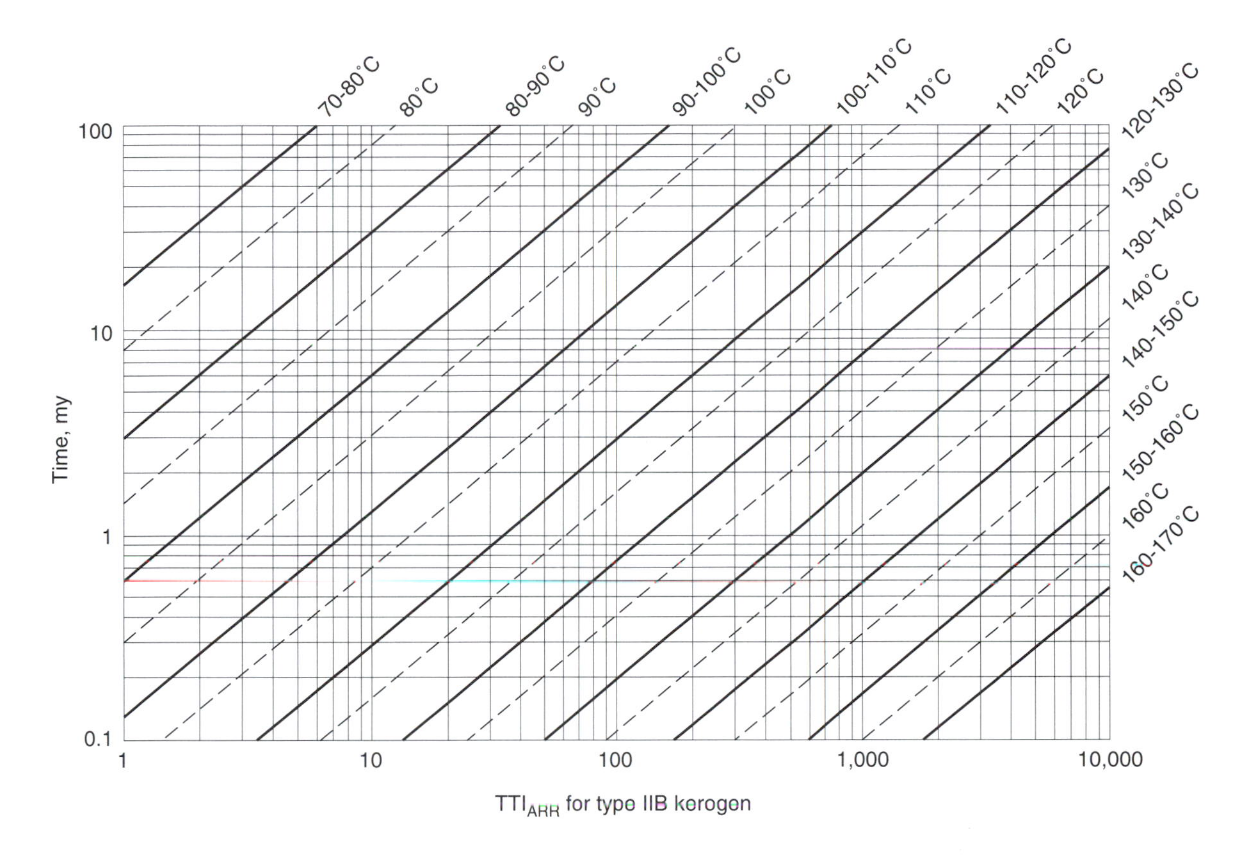

Figure 6-4

Graph showing the relationship among time, temperature, and TTI_{ARR} for kinetics of type IIB kerogen as represented by the Permian Retort Shale Member of the Phosphoria Formation of Montana. [Hunt et al. 1991]

generated from all types of kerogen actually begins well up in the oil window around 60°C (140°F), but the bulk of gas comes from cracking oil around 150 to 160°C (302 to 338°F). The Arrhenius equation kinetics shown in Figure 6-11 were derived to represent all oil cracking (Quigley et al. 1987). However, oils differ in composition, and future work is needed to determine the variability in oil-cracking kinetics. Figure 6-11 is suitable for cracking typical naphthenic 30°API gravity oils, but not necessarily for paraffinic, heavy, or high-sulfur crude oils.

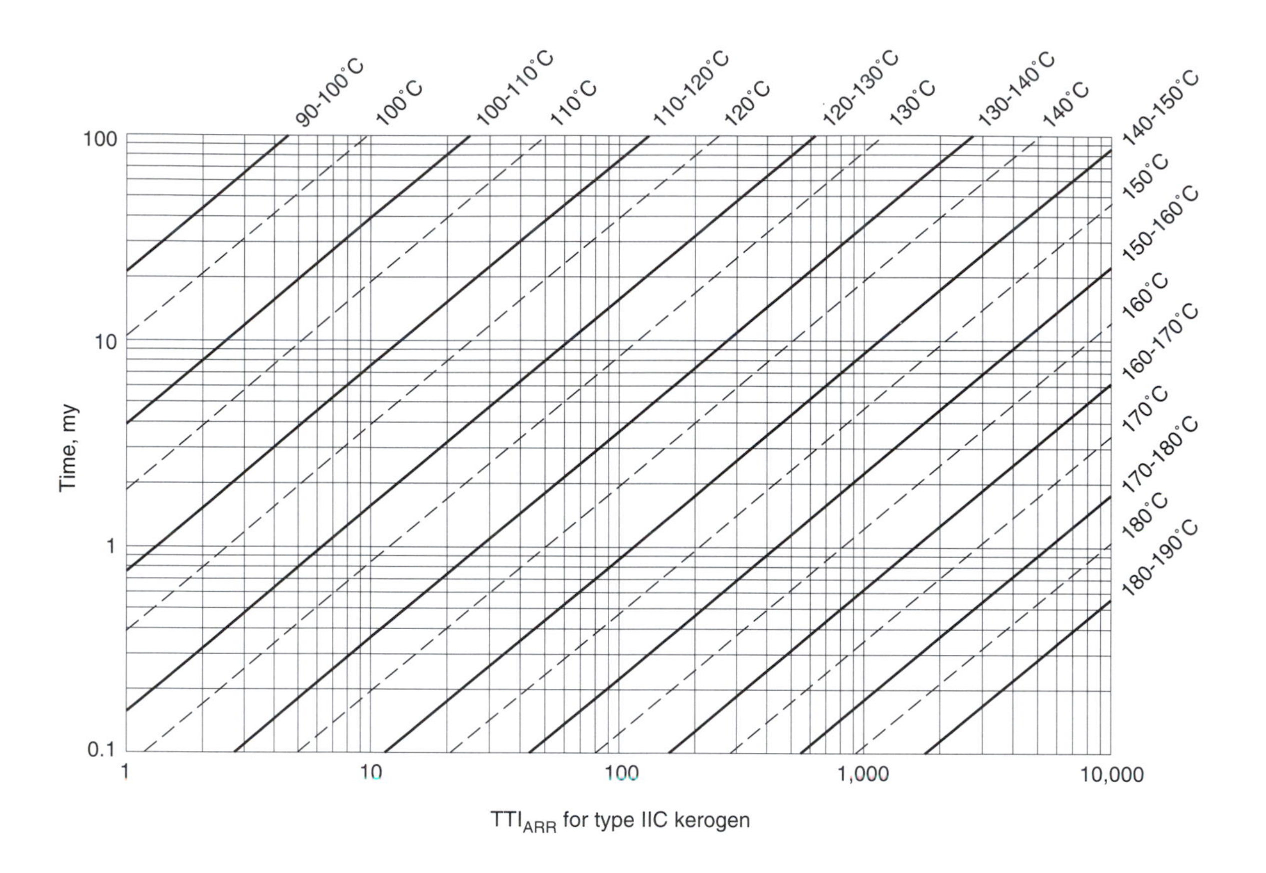

Figure 6-5

Graph showing the relationship among time, temperature, and TTI_{ARR} for kinetics of type IIC kerogen as represented by the Alum Shale of Sweden. [Hunt et al. 1991]

Addition of the TTI_{ARR} values for each 10°C interval, or period of constant temperature, in the burial history curve of a source rock gives a summation index (ΣTTI_{ARR}). This index gives the cumulative effect of time and temperature on the oil generation of a source rock. Despite the different values for the kinetic parameters and the E and A for different source rock types (Table 6-1), the ΣTTI_{ARR} values are always related to the same percentage of oil generated ($x\%$) by the expression

$$x\% = [1 - \exp(-\Sigma TTI_{ARR}/100)] \times 100 \qquad (6\text{-}4)$$

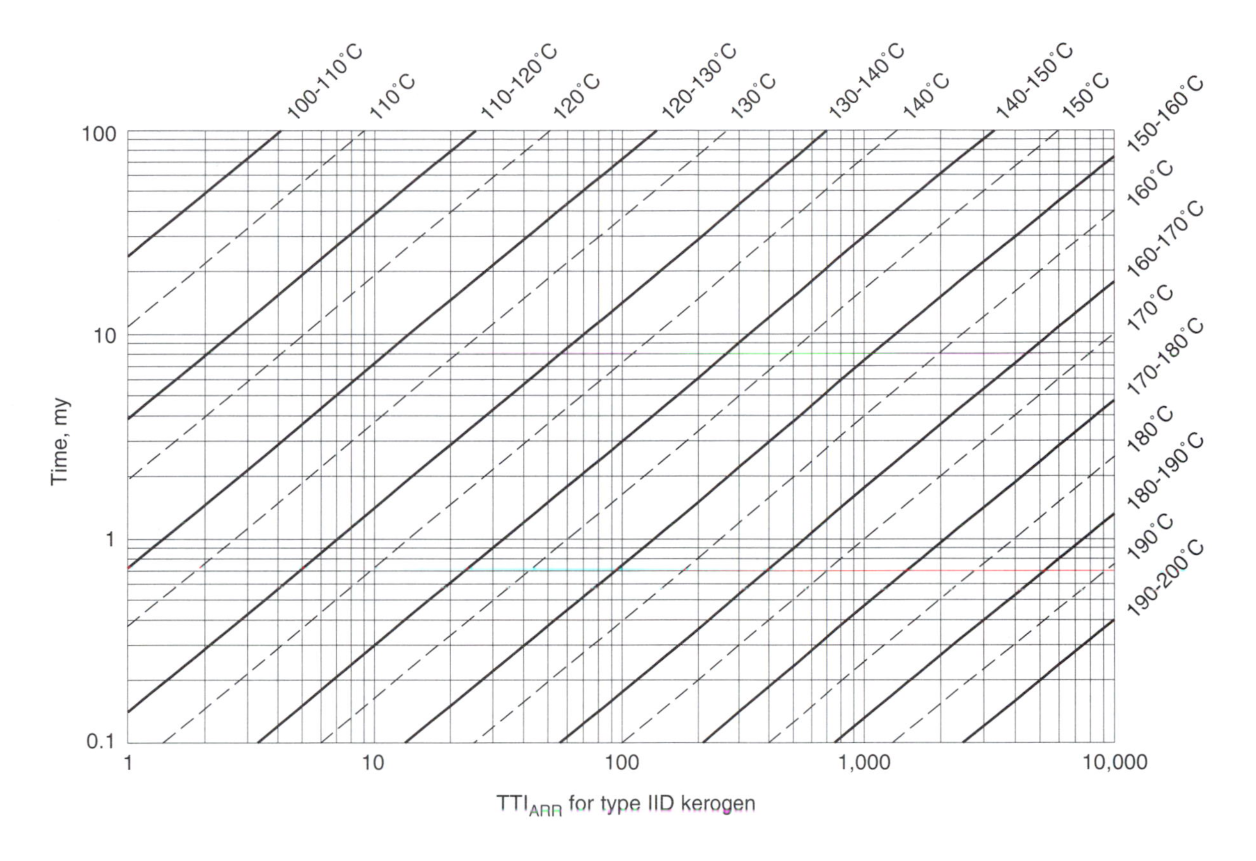

Figure 6-6

Graph showing the relationship among time, temperature, and TTI$_{ARR}$ for kinetics of type IID kerogen as represented by the Woodford Shale of Oklahoma. [Hunt et al. 1991]

This expression is shown graphically in Figure 6-12. On this graph, the beginning of oil generation may be equated to a ΣTTI_{ARR} of one and the end to a ΣTTI_{ARR} of 460. This same range also may be used for determining the beginning and end of oil cracking to gas when ΣTTI_{ARR} is being determined from the oil-to-gas graph (Figure 6-11).

The differences in the kinetics of different kerogens such as types IIA, B, C, and D in Table 6-1 are significant and must be taken into account when evaluating the timing of oil generation in a source rock, as described for Figure 6-2. This also may be demonstrated from the graphs by considering the 10°C

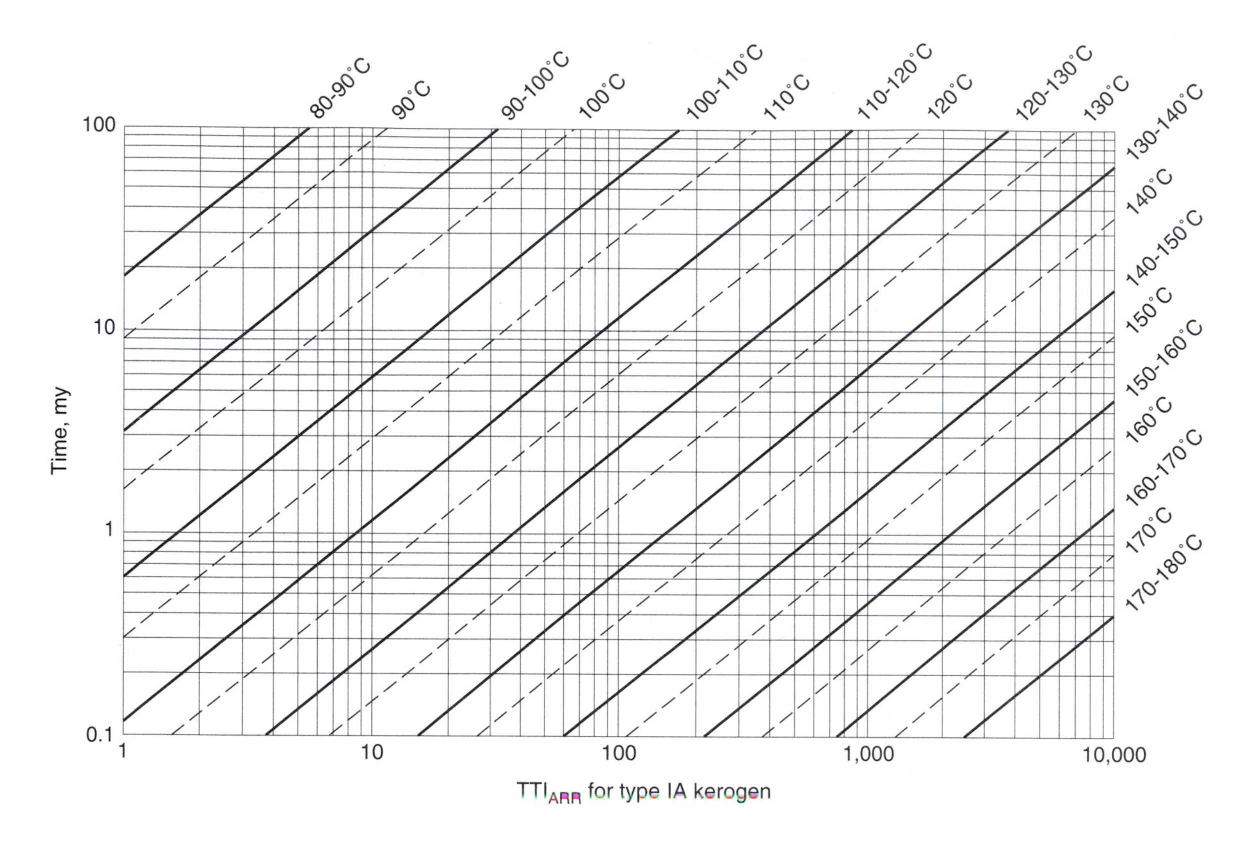

Figure 6-7

Graph showing the relationship among time, temperature, and TTI_{ARR} for kinetics of type IA kerogen of the Green River Formation at Anvil Point, Colorado. [Hunt and Hennet 1992]

temperature range required for a source rock to obtain a TTI_{ARR} of 1 in 1 million years for each set of type II kinetic parameters given in Table 6-1. The type IIA kinetics require a temperature range of 50 to 60°C (Figure 6-3), and the type IID kinetics require a temperature range near 120 to 130°C (Figure 6-6). Accordingly, the type IIB (Figure 6-4) and type IIC (Figure 6-5) kinetics require intermediate temperature ranges of 90° to 100°C and 110° to 120°C, respectively. Therefore, the use of a fixed set of kinetic parameters to describe oil generation from all source rocks bearing type II kerogens is not valid.

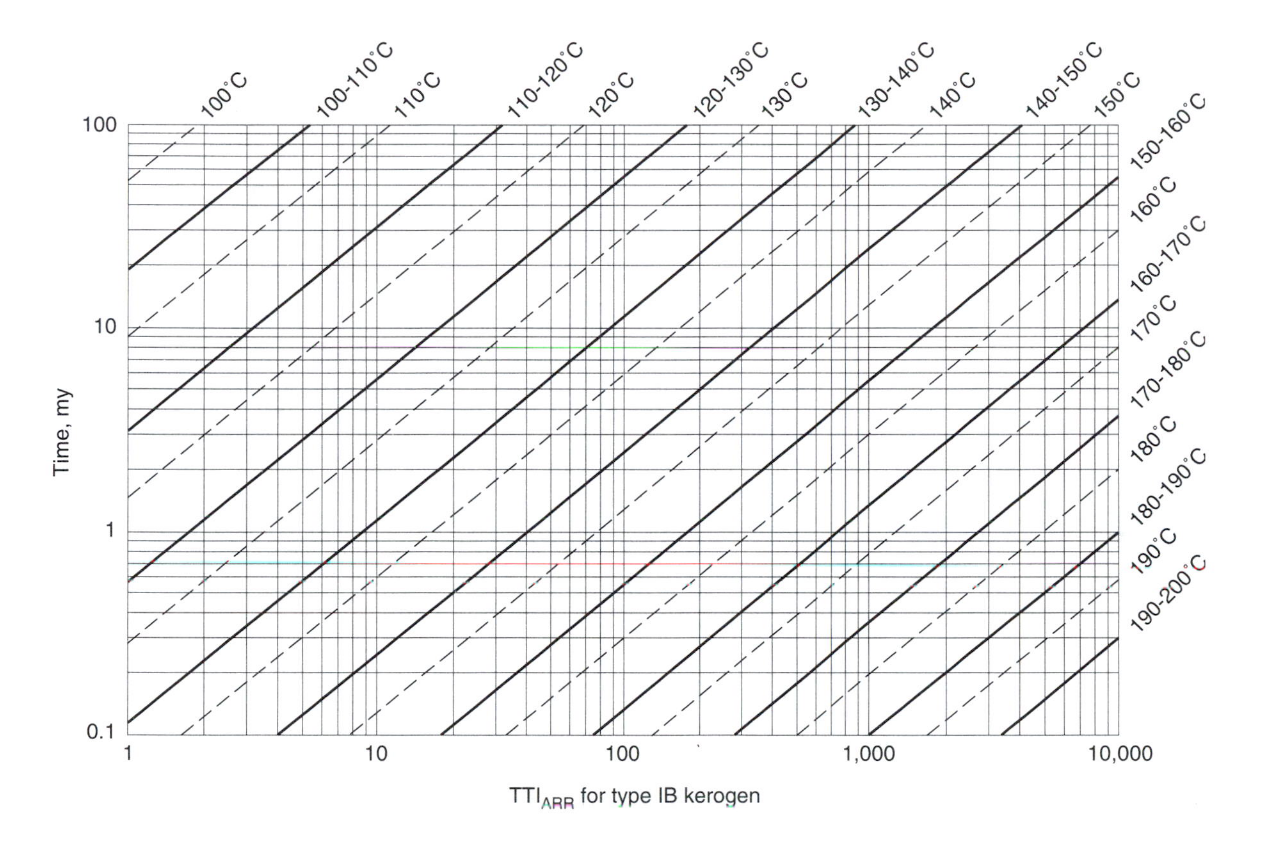

Figure 6-8

Graph showing the relationship among time, temperature, and TTI_{ARR} for kinetics of type IB kerogen of the Green River Formation in the Uinta Basin, Utah. [Hunt and Hennet 1992]

Case Studies

The following examples illustrate how the oil windows of source rocks are determined using the TTI_{ARR} graphs. These examples are based on limited data from the literature and are assumed to be reasonably accurate. The intent of these examples is to illustrate the applications of the graphical method rather than to provide rigorous exploration models for the basins or regions mentioned in the examples.

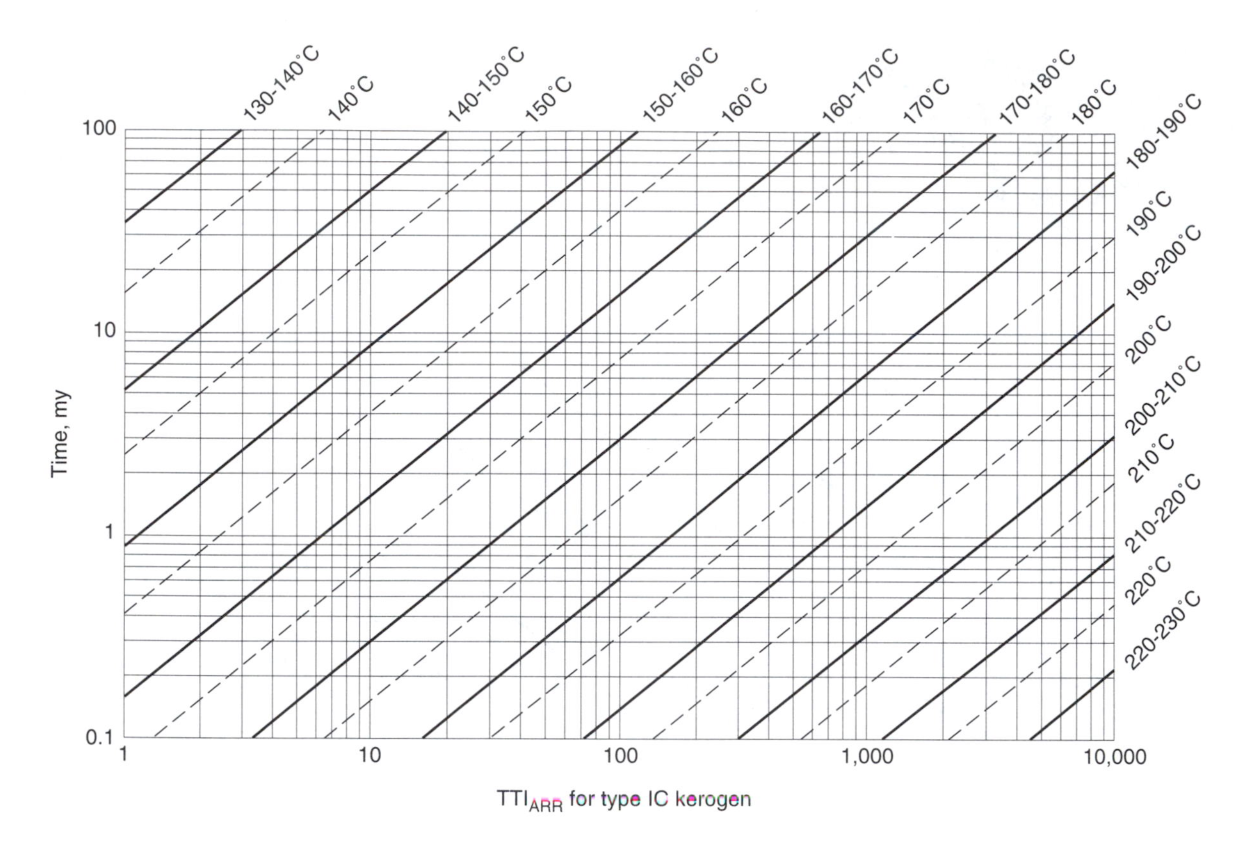

Figure 6-9

Graph showing the relationship among time, temperature, and TTI_{ARR} for kinetics of type IC kerogen of the Green River Formation (Tipton Member) in Wyoming. [Hunt and Hennet 1992]

Kerogen Type IIA Kinetics

Figure 6-13 shows the burial history curves for the Monterey Shale and shallower formations in the Point Conception COST well from Petersen and Hickey (1987). Table 6-2 contains time–temperature data plus the Arrhenius TTI and ΣTTI values for the curve representing the base of the Monterey in this well. The TTI_{ARR} values for the Monterey were obtained from the graph for type IIA kinetics in Figure 6-3. Geochemical data from this well have puzzled geochemists because some geochemical maturity indicators reported by Petersen and Hickey (1987) showed the Monterey to be in the oil window at a depth of

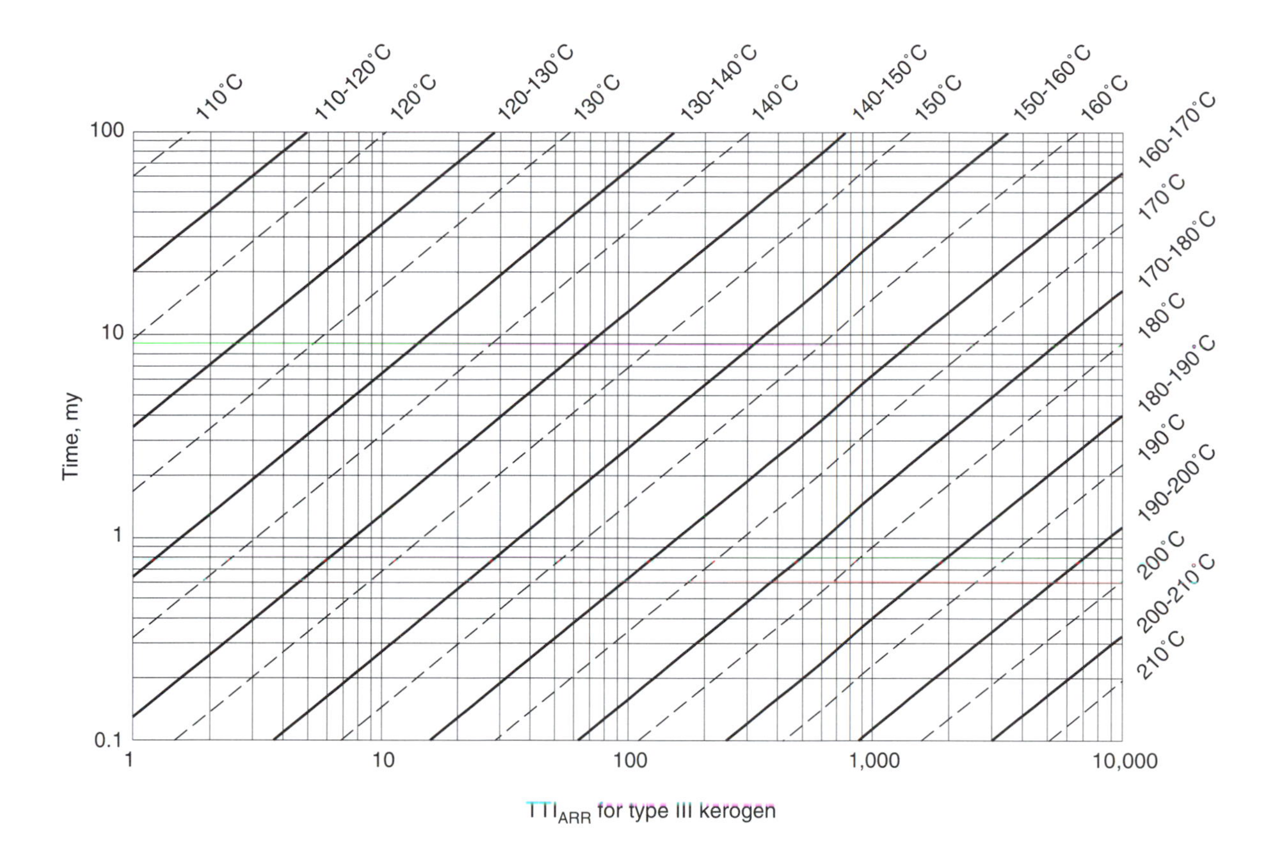

Figure 6-10

Graph showing the relationship among time, temperature, and TTI_{ARR} for kinetics of type III kerogen in the Tent Island Formation of the Beaufort–Mackenzie Basin of Canada. [Hunt and Hennet 1992]

2,134 m (7,001 ft), equivalent to a subsurface temperature of 83°C. But the calculated Lopatin TTI value of 0.3 indicated that the Monterey Shale was not in the oil window in this well.

In Figure 6-3, the intercepts of each solid diagonal temperature range line with the exposure time line from the ordinate gives the TTI_{ARR} values along the abscissa. As shown in this example, the Monterey Shale was subjected to the 70-to-80°C temperature range for 1 million years (Table 6-2), which results in a TTI_{ARR} value of 20 (Figure 6-3, Table 6-2). Summation of this value with the preceding values calculated for the lower 10°C temperature ranges results in a

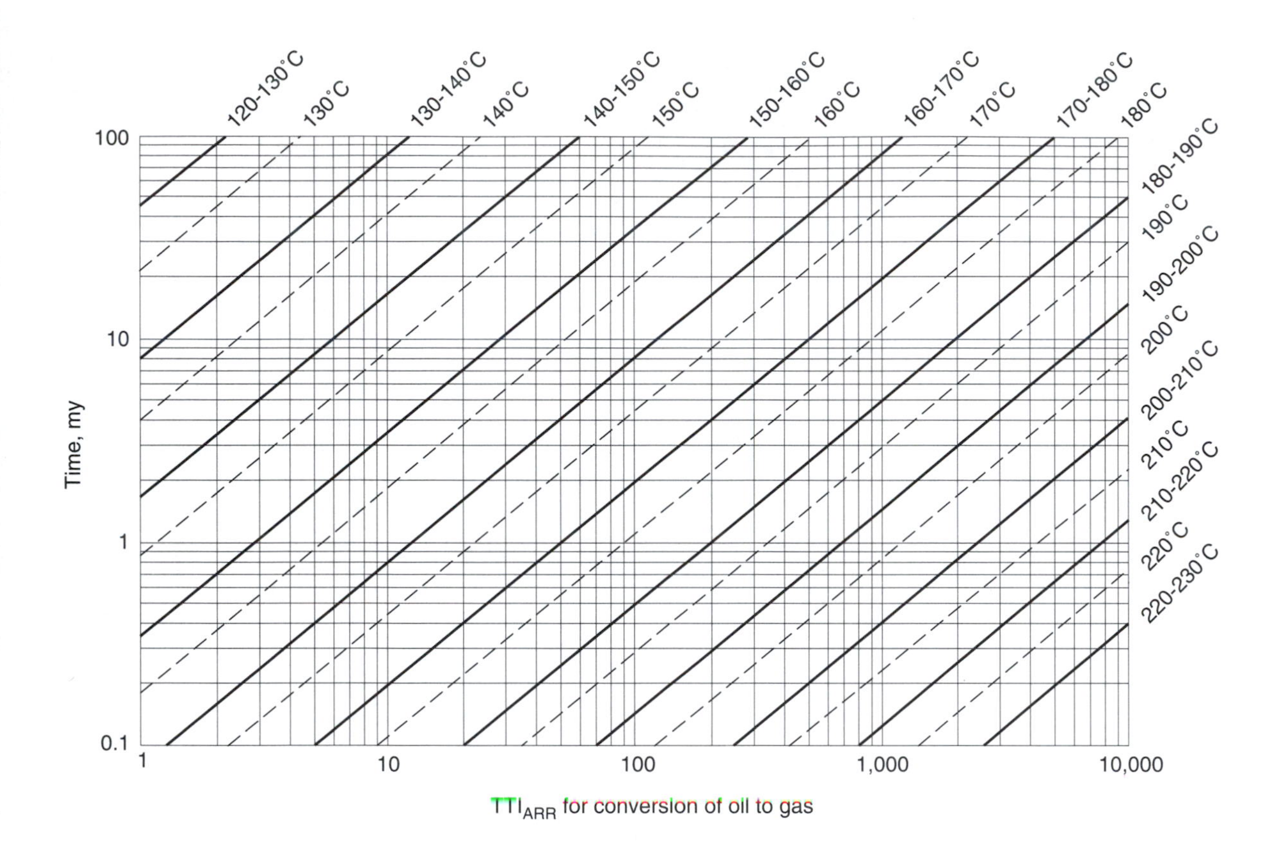

Figure 6-11

Graph showing the relationship among time, temperature, and TTI_{ARR} for the conversion of oil to gas. [Hunt and Hennet 1992] The kinetic parameters are $E = 230$ kJ/mol and $A = 3.17 \times 10^{26}$/m.y. [Quigley et al. 1987]

ΣTTI_{ARR} of 27 at this point in its burial history. Based on Figure 6-12 this summation value indicates that 26% of the oil has been generated. By the time the Monterey had been buried to a temperature range of 90–100°C, about 98% of the oil had formed. Oil generation started around 7 Ma and ended about 1 Ma, as shown in Table 6-2 and Figure 6-13.

Another example of a high-sulfur type IIA (II-S) kerogen is the Senonian bituminous rock (SBR) in the Amiaz-1 well of Israel, described by Tannenbaum and Aizenshtat (1984). The organic sulfur contents of the kerogens in this rock unit range from 10 to 12 wt%. These authors used molecular indicators of maturation, such as the isomerization of hopanes and steranes plus CPI, and the C_{15+} yield to estimate the level of maturation. Their data showed that the

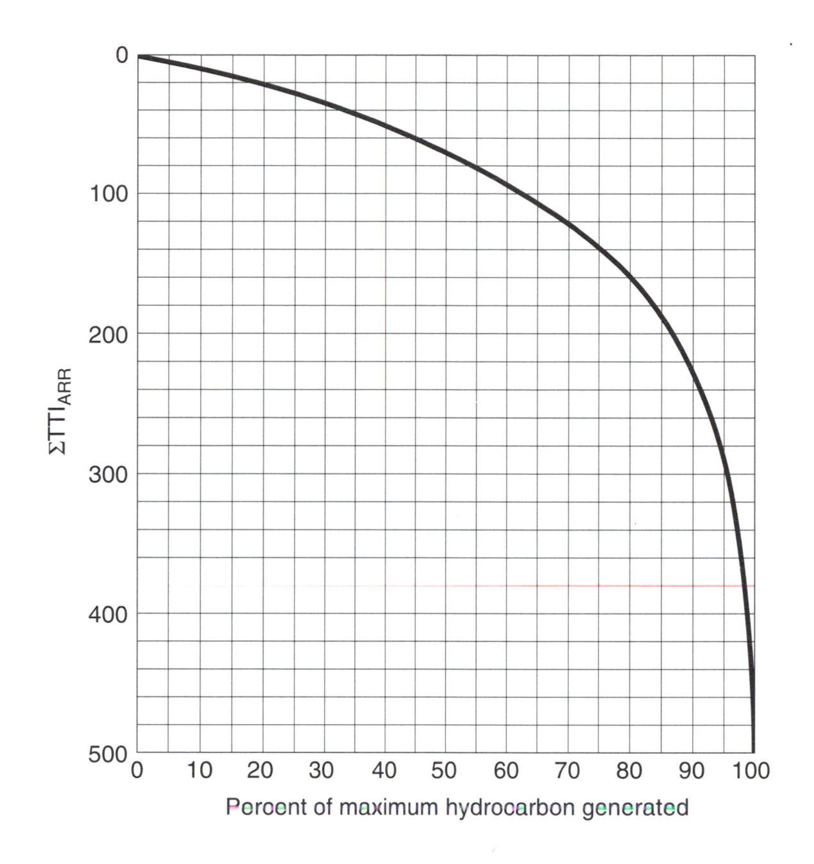

Figure 6-12

Graph showing the relationship between ΣTTI_{ARR} and the percentage of oil or gas generated, based on Equation 6-4 (see p. 154). [Hunt et al. 1991]

SBR was well into the oil window in the Amiaz well. As shown in Table 6-2, the oil generation started about 4 Ma at 50°C in this well and is continuing. The ΣTTI_{ARR} value of 134 at the bottom of the well suggests that about 73% of the oil has been generated.

Kerogen Type IIB Kinetics

The Phosphoria Formation of Montana, which is used as a standard for type IIB kinetics, was buried to a depth of about 6,000 m (19,680 ft) by the Absaroka

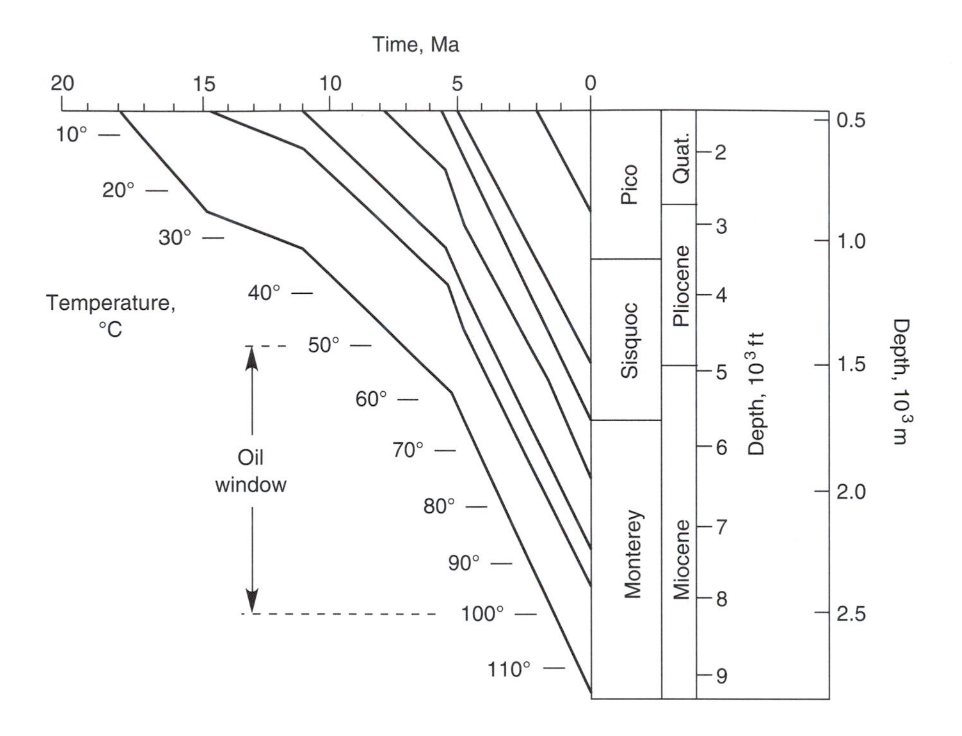

Figure 6-13

Burial history curve of the Miocene Monterey Formation in the Point Conception COST well. [Petersen and Hickey 1987]

thrust plate in Wyoming, according to Warner and Royse (1987). Their burial history curve for the top of the Phosphoria is shown in Figure 6-14. Table 6-3 contains the TTI_{ARR} data from Figure 6-4. Oil generation started about 95 Ma when burial temperatures reached 80 to 90°C and ended about 81 Ma at 130°C.

Since the burial continued, however, the oil that did not escape updip was cracked to gas. Also, since the oil generation ended, the kerogen began generating primarily gas. To determine the gas window, the first two columns in Table 6-3 are used with Figure 6-11 (oil to gas) to calculate the TTI_{ARR} values, in the far right columns of Table 6-3. The gas window started about 79 Ma and ended around 73 Ma, before uplift. Both oil and bitumen are found in the shallower sections, but the major production is gas and the Phosphoria in this area is postmature to oil, based on vitrinite reflectance (Warner and Royse 1987). Light oil and condensate generation would have occurred between the oil and gas windows in Figure 6-14.

TABLE 6-2 Arrhenius TTI Values for Source Rocks with Kerogen Type IIA Kinetics

Temperature range, °C	*Exposure time, m.y.*	*Type IIA TTI*	*Type IIA ΣTTI*	*Oil window*[a]
Monterey Shale Point Conception COST Well				
40–50	2	0	0	
50–60	1.6	2	2	B
60–70	1	5	7	
70–80	1	20	27	
80–90	1	80	107	
90–100	1	300	407	E
100–110	1	1,000	1,407	
Senonian Bituminous Rock Amiaz-1 Well				
40–50	3	0	0	
50–60	2	2	2	B
60–70	0.5	2.5	4.5	
70–80	0.5	10	14.5	
80–90	0.5	40	54.5	
90–95	0.5	80	134.5	↓

[a]B = Beginning and E = End of oil window.

The Santonian Brown Limestone (SBL) is considered by Chowdhary and Taha (1987) to be one of the main source rocks in the Gulf of Suez. Its classification as type IIB is based on the organic sulfur content of its kerogen (Eglinton et al. 1990). Also, crude oils that are known to have been generated from this rock have moderately high sulfur contents of 2.3%. Various maturity indicators presented by Chowdhary and Taha show that the SBL is well into or past peak generation in the DD83-1 well. Using their burial history curve (Figure 6-15) and the graph (Figure 6-4) for kerogen type IIB, Table 6-3 suggests that oil generation may have started as early as 9 Ma at a depth of 2,290 m (7,500 ft) within the 90-to-100°C temperature range and ended 2.5 Ma in the 130-to-140°C range. The oil window in Figure 6-15 is only for the brown limestone. If there are any source beds in the younger formations, they have different kinetics. Consequently, the same oil window cannot be drawn across all the formations, as was done when using a single time–temperature parameter, as in the Lopatin method (Waples 1985, p. 153).

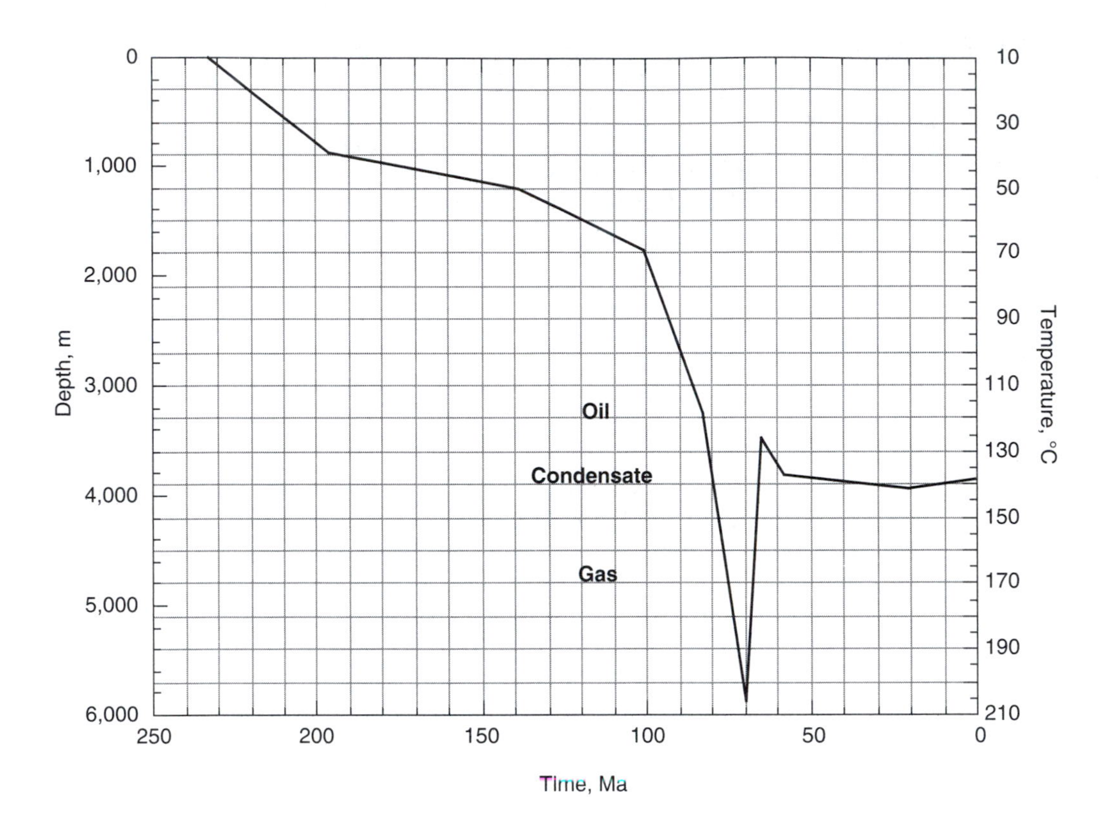

Figure 6-14

Burial history curve of the Phosphoria Formation in the Mobil 22–19G Tip Top well of southwest Wyoming. [Warner and Royce, 1987]

Kerogen Type IIC Kinetics

The giant Salym oil field in the West Siberian Basin produces oil from fractures in the Bazhenov Shale source rock. According to Lopatin (personal communication), the sulfur content of the kerogen is higher than in most clastics, due to some gypsum occurring in the Bazhenov suite. A burial history curve for Hole 184 in the Salym field starting at 2,000 m is shown in Figure 6-16 (Lopatin, personal communication). The geothermal gradient in this interval is 3.3°C/100 m. The time–temperature data in Table 6-4 were calculated from the curve using

TABLE 6-3 Arrhenius TTI Values for Source Rocks with Kerogen Type IIB and Oil-to-Gas Kinetics

Temperature range, °C	*Exposure time, m.y.*	*Type IIB TTI*	*Type IIB ΣTTI*	*Oil window*[a]	*Oil to gas TTI*	*Oil to gas ΣTTI*	*Gas window*[a]
Phosphoria Formation, Mobil 22–19B Tip Top Well, Wyoming							
70–80	3.2	0	0				
80–90	3.2	1.1	1.1	B			
90–100	3.2	5.4	6.5	|			
100–110	3.2	25	32	|			
110–120	3.2	110	142	|			
120–130	1.5	190	332	E			
130–140	1.5	700	1,032				
140–150	1.5	2,500	3,532		0	0	
150–160	1.5				4	4	B
160–170	1.5				18	22	|
170–180	1.5				75	97	|
180–190	1.5				300	397	E
190–200	1.5				1,000	1,397	
Brown Limestone, DD83-1 Well, Gulf of Suez							
80–90	2	0	0				
90–100	2	3	3	B			
100–110	1.5	12	15	|			
110–120	1.5	45	60	|			
120–130	1.5	200	260	|			
130–140	1.5	700	960	E			

[a]B = Beginning and E = End of oil or gas window.

kerogen type IIC kinetics (Figure 6-5). The Arrhenius TTI data indicate that generation started when the Bazhenov Shale reached 90-to-100°C around 100 Ma. Currently, the Bazhenov is nearing the end of oil generation.

It is interesting that the Lopatin TTI and Arrhenius TTI calculations show the same oil-window interval for the Bazhenov Shale (Hunt and Hennet 1992). These two modeling methods tend to agree with type IIC and IID kerogens when the heating rates are close to 1°C/m.y., as the Bazhenov Shale was. It was buried through an interval of 110°C in 107 m.y.

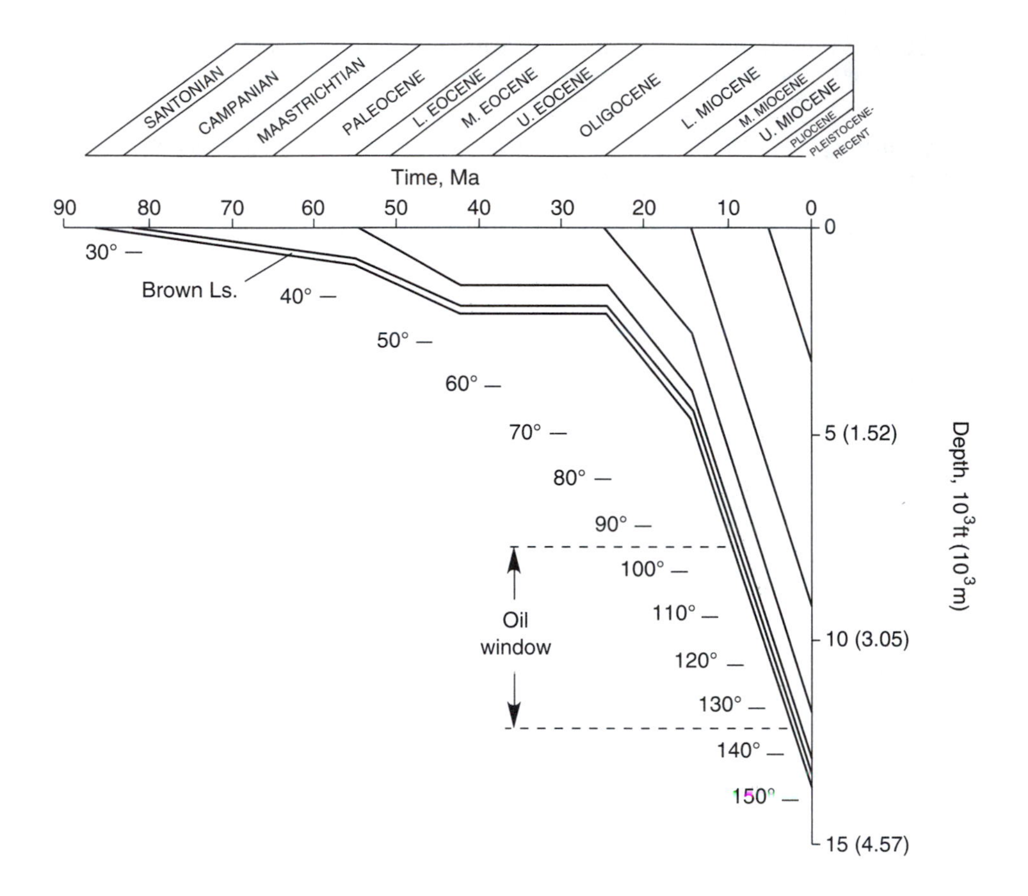

Figure 6-15

Burial history curve of the Santonian Brown Limestone in the DD83-1 well in the Gulf of Suez. [Chowdhary and Taha 1987]

The Bakken Formation of the Williston Basin of North Dakota is an extremely organic-rich source rock deposited under strongly anoxic conditions. The TOCs are as high as 15 wt% (30 volume%), and the mature interval is overpressured (Webster 1984). These characteristics are the same as for the Bazhenov Shale discussed earlier. Also, the Bakken is associated with overlying carbonates and anhydrites. All these factors indicate that it is a type IIC kerogen.

Figure 6-17 is a burial history curve of the Bakken Formation in the California Oil No. 1 Rough Creek well (Webster 1984). Table 6-4 contains the

TABLE 6-4 Arrhenius TTI Values for Source Rocks with Kerogen Type IIC Kinetics

Temperature range, °C	*Exposure time, m.y.*	*Type IIC TTI*	*Type IIC ΣTTI*	*Oil window*[a]
Bazhenov Shale of the Western Siberian Basin				
90–100	14	0.6	0.6	B
100–110	23	5.8	5.8	|
110–120	16	20	26	|
120–130	47	300	326	E
Bakken Formation, Williston Basin, North Dakota				
90–100	9	0	0	
100–110	9	2.3	2.3	B
110–120	10	14	16	|
120–130	56	360	375	E

[a]B = Beginning and E = End of oil window.

TTI_{ARR} data derived from this figure and from Figure 6-5 (type IIC kinetics). The results indicate that oil generation started when the Bakken was heated to 100°C about 75 Ma. Today, the oil generation is essentially complete, and some conversion of oil to gas has commenced. This is deduced from Figure 6-11 (oil-to-gas kinetics). It shows that a TTI of 1 occurs by exposure of a source rock to 120 to 130°C for 45 m.y. The Bakken has been at this temperature for 56 m.y. (Table 6-4).

The beginning of the oil window at 100°C is equivalent to a depth of 2,780 m (9,118 ft), which is about where Meissner (1978) observed an increase in the electrical resistivity log due to hydrocarbons filling the pore spaces in the Bakken.

Kerogen Type IID Kinetics

Table 6-5 contains time and temperature data for the Kimmeridge Shale source rock in the Central Graben of the Norwegian North Sea near the Ekofisk field (Leonard, 1989). The burial history curve is shown in Figure 6-18. According to Leonard (1989), oil generation started about 40 Ma, peaked around 15 Ma, and ended about 5 Ma The ΣTTI_{ARR} values based on type IID kinetics generally agree with this. The type IID kinetics was chosen based on analyses of the organic sulfur content of Kimmeridge Shale samples (Eglinton et al. 1990).

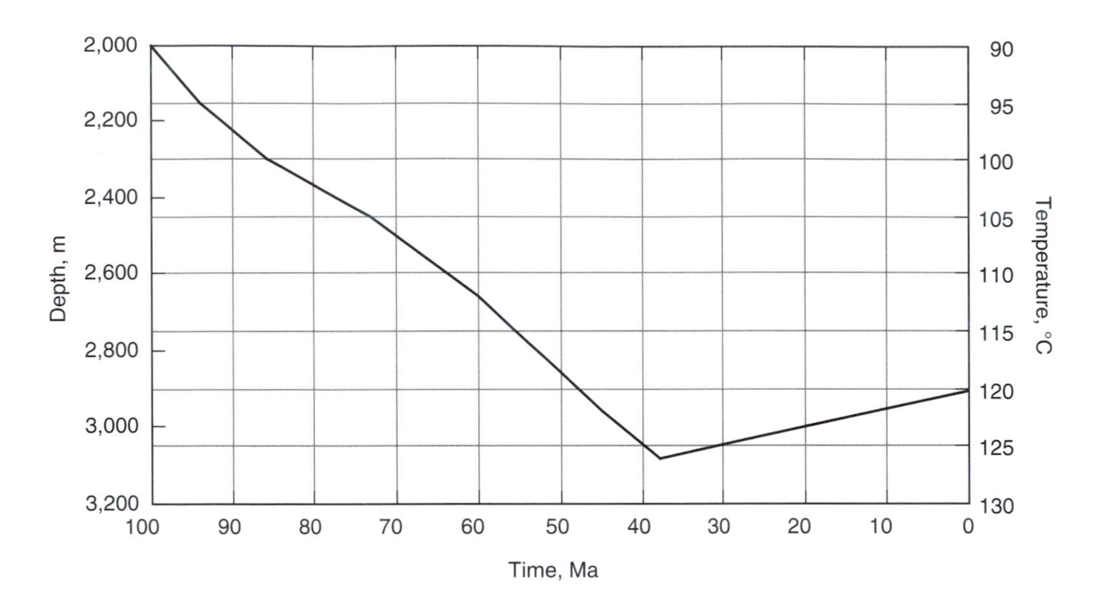

Figure 6-16

Burial history curve starting at 2,000 m of the Bazhenov shale in Hole 184, Salym Oil field, West Siberian Basin. [Lopatin, personal communication]

Table 6-5 also contains TTI_{ARR} data for the gas window on the far right. The ΣTTI_{ARR} of 287 when used on Figure 6-12 indicates that about 95% of the oil remaining in this deep part of the source rock has been converted to gas. Any oil that migrated upward, of course, would not be affected.

The Canadian Hunter Brassey field in the Deep Western Canada Basin produces 54° API gravity oil from the Triassic Artex dune sandstone, an overpressured reservoir at 2,903 m. The seal is a dolomite with pores plugged with anhydrite. The probable source rocks are the Doig–Montney Permian Shales at about 3,000 m. The burial history curve at the base of the Doig (top of the Montney) is shown in Figure 6-19 (H. von der Dick, personal communication). The Doig is believed to have been buried to about 5,500 m around 40 Ma. Subsequently, it was uplifted to 3,000 m. Table 6-6 contains time–temperature data plus Arrhenius TTIs for the oil window from Figure 6-6 and the gas window from Figure 6-11. These data indicate that oil generation started about 78 Ma when the Doig reached 110°C and ended in 45 Ma at about 150°C. Oil-to-gas cracking started about 55 Ma after peak oil generation. Approximately 40%

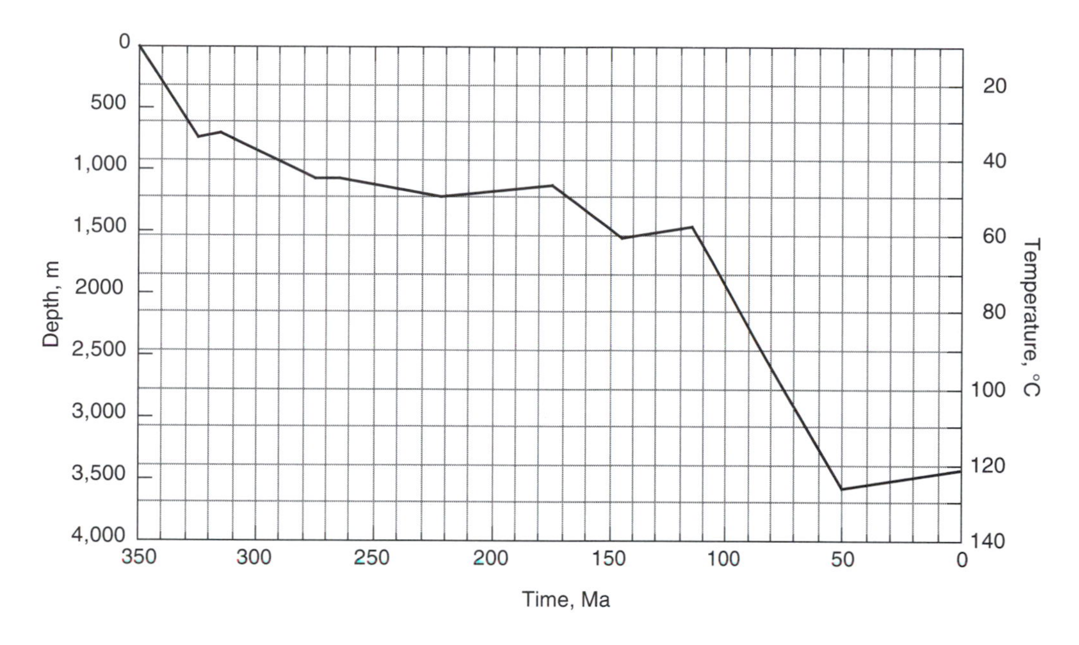

Figure 6-17

Burial history curve of the Bakken Formation in the California Oil No. I Rough Creek well in the Williston Basin, western North Dakota. [Webster 1984]

of the oil was converted to gas before cooling due to uplift caused the cracking to cease (29 Ma). If uplift had not occurred, the Brassey field would be entirely gas. Some pyrobitumen, which is probably a residue of the conversion of oil to condensate and gas, was encountered in the Artex sandstone reservoir (H. von der Dick, personal communication).

The Mowry Shale is a well-known petroleum source rock in the Rocky Mountain region of the western United States. Its low-sulfur kerogen classifies it as type IID. The time and temperature data in Table 6-6 were derived from the burial history curve (Figure 6-20) of the Amoco B.N. well in the northern Bighorn Basin (Hagen and Surdam 1984). The ΣTTI_{ARR} values from Figure 6-6 indicate that oil generation from the Mowry Shale in this part of the Bighorn Basin started around 52 Ma at a temperature of about 125°C and ended around 20 Ma at about 145°C. Similar results were obtained by Hagen and Surdam (1984) using the Lopatin model. As previously mentioned, TTI_{LOP} and TTI_{ARR} tend to agree, using type IID kinetics at moderate heating rates. The heating rate of the Mowry Shale in this basin from surface to maximum depth was about 1°C/m.y.

TABLE 6-5 Arrhenius TTI Values for the Kimmeridge Shale, Central Graben, North Sea, with Kerogen Type IID and Oil-to-Gas Kinetics

Temperature range, °C	*Exposure time, m.y.*	*Type IID TTI*	*Type IID ΣTTI*	*Oil window*[a]	*Oil to gas* TTI_{ARR}	*Oil to gas* ΣTTI_{ARR}	*Gas window*[a]
100–110	7	0	0				
110–120	8	2	2	B			
120–130	8	12	14	│			
130–140	8	56	70	│			
140–150	8	270	340	│	5	5	B
150–160	2	270	610	E	6	11	│
160–170	2				26	37	│
170–180	1				50	87	│
180–190	1				200	287	↓

[a]B = Beginning and E = End of oil or gas window.

Kerogen Type I Kinetics

TTI_{ARR} graphs based on kerogen type I kinetics from Table 6-1 are shown in Figures 6-7, 6-8, and 6-9. All three of these are from different regions of the Green River Formation (GRF) in the western United States. The oil-generation interval for the GRF in the Uinta Basin of Utah was studied in some detail by Sweeney et al. (1987). Their burial history curve for the base of the Eocene in the Shell Brotherson 1-11B4 well is shown in Figure 6-21. The time–temperature data from this curve are in Table 6-7. The TTI_{ARR} graph (Figure 6-8) was used to obtain the TTI_{ARR} values in Table 6-7. During burial of the GRF there was a long period (20 m.y.) of constant temperature at about 150°C. Consequently, the dashed line marked 150°C on Figure 6-8 was used during this time period to determine the Arrhenius TTI. The ΣTTI_{ARR} shows that oil generation started about 37 Ma at a temperature of 120°C and ended 26 Ma during the constant-temperature interval. This agrees with the model study by Sweeney et al. (1987) who estimated that the peak oil generation was about 30 Ma. The long exposure of the Green River Formation to 150°C could have converted up to 26% of the oil to gas (Table 6-7). This value may be a bit high, since the paraffinic Altamont oil would be harder to crack than the oil used to derive Figure 6-11. However, gas/oil ratios in the Altamont field are as high as 1,500 cubic feet of gas per barrel of oil.

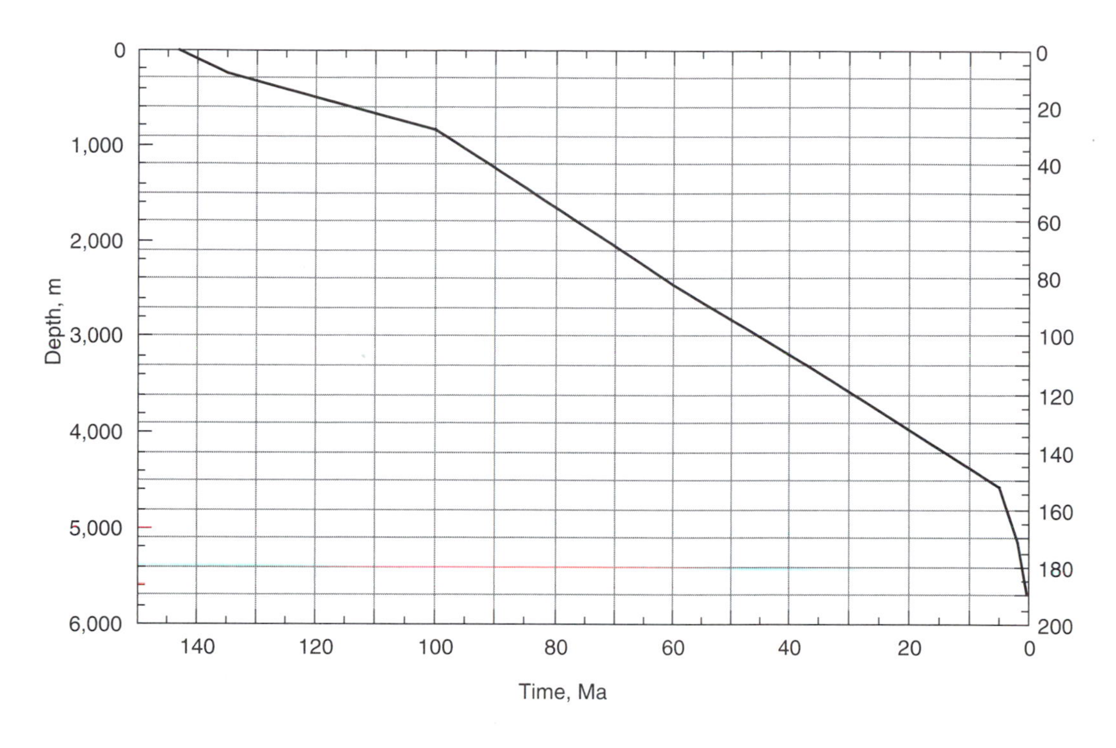

Figure 6-18

Burial history curve of the base of the Upper Jurassic Kimmeridge Shale in the Central Graben of the North Sea. [Leonard 1989]

Kerogen Type III Kinetics

Type III kerogens are slower reacting than the type IIs and some of the type Is. This is apparent from Figure 6-2, which shows only type IC having a slower reaction rate than type III. In fact, type III kinetics are not very different from the kinetics for cracking oil to gas. A comparison of these two Arrhenius graphs (Figures 6-10 and 6-11) indicates that they differ by only 10°C. This means that as the type III kerogen shifts from oil to condensate and gas generation, the oil previously formed is already beginning to crack. This suggests that source rocks with type III kerogens tend to generate petroleum with a higher condensate and gas-to-oil ratio than the other kerogen types do.

The kinetic parameters for Figure 6-10 were obtained by open pyrolysis (Issler and Snowdon 1990), which includes bitumen, oil, and gas from kerogen.

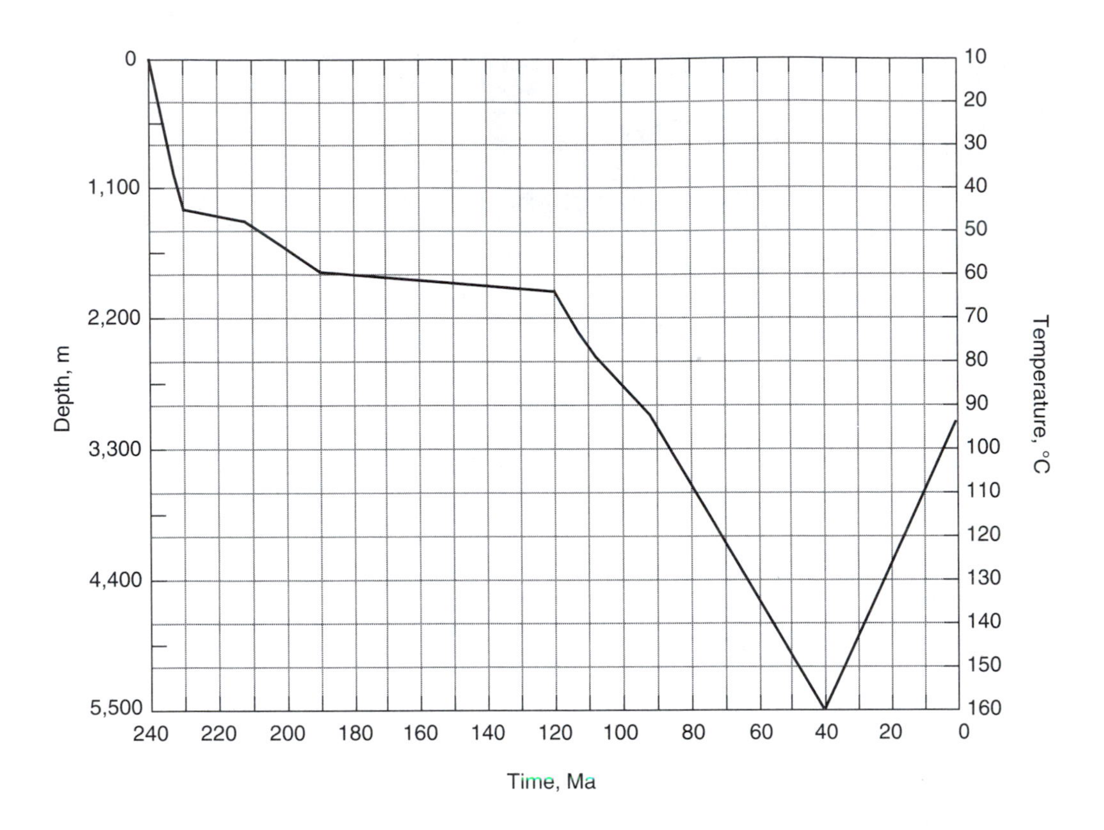

Figure 6-19

Burial history curve of the base of the Doig Shale in the Canadian Hunter Brassey field of the Deep Western Canada Basin. [H. von der Dick, personal communication]

However, over 60% of the hydrocarbon generation was within $E = 230 \pm 10$ kJ/mol, so this E value is assumed to be for the oil window. Gas generation from kerogen would continue at higher activation energies. This assumption is reasonable if the hydrogen index (HI) from open pyrolysis of an immature source rock is above 200, since this indicates some oil generation and expulsion. At HI values below 150, gas would be the main product.

Figure 6-22 is part of the burial history curve of the South Padre Island COST No. 1 well drilled off the south coast of Texas. Only the section starting at 2,100 m (6,888 ft, equivalent to 90°C) is shown, since the upper section at lower temperatures is not relevant to the TTI calculation. The geothermal gradient in this deeper section is 3.3°C/100 m. Table 6-8 contains the time–temperature data plus the TTI_{ARR} values determined from Figure 6-10. Oil generation from

TABLE 6-6 Arrhenius TTI Values for Source Rocks with Kerogen Type IID and Oil-to-Gas Kinetics

Temperature range, °C	*Exposure time, m.y.*	*Type IID TTI*	*Type IID ΣTTI*	*Oil window*[a]	*Oil to gas TTI*	*Oil to gas ΣTTI*	*Gas window*[a]
Doig Formation, Brassey Field, Deep Western Canada Basin							
100–110	7.5	0	0				
110–120	7.5	1.8	1.8	B			
120–130	7.5	11	12.8	|			
130–140	7.5	53	65.8	|	0.9	0.9	B
140–150	7.5	250	316	|	4.5	5.4	|
150–160	15	1,900	2,216	E	42	47.4	|
150–140	5.6				3.4	50.8	|
140–130	5.6				0	50.8	↓
Mowry Shale of the Bighorn Basin of Wyoming							
110–120	1.4	0	0				
120–130	13	17	17	B			
130–140	20	150	167	|			
140–150	10	340	507	E			
130–140	7	50	557				
120–130	5	7	564				

[a]B = Beginning and E = End of oil or gas window.

the type III kerogen in the Miocene source rock started about 6 Ma when the temperature exceeded 130°C at a depth of 3,384 m (11,100 ft). It ended less than 1 Ma at 4,600 m (15,090 ft), but some generation of gas from kerogen continues. Any oil that did not escape upward would be cracked to condensate and gas, starting at 3,960 m (12,990 ft), based on Figure 6-11 (oil to gas). Note that the oil window in Figure 6-22 as determined by Arrhenius kinetics matches the oil window for this well in Figure 5-8 as determined by the yield of C_6 plus C_7 hydrocarbons. It also fits with other geochemical data obtained on this well by Huc and Hunt (1980).

Although the kinetic data and the geochemical analyses clearly define an oil window in this Miocene section of the U.S. Gulf Coast, it does not mean that commercial quantities of oil were generated and expelled from the Miocene.

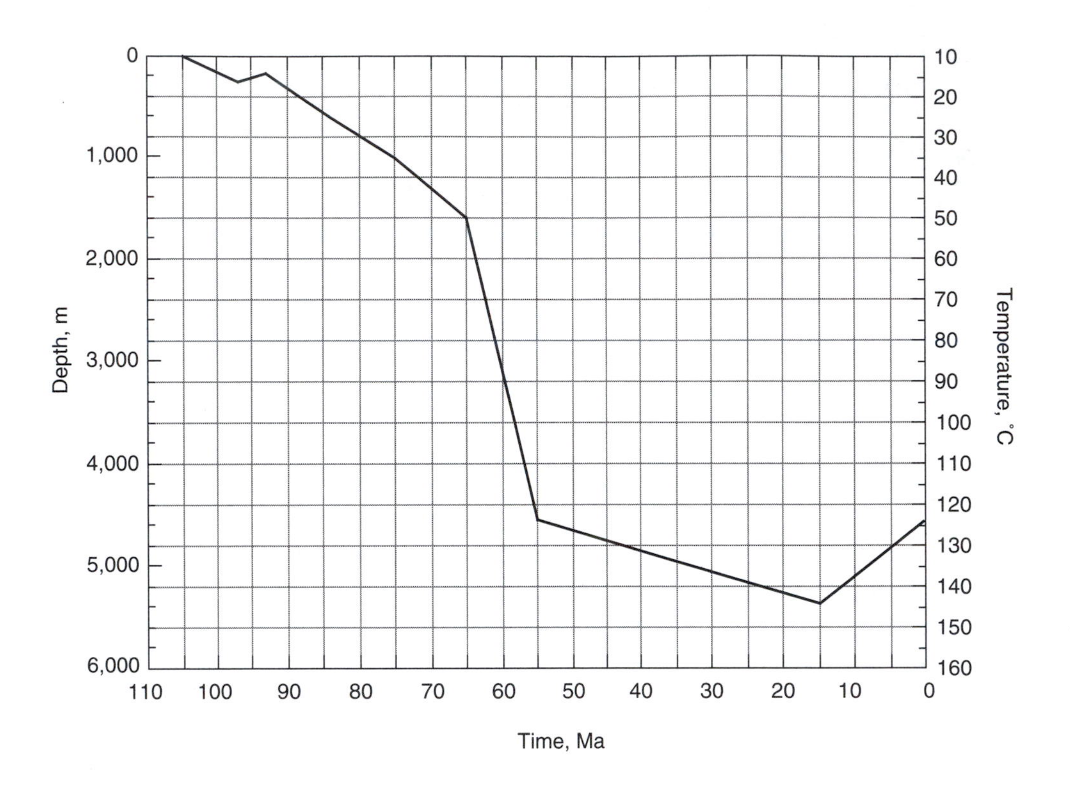

Figure 6-20

Burial history curve of the base of the Mowry Shale in the Amoco B.N. well in the Northern Bighorn Basin of Wyoming. [Hagen and Surdam 1984]

The problem is that the TOC values in the Miocene range from 0.5 to 1.5%, which may be too low for expulsion of an oil phase unless considerable amounts of free gas are present. This will be discussed further in Chapter 8.

The sources of gas in the giant Elmworth field in the Deep Western Canada Basin are the Lower Cretaceous and Jurassic coals and coaly shales (type III). Figure 6-23 shows a postulated burial history curve for the deepest major coal bed (Layer 4 in Welte et al. 1984). Precise data were not available for the thickness of eroded rock, so the maximum burial depth was estimated from vitrinite reflectance data by Welte et al. (1984).

The Arrhenius TTI data in Table 6-9 were calculated from Figure 6-10 for the Layer 4 coal bed shown in Figure 6-23. Oil generation by the coal started at around 120°C about 56 Ma and ended around 165°C about 33 Ma.

When uplifting and erosion began, nearly all of the oil and some of the gas had been generated from the coaly kerogen. Gas generation from coal probably

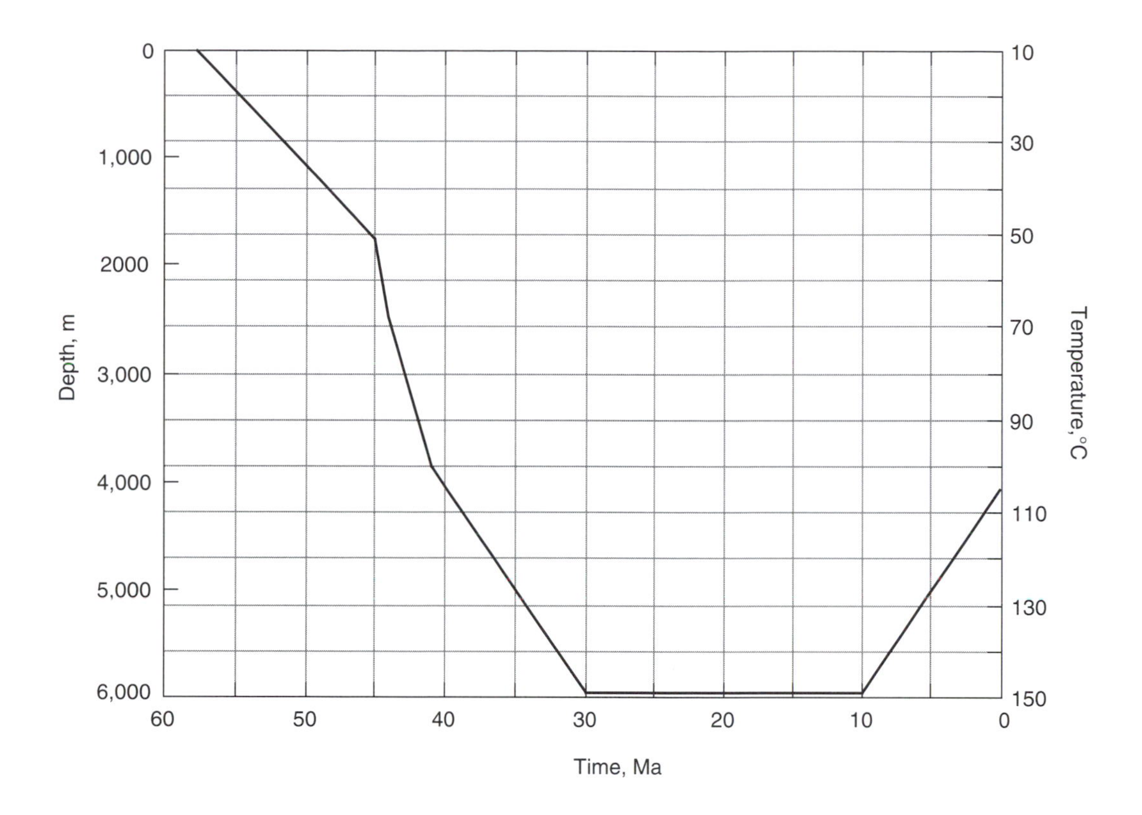

Figure 6-21

Burial history curve of the base of the Eocene Green River Formation in the Shell Brotherson 1-11B4 well, Altamont field, Uinta Basin, Utah. [Sweeney et al. 1987]

continued until about 9 Ma when temperatures dropped below 120°C. Meanwhile, about 65% of any oil trapped in the coals, coaly shales, or sandstones at the depth of coal Layer 4 would have been converted to gas (ΣTTI_{ARR} = 101, Table 6-9). The gas-from-oil interval started around 140°C about 45 Ma and ended around 140°C about 19 Ma. Based on Arrhenius kinetics, it is reasonably certain that there has been no significant gas generation from coal Layer 4 since it cooled below 120°C.

Comments on Using TTI_{ARR} Graphs

The TTI_{ARR} graphs in this chapter provide a simple and rapid method for using Arrhenius kinetics to determine the position of the oil- and gas-generating

TABLE 6-7 Arrhenius TTI Values for the Green River Formation in the Altamont Field of the Uinta Basin, Utah, with Kerogen Type IB and Oil-to-Gas Kinetics

Temperature range, °C	*Exposure time, m.y.*	*Type IB TTI*	*Type IB ΣTTI*	*Oil window*[a]	*Oil to gas TTI*	*Oil to gas ΣTTI*	*Gas window*[a]
110–120	2	0	0				
120–130	2	3.7	3.7	B			
130–140	2	18	22	│	0	0	
140–150	2	83	105	│	1.3	1.3	B
150	20	1,600	1,705	E	24	25.3	│
140–150	2				1.3	26.6	│
130–140	2				0	26.6	↓

[a]B = Beginning and E = End of oil or gas window.

windows in a basin. Their application also helps geologists develop a working understanding of oil generation kinetics. Anyone using these graphs can quickly see how changes in the time–temperature input data for a basin can change the conclusions regarding the position of the petroleum-generation window.

It is very important to recognize that there is considerable variation in the kinetic parameters of type II kerogens with different sulfur contents, as stated earlier. For example, Baskin and Peters (1992) found that petroleum generation

TABLE 6-8 Arrhenius TTI Values for the Miocene Shales in the South Padre Island COST Well, Offshore Texas, with Kerogen Type III and Oil-to-Gas Kinetics

Temperature range, °C	*Exposure time, m.y.*	*Type III TTI*	*Type III ΣTTI*	*Oil window*[a]	*Oil to gas TTI*	*Oil to gas ΣTTI*	*Gas window*[a]
120–130	1	0					
130–140	2.4	3.7	3.7	B			
140–150	1	7.8	11.5	│	0	0	
150–160	1.2	40	51.5	│	3.3	3.3	B
160–170	1.1	170	222	E	15	18.3	│
170–180	1	620	842		50	68.3	↓

[a]B = Beginning and E = End of oil or gas window.

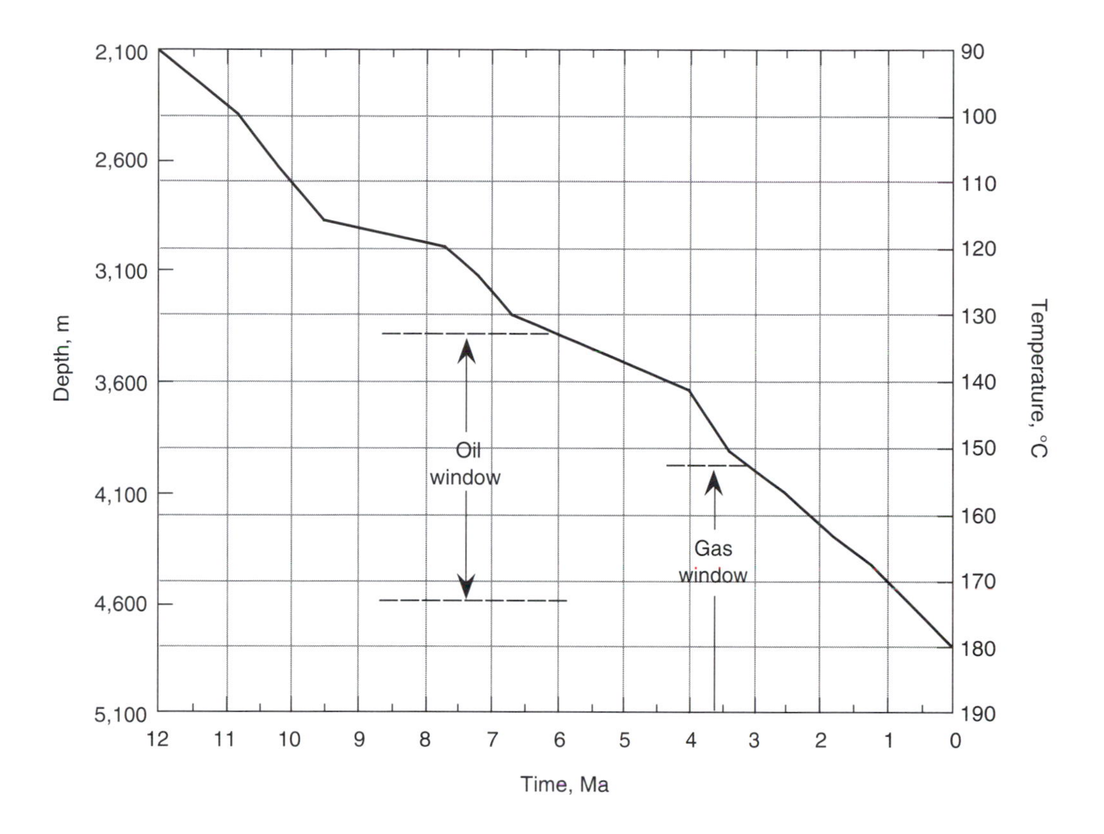

Figure 6-22

Burial history curve from 2,100 m to 5,100 m (6,800 ft to 16,728 ft) of the Lower Miocene in the South Padre Island COST well drilled off Texas, Gulf of Mexico. [Huc and Hunt 1980]

started at a vitrinite reflectance as low as 0.3% R_o in the high-sulfur Monterey Shale of California and the Upper Senonian Shale of Israel, compared with 0.75% R_o in a low-sulfur Tertiary shale of Sudan (Figure 6-24). The Sudan shale required a temperature 60°C higher than the Monterey Shale (roughly equivalent to 2,000 m of burial) to initiate petroleum generation. Their data emphasize again that you cannot use a single set of kinetic parameters for all type II kerogens.

A critical question in using these graphs is which graph to use when studying a specific source rock. The best approach is to determine experimentally the kinetic parameters of an immature sample of the source rock being considered

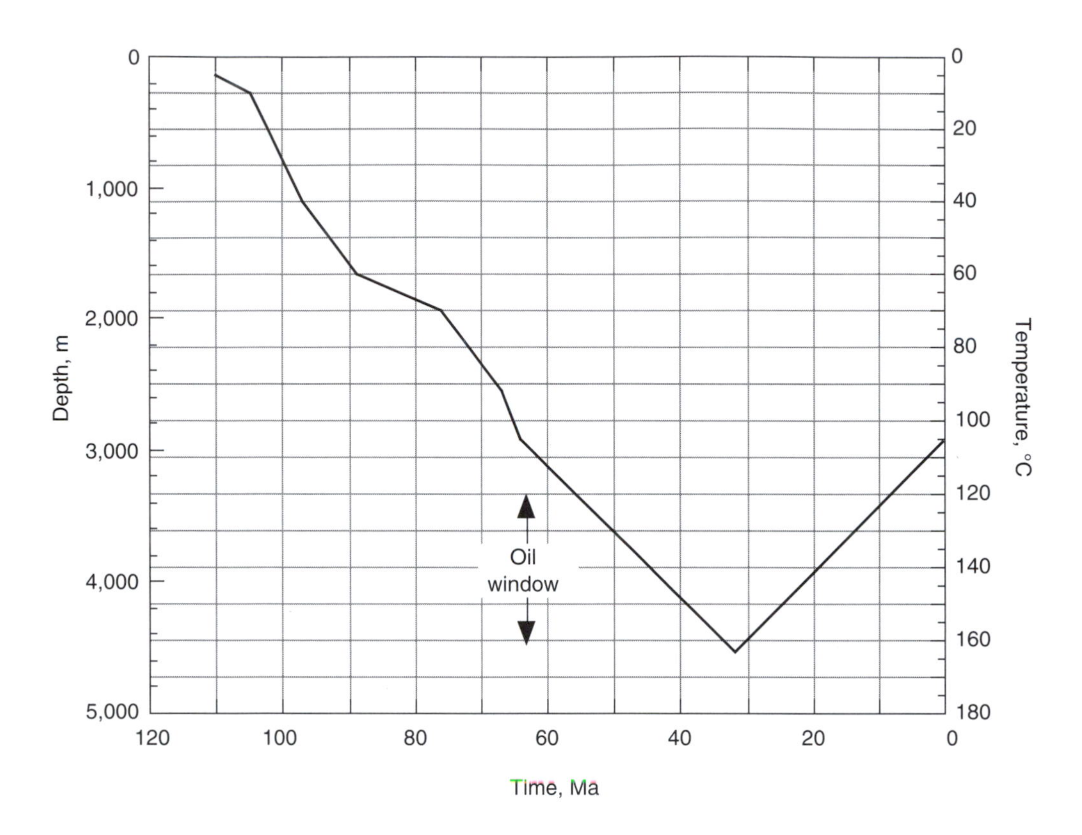

Figure 6-23

Burial history curve of Layer 4 of the Jurassic and Lower Cretaceous coals in Well 6-28-68-13W6M, Elmworth field, Deep Western Canada Basin. [Welte et al. 1984] Layer 4 is a coal seam interlayered with shaly sands.

in a basin evaluation by hydrous pyrolysis and to construct a TTI_{ARR} graph using the method described in Hunt et al. 1991. Or try using any other laboratory experiment to determine the Arrhenius kinetic parameters for constructing the TTI_{ARR} graph, with the understanding that the kinetics apply only to the specific products in the experiment, such as kerogen to expelled oil or oil to gas.

If this is not feasible, a second approach is to have a geochemical service laboratory analyze the source rock to determine the kerogen's maturity and type and to follow up by analyzing the kerogen for organic sulfur (Eglinton et al. 1990). If the kerogen is type II, then the sulfur analysis will classify it as IIA, B, C, or D, and the corresponding graphs can be used. If it is type III, then use Figure 6-10.

TABLE 6-9 Arrhenius TTI Values for Oil and Gas Generation in the Elmworth 6-28-68-13W6M Well of the Deep Western Canada Basin, Using Kerogen Type III and Oil-to-Gas Kinetics

Temperature range, °C	*Exposure time, m.y.*	*Type III TTI*	*Type III ΣTTI*	*Oil window*[a]	*Oil to gas TTI*	*Oil to gas ΣTTI*	*Gas window*[b]
110–120	5.6	0	0				
120–130	5.6	1.6	1.6	B			
130–140	5.6	8.5	10	│			
140–150	5.6	43	53	│	3.4	3.4	B
150–160	5.6	190	243	│	16	19	│
160–170	5	800	1,043	E	63	82	│
160–150	5.5	190	1,233		15	97	│
150–140	5.5	43	1,276		3.3	101	↓
140–130	5.5	8.5	1,285		0	101	
130–120	5.5	1.6	1,287				
120–110	5.5	0	1,287				
110–100	3						

[a]B = Beginning and E = End of oil generation. Gas generation from kerogen (coal) may continue.

[b]B = Beginning of conversion of oil to gas. Uplift stops conversion.

Unfortunately, there are not enough data available on the variations of type I kerogen kinetics with composition to determine which graph to use. It would be expected, however, that the type I kerogens with the most paraffin chains would react the most slowly. According to this interpretation, the type I Tipton graph (Figure 6-9) would be used for kerogen compositions with the most paraffinicity, and the type I Anvil graph (Figure 6-7) would be used for those with the least paraffinicity. The type I Uinta (Figure 6-8) would represent the mean.

It should be recognized that the composition of a kerogen may change laterally or vertically within a source rock as its depositional facies changes, so analyses of several samples may be needed. For example, Lewan (1986) noted that the organic sulfur content in type II kerogens of the Monterey Shale vary from 11.2 wt% in the distal pelagic chert–rich facies to 4.6 wt% in the more proximal clastic facies. This would classify different parts of the Monterey as types IIA, B, C, and D. Therefore, one should use caution in overgeneralizing the composition of a regional source rock.

Another approach applicable to type II kerogens is based on our observation that source rocks composed of claystones, mudstones, and siltstones

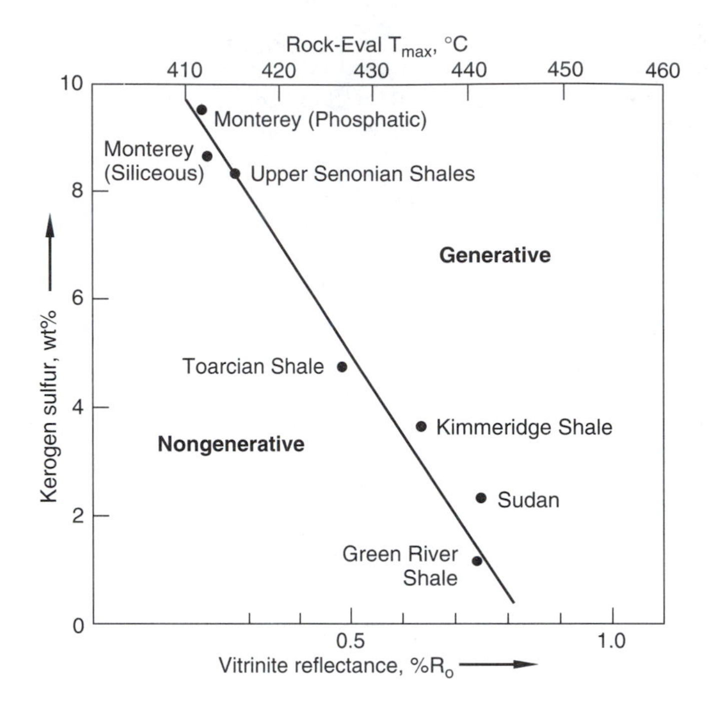

Figure 6-24

Correlation between the wt% of organically bonded sulfur in kerogen and the onset of oil generation for a variety of known source rocks. [Baskin and Peters 1992]

contain more type II kerogens with lower organic sulfur contents than do source rocks associated with evaporites or biogenic–pelagic strata such as carbonates, diatomites, and cherts. The reason for this is that sources of iron are characteristically associated with detrital sediments and react during early diagenesis to lock up sulfur as pyrite. Using this generalization, biogenic–pelagic source rocks would have oil-generation kinetics similar to that of kerogen types IIA or IIB, and clastic source rocks would have oil-generation kinetics similar to types IIC or IID. A concern here is that in a few euxinic clastic environments, the buildup of H_2S can exceed the detrital iron content of the sediment, which may result in kerogens with high organic sulfur contents.

For type II source rocks in areas where information on only crude oil composition is available, a third approach may be used. This is based on the sulfur content of the asphaltenes in oils. Orr's 1986 study of Monterey crudes showed that high-sulfur kerogens generate oils whose asphaltenes are high in sulfur. The asphaltenes of oils from type II-S kerogens, equivalent to IIA and B, have atomic S/C ratios > 0.035. Medium- and low-sulfur kerogens, equivalent to

TABLE 6-10 Approximate Temperature Range in °C Required for Different Kerogen Types to Reach a TTI_{ARR} of 1 in 1 m.y.

Kerogen type	*Temperature range, °C*	*Kerogen type*	*Temperature range, °C*
IA	100–110	IIA	50–60
IB	120–130	IIB	90–100
IC	150–160	IIC	110–120
III	130–140	IID	120–130

types IIC and IID, generate oils in which the asphaltenes have S/C ratios of 0.02 and < 0.01, respectively. The S/C ratios of crude oil asphaltenes can be determined by elemental analysis after removal of occluded resins and aromatics (Orr 1986).

The total sulfur content of a crude oil is always determined by the refinery that uses the oil. Consequently, it can be used as a rough classification guide if no other analytical data are readily available. However, in any crude oil analyses, it must be determined that the oils being compared are not biodegraded and have no secondary sulfur enrichment.

When no information is available on the kerogen or oils, the graph for a type IIC kerogen with a medium reaction rate (Figure 6-5) can give a first approximation of the oil window for type II source rocks.

The application of time–temperature relationships for oil generation has progressed from using one parameter for all oil generation (TTI_{LOP}) to using three for each of the major kerogen types (I, II, III) and now to using four within the type II group (IIA, B, C, and D) and three within the type I group. An obvious question is how many sets of kinetic parameters will ultimately be required to cover all potential source rocks.

Concerning the graphs, it is probable that six type II graphs will suffice for nearly all kerogens in this class. The TTI_{ARR} values are determined for 10°C intervals on a burial history curve. Note that the first diagonal lines in the upper left-hand corner of Figures 6-5 and 6-6 for kerogen types IIC and IID differ by only 10°C. Type IIC is 90 to 100°C, and type IID is 100 to 110°C. It is questionable whether or not a graph between these two would be a useful refinement, considering all the other uncertainties in estimates of paleotemperatures and burial history curves. Figures 6-3 and 6-4 (types IIA and IIB) have a bigger spread of 20 to 30°C. One additional graph may be needed here, plus one more between types IIB and IIC and a graph for a slower-reacting kerogen than type IID. This would mean six graphs to cover all type II kerogens, three for type I, and one for type III. Not enough information is available to describe reaction-rate differences within the type III kerogen group.

In model studies involving a Paleozoic cratonic basin and a Neogene pull-apart basin, the computer used 2,000 increments along the burial history curves to determine the ΣTTI_{ARR} with different type II kerogens. The same burial curve and kerogen types were used to determine the ΣTTI_{ARR} graphically, as described in this chapter. The ΣTTI_{ARR} values determined graphically came within ±3% of those determined by the computer.

Anyone reviewing case studies like the Phosphoria Formation (Figure 6-14, Table 6-3) might get the impression that no gas is generated until after the source rock passes through the oil window. Quite the contrary: Some gas is generated from kerogen all through the oil window, as shown in Figure 5-14 for the Aquitaine Basin. No graphs were made for the kinetics of gas from kerogen because the activation energies cover such a wide range that their application would be difficult. Also, as stated earlier, more gas comes from cracking oil than from the kerogen of conventional source rocks. Since kerogen types I and II yield more oil than type III does, they ultimately will yield more gas than type III will if their expelled oil is cracked to gas. Kerogen type III is called gas prone, not because of its gas yield, which is low, but because the gas/oil ratio is much higher in the products of type III compared with the products of type I or II.

Small amounts of gas are generated from kerogen and coal at the very earliest stages of oil generation. After peak gas generation around 150°C, some methane continues to be formed well into the meta-anthracite stage. This occurs after all the oil is cracked to gas, so gas from coal and from disseminated kerogen covers a much wider time–temperature range than does the gas window from cracking oil to gas.

The significant differences in the kinetics of the type II kerogens were previously discussed by comparing the 10°C temperature range required for a source rock to obtain a TTI_{ARR} of 1, which is the beginning of oil generation, in 1 million years. Similar data for all kerogen types listed in Table 6-1 are shown in Table 6-10. The slow-reacting type IC requires a temperature of 150 to 160°C to initiate generation. This is 100°C higher than the fast-reacting type IIA. It would be equivalent to about 3.3 km of additional depth at an average geothermal gradient of 3°C/100 m. In comparison with the kerogen decomposition, the beginning of the conversion of oil to gas in 1 m.y. would require a temperature of about 150°C.

Finally, it is beyond the scope of this book to go into detail about the problems in obtaining precise input data of the thermal and burial history of the source rock. Entire books (e.g., Lerche 1990) have been written on this subject. But the importance of such data can be seen by comparing two type III source rocks buried for 15 million years, one at 130 to 140°C and the other at 150 to 160°C. The first has generated only 25% of its oil, and the second is past 100% generation. An error of only 20°C in burial-temperature history would change the interpretation from early-oil to postoil (gas) generation.

The case studies discussed in this chapter pertain to individual wells. Basin studies generally use data from multiple wells scattered over extended areas in order to determine the hydrocarbon-generating areas through time. Such a study was made by Talukdar et al. (1986) in the Maracaibo Basin of Venezuela. Burial history curves were made for the La Luna Formation in eighty-one wells located in different parts of the basin. Lopatin's TTI index of maturity was calculated and areal maps were made at different time intervals to determine the generation kitchens through time. Since most of the La Luna kerogen is a type IIC, the TTI_{ARR} would give results similar to those for the TTI_{LOP} at moderate heating rates, as previously discussed.

Using this approach, Talukdar et al. (1986) concluded that the La Luna Formation was immature throughout the basin at the end of the Eocene, 38 Ma, except on the northeast flanks. There, it was buried deep enough to go completely through the oil and gas windows into postmaturity with respect to gas. About 12 Ma, at the end of Middle Miocene, the La Luna had just begun to enter the oil-generation window in the southwest part of the basin, as shown in Plate 2B. In most of the basin, the La Luna was still immature. By 5 Ma, the hydrocarbon-generating areas in both the southwest and the northeast had begun to spread across the basin (Plate 2C). Condensate and wet-gas generation began to increase, as shown in light red. At the present time (Plate 2D), the La Luna Formation in the south and southeast parts of the basin is in the dry-gas zone. In most of the basin, the La Luna is generating either oil or gas. The only immature part is a narrow belt bordering the entire western side of the basin.

These three plates show clearly how the oil and gas kitchens for a source rock can spread across a basin as the source rock is buried deeper through time. Similar studies were reported by Leonard in the Columbus basin off Trinidad (1983) and in the North Sea (1989) using Arrhenius kinetics. By combining such information with data on the evolution of migration pathways, traps, and structures, one can better evaluate the risk in prospecting for economic hydrocarbon accumulations.

SUMMARY

1. The time and depth of oil generation from petroleum source rocks can be determined using time–temperature index (TTI) graphs based on the Arrhenius equation.

2. Modeling oil generation requires determining the exposure time that a source rock experiences at varying temperatures with burial. Rapid burial with high geothermal gradients results in a different level of maturation than does slow burial with low geothermal gradients.

3. Modeling generation involves applying the Arrhenius equation to a burial history or a geohistory curve, which is a plot of source rock depth and temperature through time.

4. Type II oil-generating kerogens are designated as types IIA, B, C, and D, depending on their reaction rates and organic sulfur contents. The fast- and medium-fast-reacting kerogen types IIA and B (also called II-S), have higher-sulfur contents, and the medium- and slow-reacting kerogen types IIC and D,

respectively, have lower-sulfur contents. The activation energies (E) of the kerogens are inversely proportional to their sulfur contents.

5. Type I kerogens as a group are slower reacting (or more refractory) than type II, and type III in this study is slower than the medium type I. Type III kinetics are close to those for cracking oil to gas. This may be a factor causing type III to yield proportionately more gas and condensate than oil, compared with the other kerogen types.

SUPPLEMENTARY READING

Burrus, J. (ed.). 1986. *Thermal modeling in sedimentary basins.* Paris: Éditions Technip, 600 p.

Lerche, I. 1990. *Basin analysis, quantitative methods,* vols. 1 and 2. San Diego: Academic Press, vol. 1, 562 p., vol. 2, 571 p.

Naeser, N. D., and T. H. McCulloh (eds.). 1989. *Thermal history of sedimentary basins.* New York: Springer-Verlag, 319 p.

Waples, D. W. 1984. Thermal models for oil generation. In J. Brooks and D. Welte (eds.), *Advances in petroleum geochemistry,* vol. 1. London: Academic Press, pp. 7–67.

Waples, D. W. 1994. Maturity modeling: Thermal indicators, hydrocarbon generation, and oil cracking. In L. B. Magoon and W. G. Dow (eds.), *The petroleum system—From source to trap.* AAPG Memoir 60. Tulsa: American Association of Petroleum Geologists, pp. 285–306.

Chapter

7

The Origin of Natural Gas

Natural gas is the gaseous phase of petroleum. Typically, a reservoir gas contains 70 to 100% methane, 1 to 10% ethane, lower percentages of higher hydrocarbons through the hexanes, and traces up through nonanes (C_9H_{20}). The percentages of nonhydrocarbon constituents, such as carbon dioxide, nitrogen, and hydrogen sulfide, may vary from very low to 100%. Natural gas is classified in the field as *dry gas* or as *wet gas* if it has less than 0.1 or more than 0.3 gallons of condensible liquids per 1,000 ft^3 (<1.3 or >4 liters/100 m^3). The terms *sweet* and *sour* refer to gases that are low and high, respectively, in hydrogen sulfide. *Reservoir gas* may occur underground in the free gaseous state or as gas dissolved in oil or water. *Associated gas* occurs with oil as free gas, gas dissolved in oil, or liquefied gas. *Nonassociated gas* occurs alone as free gas and as gas dissolved in water. At subsurface pressures beyond 4,000 psi (28 MPa) and temperatures above 200°F (93°C), most fields containing hydrocarbon gases and oil exist as a single-phase petroleum fluid with gas dissolved in oil or vice versa (Katz et al. 1959, p. 465). Gas is highly compressible compared with oil or water. At 4,000 feet (1,220 m) under normal hydrostatic pressure, gas fills only 1% of the space it occupies at the surface. If there are two equal-size gas reservoirs at 2,000 and 10,000 ft (610 and 3,049 m), the latter reservoir will hold roughly five times as much gas as the former. Gas also is very mobile in the subsurface. It is found more widely distributed vertically and laterally than oil because it migrates more easily. Also, it is generated from source rocks that are more widely distributed than source rocks of oil.

World gas reserves have been steadily increasing as drilling goes deeper and as new production techniques such as horizontal drilling have been developed. Proven original reserves at the end of 1990 were equivalent to about 1×10^{12} barrels of oil on a BTU basis (Masters 1993). The former Soviet Union holds 38% of this, and the Middle East 30% (True 1991).

Since proven reserves have been doubling about every ten years, it is conservatively estimated that the ultimate recoverable conventional gas resources will be at least 10×10^{15} ft^3 (2.83×10^{14} m^3), equivalent to 1.65×10^{12} barrels of oil (Wyman 1985). This compares with ultimate oil resources of 1.8×10^{12} barrels (Masters et al. 1984), about the same as the amount of available gas. However, oil is being depleted about 2.6 times faster than gas. This means that gas will probably be the world's major fossil energy source during the next two centuries (Wyman 1985).

Worldwide resources of unconventional gas from tight sandstones, black shales, coal, hydrates, and the like far exceed conventional resources. In the United States and Canada alone, there are probably 2,000 TCF (trillion cubic feet) of gas in tight sandstones. The black Devonian shales of North America hold another 2,000 TCF. Coal deposits worldwide contain thousands of TCF of gas in place waiting for the economics of production to improve.

Gas is ubiquitous in petroleum basins. Zor'kin and Stadnik (1975) estimated that there are 86,000 TCF (2.4×10^{15}m^3) of gas dissolved in the pore waters of five petroliferous basins in the former Soviet Union. This is more than fifty times the amount of gas in the reservoirs of those basins. Zor'kin and Stadnik think that the active hydrocarbon-generation kitchens are distinguished by the high gas saturation of the entire section. The hydrocarbons are redistributed between gas-generating source beds and the formation waters.

Although gas source rocks extend back into the Precambrian, at least three-fourths of the world's recoverable gas, like oil, is believed to have been generated in the last 100 m.y (Klemme and Ulmishek 1991). This chapter discusses the origin of natural gas and some characteristics of its accumulations.

Sources of Natural Gases

Both the hydrocarbons and nonhydrocarbons in natural gas have multiple sources. The major sources of the hydrocarbon gases are (1) methanogenic bacteria, (2) all types of kerogens, (3) coal, and (4) oil in source and reservoir rocks. The thermal cracking of kerogen and coal to generate methane (items 2 and 3) is called *primary cracking,* whereas the cracking of oil (item 4) is called *secondary cracking.* The major nonhydrocarbon gases—CO_2, H_2S, and N_2—are formed by both organic and inorganic processes.

All known hydrocarbon gas accumulations are biogenic in origin in that they come from the decomposition of organic matter in the earth's crust. No known abiogenic methane accumulations exist based on stable isotope measurements, although minor seeps of possible abiogenic gases are occasionally reported (Schoell 1988).

Figure 7-1 is a schematic illustration of the temperature ranges in which the various gases are formed. The curves show in a very general way the relative volumes of the gases from sapropelic (types I and II) and from humic (types III and IV) organic matter when both are deposited in an aquatic environment, marine or nonmarine. Bacterial methane that formed during diagenesis is

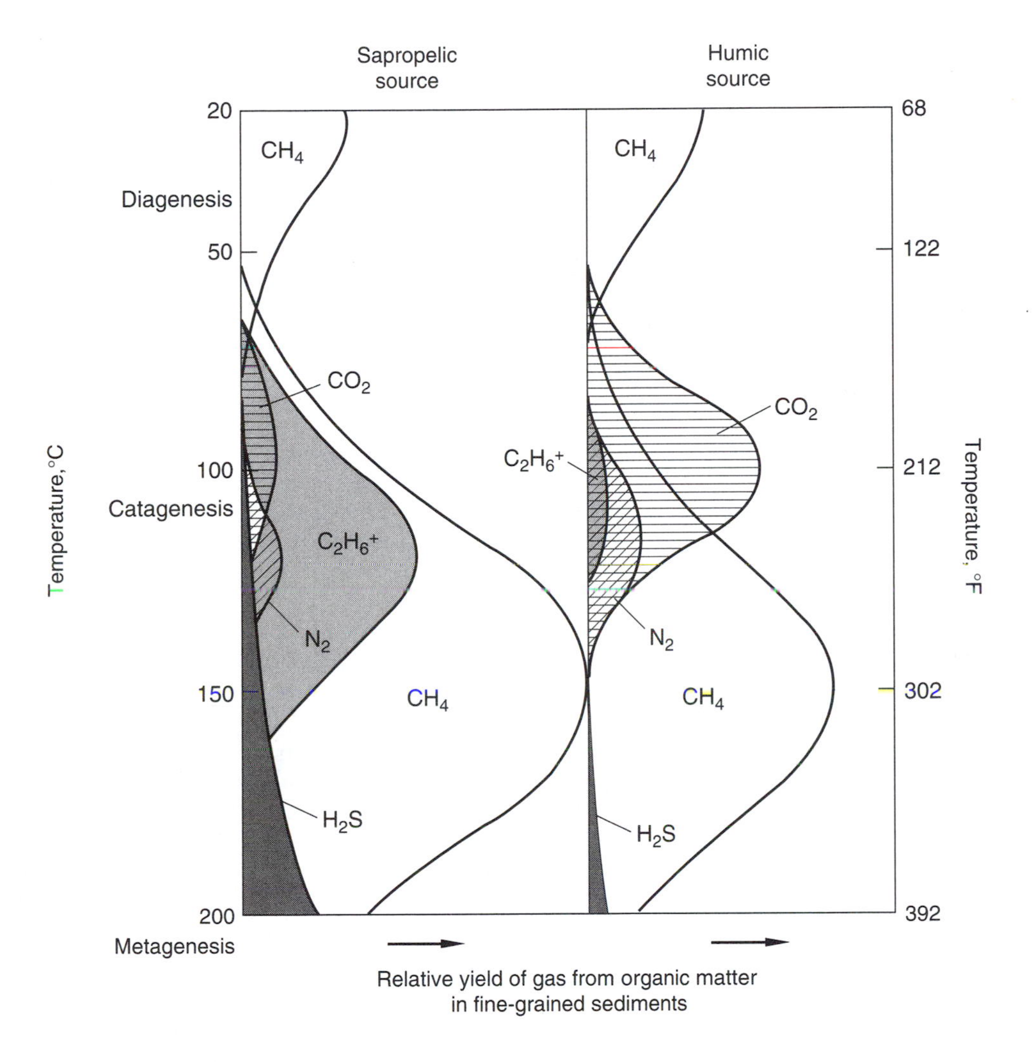

Figure 7-1

The generation of gases from organic matter with temperature. The C_{2+} represents hydrocarbon gases heavier than methane. The N_2 is generated initially as NH_3. Inorganic sources of CO_2, H_2S, and N_2 are not shown.

estimated to represent 12% of the total methane in nonreservoir rocks (Figure 5-16) and possibly 20% of the methane in conventional reservoirs (Rice 1993a). Far larger amounts were formed through geologic time, but because shallow reservoirs generally have no seal, most of this early methane was lost. (More details on the conditions for bacterial gas generation and accumulation can be found in Rice 1993a.) Most bacterial methane pools are in sand–shale sequences containing continental (type III) organic matter rather than in carbonate–evaporite sequences containing marine or lacustrine source material (Polivanova 1977; Rice 1993a). Polivanova attributes this to the suppression of methanogenic activity in evaporite deposition. But the increased CO_2 available from continental humic material also may be a factor. Figure 7-1 does not show the CO_2, H_2S, and N_2 formed during diagenesis because very little of it survives early microbial and chemical reactions (for details on this, see Hunt 1979, p. 152).

Microbial generation of methane ceases as bacterial populations decrease; temperatures rise above about 70°C (158°F); and nutrients for metabolism become less available. Somewhere in the first 1,000 m (3,280 ft) of sediment depth is the end of microbial methane generation. Meanwhile, thermally formed methane and higher hydrocarbons plus CO_2 begin to increase as temperatures exceed 50°C. The CO_2 from organic sources peaks at about 100°C (212°F), but the CO_2 from inorganic sources can occur at much higher temperatures. Figure 7-1 does not indicate the inorganic sources of nonhydrocarbon gases because they are functions of temperature, mineralogy, and pore fluid composition rather than of organic matter. For example, some CO_2 comes from the chemical and thermal decomposition of carbonates. Thermochemical sulfate reduction forms abundant H_2S in high-temperature reservoirs associated with evaporite sequences (Orr 1974).

Wet gas forms from sapropelic OM in and below the oil window and from the thermal decomposition of oil, whereas most organic CO_2 is from humic OM (Figure 7-1). The peak methane generation from OM is around 150°C with sapropelic OM, ultimately yielding two or three times as much methane as humic OM. This is because types I and II kerogens generate gas during and after generating oil in greater amounts than do types III and IV. In addition, any sapropelic source oil that is buried deeper is eventually converted to gas.

Hydrogen sulfide is generally found at depths beyond 3,000 m (9,840 ft), with more coming from sapropelic OM deposited in a marine environment, compared with OM in fluvial or lacustrine deposits. This results from the early diagenetic influence of the higher sulfate content of marine waters. No one has identified a peak in the H_2S generation window, but it is probably beyond 5 km, based on Le Tran et al. 1974.

Bacterial Source

The term *biogenic* has been widely used in the literature to describe methane formed by methanogenic bacteria in surface sediments. Biogenic, however, refers to all methane from OM. It is more correct to label this early methane as *bacterial* or *microbial*. Bacterial methane is formed by both microbial fermentation and CO_2 reduction, although most reservoired gases are formed by the latter process (for details, see Schoell 1988). The reaction is

$$4H_2 + CO_2 \rightarrow CH_4 + 2H_2O$$

Under anoxic conditions, bacteria that ferment organic matter can thrive to produce hydrogen and CO_2. The hydrogen is consumed by sulfate reduction until much of the sulfate is gone, after which methane is generated. This may occur in the water column or in the sediments. The stable isotopes of both the hydrogen and carbon of the methane are fractionated by these biological processes, so it is possible to distinguish biologically formed methane from that formed thermally during catagenesis. When bacteria form methane by the reduction of CO_2, they preferentially consume the lighter $^{12}CO_2$ rather than the heavier $^{13}CO_2$. Depending on the ^{13}C value of the carbon source, the bacteria can create methane with a $\delta^{13}C$ of –109‰ relative to the Peedee belemnite. (The PDB standard and δ are explained in Chapter 3.) Methanogenic bacteria also prefer to utilize light hydrogen rather than heavy hydrogen (known also as deuterium D). Consequently, bacterially formed methane is very deficient in deuterium with a δD of –250‰ relative to the standard mean ocean water (SMOW) (Schoell 1988).

Table 7-1 lists the ranges for carbon and hydrogen isotopes for the bacterial and thermal sources of gas. Compare this table with Figure 7-1. Dry bacterial gas is formed in surface sediments at temperatures from 20 up to about 80°C (Figure 7-1). Most of the wet thermogenic gas is formed from the cracking of sapropelic kerogen and oil, mainly in the upper part of the temperature range, from about 80 to 150°C. Small amounts of wet gas also may form from humic sources such as coal, depending on its hydrogen content. Dry thermogenic gas is formed from all kerogen types in the high-temperature range from around 150 to over 200°C.

The ^{13}C and D ranges in Table 7-1 are only guidelines. Anomalous values occur in nature outside these ranges because of variable sources of ^{13}C and D and the extensive mixing of gases caused by vertical migration. For example,

TABLE 7-1 Approximate Range in Carbon and Hydrogen Isotopes for Different Types of Petroleum Gases

Gas type	$\delta^{13}C^a$	δD^b
Dry bacterial	–110 to –60	–250 to –150
Wet thermogenic	–60 to –30	–300 to –120
Dry thermogenic	–40 to –15	–150 to –70

[a] ‰ relative to PDB.

[b] ‰ relative to SMOW.

the conversion of CO_2 to methane by bacteria causes a fractionation of approximately –70‰ in carbon isotopes (Rosenfeld and Silverman 1959). If the $\delta^{13}C$ of the CO_2 is –15‰, it will form methane with a $\delta^{13}C$ of –85‰ in the bacterial range (Table 7-1). But if the CO_2 is +30‰, it will yield –40‰, which is not in the bacterial range even though it is microbial in origin. Such CO_2, as bicarbonate, has been found in cores from the deep-sea drilling program. But this is an anomalous source. Empirical studies of hundreds of natural gases have shown the ranges in Table 7-1 to be valid for most samples.

Gases analyzed from the surface down in a thick sedimentary column exhibit an increase in the heavy isotope ^{13}C. A matching increase in the C_{2+} (wet gas) of this gas fraction also occurs with depth. Examples of both are shown in

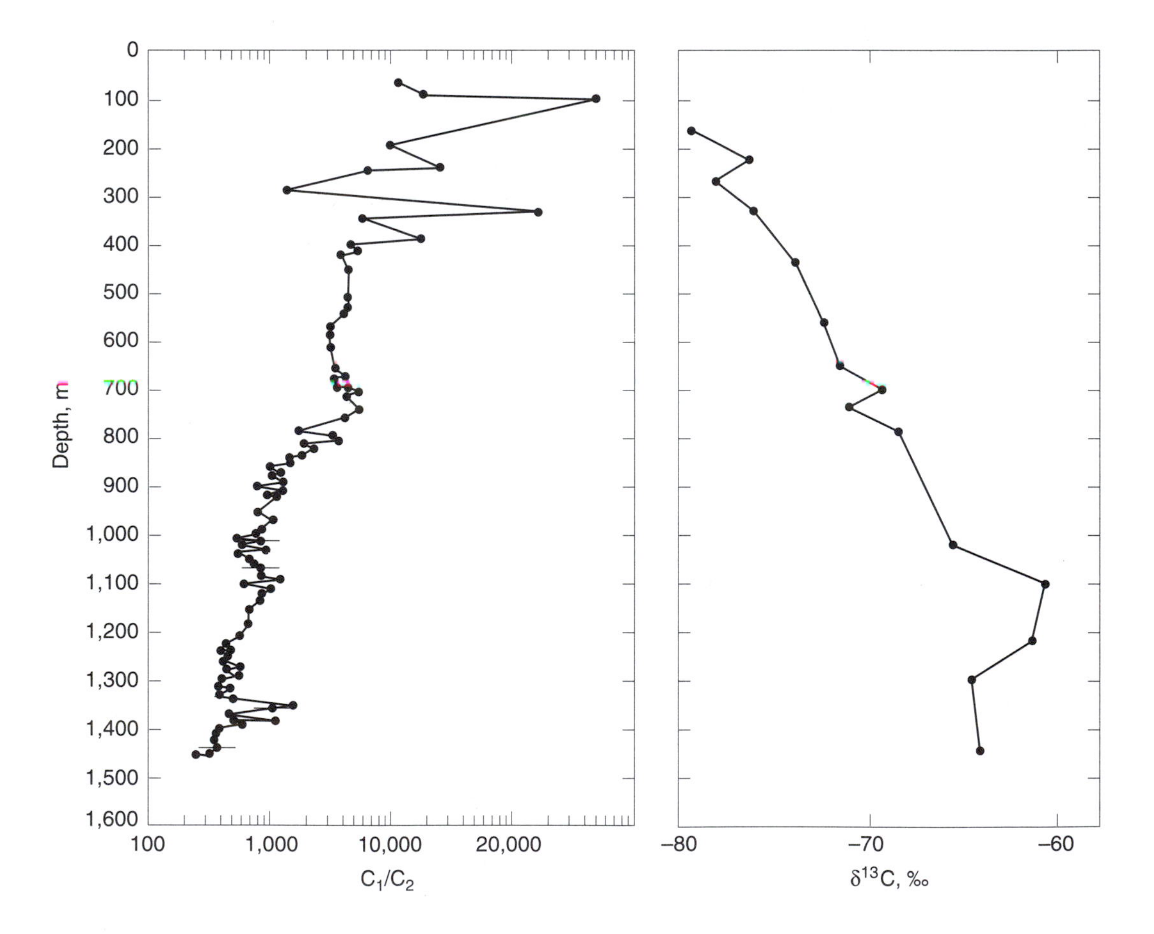

Figure 7-2

The change in $\delta^{13}C$ and log C_1/C_2 with depth in gas pockets of cores from IPOD Holes 397 and 397A. [Whelan 1979]

Figure 7-2 for gases analyzed in cores from a well drilled off the Canary Islands by the International Program of Ocean Drilling (IPOD). The $\delta^{13}C$ values in this well become more positive from –80 in the first 200 m to –62‰ about 1,200 m deeper in Cretaceous sediments (Whelan 1979). Over the same depth range, the methane/ethane ratio decreased from about 20,000 to 300.

Although all these samples are still in the bacterial range, according to Table 7-1, they do show a trend in the carbon isotopes and gas wetness that continues to much greater depths in typical petroleum basins. For example, Figure 7-3 shows the change in $\delta^{13}C$ and percentage (C_2-C_4/C_1-C_4) for head space cuttings gases coming from the Antelope Creek well in the Powder River Basin of Wyoming. A continuous change in $\delta^{13}C$ occurs with increasing depth from

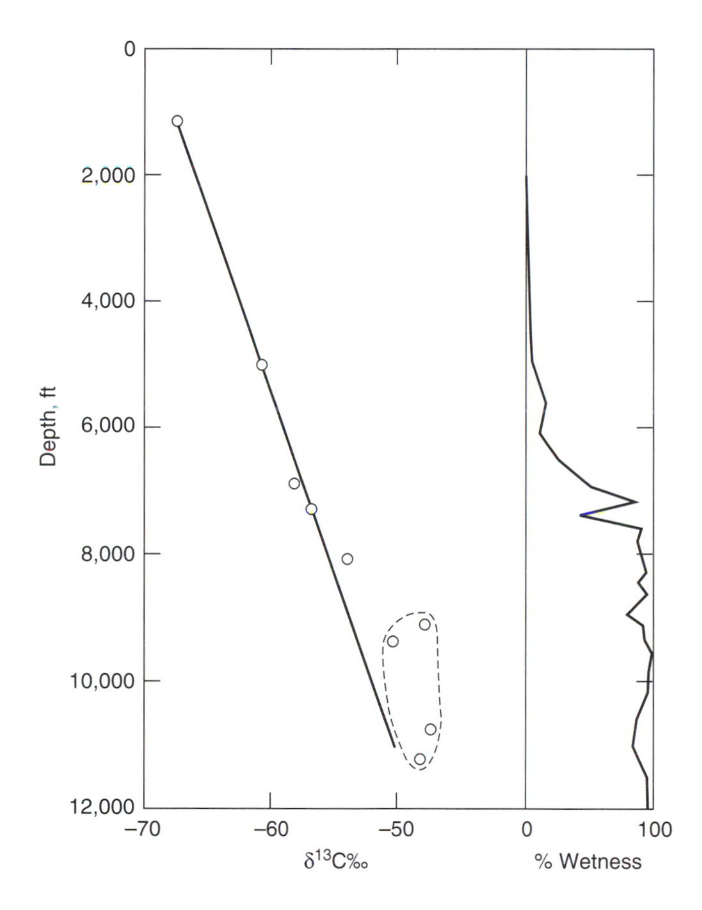

Figure 7-3

The change in $\delta^{13}C$ and the percentage of wet gas (C_2-C_4/C_1-C_4) with depth in head space gas from cuttings of the Antelope Creek well, Powder River Basin, Wyoming. Encircled data represent thermal gases migrating upward. [Reitsma et al. 1981]

about –67‰ at 1,000 ft (305 m) to –47‰ at 11,560 ft (3,523 m). The big increase in wetness in this well occurs between 6,000 and 7,000 feet (1,829 and 2,734 m), where large amounts of propanes and butanes began to appear. The latter hydrocarbons are not formed by bacteria, so the $\delta^{13}C$ values from about –55 to –47 represent wet thermogenic gases (Reitsma et al. 1981). The head space technique involves sealing wet drill cuttings in a can at the well site and analyzing them at a laboratory for free hydrocarbons (for details, see Chapter 14).

Figure 7-4 contains carbon isotope data and gas wetness on cuttings collected from well 33/6-1 in the North Sea (Schoell 1984a). There is mainly dry

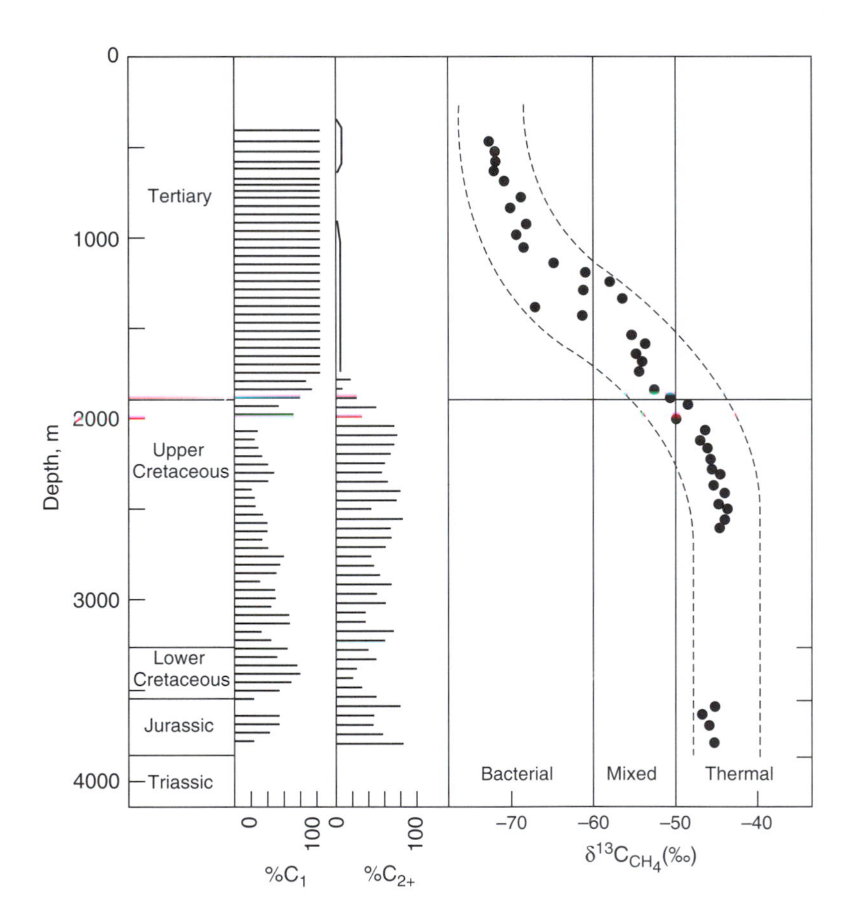

Figure 7-4

The change in $\delta^{13}C$ of methane and the percentages of methane and C_{2+} in total gas with depth in head space gas from cuttings of Well 33/6-1 in the North Sea. [Schoell 1984a]

bacterial gas down to about 1,800 m, with the $\delta^{13}C$ going from –75 to –55‰. From 1,800 to 2,100 m, the ethane plus fraction increases rapidly with a corresponding increase in the $\delta^{13}C$ from –55 to less than –50‰. This represents the beginning of significant thermogenic gas formation which continues to 3,700 m near the bottom of the hole.

These last two examples illustrate the change from dry bacterial to wet thermogenic gas with depth. Deep holes like the 20,000 ft (6,098 m) Inigok well drilled on the North Slope of Alaska show the shift from wet to dry thermogenic gas. Figure 7-5 presents the $\delta^{13}C$ and wetness values for gas samples from

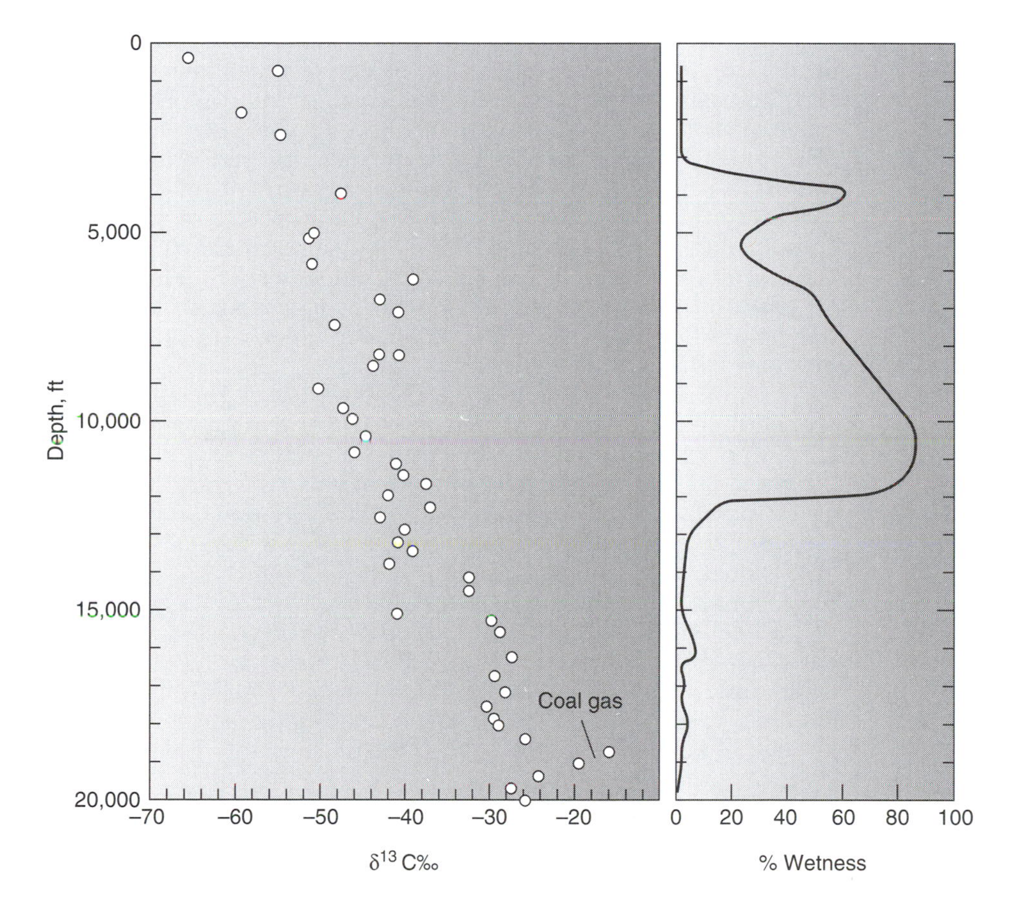

Figure 7-5

The change in $\delta^{13}C$ and gas wetness with depth for gas samples from cores and cuttings of the Inigok well, North Slope of Alaska. The wetness is defined as the content of C_2–C_4 hydrocarbons as a percentage of C_1–C_4 hydrocarbons. [L. Magoon, personal communication]

cores and cuttings in this well (L. Magoon 1992, personal communication). In this example, the isotope values and gas wetness indicate that there has been vertical migration, possibly through faults and fissures up several thousand feet into sandstones and siltstones of the Torok and Nanushuk groups. Dry bacterial gas was recovered down to about 3,000 ft, where there was a sudden increase in gas wetness along with an increase in $\delta^{13}C$ from –55 to –47‰. The latter number is in the thermogenic gas range.

Other geochemical indicators show that source rocks for this zone are immature for gas generation, indicating that any thermal gas has migrated up from below. No major source rocks were encountered before about 9,000 ft, so the $\delta^{13}C$ values of –48 to –33‰ above 9,000 ft in Figure 7-5 indicate that gas has migrated from deeper in the section and mixed with bacterial gases entrained in the shallower sediments. At about 12,000 feet there is an abrupt decrease in gas wetness, indicating the end of the oil and condensate windows and the beginning of the dry thermogenic gas interval. The $\delta^{13}C$ values in this range increase from –42 to –16‰ at 19,000 ft. The gases with these low negative $\delta^{13}C$ values appear to be coming from coals in a very advanced state of maturation (R_o equals 5% at a bottom hole temperature of 201°C, 394°F in this well).

The change with depth in the $\delta^{13}C$ and the C_{2+} of subsurface gases was combined in a plot by Bernard (1978) to distinguish bacterial and thermogenic gases from each other. Figure 7-6 (A) is a modification of Bernard's original plot for gases from vents, seeps, and sediments in various areas. The $\delta^{13}C$ values in this figure increase to the right, which corresponds to the general plot of $\delta^{13}C$ in Figure 3-13.

The four lowest solid circles in the lowermost right corner of Figure 7-6 (A) are thermogenic underwater vent gases from offshore petroleum production in the Gulf of Mexico. The other fourteen solid circles are underwater seeps. Eight are bacterial; five are mixtures of thermal and bacterial gas; and one is the seep of a thermogenic gas (Bernard et al. 1976). The two California seeps are thermogenic, and all but one of the Alaska seeps are bacterial. Faber and Stahl (1984) analyzed 350 bottom-sediment gases from the North Sea and found all of them to fall into the thermogenic gas area of Figure 7-6 (A).

Schoell (1983, 1993) developed a plot of δD versus $\delta^{13}C$ to recognize different genetic types of gases. Figure 7-6 (B) shows the approximate areas within which different gases fall at the time of their formation, based on hundreds of gas analyses. Besides distinguishing bacterial and thermal gases, this plot gives a rough indication of maturity, with the most mature gases falling in the lower right corner of dry thermogenic gas (T_D). The two plots in Figure 7-6 are only guidelines. Overlaps can occur within the designated areas. Also, gases like T_o (gas with oil) can migrate away from oil and still have the T_o signature. If there is any doubt about classifying a particular gas, a more detailed analysis of individual gas components can often clarify the problem.

The bacterial gases of the Po Basin differ from the previous examples cited in that they show no change in $\delta^{13}C$ with depth (Matavelli et al. 1983). Bacterial gases in Lower Pliocene reservoirs at depths of 991, 2,064, 3,504, and 4,467 m all have $\delta^{13}C$ values of about –70‰. The last gas at 4,467 m (14,800 ft) is the deepest bacterial gas known.

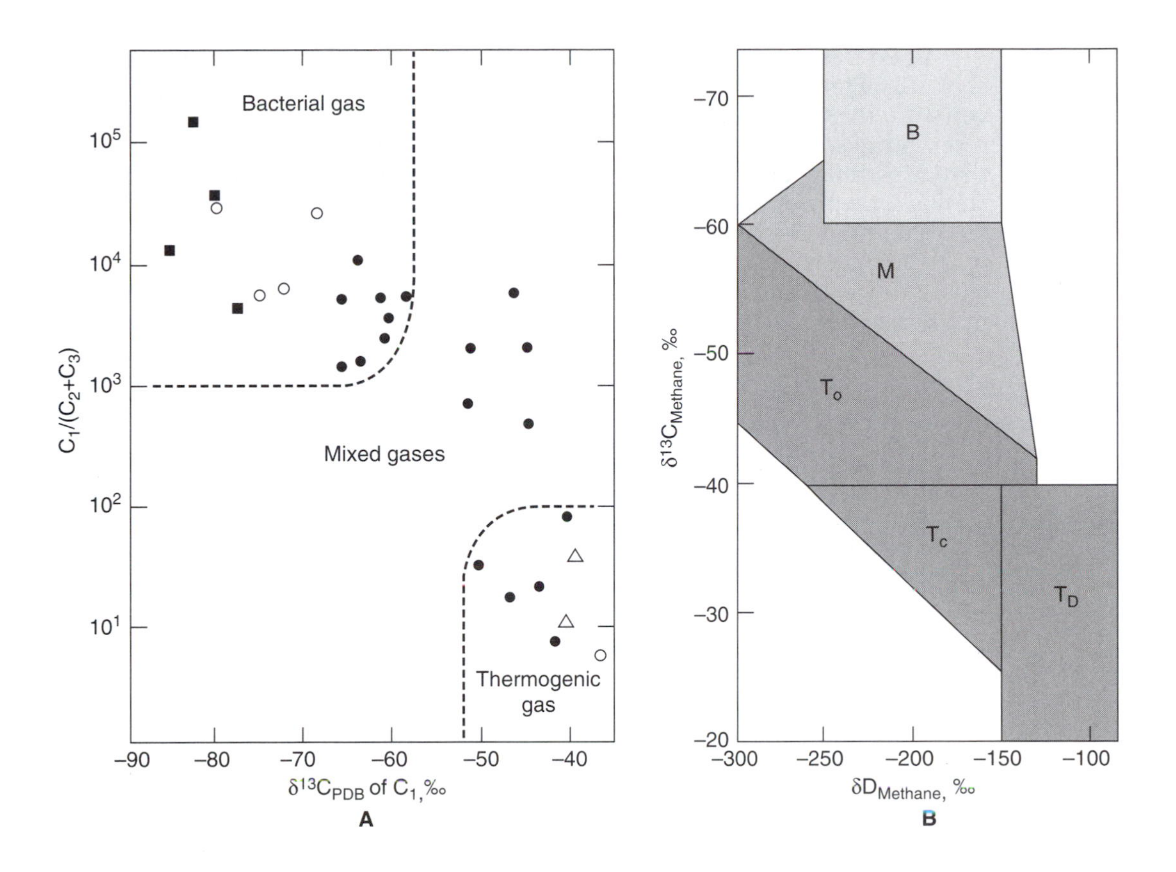

Figure 7-6

(A) Revised Bernard plot comparing gas wetness and $\delta^{13}C$ of methane for gases in vents, seeps, and sediments from various areas: ● Gulf of Mexico; ○ Norton Sound, Alaska; ■ western Gulf of Alaska; △ offshore southern California. [Claypool and Kvenvolden 1983] (B) Schoell's plot of methane δD versus $\delta^{13}C$ for characterizing the origin of natural gases. B = bacterial gas, M = mixed gases, T_O = thermogenic gas with oil, T_C = thermogenic gas with condensate, T_D = dry thermogenic gas.

Considerable mixing of the gases in the Po Basin indicates extensive vertical migration from the deepest reservoirs. At depths shallower than 2,000 m, there is a complete spread in the distribution of bacterial, mixed, and thermal gases. Messinian age reservoirs produced all three types of gases, but they were predominately of mixed origin.

All Po Basin gases do show a trend in carbon isotopes with reservoir age, as shown in Figure 7-7 (Matavelli et al. 1983). All the gases in Pre-Tertiary reservoirs are thermogenic, and all the Middle and Upper Pliocene and Pleistocene reservoir gases are bacterial. No bacterial gases were found in reservoirs older than Messinian, and no mixed gases are in Pre-Tertiary reservoirs. Figure 7-7 is similar to a previous figure by Hunt (1979, p. 177), which at that time showed no remaining recognizable bacterial gases in the world in reservoirs older than Cretaceous. Since then, Jenden et al. (1988) have reported bacterial gas in shallow (150 to 250 m depth) Carboniferous reservoirs above an unconformity in Kansas, but it is not clear when the gas formed.

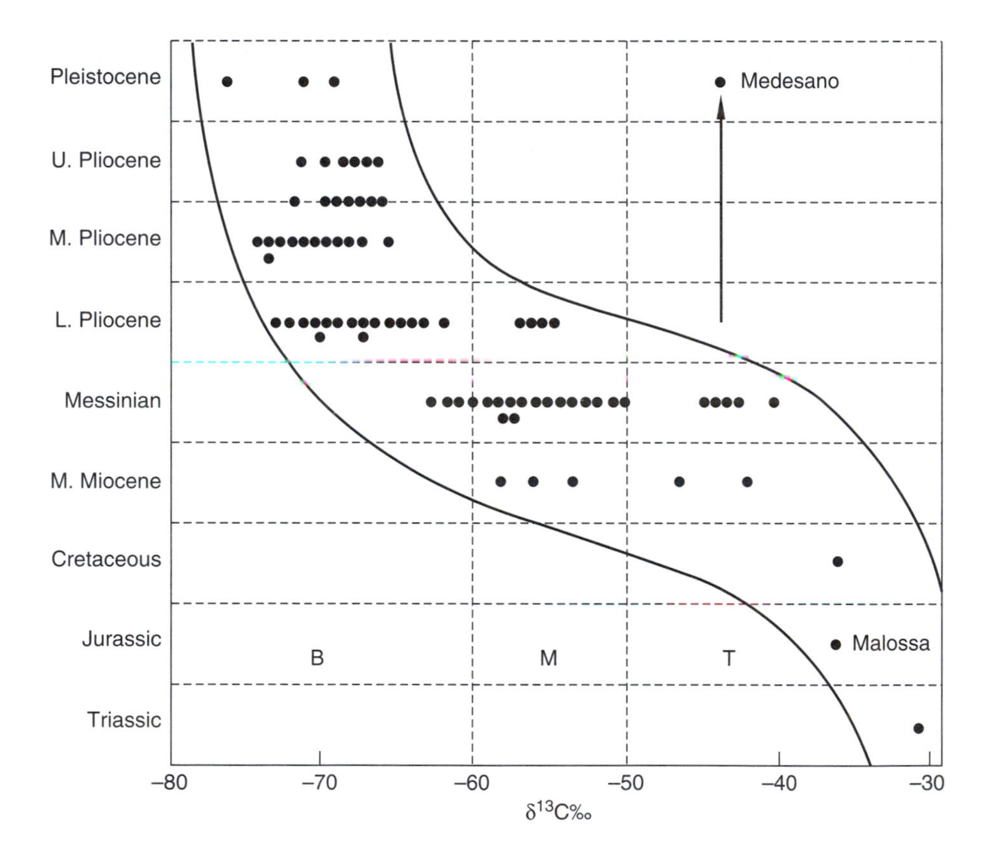

Figure 7-7

The relationship of the $\delta^{13}C$ of methane and the age of producing reservoirs for gases of the Po Basin. B, M, and T stand for bacterial, mixed, and thermal gases, respectively. [Matavelli et al. 1983]

Plots of $\delta^{13}C$ versus δD for methane have been useful in classifying bacterial gas origins such as CO_2 reduction versus fermentation and bacterial activity in various depositional environments (Schoell 1983). Also, unusual sources of methane such as from geothermal areas, the Canadian Shield, and ophiolites can be distinguished by such plots (Schoell 1988). But using these plots for all petroleum gas studies has been hindered by the fact that δD values overlap for several types of gases, as shown in Table 7-1. Also, relying on methane alone for defining gas origins requires separating out all the things that can happen to methane isotopes. For example, possible fractionation during migration through very low permeability conduits, fractionation during desorption from coal, bacterial oxidation, and the mixing of different thermal sources sometimes make interpretation a difficult process.

An example of the problems in using methane alone is shown in Table 7-2, which contains $\delta^{13}C$ values for the individual gaseous hydrocarbons from a series of stacked producing intervals in the Lena field offshore Louisiana (James 1990). All three of these gas pools ranging in depth from 4,963 to 8,521 ft (1,513 to 2,598 m) are in immature Pliocene sands. Only the shallowest gas, which is 98% methane and has a $\delta^{13}C$ of –57.2‰, seems to be bacterial. The other two have $\delta^{13}C$ values of methane in the thermogenic range. They probably migrated vertically along faults from deeper sources. The question is whether one or more deep sources contributed to these pools. The $\delta^{13}C$ analyses of the C_2 through C_4 gases shows them to be essentially identical. The variations for each hydrocarbon across the three pools are less than 1‰. Such data indicate that a single source of gas filled the deepest reservoir and mixed in varying amounts with the bacterial gas in the shallower reservoirs. But this would be difficult to determine with plots using only methane.

The thermal destruction of C_2 to C_4 gases at high maturities, however, (>2% R_o) means that more reliance would have to be placed on the $\delta^{13}C$–δD plots. Both of these isotopes become more positive at high maturities.

TABLE 7-2 Carbon Isotopic Compositions of C_1–C_4 Gases from Stacked Producing Intervals, Lena Field, Offshore Louisiana

	Compositions in ‰$\delta^{13}C_{PDB}$		
Median depth	*8,521 ft*	*8,025 ft*	*4,963 ft*
Methane	–44.1	–49.0	–57.2
Ethane	–25.8	–25.1	–25.6
Propane	–23.7	–23.4	–23.7
Isobutane	–25.7	–25.3	–25.0
n-Butane	–23.5	–23.3	–23.0

Source: Data from James 1990.

Gas Hydrates

Gas hydrates are crystalline compounds in which the ice structure of water is distorted to form cages that contain the gas molecules. The hydrates are solids resembling wet snow in appearance, and they form both above and below 0° under specific pressure–temperature conditions. The water molecules form two kinds of unit cell structures (for details, see Hitchon 1974). The smaller unit structure contains 46 water molecules, which can hold up to 8 methane molecules. Gases such as CH_4, C_2H_6, H_2S, and CO_2 can fit into this structure. The larger unit cell contains 136 molecules of water. Gases such as propane and isobutane can fit into it. These are the only gaseous hydrocarbons that form hydrates. The pentanes and *n*-butane molecules are too large. A methane hydrate in which all the ice stages are completely filled with methane contains about 172 m^3 CH_4/m^3 of hydrate at STP. In nature the more typical ratio is 40 m^3 CH_4/m^3 of hydrate. Natural hydrate units generally have six or seven H_2O molecules per CH_4 molecule.

The pressure–temperature diagram for methane and a 0.6 gravity gas ($C_1 + C_2 + C_3$) is shown in Figure 7-8 (Katz et al. 1959). Hydrates are formed by increasing pressures and are decomposed by increasing temperatures. Since the pressure required to form gas hydrates increases logarithmically as the temperatures increase linearly, it is apparent that the hydrates in most sedimentary basins decompose in the temperature range 70 to 80°F (21 to 27°C), because the pressures are inadequate to preserve them.

Makogon (1981, p. 182) determined the depth below ocean bottoms at which hydrates of gases can form. Table 7-3 shows data for two areas, the subtropical Pacific and the Arctic Ocean. Note that pure methane requires greater sediment depths to form hydrates than do any of the other hydrocarbon or nonhydrocarbon gases. Mixing methane with other gases always lowers the pressure requirements to form hydrates. Makogon (1981, pp. 20, 21) also

TABLE 7-3 Sediment Depths Below the Ocean Bottom at Which Hydrates of Individual Gases Can Form

	Depth to top of hydrate, m	
Gas	*Subtropical Pacific*	*Arctic Ocean*
H_2S	70	1
CO_2	250	70
Gas 0.6 density	300	100
CH_4	500	300

Source: Data from Makogon 1981.

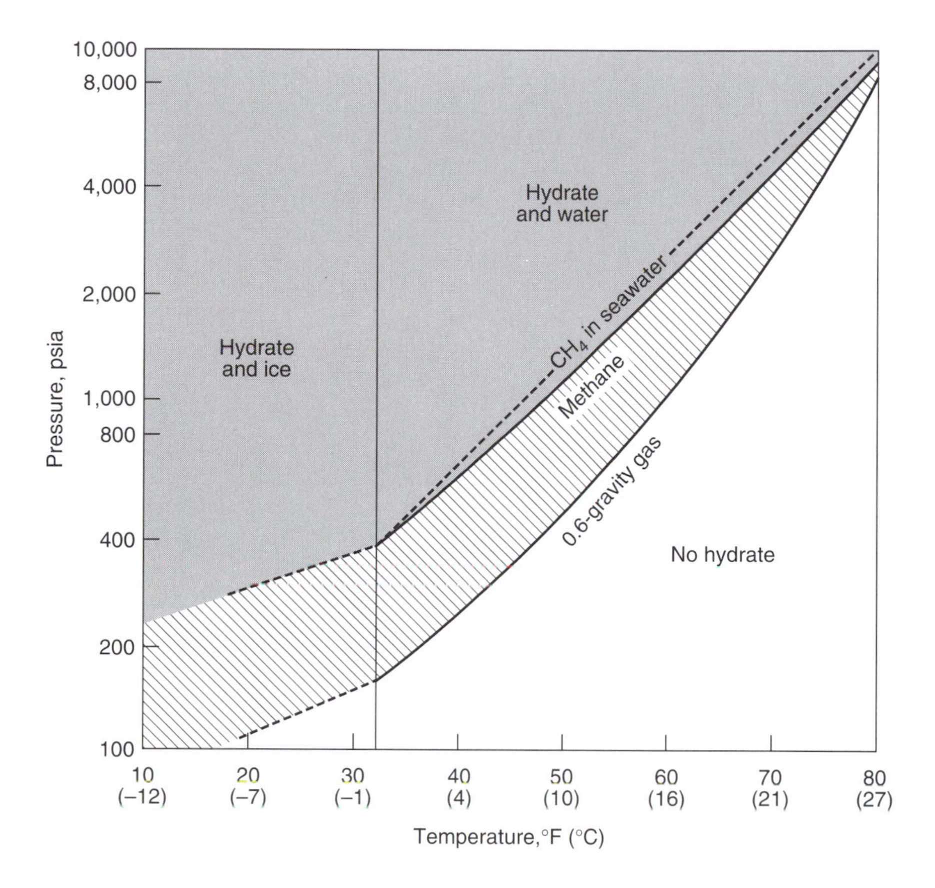

Figure 7-8

Pressure–temperature diagram for gas hydrates. [Data from Katz et al. 1959]

found that the distribution of individual gases in a hydrate is controlled mainly by the composition of the parent gas. Free gas with 1.3 and 4.5% H_2S can form hydrates with 37 and 56% H_2S, respectively.

The salinity of water lowers the temperature at which the hydrate forms, the effect being greater at high temperatures than at low temperatures. Figure 7-8 shows a dotted line for the methane seawater hydrate limit.

Katz (1971) developed a temperature–depth curve to predict the depths at which gas hydrates will occur (Figure 7-9). This shows the curve for methane and 0.6 gravity gas, assuming a hydrostatic head of 0.435 psi per foot of depth and considering salinity effects as negligible. In order to determine the hydrate zone, it is necessary to know the geothermal gradient and the depth of the permafrost when drilling in permafrost territory, or the sea-bottom temperature when drilling offshore. Assuming a constant geothermal gradient, a line can be drawn, as shown on Figure 7-9 for the Cape Simpson, Alaska, area, using Katz's data. The methane hydrate zone here would be only a few hundred feet thick, but the 0.6 gas hydrate would be about 1,900 ft (579 m) thick. Obviously, areas of high geothermal gradients have thinner hydrate zones than do areas of low geothermal gradients. The maximum thickness of the potential hydrate zone can be more accurately estimated if subsurface temperatures are available.

The temperature data from Prudhoe Bay (Holder et al. 1976) and from the Messoyakha field of Western Siberia (Makogon et al. 1971) are plotted in Figure 7-9. The permafrost layer at Prudhoe Bay extends to about 2,000 ft (610 m). The potential-hydrate depth interval overlaps the permafrost, extending from about 700 to 3,500 ft (213 to 1,067 m). For a 0.6 gravity gas, it would extend

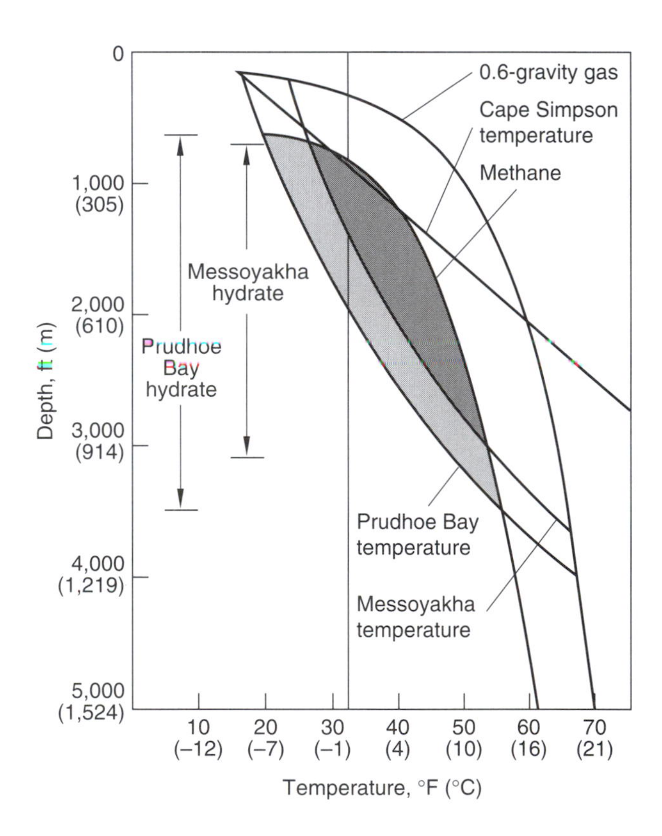

Figure 7-9

Depth–temperature curve for predicting the depth and thickness of gas hydrates. The geothermal gradient is lower in permafrost than in deeper sections of Messoyakha and Prudhoe Bay. [Data from Holder et al. 1976; Makogon et al. 1971]

down to 4,000 ft (1,219 m). At Messoyakha, the hydrate interval has been measured from about 1,148 to 2,854 ft (350 to 870 m).

Hydrates form initially from bacterial gas in near-surface unconsolidated muds containing 40 to 70% water. Once formed, the hydrates immobilize the pore water. Any methane migrating into this zone may be converted to hydrate, assuming saturation in the required temperature–pressure range. Further deposition of methane muds builds up the hydrate until the bottom of the crystallized mass reaches the decomposition temperature and is converted to a gelatinous ooze. Hedberg (1974) suggested that decomposition at the base of hydrates could cause mud diapirs, mud volcanoes, and other phenomena of overpressuring. The Russians have observed pressures of 100,000 psi from the decomposition of methane hydrates in closed experimental systems.

The cooling of the surface rocks during glacial cooling episodes could create thick sections of permafrost and gas hydrates, which would later be decomposed during warm intervals. Makogon et al. (1972) pointed out that the permafrost is now receding in both west and east Siberia. It was much more extensive in the past, and therefore they expect large accumulations of gas to form from the decomposition of hydrates.

Hydrates can be recognized in the subsurface by anomalously fast seismic data velocities and by variations in drilling rates, but sampling with a pressure core barrel is required for proof of a hydrate. Stoll et al. (1971) based their suggestion that there were gas hydrates in the deep ocean on the fact that seismic reflectors tended to follow surface contours, as would a hydrate, rather than to follow the bedding planes. Stoll formed hydrates in the laboratory and found that the seismic velocity increased from 1.85 km/sec to 2.69 km/sec as the solid crystalline hydrate was formed.

Drillers on the Deep Sea Drilling Project (DSDP) noted a marked decrease in drilling rates from less than 1 minute per meter to 5 or 6 minutes per meter when crystalline hydrates were encountered. The bottom of a hydrate interval is usually well defined on seismic records, owing to the sharp drop in velocities from around 3 km/sec to 0.5 and even 0.2 km/sec (Kvenvolden 1993). This is the well-known BSR (bottom-simulating reflector). Sometimes a "bright spot" shows up, indicating free gas to be below the hydrate. Hydrate sections are known to thin as much as 50% over salt diapirs because of the heat conductivity of the salt. This makes such sites a good place to look for free gas.

Gas hydrates have excited the interest of many production geologists because a hydrate reservoir can hold six times as much gas as free gas held in the same space. However, this enrichment factor decreases with depth owing to the compression of free gas. The ratio of gas in a hydrate to free gas in a reservoir of the same size decreases from 6 at about 900 ft (274 m) to 2 at 2,500 ft (762 m) and 1.25 at 4,000 ft (1,219 m). This is assuming a normal hydrostatic gradient and a geothermal gradient of 1.5°F/100 ft (2.7°C/100 m). At Messoyakh the ratio is 1.8.

The only way to produce gas from a hydrate is to decompose the hydrate. To date, most of the gas production from hydrate fields has been in Russia from free gas below the hydrates. Many of these fields are in the West Siberian and Timan–Pechora Basins. The top of the hydrate in West Siberia varies from 200 to 500 m in depth, with the shallowest zones along the Ob River. The depth to the base of the hydrate ranges from 400 to 1,000 m (Cherskiy et al. 1985). In the Timan–Pechora Province the hydrate base ranges from 300 to 1,030 m in depth.

When the Russians first tried to produce gas from hydrate sections, they set large (> 50 cm diameter) casings and used no blowout prevention equipment. The increase in temperature and decrease in pressure from the production operation caused several hundred meters of hydrate to decompose around the well bore and eventually to crush the casings. This caused a tremendous loss of gas. In later drilling the Russians used small insulated casings and blowout prevention equipment on all wells drilling into hydrates.

There are basically three ways to produce gas by decomposing hydrates: (1) by reducing the pressure, (2) by increasing the temperature, and (3) by adding a chemical like alcohol (Makogan 1981, pp. 195 ff). Of these three techniques, the only economical one has been to produce free gas from below the hydrate or from within the hydrate close to its base. This reduces the pressure sufficiently to cause additional hydrate to decompose.

In the Messoyakha gas field the base of the hydrate is the 10°C isotherm. Table 7-4 shows the difference in gas production of the free gas below the hydrate and the gas coming from decomposition within the hydrate. Obviously, the gas flow from perforations far above the base of the hydrate would be uneconomic.

Lost circulation is another hazard of drilling hydrates. One cubic meter of pore water can absorb up to 220 m^3 of gas in forming a hydrate. An abnormally low reservoir pressure will form if the gas–water contact is at a constant depth during hydrate formation.

Gas hydrates are found in all the world's oceans at water depths greater than about 500 m (Kvenvolden 1988). The area of sediments beneath the world's oceans that meet the thermodynamic conditions for hydrate formation is estimated to be eight times greater than the corresponding area on the continents (Makogon 1981, p. 183). Kvenvolden (1988) listed thirty-eight areas around the

TABLE 7-4 Gas Production Relative to Distance from Base of Hydrate for Five Wells in the Messoyakha Field, Russia

Depth of perforations, m	*Depth of hydrate base, m*	*Interval distance, m*	*Gas flow, m^3/day*
727	791	+64	26
794	800	+6	133
793	787	–6	413
795	766	–29	626
793	734	–59	1,000

Source: Data from Makogon 1981.

world, both oceanic and continental, that contain known or inferred gas hydrates based on the BSR or sampling. Actual hydrate samples were obtained by coring at ten of these locations. Estimates of the quantity of methane trapped in these hydrates worldwide are highly speculative, but Kvenvolden thinks that a conservative estimate is about 700 QCF (quadrillion cubic feet). This is equivalent to 116 trillion barrels of oil, or about thirty times the combined oil and gas resources presumed to exist in the world. Some of these hydrates may have large gas accumulations trapped beneath them, but the economics of recovery is highly uncertain. Most of these hydrates will probably not release their gas until the next major period of global warming.

Gas Seals

Methane is much more mobile than oil in the subsurface because of its small size (effective diameter = 0.38 nm) and greater buoyancy. The buoyancy is due to a higher hydrogen content than oil. Methane can penetrate some rocks that are effective seals for oil, but it cannot move through hydrates. The best seals for creating large gas accumulations are, in order, (1) gas hydrates, (2) evaporites, and (3) shales of low porosity and permeability. A *giant* oil field contains more than 0.5×10^9 barrels of expected ultimate recoverable oil, and a *supergiant* oil field contains more than 5×10^9 barrels of recoverable oil (Fitzgerald 1980). Giant and supergiant gas fields are defined as having more than 0.5 and 5×10^9 BOE (barrels of oil equivalent on a BTU basis; 1 barrel of oil = 6,040 ft^3 of gas). This calculates to 3 and 30 TCF (trillion cubic feet) for giant and supergiant gas fields, respectively.

According to Klemme (1983), all the supergiant gas fields in the world are overlain either directly or somewhat above the reservoir by a section composed of permafrost with attendant hydrates or by regional evaporite beds. Most of the giant Russian fields are in the former group, and those of the Middle East are in the latter. In contrast, rift basins like the North Sea are predominately areas of giant oil fields with low gas-to-oil ratios despite high geothermal gradients. This is partly due to the loss of gas through the reservoir cap rocks. Plumes of leaking gas have been observed on reflection seismic profiles of fields in rift basins like the North Sea (Thomas 1980).

Kerogen Source

Methane is generated thermally from all types of kerogen throughout the oil window. Some thermal methane starts forming in small amounts as the shallower bacterial methane is diminishing, so that the two processes overlap. Hedberg's 1974 and 1980 papers emphasized that some methane forms early, although he agreed that most of it forms after the peak in oil generation. Hedberg felt that the early thermal generation of methane was supported by

(1) the substantial quantities of methane dissolved in oil and (2) the frequency of nonassociated gas accumulations, gas caps, and high GOR (gas/oil ratio) reservoirs among oil accumulations in the upper parts of mature oil-bearing sequences. Isotope data, discussed later, support the idea that some hydrocarbon gases form at low maturities in the oil window.

Lillack et al. (1991) obtained a whole series of reaction peaks representing methane generation, by means of laboratory pyrolysis of petroleum source rocks. Activation energies for these peaks appeared to range from 126 to 262 kJ/mol for methane formation. This suggests that some methane may be generated from kerogen over a temperature range from 50°C to greater than 170°C. In fact, small amounts of thermogenic methane were observed in canned cuttings from a depth of 6,098 m (20,000 ft) at 200°C (392°F) in the Inigok well of Alaska. The cuttings contained 9 cc CH_4 at STP per 10^3 cc of sediment. The H/C ratios of the kerogen at this depth were about 0.3, and the vitrinite reflectance was 4.8% R_o, indicating a very advanced state of maturation (Claypool and Magoon 1988).

The wet gas components, C_2–C_7, are formed in the parts per billion (ppb) range during diagenesis by low-temperature carbonium ion or free-radical reactions (Figure 7-1). These reactions yield branched hydrocarbons as the dominant products. At temperatures higher than 50°C there is an exponential increase in hydrocarbons, and the straight-chain (normal) compounds become dominant. This shift from branched- to straight-chain structures occurs at the beginning of the oil and wet gas (C_2H_6+) windows, as discussed for Figure 5-8. The wet gas window ranges from about 70° to 150°C (158 to 302°F), with peak generation around 120°C (248°F). The methane window peaks around 150°C (302°F), as shown in Figure 7-1.

The distribution of light hydrocarbons with depth in the C_2H_6+ range indicates that kerogen generates different hydrocarbons in early stages of catagenesis compared with late stages. This was examined in Chapter 5, in which examples were given of hydrocarbons with tertiary carbon atoms, that is, a carbon attached to three other carbons, forming earlier than those with a quaternary carbon atom, which is a carbon attached to four other carbons (Hunt 1984). Ratios of tertiary to quaternary hydrocarbons may be rough indicators of maturity in the oil window.

At temperatures beyond 150°C, wet gas decreases exponentially to very low values (Figure 7-1). In the previously mentioned Inigok well at 6,098 m where the temperature is 200°C, there are only 0.03cc C_2–C_4 hydrocarbons /10^3cc of sediment. The hydrocarbons are 99.7% methane.

This hydrocarbon-depth distribution pattern of dry gas near the surface followed by wet gas in the oil window and dry gas below (as shown schematically in Figure 7-1) is repeated in many sedimentary basins throughout the world. The heavier gaseous hydrocarbons are formed through approximately the same temperature–depth interval as the liquid hydrocarbons; deeper than the bacterial methane but shallower than the maximum formation of thermogenic methane.

An example of this is the Beaufort Basin of Canada, which is shown in Figure 7-10 (Evans and Staplin 1971). Dry gas is in well cuttings of fine-grained

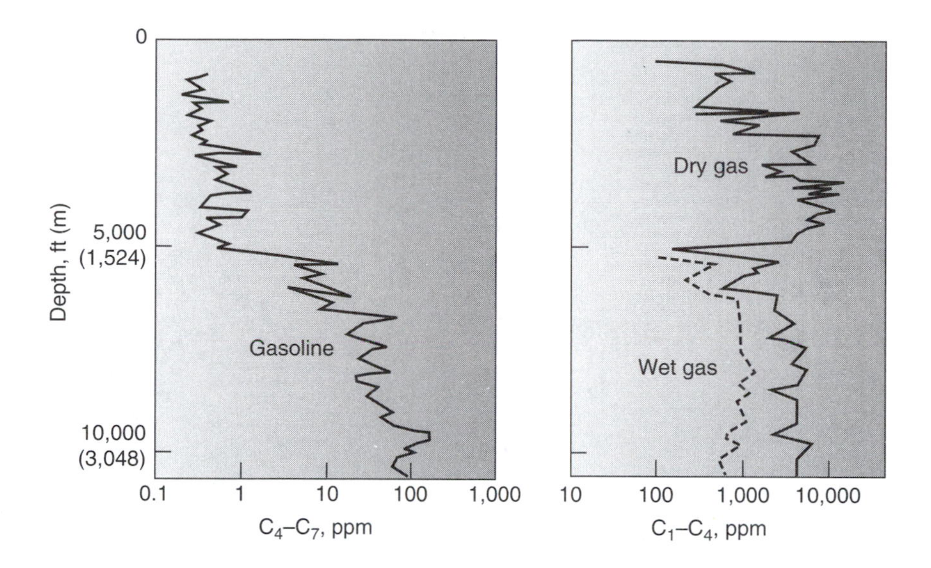

Figure 7-10

Cuttings gas analysis of a well in Beaufort Basin, Northwest Territory, Canada. Methane and ethane plus gas yields were obtained from the laboratory analysis of cuttings canned at the well site. Only methane was obtained from cuttings to a depth of 5,000 ft (1,524 m) through the diagenesis zone, whereas wet gas was obtained in the catagenesis zone to total drilling depth. [Evans and Staplin 1971]

sediments to a depth of about 5,000 ft (1,524 m), and wet gas from this depth to about 10,000 ft (3,048 m). Also, the distribution of gasoline, C_2–C_7 hydrocarbons, increases from less than 1 to 100 ppm in the same range that the wet gas appears. These authors also observed the transition from dry gas to wet gas with depth on the eastern updip side of the Western Canada Basin. In the deeper and older Devonian sediments of northern Alberta, they observed the transition from wet gas to dry gas at a depth equivalent to a paleotemperature of about 160°C (320°F). The data in Figure 7-10 are in an area that was uplifted several thousand feet and eroded, so that the transition from dry gas to wet gas was actually deeper than shown. (Other examples of the shift from dry to wet to dry gas in shale cuttings with depth in both source and reservoir rocks are shown in Hunt 1979, pp. 179, 443–444.)

An example of the dry-wet-dry gas shift in reservoirs is given in Figure 7-11 for natural gas pools in south Louisiana (Meyerhoff 1968). The methane is shown as a percentage of the total hydrocarbon gas components. Reservoirs down to about 6,000 ft (1,929 m) contain dry gas, and from there to about 16,000 ft (4,878 m), there is wet gas followed by dry gas below. The wet gas interval contains up to 30% C_{2+} gases.

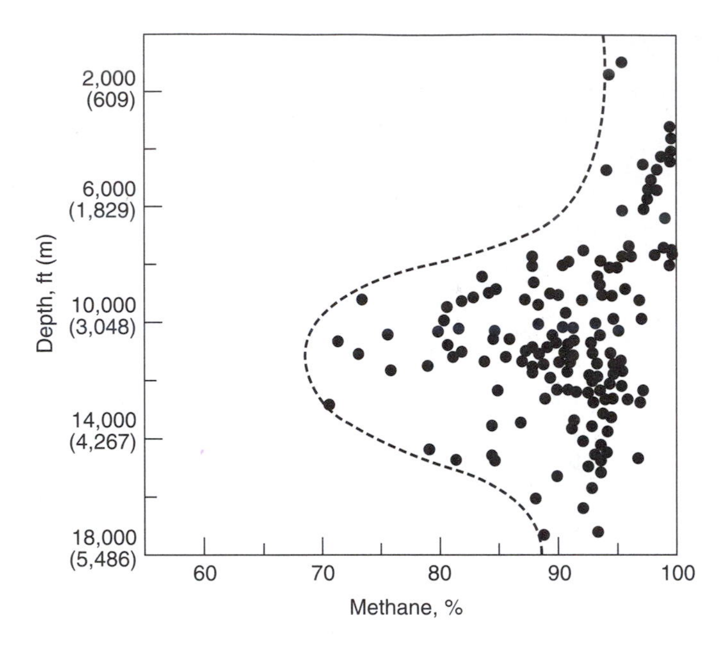

Figure 7-11

Methane as a percentage of the total hydrocarbons in the reservoir gases of south Louisiana. [Meyerhoff 1968]

The basic differences in the chemical structure of sapropelic (types I and II) and humic (types III and IV) kerogens of Figure 7-1 are illustrated in Figure 7-12. The sapropelic, oil-generating kerogens are much higher in hydrogen and have many long chains and individual ring structures that can break up to form liquid fractions of petroleum. As these chains break off, they strip hydrogen from the remaining structure, which condenses to form the compact-ring system of the gas–generating humic kerogen at the bottom of Figure 7-12. The thermal degradation of kerogen yields oil first, followed by increasing amounts of wet gas and methane and finally by only methane. The humic gas–generating kerogen has only a few short side chains plus single methyl groups (shown by one line) and a large number of condensed rings. This gas-generating material can form only a small amount of C_2–C_4 gases, so its major product is methane.

This means that a kerogen shown by pyrolysis to be gas prone is one that either was humic gas–generating to begin with or was sapropelic initially but evolved into the humic gas–generating structure after generating oil.

The importance of hydrogen in this process is shown schematically in Figure 7-13, starting with a hypothetical type II kerogen structure having an H/C ratio of 1.43. The kerogen can break down to yield small gasoline-range

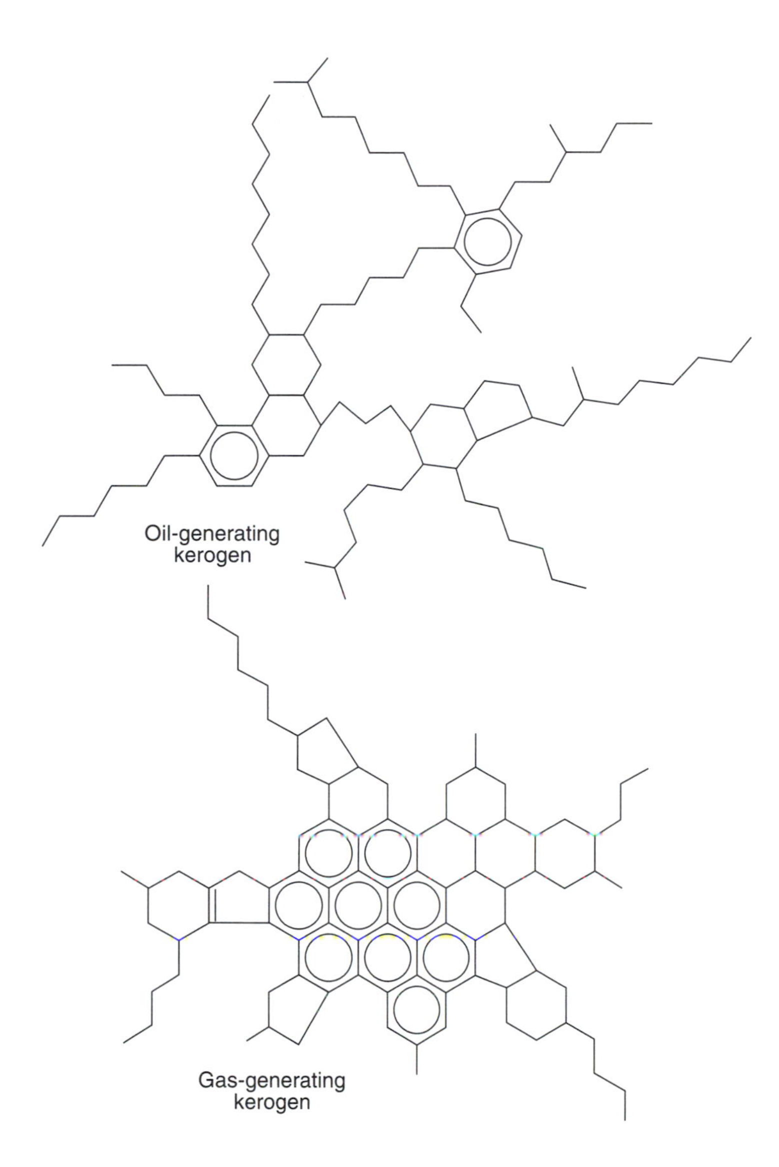

Figure 7-12

Schematic structures of oil-generating and gas-generating kerogens.

molecules such as toluene, 3-methyloctane, and 3-methylpentane, which are illustrated on the second line of this figure. The kerogen that provides hydrogen for these small molecules by condensing the larger molecules into rings is on the left side of the second line. The process continues on the third line, with

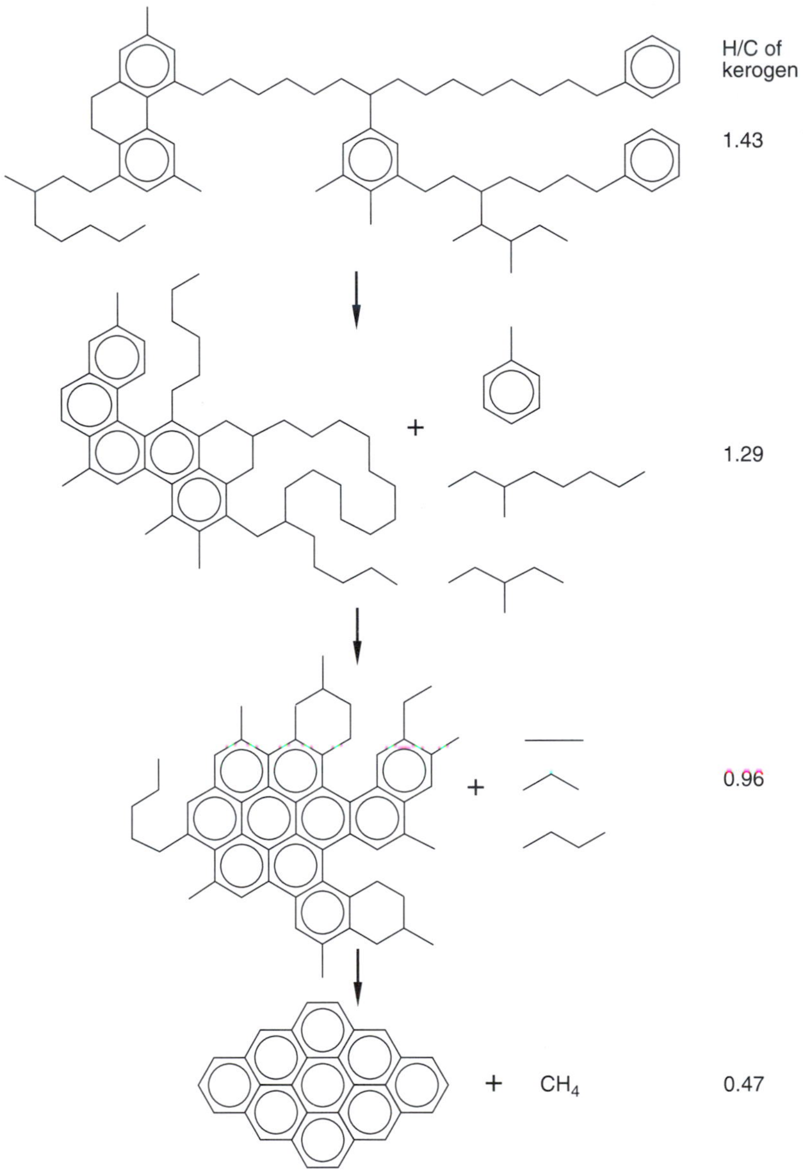

Figure 7-13

The thermal alteration of kerogen involves hydrogen disproportionation reactions in which the kerogen loses hydrogen to form gasoline, wet gas, and dry gas in succession. The hydrogen-depleted kerogen condenses and aromatizes, eventually to form graphite.

naphthene and aromatic rings forming in the kerogen, thereby releasing hydrogen to make the smaller gas molecules. Finally, at the bottom, methane is about the only hydrocarbon that can form, since mainly single-carbon methyl groups remain attached to the highly condensed ring structure on the third line. The kerogen continues to lose more hydrogen than carbon as its structure approaches that of pyrobitumen and graphite.

Although bitumen and oil may be intermediate products in the generation of gas from some types of kerogen, the overall reaction appears to be approximately as follows:

$$C_5H_5 \text{ (kerogen)} \rightarrow CH_4 \text{ (gas)} + C_4H \text{ (pyrobitumen)}$$

The decrease in the H/C ratio of the kerogen (right column of Figure 7-13) is typical of all kerogens and coals, as they are exposed to increasing temperature and time. Consequently, the H/C ratio of kerogen becomes an important maturation indicator.

In Figure 5-15 the H/C ratio of kerogen in the Phosphoria Formation of Wyoming changed from 1.26 to 0.6 through the oil window. Another case study is the Bertha Rogers well, Anadarko Basin, Oklahoma (Price et al. 1981), where the H/C ratio changed from 1.22 at 1,568 m (5,143 ft) to 0.25 at 9,311 m (30,540 ft). As mentioned in Chapter 5, a H/C ratio of 0.25 contains thirty-seven fused aromatic rings. It is incapable of generating any hydrocarbons except traces of methane. Such low H/C ratios are typical of the kerogen in the overlap between the end of metagenesis and the beginning of metamorphism. The latter stage is characterized by the greenschist facies (Hayes et al. 1983). At this stage there is essentially no hydrocarbon-generating capability left in the kerogen.

A more dramatic representation of the change in kerogen structure with the attendant loss of hydrogen in the oil window can be seen in Plates 3A, 3B, and 3C. These excellent chemical models by Behar and Vandenbrouke (1987) are based on extensive analyses of natural samples by techniques such as elemental analyses, electron microscopy, ^{13}C NMR, thermogravimetry analyses, and pyrolysis. These structures go from a high to a low H/C ratio. Each illustration shows the paraffins as white wiggly lines; the naphthenes as red, bent, five- and six-sided rings; and the aromatics as flat, yellow, hexagonal rings.

Plate 3A represents a type I kerogen from the upper Mahogany marker of the Green River Formation in the Uinta Basin, Utah. The H/C ratio is 1.64 and the O/C is 0.06, which is in the diagenesis stage before any significant petroleum generation. The white paraffin chains dominate the picture, indicating a high level of petroleum-generating capability. It is the white chains and red naphthene rings that form over 75% of most crude oils. A porphyrin structure with nitrogen atoms in blue is on the lower, slightly left side of the figure (see Figure 4-8 for comparison). On the lower right side are alcohols, esters, and other oxygen groups, with the oxygen atoms in green. The sulfur atoms, in red, connect two-ring structures in the upper left corner.

Plate 3B is based on the type II kerogen of the Toarcian shales of the Paris Basin and the Lias shales from Germany in the early stages of oil generation.

The H/C ratio is 1.25, and the O/C is 0.089. The change in H/C ratio from 1.64 to 1.25 in Plates 3A and 3B shows a shift in the dominance of the white paraffins to a dominance of naphthenes plus aromatics.

Plate 3C shows the type I Green River kerogen near the end of catagenesis. The H/C ratio is 0.83, and the O/C is 0.013. At this maturation level, most of the white paraffins and red naphthenes in Plate 3A have been converted to oil, leaving large plates of yellow condensed aromatic rings connected by a few paraffin chains. These aromatic rings have now formed parallel plates comparable to graphite.

Although these are schematic models, they provide a valid pictorial representation of the change in kerogen structures as the hydrogen content decreases. High-hydrogen kerogens have been the major generators of both oil and gas, and as the hydrogen content decreases, so does the ability to generate petroleum. A humic, hydrogen-poor type III kerogen at the end of catagenesis has a H/C ratio of about 0.67. At the H/C ratio of 0.25 observed in the bottom of the Bertha Rogers and Inigok wells previously mentioned, the figure in Plate 3C would have no white paraffin chains. Only large clusters of aromatic rings would remain, with essentially no capability of generating either oil or gas.

Coal Source

Methane is the major hydrocarbon released from the maturation of humic coals. Lewan 1993 (personal communication) found that heating coals by hydrous pyrolysis (HP) yielded 5 to 20 wt% waxy oils and 1 to 4 wt% gas. The oil yield depended on the H/C ratio of the coals, with the higher ratios yielding the most oil. These HP experiments mature the coal to near the end of the low-volatile bituminous stage.

Under natural conditions, much of the oil generated by bedded coals tends to be adsorbed in the fine pores of the coals. Eventually this trapped oil is converted to gas as the coal matures under increasing thermal stress. If a kilogram of coal generates 150 gm of oil and 40 gm of gas, the conversion of oil to gas according to the following equation yields an additional 70 gm of gas, for a grand total of 110 gm:

$$C_5H_9 \text{ (oil)} \rightarrow 2CH_4 \text{ (gas)} + C_3H \text{ (pyrobitumen)}$$

This is equivalent to 154 liters of gas per kilogram of coal, similar to the 150 liters reported by Karweil (1969), Jüntgen and Klein (1975), and others quoted in their papers for maturation into the semianthracite coal stage. An additional 50 l may be obtained under very high thermal stress, ending in the meta-anthracite stage, as in Figure 7-14(A). This shows the estimated loss of nitrogen, carbon dioxide, and methane based on the aforementioned coalification studies. Small amounts of the heavier gases ethane through butanes also are formed during coal maturation, but they are so strongly adsorbed that it is difficult to quantify their true yields. The quantity of gas trapped in coal micro-

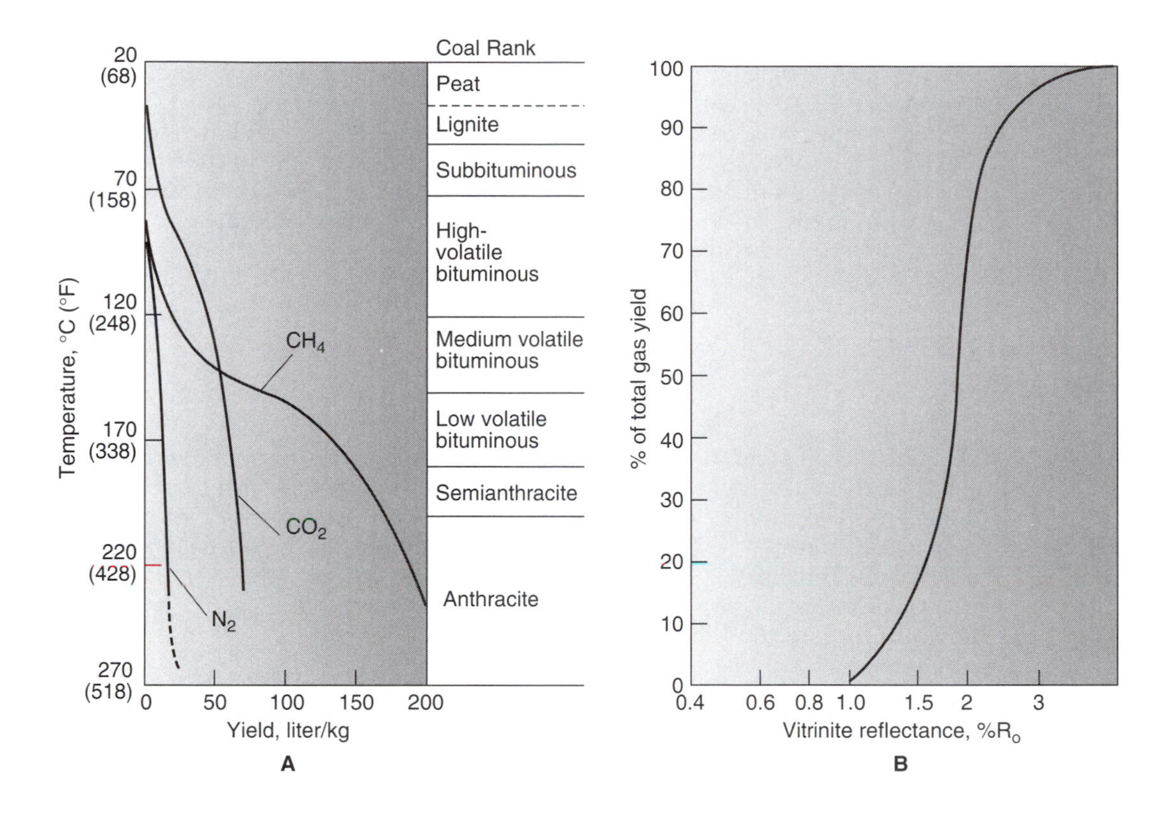

Figure 7-14

(A) Calculated curves of gases generated from coal during coalification. [Data from Karweil 1969] (B) Vitrinite reflectance versus percentage of total gas yield expelled from type III kerogen based on Shell Oil Company data files. [Smith 1994]

pores has not been evaluated in most laboratory experiments involving the generation of gas from coal. The generation of both gas and oil from coal are explored further in Chapter 11.

If this analysis is correct, it indicates that by the semianthracite stage, one-third of the gas has come directly from coal and two-thirds from the cracking of the oil generated by coal, assuming no loss of oil. By the meta-anthracite stage, the proportion is about 50% from each mechanism. However, coals with an initial low H/C ratio, which generate the smallest amounts of oil, would necessarily yield most of their gas directly from the coal. In some cases it is possible to distinguish isotopically between gas generated from either kerogen or coal and gas formed by the conversion of intermediate oil. The former gas is isotopically increasingly heavier, with greater OM maturation, whereas the latter gas shows no significant change.

When the humic components of coal such as vitrinite particles are disseminated in shales, they are classified as type III kerogens. Both coals and type III kerogens generate gas under similar conditions of thermal stress. Peak gas generation generally occurs at maturities greater than 1.3% R_o, as shown in Figure 7-14 (B). This curve, by Smith (1994) indicates that about 80% of the maximum quantity of gas available from a type III kerogen is expelled upon reaching a maturity of 2% R_o. A significant difference between gases generated by bedded coal and those from disseminated coaly particles is that the latter tend to release the generated wet gas (C_2, C_3, C_4), whereas it is absorbed by the former.

Oil Source

Much of the oil generated in sedimentary rocks remains disseminated in the fine-grained source rocks as they pass through the oil window, as pointed out in Chapter 5. The ratio of oil in nonreservoir rocks to that in reservoirs for all rocks is around 200 to 1 (Hunt 1972). In basins with large oil fields, it may range from 20 to 50 to 1. What happens to this vast amount of disseminated oil as the rocks are buried to higher temperatures? According to the equation shown in the previous section, both the nonreservoir oil and that in the reservoirs are converted to gas and pyrobitumen.

$$C_5H_9 \text{ (oil)} \rightarrow 2CH_4 \text{ (gas)} + C_3H \text{ (pyrobitumen)}$$

As the volumes of the gas phase increase, the pressures exert a powerful force causing fracture permeability and enabling fluids to expand and migrate both laterally and vertically through the adjoining rocks. Part of this gas is lost to the surface; part of it sweeps up heavier hydrocarbons and forms accumulations at shallower depths; and part of it causes the extreme pressures observed in deep-pressure compartments sealed by impermeable barriers. Essentially all the disseminated and reservoired oil is decomposed to gas and pyrobitumen when exposed to high temperatures over long time periods. However, there are quantitative constraints on the system due to the lack of sufficient hydrogen. If it is assumed that there is no source of hydrogen in the pores of a fine-grained source rock except from the kerogen and oil, then there is clearly a limit on how much gas can be formed based on the previous equation.

When describing Plates 3A through 3C, it was pointed out that although humic type III kerogen is considered gas prone, it actually yields less gas than do the sapropelic type I and II kerogens, since these contain much more hydrogen. These statements are more readily understood from a material balance of carbon and hydrogen made by Wilson Orr 1992 (personal communication), as given in Table 7-5. Orr made an estimate of the maximum TOC that can be converted to the carbon of oil and gas for the three major kerogen types. The data are based on beginning H/C ratios of 1.45 for type I, 1.25 for type II, and 0.85 for type III and an ending H/C ratio of 0.3 for all types. The numbers shown are the maximum percentages of TOC converted to carbon in petroleum

TABLE 7-5 Estimate of the Theoretical Maximum TOC Convertible to the Carbon of Hydrocarbons

	Maximum percentage of TOC converted to C in oil and gas		
Kerogen type	I	II	III
Oil	58	42	18
Gas	4.5	6.2	7.2
Total	62.5	48.2	25.2
Total if all oil goes to gas	31	26	15

Source: Data from W. Orr, personal communication, 1992.

for these kerogen types. Note that type II yields about twice as much petroleum (oil plus gas) in terms of carbon as type III. Type I yields 2.5 times as much petroleum as type III. The last line shows the maximum yields if all the previously formed oil were converted to gas. In type III, only 15% of the original TOC ends up as carbon in gas, with the remaining TOC disseminated as pyrobitumen in the rock.

Gas yields from the other kerogen types are higher, but in general, less than one-third of the carbon in the original TOC forms gas when it is the only fluid product. The limiting factor in this whole system is hydrogen. Fortunately, in most petroliferous basins, much of the oil is preserved through either uplift with erosion or vertical migration to cooler horizons where the conversion of oil to gas is negligible. Nevertheless, Orr pointed out that actual oil yields in nature rarely exceed 50% of the initial TOC and that much lower yields are more common. Some of the hydrogen and carbon in kerogen is used to convert O, S, and N to CO_2, H_2O, H_2S, and NH_3. The effect of these alternative dispositions would be to decrease hydrocarbon yields below estimates made on the basis of C and H only. Such a decrease would be moderate, however, because most O and N is lost during diagenesis and before major hydrocarbon generation.

Hydrous pyrolysis experiments that are carried out in a closed system simulating nature yield less expelled oil than the theoretical maximums based on studies by Lewan (Table 16-3). For example, the quantity of expelled oil from hydrous pyrolysis of a range of type II kerogens is between 15 and 30 wt% of the kerogen, whereas gas yields are about 3 wt% when heated to a maturation level of 1.5% R_o. Such data indicate that in most petroliferous basins, the type II source rocks would still contain between 50 and 75% of their original TOC at this maturation level.

The oil and gas yields listed in Table 16-3 are realistic guidelines for most source rocks. Experimental procedures that lead to claims of significantly higher yields than those in this table will result in overly optimistic estimates of the quantity of hydrocarbons generated in a specific basin. In the natural system oil plus gas yields of 10 to 40% are typical.

Although there is some evidence that the hydrogen of water can enter into the hydrocarbon-generation process, it is still not clear that it can combine with residual carbon to make hydrocarbons. In fact, a material balance study by Lewan et al. (1995) of the Illinois Basin petroleum system demonstrates that residual carbon increases with an increase in the maturity of the source rock.

Condensates

A *condensate pool* is a gas accumulation in which liquid hydrocarbons, mostly in the gasoline range, are dissolved in the gas phase. If the pressure of a condensate pool is reduced during production, the liquid hydrocarbons will condense within the reservoir, resulting in a much lower recovery than production in the gaseous phase. Repressuring can return part of the condensate to the gaseous phase. (More information about condensates is given in Chapters 3 and 12.)

Condensates are common at temperatures above 100°C and pressures above 5,800 psi (40 MPa). In typical basins this is equivalent to depths beyond 12,000 ft (3,660 m) and vitrinite reflectance values around 1.2 to 1.8% R_o. For example, condensates in the Persian Gulf usually start at depths of 3.5 km, in the South Caspian Basin at 3.5 km, and in the Sirte Basin of Libya and the Maracaibo Basin of Venezuela at about 4 km. This has led geochemists to consider condensates to be an intermediate maturity phase between oil and gas. However, in some basins such as Sacramento, Anadarko, and Arkoma in the United States; Cooper in Australia; and Po in Italy, condensate is produced from a variety of depths over a range of reservoir maturities. Also, many condensates occur in low-maturity reservoirs ($R_o < 0.6\%$) in deltas such as the Mississippi and Mackenzie.

At first these early condensates were explained as originating from resinites, but hydrous pyrolysis experiments with resinites showed that their pyrolysates were distinctly different from most naturally occurring condensates (Hwang and Teerman 1988; Lewan and Williams 1987). Subsequently, it was realized that these condensates came from the vertical migration of gas and light-oil fractions formed by partial vaporization or gas-stripping processes of gas-saturated oil accumulations at greater depths.

Consequently, condensates may have several modes of origin such as (1) from kerogen as an intermediate product between oil and gas, (2) from the conversion of oil to gas in a reservoir, and (3) from the gas-phase migration of hydrocarbons out of an oil pool because of physical phenomena such as vaporization or gas stripping.

The vaporization process of gas-carrying oil in the gas phase vertically through faults, fractures, and other permeable pathways and being trapped as condensate in the shallower reservoirs has been investigated by several geochemists, starting with Zhuze et al. in 1962 (for more details, see Thompson

1987). Silverman (1965) called the process *separation-migration.* More recently Thompson (1987) defined it as *evaporative fractionation* in order to emphasize that it involves a fractionation process that can form aromatic condensates. Before Thompson's experiments, it was not clearly understood why some condensates contain up to 45% aromatic hydrocarbons such as those from the Upper Devonian of Canada (Hitchon and Gawlak 1972), when crude oils typically contain only 15 to 20% aromatics.

Thompson (1987) carried out as many as eleven successive equilibration steps of oil supersaturated with methane at 6,000 psi and 127°C (260°F). Table 7-6 shows the increase in aromaticity and decrease in paraffinicity in the left and right columns, respectively, for the eleven condensates formed in his experiment. Aromaticity is measured as the ratio of toluene to *n*-heptane and paraffinicity as *n*-heptane to methylcyclohexane. Although the first two condensates in Table 7-6 are slightly less aromatic than the parent oil, succeeding condensates become highly aromatic and naphthenic (less paraffinic). Thompson (1987) called these *evaporative condensates* and showed several examples of condensates from North America and Europe that follow the same trend shown in Table 7-6.

In evaporative fractionation, the original reservoir oil may be supersaturated with methane by either thermal alteration of conventional oil, as described in Chapter 12, or methane migration through the reservoir from a deeper

TABLE 7-6 Laboratory Experiment Showing Increase in Condensate Aromaticity Due to Evaporative Fractionation

Sample	*Aromaticity, toluene/n-C_7*	*Paraffinicity, n-C_7/MCH*[a]
Original oil	1.24	0.67
Condensate 1	0.85	0.77
Condensate 2	1.14	0.69
Condensate 3	1.33	0.62
Condensate 4	1.77	0.46
Condensate 5	2.24	0.45
Condensate 6	3.18	0.35
Condensate 7	4.50	0.29
Condensate 8	6.13	0.25
Condensate 9	8.00	0.20
Condensate 10	8.50	0.21
Condensate 11	7.00	0.20

[a]Normal heptane/methylcyclohexane.

Source: Data from Thompson 1987.

source (gas stripping). As the gas with dissolved oil migrates upward, hydrocarbons liquify and drop out on the way, just as in the distillation tower in Figure 3-6. Consequently, the shallowest oils and condensates have the highest API gravity (lowest specific gravity).

In a broad sense, condensates are either thermally generated light hydrocarbons or light hydrocarbons accumulating from vaporization and migration processes involving methane as the carrier gas, or mixtures of both. Since they migrate in the gas phase, they may be found at a variety of depths extending from immature to postmature sediments. Some condensates are unusually high in aromatics–such as benzene, toluene, xylenes, and trimethylbenzenes—whereas others have hydrocarbon distributions typical of light crude oils. Some deep thermally formed condensates have been reported to contain diamondoids, the thermally stable mono-, di-, tri-, and tetracyclic hydrocarbons in the C_{10}–C_{22} range (see Chapter 12). Petrov (1984, p. 95) stated that 15 to 20% of the C_{11}–C_{14} tricyclanes in several Russian crude oils are diamondoids.

In the northern North Sea, the present-day thermally generated condensate window is between 4,500 m and 5,400 m (14,800 and 17,700 ft), according to Reeder and Scotchman (1985). This is equivalent to a temperature of about 150 to 180°C (Figure 6-18). These authors observed a strong correlation between condensates with high API gravities and the areas where the Kimmeridge Shale has been buried to over 4,000 m (13,100 ft) equivalent to temperatures $>135°C$. Their depth–temperature interval for the condensate window agrees closely with the end of the oil window and the beginning of the conversion of oil to gas based on Arrhenius kinetics for type IID kerogens, as shown in Table 6-5 in the previous chapter.

Not all high API gravity crude oils are condensates, especially if they contain little or no methane. In 1979 a 75°API gravity oil was recovered from Oligocene sands in Texas (Smith and Hilton 1980). The oil contained less than 0.1 wt% C_1+C_2, 6% C_3, 21% C_4, 18% C_5, 14% C_6, and 41% C_7^+. Over half the oil was in the C_4–C_{16} range. A reservoir fluid study confirmed that the oil existed in a liquid state under reservoir conditions at 3,390 ft (1,034 m), so it is not a condensate. A lipid-rich cannel-type coal was suggested as a possible source for this unusual oil. However, the authors hypothesized that some undefined migration and trapping process must have allowed the methane and ethane to escape but retained the propane and higher hydrocarbons.

Evaluation of Gas Source and Maturity

The carbon isotope values of the individual hydrocarbon gases, methane through *n*-butane, depend on both source and maturation. The $\delta^{13}C$ values are initially determined by source, with methane having the most negative values (less ^{13}C) and ethane, propane, and *n*-butane having increasingly positive values in the order shown. Initially, the *n*-butane may have 25‰ more ^{13}C than the methane. With increasing maturation of the source, this difference decreases to less than 10‰.

One of the first field observations of this spread in isotope values occurred with a gas that was recovered with an oil-saturated core from a hole drilled by the Deep Sea Drilling Project in the Gulf of Mexico in 1968. The gas was from a sediment depth of about 450 ft (137 m) under a water column of 11,720 ft (3,572 m). The sediment was an anhydrite cap rock that had been altered to gypsum, calcite, and sulfur. Carbon and sulfur isotope data indicated that microbial oxidation precipitated the sulfur from H_2S, and the calcite, from methane or oil. The oil was found to be immature (for details, see Hunt 1979, pp. 296–298). Carbon isotope analysis of the gases showed a wide spread in $\delta^{13}C$ in going from methane to ethane, propane, and butane. The values were –50.8, –31.0, –26.2, and –23.6‰, respectively. When these data were plotted on James's (1983) calculated maturity diagram (C in Figure 7-15), they indicated that the gas was immature, equivalent to a vitrinite reflectance (R_o) value of $< 0.6\%$.

The immaturity of both the oil and the gas supports the concept that biogenic organic matter generates hydrocarbons in deep ocean environments well beyond the continental slope. The methane for this sample plots above the calculated maturity line in Figure 7-15 because of mixing with methane formed during diagenesis by low-temperature bacterial and chemical reactions. The maturity lines in Figure 7-15 were derived on a theoretical basis and confirmed by analyses of hundreds of gases by James (1983).

The diagram plots the algebraic difference in ppt (‰) between the isotopic compositions of individual gases on the vertical axis. The scale is divided into 10‰ units. Specific $\delta^{13}C$ values are not listed, since these vary with both the source and the maturity of the gas, whereas the spread in the lines is due mainly to maturity. The maturity of a gas is determined by measuring $\delta^{13}C$ of the gas components and comparing them with the calculated lines in Figure 7-15. The best fit of the data is determined by sliding the vertical line of measured isotopic separations horizontally and vertically until they fit the calculated separations, particularly for the C_3–C_5 hydrocarbons. This indicates the maturity of the source from which most of the gas was generated. For example, sample K (Figure 7-15) is clearly very mature, whereas B is very immature. Various maturity scales are on the horizontal axis, such as vitrinite reflectance ($\%R_o$), the level of organic maturity (LOM; Hood et al. 1975), and the thermal alteration index (TAI). (Details of the construction of the plot and its applications are in James 1983 and 1990, Galimov 1988.)

The $\delta^{13}C$ separations between gas components decrease continuously with increasing maturity to an LOM of around 13 ($R_o > 1.5\%$). At greater maturities, the C_2–C_5 gases are so thermally degraded that the small amounts remaining show an increase in isotopic separation with maturity until they disappear completely. Only the methane isotope is currently measurable at maturities above about LOM 15 ($R_o > 2.3\%$). The isotope separations go through a minimum at LOM of 12 to 13, with increasingly wider separations occurring at both the low and high maturities. This is shown in Figure 7-16, which plots the generalized maturities of thirty-six gases from the Delaware–Val Verde Basin in Texas (James 1983). Figure 7-16 shows that a wide spread in isotope values by itself does not necessarily indicate an immature gas, since the spread occurs at

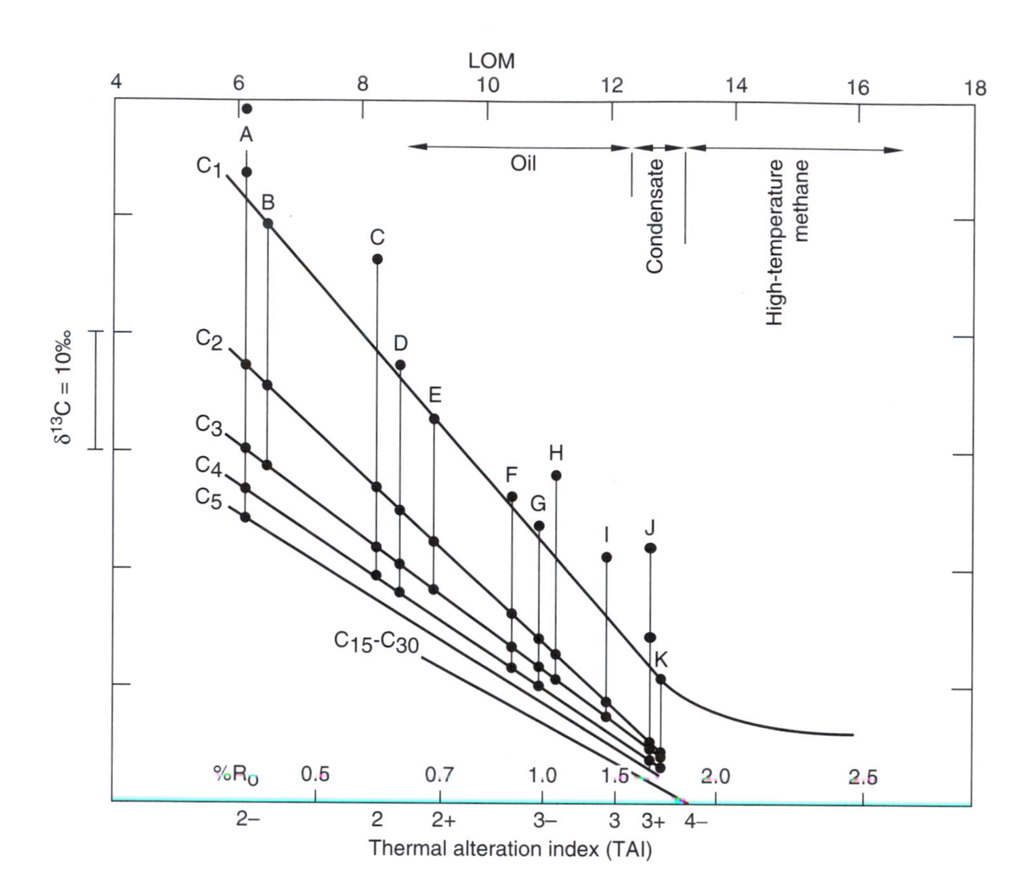

Figure 7-15

Measured carbon isotopic separations of reservoired gases plotted on a calculated maturity diagram by James (1983). Three maturation scales, $\%R_o$, TAI, and LOM, are shown. LOM is the level of organic maturity (see Chapter 10). The source of gases A through K is discussed in the text.

both very high and very low maturities. Other data such as the actual ^{13}C value, which becomes more positive at high maturities, and the relevant geological information on the sample would need to be assessed. If the data fit the profile on the right side of Figure 7-16, it is possible to make a rough estimate of the maturation level that a specific gas may have beyond an LOM of 13.

Some examples of gases at different maturation levels are shown in Figure 7-15 along with the previously discussed DSDP sample from the Gulf of Mexico. Two gas samples from the Norwegian shelf in the North Sea are plotted, one from block 2/2 (Gabrielsen et al. 1985) and the other from the

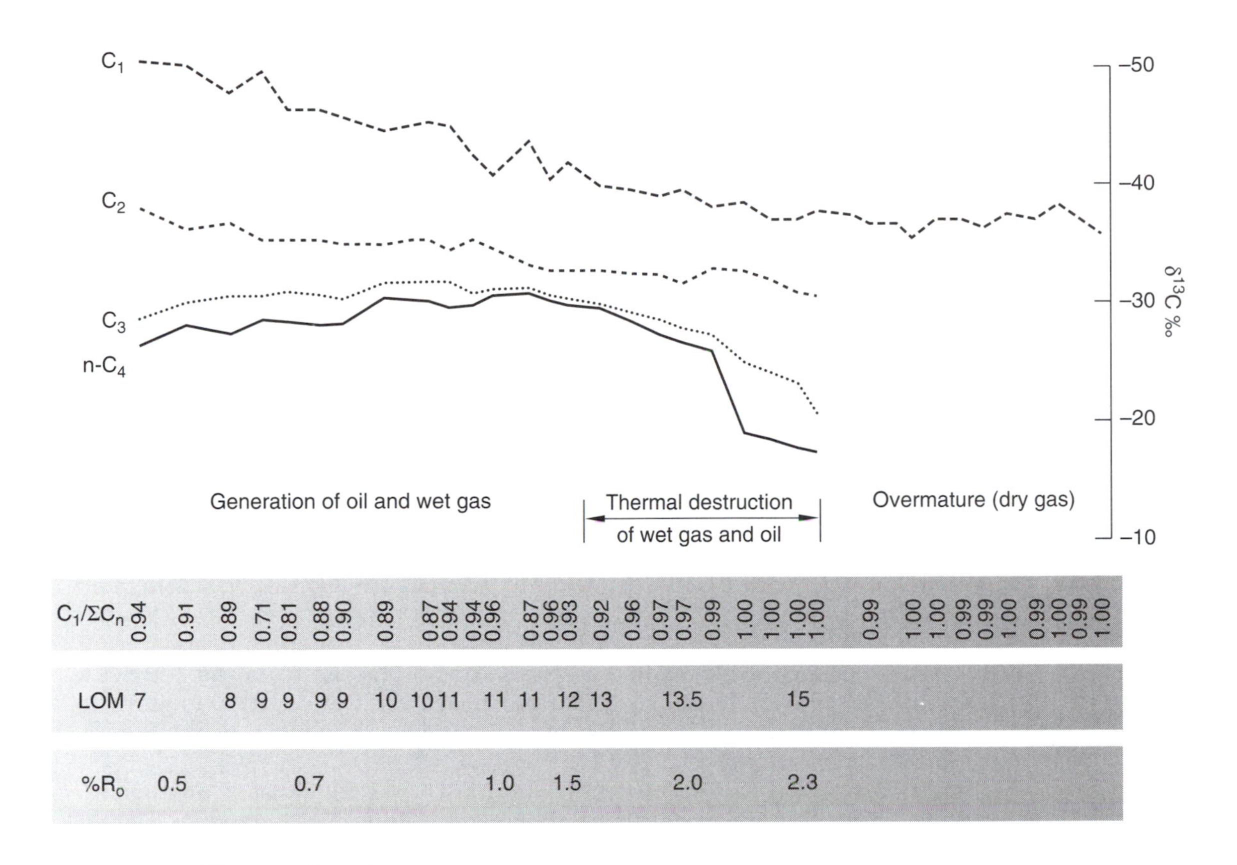

Figure 7-16

The $\delta^{13}C$ profiles of individual hydrocarbons in thirty-six gases from the Delaware–Val Verde Basin, Texas, in various stages of maturation. [James 1983]

Midgard field in the Haltenbanken area (Elvsborg et al. 1985). The gas from block 2/2 (D, Figure 7-15) has a low to moderate maturity (LOM 8–9, R_o = 0.6%), indicating early generation in the oil window. The gas is mostly thermogenic based on the $C_1/(C_2 + C_3)$ ratio of 2.5 and a $\delta^{13}C$ for methane of −48.6‰. Other geochemical data indicate an Upper Jurassic source rock with the gas migrating vertically into Oligocene reservoirs. The Midgard sample (J) is a gas condensate with an indicated maturity of LOM 13. The probable source of this gas is a Lower Jurassic coal unit buried to a subsea depth > 4 km. The gas condensate migrated vertically into Middle and Lower Jurassic sands at shallower depths.

The Marlin field, Gippsland basin, Australia, produces gas (K) from a Tertiary reservoir at 4,454 ft (1,358 m). The $\delta^{13}C$ for the gas (James 1983) plots at a maturity of LOM 13, compared with the reservoir maturity of LOM 7. The LOM values for the underlying rock sections indicate that the gas source rock is the Upper Cretaceous nonmarine Latrobe group containing type III kerogens and coals at an LOM of 13. This is an example of matching a gas with its source rock based on similar maturities.

Two gas samples from the Western Canada Basin are plotted on Figure 7-15 (for additional Canadian gas data, see James 1983 and 1990). The Medicine Hat sample (A) from a reservoir at 2,200 ft in the southeastern corner of Alberta has a very immature signature (LOM 6, $R_o < 0.5\%$). James (1983) considers the gas to be indigenous to the Cretaceous interval in which it occurs. The Leduc gas (F) at a present-day depth of 5,315 ft was generated from the Upper Devonian Duvernay Formation. The LOM of 10.5 for the gas is more mature than the Leduc reservoir LOM of 8 (James 1990). The Duvernay source rock has a matching LOM of 10.5 about 60 km downdip southwest along the reef trend. Gas migration appears to have been laterally updip from this source, which was buried to more than 12,000 ft (3,659 m) in the past.

The immature gas (B) from the Salina field in Kansas was generated along with an immature oil from Upper Ordovician source rocks, and they are currently at 3,200 ft (Jenden et al. 1988). The field is in the Salina Basin, west of the Nemaha Uplift. This gas and a Mississippian gas (E) from the Sedgwick Basin of Kansas (Jenden et al. 1988) are characteristic of the immature to marginally mature gases generated by local Paleozoic source rocks. Butane isotopic analyses are not available for these samples.

The Coyanosa gas (G) in the Delaware basin of west Texas is another example of correlating a gas with its source rock based on maturities (James 1983). The gas with an LOM of 11 is produced from a Pennsylvanian reservoir that is in a fault block protruding into the organic-rich Permian–Wolfcampian shales. These shales with 2 to 4% TOC have an LOM of 11 corresponding to the gas (Figure 7-15). Deeper possible source rocks in the areas have maturities from LOM 12 to 14, higher than the gas. Also, gases from deeper reservoirs are drier in composition and isotopically heavier than the Pennsylvanian reservoired gas.

Gas samples B, D, E, F, and G in Figure 7-15 all are from kerogen type II source rocks. As such, they tend to match the theoretical maturity lines in Figures 7-15 more closely than the Sleipner gases (H and I) of the North Sea generated by a type III Jurassic kerogen or III–II mixture (James 1990). The butane data are not shown for the Sleipner gases (H and I) because they would plot between the ethane and propane lines in Figure 7-15. The values are reversed from the theoretical isotopic separations. Such reversals indicate that the source effect is overriding the maturity effect on which the James plot is based.

According to James, these reversals occur more frequently with (1) the heavier hydrocarbons, butane and pentane; (2) branched hydrocarbons like isobutane and isopentane; (3) gases generated by Paleozoic and Mesozoic rocks that have not been exposed to high subsurface temperatures; and (4) gases from type III kerogens. The different coal macerals in type III, such as liptinite and vitrinite, appear to generate gases of different isotopic composition at the same maturation level. Gases from vitrinite are more enriched in ^{13}C than gases from liptinite. Consequently, gases formed by different layers of lacustrine and fluvial nonmarine strata have many more reversals and odd isotope distributions than gases from predominately marine type II sources, which are more uniform and consist mostly of liptinite. Where gases from type II sources show reversals, the mixing of gases is usually indicated. Where reversals occur with type III source gases, comparisons with other gases in the area are needed to determine whether the reversals are due to mixing, multiple sources, or a strong source control. If it is a strong source control, the carbon isotopic separation of ethane and propane will give the most reliable maturity estimates.

Where the source influence is strong, the wet gas components are particularly useful for correlating gases with one another and with crude oils containing dissolved wet gases. Gases can be correlated with one another simply by comparing the molecular composition and the $\delta^{13}C$ values of the individual gas components. Comparisons of gas families are most reliable if isotope data for the ethane through *n*-butane are available.

Microbial biodegradation can severely alter the carbon isotopic composition of gases by causing selective enrichment of ^{13}C in hydrocarbons such as propane and *n*-butane (James and Burns 1984). This would invalidate most maturation and correlation studies. Biodegradation is discussed further in Chapter 12.

Nonhydrocarbon Gases

Carbon Dioxide

The often erratic distribution of carbon dioxide (CO_2) in reservoirs is caused by a variety of factors, such as multiple sources, high solubility in formation fluids, and high reactivity. The solubility of CO_2 in water at STP is about 1 volume of CO_2 to 1 of water, a volume ratio of 1/1. At 7,000 psi (48 MPa), equivalent to a depth of about 15,000 ft (4,573 m), the volume ratio of dissolved CO_2 in a typical formation water is about 30/1. At the same pressure, the volume ratio of CO_2 dissolved in a 40° API gravity oil is about 170/1, or more than five times greater than its solubility in water. The critical temperature of CO_2 is 87.8°F (31°C). This means that any undissolved CO_2 in reservoirs at temperatures above this will exist as a gas, regardless of the pressure.

There are three important sources of CO_2, one organic and two inorganic. The first source is from the thermal degradation of organic matter, which occurs during diagenesis and catagenesis and is mostly completed at the end of the oil window. Continental derived humic materials contribute most of the organic source CO_2. Up to 75 liters of CO_2 per kilogram of coal (2,700 ft^3/ton) are released during the maturation of coal from the lignite to the anthracite stage (Karweil 1969). The CO_2 peaks in Figure 7-1 are mostly due to the decomposition of carbonyl (C=O), methoxyl ($-OCH_3$), phenolic hydroxyl (–OH), and other oxygen groups in the catagenetic stage. This organic source of CO_2 seems to predominate in and above the oil window, whereas inorganic sources predominate at greater depths.

The second important source is described by Hutcheon and Abercrombie (1989, and earlier papers referenced therein). It involves a reaction between kaolinite and carbonates to produce chlorite and CO_2, as follows:

$$5FeCO_3 + SiO_2 + Al_2Si_2O_5(OH)_4 + 2H_2 \leftrightarrow Fe_5Al_2Si_3O_{10}(OH)_8 + 5CO_2$$

This reaction appears to occur at temperatures above 100°C and reaches equilibrium at 160°C. Hutcheon and Abercrombie (1989) think that this is the source of large volumes of CO_2 produced during steam-assisted recovery of heavy oil at shallow depths but high temperatures. They also think it is the source of CO_2 in gases of the Venture field at Sable Island off the east coast of Canada. Illite also can react with carbonates at high temperatures to yield CO_2 and chlorite or feldspar.

Similar reactions may be the cause of the large increase in CO_2 in overpressured compartments of the U.S. Gulf Coast. In the Miocene through Jurassic reservoirs of the Texas Gulf Coast, the CO_2 increases from <1 mole% in reservoirs at 7,000 ft (2,134 m) to 7 mole% in reservoirs at 12,000 ft (3,659 m), according to Franks and Forester (1984). Land and Macpherson (1992) found that the CO_2 as dissolved bicarbonate in aqueous solutions throughout the Gulf becomes isotopically heavier with increasing depth. This indicates that the increase in CO_2 is from an inorganic source.

Another area showing an increase in inorganically sourced CO_2 with reservoir depth is the Greater Sleipner field area offshore Norway (James 1990). Gas samples shallower than 2,700 m (8,859 ft) subsea depth contain <1 mole% CO_2, whereas some samples at 3,900 m (12,800 ft) contain up to 36 mole% CO_2. The CO_2 in shallow reservoirs at 2,400 m (7,870 ft) is light, with a $\delta^{13}C$ of –12‰, indicating an organic source. Deeper CO_2 at 3,600 m (11,800 ft) is heavy ($\delta^{13}C = -3$‰), typical of an inorganic carbonate source.

The third source of CO_2 is volcanic activity. When high temperature magma penetrates carbonate rocks, the carbonate decomposes, giving off CO_2. The CO_2 migrates as a gas and can be trapped in reservoir rocks of nearby structures. This relationship between CO_2 generation and high-temperature igneous intrusions was first discussed by Holmquest (1965) for the Delaware–Val Verde basins of New Mexico and Texas. Figure 7-17 shows the change in CO_2 concentration of gases from the Ellenberger Dolomite of Cambro-Ordovician age. Ellenberger gases in the regions closest to the igneous intrusion of the Diablo Platform and Marathon Uplift contain the highest percentages of CO_2. The CO_2 concentrations decrease from the igneous intrusion toward the central basin platform, where CO_2 concentrations are zero. There also is an upward decrease in CO_2 from the Ellenberger as well as the horizontal increase toward the intrusion.

In Colorado, a large igneous intrusion forms the core of the Sleeping Ute Mountain Range. In forming the mountain, the intrusion passed through the

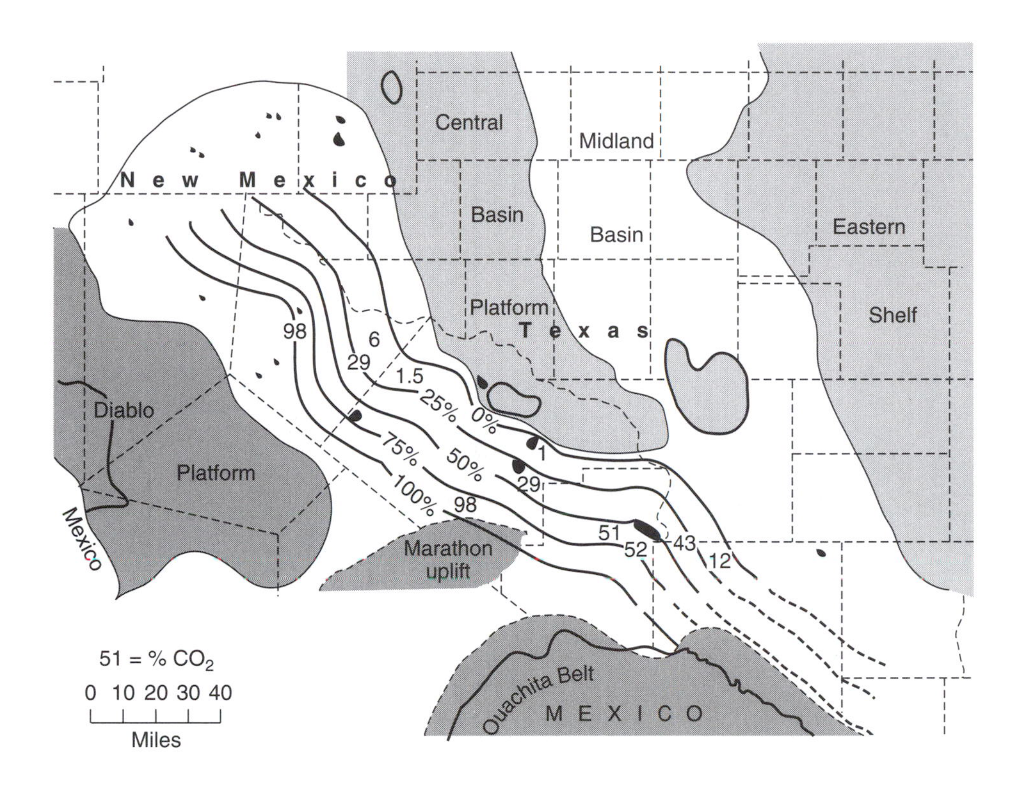

Figure 7-17

Contours of the percentage of CO_2 in gases from the Ellenburger Formation in the Delaware–Val Verde basins of New Mexico and Texas. [Holmquest 1965]

Mississippian Leadville limestones and dolomites, thermally decomposing the nearby carbonates. The released CO_2 moved updip and accumulated within the McElmo Dome, as shown in Figure 7-18. The reservoir is estimated to contain 10 trillion cubic feet (TCF) of essentially pure CO_2. A similar situation exists in New Mexico's Bravo Dome, where a huge CO_2 reserve is trapped in Permian sandstones. Another example is the Kevin Sunburst Dome on the Sweetgrass Arch area of northwestern Montana. All three of these examples are gas accumulations in large structural domes, containing more than 90% CO_2 that was formed by igneous interactions occurring with deep sedimentary carbonates downdip or on the flanks of the dome. These large reserves represent only trapped CO_2. Very little is known about the quantity of CO_2 that is formed in these sedimentary basins from the interaction of carbonates and igneous intrusions. All three of the aforementioned CO_2 reserves are being produced for tertiary recovery of oil from nearby oil fields.

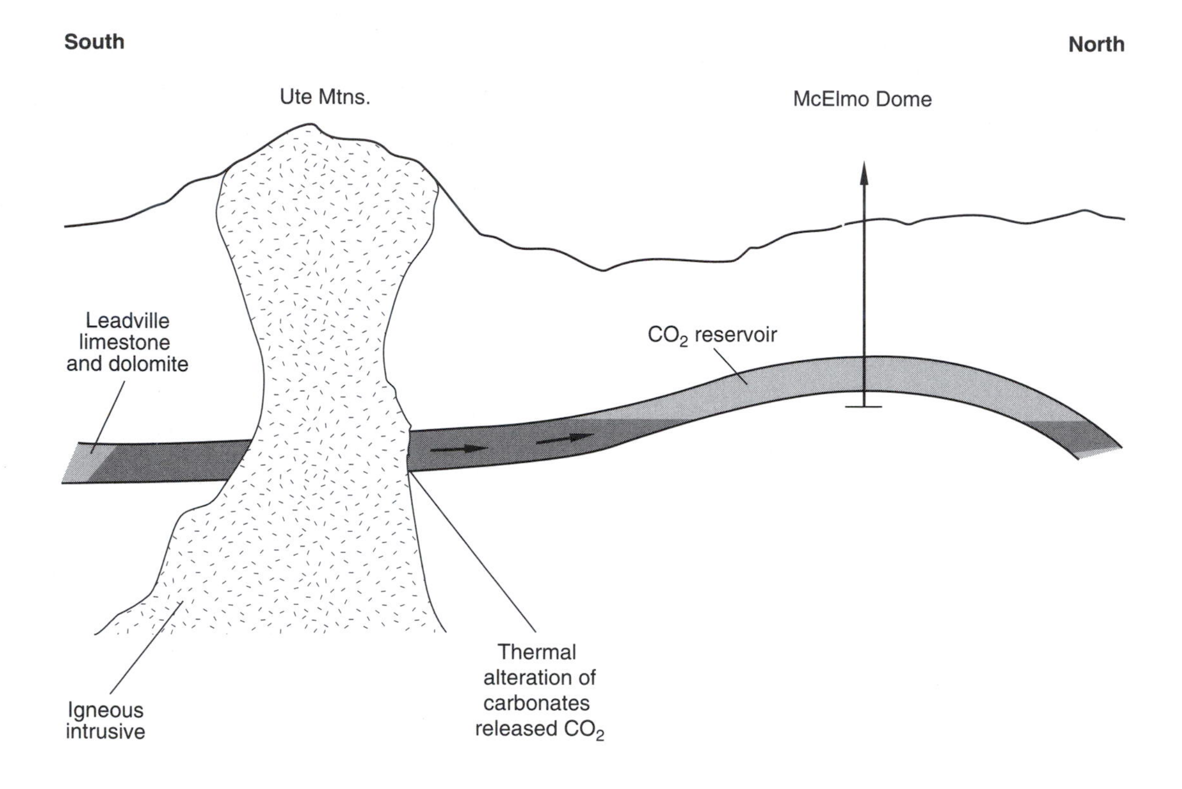

Figure 7-18

Origin and accumulation of CO_2 at McElmo Dome, Colorado. The volcanic activity that generated CO_2 and expelled it into McElmo Dome also formed the Sleeping Ute Mountain Range.

The different sources of carbon dioxide cause different $\delta^{13}C$ values of the carbon, as shown in Table 7-7. Thermogenic CO_2 from OM has a more negative $\delta^{13}C$ than that from the decomposition of carbonates. Bacterial oxidation of CH_4 to CO_2 can result in a wide range of $\delta^{13}C$ values for the CO_2, depending on the $\delta^{13}C$ of the original CH_4. In comparison, volcanic and atmospheric CO_2 are essentially identical. The variations in Table 7-7 can be used to evaluate multiple CO_2 sources, as James (1990) did in the North Sea. His data indicate that CO_2 from deep in the Viking Graben has an inorganic source ($\delta^{13}C$ = –3 to

TABLE 7-7 Variation in $\delta^{13}C$ of CO_2 from Different Sources

Source	$\delta^{13}C$‰
1. Thermal degradation of organic matter	–8 to –12
2. Thermal destruction of carbonates	+4 to –5
3. Bacterial oxidation of methane	–20 to –59
4. Volcanic degassing	–8
5. Atmospheric CO_2	–8

–5‰), whereas that from reservoirs high on the Norwegian shelf to the east has an organic source ($\delta^{13}C$ = –8 to –12‰).

Hydrogen Sulfide

Hydrogen sulfide (H_2S) is the deadliest gas produced in large quantities in nature. As little as one part per thousand of H_2S in air causes respiratory paralysis and sudden but agonizing death from asphyxiation. There is a misconception that the foul odor gives a warning, but in nonlethal concentrations below 0.1 ppt, H_2S rapidly dulls the sense of smell, so increasing exposure unknowingly becomes fatal. In 1975 nine people were killed in Denver City, Texas, when a mixture of H_2S and CO_2 escaped from a pipeline leak 500 ft (150 m) from their house. In 1950 when H_2S escaped from a gas treatment plant 130 miles northeast of Mexico City, 22 persons were killed and 320 were hospitalized. Such accidents occur because a combination of H_2S, CO_2, and water causes most metals to become brittle and crack–a process called *hydrogen embrittlement*. Specially fabricated high-carbon steels are therefore required for transporting or drilling into high H_2S gases.

In addition, H_2S is so reactive that much of it is converted in sediments to elemental sulfur, metallic sulfides, or organic sulfur compounds. The solubility of H_2S is more than twice that of CO_2, but its critical temperature of 100.4°C (213°F) is much higher than for CO_2. Above this temperature, any undissolved H_2S is in the gas phase. Figure 7-1 shows H_2S starting to form from organic matter during catagenesis, with generation increasing with increasing depth and temperature. Not shown are the large amounts of H_2S formed by microbial sulfate reduction during diagenesis, because nearly all of this is oxidized to sulfur. The huge sulfur deposits associated with salt domes in the U.S. Gulf coast are the result of bacterial H_2S being oxidized to sulfur (Thode et al. 1954). The amount of bacterial H_2S that becomes a constituent of natural gas is negligible compared with the H_2S from other sources.

The H_2S in the subsurface originates from both organic and inorganic sources. Le Tran et al. (1974) found that most of the H_2S disseminated in fine-grained rocks appears to be coming from the thermal decomposition of the kerogen. Le Tran analyzed the adsorbed hydrogen sulfide and methane in sediment samples with depth in three wells of the Aquitaine Basin. He also analyzed for C_{15+} liquid hydrocarbons in order to determine the zones of heavy hydrocarbon generation. Figure 7-19 presents the results for the Jurassic Mano Formation. Depth is plotted linearly, with the gas yield in ml/g of TOC plotted logarithmically. The location of the C_{15+} hydrocarbon yield peak and approximate present-day temperatures also are indicated. Similar results were obtained in a study of the Lower Cretaceous Barremian–Neocomian Formation.

The data indicate that organic sulfur structures in the dispersed kerogen crack thermally to form H_2S. This process occurs later than the generation of methane. The largest quantities of both gases are formed at temperatures greater than 120°C (248°F). This temperature is minimal, since the geological evidence indicates that paleotemperatures were higher. The oil-generation peak is about 90°C (194°F) in Figure 7-19. The peak in methane generation is in the

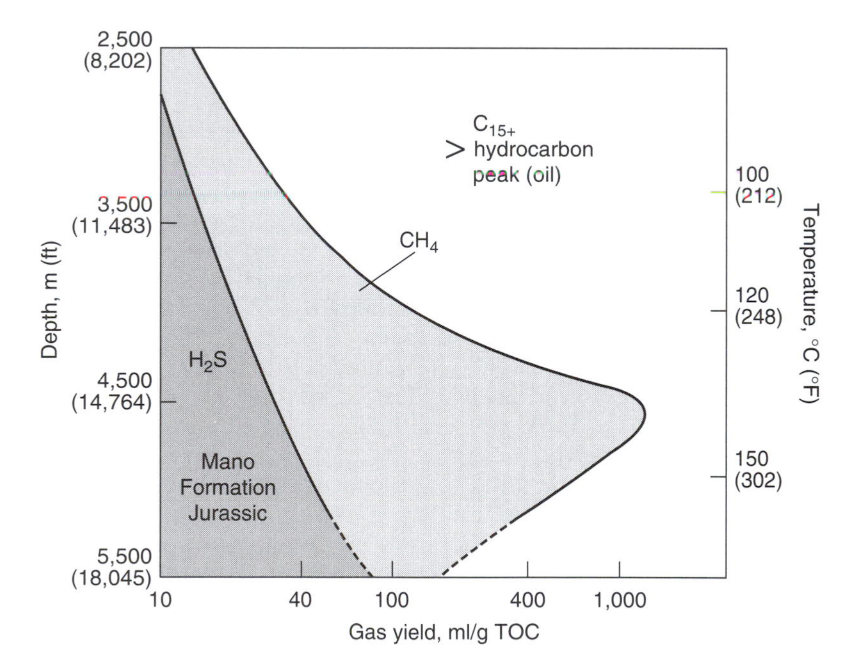

Figure 7-19

Generation of CH_4 and H_2S from the thermal decomposition of organic matter in carbonate source rocks of the Aquitaine Basin, France. Oil generation (C_{15+} hydrocarbons) peaks before H_2S generation. [Le Tran et al. 1974]

range of 140°C (284°F), and the H_2S generation does not peak at the maximum depth analyzed, equivalent to almost 170°C (338°F).

Le Tran (1972) also analyzed for adsorbed H_2S in cores and cuttings of wells from other basins. He found that the increase in H_2S always paralleled the increase in CH_4 in the deepest part of the basin. Also, the formation of H_2S was always related to the dispersed organic matter. Where there was little or no OM in the fine-grained rocks, there was no H_2S, and zones high in OM were high in H_2S, other things being equal. The yields of H_2S always increased with depth of burial, with the largest increase occurring after the oil generation peaked.

Probably only a small part of the H_2S in reservoir gases comes from nearby source rocks. Because of its high reactivity, it is difficult for H_2S from a source rock to migrate to a distant reservoir, whereas other sources already in the reservoir include the thermal decomposition of sulfur-containing oil and the reduction of sulfate in the interstitial water. At a temperature of 104°C (220°F) and a pressure of 2,000 psi (14 MPa), H_2S is forty times more soluble in water than at STP. Any H_2S formed by the kerogen is rapidly disseminated by fluid migration and precipitated as pyrite on contact with iron. In areas of low reactive iron content, as in certain carbonate sequences, some H_2S from kerogen may enter the reservoirs, but it usually is surpassed in volume by H_2S derived from the reservoir oil and from the sulfate in the pore waters. The evidence for this is the fact that the $\delta^{34}S$ of H_2S starts out near that of the reservoir oil and presumed source rock in immature accumulations but gradually approaches the $\delta^{34}S$ of the pore water sulfate with increasing maturity (Orr 1977). Initially, the $\delta^{34}S$ of the H_2S is 10 to 15‰ lighter (more negative) than that of the sulfate.

Even oil degradation cannot form high H_2S gases. Laboratory experiments have indicated that when oil is thermally degraded, about half of its sulfur forms H_2S, and the other half remains as a pyrobitumen residue on the rock (Orr 1977). Orr showed that when a typical high sulfur oil (3.2% S) is degraded to methane, H_2S, and a bitumen residue with a H/C of 1, the maximum volume of H_2S in the produced gas is 3%, as shown. This is assuming a closed reservoir system with all the sulfur coming from the oil.

$$\underset{\text{oil}}{C_{100}H_{170}S_{1.4}} \longrightarrow 23CH_4 + 0.7H_2S + \underset{\text{pyrobitumen}}{C_{77}H_{77}S_{0.7}}$$

$$\text{Volume of } H_2S \text{ in the gas} = (0.7)(100)/23.7 = 3.0\%.$$

This thermal reaction does not cause significant fractionation of the sulfur isotopes (Orr 1977).

Orr examined seventy-one gases from four sour gas provinces for H_2S content. These were the Devonian of Alberta, the Jurassic of the southeastern United States, the Permian of northwest Germany, and the Upper Jurassic of southwest France. He found that about 65% of the gases contained more than 3% by volume of H_2S. This could not be explained by only degradation of the reservoir oil based on the reaction just given.

Orr's earlier 1974 paper was the first to propose that the reduction of the sulfate in the reservoir pore water could account for these high percentages. He called the reaction *thermochemical sulfate reduction* (*TSR*). It involves the reaction of H_2S with sulfate to form elemental sulfur and polysulfides, followed by the reaction of sulfur with hydrocarbons to form H_2S and CO_2. The net reaction is as follows:

$$SO_4^{=} + 1.33(CH_2) + 0.66\ H_2O \rightarrow H_2S + 1.33CO_2 + 2OH^-$$

The end result is that the H_2S is from the reduction of the sulfate and the CO_2 is from the oxidation of the oil. Consequently, the $\delta^{34}S$ of the H_2S approaches that of the sulfate, and the $\delta^{13}C$ of the CO_2 approaches that of the oil. The process is autocatalytic, since H_2S is a catalyst as well as a product of the reactions.

Toland (1960) demonstrated that these reactions take place in a few hours at temperatures above 300°C (570°F). Orr (1974) carried out similar reactions at lower temperatures and established with certainty that these reactions can occur at geologically significant rates in the temperature range of 77 to 121°C (170 to 250°F). Orr (1977) also found that the shift in the $\delta^{34}S$ toward that of the reservoir sulfate with increasing maturity appears to be common in reservoirs where sulfate is abundant and temperatures exceed 77°C (170°F). For example, in the Lacq gas field, Aquitaine Basin, France, the $\delta^{34}S$ for the H_2S and reservoir anhydrite are both +16‰ at a depth of 4,000 m (13,120 ft) and a reservoir temperature of about 150°C (302°F). This similarity indicates that most of the H_2S was formed by TSR of the sulphate.

Another factor supporting the role of sulfates is the association of high H_2S gases with carbonate–evaporite sequences. Table 7-8 lists gas deposits with more than 5% H_2S (Le Tran 1972). These all are carbonate reservoirs with either interbedded anhydrite or anhydrite above or below the reservoirs.

In all these examples, the amount of iron in the rock is insufficient to eliminate the H_2S as pyrite. The ratio of iron in shales to iron in carbonates is about 12/1 and that of iron in sandstones to iron in carbonates is 3/1. This means that up to twelve times more H_2S can be changed to iron sulfides in shales than in carbonates. In addition, there is more sulfur initially in marine kerogen and pore fluids during carbonate deposition than in nonmarine kerogen and fluids during the deposition of clastics. It is common to find H_2S in Paleozoic limestones in areas such as Ohio, Indiana, Illinois, Michigan, and western Kentucky. Although Kansas and Oklahoma have almost no H_2S in younger formations, there are large amounts in the Cambro-Ordovician carbonates of both states.

More H_2S is being found throughout the world as wells are drilled deeper. The giant Tengiz field northeast of the Caspian Sea in Kazakhstan contains 25 billion barrels of very sour oil along with 40 TCF of high H_2S gas at depths between 3,660 and 7,012 m (12,000 and 23,000 ft). Gas fields with more than 10% H_2S also are common in Jurassic Smackover reservoirs of Alabama, Mississippi, and Florida at depths between 4,000 and 5,800 m (13,120 and 19,000 ft). Mobil's Mary Ann field, a Jurassic Norphlet reservoir offshore of Alabama, has 9% H_2S. As the gas is produced, the H_2S is converted to more than 200 tons of sulfur per day, which is sold at a profit.

TABLE 7-8 Natural Gas Deposits with High H_2S Content

Region	*Reservoir age*	*Lithology*	*Depth (m)*	*%H_2S in total gas*
Lacq, France	Upper Jurassic and Late Cretaceous	Dolomite and limestone	3,100–4,500	15
Pont d'As-Meillon, France	Upper Jurassic	Dolomite	4,300–5,000	6
Weser-Ems, Germany	Permian (Zechstein)	Dolomite	3,800	10
Asmari–Bandar, Shahpur, Iran	Jurassic	Limestone	3,600–4,800	26
Urals–Volga, Russia	Late Carboniferous	Limestone	1,500–2,000	6
Irkutsk, Russia	Late Cambrian	Dolomite	2,540	42
Alberta, Canada	Mississippian	Limestone	3,506	13
	Devonian	Limestone	3,800	87
South Texas	Late Cretaceous (Edwards)	Limestone	3,354	8
	Upper Jurassic (Smackover)	Limestone	5,793–6,098	98
East Texas	Upper Jurassic (Smackover)	Limestone	3,683–3,757	14
Mississippi	Upper Jurassic (Smackover)	Limestone	5,793–6,098	78
Wyoming	Permian (Embar)	Limestone	3,049	42

Molten Sulfur

Sulfur melts between 113 and 120°C (235 to 248°F), so any reservoirs above that temperature may contain molten sulfur. Molten sulfur is twice as heavy as water, so it tends to flow into synclines rather than anticlines.

Molten sulfur and associated H_2S are generally restricted to deep stratigraphic zones that consist mainly of carbonate and sulfate rocks. Wells that have produced molten sulfur on a drill-stem test have never produced economic quantities of hydrocarbons from the same depths as the molten sulfur. Some examples reported by Les Magoon in 1992 (personal communication) are given next.

In the United States, large quantities of molten sulfur were recovered from the Cambro-Ordovician Arbuckle limestone–dolomite at 31,441 ft (9,586 m) in the Bertha Rogers well, Anadarko Basin, Oklahoma. Molten sulfur also was recovered from the Jurassic Smackover carbonate–evaporite sequence at 20,000 ft (6,098 m) in a well in southern Mississippi along the U.S. Gulf Coast.

A well drilled by Pemex in northeastern Mexico recovered molten sulfur in a Mesozoic carbonate sequence at about 12,000 ft (3,659 m). Molten sulfur also was recovered from Devonian carbonates at a similar depth in the Deep Western Canada Basin.

The Permian Zechstein dolomite has yielded molten sulfur in several wells drilled at depths ranging from 10,000 to 14,000 ft (3,049 to 4,268 m) in western Germany.

The source of this sulfur is not always clearly defined. The Inigok well on the North Slope of Alaska encountered a heavy flow of H_2S gas at 17,570 ft (5,357 m) in the Mississippian–Lisburne Carbonate. Subsequently, elemental sulfur was recovered with shale cuttings from this interval. The S and H_2S had $\delta^{34}S$ values of +3.3 and +3.9‰, respectively. No anhydrite was reported in the section, and only traces of sulfate were found in the mud samples. Thus the evidence suggests that the sulfur came from the downhole oxidation of the H_2S (L. Magoon 1992, personal communication). H_2S itself may have come from the organic matter of underlying shales or coals. The H_2S is known to come from the decomposition of some coals, starting at around 150°C (Klein and Jüntgen 1972). There are numerous coal beds in the Inigok well between 17,570 ft (5,357 m) where the strong H_2S flow was encountered and the total depth (TD) of the well is at 20,102 ft (6,127 m). The present-day temperatures in this interval are 175 to 200°C (347 to 392°F).

Elemental sulfur reacts with H_2S to form polysulfides, which are powerful oxidizing agents. They can convert saturated hydrocarbons like hexanes to thiophenes, thiols, and sulfides. At elevated temperatures in the presence of H_2S, pentanes, butanes, propane, ethane, and, finally, methane are oxidized to CO_2, forming H_2S as a by-product. Consequently, it is not surprising that wherever molten sulfur is recovered on a drill-stem test, no appreciable quantities of methane or higher hydrocarbons are present.

Nitrogen, Hydrogen, and Helium

Ammonia (NH_3) is a common constituent of pore waters in sedimentary basins. The Russians have reported ammonia concentrations in oil-field waters to range from 50 to more than several hundred mg/l (Hunt 1979, p. 470, and references therein). For example, in the Volga–Urals region, it ranges from 600 to 1,625 mg/l (Kudel'skiy 1973). Coal beds can give off as much as 20 liters N_2 (as NH_3)/kg of coal during maturation from the bituminous to anthracite stages (Figure 7-14). This is about one-tenth the yield of methane. The pyrolysis of coals by Klein and Jüntgen (1972) indicated that nitrogen was released in two peaks, the first around 100°C (212°F) and the second about 200°C (392°F).

Most of the ammonia dissolved in pore waters is oxidized to nitrogen through contact with heavy metal oxides or meteoric waters containing oxygen.

Many high-nitrogen gases are associated with red beds. Guseva and Fayngersh (1973) described the presence of gases containing more than 90% nitrogen in red beds of the Central European and Chu-Sarysuy oil and gas basins of Russia. Nitrogen gases in both basins occur mainly in red beds of Carboniferous, Permian, or Triassic strata overlain by thick salt beds. The authors think that the nitrogen is formed by the reaction of ferric oxide with ammonia or nitrogen-bearing organic compounds.

In the North Sea off the coast of Germany, high-nitrogen gases are found in red beds of the Rotliegendes formation under a thick cover of Zechstein salt. Lutz et al. (1975) attributed the nitrogen to the thermal alteration of underlying Carboniferous coal measures and the dispersed OM in shales. When it contacts the red beds, the ammonia formed oxidizes to nitrogen.

Nitrogen may also have an atmospheric origin or come from mantle outgassing. Many gases occluded in igneous rocks are high in nitrogen. Petersil'ye et al. (1970) found 24 to 40% N_2 and 0.6 to 3.7% helium in methane gas issuing from holes drilled in ultramafic rocks of the Kola peninsula. The frequently observed correlation of nitrogen and helium in natural gas has indicated a deep-seated source for some nitrogen. Coveney et al. (1987) list several igneous rocks containing gas seeps with up to 43 mole% N_2.

In Kansas near the Mid-Continent rift system is a gas accumulation containing about 35 mole% hydrogen, the remainder being nitrogen with only traces of hydrocarbons (Coveney et al. 1987). The hydrogen was analyzed for hydrogen isotopes to evaluate its origin. The results were consistent with a hydrogen origin from reactions in the crust involving ferrous iron oxidation rather than from mantle outgassing. Organic sources were discounted because only traces of CH_4 and CO_2, the expected by-products of biogenic activity, were found in the gas. Hydrogen is a common constituent of many well gases and subsurface waters, but it is rarely reported in the literature because it is not included in conventional gas analyses. Zinger (1962) found up to 43% hydrogen in the gas dissolved in oil-field waters of Paleozoic rocks of the lower Volga region of Russia. Samples were taken in new wells directly after perforation of the casing and after the water attained constant chemical composition, so the hydrogen could not be attributed to acid reactions with the casing. Zinger's paper also reports other Russian data on hydrogen in natural gases and in coal gases. In western Siberia, Nechayeva (1968) found hydrogen in about 15% of all gas samples analyzed. About 60% of the samples with hydrogen had under 1%. Peak hydrogen concentrations were 0.9% for gas fields, 6% for gas condensate fields, and 11% for oil fields. The highest concentrations were in gases from organic-rich Jurassic sediments, suggesting that the hydrogen was coming from the thermal decomposition of the organic matter. Analyses of both well gases and sedimentary rocks from the North American continent have also shown hydrogen to be relatively common, with up to 15% in some natural gases.

Hydrogen is so mobile and reactive that it cannot be permanently retained in a geological trap. Its presence thus indicates that it is either being actively generated from reactions in the reservoir or in adjacent source beds or is diffusing up from deeper sources.

Helium has two isotopes, ^3He and ^4He, whose relative abundances indicate the source of helium. The ^4He is generated almost exclusively by the disintegration of radioactive elements in the earth's sedimentary rocks. The highest values are associated with uranium ore deposits. The ^3He has a mantle origin. Consequently, low $^3He/^4He$ ratios of about 10^{-8} indicate a sedimentary origin, and ratios of 10^{-7} to 10^{-5} indicate a mantle source. Furthermore, the ratio of $^4He/^{40}Ar$ (radiogenic argon) is around 10 to 20 for a sedimentary origin and 1 to 2 for a mantle origin. The $^4He/^{40}Ar$ ratios for most natural gas pools cluster around 12 and go up to 170, indicating a sedimentary origin for most of the associated helium (Krouse 1979, and references therein).

Helium has an effective diameter of 0.2 nm (10^{-9}m) compared with 0.23 nm for hydrogen, 0.33 nm for CO_2, 0.34 nm for N_2, and 0.38 nm for CH_4. Consequently, helium migrates more readily than the other gases through most rocks. Few cap rocks retain helium, so its presence in a reservoir indicates a currently active source and dynamic processes.

Many helium occurrences are associated with down-to-basement faults. A survey of soil gases along the San Andreas fault in California showed helium values as high as 430 ppm associated with hydrogen at 50 ppm and methane at 3 ppm. The presence of only traces of methane in gases coming from down-to-basement faults and areas like the East Pacific rise and Mid-Continent rift system (Coveney et al. 1987) is further evidence that abiogenic methane is not a significant potential source of methane from the mantle. There has been no reported abiogenic methane associated with helium in commercial petroleum reservoirs anywhere in the world based on carbon isotope measurements.

Large natural gas fields in coastal zones and offshore areas such as in southern Mexico, northern Alaska, the Persian Gulf, and the U.S. Gulf Coast commonly have less than 0.0007% helium (Cook 1979). The average helium content of all natural gas reserves in the United States that contain helium is 0.068%.

Gas Distribution in Basins

The discussion of Figure 7-1 leads to the conclusion that dry bacterial methane can be found in the immature organic facies of a basin, methane plus wet gases in the deeper mature facies, underlain by dry thermal methane in the postmature facies. This hydrocarbon-depth distribution pattern is repeated in many sedimentary basins throughout the world. The heavier gaseous hydrocarbons are formed through the same temperature–depth interval as the liquid hydrocarbons. Examples of this are shown in Figures 7-10 and 7-11.

Vertical migration in some areas distorts this pattern, but the general relationship is the same everywhere. Figure 7-20 illustrates this graphically for the Western Canada Basin. Added to this is the distribution of nonhydrocarbon gases, as observed in Canada and in many other sedimentary basins. The immature facies (equivalent to the diagenesis stage in this text) contains kerogen with light yellow spore and pollen particles, indicating that it has never

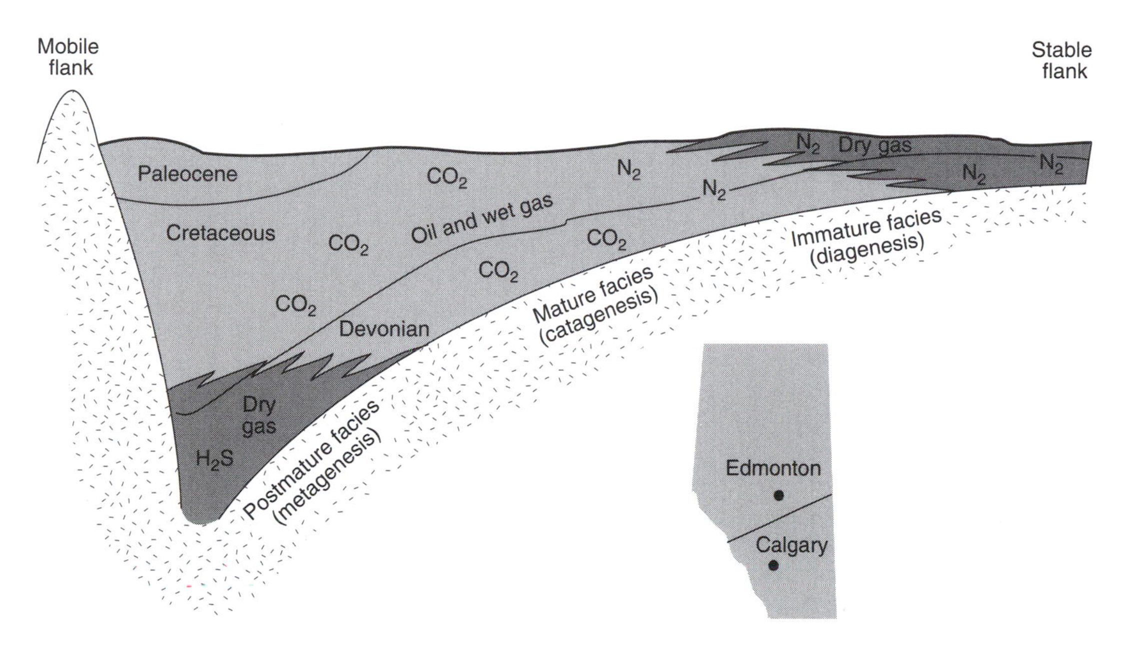

Figure 7-20

The distribution of gases in different maturation facies of the Western Canada Basin. [Evans and Staplin 1971]

been thermally altered. The dominant gases are bacterial methane and nitrogen. The deeper, mature facies (equivalent to the catagenesis stage) contain spores and pollen ranging from amber to dark brown in color, indicating that the increasing temperatures have begun to cook the kerogen. Wet gas and oil are in the upper part of this range, with thermal condensate in the lower part and carbon dioxide throughout.

Near the end of catagenesis and the beginning of metagenesis, the spores and pollen in the kerogen go from dark brown to black, indicating that the cooking process has eliminated most of the heavier hydrocarbons. The gases in this zone are methane, CO_2, and hydrogen sulfide. In all these facies, methane is usually the dominant reservoir gas.

The organic facies changes shown in Figure 7-20 are due to the increase in temperature with depth. The temperature also may increase laterally because of different geothermal gradients within the same basin. In the Rainbow area of Alberta, Canada, the Middle Devonian carbonates in the east are rich in wet gas, but moving west there is a change to dry gas (Evans and Staplin 1971). Middle Devonian temperatures at the same depth increase westward more than 10°C (50°F) from the wet gas to the dry gas areas. In the Dnieper–Donets depression of Russia there are oil and gas pools to the northwest, but only gas and condensate to the southeast within the same Paleozoic horizons (Kravets 1974). The depression is a graben formed along a system of faults in the Precambrian crystalline basement and filled by 2.5 to 11 km (8,200 to 36,090 ft) of Devonian through Cenozoic sediments. Subsurface temperature measurements at the northwest edge of the oil-producing area show a geothermal gradient of 2°C/100 m (1.1°F/100 ft), whereas at the southeast edge of the gas-producing area, the gradient is 3°C/100 m (1.7°F/100 ft). Clearly, the changes in maturity, such as wet gas to dry gas, may occur if the temperature increases laterally in a formation. It has the same effect as does a temperature increase vertically.

Among the nonhydrocarbon gases, nitrogen is found in the highest concentrations on the stable shelves of basins, as shown in Figure 7-20. This distribution is partly related to diagenetic ammonia being oxidized in near-surface sediments and partly to the ease with which nitrogen migrates upward. Among natural gas constituents, nitrogen is the smallest molecule after helium.

Zor'kin and Stadnik (1975) studied the composition of gases in pore waters of Mesozoic and Paleozoic rocks of the Caspian depression. They found that the dissolved gases are primarily nitrogen in the eastern, shallower part of the depression and along its margin. Down the section, toward the central, deeper part, the concentration of methane and wet gases increases. Zor'kin et al. (1972) stated that subsurface waters of the marginal areas of the Baltic Basin, the Mid-Russian Basin, and the Volga–Urals oil and gas province all are high in nitrogen. Devonian formation waters on the south flank of the Baltic shield are fresh to brackish and contain nitrogen-rich gas. Going basinward, the methane content increases, and the nitrogen content decreases. On the Russian Platform, the gases become richer in hydrocarbons and poorer in nitrogen going down toward the deeper parts of the L'vov, the Baltic and Moscow basins, and the Yarensk trough.

An aerial view of the nonhydrocarbon gas distribution relative to the "hot lines" of the Western Canada Basin is shown in Figure 7-21. The hot lines were defined by Evans and Staplin (1971) as the transition zone from the wet gas to the dry gas facies. This zone represents, in a general way, the westerly limit of oil exploration for each of a sequence of reservoir horizons. The limit is caused by the deepening of the basin and increased temperatures going westward. Line A is the westerly limit of wet gas and oil for the Mississippian; line B for the Upper Devonian Wabamun; line C, Upper Devonian Woodbend–Winterburn; and line D, Middle Devonian Elk Point. Dry gas tends to be west of these hot lines, and wet gas and oil generally are to the east. Areas of gas accumulations with high N_2, CO_2, and H_2S concentrations are based on Hitchon's (1963) data. The nonhydrocarbon gases tend to follow these maturity lines in a very general way, as shown in Figure 7-20. Since H_2S forms last and therefore deepest among the natural gas components, it is found only in the deepest, highest-temperature regimes, generally west of the hot lines in each formation of Figure 7-21. To some extent, of course, the gas distributions are affected by migration within the formations away from their source areas. The profile does clarify the general relationship of nitrogen on the shallow, stable flank to the east, with CO_2 in the deeper transition zone and H_2S in the very deep high-temperature sedimentary rocks on the mobile deformed flank to the west.

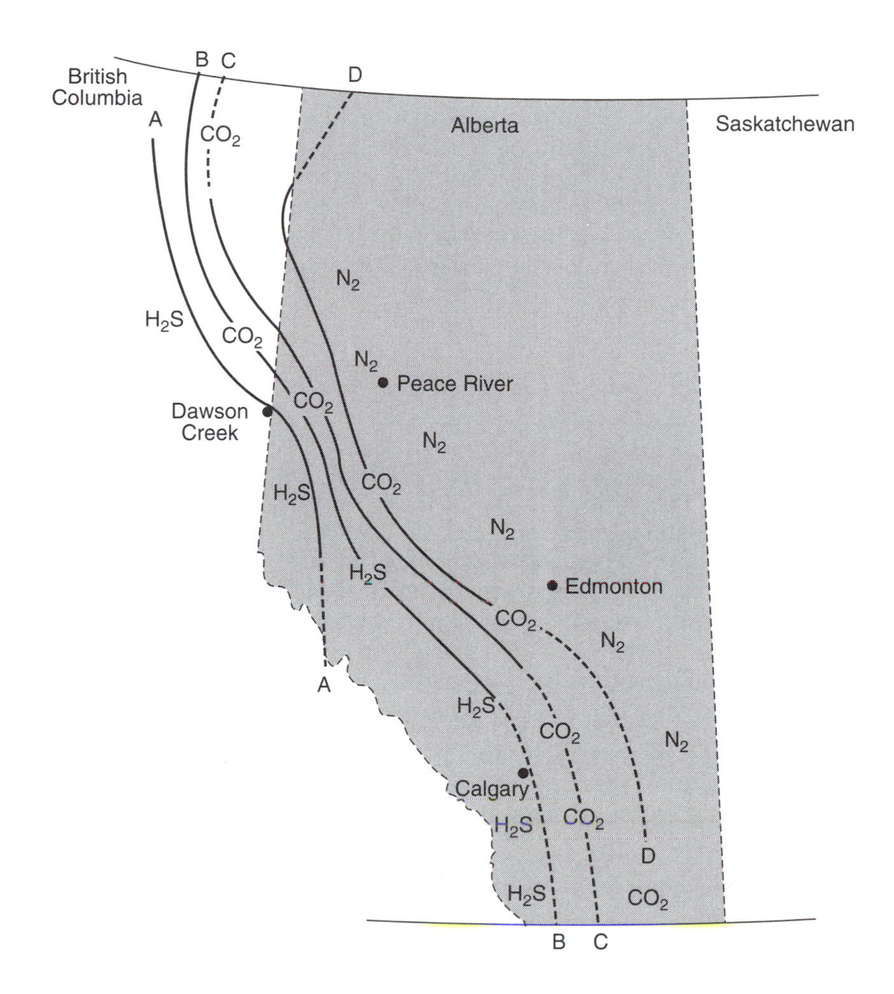

Figure 7-21

The distribution of N_2, CO_2, and H_2S relative to the hot lines of the Western Canada Basin. The change from oil to gas going west occurs at line D in the Devonian Elk Point Formation and at line A in the Mississippian. [Evans and Staplin 1971]

The Lower Cretaceous Mannville Shale of Alberta contains a type III humic kerogen that is a major generator of CO_2. McIver (1967) calculated that one m^3 ($35\ ft^3$) of this shale would generate $7\ m^3$ ($247\ ft^3$) of CO_2 at STP during thermal alteration through a 2,000 m (6,562 ft) depth interval. Carbon dioxide is common in Mannville reservoir gases paralleling the hot-line zones from Calgary to British Columbia.

The older Appalachian Basin in the eastern United States shows the same kind of trend in hydrocarbon and nonhydrocarbon gases illustrated in Figure 7-20. In the deep, deformed mobile side of the basin along the Allegheny front, dry methane predominates. Moving northwest into the central basin, there is wet gas; continuing onto the shelf area extending from New York through Ohio into Kentucky, the gas is dry again. Although noncombustible nitrogen is a minor problem in the Appalachian areas, the highest nitrogen contents are on the stable, shallow shelf of Ontario, Ohio, and Kentucky.

SUMMARY

1. The major sources of the hydrocarbon gases are (1) methanogenic bacteria, (2) all types of kerogens, (3) coal, and (4) oil in source and reservoir rocks. The major nonhydrocarbon gases—CO_2, H_2S, and N_2—are formed by both organic and inorganic processes. All known hydrocarbon gas accumulations are biogenic in origin in that they come from the decomposition of organic matter in the earth's crust. No known abiogenic methane accumulations exist based on stable isotope measurements.

2. About 20% of the methane in conventional reservoirs worldwide is bacterial in origin. From 40 to 55% is thought to come from the thermal decomposition of oil in reservoir and nonreservoir rocks and in coal. The remaining 25 to 40% is directly from the thermal decomposition of kerogen.

3. The wet gases, ethane, propane, and the butanes are formed primarily in the oil window, whereas methane forms throughout the entire sedimentary column, bacterially during diagenesis and thermally during catagenesis and metagenesis. This difference in the temperature of formation of methane and the wet gases causes a vertical distribution in many sedimentary basins of shallow dry gas on the stable shelf in the diagenesis zone and wet gas in the deeper catagenesis zone, underlain by deep dry gas in the metagenesis zone in the deepest part of the basin.

4. The source and maturity of reservoired gases can be evaluated by the carbon isotopic and molecular compositions of the individual gases. Bacterial gases generally are >95%, methane with the $\delta^{13}C$ of the methane ranging between –60 and –80‰. Wet gases formed during early catagenesis at the very beginning of the oil window have a wide spread in $\delta^{13}C$ of more than 20‰ between methane and *n*-butane. At the postmature end of the oil window, this difference shrinks to less than 10‰. Also, at higher maturities the gases become heavier with increasing amounts of ^{13}C.

5. Condensates are reservoired gases that condense to liquids at the surface. They originate from either the thermal decomposition of oil in the reservoir or the evaporative fractionation and gas-stripping processes involving vertical

migration. Condensates formed by the latter processes may be reservoired at a variety of depths extending from postmature to immature sediments.

6. Carbon dioxide has three important sources: (1) from the thermal decomposition of kerogen in and above the oil window, (2) from the chemical reaction of kaolinite and carbonates at temperatures >100°C, and (3) from the very high temperature decomposition of carbonates heated by magma.

7. Hydrogen sulfide in reservoirs has two major sources: (1) the thermal decomposition of high-sulfur oils deeper than the oil window and (2) the thermochemical reduction of sulfate in pore waters accompanied by the oxidation of reservoired hydrocarbons. Some H_2S also comes from the thermal degradation of kerogen and coal.

8. Nitrogen has three sources: (1) the oxidation of ammonia in the pore waters of sedimentary basins (ammonia comes from the maturation of organic matter, particularly coals); (2) the atmosphere; and (3) mantle outgassing. Nitrogen, hydrogen, and helium form independently of petroleum, but because of their small molecular size and migrant behavior, they can be indicators of fluid migration pathways in the subsurface.

9. The typical pattern of gas compositions in many sedimentary basins is nitrogen plus dry gas on the shallow stable shelf; CO_2 plus wet gas within the oil-generation window; and CO_2, H_2S, and dry gas only in the deepest part of the basin against the mobile flank.

SUPPLEMENTARY READING

Howell, D. G. (ed.). 1993. The future of energy gases. *U.S. Geological Survey Professional Paper* 1570. Washington, DC: U.S. Government Printing Office, 890 p.

Klemme, H. D. 1983. The geologic setting of giant gas fields. In C. Delahaye and M. Grenon (eds.), *Conventional and unconventional world natural gas resources.* Laxenburg, Austria: International Institute for Applied Systems Analysis, pp. 133–160.

Makogon, Y. F. 1981. *Hydrates of natural gas.* Tulsa: PennWell, 237 p.

Masters, J. A. (ed.). 1984. *Elmworth: Case study of a deep basin gas field.* AAPG Memoir 38. Tulsa: American Association of Petroleum Geologists, 316 p.

Chapter

8

Migration and Accumulation

The process of petroleum migration has been divided into two parts since the time of V. C. Illing (1933). Primary migration has been considered to be the movement of oil and gas within and out of the nonreservoir source rocks into the permeable reservoir rocks, whereas secondary migration has been described as the movement within carrier rocks and reservoir type rocks, leading to a petroleum accumulation. Today it is recognized that the process is more complicated because there is extensive movement through many fractured source rocks, which can act in some cases as both source and reservoir. Examples are the Altamont field in fractured Green River shales of the Uinta basin, Utah; the Florence field in fractured Pierre shale of Colorado; the Austin Chalk production in Texas; and the Big Sandy production from the fractured Devonian brown shales of Kentucky and West Virginia. Also, migration occurs along faults and joints that are neither source nor reservoir rocks. Consequently, *primary migration* is more correctly defined as any movement within the fine-grained portion of the mature source rock, and *secondary migration* is any movement outside of it. *Tertiary migration* is the movement of a previously formed accumulation.

Primary oil migration in source rocks with 2% or more TOC occurs initially as a bitumen that absorbs water and expands into the micropores and bedding plane partings to form a water-saturated bitumen network (Lewan 1993b). As the bitumen partially decomposes to oil, the latter mixes with gas (mainly $CH_4 + CO_2$) from the kerogen and migrates out as an oil–gas mixture. This may be as a single or separate phase, depending on the pressure–temperature conditions. The generation causes the migration. The generation and migration of light oil, condensate, and gas also may come from low, < 2%, TOC source rocks without an apparent bitumen intermediate, particularly in deltaic sediments. Finally, migration in solution or by diffusion also occurs with the smallest and most soluble molecules, such as methane, ethane, benzene, and toluene. These latter processes are not important for migrating an entire petroleum, but they can transport large volumes of gas.

Although most petroleum migrates as a separate phase from water, the water is still an important factor influencing the direction, distance, and areal extent of petroleum migration. As A. I. Levorsen (1967) once stated, "The geology of petroleum is essentially the geology of fluids." What has changed since Levorsen's time is the recognition that the generation of oil and gas is a more significant factor in the expulsion of petroleum from the source rock than mechanical compaction. Compaction, enough to influence primary migration, does not occur in many shales by the time the depths of oil generation are reached, as will be explained later (Hunt et al. 1994). Also, oil and gas are expelled from 2 cm chunks of source rocks in hydrous pyrolysis experiments in which compaction plays no role (Lewan 1993b). At the depths of the oil-generation window, the very localized high-pore pressures caused by the generation of liquid and gaseous hydrocarbons plus CO_2 force the fluids along bedding planes and through hydraulic fracture systems until they reach pre-existing joints, faults, and fractures where secondary migration can proceed. The dominant factors controlling the direction of fluid movement beyond the source rock are the existing or developing fracture patterns plus chemical changes that influence permeability (pressure solution, dissolution, recrystallization, and cementation), thereby creating pathways and seals. Mechanical compaction has essentially ended in many source rocks at the depths where temperatures exceed 93°C (200°F). Primary migration is driven mainly by petroleum generation and the buoyancy of the hydrocarbon products.

Water

Water (H_2O) is a liquid at standard temperature and pressure (STP). In contrast, other major elements in the life cycle–carbon, hydrogen, nitrogen, and sulfur–form gaseous hydrides, namely, CH_4, H_2, NH_3, and H_2S. Why is water a liquid? The reason is that the hydrogen of a water molecule has such a powerful affinity for oxygen that it forms temporary hydrogen bonds with the oxygens of neighboring water molecules. This causes the water to be a liquid polymer at STP.

The polymerization of water molecules causes liquid water to have many anomalous physical properties, compared with other substances. For example, it has the highest heat capacity of all solids and liquids at STP. It conducts heat more readily than any other liquid. Its surface tension, dielectric constant, and latent heat of evaporation are the highest of all liquids. In general, water dissolves more substances and in greater quantities than any other liquid. Its maximum density occurs at temperatures above the freezing point. These

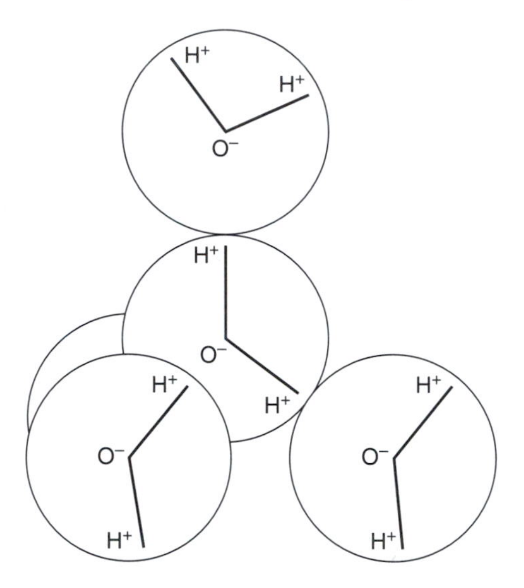

Figure 8-1

The tetrahedral structure of water molecules, forming a polymer.

anomalous properties of water compared with those of other liquids are important to chemical, biological, and geological processes in nature. For example, the high heat capacity results in a large transfer of heat by water movement, which prevents extreme ranges in temperature. Our uniform body temperature is maintained through the heat transfer of fluids that are mostly water.

Water is mostly monomeric in steam, the gaseous form. In its solid form, ice, the water molecules are arranged in a tetrahedral structure, as shown in Figure 8-1. When ice melts, only about 15% of the hydrogen bonds are broken. Most of the water remains structured as tetrahedrae. For example, the dielectric constant, which is a property highly dependent on the spatial disposition of atoms and charge sites, changes very little. It is about 74 in ice and 88 in water at 0°C (Dorsey 1940). Other physical properties, such as the X-ray radial distribution curve for water and the densities of water and ice, indicate that the tetrahedral arrangement still exists in the water phase. The affinity of hydrogen for oxygen also expresses itself in the structuring of water next to mineral surfaces. The hydrogens of water bond to the oxygens of the mineral surface. This causes the viscosity of water to increase markedly within a nanometer (10^{-9} m) of the mineral surface. (For more information on structured water at mineral contacts, see Hunt 1979, pp. 188–190.)

Oil-Field Waters

Water occupies all the pore spaces in sedimentary rocks below the water table except those containing oil, gas, or bitumen. The water contained in the small spaces between the mineral grains is called *interstitial* or *pore water*. Interstitial waters contain ions in solution in varying concentrations depending on their source and environment of deposition. As Table 8-1 shows, the bulk of the ions are contributed by seawater. River, lake, and rain waters act primarily to dilute the seawater.

The salinities of surface and subsurface waters are generally expressed in terms of milligrams per liter (mg/l) or grams per liter (g/l) of total dissolved solids (TDS.) Fresh water has < 1 g/l, brackish water 1 to 10 g/l, saline water 10 to 100 g/l, and brine > 100 g/l (Carpenter 1978). For comparison, seawater is about 36 g/l (Table 8-1).

Surface salinities much higher than seawater are found in interior basins wherever the rate of evaporation exceeds the rate of water input. According to Hite and Anders (1991), *vita* and *metasalinities,* in which carbonates form, range from 35 to 142 g/l of TDS. Anhydrites form in *penesaline* environments (142 to 250 g/l), anhydrite and halite in *saline* environments (250 to 350 g/l), and halite and potash in *supersaline* environments (350+ g/l). Subsurface waters are divided into three major classes based on their history. *Meteoric* waters include waters that are currently part of the hydrogeologic cycle, that is, surface and ground waters moving down from the surface through permeable strata.

TABLE 8-1 Chemical Composition of Various Waters, g/l

			Mud pore water[a]		*Oil-field waters*[b]	
	River water	*Seawater*	*9.5 m (31 ft)*	*335 m (1,099 ft)*	*1,570 m (5,151 ft)*	*1,814 m (5,951 ft)*
Cations						
Na^+	0.006	10.8	10.5	7.8	53.9	57.0
K^+	0.002	0.4	0.4	0.3	–	–
Mg^{2+}	0.004	1.3	1.3	0.4	2.1	2.2
Ca^{2+}	0.015	0.4	0.4	2.7	15.0	18.0
Anions						
Cl	0.008	19.4	19.6	23.4	115.9	126.0
SO_4^{2-}	0.011	2.7	2.8	2.8	0.1	0.07
HCO_3^-	0.059	1.4	0.1	0.05	0.05	0.06
Total	0.105	36.4	35.1	37.4	187	203

[a]Interstitial water in deep-sea drilling carbonate mud samples obtained in Hole 292 east of the Philippines. Sediment depth below seafloor shown in meters and feet. Sodium by difference (White 1975).

[b]Pennsylvanian sands of the Tonkawa and Morrow formations in Texas and Oklahoma (Dickey and Soto 1974). Samples were not analyzed for potassium.

Connate waters are fossil reservoir fluids in closed hydraulic systems that have not been in contact with the surface since deposition. They are not fossil seawater. They almost never have the chemistry of the original seawater, since they undergo considerable chemical modification and migration after burial. Instead, the term is used mainly to characterize fossil waters that have neither a meteoric nor a juvenile signature. *Juvenile* waters are those still coming from the mantle, but they are harder to identify with assurance, since they may be recycled meteoric waters.

The pore waters of clastic sediments undergo several changes in chemical composition during the first few hundred meters of burial. As listed in Table 8-1 under mud-pore water, the ratio of magnesium to calcium changes from 3 at 9.5 m to 1/7 at 335 m. The largest change in the composition of pore waters is the increase in salinity in reservoir sandstones with depth, which can range from 2 to 50 g/l per thousand feet (305 m) of increasing depth. This increase for various areas is shown in Figure 8-2 (Dickey 1969). The probable cause of this increase in salinity is the dissolution of salt from either salt diapirs or evaporite beds originally deposited in the basin. A less plausible explanation is the

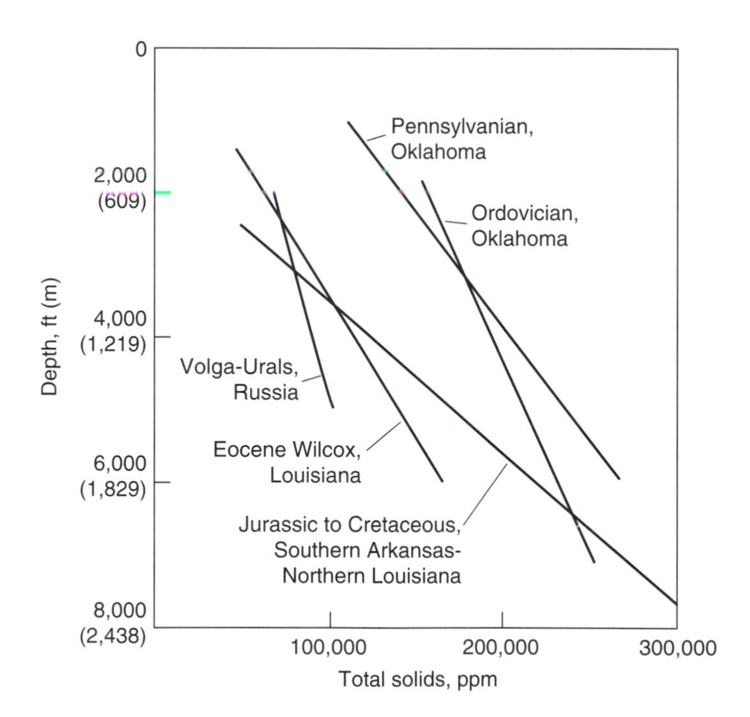

Figure 8-2

Change in salinity of reservoir waters with depth. [Dickey 1969]

membrane filtration hypothesis, which requires substantial hydraulic gradients to force water molecules to migrate through the semipermeable membranes of shales. After the first few thousand feet of burial, water tends to leave the shales via numerous vertical regional fracture systems, which provide a far more permeable pathway than the shale matrices (Gretener 1986). Although there is probably some membrane filtration, it does not appear to be the dominant mechanism increasing pore-water salinities in most basins.

In regard to the U.S. Gulf Coast, Hanor (1984) noted that the increase in salinity shown in Figure 8-2 was primarily due to increases in Na and Cl concentrations with depth. Other ions such as magnesium, bicarbonate, and sulfate did not show the linear trend of Na and Cl. Hanor (1984) concluded from his studies that molecular diffusion was the principal mechanism for transporting dissolved salts upward through the basin. He made first-order estimates of the magnitude of this diffusion from mid-depth conditions (50°C, 120°F) and concluded that the vertical velocity of the dissolved salt was approximately 9 mm/yr (0.35 in/yr). He estimated it would take only 34,000 years to migrate dissolved NaCl from the top of the Jurassic Louann salt to the base of the freshwater zone in the northern Louisiana and southern Arkansas areas of the U.S. Gulf Coast.

Jensenius and Munksgaard (1989) used fluid inclusion data along with carbon and oxygen isotope compositions of a chalk reservoir in the Danish central trough to evaluate the effects of hot-water flushing of the reservoirs. This flushing system existed around salt diapirs and was related to the expulsion of overpressured fluids from the surrounding sediments. Their fluid inclusion salinities indicated that flushing waters caused considerable salt dissolution, thereby enriching the ascending waters with both sulfate and NaCl. Similar salinity anomalies have been observed in the Tertiary section of the Louisiana Gulf Coast. Hanor and Bailey (1983) found that the saltiest waters in this section were in thick, sandy sequences in proximity to salt domes. They traced plumes of elevated salinity upward and laterally away from some of these domes, indicating that dissolution and mass transport were currently under way. Several investigators have reported shale salinities to be much less than that of the adjacent sands (see Hunt 1979, p. 191, for details). For example, the Paleozoic Bartlesville shale of Oklahoma contains 41.9 g/l TDS, whereas the associated sand contains 167 g/l. This implies that the diffusion of salts in the subsurface is not penetrating the matrix of the shales.

Halbouty et al. (1970) compared the oil-field waters of 260 giant oil and gas fields in the world. About half these fields contain waters with salinities less than 40 g/l; a fourth have salinities greater than 150 g/l; and the remaining fourth are in the intermediate range. The mean salinity in sandstone reservoirs was about 27 g/l, compared with around 90 g/l for carbonate reservoirs.

Salinities in high-pressure compartments may be lower or higher than salinities in overlying normally pressured sediments. In the Tertiary of south Louisiana, formation waters contain an average of 80 g/l TDS at 4,000 ft (1,220 m) in the normally pressured section and 40 g/l in the overpressured compartment below 10,000 ft (3,049 m). In contrast, in the Upper Jurassic carbonate reservoirs in many areas of southeast Turkmenia, high-pressure brines between salt beds contain from 348 to 546 g/l TDS at depths ranging from 2,594 to 3,720 m (8,508 to 12,202 ft) (Novokshchenov 1982). The high salinities in the small high-pressure compartments of Turkmenia are comparable to those in the Jurassic Smackover of the U.S. Gulf Coast (Forgotson 1979). Evaporites flow under overburden pressure, creating an isolated sealed-in reservoir compartment. As the evaporites dissolve and raise the salinity, more flows into the pores and so raises the pressure.

Morton and Land (1987) observed salinities from 8 to 250 g/l TDS in the overpressured Oligocene sands of the Texas Gulf Coast. Ionic and isotopic analysis indicated that the high salinities are generated by the dissolution of salt. This was mainly either the local dissolution of diapirs that penetrated the Tertiary section or the vertical transport of brines from deep-seated salt dissolution. Low salinities indicated dilution with water from the smectite to illite transformation and mineral dehydration reactions. Low salinities also were associated with areas where diapiric salt structures were uncommon. Thermal convection within the overpressured zone appeared to concentrate the less dense, low-salinity water and oil in zones of secondary porosity just below the top of the overpressured zone. Nevertheless, salinity differences across the boundary between the geopressured and the hydropressured zones often exceeded 100 g/l TDS.

Porosity of Shales

Rock porosities decrease with depth owing to compaction and diagenesis. Mechanical compaction is the reduction in the bulk volume of sediments, mainly as a result of the compressional stresses caused by the overburden load. Chemical compaction and diagenesis involve the reduction of rock volume due to reactions such as intergranular pressure solution (IPS,) stylolitization, and cementation. Rock porosity is defined as the ratio of the interstitial volume to the total rock volume. The porosities of both argillaceous and calcareous rocks decrease mainly because of mechanical compaction during about the first 2 or 3 km of burial. After this, mechanical compaction phases out, and chemical reactions become increasingly important.

Two kinds of porosity-depth curves have been published over the past several decades. Composites of porosity-depth data from several wells over a large area generally show an exponential curve of porosity decreasing very rapidly at the surface and more slowly with increasing depth. Several examples, along with a discussion of their variability, were published by Rieke and Chilingarian in 1974 (p. 42) and Hunt in 1979 (p. 202). The second type of curve that shows up in many single-well analyses indicates that porosity decreases exponentially down to a porosity of 20 to 30%, after which it tends to follow a straight-line decrease to a porosity of 3 to 15%. At greater depths there frequently is no further decrease in porosity, indicating no compaction, particularly at porosities < 10%.

Hedberg (1936) was the first to observe this latter phenomenon in the Eastern Venezuelan Basin (Figure 8-3). Most of his data were from a single well supplemented with a few shallow and deep data points from two other wells in the same area. Hedberg explained the lines in Figure 8-3 as follows: Mechanical compaction and dewatering reduce the initial 70% porosity to about 34%. During the next leg, from 34 to 10% porosity, mechanical effects such as particle deformation and expulsion of bound water from mineral lattices gradually decrease, and chemical diagenesis becomes more important. At depths where primary porosities reach 10% (around 3 km) depth, mechanical compaction has ceased. Further increases or decreases in porosity come from pressure solution, cementation, and hydraulic and tectonic fracturing.

In 1959, Storer published dry-bulk density–depth curves for several wells in the Po Basin of Italy. They showed some of the same linear changes reported by Hedberg. Linear shale compaction trends also were reported by Korvin (1984) and Wells (1990).

During the 1970s and 1980s the Amoco Production Company measured dry-bulk densities and porosities directly on shale cuttings and cores from several hundred wells in the U.S. Gulf Coast Basin. An example of their work is shown in Figure 8-4. It contains the shale dry-bulk density and porosity measurements for a normally pressured well in Frio County, Texas. Note that

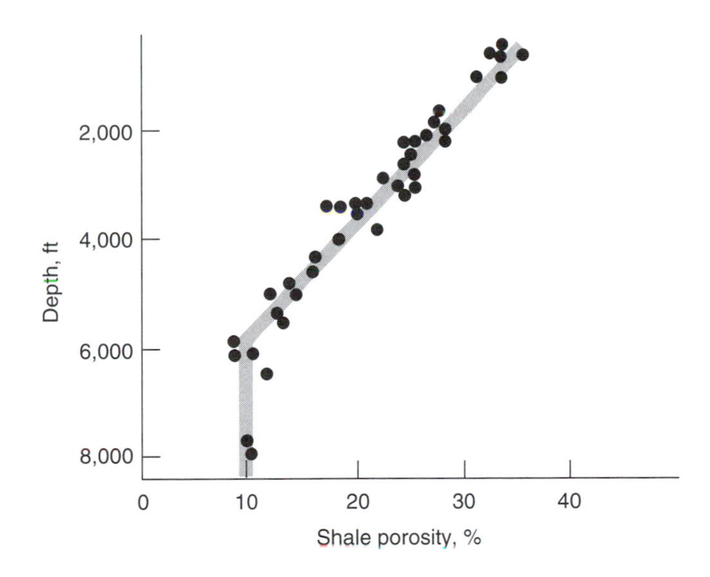

Figure 8-3

Graph illustrating a two-stage linear decrease in porosity with depth for shales from the Eastern Venezuela Basin. [Hedberg 1936]

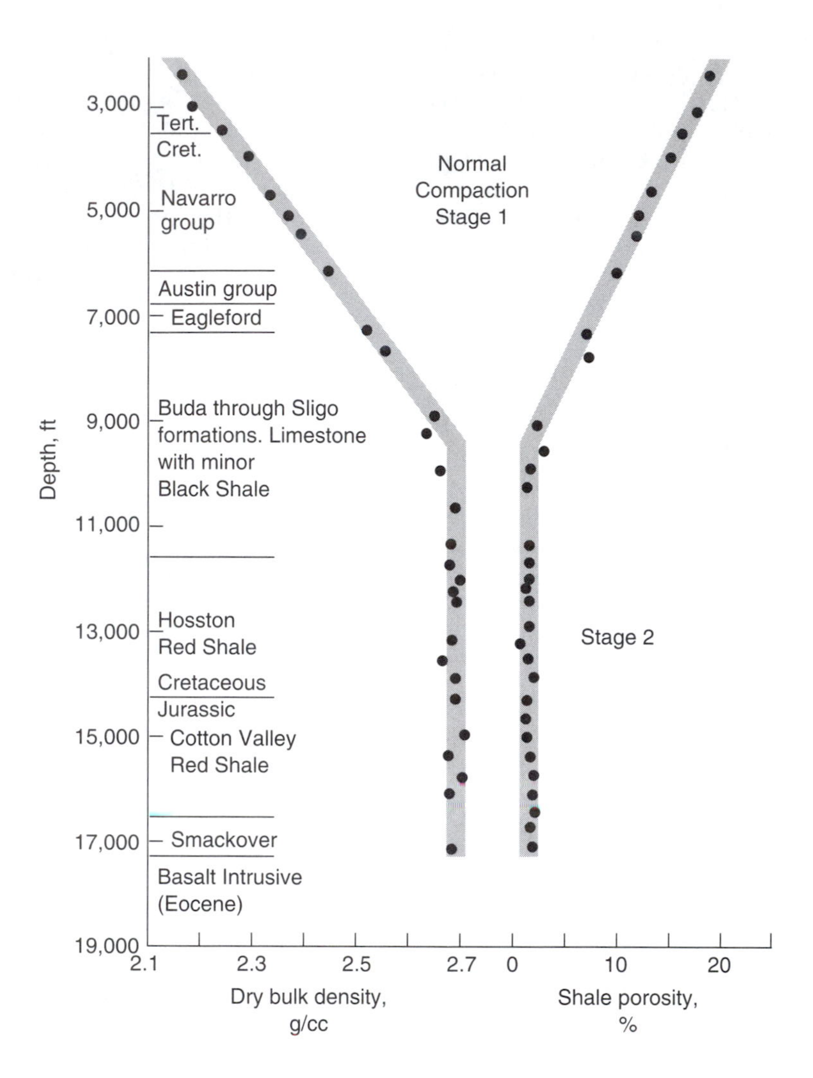

Figure 8-4

Shale dry-bulk density and porosity versus depth shown as two linear segments in angular end-to-end contact for a normally pressured well in Frio County, Texas. There is no slope in the Stage 2 segments, indicating no compaction. [Powley 1993]

starting at a depth of 2,000 ft, the density increases and the porosity decreases along straight-line segments until they reach relatively constant values of 3% porosity and 2.7 g/cc density at a depth of about 9,500 ft. From here to a total depth at 17,500 ft in a basalt intrusive, there is no systematic decrease in porosity, indicating that no compaction is occurring. This two-stage compaction model is defined as the normal compaction curve for this well. The low porosities are thought to be due to the rocks containing mainly carbonates and red shales with kaolinite, both of which have very low mineral surface areas. If this well contained water-adsorbing mixed-layer clays, the two straight-line segments would still occur, but Stage 2 would be displaced to a higher porosity, depending on the content of such clays.

Some geologists have speculated that the Stage 2 density and porosity lines in Figure 8-4 represent undercompaction. However, undercompacted shales are universally within overpressured compartments, and there is no evidence that this well was ever overpressured (Powley 1993). Overpressures can be recognized by drill-stem tests, mud weights, and resistivity logs. This Texas well was drilled with 9 lb/gal mud to total depth. There was persistent lost circulation in the well and no evidence of a shift from normal to low resistivities. All these observations indicated normal hydrostatic pressure through the entire well.

The depth below which there is no further compaction appears to vary with the internal surface areas of the shale (clay mineral content). Shales with small internal surface areas (e.g., those composed of very fine-grained quartz, carbonates, or kaolinite) stop compacting at porosities of ~3%. Shales with 20% mixed-layer water-adsorbing clays stop compacting at porosities of ~10%. The cessation of compaction is not related to overpressuring. This phenomenon occurs with normally pressured shales. The two-stage, linear, composition-dependent compaction is thus a "normal" compaction trend. In Stage 2, the cessation of compaction may be due to the development of a rigid, silicified framework in the shales, partial cementation and recrystallization, or the possibility that the very small pores contain immobile, structured water that is not released except under extreme pressures.

The depth of the change from Stage 1 to Stage 2 can be ascertained only from individual well data or closely spaced wells in the same pool. If data from several oil fields were lumped together, as in computer composites of data, the resulting curve would not exhibit the straight-line detail of single well profiles and would not be useful for determining the depth of the shift from a linear decrease in porosity to no decrease (no compaction).

A statistical analysis of porosity data for shales from twenty wells throughout the Texas and Louisiana Gulf Coast was carried out to determine whether the shale porosity data fit a two-stage linear curve better than a one-stage exponential curve. Correlation coefficients (R) and coefficients of determination (R^2) were obtained for both curve fits. If the porosity is decreasing exponentially with depth rather than in two linear stages, one would expect to see a better coefficient of determination (R^2) for the former, compared with the latter. However, the two-stage linear plot fits the data much better than the commonly used one-stage exponential plot in all twenty wells. For example, the R^2 one-stage exponential coefficient for a well in DeWitt County, Texas, is 0.834, compared with the combined two-stage linear coefficient of 0.955 (Hunt et al. in press).

In addition, it was determined that there is no slope in the Stage 2 porosity/depth profile within the 95% confidence level for nineteen of the twenty wells. This indicates that no compaction is occurring in Stage 2.

The top of Stage 2 shows no relation to the lithology in Figure 8-4. In Figure 8-5, however, the top coincides with the top of the Jurassic in the normally pressured COST B-2 well offshore New Jersey. Within Stage 2, the shale porosities stay relatively constant in the 4-to-5% range to total depth (Hunt et al. in press).

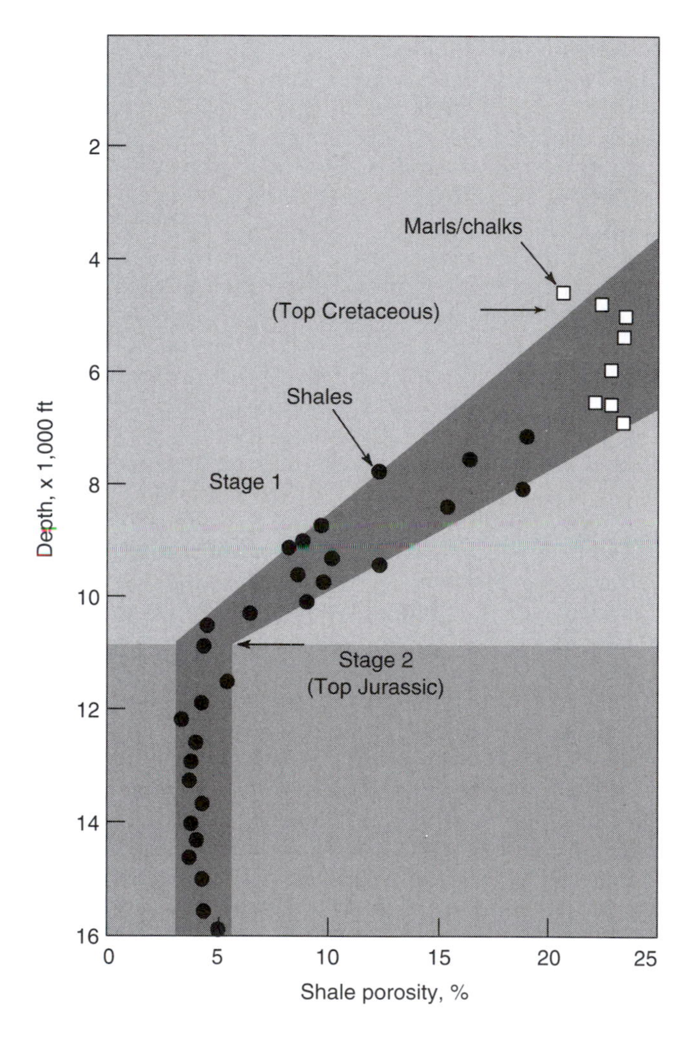

Figure 8-5

Two-stage porosity profile with depth for the normally pressured COST B-2 well offshore New Jersey. Stage 2 shows no systematic decrease in porosity with depth, indicating no compaction in Jurassic and older rocks. [Hunt et al. in press]

The correlation between the minimum porosities reached in Stage 2 and the mineralogy of the shale are evident when comparing shales from different areas, but they are not readily apparent when comparing shales in the same well. For example, Figure 8-6 compares the porosity and smectite–illite content of shales from two COST wells, the Baltimore Canyon well and a well offshore Texas, both of which are normally compacted. Clearly, the B-2 well has an average lower shale porosity and correspondingly lower smectite–illite content than the Texas well, even though the individual data are somewhat scattered.

Chilingar and Knight (1960) demonstrated this relationship in the laboratory, by subjecting pure clay minerals to 200,000 psi (1,379 MPa) pressure. Kaolinite with a small surface area reached a minimum porosity of 8%, whereas smectite with a large surface area stopped at 18% porosity.

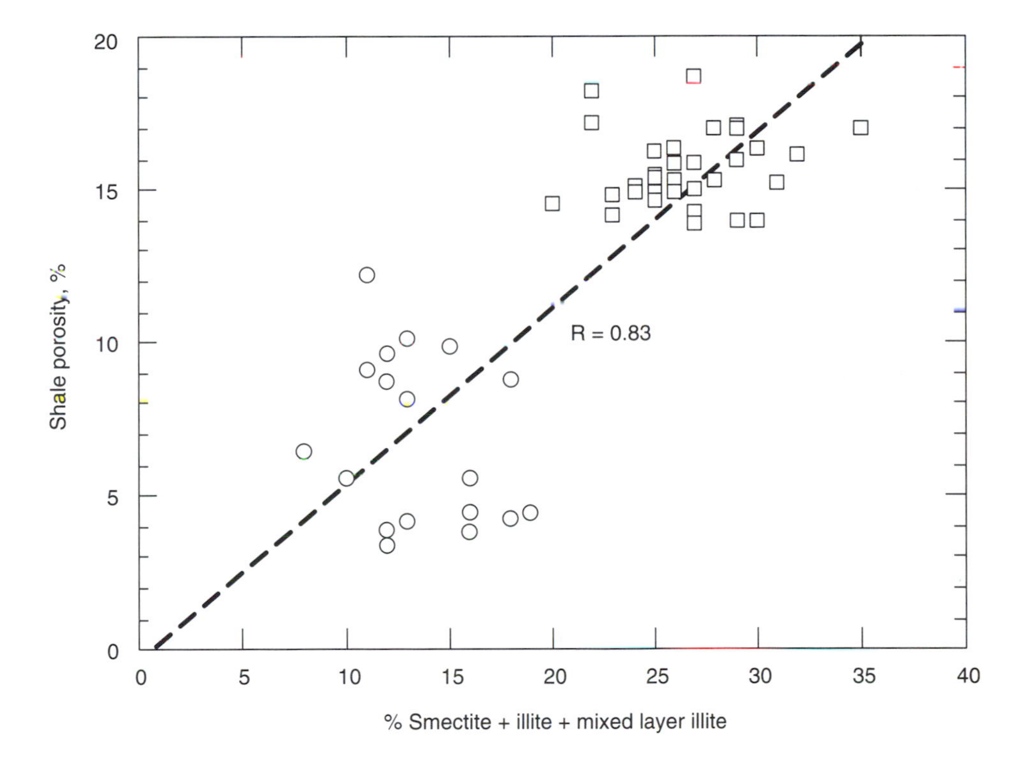

Figure 8-6

Correlation of shale porosity and smectite–illite content in Stage 2 shales that are normally pressured. The circles are from the COST B-2 well in Figure 8-5. The squares are from a well offshore Texas. [Hunt et al. in press]

The shift from Stage 1 to Stage 2 occurs within a narrow temperature range, with the median around 93°C (200°F). Figure 8-7 shows the temperatures at the top of Stage 2 for sixty-five wells in the Gulf Coast (Hinch 1980). The temperatures are higher in young shales and somewhat lower in older shales. At depths where temperatures are greater than those shown, there is no significant decrease in porosity or increase in density in most normally compacted shales. Compressibility would cause a porosity loss of less than 1% between about 3 and 6 km depth (Bradley 1986). Such a loss could not be measured, considering the precision of most porosity measurements. Figure 8-7 also suggests that a kinetic energy reaction may be involved in the changeover from compaction Stage 1 to Stage 2. This occurs at temperatures as high as 230°F (110°C) over short geologic times and as low as 181°F (83°C) over long geologic times.

The compaction profiles shown in Figures 8-3, 8-4, and 8-5 represent normal compaction in various areas. (Additional examples of porosity–depth profiles are in Hunt et al. in press.) Undercompaction due to rapid burial without adequate drainage can cause a bulge in higher porosities and lower densities, starting in Stage 1 or 2 and continuing until the shales dewater. Undercompaction is discussed in connection with overpressures in Chapter 9. These patterns develop temporary variations from the normal compaction curves, but most of them tend to disappear with time and further burial.

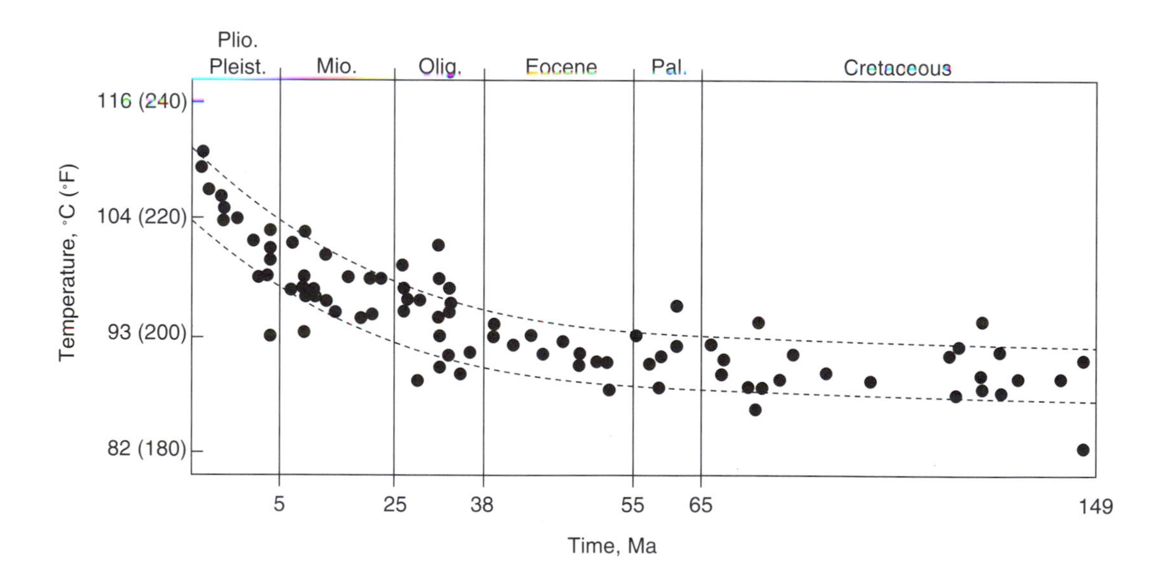

Figure 8-7

Temperatures at the top of Stage 2 where mechanical compaction ceased for sixty-five shales in the Gulf Coast province. [Hinch 1980]

Secondary Porosity

The decrease in primary porosity caused by mechanical compaction tends to be regional, whereas secondary porosity is generally localized. For example, at around 5,000 m (16,400 ft) in the overpressured section of the Venture field in the Scotian Basin off Nova Scotia, the shales are normally compacted with porosities of 0.9 to 9% (Katsube et al. 1991). Two highly overpressured gas fields near the Venture field are Arcadia and South Venture. They both are normally compacted with about 8 to 9% porosity and pressure–depth ratios of about 0.9 psi/ft between 5,600 and 5,800 m (18,373 and 19,024 ft) (Drummond 1992). A detailed petrographic study by Jansa and Noguera Urrea (1990) also confirmed that the calcareous shales and limestones in the Venture area are highly indurated and fully compacted. Illite crystallinity measurements show an advanced late stage of shale diagenesis. There are stylolites in some of the carbonates, indicating that there was considerable overburden compaction before the advent of overpressures. Also, the sandstones that did not develop secondary porosity are normally compacted. They show grain-to-grain solution and suturing between grains. All these are characteristics of well-compacted rocks.

However, sandstone porosities within the Venture field itself at 5,000 m are 9 to 28%. This is much higher than those of the sandstones and shales in the adjacent areas just mentioned. Thin-section analyses of Venture sandstones by Jansa and Noguera Urrea (1990) revealed that these sandstones originally were probably normally compacted with the same remnant primary porosity of about 9% observed in the Arcadia and South Venture wells. The additional porosity, raising it from 9 to 28%, was all secondary. This localized secondary porosity at Venture was formed by the dissolution of both ferroan and nonferroan cements, sparry calcite cement, feldspars, and other mineral alterations. The fact that such high porosities are localized in selected sandstone bodies indicates that these are pathways of greater permeability subject to chemical diagenesis.

The Venture field area is an example of normal compaction occurring regionally to about 8 to 9% porosity followed by or accompanied by localized chemical diagenesis. Dissolution increased the sandstone porosity to 28% in some areas, and cementation and mechanical compaction changed it to zero in others. Several sandstones in the area have essentially no porosity, because of pressure solution and silica cementation. The highly overpressured section in the Scotian Basin field area is comparable to most overpressured sedimentary sections in the world that are not mechanically undercompacted (Powley, personal communication).

Proshlyakov's (1960) study of rock porosities in the Jurassic Cis-Caucasus of Russia showed the same difference between shales and sandstones as observed at Venture field. The shales were normally compacted to the total depth drilled. The adjacent sandstones, however, were normally compacted only to about 10,000 ft (3,049 m), after which the porosities varied widely because of chemical reactions causing both cementation and dissolution locally. Shale porosities ranged from 4 to 7% at 3,354 m (11,000 ft), whereas sandstones at the same depth had porosities of 4 to 28%. The wide spread in sandstone porosities compared with the shale porosities at the same depth is further evidence that fluid circulation is causing more chemical alterations in the sandstones than in the shales. Fluids move in compacted shales primarily through fractures. Volkmar Schmidt observed secondary porosity in silty shales of the Beaufort–Mackenzie Basin. Porosity was created mainly by leaching along microfractures.

Reservoir sandstones at great depths generally show greater scatter in individual porosities than shales because they represent major channels of fluid migration. Loucks et al. (1984) found sandstone porosities ranging from 3 to 30% at depths of 8,000, 11,000, and 15,000 feet (2,439, 3,354, and 4,573 m) in the Lower Tertiary of the Texas Gulf Coast. Over half the total porosity below 10,000 feet (3,049 m) was secondary, owing to the chemical dissolution of framework minerals and cements and also owing to mechanical stresses that promote fractures.

Franks and Forester (1984) observed a strong correlation between the carbon dioxide content of Wilcox reservoir gases and the occurrence of secondary porosity in the sandstone reservoirs, as illustrated in Figure 8-8. Secondary porosity increased until it reached 100% of total porosity at 16,000 ft (4,878 m). The most rapid increase in secondary porosity occurred between 6,000 and 8,000 ft (1,829 and 2,439 m). This is the same depth interval in which there is a rapid increase in mole% CO_2 in reservoir gases. This suggests that CO_2 is causing the increase in porosity from the dissolution of carbonate cements and feldspar. The source of the CO_2 is probably both organic and inorganic, the latter due to deep high-temperature reactions, as discussed in Chapter 7.

Porosity of Carbonates and Evaporites

Carbonate rocks lose porosity with burial by a variety of processes, including mechanical compaction, intergranular pressure solution, stylolitization, cementation, recrystallization, dissolution, and fracturing. Mechanical compaction reduces porosity to about 30%, at which point it becomes less effective because of the high differential stresses required to break the grain-supported fabric. Chalks lose porosity faster than do shallow water carbonates. Chalks from the North Sea, the U.S. Gulf Coast, and the U.S. interior basins reach porosities of 10% at about 2 km sediment depth, whereas South Florida platform limestones required about 3 km of burial to reach 10% porosity (Scholle and Halley 1985).

Chemical compaction causes most of the porosity changes in carbonates at porosities below 30%. Pressure solution is a sequential process of dissolution at grain contacts with propagation along contacts and removal of dissolved ions by fluid flow and diffusion followed by reprecipitation. This process can considerably reduce porosity in a variety of carbonate rocks and sandstones. It is inhibited by clay minerals and the cementation of grain contacts by silica or calcite.

Because of such chemical changes, the porosity of carbonates does not change uniformly with depth and does not exhibit the no-porosity decrease of

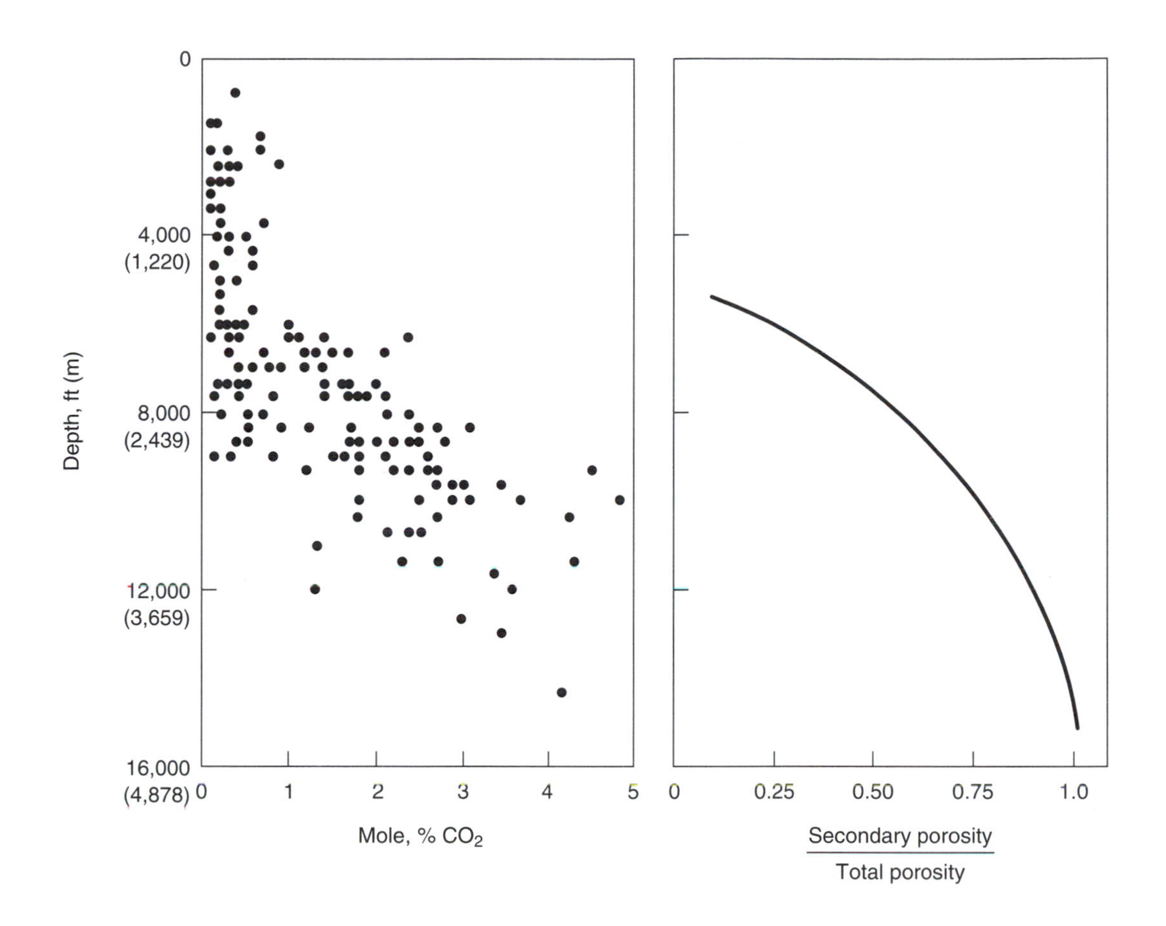

Figure 8-8

The relation between the secondary porosity of Wilcox reservoir sandstones, U.S. Gulf Coast, and CO_2 in reservoir gases. [Franks and Forester 1984]

shales (Stage 2, Figures 8-4, 8-5, and 8-6). No compaction stages can be defined for carbonates because of these irregularities.

Limestones generally have a lower porosity than do interbedded shales. For example, Hunt et al. (in press) reported that the porosities of shales below 12,000 ft (3,659 m) in a well in Lee County, Texas, varied between 10 and 16%. Interbedded limestones at the same depths had porosities of 2 to 7%.

Dunnington (1967a) estimated that 2,000 to 3,000 ft (610 to 914 m) of overburden is required to initiate the pressure solution of carbonates to form stylolites. Stylolites commonly cause a 10% reduction in rock volume after lithification. The more extensive stylolites of large amplitude tend to form during late stages of diagenesis. All of them concentrate the organic matter and insoluble minerals that were present in the dissolved carbonates.

Chemical diagenesis controls essentially all porosity changes in carbonate rocks well before the depth of the oil-generation window is reached. Carbonate porosities at these depths (°3 km) are generally less than 20% (Scholle and Halley 1985). Cementation becomes important in these deeper zones, but so does pressure solution, because of the presence of increasing amounts of inorganic H_2S and CO_2. Both gases dissolve in the pore fluids, creating acidic solutions that cause increased dissolution of carbonate rocks.

Coogan (1970) noted that 50 to 90% of the primary porosity in the Jurassic Smackover of the Mississippi salt basin had been eliminated where the Smackover is buried to 10,000 ft (3,049 m). At greater depths, secondary porosity increases locally, probably due to variations in the influx of CO_2 and H_2S. This increase in secondary porosity causes a higher total porosity in the deep Smackover, compared with platform carbonates in areas such as south Florida. Smackover reservoir carbonates range from 9 to 23% porosity at about 12,000 ft (3,659 m) and 10 to 21% porosity at 18,000 ft (5,488 m). South Florida carbonates at the same depth-interval range from about 2 to 15% (Scholle and Halley 1985).

Evaporites lose porosity more rapidly with depth than do shales or carbonates. Halite has initial porosities of 30 to 50%. This diminishes to about 10% in the first 40 m (131 ft) of burial. Gypsum loses up to 38% of its initial volume and converts to anhydrite in the first 600 m (1,968 ft) of burial (Schreiber 1988).

Pore Diameters and Oil Expulsion

Rock and coal pores are classified by width according to the 1962 IUPAC classification. Macropores have pore widths greater than 50 nm (nanometers), mesopores 2 to 50 nm, micropores 0.8 to 2 nm, and ultramicropores less than 0.8 nm (Walker 1981).

Any evaluation of primary migration mechanisms needs some understanding of source rock–pore size distributions. These can be measured in the range between 2 and 60 nm by nitrogen desorption and between 3 and several hundred nm by mercury intrusion porosimetry. A third technique for confirming results of the others is small-angle neutron scattering (SANS), which covers the range from about 1 to 100 nm. Figure 8-9 shows cumulative pore-width distributions using these three techniques for the Cherokee Shale of Oklahoma (Hall et al. 1986). This is a Pennsylvanian-age source rock that entered the oil-generation window before being uplifted to its present depth of 2,921 ft (890 m). It has a porosity of 5.2% and is the source of the nearby Burbank oil field. It is well into the oil-generation window, so the size of the pores can provide insight into the ability of hydrocarbon molecules to migrate out. All three techniques agree in showing that 30% of the pore volume is in pores larger than

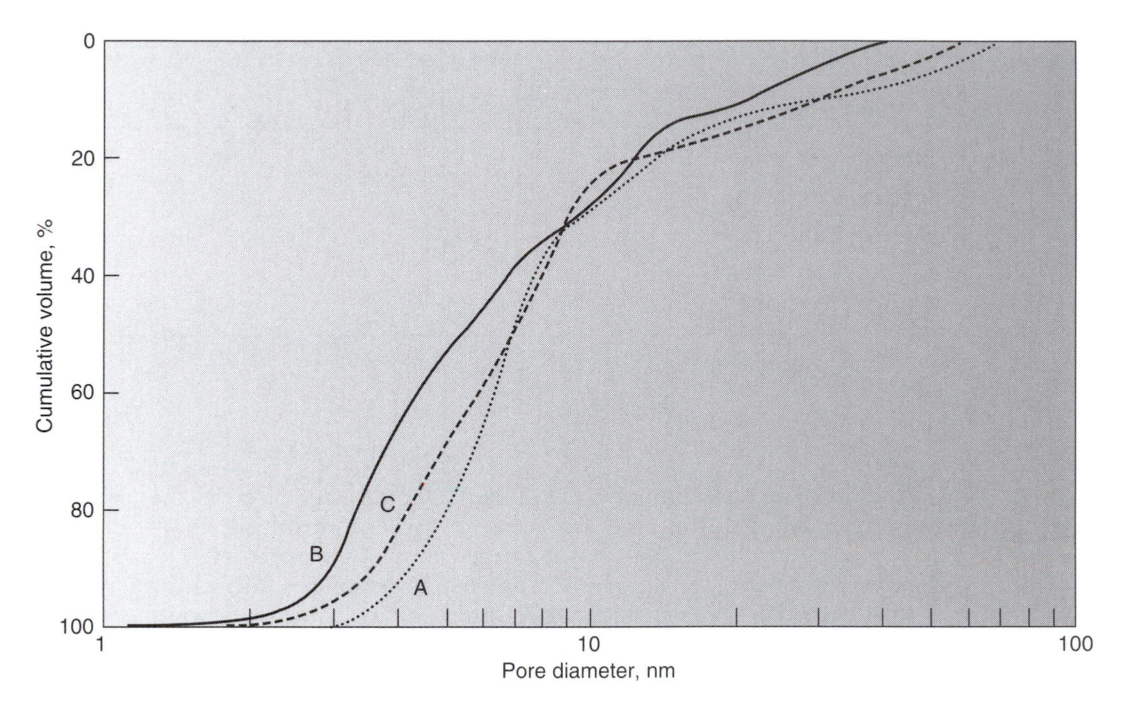

Figure 8-9

Cumulative pore width distribution for Cherokee Shale, Oklahoma. Curve A, SANS data using Vonk's method [1976]; curve B, nitrogen desorption; and curve C, mercury porosimetry. [Hall et al. 1986]

10 nm in diameter and that 70% are smaller. There is more variation among techniques in small diameters, but on the average, 20% of the pore volume is smaller than 4.5 nm, and 80% is larger.

Table 8-2 shows the median pore diameters and porosities of source rocks measured by the various techniques just described. Hall et al. (1986) found that shales with measured porosities of 5 to 10%, typical of the oil-generation window, generally have median pore widths > 5 nm. However, an overall evaluation of measurements by the different techniques indicated that a significant amount of the total porosity is present in micropores and ultramicropores (< 2 nm width) that are below the range of most analytical techniques. Also, some large pores in shales may be artifacts introduced during sample preparation (Borst and Smith 1982). This would agree with Momper's (1978) conclusion that in a typical compact shale of low porosity, more than half the pores are less than 3 nm diameter and the remainder may extend up to several

TABLE 8-2 Median Pore Diameters and Porosities of Shale Source Rocks

Shale	*Diameter, nm*	*Porosity, %*
Bakken, North Dakota[a]	5	4.3
Cherokee, Oklahoma	7	5.2
Monterey, California[b]	10	8.5
Monterey, California[b]	16	12.7
Tertiary, U.S. Gulf Coast[b]	20	15

Sources: [a]Hall et al. 1986; [b]J. Popek, personal communication.

hundred nm. The pore diameters in Table 8-2 would need to be further reduced by about 2 nm if structured water were present, resulting in median effective diameters in the 3 to 8 nm range below 10% porosity. This is without considering the percentage of unmeasurable micro- and ultramicropore (< 2 nm) porosity. Also, organic matter could occupy some of the pore space.

Table 8-3 lists the effective molecular diameters of some of the petroleum molecules that would need to migrate through the source rocks. The small hydrocarbon molecules could migrate through all but the smallest micropores. Large complex molecules would be trapped in half or more of the pores. This means that the shales would act like molecular sieves, releasing fewer of the large molecules and more of the small ones. Numerous studies have verified that the small paraffinic and naphthenic molecules migrate from shales to sandstones more easily than do the large aromatic–asphaltic molecules. Hunt (1979, pp. 499–503) cited several comparisons of oils extracted from source

TABLE 8-3 Effective Molecular Diameters of Reservoir Fluids

Molecule	*Diameter, nm*
Water	~0.3
Methane	0.38
n-Alkanes	0.47
Cyclohexane	0.48
Complex ring structures	1–3
Asphaltene molecules	5–10

Source: Welte 1972.

rocks with associated reservoir oils. These show the latter to contain twice as many paraffins and naphthenes (saturates) and less than half the asphaltic (NSO) compounds than the former. Part of this is due to preferential adsorption on the kerogen, but part is a filtration process.

More recently, Leythaeuser et al. (1987) reported a gradational change in extractable oil composition through a 15 m sampled interval of the Kimmeridge Shale source rock of the North Sea as an underlying sandstone was approached. A shale sample 15 m above the sandstone contained a C_{15+} extract with a saturate/NSO ratio of 0.5, whereas a shale next to the sandstone had a ratio of 0.2, much more depleted in saturated hydrocarbons than the shale farther away. The saturate hydrocarbons in this adjacent shale had moved prefentially into the sandstone, compared with the NSOs. The oil in the sandstone recovered on a DST had a saturate/NSO ratio of 3.6.

Leythaeuser et al. (1984) estimated expulsion efficiencies of the C_{15+} alkane fraction of oil as it migrated from thin shales to adjacent sandstones in an uplifted section of Paleocene, Firkanten Formation, Svalbard, Norway. Efficiencies were estimated by expressing concentrations of the migrated hydrocarbons as percentages of the amounts found in the center of a thick unmodified part of the shale. Efficiencies for individual normal and isoprenoid hydrocarbon expulsions from a shale only 5 cm thick into an adjacent sandstone are given in Figure 8-10. The expulsion efficiency for C_{15} is about 85%. Efficiencies for the smaller molecules would be expected to be even greater, since the peak in efficiencies is trending upward to the left of Figure 8-10. The data indicate that all of the alkanes through the diesel-fuel range (C_{18}) had no problem migrating. But large molecules in and above the lubricating-oil range ($> C_{26}$) show an expulsion efficiency of only about 20%. Considering the data on a relative basis, there is no doubt that small molecules migrate much more easily from shale to sandstone than do large ones.

Mechanisms of Primary Migration

The mechanisms of hydrocarbon migration inside the matrix of a fine-grained source rock are diffusion, solution, and as an oil–gas phase. Most primary migration occurs in an oil–gas phase, with diffusion and solution relevant to only the smallest and most soluble hydrocarbon molecules.

Diffusion

Diffusion is a spontaneous, irreversible process in which hydrocarbons move in the direction of lower concentration. Diffusion tends to disperse hydrocarbons rather than concentrate them, and it is an exceedingly slow process. Methane, which moves further by diffusion than do all the other hydrocarbons, still takes 140 million years to move vertically through 1,740 m (5,709 ft) of

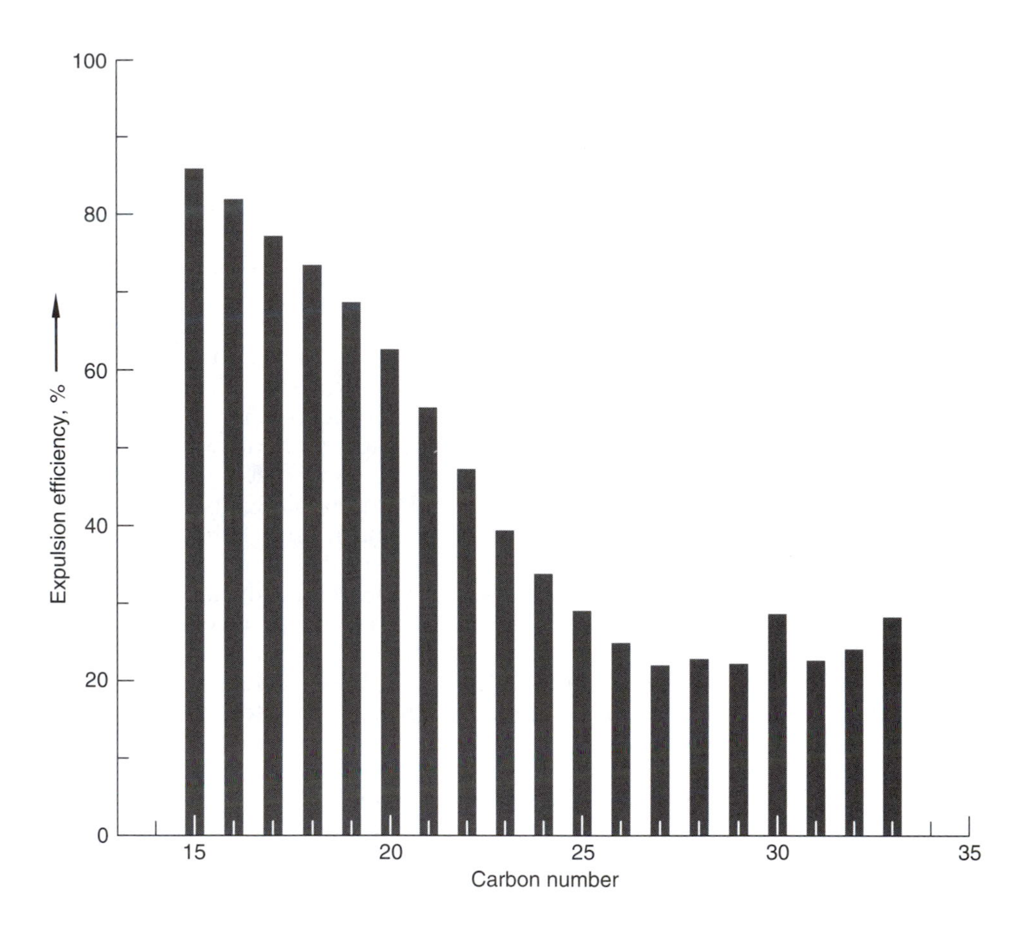

Figure 8-10

Expulsion efficiencies of individual normal and isoprenoid alkanes from thin shale to adjacent sandstone in Paleocene Firkanten Formation, Svalbard, Norway. [Leythaeuser et al. 1984]

rock matrix, based on calculations by Smith et al. 1971. Most of the rapid vertical migration observed in the field is occurring with pore-water movement along faults, joints, and similar high-permeability pathways.

Jasper et al. (1984) compared the migration by diffusion of the hydrocarbons methane through *n*-butane formed in alternating black and gray shales of Cretaceous age at a sediment depth of about 1,050 m (3,444 ft) in a DSDP core. A small amount of all four of these hydrocarbons had been generated in the black and gray shales but not in the interbedded green shales. The migration of methane and ethane by diffusion had distributed them through all the lithologies, whereas the propane and *n*-butane were found in only the black and gray shales where they originated. These larger molecules had not migrated vertically by diffusion into the green shales, even over distances of a few centimeters.

Reitsema et al. (1981) determined the carbon isotope values for methane through butanes in head space gases from shale cuttings of three wells in Wyoming and Colorado. The gas analyses extended from the surface to about 13,000 ft through sedimentary ages from Tertiary into Jurassic. If diffusion had been the main process involved in the vertical migration of the hydrocarbons, large isotopic fractionations of the carbon would exist in the shales. No significant fractionation was observed with ethane, propane, or butane, indicating that diffusion was not an important mechanism for moving the hydrocarbons.

The Kingak Shale of Jurassic age is thought to be a major source of the giant oil accumulations on the North Slope of Alaska (Magoon and Claypool 1983). Figure 8-11 shows the distribution of methane, ethane, and four gasoline-range hydrocarbons in shale cuttings from the Inigok No. 1 well drilled on the North Slope (Hunt 1987). The liquid hydrocarbons include a paraffin, two naphthenes, and an aromatic. The highest concentrations of these hydrocarbons, except for methane, are within the Kingak Shale where they were generated. Above the lower part of the overlying Torok Formation and below the Kingak Shale, the concentration of these hydrocarbons, except for methane, ethane, and toluene, is essentially zero. Toluene is the most soluble of the liquid hydrocarbons shown, but even most of its concentrations are close to zero. These data indicate that neither diffusion nor solution is causing the migration of these hydrocarbons very far through the matrix of the shale sections above and below the Kingak Shale.

There is a suggestion that solution–diffusion processes are raising the concentrations of all hydrocarbons in the first 400 m (1,312 ft) above the Kingak Shale. None of the liquid hydrocarbons is migrating downward, as would be expected. Methane concentrations are highest below the Kingak where the sediments are within the dry gas-generation window. But methane is distributed irregularly throughout the entire sedimentary column. Ethane shows somewhat higher concentrations than the liquid hydrocarbons above the Kingak Shale, but most ethane analyses are close to zero. This further supports the observations that only methane and, to a lesser extent, ethane move significantly by diffusion.

Magoon and Claypool (1983) did observe low concentrations of migrated wet gas in a sandstone at about 3,500 ft (1,067 m) in this well. Such gas could have migrated up with pore fluids traveling along a joint or a fault from a deeper source. Vertical migration along faults into shallower sandstones is a common pathway leading to hydrocarbon accumulations in deltas and foreland basins. Hydrocarbons do not migrate vertically through the matrix permeability of shales when fault and fracture permeabilities provide a much easier pathway.

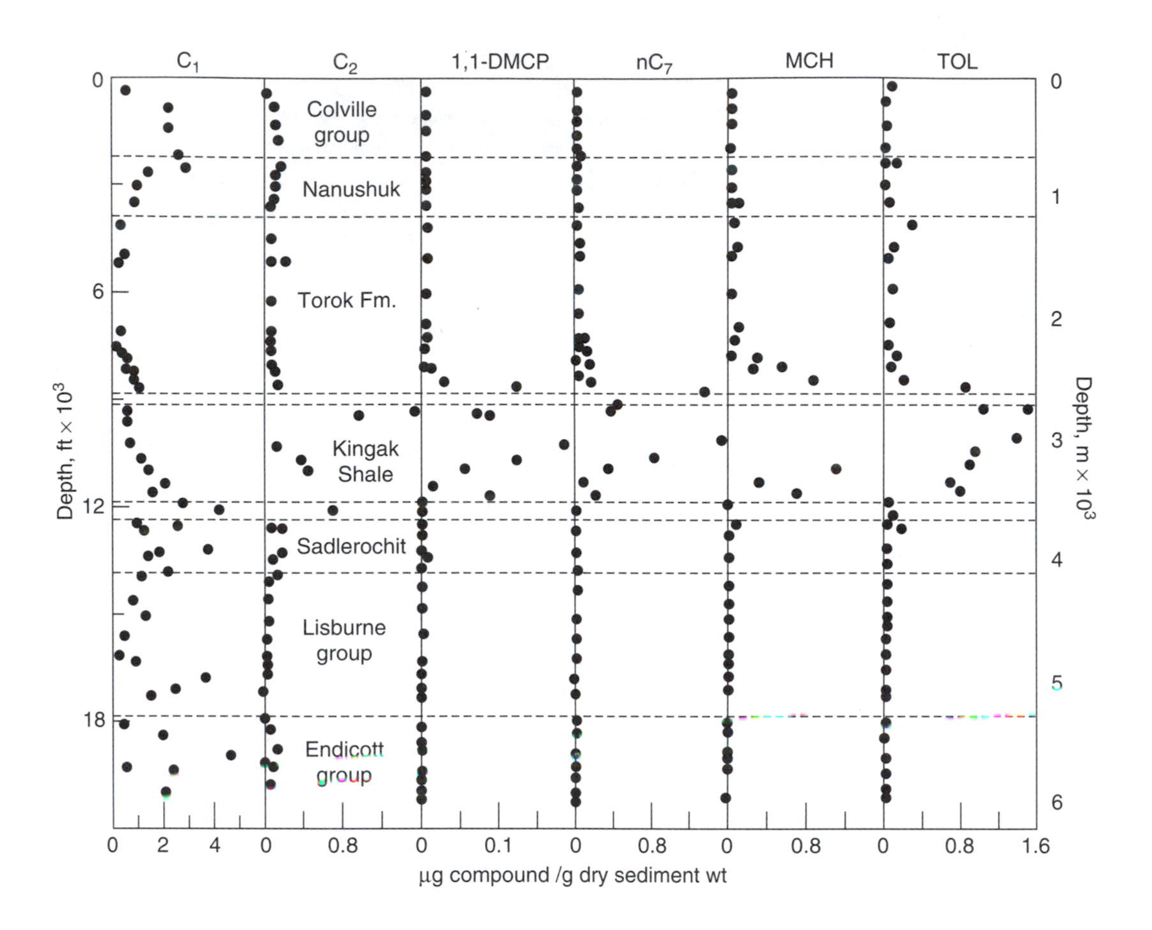

Figure 8-11

Distribution of four C_7 hydrocarbons: 1,1-dimethylcyclopentane, normal heptane, methylcyclohexane, and toluene compared with methane and ethane in the Inigok No. 1 well drilled in the Ikpikpuk Basin, North Slope of Alaska. [Hunt 1987]

Migration in Solution

Methane is widely distributed in the subsurface because of its solubility in pore fluids and its ease of migration in the gas phase. Methane shows up in mud logs from almost all areas and stratigraphic sections where methane is forming and migrating. It is prominent in the oil- and gas-generation windows, where its solubility is about 20 ft³/bbl pore water or 2,500 ppm (Blount et al. 1980), assuming a temperature of 100°C (212°F), a pressure of 7,000 psi (48MP), and a salinity of 150,000 ppm TDS (total dissolved solids). The solubility of most of the larger liquid hydrocarbons in pore fluids in the petroleum-generation window, however, is less than 50 ppm, which is insufficient to be a significant factor in moving whole oil from the source to the reservoir.

Although methane is relatively insoluble in fresh water at STP (25 ppm or 0.2 ft^3/bbl), its solubility increases rapidly with increasing pressures and temperatures. This is offset somewhat by the greater salinity with depth. At about 8,000 ft (2,438 m) the solubility of methane is 100 times greater than at the surface, and at 20,000 ft (6,096 m) it is about 300 times greater (Culberson and McKetta 1951). This increasing solubility with increasing pressure and temperature means that migrating pore fluids readily pick up methane from source beds and release it during any vertical migration to regimes of lower temperature and pressure. The effect would be even more pronounced in methane released from abnormal pressure compartments, where the solubility increases faster than in normally pressured sediments, owing to higher temperatures and pressures.

Table 8-4 lists the solubility of methane with depth at varying pressures, salinities, and temperatures characteristic of the depths listed. Assuming no overpressures, solubilities would increase steadily with depth despite increasing salinities, because the pressure and temperature effects overcome this. In overpressured compartments of basins, these solubility numbers would be exceeded because of the effects of increasing pressure and temperature and sometimes decreasing salinities. A 15,000 ft well encountering pore waters at 13,000 psi and a temperature of 340°F (171°C) could dissolve 39 ft^3/bbl of methane if the salinity were 150,000 ppm TDS. At 50,000 ppm TDS, about 60 ft^3/bbl of methane would be in solution.

Temperature seems to have the most pronounced effect on solubility. If the temperature at the 10,000 ft depth in Table 8-4 were doubled, the solubility would rise from 15 to 33 ft^3/bbl. If the salinity were cut in half, the

TABLE 8-4 Solubility of Methane with Depth as a Function of Pressure, Salinity, and Temperature

Depth, ft (m)	*Pressure, psi*	*Salinity, ppm TDS*	*Temperature, °F (°C)*	*Solubility, ft^3/bbl[a]*
5,000 (1,524)	2,325	150,000	145 (63)	12
10,000 (3,049)	4,650	200,000	230 (110)	15
15,000 (4,573)	7,050	250,000	315 (157)	19
20,000 (6,098)	9,400	300,000	400 (204)	23
25,000 (7,622)	12,200	350,000	485 (252)	27

[a]Estimated from data of Blount et al. 1980. To convert ft^3/bbl to mol fraction methane, divide by 7,370.

solubility would increase from 15 to 21 ft^3/bbl, and if the pressure were doubled, it would climb from 15 to 19 ft^3/bbl (Blount et al. 1980).

Any methane-saturated pore waters moving vertically up along faults would release free gas. The decreasing temperature and pressure changes would have more effect on solubility than any decrease in salinities that might occur.

The solubilities of some heavier gaseous and liquid hydrocarbons are shown in Table 8-5. McAuliffe (1966) measured the solubilities of sixty-five hydrocarbons in fresh water at room temperature. He found that for each homologous series of hydrocarbons, the logarithm of the solubility in water was a linear function of the hydrocarbon molar volume. In other words, the small molecules dissolve much more readily than the large ones. McAuliffe also found that forming a hydrocarbon ring from a chain increased the water solubility and that increasing the unsaturation of the ring further increased the solubility. This means that the order of increasing solubility for the major hydrocarbon groups is paraffins → naphthenes → aromatics. The mono-olefins have solubilities comparable to the naphthenes (cycloparaffins). Di- and triolefins have proportionately greater solubilities.

Zhuze et al. (1971) quoted the work of earlier Russian scientists showing that benzene, toluene, and methylcyclohexane solubilities all increase with rising temperature. Thus the solubility of methylcyclohexane more than doubles between room temperature and 115°C (239°F). They also found that water saturated with gas reduced the solubility of liquid hydrocarbons, depending on the composition of the hydrocarbons, and the temperature–pressure condi-

TABLE 8-5 Solubility of Gasoline-Range Hydrocarbons in Pure Water at STP

Hydrocarbon	*Solubility, ppm*
n-Pentane	40
n-Hexane	9.5
n-Heptane	2.2
n-Octane	0.4
n-Nonane	0.12
Cyclopentane	160
Cyclohexane	67
Methylcyclohexane	16
Benzene	1,740
Toluene	554
Orthoxylene	167

tions. The solubility of a mixture of hydrocarbons was found to be about 50% lower than the sum of the solubilities of the individual hydrocarbons, indicating that hydrocarbons tend to displace one another in solution.

Price (1976) made a detailed study of the effect of temperature on the solubilities of hydrocarbons and crude oil fractions. He found that the aqueous solubility of hydrocarbons increases gradually to around 100°C (212°F), at which the increase accelerates because of a change in the solution mechanism. Price (1973) postulated that the change in slope of the solubility line could be caused by the breakup of aggregates of molecules to a true molecular solution at the higher temperatures. It also is possible that clustered water structures, which are prevalent at lower temperatures, would be broken up at higher temperatures, resulting in more unbonded water available to dissolve the hydrocarbons.

Zhuze et al. (1971) found that hydrocarbons such as benzene, toluene, and methylcyclohexane are less soluble in water saturated with gases such as nitrogen, helium, carbon dioxide, and methane than in pure water. This effect was noted over the temperature range from about 50 to 150°C (122 to 302°F). This means that the water saturated with almost any subsurface gas would cause more exosolution of hydrocarbons than would waters relatively free of gas.

Hydrocarbons dissolved in pore waters come out of solution because of the following changes: increase in salinity, decrease in pressure and temperature, partitioning with an oil–gas phase, and increase in gas saturation. The decrease in solubility with increasing salinity is quite dramatic. Price (1976) noted that *n*-pentane and methylcyclohexane solubilities decreased from 40 ppm at STP in fresh water to less than 3 ppm in water containing 300,000 ppm TDS. High salinities in the oil-generation window tend to counteract the pressure effects trying to keep hydrocarbons in solution. In summary, migration in solution is important only to methane and possibly to ethane and the small aromatic hydrocarbons benzene and toluene. It cannot explain migration of an entire oil.

Gas-Phase Migration

Compressed gas can dissolve increasing amounts of heavy liquid hydrocarbons as the pressure and temperature increase. Sokolov et al. (1963) and Sokolov and Mirnov (1962) demonstrated that subsurface gases dissolve large amounts of liquid hydrocarbons under temperature–pressure conditions corresponding to depth ranges of 6,000 to 10,000 ft (1,829 to 3,048 m). Rzasa and Katz (1950) studied the critical pressure for binary systems with methane and found that methane–decane mixtures are single phase at pressures above 5,400 psi (37 MPa). Neglia (1979) reported that the Malossa field in Italy is condensate with heavy liquid components dissolved in the gas phase. This field produces from a Triassic dolomite at a depth of about 20,000 ft (6,096 m). The reservoir pressure is 15,431 psi (106 MPa), and the temperature is 307°F (153°C). Rzasa and Katz's data (1950) indicate that hydrocarbons through C_{18} would be dissolved in the gas phase under these conditions.

Neglia (1979) thinks that gases generated in source rocks migrate vertically through microfractures while dissolving oil from adjacent pores. Molecular distillation of oil occurs because its vapor pressure in the liquid phase is higher than the vapor pressure in the gaseous phase. Eventually, however, the migrating gases reach a level at which the reduced pressure and temperature result in retrograde condensation with the formation of an oil phase.

Figure 8-12 shows data published by Neglia (1979) indicating that microfractures may be forming in the oil-generating depth range. This plot of Tertiary shale permeability with depth shows a decrease in permeability to about

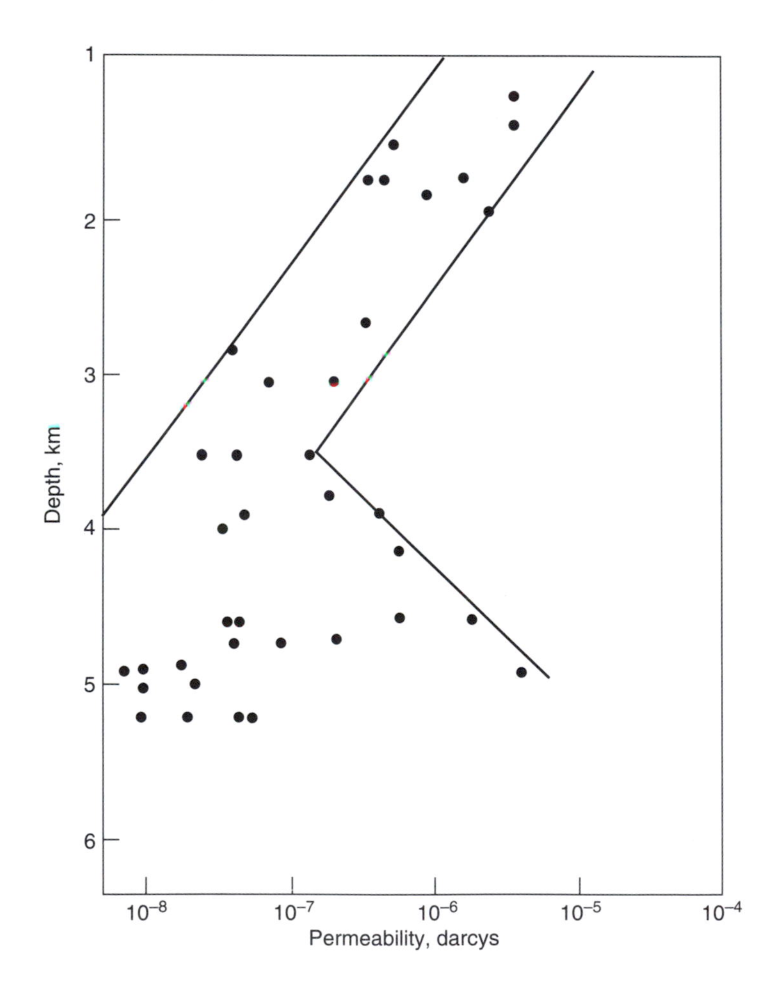

Figure 8-12

Semilog plot of permeability of Tertiary shales versus depth. [Neglia 1979]

3,500 m (11,483 ft), after which there is an increase. Assuming a geothermal gradient of 2.7°C/100 m (1.5°F/100 ft), this increase starts around a temperature of 114°C (237°F), which is within the oil-generating range for these Tertiary sediments. The increase in permeability is attributed to the development of microfractures.

Rumeau and Sourisse (1973) concluded that the hydrocarbons in the carbonate reservoirs of the southwest Aquitaine Basin probably entered in the gas phase because the flow of fluid from compaction ceased very early.

Gas-phase migration along vertical faults and through permeable Tertiary sandstones may explain the presence of aromatic-rich condensates in the Pleistocene of the Gulf Coast. When gas is passed through oil, it tends to concentrate more of the light aromatics in the gas phase. Aromatic condensates also are found in the Western Canada Basin (Hitchon and Gawlak 1972) and in East Turkmen (Gavrilov and Dragunskaya 1963).

Gas-phase migration cannot account for giant oil accumulations such as those of the Middle East unless it is assumed that huge volumes of gas have been lost. This is unlikely in the Arabian–Iranian Basin, since it has multiple evaporite seals. Also, it does not account for oil accumulations formed by source rocks at an early stage of generation like the Monterey Shale at Point Conception, California. Nevertheless, it may be the most reasonable explanation for some of the accumulations of the U.S. Gulf Coast, the Niger Delta, the Mackenzie Delta, the Mahakam Delta, and the Po Basin where microfractures could direct the gases into preexisting major vertical fault systems. With upward migration, the reduction in pressure and temperature would cause retrograde condensation of the oil in reservoir type rocks along the fault zones. Also, the type III kerogens of these deltas tend to generate more gas and condensate than oil.

Oil-Phase Migration

Lewan (1987) combined detailed petrographic analyses with hydrous pyrolysis on cores of the Woodford Shale and its stratigraphic equivalents from the midcontinent region of the United States. He examined different maturity levels in order to follow the generation and migration of the oil. He found that increasing the thermal stress generated a continuous bitumen network within the shale groundmass. As temperatures went up, the bitumen formed an oil that impregnated micropores in the shale and then expelled the excess into adjacent fractures. Any remaining bitumen and retained oil were ultimately carbonized to a pyrobitumen, which was visible as an opaque groundmass. The decrease in density of the bitumen and oil products, compared with the original kerogen, caused a net volume increase of the total organic matter within the matrix permeability of the shale. This caused the expulsion of oil from the rock samples.

Lewan determined from this and subsequent studies that the minimum TOC required to form a continuous bitumen network ranged between >1.5% for cherts and 2.4% for claystones. Source rocks that contained organic matter concentrated along laminae appeared to have lower minimum TOC requirements than did those containing the organic matter dispersed throughout the shale groundmass.

A bitumen and oil network resulting in oil-wet shales generally shows abnormally high resistivities compared with typical water-wet shales. Meissner (1978) used high resistivities to identify the oil-generation zone in the Bakken shale of the Williston Basin. Talukdar et al. (1986) made similar observations in the La Luna source rock of the Maracaibo Basin, Venezuela. The petrographic examination of the La Luna showing the pores and microfractures to be filled with bitumen and oil was similar to Lewan's observations of the Woodford Shale.

A bitumen network tends to increase the oil wettability of a shale, resulting in a mixed oil–water wettability that would facilitate primary migration of the oil phase. Experiments with water-wet sandstones indicate that oil will move along with the water only if it occupies about 20% or more of the pore volume. In sandstone cores of mixed oil and water wettability, oil apparently flows at residual oil saturations as low as 10% (Salathiel 1973).

All these observations demonstrate that an oil-phase migration through the generative kitchen of a type II source rock with > 2% TOC is occurring. Furthermore, this process would be enhanced in leaner type III source rocks by the presence of methane and CO_2 derived from organic and inorganic sources. Inorganic CO_2 (Figure 8-8) would lower both the viscosity and the interfacial tension of the oil. McKirdy and Chivas (1992) recovered a small amount of condensate in the Otway Basin of Australia that apparently was flushed out during CO_2 production. The authors concluded that the CO_2 stripped the condensate from poor-quality types III and IV kerogens in the rock.

Even the bitumen formed initially in a source rock with type II kerogen will migrate more readily if CO_2 enters the pores. At a subsurface temperature of 90°C (195°F) and a pressure of 4,000 psi (28 MPa), about 500 SCF of CO_2 will dissolve in a barrel of 10° API oil, thereby reducing its viscosity by a factor of 30 (Murtada and Hofling 1987).

Methane generated from kerogen within a source rock or from the partial conversion of oil to gas can dissolve in oil and lower both its viscosity and density. A 30° API oil with no gas would have a density of 0.88 gm/cc, compared with a density of 0.55 for a dissolved gas–oil ratio of 1,800 ft^3 gas/bbl of oil.

Momper (1978) estimated that at peak oil generation, the conversion of organic matter to liquids and gases can cause a net volume increase of up to 25% over the original organic volume (Figure 8-13). In the restricted pore space of a fine-grained source rock, this would create a localized pressure buildup, causing the opening of existing microfractures or formation of new ones and the expulsion of oil. Overpressures of 0.6 to 0.7 psi/ft (13.5 to 15.8 kPa/m) would be sufficient to reopen closed vertical fractures (Momper 1981). After the oil is expelled, the fractures would close until the pressure builds up from subsequent generation. This results in the pulsed expulsion of oil until the generating system runs down. Generation causes migration. When generation stops, the primary migration stops.

Squeezing oil from shales by mechanical compaction is an alternative process that has been used to explain primary oil migration. Recent studies

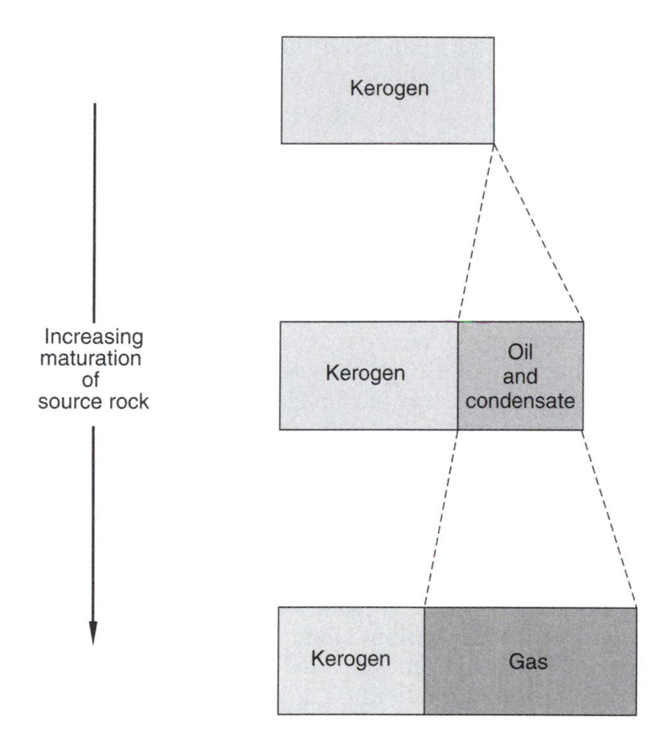

Figure 8-13

Increase in net volume of kerogen plus generated fluids with increasing maturation of source rock.

have shown, however, that many Gulf Coast shales do not mechanically compact within the oil-generation window, as will be discussed in Chapter 9.

Finally, the discussion as to whether hydrocarbons migrate in an oil or gas phase becomes academic in very deep formations where pressures exceed the critical point. This is the point at which the densities, viscosities, and surface tensions of oil and gas are so similar that they are considered to be in a single phase. Katz et al. (1959, p. 465) think that most pools consisting of gas and oil at pressures above 4,000 psi (28 MPa) and temperatures above 200°F (93°C) exist as single-phase fluids. In typical oil basins, this would correspond to depths greater than 8,000 ft (2,439 m) at hydrostatic pressure. Pressures would be much higher than hydrostatic within restricted shale pores as oil and gas are generated. This suggests that primary migration as a single-phase fluid may be the norm in all deep wells and may even occur in some formations as shallow as 5,000 ft (1,524 m). The reduction in pressure on entering a shallower reservoir rock could cause retrograde condensation, that is, the formation of separate oil and gas phases. If the reservoir rock also were overpressured, secondary migration might occur as a single-phase fluid.

Permeability of Source Rocks

The matrix permeability of petroleum source rocks ranges from 10^{-3} to 10^{-11} darcies. The fracture permeability is proportional to the cube of the width of the fracture (Nelson 1985, p. 251). Consequently, it can be several orders of magnitude larger than the matrix permeability, even for microfractures. R. W. Jones, formerly of Chevron, posed an interesting question at the 1986 AAPG meeting: "Suppose you have a shale slab one mile square with a permeability of 10^{-8} darcies that is cut by a fracture *W* centimeters wide. How wide is the fracture when the quantity of fluid passing through the square mile of shale is equal to the quantity passing through the fracture?" The answer is 6×10^{-4} cms, or 6 μm. These comparisons make it clear that flow of fluid through fine-grained source rocks is through the fracture permeability and not the matrix permeability.

Plates 3D, 4A, B, C, D, and 5A are photomicrographs of thin sections showing petroleum migration along microfractures. The microcracks or microfractures in this discussion are those with fracture widths below 100 μm. Macrofractures have widths greater than 100 μm. The Upper Cretaceous Querecual Formation is believed to be the major source of petroleum in the Eastern Venezuela Basin. Plate 3D, courtesy of S. C. Talukdar, shows fluorescent oil moving through 5 to 20 μm wide fractures in the Querecual Formation. This is a calcareous shale in which the vitrinite reflectance, R_o = 0.93%, indicates that it is at peak maturity in the oil window. Maximum burial depth was about 5,000 m. The fractures are either hydraulic or regional, with Mode I fracture propagation brought about by a tensile stress acting perpendicular to both the fracture plane and the propagation direction (Nelson 1985, p. 117).

Plate 4A depicts a thin section from a core at 10,911 ft (3,326 m) in the Lower Wolfcamp source rock of the Permian Basin (courtesy of Morad Malek-Aslani). Fluorescent oil is moving along vertical microfractures in the rock matrix. The Wolfcamp, which is the major source of the oil in the Permian Basin, is a slightly dolomitic argillaceous limestone. The fracture on the left of Plate 4A has the straight-line characteristics of Mode II fractures caused by shear stresses acting within the fracture plane and parallel to the propagation direction. These are tectonic fractures that go through grains rather than around them indicating a higher stress level.

Talukdar et al. (1986) carried out petrographic studies of thin sections and polished slabs of the La Luna Formation, which is the major source of oil in the Maracaibo Basin of Venezuela. It is the stratigraphic equivalent of the Querecual. Samples from the oil window show fluorescent bitumen filling both parallel and oblique fractures plus the cavities of foraminifera. Bitumen also occurs as regular, very thin layers parallel to bedding stratigraphy and in a diffused condition associated with kerogen throughout the rock groundmass. Plate 4B displays a fracture filled with fluorescent bitumen C cross-cutting fluorescent kerogen K and the mineral matrix M. The irregular pathway of fracture C is characteristic of hydraulic tension fractures (Mode I), probably formed by an increase in pore-fluid pressure. Talukdar et al. (1986) think that the generation of oil opened existing fractures and created new ones through the volume increase of the pore fluid resulting from the partial thermal transformation of kerogen to bitumen, oil, and gas. The La Luna source rock is thought to have contained very little water during the bitumen oil-generation stage.

Plate 4C is a fluorescent photomicrograph taken by Morad Malek-Aslani of a sample from the Smackover Formation at a depth of 15,178 ft (4,627 m) at the eastern end of the Interior Salt Basin in Santa Rosa County, Florida. The slide shows a black band of bitumen expelling fluorescent oil into fractures in the matrix of the Jurassic Smackover source rocks. The algal carbonate rocks of this formation are the main source of oil and condensate in this basin.

The Posidonia Shale of Lower Jurassic age is a well-known source rock for many oils in western Europe. Littke et al. (1988) found that numerous micro- and macrofractures filled with bitumen occur in the mature Posidonia Shale, where vitrinite reflectance values exceed 0.68% R_0. Immature sections of the Posidonia contained only a few vertical fractures. In addition, selected samples of the shales overlying and underlying the Posidonia at similar levels of maturation exhibited no fractures. This suggests that the fractures were caused by the generation of petroleum. Horizontal fractures were described as being interconnected by thin, irregular vertical fractures. These would be characteristic of hydraulic fractures formed by pore-fluid pressures.

In their study (1966) of the Maikop Shales from the West Kuban downwarp, Teslenko and Korotkov noted that the calcareous shales at a depth of about 3,200 m (10,499 ft) contain a network of vertical microfractures filled with calcite. The migration of bitumens along the microfractures was evident from luminescence tests. The adjacent layers of noncarbonate shales contain no visible microfractures.

These examples illustrate that many fine-grained shale source rocks contain microfractures. But fine-grained shales are often thought of as containing a high percentage of clay minerals such as smectites and illite (Shaw and Weaver 1965). Such minerals tend to make a rock ductile rather than brittle. Ductile shales are caprocks for many large oil accumulations. The mineralogical composition of many shales that are petroleum source rocks, however, are very different from the usual conception of shales having a high clay mineral content. Most world-class source rocks are low in clay minerals and are brittle rather than ductile.

Mineralogy of Shale Source Rocks

The term *shale* is properly defined by its particle size rather than its mineralogy. Twenhofel (1932, p. 240) defined it as having particles less than 1/256 mm or about 4 μm in diameter. *Siltstone* was defined as having grain sizes between

1/16 and 1/256 mm, and *sandstone* between about 1 and 1/16 mm. More recently, Lewan (1978) defined a shale as containing more than 65% by volume of material smaller than 5 μm.

Generally, this < 5μm fraction is thought of as containing mainly clay minerals, but this is not always the case. For example, Bradley (1993, personal communication) evaluated X-ray and elemental analyses of several hundred shales from the U.S. Gulf Coast. The shale sections in the cores were defined as shale by electric logs and visual inspection. The clay-size fraction (< 4 μm) consists of 74% quartz and 26% clay minerals. The average particle size of the quartz was 2 μm, and that of the clays, 0.1 μm. The average Tertiary Gulf Coast shale based on this study contains 67% quartz, 8% feldspar, 20% clay, 4% carbonate, and 1% organics and other minerals.

Among well-known source rocks, the Green River Oil Shale of the Uinta Basin, Utah, is composed mainly of calcite, dolomite, and quartz. Only 25% of the samples analyzed contain clay minerals, and these have less than 10% (Hunt et al. 1954). The Woodford Shale of Oklahoma is 85 to 95% fine-grained quartz (Lewan 1987). The Monterey Shale of California is mostly calcareous and dolomitic porcelanite and microquartz (Isaacs 1987). The Draupne and Heather Shales of the northern North Sea contain 53 to 57% quartz and < 5% smectites plus illite (T. Barth, personal communication). The Nordegg Shale of the Western Canada Basin and the Bakken Shale of the Williston Basin in the United States have less than 20% clay minerals. The La Luna Shale of the Maracaibo Basin, Venezuela, is 50 to 90% carbonate. The New Albany Shale of the Illinois Basin has up to 68% quartz. Many black, fissile rocks are 60 to 80% fine-grained quartz. Fracture intensity is greatest in the finest-grained shales containing the most brittle minerals. These include the various forms of quartz, feldspars, dolomite, and calcite (Nelson 1985, p. 153).

Comer and Hinch (1987) found that parts of the mature Woodford Shale source rock of Oklahoma contained oil-filled stylolites and a fine polygonal fracture network with vertical tension fractures extending from the bedding planes. The fractures appeared to be tectonic in origin. Some oil-filled fractures were smaller than 10 μm in width.

Many shale fractures are invisible except by microscopic examination of thin sections. The shale particles are so small in these source rocks that there is nothing to prop open the fractures. It is like many of the shale caprocks in the world. Some shale seals over oil, and gas fields may contain fractures, but they are in a stress phase that keeps the fractures tightly closed until excess pressures develop. For example, some gas storage reservoirs in anticlines in the American Midwest have threshold values of pressures beyond which leakage occurs. If the injected gas raises pressures beyond this, the fractures will open, and gas will leak out; with leakage the pressure will drop and the fractures will close. The reservoir caprock acts like a pressure-release valve.

The Devonian black shales in the Appalachian Basin produce hydrocarbons from shale fractures at 10,000 ft (3,049 m) because the fractures contain dolomite crystals that keep them propped open.

Many source rocks also contain regional fractures and joints that formed before any hydrocarbon generation. The Big Sandy field of Kentucky and West

Virginia produces gas from fractured, Devonian brown shales that contain a mixture of regional orthogonal fractures with localized tectonic fractures (Nelson 1985). Similar mixtures of fracture systems are found in the brittle, calcareous Niobrara shale source rocks of the Powder River Basin in Wyoming. The La Luna source rock of the Maracaibo Basin in Venezuela has regional orthogonal fractures as well as localized hydraulic fractures. The Querecual source rock of eastern Venezuela also contains regional fractures. Consequently, the primary migration of hydrocarbons can occur through localized microfractures in source rocks and continue through preexisting tectonic and regional fractures, faults, joints, and the like, which provide a pathway out of the source rocks to more permeable reservoir type carrier beds.

Secondary Migration

The main driving force in secondary petroleum migration is the buoyancy of the hydrocarbons. The resisting force is the capillary pressure of the rock–water system. Modifying forces include hydrodynamic fluid flow and isolated abnormal pressure compartments. The buoyant force is the difference in density between the hydrocarbon phase and the water phase. The greater this difference is, the greater the buoyant force will be for a given hydrocarbon column. Water densities range from 1 to 1.2 gm/cc; oil densities are 0.5 to 1 gm/cc; and gas densities are < 0.5 gm/cc. This results in oil–water buoyancy gradients ranging from 0 to 0.3 psi/ft (6.8 kPa/m). Gas–water buoyancy gradients in the subsurface range from about 0.2 to 0.5 psi/ft (4.5 to 11 kPa/m) (Schowalter 1979).

Zieglar (1992) compared hydrocarbon columns, buoyancy pressures, and seal efficiencies for more than 800 oil and gas pools in California and the Rocky Mountains. He concluded that the average values for converting column lengths to buoyancy pressure in these areas were 0.1 psi/ft (2.3 kPa/m) for oil and 0.35 psi/ft (7.9 kPa/m) for gas. This means that a gas column has more than three times the buoyant pressure of an oil column of the same length.

In most situations, gas forces its way into the shallowest part of a trap above the oil or water, but there are anomalous exceptions. In the Deep Western Canada Basin, gas is trapped downdip from water in a very low permeability rock with an updip gas–water contact (Gies 1984). Gies considers the gas column to be in a dynamic state. High gas pressures presumably caused by slow gas generation from below are causing gas to move slowly upward where it gradually escapes across the gas–water contact. The accumulation persists as long as gas is feeding in faster than it is escaping.

The *capillary pressure* is the pressure required for the oil or gas to displace the water from the rock it is trying to penetrate. Purcell (1949) expressed this as follows:

$$Pd = \frac{2\gamma \cos\theta}{R}$$

where Pd = the hydrocarbon–water displacement pressure in dynes/cm^2, γ = interfacial tension (dynes/cm), θ = wettability expressed as the contact angle between the rock and hydrocarbon (Figure 8-14), and R = the radius of the largest connected pore throats (cm). Schowalter (1979) pointed out that the wettability term equals 1 for water-wet rocks and is not significantly different for rocks having as much as 25% oil-wet surfaces. This reduces the displacement pressure equation to $Pd = 2\gamma/R$.

Interfacial tensions for oil–water systems range from 5 to 35 dynes/cm and for gas from 30 to 70 dynes/cm at STP (Schowalter 1979). Gas–water interfacial tensions decrease with both increasing temperatures and pressures. Schowalter estimated that in deep reservoirs the interfacial tensions would be around 22 to 25 dynes/cm. Oil–water interfacial tensions decrease mainly with increasing temperature, reaching values around 10 to 20 dynes/cm. This means that oil migrates more easily than gas through water-wet rocks, since a lower interfacial tension results in a lower Pd.

At even greater depths, however, the interfacial tensions tend to approach each other as the oil changes to condensate. Also, as previously mentioned, at pressures above 4,000 psi (28 MPa) and at temperatures above 200°F (93°C), many petroleum accumulations are single-phase fluids. This would be near a depth of 9,000 ft (2,744 m) in some areas.

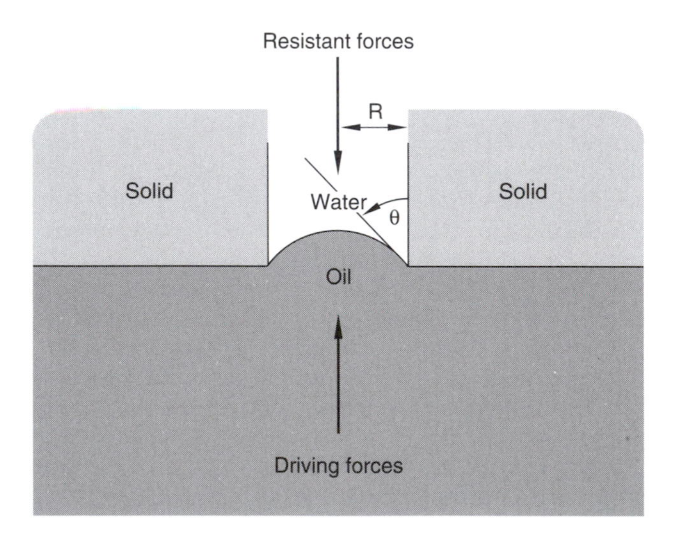

Figure 8-14

For migration to occur, work is required to increase the surface area of the oil–water interface to force the oil phase through the smaller pore throats of a confining bed. The minimum pressure required to force hydrocarbons into the largest connected pore throats of a rock is defined as the *displacement pressure*. [Schowalter 1979]

The pore-throat radius is the controlling factor in the displacement pressure equation. A trap will occur at a change in permeability if the buoyancy of the oil or gas column does not exceed the capillary displacement pressure. The concept of a capillary barrier was demonstrated long ago in experiments by Illing (1933). He placed alternate layers of coarse and fine sands in two tubes, one wet with oil and the other wet with water. When an oil–water mixture was passed through the coarse–fine layers of water-wet sands, the oil was segregated in the coarse layers and the water in the fine layers. When the oil–water mixture was passed through oil-wet sand layers, the water segregated in the coarse layers and the oil in the fine layers. This clearly indicated the effect of a capillary barrier. A small opening wet with water rejects oil, whereas a small opening wet with oil rejects water. Because the majority of rocks in the subsurface are water wet, water can pass uninhibited from a sandstone to a shale, but oil is trapped by the capillary barrier of the fine, water-wet pore openings.

The pressures developed by capillary barriers are enormous. Hubbert (1953) calculated that the capillary displacement pressure in a shale with a particle diameter of 10^{-4} mm would be about 40 atmospheres, compared with less than 0.1 atmosphere for a sandstone. This means that a globule of oil attempting to enter a shale from a sand would be rejected from the shale by a pressure on the order of tens of atmospheres. The capillary pressure in a siltstone can be several times greater than that in the sandstone. Almost any change from a coarse to fine particle size results in some accumulation of oil.

Migration Pathways

Capillary displacement pressures in nature of several hundred psi are not necessarily required to move oil and gas through a shale caprock. Fluid pressures in overpressured compartments can force open preexisting fractures in the shale long before the pressure required to penetrate the shale groundmass is reached. This is why many gas storage reservoirs leak (as previously stated).

Plate 4D is a photomicrograph of a thin section of the Woodford Shale source rock of Oklahoma (Olson 1982). Its TOC is 2.6%, and the vitrinite reflectance is 0.4% R_o, indicating that it is too immature to have generated oil. Bitumen from a deeper, more mature oil-generating section of the Woodford has moved up and forced open preexisting bedding plane fractures. The curved, bitumen-filled fracture between the bedding planes is probably a natural hydraulic fracture caused by pore pressure, since it terminates in both the top and bottom bitumen layers. The straight, knife-edged fractures indicated by arrows above and below the bitumen layers are tectonic shear fractures caused by faulting or uplift. A Woodford sample (Plate 5A) in the early stages of oil generation (R_o = 0.5%) has a stylolite and fine polygonal fracture network all filled with bitumen (Comer and Hinch 1987). The TOC is 4.5 wt%. The fractures are believed to have formed during the Late Paleozoic and to have filled with bitumen on uplift.

Lewan (1987) observed bitumen-filled fractures parallel to bedding in mature samples of the Woodford shale and other source rocks. He believes that these were formed by the pressure increase caused by hydrocarbon generation. Hedberg (1931) found bitumen distributed uniformly through the groundmass of the La Luna source rock, but the oil in the underlying nonsource Cogollo Limestone was confined to microfractures.

Joints, Faults, and Lineaments

Joints, which are the most ubiquitous structures in some sedimentary rocks, are tension fractures with no visible offset. Internal fluid pressures slightly in excess of the least compressive stress are believed to be the driving force forming joints. Joints tend to be oriented perpendicular to this stress. The distances between joints in granodiorite mapped in the Sierra Nevada of California range from about 20 cm to nearly 25 m (Pollard and Aydin 1988).

Rogers and Anderson (1984) measured methane concentrations and fracture distributions in the vicinity of a major northwest trending lineament that extends across the valley and ridge province of Pennsylvania. Bore holes drilled on the lineament and within 1 km of it had methane concentrations at least ten times larger than wells farther away from the lineament. The methane comes from the Devonian brown shale source rock about 1,000 ft (305 m) below. The methane is concentrated along the lineament because it represents a higher permeability pathway to the surface. The authors also reported a general increase in fracture density or spacing approaching the lineament.

As previously mentioned, fracturing and fissuring are more common in brittle rocks such as carbonates and calcareous shales. Radchenko et al. (1951) cited many examples of smears and seeps of liquid hydrocarbons and impregnations of solid asphalts along bedding planes of fissures and faults in carbonate rocks, sandstones, and some metamorphosed rocks. Goldberg (1973) found several types of bitumens in a dense network of nearly vertical fractures occurring through the argillaceous carbonate Ordovician cover of Cambrian oil reservoirs in the Baltic region. The highest concentration is in the lower 20 to 25 m (66 to 82 ft) of the Ordovician section. The vertical fractures range in thickness from less than 1 mm (0.04 in) to 3 mm (0.12 in). The paleopermeability caused by the fractures was determined to be from 85 to 110 md (millidarcies) before the filling of the fractures with bitumens. It was higher in some mineral-filled fractures. This situation is analogous to the vertical bitumen-filled joints in the dolomitic shales of the Uinta Basin reported by Hunt et al. (1954). In this study, the vein thicknesses changed from about 1 to 50 mm (0.04 to 2 in) over a vertical distance of about 150 m (492 ft).

Tension fractures at great depths have been discounted in the past on the basis that absolute tension is impossible very deep in the earth's crust. However, these calculations did not take into consideration the role of fluid pressure in the tension-fracturing process. Secor (1965) demonstrated from theoretical considerations that tension fractures can develop at increasingly greater depth in the earth as the ratio of fluid pressure to overburden pressure approaches 1.

Secor also noted that previously formed fractures can be opened up by fluid pressure at great depths.

The splitting of shear waves in the earth's crust has been attributed to the presence of parallel, vertical, water-filled microfractures that occur everywhere in the crust (Muir-Wood 1988). These are called *extensive–dilatancy–anisotropy,* or *EDA,* fractures. They are thought to contain hot fluids and exist in a range of sizes from a few micrometers up. The existence of these water-filled fractures has been confirmed at depths of 12,000 m (39,360 ft) in the Kola well drilled in the Arctic Circle of Russia (Kozlovsky 1984).

Kushnareva (1971) investigated a fault extending from carbonate rocks of Upper Devonian to Upper Permian in the West Soplyas area at the southeast end of the Pechora–Kozhva arch of Russia. Oil accumulations occur wherever there are good reservoirs throughout this interval. In the fine-grained rock, regardless of composition, oil occurs in discordant microjoints, stylolites, and hairline fractures. Secondary mineral accumulations of calcite, metallic sulfides, and fluorite were also noted. Kushnareva concluded that the clear evidence of oil movement along discordant joints suggested a wide-front, upward vertical migration of fluid. The fault was the main channel for the percolation of hydrothermal fluids and then of the oil.

The movement of fluids along vertical fractures has been observed in several areas of the U.S. Gulf Coast. Forgotson (1969) reported a Louisiana well in which a fault became a passageway for the upward movement of high-pressure fluids from a sandstone around 13,000 ft (3,963 m). The fluids fractured the formations above them on the downthrown side of the fault until the containing pressure of the formation and the diminishing pressure of the invading fluids were equalized.

Reverse faults generally have lower permeabilities than do normal faults. Also, fluids travel more readily along active moving faults in deltas. If a fault stops adjusting to subsidence, it can seal and prevent further migration. In the Gulf Coast and Nigeria the faults that continue moving are the upwardly concave growth faults, the so-called rollover faults. They are analogous to landslides caused by gravity and are arcuate in shape. Most of them are visible on seismic sections because the section increases across the fault and expands on the downthrown side.

The large-scale vertical migration of formation waters and crude oil in the Louisiana Gulf Coast was documented by Hanor and Sassen (1990). They concluded that the vertical upward flow occurred through fault and fracture permeabilities rather than the matrix permeability of the shale. Their reasoning is that if water were moving through the shale groundmass, the shallower water should be lower in salinity because of the membrane filtration effect discussed earlier. Dissolved anions should be retarded by the negatively charged clay particles, producing fresher water moving up section. Instead, they found that salinities increase above the overpressured zone. The higher salinities are localized in the same areas as hydrocarbon production over salt domes. This indicates dissolution of the salt as hot fluids from the geopressured zone move up along it. Crude oils in reservoirs stacked above one another from 2,521 m (8,269 ft) to 433 m (1,420 ft) are reservoired in immature sandstones at temperatures < 65°C (149°F). Clearly, they had to come from source rocks much deeper in the section. The chemical similarity of the oils indicates a single source.

Fissures and fractures associated with faulting induces secondary permeability and porosity that favor vertical fluid migration. Faults may act as conduits or seals, depending on a variety of factors (Eremenko and Michailov 1974). The normal stress across a fault and the nature of the fault surface and of the strata cut by the fault are among the critical factors.

Link (1952) cited several cases of vertical migration of hydrocarbons along faults. Seeps associated with the Rothwell field of Ontario, Canada, leak from a fault interrupting a series of flat beds from Ordovician to Devonian in age. The Norman Wells field in Canada is a leaking reef limestone in which oil rises up along the fault zones on the edge of the reef. The faulting is caused by draping over the reef itself and is a classic example of the loss of oil from a stratigraphic trap through extension faults without folding.

The Russian literature identifies many examples of bitumen along fault planes and of small accumulations associated with the updip end of fault planes. For example, Kudryatseva et al. (1974) cited migration along a fault zone in the Savan Hot Springs region on the eastern margin of the Golygino Basin in southwestern Kamchatka. The fault, which is more than 2 km (6,562 ft) long, has 100 to 120 m (328 to 394 ft) of visible throw. The fault zone is up to 20 m (66 ft) wide, and the rock throughout it is saturated with methane. Gas escapes through two large vents at the rate of about 30 m^3/day (1,059 ft^3/day). There also are several smaller vents in the fault zone. Core holes drilled on the site fill with water containing small amounts of kerosine. The kerosine consists almost entirely of condensed napthenic or aromatic hydrocarbons.

On the other hand, Weeks (1958) looked at several examples in which faults act as seals rather than conduits for migration, such as the Velasquez field of the Magdalena Valley, Colombia. He noted that there were no seepages associated with the lower shelf and hinge belt of the Eastern Venezuela Basin, which has a maze of faults serving as barriers to trap oil in a large number of fields.

Levorsen (1967, p. 260) discovered many examples of reservoir traps caused by tightly sealed normal, reverse, and thrust type faults. Dickey and Hunt (1972) pointed out that hundreds of normal faults in the La Brea–Parinas field of Peru act as seals, subdividing the field into separate reservoirs.

It is clear that both leaking and sealing faults exist and that no single general statement can be made regarding vertical migration along faults, simply because there is such a wide variation in the level and maintenance of continuous permeability.

Unconformities, Sheet, and Channel Sandstones

Unconformities are generally thought of as conducive to harboring traps. North (1985, p. 319) believes that the most important structural phenomena involved in trapping oil and gas on a worldwide basis are the sub-Cretaceous and intra-Cretaceous unconformities. But unconformities also represent more permeable migration pathways, particularly when they are overlain by a massive sheet sandstone, as in the Western Canada Basin.

Many oil fields are related to unconformities, possibly because the weathered surface or the overlying transgressive sandstones form a permeable zone through which water and oil can migrate. Many seeps occur at the outcrop of an unconformity if the unconformity is a controlling factor in oil migration. The post-Mississippian, pre-Pennsylvanian unconformity in eastern Oklahoma has considerable control over the location of oil accumulations, and seeps occur wherever it crops out. In the Uinta Basin, asphalt-impregnated sandstones occur at the outcrop of the unconformity overlying the Wasatch formation. The asphalt lies close to the Cretaceous–Tertiary contact.

Much of the oil on the Barrow Arch, North Slope of Alaska, was sourced downdip in the Kingak Shale and the Shublik Formation immediately underlying the Lower Cretaceous unconformity. The oil moved updip along the unconformity into subunconformity Permo-Triassic sandstones or younger (Cretaceous) sandstone reservoirs.

A spectacular series of seeps ring the Maracaibo Basin in Venezuela (Dickey and Hunt 1972). All the seeps are within a few hundred meters of the outcrop of the pre-Miocene unconformity. Most of the large productive oil reservoirs of the area are sandstones either just above or just below the unconformity surface. Shows of oil are encountered at the unconformity almost everywhere it is penetrated by wells. Obviously, the unconformity-surface zone served as a channel along which oil migrated, saturating porous beds above and below it. Two seeps, Inciarte and Mene Grande, are particularly large and active, indicating that the secondary migration of petroleum is still occurring.

The two largest accumulations of heavy oil in the world are in the Eastern Venezuela and Western Canada Basins. Both occur at unconformity outcrops. Demaison (1977) pointed out that both of these giant accumulations are reservoired in immature sediments. The sources in both cases occur around the downdip extension of the unconformities where the source rocks were originally buried beyond 10,000 ft (3,049 m).

Petroleum Traps and Seals

A *trap* is any combination of permeable rocks that holds commercial quantities of petroleum. The *closure* of a trap is the vertical distance from the highest point or crest to the *spill plane,* that is, the level at which the oil spills below the trap into adjacent permeable beds. The gross-pay zone in a trap is the distance from the top of the petroleum accumulation to the lowest point of the oil–water contact. The net-pay zone is the commercially productive part of this interval. The three most important factors in a trap are (1) its proximity to a hydrocarbon migration pathway, (2) the permeability of its seal, and (3) the height of the closure (trap size). If the trap is not in a hydrocarbon generation–migration fairway, it will not accumulate oil, regardless of its size and seal. If the seal leaks, the trap will contain oil only if the rate of accumulation exceeds the rate of loss. If the first two factors are favorable, then the height of the closure will determine the size of the accumulation.

The best petroleum seals are gas hydrates. Following close behind are evaporites, especially salt. In Chapter 7 it was stated that all the supergiant gas fields in the world have seals of either permafrost with hydrates (as in West Siberia) or regional evaporite beds (as over the Hugoton field in the United States). Also, the importance of drilling below gas hydrates to produce free trapped gas was mentioned.

Evaporites represent the second-tightest barrier to hydrocarbon migration existing in sedimentary basins. For many years, L. G. Weeks (1958, 1961) emphasized the importance of evaporites as caprocks for oil accumulatons. He pointed out that many cycles of deposition involve organic-rich carbonate marls or muds that end with evaporites. The evaporites act as an excellent seal that effectively traps most of the hydrocarbons generated during preevaporite sedimentation. Multiple evaporite seals, including the Hith Anhydrite, have trapped most of the major oil and gas fields in the Central Arabian Gulf (Murris 1980).

The ability of evaporites to prevent the vertical migration of hydrocarbons is caused by the capability of salt to undergo plastic flow at elevated temperatures and the small size of the crystal lattice of sodium chloride. The distance between NaCl lattice units is 2.8×10^{-10} m, whereas the smallest hydrocarbon, methane, has a molecular diameter of about 4×10^{-10} m. The only way that hydrocarbons can migrate vertically through salt crystals, even by diffusion, is if the salt contains impurities that render it brittle and capable of fracturing under tectonic stress. Antonov et al. (1958) examined four specimens of rock salts ranging from Devonian to Permian in age and found that they did have a mosaic of very fine fractures that permitted some vertical diffusion of methane. As a rule, however, evaporites may be considered the strongest mineralogical barriers to the vertical migration of hydrocarbons.

It is not the matrix permeability but the fracture permeability of a seal that is critical. Downey (1984) calculated that a single fracture, 0.001 inch (0.025 mm) wide, in a seal overlying a 500 ft oil column would leak oil at a rate exceeding 150 million barrels per thousand years. Fractures provide permeability pathways for primary migration from brittle, fine-grained source rocks. They also can provide pathways through the top seals of traps. Downey (1984) compared seals in terms of ductility, since the more ductile rocks tend to flow under deformation, whereas the brittle ones break. His order of decreasing ductility and increasing brittleness is (1) salt, (2) anhydrite, (3) clay mineral-rich shales, (4) silty shales, (5) calcareous shales and carbonates, and (6) cherts. As previously mentioned, many shales tend to be brittle; consequently, many shales are a poor seal for oil and gas accumulations. Only a small percentage of shales have clay mineral contents above 40%, which would make them ductile. High kerogen contents can make shales—but not argillaceous carbonates—ductile. The Green River oil shale with TOCs above 10% is quite brittle.

The risk of retaining an oil accumulation differs for structural traps compared with stratigraphic traps. In a simple anticlinal closure (Figure 8-15), the buoyant force of the oil column is directed vertically upward perpendicular to the bedding. In a typical clastic sequence there may be a series of beds, as shown with displacement pressures (Pd) ranging from 0.2 to 200 psi. It takes only one

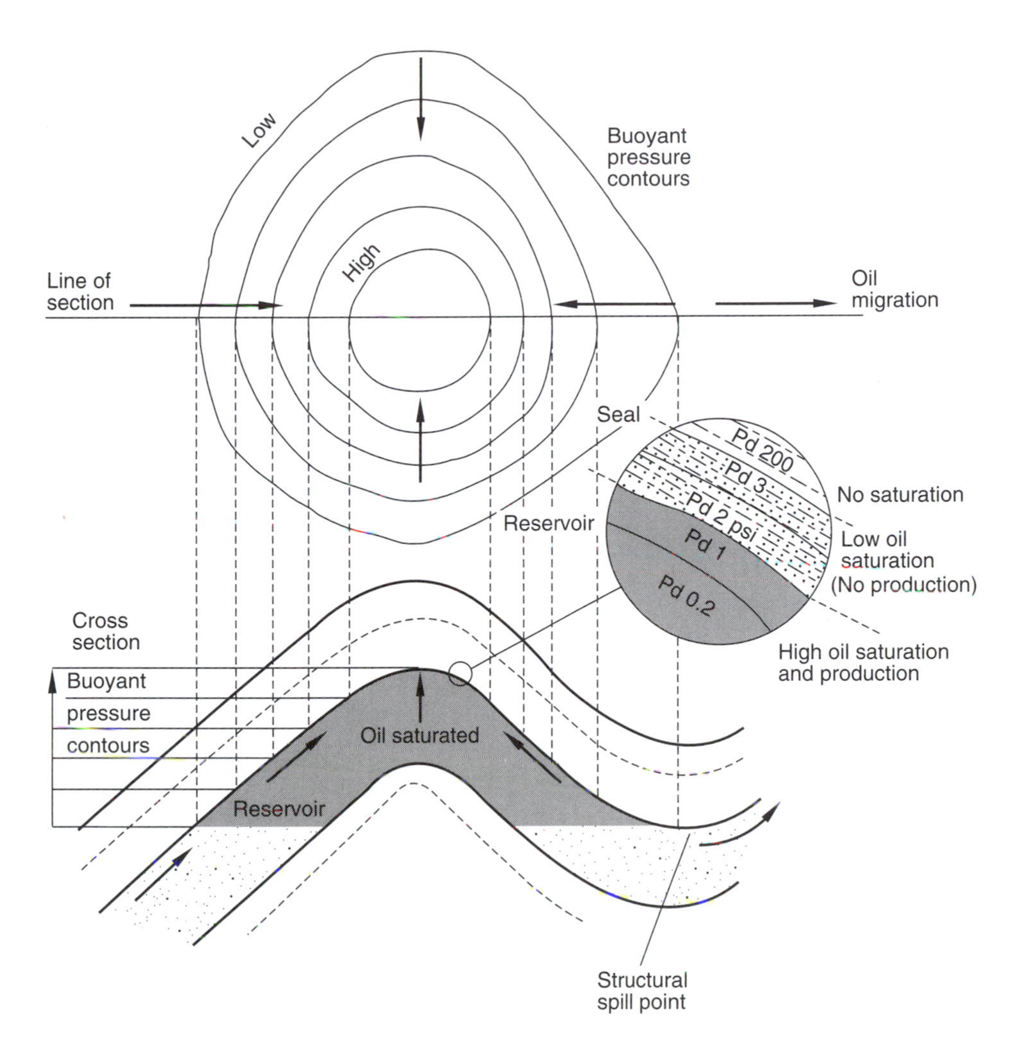

Figure 8-15

Map view of structural type trap. [Schowalter 1979]

high Pd ductile bed to trap the oil vertically and laterally down to the spill point, assuming that the bed covers the entire structure in three dimensions.

In contrast, a stratigraphic trap (Figure 8-16) in which the oil column can move parallel to the bedding plane would leak if there were only one small silt bed with a low displacement pressure, such as the 2 psi shown. Since the buoyant force of the oil is directed updip and parallel to the bedding, rather than perpendicular, it forces its way into and through any bed with a low Pd. Consequently, the risk of not finding hydrocarbons in a stratigraphic trap is greater than for an anticlinal trap (Schowalter 1979).

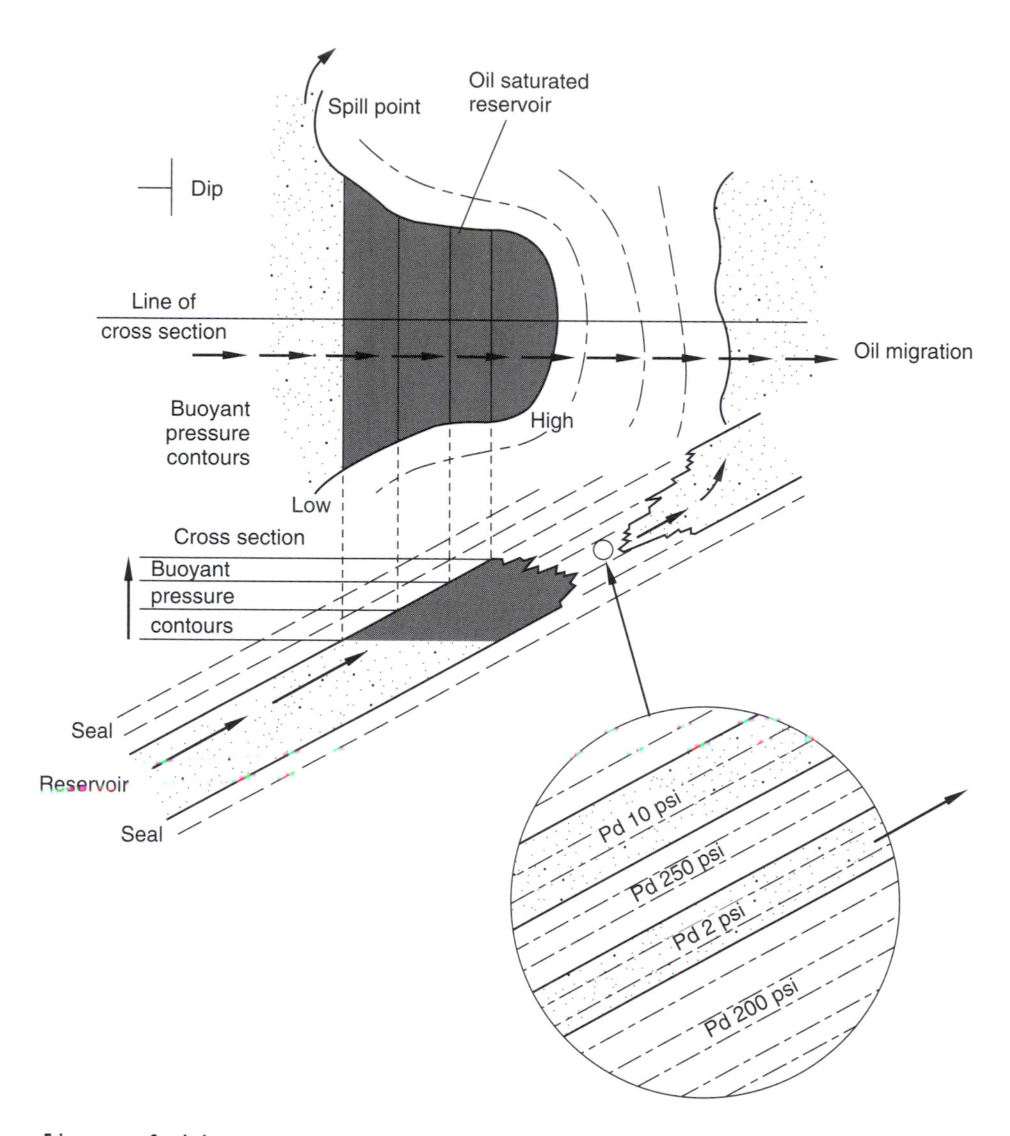

Figure 8-16
Map view of stratigraphic type trap. [Schowalter 1979]

As mentioned previously, both unconformities and faults can act as either seals or pathways. Downey (1990) considers fault closure to be inherently more risky than anticlinal traps because of our failure to understand which faults act as seals and which permit leakage. In an active compressional setting, a fault plane is rarely a pathway. But in tensional settings at shallow depths and in deep geopressured compartments, faults can be active migration pathways. Downey (1990) recommends mapping fault traps in three dimensions to evaluate the seal risk on a prospect.

Capillary barriers at coarse–fine-grained sediment interfaces represent the most common type of barrier to petroleum migration. Other types of barriers include such things as secondary mineralization and asphalt accumulation, especially at very shallow depths.

The solubility of SiO_2 in pore waters decreases from about 125 ppm at 150°C (302°F) to 50 ppm at 100°C (212°F). Consequently, any fluids saturated with SiO_2 and moving surfaceward would precipitate silica. Quartz cementation obliterates the intergranular porosity in some of the top seals in abnormally pressured compartments of the Venture gas field off Nova Scotia (Jansa and Noguera Urrea 1990). Both silica and calcite cementation contribute to the formation of pressure seals in the North Sea (Hunt 1990).

Hinch (1993, personal communication) observed that microfractures formed naturally in shales tend to be healed by the precipitation of minerals; quartz precipitates first, followed by calcite in the fracture. Schmidt and McDonald (1979) observed calcite cementation in sandstones at a vitrinite reflectance of R_o = 0.4 to 0.5%. This would be below 100°C (212°F) in most basins.

Oil accumulations in carbonate reservoirs can be retained by the postaccumulation formation of stylolites. Dunnington (1967a) cited examples of Middle East reservoirs where the late formation of stylolites in the porous and permeable part of the reservoir has sealed off any further petroleum migration.

Asphalt can act as a barrier to oil and gas movement. Levorsen (1967, p. 336) describes several examples of fields where surface outcrops of asphalt prevent the escape of downdip accumulations of oil and gas. The asphalt forms from the weathering and bacterial degradation of seepages. Most notable are the Bolivar coastal fields of Maracaibo, Venezuela, where asphalt and heavy degraded oil is updip and lighter oil is downdip in the same reservoirs (Dickey and Hunt 1972). Subsurface deasphalting of oil and degradation through the agent of meteoric waters may cause asphalt barriers at oil–water contacts. Such asphalt barriers have been reported from Hawkins field (Texas), Prudhoe Bay (Alaska), Frannie (Wyoming), Ghawar (Saudi Arabia), Oseberg (North Sea), and Burgan (Kuwait). In some fields they may be the cause of inclined oil–water contacts.

Direction and Distances of Secondary Migration

Klemme and Ulmishek's (1991) study of effective petroleum source rocks in the world indicates that 60% of the world's conventional oil moved vertically from source to reservoir and 40% moved laterally. But 43% of these are both lateral and vertical within the same stratigraphic sequence as the source. For example, the Upper Jurassic in the Middle East central Gulf area contains both the Hanifa source and the Arab D reservoir rocks. Vertical migration was defined in their study as crossing a stratigraphic boundary, which means that migration could be either up or down (stratigraphically, not physically). Lateral migration was defined as > 25 mi (40 km).

Migration directions and distances from source to reservoir rock depend on basin size and configuration. Klemme (1992) considers deltas, small interior rift basins, and divergent margin and active margin basins to have dominantly short-distance vertical migration with minor lateral migration. Examples of these are shown in Table 8-6. Foreland basins and fold belts have dominantly lateral migration unless the foreland ramp is small or deformed by the fold belt. Interior cratonic sag basins have both long-distance lateral and short-distance vertical migration. Large interior rift basins with several rifts and one sag, like the West Siberia Basin of Russia and the Great Artesian Basin of Australia, have considerable lateral migration, along with primarily vertical migration.

Interior cratonic sag basins such as the Illinois and Williston basins are generally less than 320 km (200 mi) in diameter. The petroleum migration distances are intermediate between those in interior rift and foreland basins. In the Williston Basin there is vertical migration into the Nesson anticline and lateral migration of 100 mi (161 km) or more to the oil fields of Canada. In the Illinois Basin, Demaison (1984) reported most of the oil fields to be 30 mi (48 km) updip from the mature New Albany Shale source rock.

An example of a small interior rift basin is the North Sea, where some Upper Jurassic shales and overlying sandstones act as source and reservoir (lateral). But other oil from the same source moves higher into Cretaceous reservoirs (vertical). Vertical migration along extensional faults and fracture zones is common in rift basins like the North Sea and the Sirte Basin. Such basins are generally less than 160 km (100 mi) in width, and vertical migration distances are relatively short. In some parts of the Viking and Central Grabens in the North Sea, oil migrated from the Jurassic Kimmeridgian source rock up 2,000 ft (610 m) or so into the Cretaceous Chalk. The correlation of the reservoir and source oils was established with biomarkers (Van den Bark and Thomas 1981).

Vertical migration in the North Sea depends on faults to carry the oil from source to reservoir. For example, Leonard (1989) determined the migration and trapping efficiencies for a number of structures in the Central Graben of the Norwegian North Sea. The efficiencies were calculated by dividing the oil in place in the reservoirs by the quantity of oil generated and expelled by the source rock within the drainage area (kitchen) of that structure. Figure 8-17 shows his interpretation for migration into the Tor and Southeast Tor fields in the North Sea. These two structures are next to each other at about the same depth of burial with the same reservoirs and source rocks, but the migration and trapping efficiencies are very different. Tor trapped 11% of the oil expelled, but Southeast Tor trapped only 1.5%. The difference is that the faults under the Tor field lead directly from the Upper Jurassic source shales to the reservoir rock, whereas at Southeast Tor a salt dome partly blocks the passage of oil from the source rock to the faults leading to the reservoir. Some oil probably leaks through small faults, bypassing the salt at Southeast Tor, but the migration pathway is much more impeded than at Tor.

TABLE 8-6 Distribution of Petroleum by Basin Type

Basin classes		*Percentage of world's petroleum reserves (BOE)*
1. Interior cratonic sag		0.7
Denver	Williston	
Illinois	Upper Amazon	
Michigan		
2. Interior rift		23.2
Cuyo	San Jorge	
Erg Oriental	Sirte	
Gippsland	Songliao	
Great Artesian	Suez	
North Sea	West Siberia	
Reconcavo		
3. Divergent margin		8.1
Baku	North Caspian	
Bombay	Northwest Shelf (Australia)	
Campos	Potiguar–Ceara	
Chicontopec	Reforma–Campeche	
Congo	Scotia	
Gabon	Sergipe–Alagoas	
Gulf Coast	Tampico	
Ivory Coast		
4. Active margin		6.3
Los Angeles	Sacramento–San Joaquin	
Magdalena	Santa Maria	
Maracaibo	Ventura–Santa Barbara	
Progreso–Talara	Vienna	
5. Foreland basins and fold belts		56.2
Alberta	Llanos–Barinas	
Appalachian	Magallenes	
Arabian–Iranian	North Slope of Alaska	
Big Horn	Powder River	
Eastern Venezuela	Wind River	
6. Deltas		5.5
Mackenzie	Niger	
Mahakam	Nile	
Mississippi	Po	

Source: Classifications and percentages from H. D. Klemme 1992.

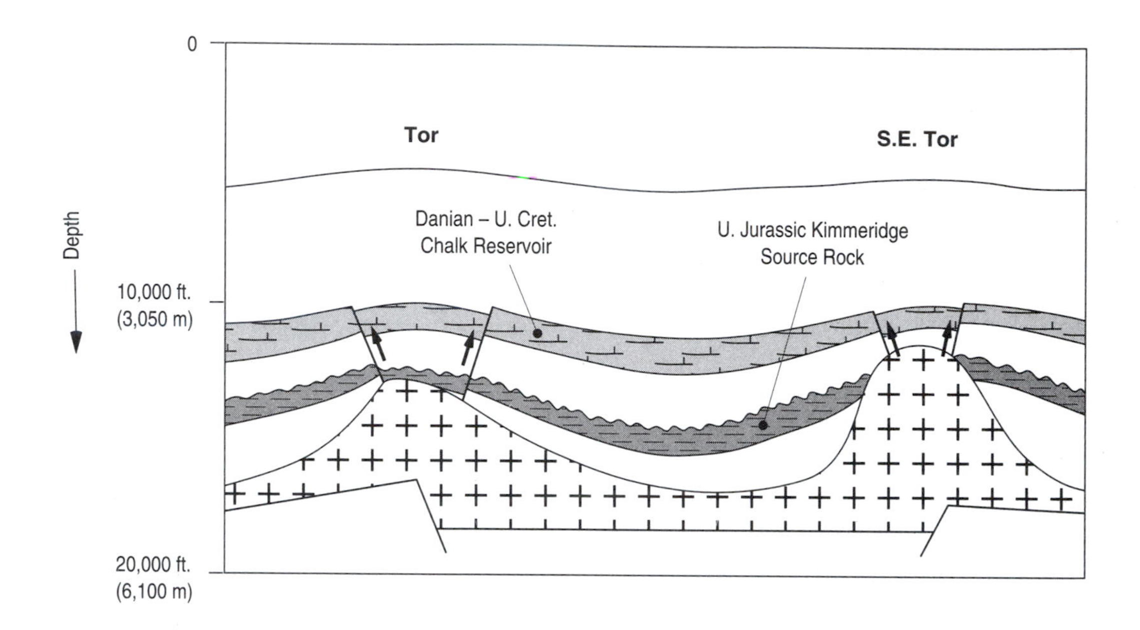

Figure 8-17

Vertical migration at the Tor and Southeast Tor fields in the Central Graben of the southern Norwegian North Sea. [Courtesy of R. C. Leonard]

Stacked reservoirs with the same genetic type of oil are found in many vertically drained petroleum systems, particularly deltas. In the Columbus Basin off Trinidad, the Naparima Hill and Gautier Cretaceous source rocks are deeper than 12,000 ft (3,659 m). Oil from these rocks migrated vertically up a complex network of normal faults to reservoirs that extend from 10,000 to 4,200 ft (3,049 to 1,280 m) in depth (Heppard et al. 1990). The generation and migration of these oils both occurred within 3 million years. Similar vertical stacking is observed in most deltas.

In contrast, the long-distance lateral migration of several hundred miles is typical of foreland basins, as shown in Figure 8-18. The Eastern Venezuelan Basin contains an estimated 1.5 trillion barrels of heavy oil generated by the organic-rich Cretaceous Querecual source rock which is now buried to more than 30,000 ft (9,140 m) in the Maturin depocenter. The oil probably migrated through fractures updip into the Oligocene–Cretaceous unconformity where it continued over a distance of about 275 km (172 mi) to the Orinoco heavy oil belt. Migration was along a gently dipping, featureless homoclinal slope, possibly assisted by hydrodynamics due to a substantial hydraulic head on the faulted northern mobile side of the basin. Low-amplitude faulting on the foreland plate trapped some of the oil at its original 30 to 40° API gravity, as shown in Figure 8-18. Hot pore waters moving up from the depocenter with the oil probably contributed to the higher geothermal gradient observed on the foreland plate, compared with the depocenter. Roadifer (1987) estimated that at least 25 trillion barrels of oil were generated within the depocenter during the subsidence of the Querecual source rock. The heavy oil belt itself is more than 640 km (400 mi) long and about 60 km (40 mi) wide. Most of the conventional oil and all the heavy oil are now preserved in the immature, nongenerating

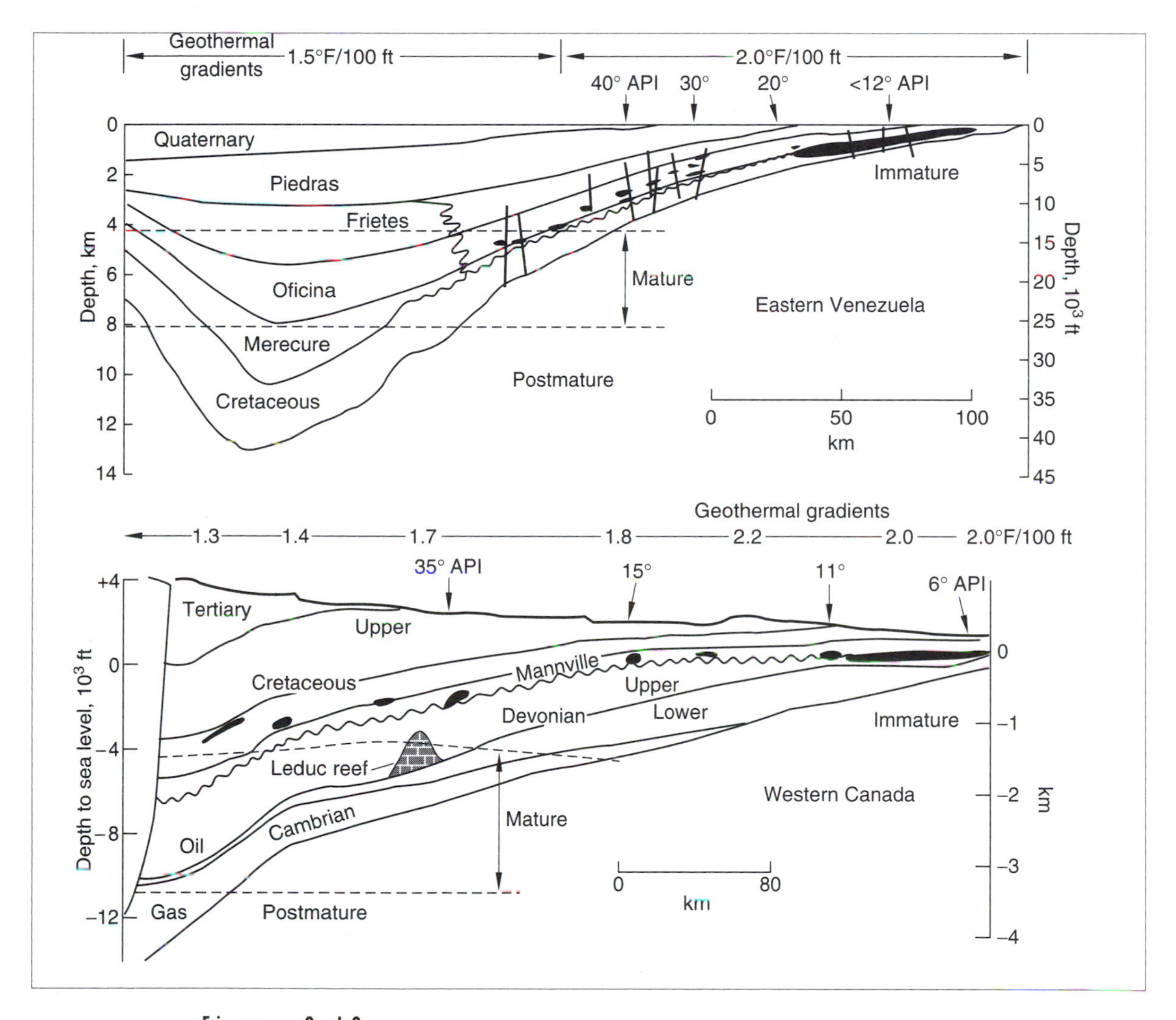

Figure 8-18

Long-distance lateral migration in the Eastern Venezuelan and Western Canada Basins. The oil-generation windows are labeled mature. [Demaison 1977; Roadifer 1987]

foreland plate. Most of the Querecual source rock is now capable of generating only gas.

The generation–migration process for the Western Canada Basin (Figure 8-18) was similar except that there were as many as ten source rocks, from the Devonian, Mississippian, Triassic, Jurassic, and Cretaceous formations, feeding oil into the Devonian–Cretaceous unconformity and the massive Mannville sandstone above it (Creaney and Allan 1990). Most of the 1.2 trillion to 1.7 trillion barrels of heavy oil lie at the unconformity outcrop analogous to the Tar Belts in eastern Venezuela (Roadifer 1987). The maximum migration distance from source to reservoir is around 400 km (250 mi). Creaney and Allan (1990) estimated that the peak generation for some of the source rocks was about 40 million years ago. This means the oil had to move 400 km in 40 million years or less, which amounts to a minimum migration rate of 1 cm/year.

Demaison (1977) noted that the API gravities of the oils in both the Western Canada and Eastern Venezuela Basins decrease as they approach the outcrop (Figure 8-18). The original gravity of the unaltered oils probably was 25 to 35° API. As the oils migrated toward the unconformity outcrop, which had no seal, they were water washed and biodegraded by invading meteoric waters. The API gravity was thus lowered by the removal of most of the gasoline and kerosene. If evaporite covers had existed in these basins, plus 250 mi (400 km) anticlines, as they did in the Middle East, the oil accumulations might have far exceeded in volume those in the Arabian Gulf.

Klemme (1992) determined the percentage of the world's conventional petroleum reserves distributed through six general basin types, as shown earlier in Table 8-6. This table is not intended to show all the basins in the world but, rather, to emphasize significant differences in the reserves of various basin types. These differences are partly related to the drainage areas of the source rocks. For example, note that 56.2% of the world's conventional reserves are in foreland basins and fold belts, the largest of which have huge drainage areas and long-distance migration. The Alberta, Arabian–Iranian, and Eastern Venezuela Basins plus the North Slope of Alaska account for most of this oil. If the unconventional heavy oil were added, the percentage would be even higher. In contrast, deltas account for only 5.5% of the oil.

Part of the reason for this difference is that deltas have comparatively small drainage areas for individual structures with narrowly focused vertical migration. The foreland basins have huge drainage areas with broad, laterally focused migration moving along unconformities, channels, and sheet sands filling the available structures and traps over a vast area. Vertical migration is more efficient than lateral in that it can trap up to 35% of the oil from a single drainage area, compared with less than 5% for lateral. But long-distance lateral migration moves oil out of a far larger volume of source rock.

The fact that hydrocarbons can migrate with pore fluids over long distances through permeable beds means that geochemists must be cautious in condemning a section because it is too immature to generate hydrocarbons. If this nongenerating section is connected to a more deeply buried source rock through continuous sandstones, unconformities, fracture–fault systems, or continental facies, it can contain commercial petroleum accumulations.

SUMMARY

1. Primary migration of oil and gas is movement within the fine-grained portion of the mature source rock, and secondary migration is any movement outside of it. Tertiary migration is the movement of a previously formed accumulation.

2. Primary oil migration within a fine-grained mature source rock with > 2% TOC occurs initially as a bitumen that decomposes to oil and gas and migrates out in a hydrocarbon phase. Generation causes migration. Generation and migration (expulsion) of light oil, condensate, and gas also may occur from low, < 2%, TOC source rocks without an apparent bitumen intermediate, particularly from type III kerogen. Migration in solution is important only with the smallest and most soluble hydrocarbon molecules such as methane, ethane, benzene, and toluene. Migration by diffusion is not significant.

3. The salinity of pore waters in reservoir sands increases by 2 to 50 g/l per 1,000 ft (305 m) of increasing depth. The increase in salinity is caused primarily by salt dissolution and secondarily by membrane filtration.

4. Under hydrostatic conditions, the porosities of many shales in the Gulf Coast tend to decrease below 30% in two linear stages. In Stage 1, the porosity decreases because of mechanical compaction down to a subsurface temperature of 88 to 110°C (190 to 230°F). In Stage 2, there is no further systematic decrease in porosity or increase in density, indicating no compaction.

5. The minimum matrix porosity reached at the top of Stage 2 owing to mechanical compaction ranges from 3 to 15%, depending on the percentage of smectite and illite in the shale. Low minimum porosities have low contents of smectite and illite, and vice versa. Any increases or decreases in porosity in Stage 2 come from pressure solution, cementation, and hydraulic and tectonic fracturing, not from mechanical compaction.

6. Porosities of sandstones and carbonates at depths > 3 km show much greater variability than shales, primarily due to chemical diagenesis, cementation, and dissolution.

7. Shale source rocks act like sieves during primary migration, by releasing more of the small paraffinic and naphthenic molecules and retaining more of the large aromatic and asphaltic molecules.

8. World-class source rocks are brittle because they are composed mainly of detrital quartz (< 4 μm diam.) feldspars, dolomite, and calcite with less than 25% smectite, kaolinite, and illite. Consequently, they tend to develop regional orthogonal fractures and localized tectonic and hydraulic fractures, including microfractures that provide pathways out of the source rock.

9. The main driving force in secondary petroleum migration is the buoyancy of the hydrocarbons. The resisting force is the capillary pressure of the rock–water system. Modifying forces include hydrodynamic fluid flow and abnormal pressure compartments.

10. Under average subsurface conditions, gas has about twice the migration potential of oil because its buoyancy more than overcomes the higher interfacial tension of gas–water systems, relative to oil–water systems.

11. Joints, faults, and lineaments may act as fluid conduits or seals, depending on a variety of factors such as the normal stress across a fault, the nature of the fault surface, and the strata cut by the fault. Unconformities also may act as pathways or seals.

12. The three critical factors for a petroleum trap, in order of importance, are (1) its proximity to a hydrocarbon migration pathway, (2) the permeability of its seal, and (3) the trap size.

13. Vertical migration pathways include joints, faults, lineaments, fractures, and diapirs. Lateral pathways include unconformities and regional permeable beds such as channel sandstones.

14. The best petroleum seals, in order of importance, are (1) gas hydrates, (2) evaporites, (3) ductile shales rich in clay minerals, (4) low-permeability carbonates, (5) faults, and (6) asphalt.

15. Vertical migration is more efficient than lateral migration, but less petroleum is collected because individual structures drain only the relatively small areas directly below them. Lateral migration can drain petroleum from a far larger volume of source rock. For example, foreland basins and fold belts with huge drainage areas have trapped over half of the world's conventional petroleum reserves. In contrast, deltas with dominantly vertical migration account for only 5%.

SUPPLEMENTARY READING

Doligez, B. (ed.). 1987. *Migration of hydrocarbons in sedimentary basins.* 2nd IFP Exploration Research Conference, Carcan, France, June 15–19, 1987. Paris: Éditions Technip, 681 p.

Magara, K. 1978. *Compaction and fluid migration: Practical petroleum geology.* Vol. 9 of *Developments in Petroleum Science.* New York: Elsevier Science Publishing, 319 p.

Roberts, W. H. III, and R. J. Cordell (eds.). 1980. *Problems of petroleum migration.* AAPG Studies in Geology 10. Tulsa: American Association of Petroleum Geologists, 273 p.

Chapter

9

Abnormal Pressures

Abnormal pressures include both under- and overpressures. The *hydrostatic gradient* is the pressure increase with depth of a liquid in contact with the surface. The gradient for fresh water is about 9.8 kPa/m (kilopascals per meter), or 0.433 psi/ft. The *lithostatic gradient* is the total pressure increase caused by rock grains plus water. It averages 24.4 kPa/m, or about 1.08 psi/ft. If a hole were drilled through a highly permeable sand, extending from the surface to 20,000 ft (6,096 m), the pressure at all depths would be hydrostatic, following line A in Figure 9-1 for 10.4 kPa/m (0.46 psi/ft). This is the gradient for most reservoirs. An increase in salinity with depth causes the pressure gradient to increase 0.098 kPa/m (0.0043 psi/ft) for each increase of 0.01 g/cc in fluid density (Levorsen 1967, p. 685). Any departure from the commonplace hydrostatic pressure is an *abnormal pressure.* This includes *overpressures* generally ranging above 12 kPa/m (0.53 psi/ft) and *underpressures* below 9.8 kPa/m (0.43 psi/ft).

Abnormal overpressures (> 12kPa/m, or > 0.53 psi/ft) are created and maintained by the inability of pore fluids to migrate within a reasonable geologic time period when subjected to stresses causing increased fluid pressure. Several types of stresses can increase fluid pressure such as (1) rapid loading which may cause compaction disequilibrium, (2) thermal expansion of fluids, (3) compression by tectonic forces, and (4) generation of oil and gas from organic matter in the rock matrix. There are probably other forms of stress in addition to the four listed above, but all of them point to the inability of the fluids to migrate as the cause of abnormal pressure.

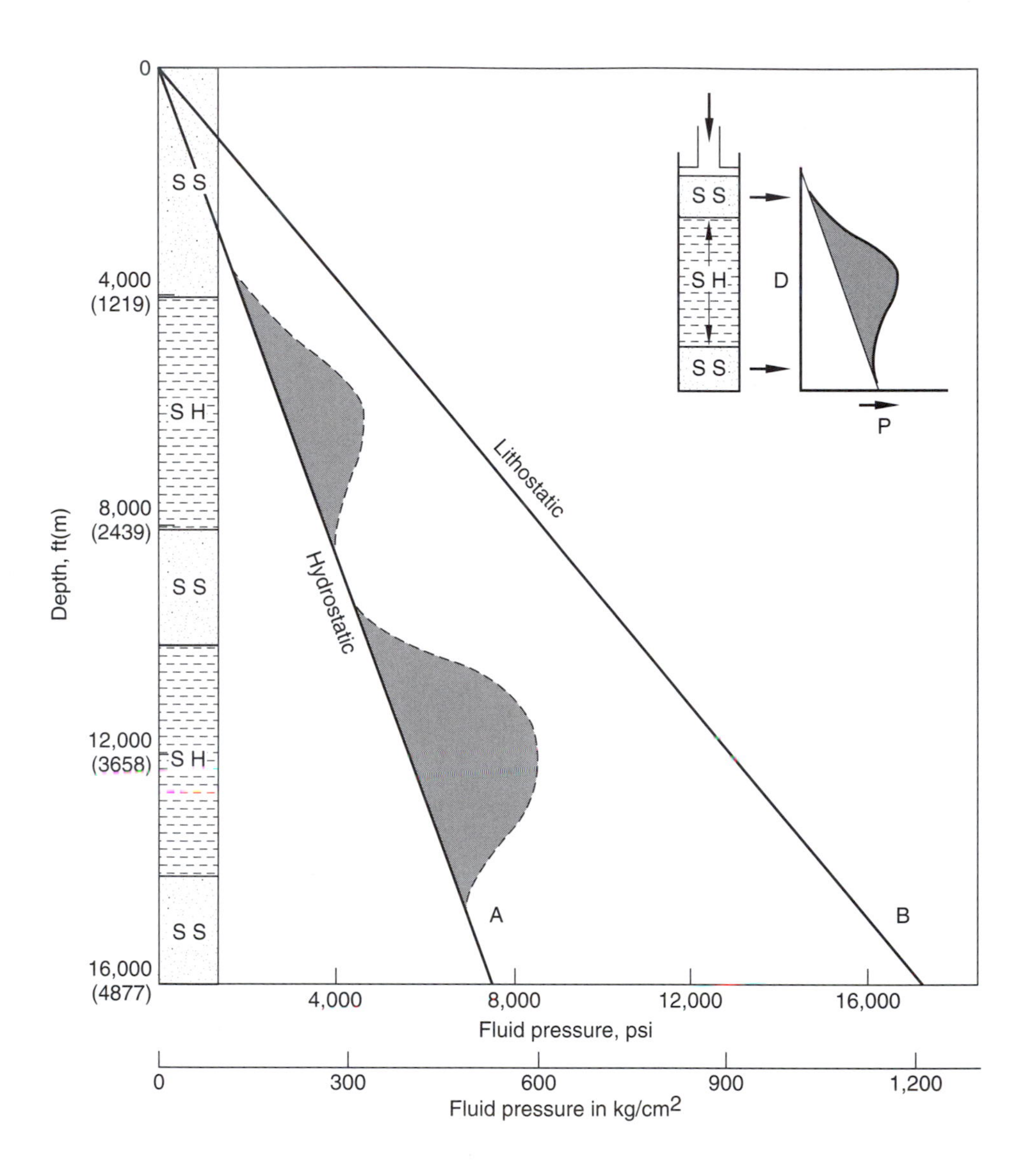

Figure 9-1

Increase in fluid pressure with depth. Line A is the hydrostatic gradient for most petroleum basins, 0.46 psi/ft (10.4 kPa/m). Line B is the lithostatic gradient, 1.08 psi/ft (24.4 kPa/m). The dashed line shows fluid pressures in the sandstone–shale sequence to the left. The diagram in the upper right shows the change in fluid pressure when pressure is applied to layers of sand and clay mud. The center of the shale is a pressure barrier to fluid movement. Fluids are escaping upward and downward to sand beds which are in contact with the surface.

Abnormal pressures are found throughout the world. Fertl et al. (1994 p. 1–15) cites examples in the Middle East, Europe, Russia, Africa, China, Australia, Indonesia and North and South America. Overpressures may be shallow or deep. For example in the Paleogene Formation in the Tadzhik Depression (Russia) the pressure is 11 MPa (1,595 psi) at a depth of 480 m (1,575 ft) (Kalomazov and Vakhitov 1975). This is about twice the hydrostatic pressure. The oil in these high-pressure regions is lighter than those in normal-pressure regions of the Tadzhik petroleum basin. In the Beshtentyak field, the Paleogene Shales at a subsurface depth between 480 and 2,787 m (1,575 and 9,141 ft) have a mean formation pressure 1.8 times hydrostatic. At about 2,700 m (8,858 ft), the fluid pressure was reported to be 2.3 times hydrostatic, which is about equivalent to the lithostatic pressure.

An important phenomenon of some, but not all, overpressured formations is the increase in the temperature gradient. Well flowline temperature gradients in areas such as parts of the Gulf Coast, the North Sea, and the South China Sea have been observed to increase appreciably prior to or when entering a high-pressure zone. This temperature increase enhances hydrocarbon generation.

Leach (1994) reported an increase in temperature with pressure in 60 Gulf Coast reservoirs. All of the reservoirs at temperatures below 93°C (200°F) have pore pressures lower than 13.5 kPa/m (0.6 psi/ft). Above 93°C (200°F), particularly in the range from about 112 to 140°C (234 to 284°F), most of the 60 Gulf Coast reservoirs are overpressured. Interestingly, most hydrocarbons form at temperatures above 93°C. For example, a steady-state catalytic model for the generation of light hydrocarbons suggests that nearly all C_7 alkanes are formed between 95 and 135°C (203 and 275°F)(BeMent et al. 1994. See Chapter 15). This is almost the same range as reported by Leach.

High geothermal gradients in overpressured shales have been attributed to the low thermal conductivity of such shales. They seem to act as heat insulators causing abnormal temperature gradients to exist across them. For more examples of high gradients see Hunt 1979, pp. 240–245.

Two Types of Overpressures

Overpressures develop in basins under two sets of conditions (Gretener and Bloch 1992). These are (1) compaction disequilibrium in which there is unrestricted lateral flow and restricted vertical flow and (2) sealed compartments in which there is restricted lateral and vertical flow. An example of compaction disequilibrium is shown in the upper right corner of Figure 9-1. Sand, clay mud, and sand are shown in alternate layers in a container. If pressure is applied on top of these beds and fluids are allowed to escape from the top and bottom sands, differential compaction will occur within the clay mud. The edges of the mud near the sands compact more readily than does the center of the mud.

This means that fluid pressures are higher in the center of the mud. This differential fluid pressure can become much greater than hydrostatic, as seen in the example at the left of Figure 9-1. A schematic indication of the variability in pressure for this sedimentary column is shown as a dashed line between lines A and B. This dashed line assumes a uniform rate of compaction for the section and continuity of sand permeability to the surface. Under these conditions, all three sands would be normally pressured (equilibrium), with the shales overpressured (disequilibrium) to varying degrees, depending on their level of hydraulic conductivity. Time is also a factor. In this example, if the pressure were increased very slowly over millions of years, as in a slowly compacting basin, fluids would have more time to escape, so pressures would be only slightly above equilibrium.

Compaction disequilibrium generally occurs in thick > 500 m (1,640 ft) ductile shales containing over 30% water-absorbing clay minerals such as smectite and illite. The center of such thick shales is a pressure barrier to vertical fluid movement through the entire section. Those fluids escaping compaction in Figure 9-1 must move down in the lower half of the shales and up in the upper half to enter the adjoining sands. The center of such thick shales is *undercompacted* because its porosity is higher in the center than that of the shale sections above and below it that are contacting the sandstones. An example of an undercompacted shale is shown in Figure 9-2, which plots porosity versus depth for a well at Mustang Island, Texas. Note that the highest porosity is at about 7,500 ft (2,287 m). This is the center of a thick shale section, with somewhat more permeable sandy or silty sediments above and below where the porosities are lower.

The second type of overpressure occurs in deep (>10,000 ft, 3,049 m) fluid compartments that are sealed on all sides, thereby restricting both lateral and vertical fluid flow (Bradley 1975). The lateral seals may be sealing faults, facies changes, or salt ridges. An example of this is the pressure/depth plot of the Morganza field step-out, Pointe Coupee Parish, Louisiana, shown in Figure 9-3. Pore-pressure measurements, made in the well with the Schlumberger repeat formation tester (RFT), show normal hydrostatic pressures down to a depth of about 18,630 ft (5,680 m), after which the pressure is increased 5,075 psi (35 MPa) across 210 ft (64 m) of lenticular sandstones interbedded with shales. This pressure increase is followed by a second large increase in pressure about 500 ft (153 m) deeper. At the bottom of the hole the overpressure gradient is 0.85 psi/ft (19 kPa/m), which is close to rock fracture pressure. The seals are essentially planar on top. They cut across both time and lithostratigraphic boundaries.

The pressure/depth profile in Figure 9-3 shows the stair-step change in pressure that is characteristic of multiple stacked-pressure compartments. The source rock is the fractured Lower Tuscaloosa Shale in Compartment 2, and most of the pay section is in Compartment 1. The generation of hydrocarbons in Compartment 2 led eventually to the fracturing of the seal, which enabled the hydrocarbons to migrate vertically and accumulate in the first compartment. Subsequent fracturing of its seal led to small accumulations in the normally pressured section above.

This second type of overpressure is due to the development of both vertical and lateral seals, which are essentially impermeable to all fluid movement of water, oil, and gas. Many such seals form at depths around 10,000 ft (3,049 m) where chemical diagenesis is active, as previously pointed out. The seals form through precipitation of minerals such as quartz and carbonates vertically in

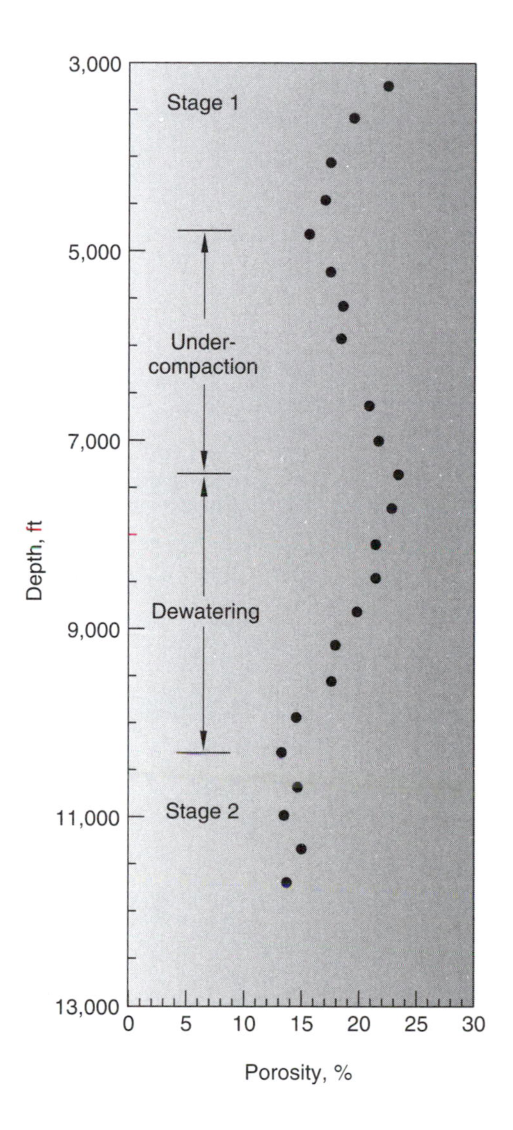

Figure 9-2

Porosity/depth profile showing undercompaction in a thick shale at Mustang Island, at 27°8′S, 96°8′W offshore the south Texas Gulf Coast.

faults and fractures and horizontally in sandstones, siltstones, and carbonates along a thermocline. The majority of these deep (> 3 km) overpressured compartments do not exhibit the compaction disequilibrium of the first type of overpressure. Rather, they are normally compacted, possibly because many of them are composed of fine-grained quartz and carbonates with generally < 25% water-absorbing clay minerals. Being more brittle than rocks in the first type of overpressure, the strata tend to fracture under high stress and to dewater normally during compaction. Extensive fluid movement through fractures and faults instead of through the rock groundmass results in cementation creating the sealed compartment that leads to overpressures.

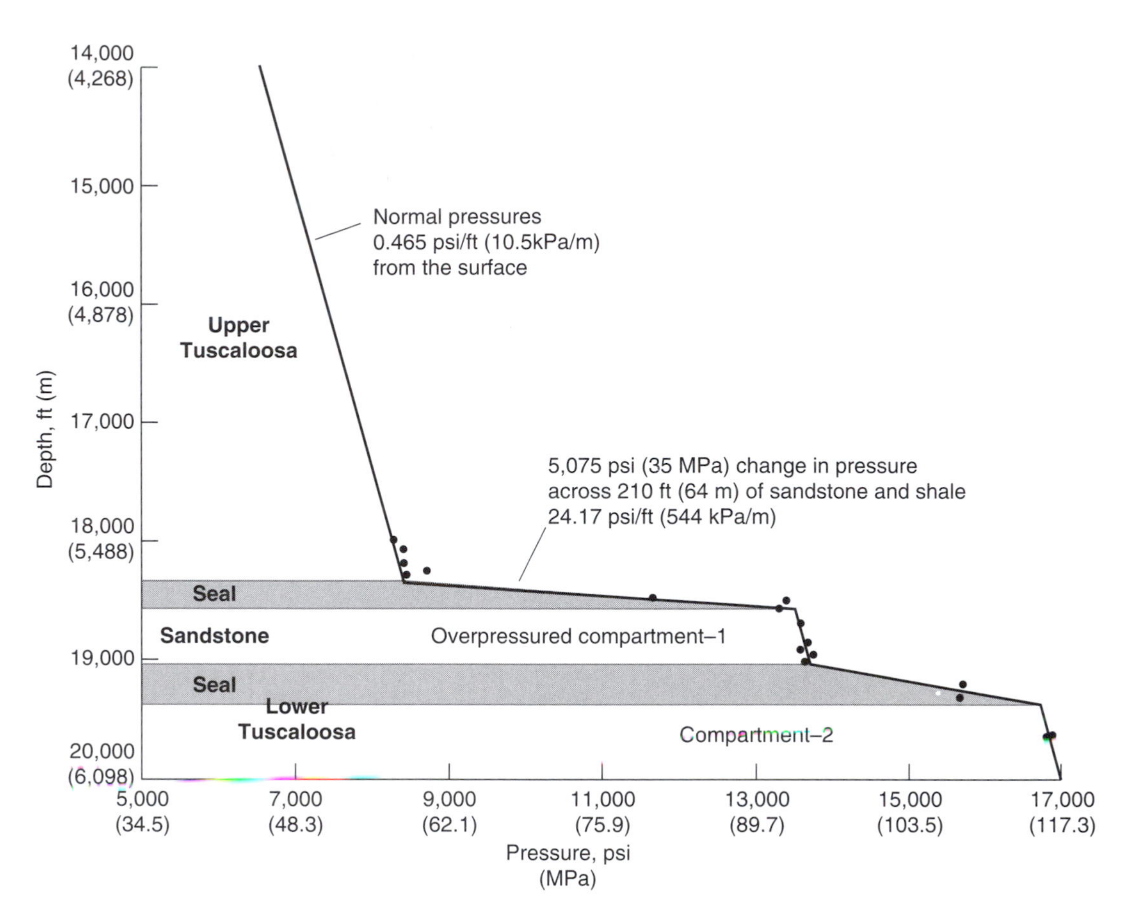

Figure 9-3

Overpressured fluid compartments in the Morganza field, Pointe Coupee Parish, Louisiana. [Courtesy of D. E. Powley]

Characteristics of Pressure Compartments

Pressures in a confined system were first explained by the seventeenth-century French philosopher Blaise Pascal, for whom the pressure unit is named in the metric system. He stated that if an external pressure is applied to a confined fluid at rest, the pressure at every point in the fluid will be increased by the amount of the external pressure. Pore fluids are an example of fluids confined in a sealed compartment. The concept of fluid- and matrix-supported seals is illustrated by the cartoons in Figure 9-4. For example, suppose that the lower part of a rock section is completely enclosed by a seal, as shown. If the entire

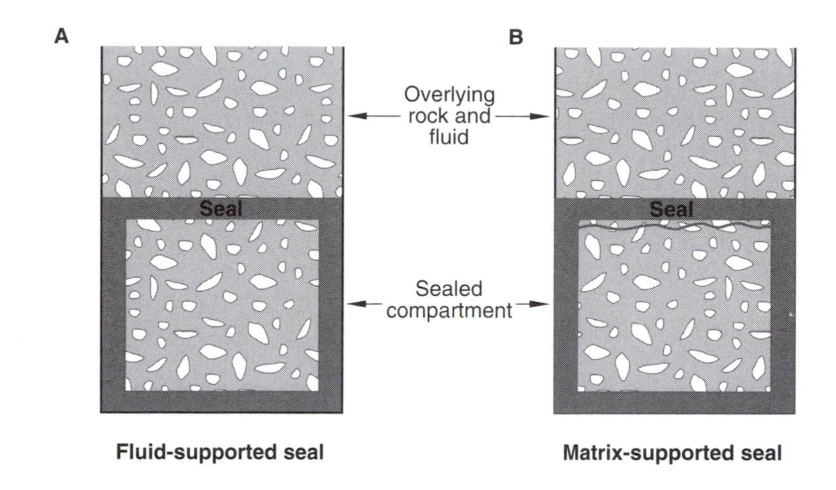

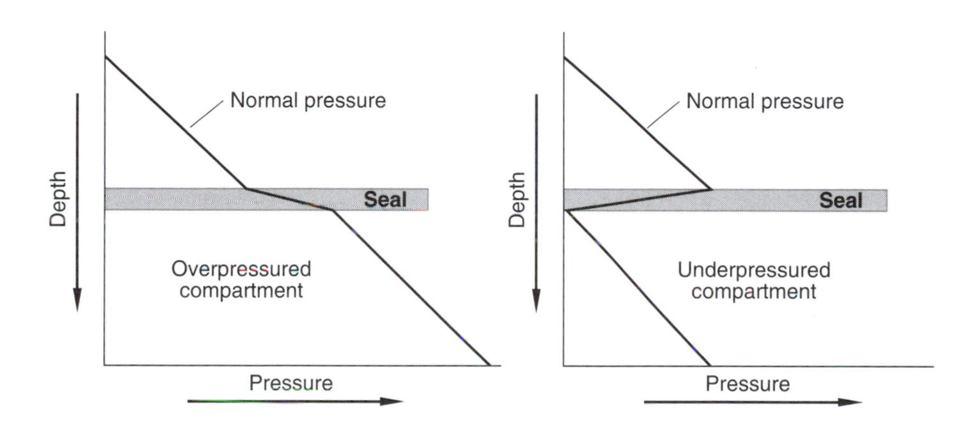

Figure 9-4

(A) Pressure/depth gradient from normal pressure to overpressure across the hydraulic seal surrounding the overpressured rock section. (B) Pressure/depth gradient from normal to underpressure across the hydraulic seal surrounding the underpressured rock section.

section is buried deeper after sealing, there will be thermal expansion of fluids in the sealed compartment, possible generation of liquid and gaseous hydrocarbons, and, in some basins, mechanical collapse or flow of the rock matrix, as with salt flows. Each of these processes increases the pressure, driving the pressure/depth gradient (P/D) to greater values (Figure 9-4 [A]). The slope of the P/D line in the fluid compartment remains the same as that in the normally pressured rocks above the seal. This assumes that the density of the fluid above and below the seal is the same and that there is internal hydraulic communication in the sealed-off rock mass. In Figure 9-4 (A) the fluid below the seal supports the weight of the seal and part of the overlying rock and fluid load. The magnitude of the increase in pressure depends on how much of the weight of the overlying rock column is borne by the compartment fluids, compared with what the rock matrix is supporting in the compartment.

Figure 9-4 (B) represents the opposite case, in which the sedimentary system is uplifted and eroded, thereby cooling the rock units after a seal has been formed. Shrinkage of the fluids with cooling is the major cause of the drop in pressure. In addition, studies of fluid withdrawal from petroleum reservoirs have found that the pore volume increases by a factor of about 7×10^{-6} for each psi (6.9 kPa) drop in pressure (Dickey 1972). These combined changes may cause the fluid level in the sealed compartment to drop slightly below the seal, leaving the rock matrix to hold up the seal and overlying rock and fluid load (Figure 9-4 [B]). The magnitude of the pressure drop depends on how much of the overlying load's weight is shifted from the pore fluid to the rock matrix below the seal. In underpressured compartments, the P/D line in the lower part of Figure 9-4 (B) may intersect the depth axis below, at, or above the base of the seal. It is common to see it above and rare to see it below. Occasionally it occurs at the seal base, as at Keyes field, discussed next.

Case Studies: Underpressures

Keyes Field, Oklahoma

In the area of the Amarillo Uplift where the Texas and Oklahoma panhandles converge, the sedimentary section has been uplifted and eroded about 1,500 m (4,920 ft). The Keyes field in the Oklahoma panhandle has been producing gas from a depth of about 1,460 m (4,790 ft) for about thirty years. The pressure/depth gradient is shown in Figure 9-5. The pressures are normal down to the Blaine Anhydrite, which acts as a seal. Below the anhydrite are two pressure compartments. In the upper compartment (1), pressures start at zero and increase down to the Wellington Salt at about 853 m (2,800 ft), which is the second seal. Pressures drop to near zero in the lower compartment (2) under this seal and then continue to increase along a gradient of 10.5 kPa/m (0.465 psi/ft) to the Pre-Cambrian basement. The Keyes field is underpressured by about 8.97 MPa (1,300 psi). The rock matrix at the base of each of the two seals in Figure 9-5 bears the entire weight of the overburden, so the fluid pressures start at about zero below these seals. Keyes field is a good example of the matrix-supported seal depicted in Figure 9-4 (B).

Underpressured reservoirs are common in the United States in rocks that have been uplifted, eroded, and cooled during the Laramide orogeny and later

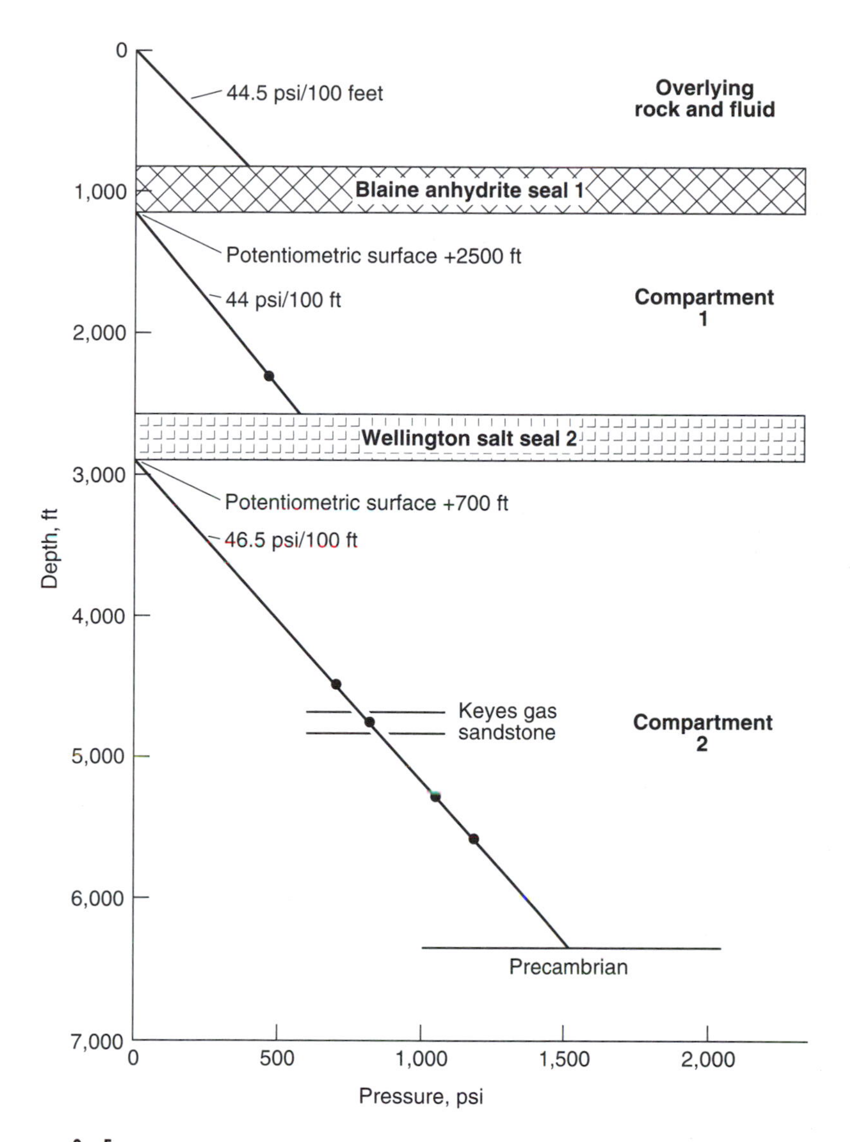

Figure 9-5

Pressure/depth gradient, Keyes field, Cimarron and Texas counties, Oklahoma, based on information from Dwight Energy Data Company. [Powley 1980]

(Russell 1972). Many fields in the uplifted western part of the Alberta Basin also are underpressured. All large gas fields in the United States, like San Juan, Hugoton, Wattenberg, and the old Seminole producing area in Oklahoma, are uplifted, eroded, cooled, and underpressured.

Underpressured reservoirs sometimes overlie overpressured reservoirs in areas of uplift. Canadian Hunter's Brassey field in the Deep Western Canada Basin produces 54°API gravity oil from the Artex Formation in a highly overpressured, normally compacted compartment at 2,903 m (9,522 ft). The top seal is a dolomite with the pores plugged with anhydrite. The source rock below the Artex Formation reached its greatest depth at 5,350 m (17,548 ft). Since then it was uplifted (with erosional cooling) to 2,903 m (9,522 ft). The Cadotte sandstone higher up at 1,376 m (4,513 ft) produces gas from an underpressured compartment (H. von der Dick 1990, personal communication).

In the Mill Creek graben of Oklahoma, south of the Arbuckle Mountains, the South Eola, Northeast Purdy, and West Criner–Payne fields are underpressured below the Marmaton–Morrow fluid pressure seal. The potentiometric surface of all the producing formations in this compartment, based on DST data, is 488 m (1,600 ft) below sea level (Powley 1985). Several thousand feet below this is an overpressured compartment with a planar top seal that cuts across the graben through several Paleozoic formations.

Case Studies: Overpressures

U.S. Gulf Coast

This area has the typical layered arrangement of two or more hydraulic systems, as pictured in Figure 9-3. The normally pressured, free-flow hydraulic system described by Hubbert (1953) is basinwide and extends from the surface down to 3,048 m (10,000 ft) or more. At greater depths, sealed, overpressured fluid compartments are common. Pressures build up in the compartments as the basin sinks, because of a combination of factors, including the thermal expansion of pore fluids and the generation of liquid and gaseous hydrocarbons from kerogen. Eventually, as the pressures approach lithostatic, the top seal breaks at the shallowest point, and the oil, gas, and other pore fluids rush out of the compartment, causing an abrupt drop in pressure. Gradually the break reseals, and the pressures begin to build again toward a later breakout.

The release of fluids is believed to be episodic, comparable to the pulsed dewatering mechanism proposed for Mississippi Valley type lead–zinc deposits by Cathles and Smith (1983). Their model studies showed that episodic dewatering was the only process that could adequately explain the color banding of sphalerite, the cycles of sulfide precipitation, and the dissolution and other distinctive local tectonic features associated with the deposits. An episodic expulsion of pore fluids could explain some of the large oil accumulations just above the seals in deltas such as the Mississippi Delta and in rift grabens such as the North Sea. Both of these are cases of vertical migration in which hydrocarbons are blown out of the pressurized compartments when their seals rupture.

Whelan et al. (1994) monitored this process in considerable detail in Eugene Island Block 330 (EI-330) offshore Louisiana. They called the process *dynamic fluid injection* and concluded from geochemical data that episodic expulsions of pore fluids were occurring up faults on short time scales, possibly as short as years. They found that abnormally high concentrations of C_3 to C_9 light hydrocarbons occur in the EI-330 oils, with the maximum carbon number increasing regularly with increasing depth. This would be characteristic of fractionation during vertical migration. In shallower reservoirs at depths around 4,500 ft (1,372 m), the light *n*-alkanes are superimposed on a background of biodegraded oils indicative of recent oil remigration. They also found anomalously high vitrinite reflectance values along a fault thought to be a carrier of hot fluids from depth. Finally, the EI-330 reservoirs have a history of anomalous overproduction of petroleum, compared with surrounding reservoirs, that dates back to 1972. This suggests that the reservoirs are being periodically refilled by relatively recent episodic injections of petroleum and related pore fluids.

Vermilion Area, Offshore Louisiana

Some of the thick, >300 m (>984 ft) seals in the deep overpressured compartments of the U.S. Gulf Coast produce oil from the permeable layers within the seals. This also occurs at Cook Inlet in Alaska, the Sacramento Valley of California, and other areas (Hunt 1990). In the Vermilion area offshore Louisiana (Figure 9-6), oil is produced within and above the seal. Pressure measurements show the characteristically abrupt discontinuity in pressure at about 4,000 m (13,100 ft). The pressure increase is 41.4 MPa (6,000 psi) across the entire seal. The seal itself is layered, with permeable sands sandwiched between calcite-filled shales with essentially no porosity or permeability. The pressure changes linearly with increasing depth to, across, and below these alternating permeable and impermeable beds. The oil and gas generated in the compartment below breaks through successive impermeable levels of the seal. It fills some sands in the seal before breaking through to the normal hydrostatic pressures above. The upper part of the seal at Vermilion contains about 30 m (100 ft) of sandstone, which has been producing oil for at least thirty years. This indicates that the reservoir is drawing on a tremendous volume or that recharging is occurring. A smaller amount of oil has been produced above the seal. Below the seal the shales are less dense and show both a higher geothermal gradient and lower resistivity. This is characteristic of most wells offshore Louisiana. In contrast, many onshore overpressured wells of Texas and Oklahoma show no change in shale porosity or density in the overpressured compartments. These onshore compartments are normally compacted. Also, in some overpressured North Sea fields and at the Malossa field in Italy, the overpressured fluid compartments are not undercompacted.

Geothermal gradients sometimes increase in overpressured zones, as seen in Figure 9-6. I called these *dogleg geothermal gradients* (1979, p. 242) and

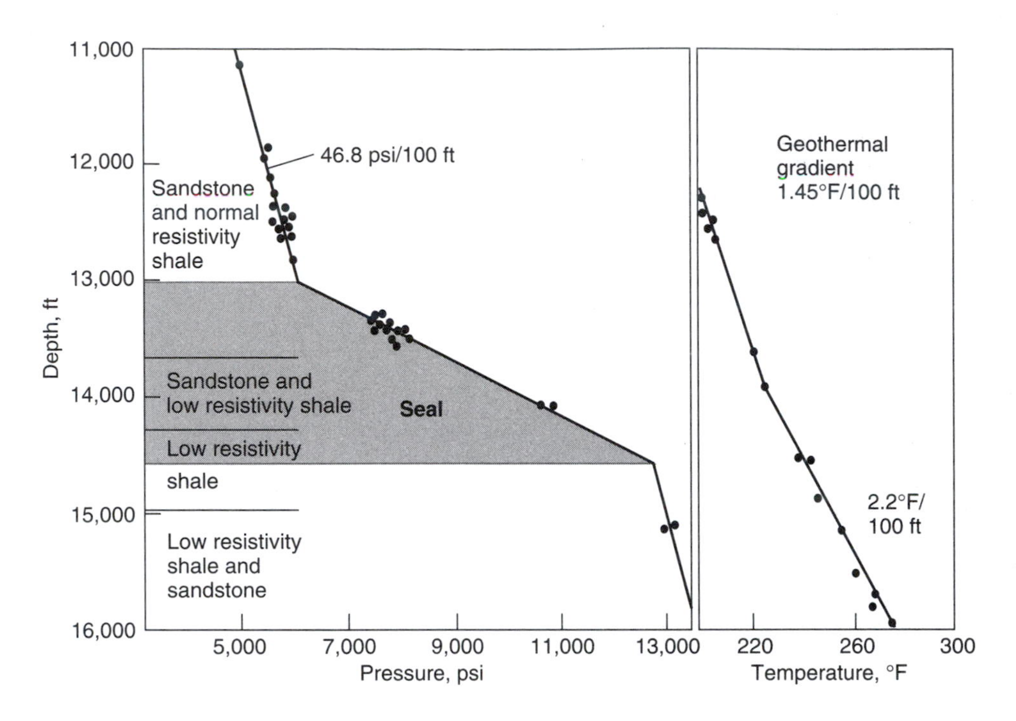

Figure 9-6

Pressure/depth gradient in Lower Miocene formations, Vermilion area, offshore Louisiana. Data points are DST measurement from six wells in Blocks 14 and 15. Geothermal gradients decrease to 1.8°C/100 m (1.0°F/100 ft) in shallower beds located above the areas shown. [Powley 1985]

pointed out that their effect on vitrinite could be used to identify paleo high-pressure zones that existed in the past. The change in slope of the gradient would cause a similar change in the vitrinite reflectance profiles, which then would be permanently recorded in the sediments. Law et al. (1989) recognized profiles with two or more changes in slope. They named them *kinky reflectance profiles* and found them in several basins of the Rocky Mountain region of the United States and Canada (see Chapter 10 for more details).

The False River field in the same Tuscaloosa trend as the Morganza field in Figure 9-3 shows an increase in geothermal gradient from 1.8°C/100 m (0.9°F/100 ft) to 4.5°C/100 m (2.5°F/100 ft) when going from normal to overpressures. Jones and Wallace (1974) found that geothermal gradients in a six-county area of the Texas Gulf Coast range from 2.5 to 10°C/100 m (1.4 to 5.5°F/100 ft). Matviyenko (1975) found overpressured Paleogene and Upper

Cretaceous shales of West Siberia to have geothermal gradients of 3.6 to 5.3°C/100 m (2 to 3°F/100 ft). Below these shales there was a normally pressured section of alternating sands and shales of Mesozoic age, with geothermal gradients of 2.8 to 3.8°C/100 m (1.5 to 2.1°F/100 ft).

Geothermal gradients are about one and a half times greater in rocks of many overpressured compartments than in normally pressured rocks of similar depth and lithology. Some compartments show no change in the geothermal gradient, such as Venture field, Canada, discussed later.

Studies in the U.S. Gulf Coast have shown that after a seal is broken, the compartment fluid pressure drops to a pressure/depth gradient of about 13.5 kPa/m (0.6 psi/ft), but not all the way back to hydrostatic. The fracture openings appear to reseal while moderate overpressures still exist in the compartments. This observation is based on pressure measurements in gas pools released by the Federal Power Commission (1973). There are almost no overpressures between 13.5 kPa/m (0.6 psi/ft) and normal hydrostatic pressure.

Long-term production from a high-porosity bed in a seal as at Vermilion is not unusual. The Willows–Beehive Bend gas field in the Sacramento Valley of California has a 300 m (1,000 ft) thick seal at a depth of about 1,300 m (4,264 ft). It consists of multiple thin impermeable layers of mineralization going across a massive sandstone. Gas has been produced for decades from an 18 m (60 ft) unmineralized section in the seal. Again, such long-term production indicates that these porosity windows within seals can be very extensive in area and that thick seals can have alternating layers of permeability and impermeability. The source rock for the gas is the Forbes Shale just below the seal (Hunt 1990).

Iran

The Alborz discovery in Central Iran described by Gretener (1982) is a spectacular example of major oil production from a reservoir almost at lithostatic pressure. The discovery well was drilled through 400 m (1,300 ft) of evaporite and penetrated the fractured Qum Limestone of Oligo-Miocene age by a mere 5 cm (2 in). At that point the entire drill string and mud column were blown out of the hole. The mud pressure was 8,000 psi (55 MPa) at a reservoir depth of about 2,700 m (8,800 ft). This was a pressure depth ratio of 0.91 psi/ft (20.5 kPa/m), essentially fracture pressure. The well "produced" 5 million barrels of oil and an unknown large volume of gas for eighty-two days before it bridged itself and the flow died. The surface temperature of the flowing oil was 115°C (240°F). The presumed source rocks downdip were at 230°C (446°F), indicating they had gone clear through the oil-generating stage to the gas-generating stage. Gretener (1982) cited this as an example of a very high pressure, commercially attractive petroleum accumulation with high reservoir permeability and a strong gas drive. He concluded that the conversion of kerogen in the Qum marls to liquid and gaseous hydrocarbons caused the high reservoir pressure and saturated gas content of the oil.

The highest pressures encountered in fluid compartments are found in carbonates, especially those with evaporite seals. Such compartments can be huge (Alborz was 12 by 50 km (7 by 30 mi). They also can be very small. An example of a miniature high-pressure compartment was encountered at 11,822 ft (3,603 m) in the Upper Jurassic Cotton Valley limestone of east Texas (Forgotson 1979). The pressure gauge exceeded 12,000 psi before the well blew out, indicating a P/D gradient of 0.99 psi/ft (22.3 kPa/m). The excessive pressure dissipated rapidly, and an offset well at the same depth registered only normal pressures. These examples demonstrate that the seals for carbonate compartments can be extremely tight, and the compartments themselves, small in volume. Also, such pressures are almost certainly caused by the generation of gas.

North Sea

At Ekofisk and nearby fields of the North Sea Basin there is a seal at about 3,300 m (10,800 ft). This seal is planar topped in that it is found at about the same depth extending northeast along the Central Graben for over 100 km. The seal is deeper north and south of Ekofisk, as shown in Plate 5B (Leonard, personal communication). It is following a thermocline because the geothermal gradient is >2.0°F/100 ft (3.7°C/100 m) at Ekofisk, and it decreases north and south to 1.8°F/100 ft (3.4°C/100 m). The seal does not follow stratigraphic horizons (Plate 5B). This also is characteristic of other areas such as Cook Inlet, Alaska (Powley 1980). Below the seal at varying depths is the organic-rich Kimmeridge Shale which has been identified as the source rock for the Ekofisk oil and many of the oils along the Central and Viking Grabens (Demaison 1984). A shallow seal also exists at about 1,830 m (6,000 ft), creating both an upper and lower Ekofisk fluid-pressure compartment (Figure 9-7). Almost all the oil and gas in the southern North Sea Central Graben is in the Upper Ekofisk compartment just above the seal. This includes large fields such as Ekofisk, Eldfisk, and Tor. The vertical distribution of the petroleum accumulations suggests that as the North Sea Basin sank, the increasing temperatures caused the thermal expansion of fluids and the increasing generation of oil and gas from the Kimmeridge Shale source rock. These increases contributed to higher pressures until the fluids eventually broke through the top seals of the lower compartment and accumulated in the nearest overlying chalk reservoirs. These examples show that the petroleum geology of fluid compartments is really the geology of seals.

The release of only a small amount of fluid from a sealed compartment causes a marked drop in pressure. Consequently, many episodic breaks are needed to create a large accumulation of oil. However, if the fluid released is mostly oil and gas that have accumulated directly beneath the seal, fewer breakthroughs are required to build a petroleum accumulation than if the effluent is mostly water.

Once there has been a breakthrough, the rate of vertical movement may be extremely rapid. A very deep test well was drilled in Iran below the giant

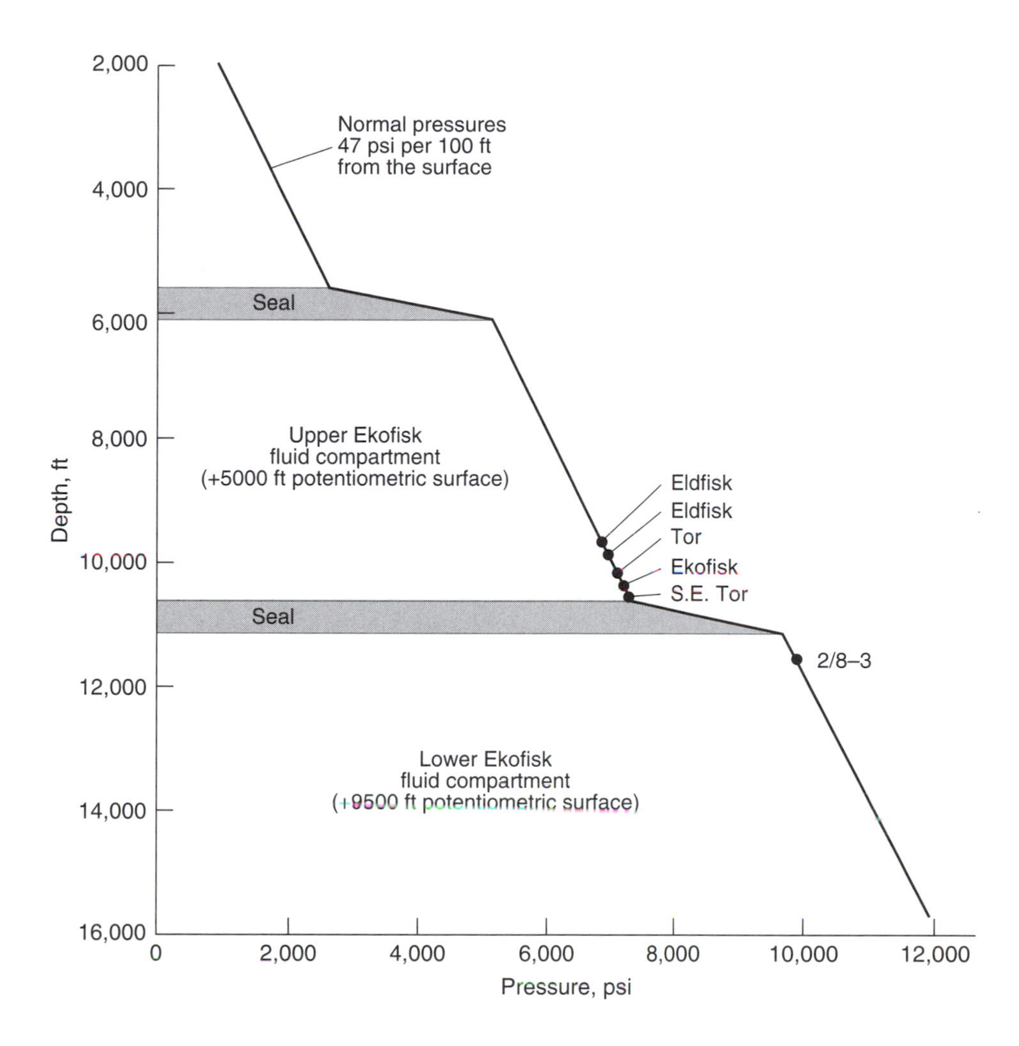

Figure 9-7

Pressure/depth gradient for Ekofisk and nearby fields in Central Graben, North Sea Basin. [Powley 1986]

Masjid-i-Suleiman Asmari reservoir of Tertiary age into the basinal facies of the Jurassic and Lower Cretaceous where fractures during folding would be unlikely to form (Dunnington 1967b). The well was subsequently cemented and abandoned, but the cement job failed. The artificial fracture so produced was able to repressure the huge producing Asmari reservoir above at a rate of about 69 kPa (10 psi)/month. This represented an enormous transfer of gas from a deep, high-pressure zone to a shallow, low-pressure zone (depleted by production) over a very short geological time scale.

Some horizontal seals in the North Sea are heavily mineralized from top to bottom with calcite and quartz. This mineralization causes drilling rates to be drastically reduced across a seal. For example, the Cretaceous chalk normally cores very rapidly, but in the Shell–Esso 30/6-2 well it took twenty-four hours to cut an 18 m (60 ft) core in a silica-enriched seal. The pores in the chalk were filled solidly with quartz from top to bottom. Several well log interpretation techniques have been developed to recognize the mineralization associated with seals as well as the changes in pressures across seals.

Venture Field, Scotian Shelf, Nova Scotia, Canada

In the North Sea cross section (Plate 5B), the pressure seals occur along thermoclines that are independent of structure and stratigraphy. The Venture field has two overpressured compartments, as does the North Sea, but the seals of the former appear to be related to lithology as well as temperature. The top seal of the first overpressured compartment includes some sandstones that have lost essentially all their primary porosity because of quartz overgrowth and massive welding of the quartz framework. Plate 8D is a photomicrograph taken by V. H. Noguera Urrea of a nonporous sandstone at 14,500 ft (4,436 m), which is at the top of the first seal. This is only about 200 ft (61 m) above the onset of low overpressures. Also, as part of the top seal, there are 70 to 80 ft (21 to 24 m) thick carbonate stringers overlain by shales. The first carbonate seal is at about 15,400 ft (4,695 m) near the Mississauga–Mic Mac formation boundary (Figure 9-8). The normal pressures above have a gradient of 0.49 psi/ft (11 kPa/m), whereas in the first pressure compartment below, the gradient ranges from 0.73 to 0.76 psi/ft (16.4 to 17.1 kPa/m). Below the second seal, starting around 16,750 ft (5,106 m), the pressure gradients range from 0.83 to 0.85 psi/ft (18.7 to 19.4 kPa/m). The DSTs taken in the upper pressure compartment indicate reservoir communication between sands in the compartment. Light oils and condensates in the 41 to 55°API gravity range were recovered with gas on several DSTs. The normally pressured Early Cretaceous Mississauga formation also is productive (Ward 1988).

Although this is a thick prograding overpressured delta complex, there is no evidence of undercompaction based on Ward Hydrodynamic's 1988 report and the previously discussed paper by Jansa and Noguera Urrea (1990). Also, the geothermal gradient did not show the dogleg in Figure 9-6. Instead, it was linear at 2.8°C/100 m through the entire section, from normal to maximum overpressures. Likewise, there were no kinks in the plot of vitrinite reflectance with depth, as was observed in the Rocky Mountains by Law et al. (1989).

The vitrinite reflectance data of Jansa and Noguera Urrea (1990) for Venture field show the R_o at about 0.85% in the first compartment and 1 to 1.3% in the second compartment. This would cover the range from the peak of oil generation through condensate and into dry gas generation. Their petrogra-

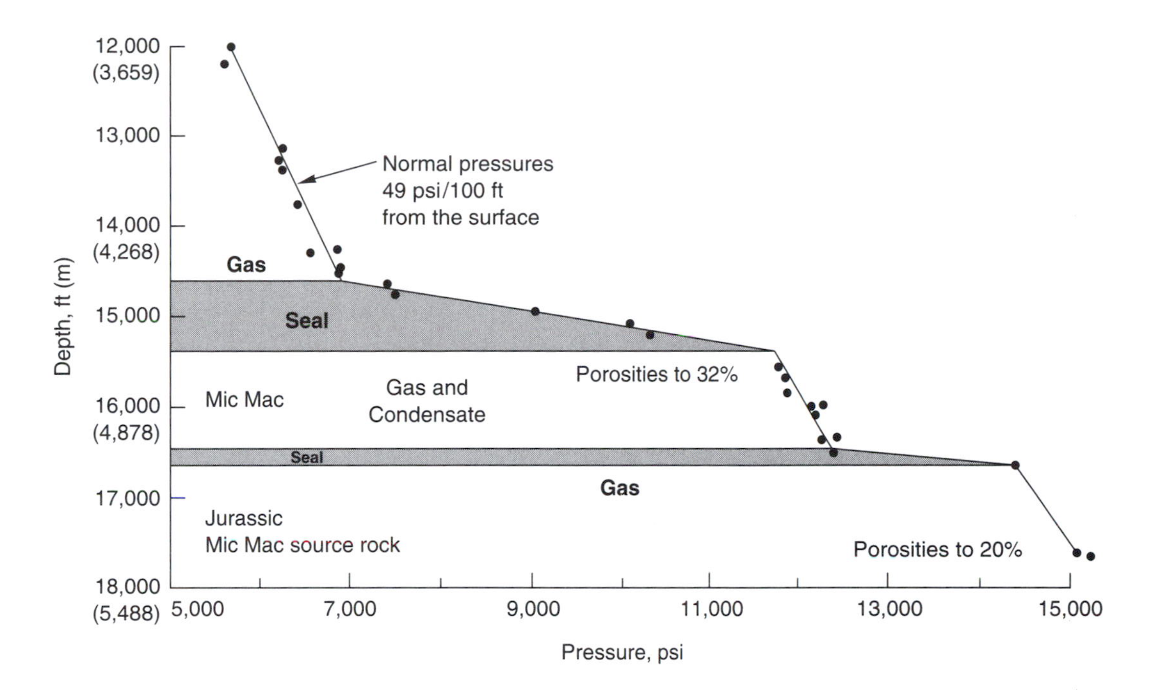

Figure 9-8

Pressure/depth gradient for Venture field, Scotian Shelf, Nova Scotia, Canada. [Data from Ward 1988]

phic studies indicated that the overpressures and the secondary porosity in the sandstones developed simultaneously. Jansa and Noguera Urrea think the main driving force for the overpressures was the generation and expulsion of gas and condensate.

Fluid Pressure Compartments of Organic-Rich Rocks

Some pressure compartments do not show the stair-step increase in pressure with depth, as illustrated in Figures 9-3, 9-6, 9-7, and 9-8. Instead, they show a highly localized pressure regime that is limited to the hydrocarbon-generating interval of an organic-rich source rock. Typical examples are the Bakken Formation of the Williston Basin, the Black Shale facies of the Green River Formation of the Uinta Basin, and the Bazhenov Formation of West Siberia. All three source rocks are in highly overpressured compartments that do not extend beyond the limits of the hydrocarbon-generating kitchens. Pressures increase to a peak in the oil-generating window and decrease below it. Furthermore, bitumen and oil fill all the large pores causing high electrical resistivities in the overpressured compartments. This contrasts with a decrease in electrical resistivities in Gulf Coast deltaic pressure compartments.

At the Antelope field in North Dakota, a highly overpressured fractured siltstone is sandwiched between two layers of dense organic-rich (10% TOC) Bakken Shale source rock (Meissner 1978). The 70 ft (21 m) Bakken interval has a P/D ratio of 0.73 psi/ft (16.4 kPa/m), compared with 0.45 psi/ft (10.4 kPa/m) in the formations immediately above and below the Bakken Shale (Figure 9-9). Pressures to the east and west in the immature, nongenerating, shallower section of the Bakken also are normal (Meissner 1978). Electrical resistivities are high in the oil-generating kitchen of the Bakken and low outside it.

A similar situation occurs in the Altamont field, Uinta Basin of Utah, where a giant, pod-shaped overpressured compartment contains 2 billion barrels of oil in the fractured Green River (Black Shale) source rock (Figure 9-10). The P/D profile differs from all previous examples in showing a gradual buildup of pressure to nearly 0.8 psi/ft (18 kPa/m) at a depth of about 13,000 ft (3,963 m) (Spencer 1987). Parts of the fractured producing section have a geothermal gradient of 7.2°C/100 m (4°F/100 ft) and a temperature of 111°C (232°F). The reason for the curve in the P/D profile (Figure 9-10) is the change in density of oil and bitumen that fill all the pores. Altamont has gas at the top of the reservoir that changes to 35°API gravity oil and then grades deeper into heavy oil (8°API) and bitumen. The overpressured interval is confined to the oil-generating kitchen, with normal pressures in the immature Green River Formation above and the lower part of the nonsource North Horn Formation below.

The Bazhenov is a calcareous, siliceous shale containing 10 to 20% TOC (Dikenshteyn et al. 1986). It is analogous to the Green River and Bakken Formations in that oil is produced from oil-wet, fractured source rocks, with all the large pores filled with oil. No water is recovered on production, and electrical resistivities in the overpressured section are very high. Overpressures of 0.73 psi/ft (16.4 kPa/m) occur within the hydrocarbon-generating interval of the Bazhenov, whereas the underlying and overlying rocks are normally pressured (Braduchan et al. 1990). Twenty pools produce directly out of the fractured Bazhenov source rock, including the giant Salym field. The oils' chemical composition is very similar to that of the bitumen extracts of the Bazhenov source rocks and different from that of the extracts of the underlying rocks (Braduchan et al. 1986). Temperatures in the Salym field are 120 to 135°C (248 to 275°F). No oil has been produced in the stratigraphically equivalent Mulym'in, Tutleym, and Yanovstan Formations, which is further evidence that the generation is confined to the Bazhenhov. Dikenshteyn and others (1983) think that the conversion of kerogen to oil and gas led to a decrease in density (increase in volume) of the total organic phase. In the sealed Bazhenov compartment, this led to an increase in pore pressures, causing extensive natural hydraulic fracturing of the source rock.

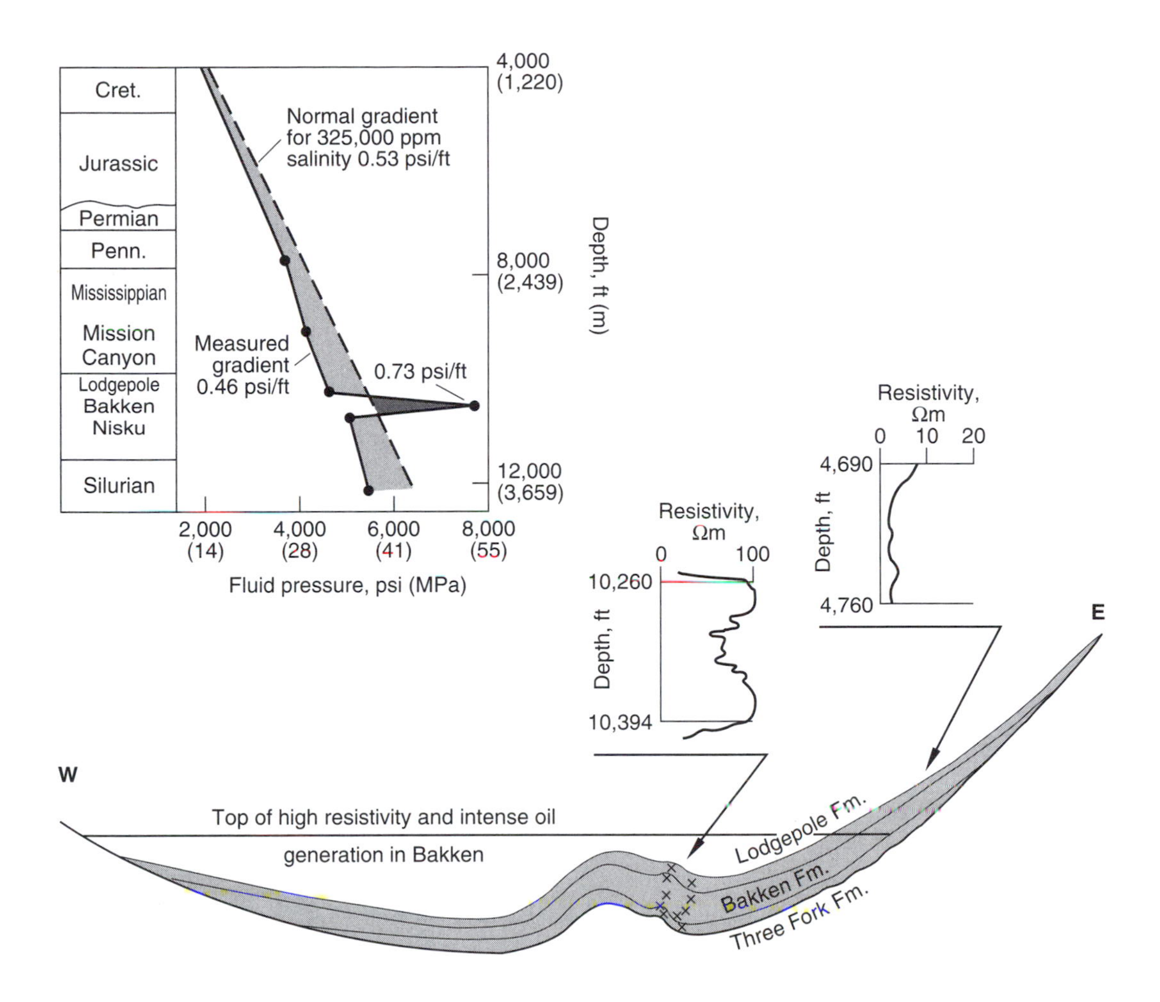

Figure 9-9

Relation of shale resistivities and fluid pressures to oil generation, Antelope field, Williston Basin, North Dakota. [Meissner 1978]

Hydraulic fracturing of the source rock because of oil and gas generation develops the fracture permeability needed to move oil and gas out of shales in abnormally pressured fluid compartments (Hunt 1990, 1991a). Snarskii (1964, 1970) was the first geologist to propose that pore-pressure increases caused by the generation of oil and gas could locally induce fractures in pores of low-permeability rocks like shales. Since then, several writers cited by Gretener and Feng (1985) have discussed the development of fractures by hydrocarbon generation in overpressured zones.

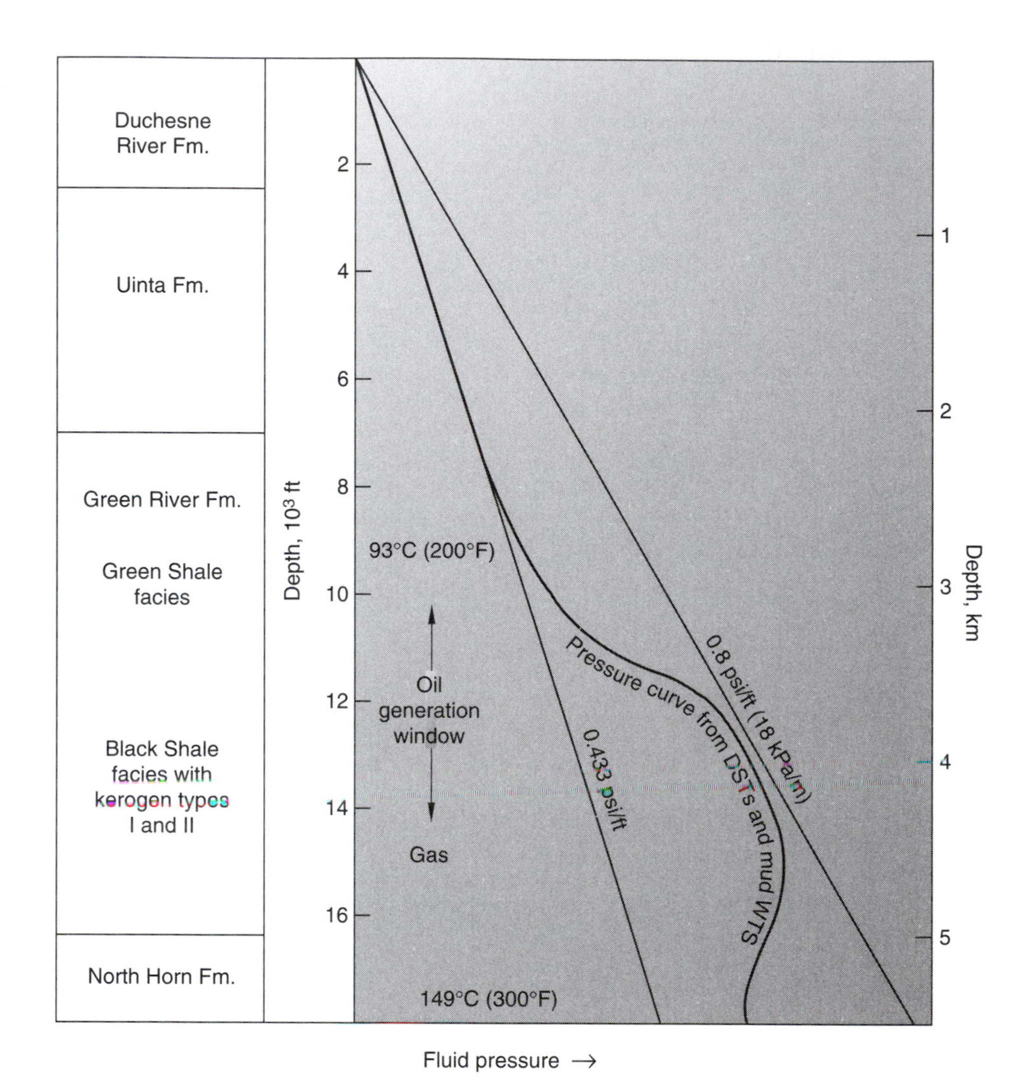

Figure 9-10

Relation of hydrocarbon generation to fluid pressures in the Shell 1-11B4 Brevorson well, Greater Altamont field, Uinta Basin, Utah. [Spencer 1987]

The Salym, Altamont, and Antelope fields are not the only ones producing petroleum from overpressured compartments. The Villa Fortuna field in the Po Basin of Italy at a depth of 6,200 m (20,336 ft) represents one of the deepest overpressured oil-producing fields in the world (Mattavelli and Novelli 1990). The nearby Malossa field also produces from within the overpressured fluid compartment as well as far above it. In the North Sea both the Gert and Lulu fields off Denmark produce from compartments of fractured overpressured Jurassic shales. Other fields producing from overpressured fluid compartments include Matagorda and Vermilion in the U.S. Gulf Coast and Samaan in the Columbus Basin off Trinidad (Hunt 1990).

Fluid Compartment Seals

Fluid seals have been identified in about 200 basins (see Hunt 1990 for a partial list of basins with abnormal pressures). Underpressured compartments are important because they prevent the water washing and biodegradation of accumulated oils. The seal in the Permian Basin compartment of west Texas has kept out the surface water. The oils are not degraded at shallow depths except where they are above the seal.

The presence of seals in the Late Pleistocene of the Gulf Coast indicates that they can form in a few thousand years. Most fluid seals are in evaporites, shales, or sands. The Zechstein Salt of Germany and Holland is a seal covering one big fluid compartment. Carbonates with secondary cementation like the Cretaceous chalk of the North Sea are less common. When carbonates act as seals, many contain pores filled with secondary quartz, whereas clastics are more likely to be cemented by calcite. For example, the Tuscaloosa Formation of the U.S. Gulf Coast is a massive overpressured sandstone with interbedded shales. The sandstone has layers of calcite cementation cutting through it to form a seal. The cementation of rocks to form seals appears to occur initially on the low-pressure side of the seal. Cores taken of seals in the U.S. Gulf Coast sometimes reveal fractures filled with calcite. Many cores taken below the seal in the interior volume of an overpressured compartment contain fractures with no cementation.

Most seals in clastic sections are about 600 ft (183 m) thick, in a range from 150 ft (46 m) to more than 3,000 ft (915 m). Most carbonate–evaporite sections have thinner seals, some as thin as 10 ft (3 m). Lateral seals appear to be nearly vertical and tend to follow fault trends (Powley 1985). Where penetrated, they have been found to contain fractures infilled with calcite or quartz.

Top seals in young Tertiary clastics such as in the Gulf Coast and Niger Delta tend to follow the irregular contact between the top of massive shales and the base of overlying sandstones. Most of the petroleum accumulations are found in these sandstones immediately overlying the shales. Fertl and Leach (1990) found that the maximum concentration of oil and gas was between 10,000 and 12,000 ft (3,049 and 3,659 m), just above the tops of overpressured compartments in 33,000 wells in the Tertiary of south Louisiana.

In contrast, horizontal or gently dipping planar seals cut indiscriminately across structures, facies, formations, and geological time horizons in clastic sediments of the Anadarko Basin, North Sea Basin, Western Canada Basin, North Slope and Cook Inlet basins of Alaska, and many Rocky Mountain basins. Most of these planar seals are in Cretaceous and older sedimentary rocks, suggesting that irregular top seals smooth themselves out over time so as to approximate a thermocline.

In areas containing carbonates and evaporites, as at Keyes field, Alborz field, and Venture field, discussed previously, the top seals usually follow stratigraphy such as carbonate or evaporite beds.

Where hot fluids break through the seals and move up into a normally pressured column, they cause local temperature anomalies comparable to the anomalies caused by hot holes in rift valleys like the Red Sea (Hunt et al. 1967). This creates hot spots that can survive for several years. Localized temperature anomalies within 1,524 m (5,000 ft) of the surface have been recognized in the U.S. Gulf Coast, when there is drilling in the vicinity of hot fluids moving out of the deep, overpressured zone. Such breakouts also temporarily alter the pressure/depth gradients. However, for a high-volume fluid release, the volume of porous and permeable beds above the seal must be great enough to accommodate the fluids ejected in pulses from overpressured zones.

Anderson et al. (1991) constructed a three-dimensional model of fluid flow history in the Eugene Island area offshore the Louisiana Gulf Coast. They found that convective heat flow anomalies at the surface indicate the movement of hot fluids upward along growth faults from the overpressured compartments below.

Many large fluid compartments consist of a series of small fluid compartments with a single top seal and several side seals. In the Tertiary of the U.S. Gulf Coast, every few kilometers there is a fault separating fluid compartments with different pressure systems. In the Jurassic Smackover of the Mississippi Salt Basin, a boxwork of large fluid compartments has internal pressures ranging from normal to 69 MPa (10,000 psi) above hydrostatic. A series of narrow fault blocks cross the Central and Viking Grabens in the North Sea. Each block has its own separate pressure system, with each fault acting as a side seal. Such a boxwork almost certainly limits the lateral migration of oil, but vertical migration can occur as pressures increase to lithostatic and temporarily open the faults.

In summary, compartment seals have close to zero permeability across the seals that completely enclose each compartment. They do not necessarily have a unique lithology, and they may not conform to geological age, lithology, facies, or structure. The abundance of seals in deep basins worldwide indicates that the formation and maintenance of seals are a normal part of basin development.

Petroleum Source Rocks in Fluid Compartments

The top seals of deep compartments in many ordinary basins are at an average depth around 3,048 m (10,000 ft) during the basin's sinking phase (Hunt 1990). This depth is equivalent to a temperature of 95°C (203°F), for a geothermal gradient of 2.4°C/100 m (1.33°F/100 ft). Time–temperature modeling based on the Arrhenius equation has shown that although petroleum generation may start at temperatures as low as 60°C (140°F), most oil and gas have been formed at temperatures higher than 95°C (203°F) (Mackenzie and Quigley 1988; Wood 1988). The higher-generation temperatures over short time periods are particularly characteristic of rapidly sinking basins. Thus, *in basins with overpressures, most of the oil and gas generation occurs in the deep, overpressured fluid compartments.*

In the North Sea, the Upper Jurassic Kimmeridgian–Volgian source rocks are actively generating and expelling oil in a fluid compartment (Plate 5B) at depths below 3,048 m (10,000 ft) and at temperatures above 93°C (200°F) (Demaison 1984). In the U.S. Gulf Coast, the source of the oil and gas is generally agreed to be deeper than the top of the overpressure, in Oligocene, Cretaceous, or Jurassic source rocks at temperatures well above 100°C (212°F). Hydraulic-flow directions in the deep, geopressured sediments of the Louisiana coastal parishes are nearly vertically upward. The geochemical characterization of oils in thermally immature Tertiary reservoirs indicates that the oils have been emplaced by vertical migration from deeper source rocks in the pressure compartments (Hanor and Sassen 1988).

The same situation exists in the Niger and Mahakam Deltas. In the Niger Delta, the Agbada (Eocene) shale source rocks in the overpressured compartments are generating petroleum that is breaking through the top seals and entering the shallow, normally pressured reservoirs above (Egbogah and Lambert-Aikhionbare 1980). In the Mahakam Delta, Indonesia, the top of the overpressures in the Handil field is at about 3,000 m (9,840 ft) (Vandenbroucke et al. 1983). At the Badak field it starts around 3,293 m (10,800 ft) (Huffington and Helmig 1990). Oil and gas are generated in both these fields in the shaly highly overpressured prodelta sediments where pressures reach 20.2 kPa/m (0.9 psi/ft) and temperatures exceed 108°C (227°F). Where the top seal fractures in each area, the oil and gas move upward into normally pressured reservoirs, ranging in depth from 450 to 2,900 m (1,476 ft to 9,512 ft) at Handil field and 1,067 m to 3,872 m (3,500 ft to 12,700 ft) at Badak field.

An interesting observation at Handil field is that the oils become lighter (API gravity increases) going upward through the 150 reservoirs in the normally pressured section. This is exactly what would occur in *evaporative fractionation,* a process characteristic of vertical migration in deltas (Silverman 1965; Thompson 1988). This process produces a sequence of increasingly lighter oils during upward migration. Furthermore, it leaves behind a heavy oil residuum that, being unable to migrate, is cracked to gas and pyrobitumen at the high temperatures of the overpressured compartments. Vandenbroucke et al. (1983) identified abundant pyrobitumen in the high-pressure fluid compartment at Handil. Mineralogical analyses of the overpressured shales by Vandenbroucke et al. also found that the dominant clay mineral is kaolinite formed by hydrothermal diagenesis. This demonstrated that fluids are circulating inside the overpressured compartments.

In the Columbus Basin off Trinidad, the Upper Cretaceous Naparima Hill and Gautier Formations generated petroleum in an overpressured fluid compartment at temperatures well above 95°C (203°F). Oil and gas have broken through the top seal and become trapped in the normally pressured Pliocene and Pleistocene sediments above (Heppard et al. 1990). Furthermore, the oils become lighter going upward. The same change has been observed in the Handil field of Indonesia. The API gravities of the oils are 32° at 3,140 m (10,300 ft) and 40° at 1,893 m (6,210 ft). Vertical migration altered both the gravity and the molecular composition of the oils.

The northwestern area of the Po Basin of Italy contains Middle to Late Triassic source rocks generating oil, condensate, and gas in a highly overpressured fluid compartment, starting around 3,800 m (12,464 ft). Some oil reservoirs are in the overpressured Triassic and Cretaceous formations, but seal breakthrough also has distributed hydrocarbons vertically through the entire Tertiary section (Novelli et al. 1987).

At the Lost Hills field in the San Joaquin Valley of California, the Monterey source rock is in an overpressured fluid compartment. Oil has broken through the seal and accumulated in the normally pressured Etchegoin (Pliocene) sands above. Other overpressured source rocks in California basins include the Cretaceous Forbes Formation, the Eocene Kreyenhagen shale, and the Upper Miocene McClure Formation.

Processes Generating Deep Overpressures

Mechanical Compaction

Compaction disequilibrium that results in overpressured, undercompacted shales is undoubtedly due to rapid loading. Since there is unrestricted lateral flow, however, the undercompacted shales eventually dewater to normal compaction, as depicted in Figure 9-2. But the rate of dewatering depends on the extent to which the unrestricted lateral flow is restricted with burial. As mentioned earlier, about one-third of the Gulf Coast area still contains overpressured, undercompacted shales. Some undercompaction is found below normal compaction, as reported at South Pecan Lake, Louisiana, by Hinch (1980). But even this undercompacted section returns to normal compaction at greater depths.

As previously discussed, mechanical compaction in many areas occurs only down to a subsurface temperature of about 200°F (93°C), as shown in Figures 8-4, 8-5, and 8-7. These figures show that compaction ends before significant oil generation begins. The unrestricted lateral flow typical of compaction disequilibrium becomes restricted with deep burial, resulting in sealed compartments. In the Gulf Coast the tops of > 3 km (10,000 ft) overpressured fluid compartments are most often found in the interval of no porosity decrease, indicating that compaction cannot play a significant role in creating overpressures in these deep, high-pressure compartments.

For example, Figure 9-11 shows a pressure/depth plot and a density/porosity profile for the Sheridan field in Colorado County, Texas (Powley 1993). This well exhibits the characteristic straight-line increase in dry-bulk density and decrease in shale porosity shown in Figure 8-4 down to about 8,700 ft (2,652 m) where Stage 2 starts. An overpressured compartment containing about 5,000 psi excess pressure starts at around 12,000 ft (3,659 m), which is more than 3,000 ft (914 m) below the beginning of the constant porosity Stage 2. Consequently, it is difficult to see how compaction can play any significant role in the development of overpressures in this well, since there is no further compaction starting around 8,700 ft. The pressure measurements (solid circles) in the profile on the left of Figure 9-11 are from drill-stem tests. There is a noticeable increase in dry-bulk density and a corresponding decrease in porosity right at the seal, as might be expected.

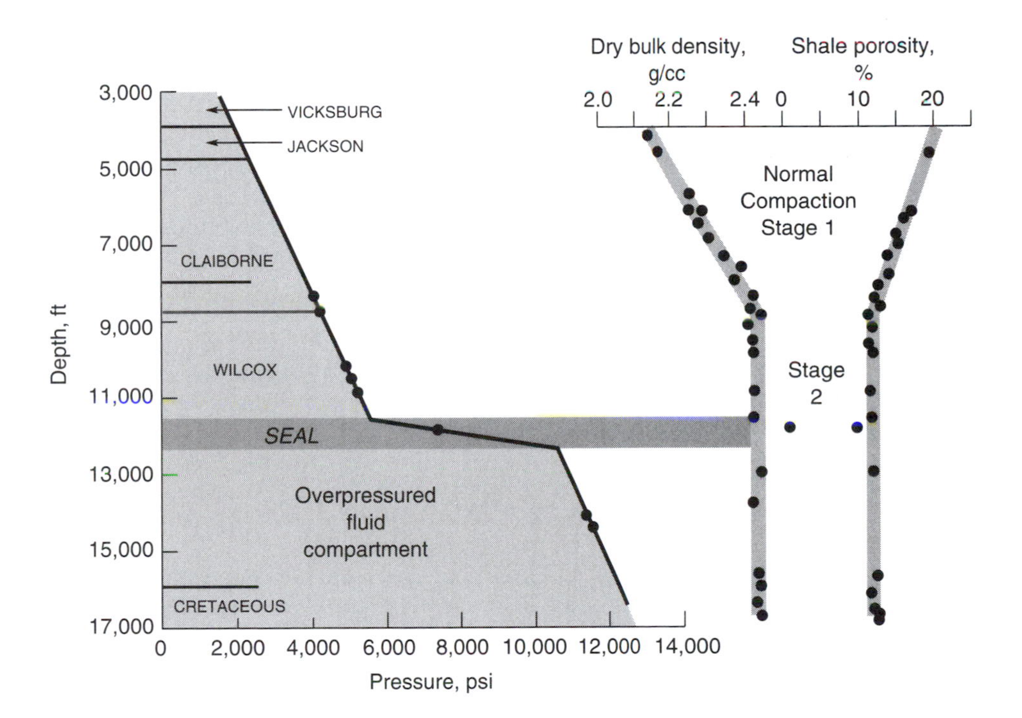

Figure 9-11

Pressure/depth gradient plus shale porosity and density versus depth for the Sheridan field in Colorado County, Texas. Pressure data points are from drill-stem tests (DSTs). The overpressure starts 3,000 ft (914 m) below the top of the constant porosity Stage 2. [Powley 1993]

Aquathermal Pressuring

The concept of excess pressures developing from the thermal expansion of pore fluids in a sealed compartment was first described by Versluys (1932) and later quantified by Barker (1972). Barker showed that for any geothermal gradient greater than about 1.5°C/100 m (0.83°F/100 ft), the pressure in an isolated volume increases with increasing temperature more rapidly than it does in the surrounding fluids. He called the process *aquathermal pressuring,* and he considered it to operate wherever an isolated volume moves down a geothermal gradient. Barker's concept does not apply to the first type of overpressures, compaction disequilibrium, in which there is an unrestricted lateral flow. But as mechanical compaction ceases and both lateral and vertical flow are restricted by the formation of sealed compartments, aquathermal pressuring can become a contributor to deep overpressures.

In the depth range of sealed compartments, 2,500 to 4,500 m (8,200 to 14,800 ft), the pressure changes approximately 862 kPA per °C (125 psi per °F) of temperature change. The higher geothermal gradients observed in many deep, overpressured compartments are believed to enhance the contribution of aquathermal pressuring to the total overpressure. Thus, the geothermal gradient shown for the Vermilion field in Figure 9-6 could result in a fluid pressure increase of 62 kPa/m of burial, or 2.75 psi/ft in the overpressured compartment.

The principal criticism of aquathermal pressuring is that even a very small leak allows fluid to drain off faster than the pressure builds up. There is considerable evidence today, however, that Bradley (1975) was right in describing compartment seals as having essentially zero permeability until pressures build up sufficiently to force open preexisting faults and fractures. Aquathermal pressuring, like all pressurizing mechanisms, could be continually repressuring compartments during periods of rapid basin sinking. (A more detailed discussion of aquathermal pressuring can be found in Magara 1978.)

Hydrocarbon Generation as a Cause of Overpressures

The most compelling evidence that hydrocarbon generation creates overpressures is from the examples of the Bakken, Green River, and Bazinov source rocks described earlier. The only indigenous fluids producible in large amounts from all three of these fractured source rocks are oil and gas. Water, where present, is the discontinuous fluid occupying only the very small pores. The rise in fluid volume as the immature kerogen is converted to mature kerogen plus oil and gas results in a substantial increase in pressure in a sealed compartment of essentially fixed volume.

Several investigators have found a correlation between the generation of hydrocarbons, particularly gas, and the development of overpressures (Dahl and Yukler 1991; Law 1984; Momper 1981; Spencer 1987). Much of the gas production from the Norphlet Formation in the Jurassic deep-gas trend of the Gulf Coast is highly overpressured, but some of it is normally pressured. There are great pressure variations among different sandstones, and there is no top seal that is equivalent for all the overpressured formations. This variability suggests that the pressures are caused by gas generation, probably from oil cracking to gas, as discussed in Chapter 12.

Model studies frequently demonstrate a close correlation between hydrocarbon generation and overpressures in rock intervals where there is no decrease in shale porosity (no compaction). For example, in the DeWitt County, Texas, well seen in Figure 9-12, the top of Stage 2 is at about 9,600 ft. The shut-in pressure at 10,500 ft is about 8,000 psi, and at TD it is around 12,000 psi. These numbers are equivalent to pressure/depth gradients of 0.76 and 0.83 psi/ft, respectively. These overpressures start and increase in the interval where there is no systematic decrease in porosity, indicating no compaction.

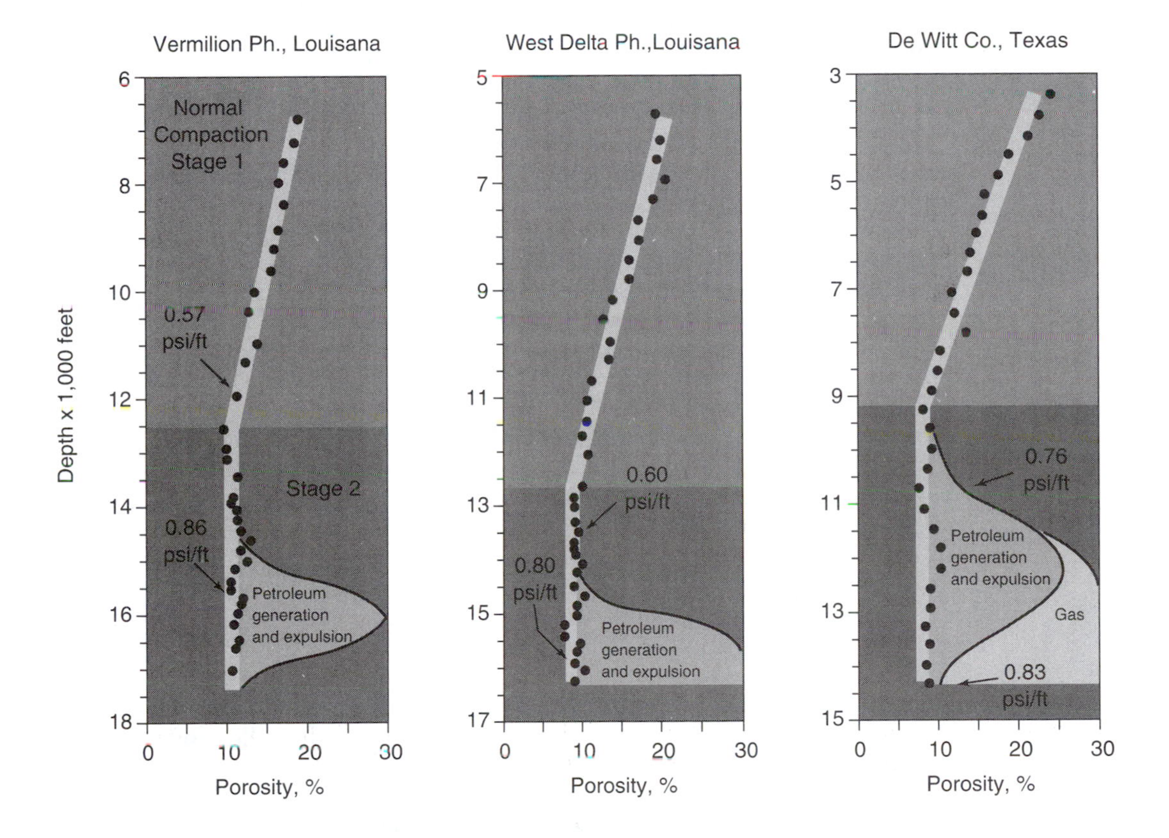

Figure 9-12

The correlation between the generation of hydrocarbons and the development of overpressures for three wells in the U.S. Gulf Coast. [Hunt et al. 1994]

The petroleum generation and expulsion curve for the DeWitt well is based on a quantitative basin analysis model by Yukler and Dow (1990) on a nearby well. Their model showed that the oil-generation window for mixed types II and III kerogen began at a depth around 10,000 ft and peaked at a depth around 11,800 ft. The oil generation phased out around 14,700 ft. However, active gas generation continued down to total depth. The increase in overpressure with depth in this well correlates directly with the increase in gas generation computed by their model. This suggests that gas generation is causing the overpressure.

Similar results were obtained in determining the petroleum-generation intervals in the West Delta and Vermilion Parish, Louisiana, wells in Figure 9-12. Both intervals were calculated using Arrhenius kinetics for type III (terrestrial) kerogen, which is typical of the Gulf Coast Tertiary. In both cases the peak in hydrocarbon generation coincided with the highest-pressure depth gradients measured in the wells, thereby indicating that gas generation is causing the overpressures. In neither of these overpressured intervals was there any systematic decrease in porosity, so compaction could not be the cause of the overpressures.

In addition, Barker (1990) calculated that if the conversion from oil to gas started at a depth of 3,659 m (12,000 ft), only about a 1% conversion would cause the pressure to exceed lithostatic (Figure 9-13). This figure also shows the completion of the conversion from oil to gas at about 17,000 ft (5,183 m), equivalent to 149°C (300°F) in Barker's model. Consequently, there is 5,000 ft (1,525 m) of rock in which gas generation is constantly recharging the pressure compartments. Most deep-pressure compartments in the world extend through this interval. Although Barker's model applies to reservoirs, large amounts of residual oil in most source rocks and associated clastics crack to gas in the depth intervals shown in Figure 9-13.

Hedberg (1974) emphasized the role of methane generation as a major cause of the overpressures that create mud volcanoes (shale diapirs). He pointed out that all mud diapirs around the world are associated with quietly or explosively escaping methane gas. Near Baku in the Caspian Sea, there are 220 mud diapirs in a trend some 200 km long. In 1964, off the south coast of Trinidad, a mud island rose from the sea accompanied by violent gas eruptions. It covered an area of ten acres about 25 ft above sea level. It disappeared eight months later. Gretener and Feng (1985) made a literature survey of high overpressures and found that some gas flow was almost universally present in high-pressure zones.

The literature is replete with examples of overpressures that are believed to be caused by hydrocarbon generation. V. C. Illing wrote the following prescient lines in 1938 (p. 233):

> It is reasonable to suppose that if oil generation continues after compaction has attained its maximum there will be a considerable generation of pressure in the formations as a result of the volume increase which takes place when organic matter is changed into oil and gas. If the process of oil generation takes place in a confined space and the rocks are sufficiently

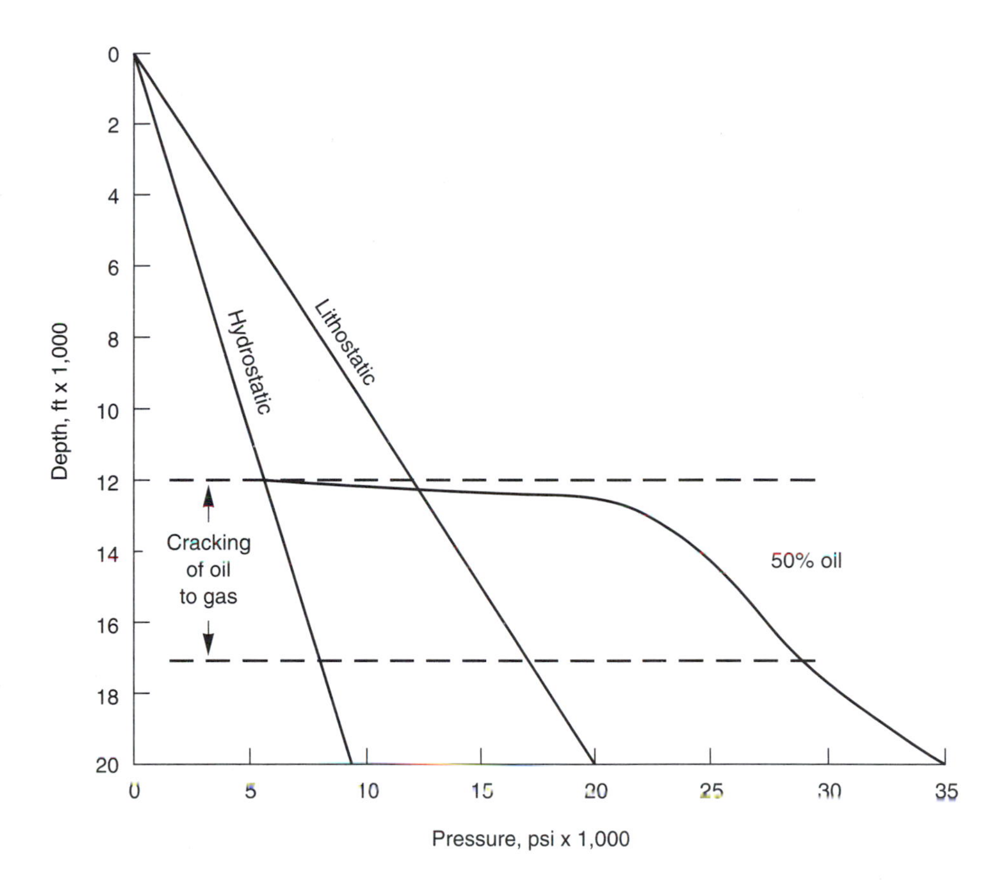

Figure 9-13

Pressure/depth profiles showing the potential pressure trend that develops when oil cracks to gas with a chemical conversion factor of 3,000 ft^3 bbl (85 m^3 bbl). The reservoir initially contains 50 volume% oil. [Barker 1990]

> strong to support it, there will be an automatic increase in volume. Such pressures could only be preserved in a medium which was surrounded by an impervious envelope.

Today we know that oil and gas are generated in a sealed compartment after mechanical compaction has ceased (Figure 9-12). Vast amounts of methane are generated by the thermal decomposition of oil and kerogen in the time–temperature range of deep (>3 km) fluid compartments. Stainforth (1984) described overpressuring in the Upper Cretaceous Campanian in the Gippsland Basin off Australia as due to hydrocarbon generation. The source rocks are considered to be the carbonaceous shales and coals of the Latrobe group, which generated large amounts of gas. Horstman (1988) found overpressures caused by compaction disequilibrium between 1,300 and 2,500 m (4,264 and 8,200 ft) in the northwest shelf of Australia. Overpressured compartments found at depths below 2,500 m closely correlate with the generation of oil and gas, leading Horstman to conclude that the deep overpressures were caused by hydrocarbon generation. Thus, both types of overpressures discussed in Figures 9-1 and 9-3 exist in this basin.

Spencer (1987) found that the tops of many overpressured reservoirs and source rocks of Rocky Mountain basins in Wyoming and surrounding states have temperatures of approximately 200°F (93°C) or higher. This is the same temperature range reported by Hunt (1990) for pressure tops in many basins worldwide and reported by Bradley (1976) and Hinch (1980) for the top of Amoco's Stage 2 (Figure 8-7). At higher temperatures (greater depths), Bradley reported no further decrease in shale porosities, indicating that compaction could not have caused the overpressures. Spencer noted that if the sediments are immature, with vitrinite reflectance (R_o) values less than 0.6%, there was no overpressuring. Rather, it occurs only where R_o values are 0.8% or higher. This indicates that the overpressuring is caused by active hydrocarbon generation.

In the deep basin of Western Canada, Canadian Hunter drilled a 10,000 ft (3,049 m) well that penetrated a coal bed. The pressure/depth ratio of shale–sand beds at about the same depth as the coal was 0.33 psi/ft (7.4 kPa/m). The P/D ratio in the coal bed was 0.6 psi/ft (13.5 kPa/m), almost twice as high as the shale. The coal is actively generating gas (R_o = 1.3 to 1.4%), which created the localized overpressure in the coal.

The largest gas fields in the Rocky Mountain region in the United States are in both over- and underpressured reservoirs. According to Law and Dickinson (1985), the tops of overpressured rocks are not related to structure or stratigraphy in basins like the greater Green River, but they do seem to correlate with the generation of gas. They noted that the largest gas fields commonly are found in low-permeability reservoirs downdip from water-bearing rocks in a basin center position. This is similar to the downdip gas accumulations of the Deep Western Canada Basin previously discussed. The conceptual model developed by Law and Dickinson indicated that overpressured accumulations of gas initially formed from the thermal generation of gas in low-permeability rocks where the rates of gas accumulation were higher than the rates of gas lost. Later in basin history, if the geothermal gradient decreased or if there were regional or local structural uplift, the generation would be retarded or stop altogether. Consequently, gas would be lost from the system faster than it would be replaced, resulting in an underpressure. In the northeastern part of the Green River Basin, Law and Dickinson estimated that approximately 60% of the gas must be lost from the overpressured reservoirs in order to evolve into underpressured reservoirs.

The vicissitudes of gas generation and expulsion or leakage from sealed compartments produce a variety of over- and underpressured compartments in the subsurface, not only with thermally generated gas, but also with bacterial gas. Barker (1987) pointed out that the burial of reservoirs containing bacterial gas may result in pressures that are above, equal to, or below hydrostatic, depending on how the gas accumulation is connected to, or isolated from, its surroundings.

Gas Chimneys

Gas chimneys, or clouds, are features over many reservoirs in overpressured compartments, as shown in Plate 5B. These clouds are caused by episodic increases in reservoir pressure owing to the periodic injection of high-pressured gases from the deeper-pressure compartments. For example, hydrocarbons breaking through the lower seal in Plate 5B are recharging Ekofisk and nearby reservoirs, analogous to the inadvertent recharging of the Asmari reservoir previously discussed. The increased reservoir pressures cause cap-rock fractures to open, discharging gas into the sediments above. When the recharging stops and pressures drop because of leakage, the fractures close until the next repressuring cycle. Such gas chimneys are not observed over most normal or underpressured reservoirs or above overpressured reservoirs that are no longer recharging.

The gas chimney shown over Ekofisk was seen by Vandenbark and Thomas (1981) on seismic profiles. A similar gas-charged cloud was observed over the highly overpressured Valhall Chalk reservoir by Munns (1985). Gas chimneys are common features throughout the overpressured central part of the Viking and Central Grabens. They also have been observed on the graben rims such as in the East Frigg area (Buhrig 1989). Gas generation was interpreted by Buhrig to be the major overpressuring mechanism in the North Sea for pore-pressure gradients 50% or more higher than hydrostatic.

SUMMARY

1. Abnormal pressures include both underpressures and overpressures. Underpressure gradients are below 9.8 kPa/m (0.43 psi/ft), and overpressure gradients exceed 12 kPa/m (0.53 psi/ft).

2. Overpressures develop in basins under two sets of conditions: (1) compaction disequilibrium in which there is unrestricted lateral flow and restricted vertical flow and (2) sealed compartments in which there is restricted lateral and vertical flow. Both types of overpressures can occur in the same area.

3. In many currently sinking basins with average geothermal gradients, a compartmented fluid pressure system exists, starting at depths of about 3,048 m

(10,000 ft). These deep systems are mostly overpressured, whereas those compartments undergoing erosion and uplift tend to be underpressured.

4. In basins with overpressures, most of the oil and gas generation occurs in the deep, overpressured fluid compartments. The generation of petroleum in the compartments plus the thermal expansion of pore fluids eventually causes fracturing of the top compartment seal during periods of basin sinking. Hydrocarbons move cross-stratally into the overlying lower-pressured rocks and accumulate in the nearest structural and stratigraphic traps. Eventually the compartment reseals and pressure builds up toward another breakout. This episodic process continues, with resealing and breakout cycles occurring in geologically short time intervals in rapidly sinking basins.

5. The process generating pore pressures at great depths is primarily the formation of gas from the decomposition of residual oil and kerogen in both source and reservoir type rocks. Aquathermal pressuring is a contributing process. But compaction is not a significant factor at great depths, since many shales going from several thousand feet above the overpressured compartments and down through them show no systematic decrease in porosity with depth, indicating no compaction.

SUPPLEMENTARY READING

Fertl, W. H., R. E. Chapman, and R. F. Hotz (eds.). Studies in abnormal pressures. *Developments in Petroleum Science,* vol. 38. Amsterdam: Elsevier, 454 p.

Orteleva, P., and Z. Al-Shieb (eds.). 1994. *Basin compartments and seals.* AAPG Memoir 61. Tulsa: American Association of Petroleum Geologists.

PART THREE

HABITAT

Chapter

10

The Source Rock

What is a petroleum source rock? This question has intrigued geologists since the earliest exploration for oil and gas. Newberry (1860) described the oil production from the Berea sand near Mecca, Ohio, as being formed by the low-temperature heating of organic matter in the Hamilton bituminous shale. Later the "Ohio black shale" became established as the source of oil and gas in Ohio and Kentucky. The early geologists believed that Kentucky's best prospects for oil fields were where the Cumberland Sandstone was overlain directly by the black shale.

Decades later, Snider summarized the general opinion of petroleum geologists:

> There seems to be a very nearly universal agreement that these organic materials are buried principally in argillaceous mud and to a lesser extent in calcareous muds and marls and in sandy muds. Coarse sands and gravels and very pure calcareous deposits are generally without any notable content of organic material. Consequently, shales and bituminous limestones consolidated from muds and marls are generally regarded as source rocks for oil and natural gas. (1934, p. 51)

Snider went on to point out that some shales and carbonate sequences containing porous and permeable sections can act as both source and reservoir.

In their search for oil, petroleum geologists continued to emphasize the location of a proper structure, the presence of suitable reservoir permeability and porosity, and some type of caprock. When all these conditions were present but no oil or gas was found, it was generally attributed to the lack of source rocks. This conclusion was based on intuitive reasoning rather than on any hard data on the characteristics of the source rock, because these characteristics were only vaguely understood.

A *petroleum source rock* is defined as any rock that has the capability to generate and expel enough hydrocarbons to form an accumulation of oil or gas. Definitions that do not include migration and accumulation are too general, because in a sense practically all rocks containing organic matter (OM) form some hydrocarbons. A *potential source rock* is one that is too immature to generate petroleum in its natural setting but will form significant quantities of petroleum when heated in the laboratory or during deep burial. An *effective source rock* is one that has already formed and expelled petroleum to a reservoir. It may be active (currently expelling) or inactive (e.g., because of uplift with erosion and cooling). Techniques for recognizing source rocks have been developed from case studies of the quantity, quality, and maturation level of organic matter in rocks associated with petroleum production. Also, crude oil–source rock correlations have helped establish that a particular rock has yielded oil to a particular reservoir.

The term *kerogen* originally referred to the organic matter in oil shales that yielded oil upon heating. Subsequently, the term was defined as all the disseminated organic matter of sedimentary rocks that is insoluble in nonoxidizing acids, bases, and organic solvents (Hunt and Jamieson 1956). Kerogen in rocks has four principal sources: marine, lacustrine, terrestrial, and recycled. Most of the world's oil has formed from marine and lacustrine kerogen, whereas most coal is from terrestrial plants, and the recycled kerogen is largely inert. The relative ability of a source rock to generate petroleum is defined by its kerogen quantity (TOC) and quality (high or low in hydrogen). Whether or not it has generated petroleum is defined by its state of maturation (immature, mature, or postmature with respect to oil).

Quantity of Organic Matter

The quantity of organic matter usually is expressed as *total organic carbon* (*TOC*). The overall efficiency of converting organic carbon in the source rock to the carbon in commercial petroleum accumulations is low, generally less than 15 wt% (Hunt 1979, p. 263). Magoon and Valin (1994) tabulated the generation–accumulation efficiencies (percentage of in-place petroleum to that generated) for sixteen petroleum systems around the world. All but one ranged between 0.3 and 14 wt%.

Klemme (1993) estimated that the recoverable oil and gas accumulated by petroleum systems such as those in the North Sea and the Middle East represent less than 5% of the hydrocarbons generated by the source rocks. Although some very localized hydrocarbon kitchens such as in the North Sea (Table 16-5) may expel and accumulate up to 30% of the oil generated, the average is much lower for an entire petroleum system. Since the overall system is inefficient, it becomes important to determine the minimum TOCs observed in rocks of varying lithologies that may generate and expel oil and gas. (See Chapter 16 for more details on petroleum systems.)

Table 10-1 lists the TOCs of rocks from different areas. The color of a rock is a rough but not always reliable indicator of its TOC content. Most sandstones

TABLE 10-1 Quantity of Total Organic Carbon (TOC) in Rocks of Various Lithologies

	Wt% TOC
Sandstone	0.03
Red shales	
Chugwater, Colorado	0.04
Big Snowy, Montana	0.04
Green shales	
Ireton, Alberta	0.11
Cherokee, Kansas	0.30
Tertiary, Colombia	0.54
Gray shales	
Frontier, Wyoming	1.2
Cherokee, Kansas	1.6
Mowry, Wyoming	3.0
Black shales	
Woodford, Oklahoma	7.0
Cherokee, Kansas	8.0
Bakken, North Dakota	11.0
Limestones and dolomites	
Cherokee, Kansas	0.2
Charles, Montana	0.3
Sunniland, Florida	1.9
Austin Chalk, Texas	2.1
Niobrara Chalk, Colorado	3.2
Calcareous shales and argillaceous limestones	
Alcanar, Spain	4.0
Antrim, Michigan	6.7
La Luna, Venezuela	7.7
Duvernay, Alberta	7.9
Toolebuc Ls, Australia	10.8
Nordegg, Alberta	12.6
Bazhenov, West Siberia	> 10
Green River, Wyoming	18

Sources: Data from Baker 1962; Huc 1988; Hunt 1961; Palacas 1984.

and red beds have very low TOCs because the organic matter has been destroyed by oxidation. TOCs generally increase in shales as the color goes from red to variegated, to green, gray, and finally to black. Baker (1972) made a detailed study of the Cherokee Shales (Pennsylvanian age) of Kansas and Oklahoma. He found TOCs as low as 0.1% in greenish gray shales and as high as 17.5% in one black shale unit. The Cherokee values in Table 10-1 are Baker's averages for a large number of samples.

Snider recognized in 1934 that in regard to TOC content, pure calcareous deposits are comparable to sands. Many white limestones have less than 0.2% TOC, whereas brown limestones may have values comparable to gray and black shales. In contrast, calcareous and dolomitic shales and argillaceous limestones have long been recognized as having the highest TOCs (Hunt 1961). All the eight source rocks listed under calcareous shales in Table 10-1 have varying amounts of calcite and dolomite mixed with fine-grained quartz and minor amounts of clay minerals. Snider's 1934 comment on some of them acting as both source and reservoir is true today, as seen in the Bakken Shale of the Williston Basin, the Bazenhov Shale of West Siberia, the Green River Shale of Wyoming, and the Monterey diatomaceous cherts and shales of coastal California.

The use of color as a rough TOC indicator should always be backed up by analytical data. Some grayish black limestones from Nevada with a petroliferous (H_2S) odor were found to have only 0.1% TOC. The color was due to manganese oxide.

The quantity of TOC in rocks is closely related to sediment particle size. A sample of Viking Shale from Alberta, Canada, was disaggregated, dispersed in water, separated by centrifuging, and analyzed for TOC (Hunt 1963). The TOC content of the siltstone size was 1.47%; clay size 2 to 4 μm, 1.70%; and clay size less than 2 μm, 5.32%. A similar study of carbonates showed the highest concentration of TOC in the lime muds and the lowest values in the skeletal grains (Gehman 1962). High TOCs in sediments are due to the preservation and transport of organic matter, not organic productivity. Lowest in preservation are the high-energy parts of coastal areas and inland seas where productivity in the water column is adequate, but strong currents and the high oxygen content of the waters intensify both the biological and the chemical degradation of the TOC. The preservation of TOC generally increases from high-energy to low-energy sediments. Inland seas such as the Caspian and silled basins such as Lake Maracaibo both show their highest TOCs in the fine-grained sediments and their lowest TOCs in the coarse sediments.

The increase in TOC with a decrease in sediment grain size was first recognized by Trask et al. (1932) in the Channel Islands region of California, where he found that clays with a median particle diameter of less than 5 μm had twice the TOC of silts with a diameter between 5 and 50 μm and four times that of fine sand whose median diameter was 50 to 250 μm. Gorskaya (1950), in a study of Recent clastic sediments, reported the following TOC contents in weight percent: sands 0.70, silts 1.0, and clay muds 1.6. Later, Emery (1960) showed that southern California shelf and beach sediments with a median grain diameter of more than 100 μm had less than 0.2% TOC compared with sediments from the offshore basins, which had particle sizes between 3 and 9 μm and TOCs in the range of 5 to 9%. Bordovskiy (1965) cited several examples in Russia of TOCs increasing with a decrease in sediment grain size. For example, TOC values in Bering Sea silts increase uniformly as particle size decreases. Bordovskiy also cited Strakhov and others as demonstrating that the accumulation of organic matter in sediments is affected by the morphological features of a basin, such as width, depth, and bottom relief.

Low-energy coastal areas and inland sedimentary basins where fine-grained clay and carbonate muds are deposited generally contain 0.5 to 5% TOC, which is in the range of most oil-forming rocks. Shallow inland seas, narrow seaways between continents, and restricted areas are the typical depositional environments for source beds of petroleum.

Even larger quantities of organic matter are preserved in areas where oxygen is eliminated and benthic organisms are suppressed. Sediments with TOC contents exceeding 10% are found in stagnant, silled basins, like the Norwegian fjords and the Black Sea, where hydrogen sulfide in the bottom water eliminates all life except anaerobes, such as sulfate reducers. The lack of oxygen restricts decomposition to reducing processes, and the poisonous effect of hydrogen sulfide kills all biota venturing into the area.

Plate 5C shows in color the clear relationship between sediment grain size and TOC in the Black Sea (Huc 1988). The dark blue in the upper left picture shows that primary production is highest in the western and lowest in the eastern part of the Black Sea. This shows no relation to the TOC values, which are highest in the center of the two major lobes of the Black Sea, as shown in the lower right figure of Plate 5C. Also, there is no direct relation between anoxicity (lower left) and TOC (lower right). The only clear relation is between grain size distribution and TOC (upper and lower right figures in Plate 5D). The low-density organic particles are deposited with the less than 2 μm mineral grains in the hydrodynamically quiescent regions in the centers of the two lobes of the basin.

Huc (1988) also reported other areas such as the Caspian Sea, the Paris Basin, the North Sea, and Western Siberia where the low-density organic particles are progressively concentrated toward the deepest, most quiescent areas of the basins. Preservation and transport mechanisms cause the highest TOCs. Such data also show that TOC values can vary widely both laterally and vertically within the same formation. Consequently, TOCs need to be measured continually in wildcat wells to define such variability.

How much organic carbon is required for a good source rock? Ronov (1958) analyzed several hundred samples of Upper Devonian shale throughout the Russian Platform from Kiev in the west to Ufa in the east. The results are shown in Figure 10-1. Although the Russian Platform has structures and interbedded sandstones with good porosity and permeability throughout this area, all the oil fields are concentrated in the area near Kuibyshev and Ufa, where the TOC ranges between 0.5 and 5 wt%. In the Saratov and Kiev areas to the south, where TOCs are around 0.5%, some gas has been found. No Devonian oil or gas has been found in the northern part of the Russian Platform, extending from the oil fields in the east to the western boundary of Russia,

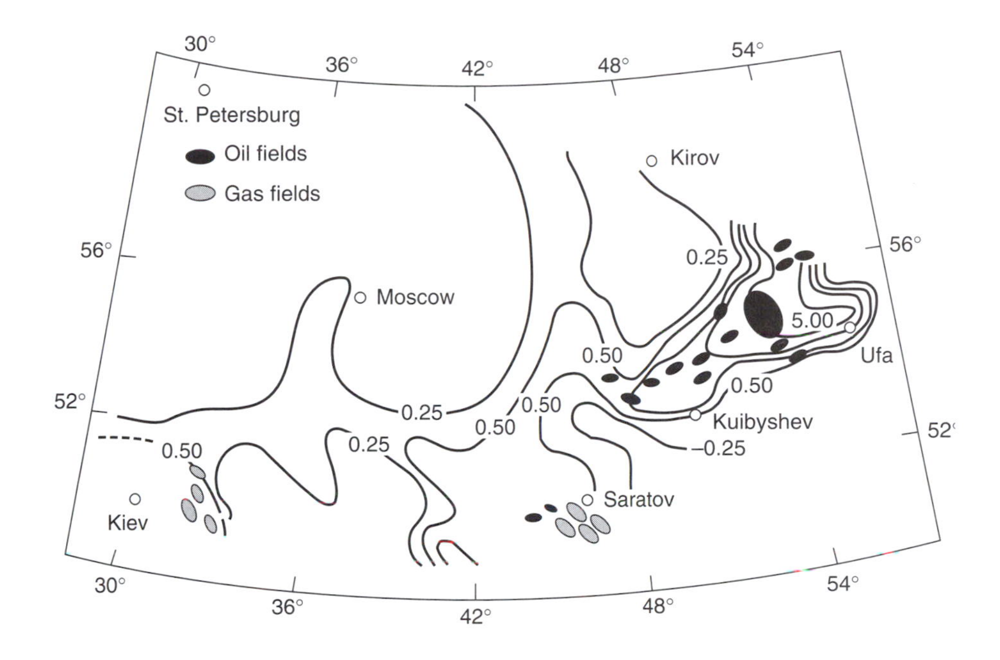

Figure 10-1

Ronov's 1958 total organic carbon (TOC) data for Upper Devonian sediments on the Russian Platform. Contours outline more than 1,000 analyses of well samples. [Figure from Hunt 1979, p. 268]

where TOC values are generally less than 0.25%. Ronov found that the mean TOC content of fine-grained rocks in petroliferous areas was 1.37% for shales and 0.5% for carbonates. In nonpetroliferous areas, the means were 0.4% for shales and 0.16% for carbonates.

Ronov (1958) prepared lithogeochemical maps relating the TOC distribution to the depositional environment and its oxidized state. He found that the northwestern and western areas of the Russian Platform represented continental and lagoonal environments where the sediments were more oxidized than in the more marine areas to the east. Ronov determined the ratios of ferric to ferrous iron in the sediments, a measure of the state of oxidation or reduction. The higher this ratio is, the more strongly oxidizing the environment will be, and the lower the ratio is, the more strongly reducing the environment will be. Ronov found that the Fe_2O_3/FeO ratios in the west were 10 where the TOC values were less than 0.25%. Going east, there was a gradual decrease in the ratio from 10 to 5 toward the central part of the platform and from 5 to 1 to the east and southeast. In the Volga–Urals petroleum province in the east, the Fe_2O_3/FeO ratio fell below 1, defining this as the most highly reducing environment on the Russian platform. Within the petroliferous provinces in the east, the highest organic carbon contents were in shales deposited in coastal-marine sediment facies. From a statistical evaluation of all his data, Ronov concluded:

> These figures indicate an existence of a certain minimum of organic substance in the major sedimentary complexes below which the transformations of disseminated organic carbon cannot be conducive to the development of economic accumulations of petroleum. This critical level lies somewhere between the organic carbon averages for the petroliferous and the nonpetroliferous areas, that is, between 1.4 and 0.4 percent, and it is probably closer to the first one of these two figures. (1959, p. 522)

Interestingly, as mentioned in Chapter 8, Lewan (1987) found that the minimum TOC required to expel generated oil in hydrous pyrolysis experiments was between 1.5 and 2%. Gas appears to be expelled down to possibly 0.5% TOC.

Much of the oil generated by source rocks never reaches a reservoir. It remains disseminated throughout the source rock matrix, is dispersed along secondary migration pathways, or is lost to the surface. Table 10-2 gives the estimated volume of oil generated in source rocks in the Powder River Basin of Wyoming, compared with the oil known to be in place in reservoirs. The major oil source rocks in this basin are the Mowry and Niobrara shales which, along with lesser source rocks, have generated about 226 billion barrels of oil (Momper and Williams 1984). Their immature TOCs are 3 and 2.5%, respectively. The Steel–Pierre is mainly a gas source. To date, about 9 billion barrels of oil have been discovered in place in Cretaceous formations. Consequently, dividing the reservoir oil by source oil in Table 10-2 gives an overall efficiency

TABLE 10-2 Oil in Cretaceous Formations, Powder River Basin, Wyoming: Effective Source Area of 10,500 square miles (27,195 km²)

Formations	*Oil generated by source rock (10^6 bbl)*	*Oil accumulated in reservoir rock (10^6 bbl)*
Steele–Pierre	1,000	30
Niobrara–Carlile	75,000	2,500
Mowry–Muddy–Newcastle–Frontier–Skull Creek	150,000	6,500
	226,000	~9,030

Sources: Revised from Hunt 1961; Momper and Williams 1984.

of 4% from generation to trapping. Similar calculations by Coneybeare (1965) for Jurassic formations in the Surat Basin in Australia give an efficiency of 5%. These numbers are similar to the previously reported estimates by Trask (1936) and Klemme (1993), indicating that overall efficiencies in general are low. Most of the dispersed and the disseminated oil that is not in reservoirs is converted to gas and pyrobitumen as the sediments are buried deeper.

Quality of Organic Matter

The most important factor controlling the generation of oil and gas is the hydrogen content of the organic matter (OM). The quantity of petroleum generated and expelled increases as the atomic hydrogen-to-carbon (H/C) ratio of the OM increases. Nearly all OM may be classified into two major types, *sapropelic* and *humic* (Potonie 1908; also see Table 10-3). The term *sapropelic* refers to the decomposition and polymerization products of fatty, lipid organic materials such as spores and planktonic algae deposited in subaquatic muds (marine or lacustrine), usually under oxygen-restricted conditions. Sapropelic OM such as fats, oils, resins, and waxes shows high atomic (H/C) ratios in

TABLE 10-3 Classification of Organic Matter in Coals and Sedimentary Rocks

	Sapropelic		*Humic*	
Coal maceral groups	Liptinite (exinite)		Vitrinite	Inertinite
Coal macerals	Alginite Cerinite[a] Sporinite Cutinite Resinite Liptodetrinite		Telinite Telocollinite Desmocollinite Vitrodetrinite	Fusinite Inertodetrinite Sclerotinite Macrinite
	Fluorescent Amorphous		Nonfluorescent Amorphous	
Kerogen Types	I	II	III	IV
H/C	1.9 to 1.0	1.5 to 0.8	1.0 to 0.5	0.6 to 0.1
O/C	0.1 to 0.02	0.2 to 0.02	0.4 to 0.02	0.3 to 0.01
Source	Marine, Lacustrine, Terrestrial		Terrestrial and Recycled	

[a]Wax.

the range from 1.3 to 2.0. Organic-rich sapropelic deposits undergo maturation to form boghead coals and oil shales. A modern freshwater, fat-secreting planktonic green alga, *Botryococcus braunii,* is the source of Australia's cooron-gite, a boghead peat. The Carboniferous equivalent of *Botrycoccus b.* is *Pila,* which is concentrated in Scottish oil shales (torbanites). The Permian oil shale "tasmanite" in Tasmania is formed from a single-celled green alga called *Tasmanites.* Some boghead coals of the Siberian Jurassic contain only amorphous liptinite (Stach et al. 1982). Amorphous OM is believed to be formed by the microbial reworking of organic debris and by the precipitation of colloidal OM such as humic acids. It can be either sapropelic or humic.

The term *humic* refers to products of peat formation, mainly land plant material deposited in swamps in the presence of oxygen (Table 10-3). Peat has an H/C ratio around 0.9, which is borderline for being an important progenitor of oil. Humic OM is derived from plant cell and wall material, which is composed mainly of lignin and cellulose plus the aromatic tannins, which have a high resistance to rotting. The humic category also includes carbonized (fusinitized) OM, such as charcoal from fires and other oxidized plant remains. Humification is accelerated by the presence of oxygen and heat in tropical climates.

When roots, bark, and wood from trees are deposited in forest and reed swamps and swamp lakes, they undergo bacterial and chemical changes that lead to the formation of peat. As the peat is buried deeper, it changes with time and temperature (coalification) to brown coal (lignite), bituminous coal, and finally anthracite under tectonic stresses. These are the major humic coals. During coalification (maturation) the atomic hydrogen-to-carbon (H/C) and oxygen-to-carbon (O/C) ratios of both the sapropelic and the humic material decrease, as indicated in Table 10-3. For example, the H/C ratio of vitrinite changes from about 1.0 to 0.3 between lignite and anthracite.

If these same particles of roots, bark, and wood are carried out to sea and deposited in a sedimentary basin as the 1 or 2% TOC of a shale, they will contribute to the humic part of the kerogen. Likewise, any spore, pollen, or algal material carried by wind or water into the marine or lacustrine environment contributes to the sapropelic part of kerogen unless it is oxidized in transit. OM from marine and lacustrine organisms is added to the disseminated land-derived humic and sapropelic material to form the kerogen of marine sediments.

The first detailed classification of the organic matter of sedimentary rocks was made by coal petrologists, who classified coal components based on their appearance under microscopic examination with reflected light, using oil-immersion objectives with 25 to 50 times magnification. Coals are composed of macerals comparable to the mineral components of rocks, except that macerals are not crystalline and they vary more widely in chemical composition than minerals (for a complete discussion of coal petrology, see Stach et al. 1982). The three major maceral groups in coals are *liptinite* (also called *exinite*), *vitrinite,* and *inertinite* (Table 10-3). Humic coals usually contain over 70% vitrinite, whereas boghead coals often have over 70% liptinite, a high proportion of which is the maceral alginite, derived from algae. These same coal macerals are recognized in disseminated form as part of the kerogen in sedimentary rocks, and petroleum geochemists use these same terms to describe them.

The sources of the liptinite macerals in Table 10-3 are defined by their names. Thus, alga, spores, cuticles, and resins form four of the macerals. *Liptodetrinite* represents the finely divided degradation remains of all the other primary liptinite macerals listed. *Amorphous* is not a maceral but, rather, describes all the unstructured, unrecognizable OM, which can be either sapropelic or humic.

Wax, which is called "cerinite" by van Krevelen (1961), is primarily of terrestrial origin, as discussed in Chapter 4. It has the highest atomic H/C ratio of all the constituents of coal. In the vitrinite group, *telinite* is the structured cell-wall material of land plants, and *collinite* is the unstructured substance that fills the cell cavities. The term *vitrodetrinite* is used for detrital vitrinitic particles (Stach et al. 1982).

Among the inertinites, *fusinite* has the lowest atomic H/C ratio of all the coal macerals. It partly originates from charcoal. *Sclerotinite* comes from fungal remains, and *macrinite* is an amorphous groundmass of high reflectance with almost no structure. *Inertodetrinite* is composed of the $< 30\ \mu m$ size fragments or other remains of all the other inertinites. All the inertinite macerals have very low H/C ratios.

Although the coal macerals in Table 10-3 are often readily identified in sedimentary rocks, amorphous kerogen is not well defined in rocks, and it has no clear counterpart in coal. The problem is that the initial bacterial degradation of kerogen, plus its dilution and dispersion during sedimentation, destroys most of the original OM structures. Even though coal beds may contain 90% recognizable macerals and only 10% amorphous OM, the kerogen of shales is typically more than 50% amorphous (Whelan and Thompson-Rizer 1993). The only clue to the origin of this amorphous mass is the small amount of associated structural kerogen whose relationship to the amorphous kerogen is not always clear.

Nevertheless, it is possible to make some classification of amorphous kerogen by viewing the sample under the microscope with the three most common microscopic illumination conditions: transmitted light, reflected light, and fluorescence using ultraviolet (UV) or incident blue light. Thompson and Dembicki (1986) defined four amorphous kerogen types in this way. Their types A and D are oil prone, and their types B and C are more gas prone.

The term *amorphous* was once mistakenly limited to the sapropelic liptinites of kerogen, which are considered to be the major oil generators. But it is now recognized that amorphous kerogen occurs throughout the entire range of sapropelic and humic kerogens. The amorphous kerogen that is potentially oil generating is what fluoresces under ultraviolet or blue light. Nonfluorescing amorphous kerogen usually does not have oil-generating potential. Senftle et al. (1987) tentatively defined fluorescent kerogens as *fluoramorphinite.* Their kerogens that had the most amorphous fluorescent material also had the highest atomic H/C ratios, indicating a high potential to generate oil.

Maturation of Organic Matter

Van Krevelen Diagrams

The atomic H/C-versus-O/C diagram was developed by van Krevelen (1961) as a simple and rapid method for following the chemical processes that occur during coal maturation (coalification). Since the main elements of coal are carbon, hydrogen, and oxygen, changes in coal composition during burial can be recognized by plotting atomic H/C-versus-O/C ratios. Figure 10-2 shows the natural maturation track for humic coal based on van Krevelen's original diagrams (solid line). This track shows how coalification changes the elemental composition of coals and also provides some insight into the products formed by this process. Moving from the right to the left across the diagram along the coalification line represents the loss of oxygen relative to carbon, which occurs with the formation of CO_2 or H_2O. Moving from the top to the bottom of the diagram represents the loss of hydrogen relative to carbon. This occurs because of the formation of oil and gas, which have higher H/C ratios than either kerogen or coal. For example, wax (1 in Figure 10-2) can be converted to the largest quantity of hydrocarbons, since it has the highest initial H/C ratio. The inertinite (9) and fusinite (10) would produce some CO_2 and water on maturation but very few hydrocarbons, since they have low H/C ratios. The end point of all these reactions is in the lower left corner of Figure 10-2, where all residual OM ends up as graphite.

Some of the sapropelic macerals of coal also are plotted in Figure 10-2 (3, 4, 5, 6, 7, and 8). When immature, they all have H/C ratios greater than 0.9. They generate the most petroleum, whether they are in coal or in kerogen disseminated in sedimentary rocks. During maturation they follow the trend of the dashed lines in Figure 10-2, losing a lot of hydrogen to make oil and gas but not much CO_2, since they start with a lower O/C atomic ratio than do the humic macerals. The boghead coals (2 in Figure 10-2) are similar to oil shales in composition, as they contain high percentages of alginite (3).

Saxby et al. (1986) simulated the maturation of a boghead coal (torbanite, 17 in Figure 10-2) and a typical brown coal by heating them in sealed tubes in the laboratory for as long as six years. They started with immature samples and gradually raised the temperature to a maximum of 400°C. The brown coal followed the natural coalification line (22, 23, 24, and 25 in Figure 10-2), whereas the torbanite, which had less than 2% oxygen to begin with, followed a straight line down the vertical axis, representing a continual decrease in the H/C ratio (18, 19, 20, and 21). The torbanite yielded the most oil, and the brown coal gave off the most CO_2, as would be expected. The two maturation lines came close together after five years when the atomic H/C ratios of both reached about 0.5 (20, 24). After six years, the aromaticity (*fa*) of the pyrolysis residues of torbanite changed from 0.16 to 0.9 through the H/C range from 1.6 to 0.35.

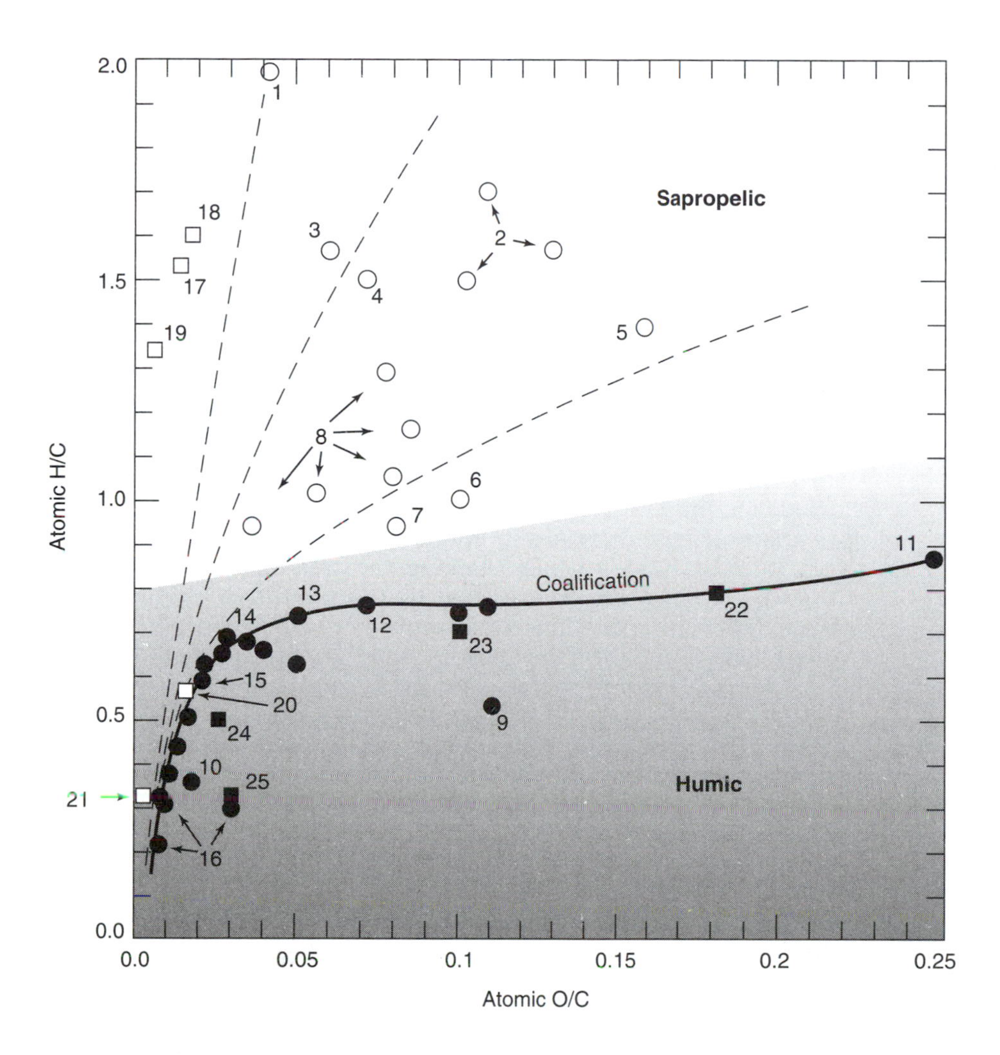

Figure 10-2

Van Krevelen maturation diagram for coals and coal macerals: (1) wax (cerinite); (2) boghead coals; (3) alginite; (4) resinite; (5) cutinite; (6) sporinite; (7) suberinite; (8) exinites; (9) inertinite; (10) fusinite; (11) lignite; (12, 13, 14) high-, medium-, and low-volatile bituminous coals; (15) semianthracite; (16) anthracites; (17) torbanite; (18, 19, 20, 21) laboratory maturation of torbanite; (22, 23, 24, 25) laboratory maturation of brown coal; (open symbols) sapropelic OM; (solid symbols) humic OM. The maturity of all samples increases toward the lower left corner of the plot, as indicated by the dashed and solid curves. [Data from Hunt 1979; Johnston 1990; Saxby et al. 1986; Stach et al. 1982; van Krevelen 1961]

Since *fa* is the fraction of aromatic TOC, it means that 90% of the carbon atoms were in aromatic rings when the experiment ended.

In 1974, Tissot et al. adopted the van Krevelen diagram to follow chemical changes in kerogen with depth and increasing thermal stress. They analyzed kerogens from shale samples taken at different depths for C, H, and O and calculated the atomic H/C and O/C ratios. When these were plotted on the van Krevelen diagram, they showed three distinct maturation pathways. The kerogens in these pathways were defined as types I, II, and III. The standard ultimately chosen for the type I was the Eocene Green River Shale of the United States; for the type II, the Toarcian shale of France; and for the type III, the Upper Cretaceous shale of the Douala Basin in Cameroon. Later, a type IV was added to include kerogens with very low H/C ratios and high O/C ratios.

Subsequently, variations of the van Krevelen plot were published, some with wider maturation pathways and the addition of vitrinite reflectance lines to define maturities (Jones 1987). Others added the positions of frequently used coal macerals to the plots (Peters and Cassa 1994). Figure 10-3 is a modification of these diagrams. It contains wider maturation track areas comparable to Jones (1987) while retaining the well-established four kerogen types. Figure 10-3 also emphasizes that these maturation pathways are only guidelines. There is a continuous range in kerogen composition, as is clear from Jones (1987) and from the kerogen subtypes discussed in Chapter 6.

Natural maturation tracks for the kerogens in three oil source rocks (types I and II) and a gas source rock (type III) are shown in Figure 10-3 (Peters 1986). The maturation lines initially show a greater decrease in O/C than H/C down to an O/C ratio of about 0.05. Under greater thermal stress, the decrease in the H/C ratio accelerates as the generation and expulsion of hydrocarbons exceed the release of CO_2 and H_2O. At H/C ratios of around 0.7, some samples show a moderate increase in the O/C ratio, because more carbon is being removed as hydrocarbons than oxygen as CO_2 and H_2O. Tissot et al. (1974) also heated, at increasing temperatures in the laboratory, samples of immature Toarcian shales from the Paris basin. The kerogen composition followed the same pathway as the natural evolution.

The dashed lines in Figure 10-3 represent thermal maturity as determined by vitrinite reflectance ($\%R_o$). All the data points to the right of the 0.5 line are of immature kerogens that have not yet generated petroleum. Moving left along the pathways causes kerogen to enter oil generation around 0.5 to 0.6$\%R_o$. As oil generation approaches a peak around 0.9 to 1.0$\%R_o$, the condensate and gas generation that started earlier begin to dominate. As gas is formed, part of the remaining oil and condensate are expelled from the source rock. Any hydrocarbons still remaining are cracked to gas, so that by the time the 2.0$\%R_o$ line is reached, the main product is methane gas. Thus the van Krevelen diagram shows not only the differences in the hydrogen content of the immature starting materials. It shows also how the loss of hydrogen to make hydrocarbons causes the oil-prone types I and II kerogens to change to the gas-prone type III kerogens that evolve toward a graphite residue. It should be noted that type II kerogen actually generates more hydrocarbon gas than does type III, but oil is

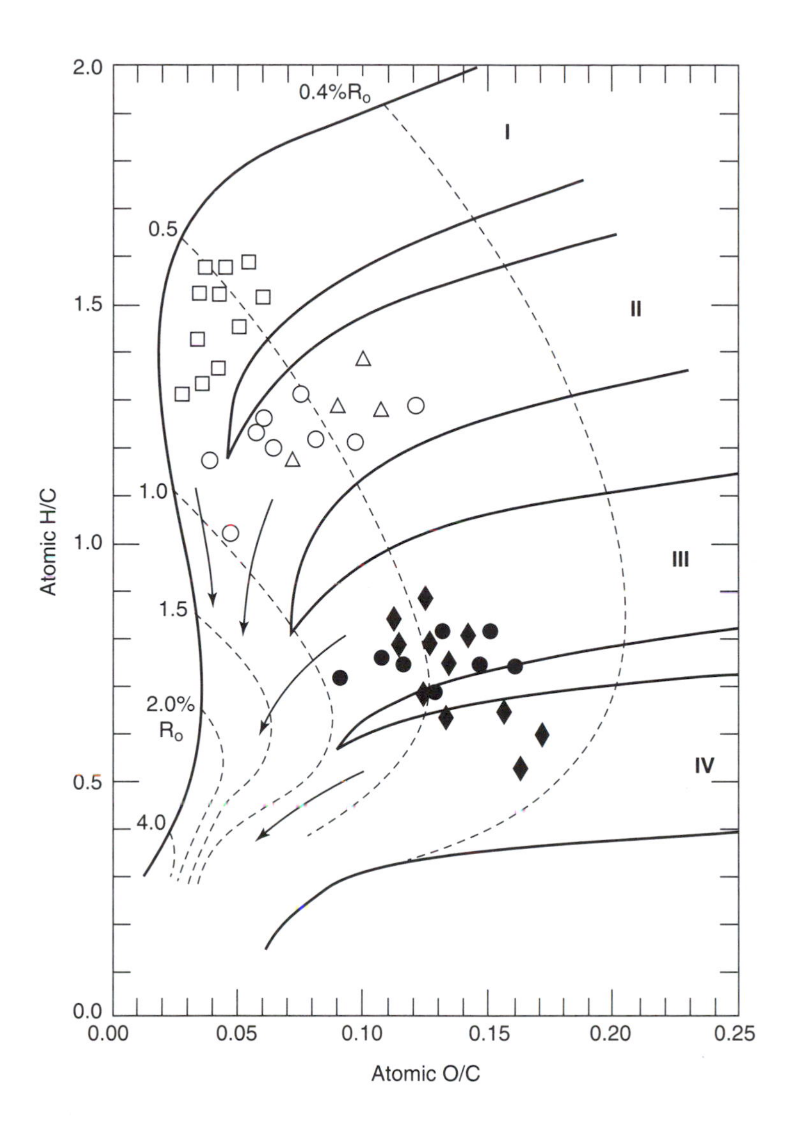

Figure 10-3

Van Krevelen diagram showing four types of kerogen at different maturity levels. Kerogens subjected to increasing thermal stress move toward the lower left corner, comparable to the movement in Figure 10-2. Symbols are □ type I, Eocene Green River Shale, United States; △ type II, Jurassic of Saudi Arabia and ○ Toarcian Shale of France; ● type III, Tertiary of Greenland; and ◆ type IV, Upper Tertiary, Gulf of Alaska. The dashed lines are isorank lines based on vitrinite reflectance, $\%R_o$. The arrows show the direction of increasing maturity. [Data from Jones 1987; Peters 1986]

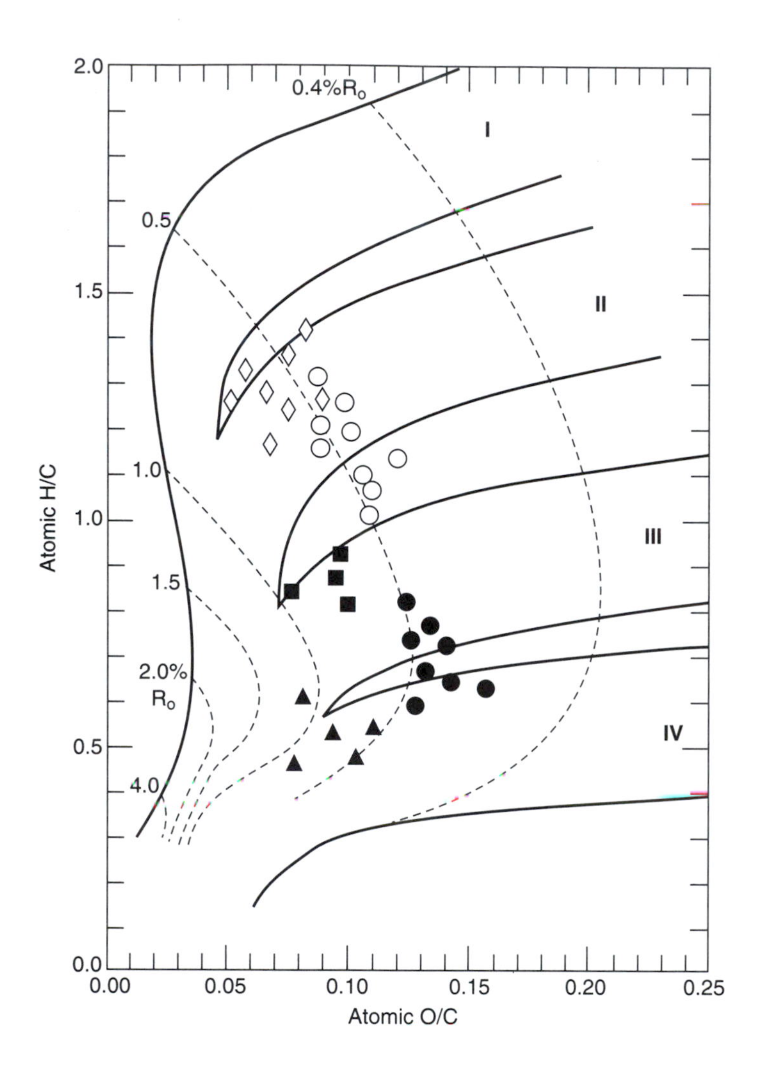

Figure 10-4

Van Krevelen diagram showing kerogen types in cuttings of a composite well section from the northern North Sea; ◇ Kimmeridge Shale, restricted marine; ○ Tertiary, mixed marine; ■ Middle Jurassic, coaly shale; ● Cretaceous, open marine; ▲ Triassic, red bed. [Data from Jones 1987]

its dominant product. Also, type III generates some waxy oil and considerable condensate, but gas is its dominant product.

Figure 10-3 also shows a type IV kerogen from Alaska, which is mainly reworked and highly oxidized organic matter. Jones (1987) classified it as being in his type D organic facies. Type IV kerogen does not generate hydrocarbons, except possibly small amounts of methane. Some CO_2 and H_2O are formed, however, as the maturation line moves to the left.

Figure 10-4 contains the entire range of kerogen types from I to IV from a composite well section in the northern North Sea (Jones 1987). It illustrates the key concept that *hydrogen preservation* with the early depletion of oxygen during deposition is required to make types I and II kerogen. The Kimmeridge Shale was deposited in a restricted marine environment, and it is the best source rock in Figure 10-4. The Triassic kerogen (type IV) was deposited in red beds and the Cretaceous kerogen (type III), in an oxidized open marine environment. They have the least hydrocarbon-generating potential of the samples shown.

Organic-rich Cretaceous black shales from Cape Verde Rise in the eastern Atlantic Ocean were penetrated by hot diabase sills during the Miocene. Cores were taken through this interval during the drilling of Site 368 of the Deep Sea Drilling Project (DSDP). Peters et al. (1983) analyzed the kerogen from several cores above and below the sills and plotted the atomic H/C and O/C ratios on a van Krevelen diagram (Figure 10-5). The differences shown are not due to variations in environmental deposition. They are due to thermal maturity caused by the intense heat of the sills. All the kerogens were probably type II with H/C ratios above 1 before the Miocene. Cracking the kerogen to yield petroleum left residues of successively lower H/C ratios approaching the sills from above or below. Today, the H/C ratio is 0.1, and the vitrinite reflectance is 4.8%R_o, at a distance of 0.3 m from one of the sills. This is a clear example of a thermal effect independent of any differences in the initial kerogen quality.

The atomic H/C ratio in the van Krevelen diagram provides some insight into the chemical structure of kerogen. For example, the ratio for the C_{10} paraffin decane is 2.2; for the corresponding cycloparaffin decalin, 1.8; and for the C_{10} aromatic naphthalene, 0.8. Consequently, when moving down the vertical axis of Figures 10-3, 10-4, and 10-5, the kerogen decreases in paraffinicity and increases in aromaticity. This was demonstrated in the kerogen structures described in Plates 3A, 3B, and 3C. Plate 3A contains mostly paraffins, shown as white wiggly lines, and Plate 3C has mostly aromatic rings, shown as fused, yellow hexagons. The H/C ratios for the kerogens in Plates 3A, 3B, and 3C are 1.64, 1.25, and 0.83, respectively.

When the Green River oil shale kerogen was made the standard for type I, it caused some geochemists to believe that lacustrine environments and oil shales in general have predominantly type I kerogen. Actually, oil shale kerogens are a mixture with as much or more type II as type I. In addition, lacustrine environments such as the Green River of the Uinta Basin, Utah, and the Cretaceous of the Songliao Basin of China have a complete range of kerogen types from I to IV (Anders and Gerrild 1984; Yang et al. 1985).

Data for the Songliao Basin are shown in Figure 10-6. The immature data points to the right of the vitrinite reflectance 0.5%R_o line reveal that there was a continuum of original kerogen ranging from type I to type IV deposited in the Songliao Basin. Type II is actually more dominant than type I. Consequently, Yang et al. (1985) subdivided these into types IIA and IIB. Jones (1987) went further, classifying all kerogens as being formed in seven organic facies designated A, A-B, B, B-C, C, C-D, and D. The dominant kerogen in the Songliao

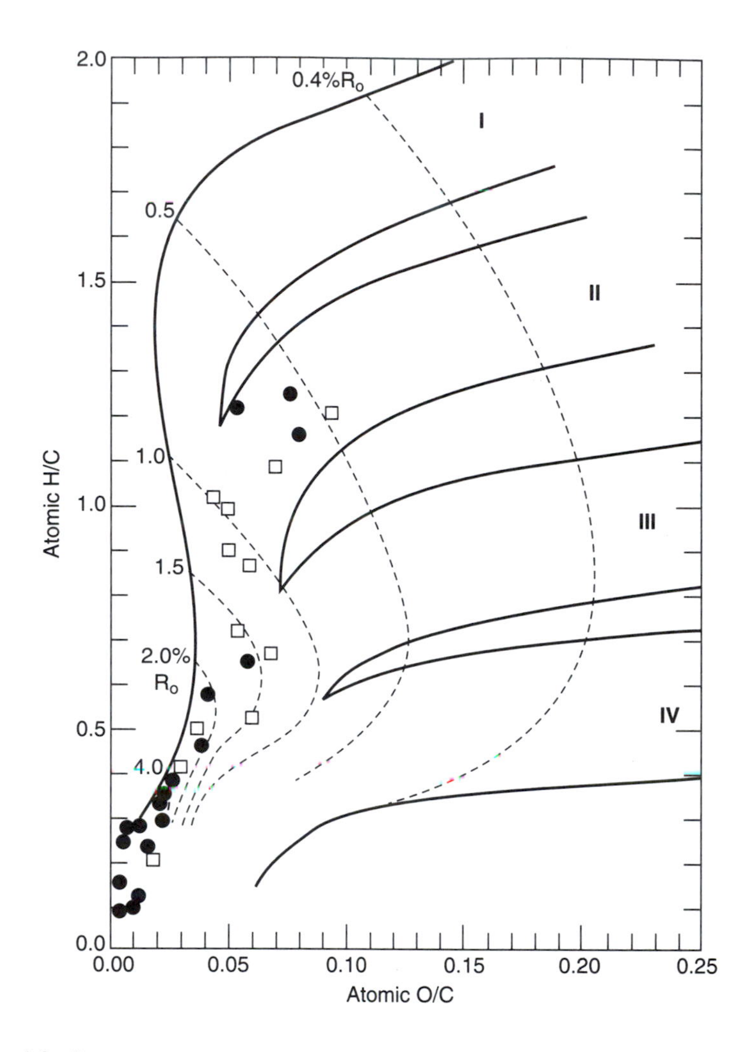

Figure 10-5

Van Krevelen plot showing the extreme thermal maturation of Cretaceous black shales in the eastern Atlantic Ocean due to penetration by diabase sills; ● above sill; □ below sill. [Data from Peters et al. 1983]

Basin based on the data in Figure 10-6 is Yang's type IIA (Jones's organic facies B) rather than the type I (organic facies A) generally attributed to lacustrine source beds. Whether type I or II is dominant in a particular lacustrine environment also might be determined from the wax content of the oil generated. According to Figure 10-2, wax is the only major starting material in nature with

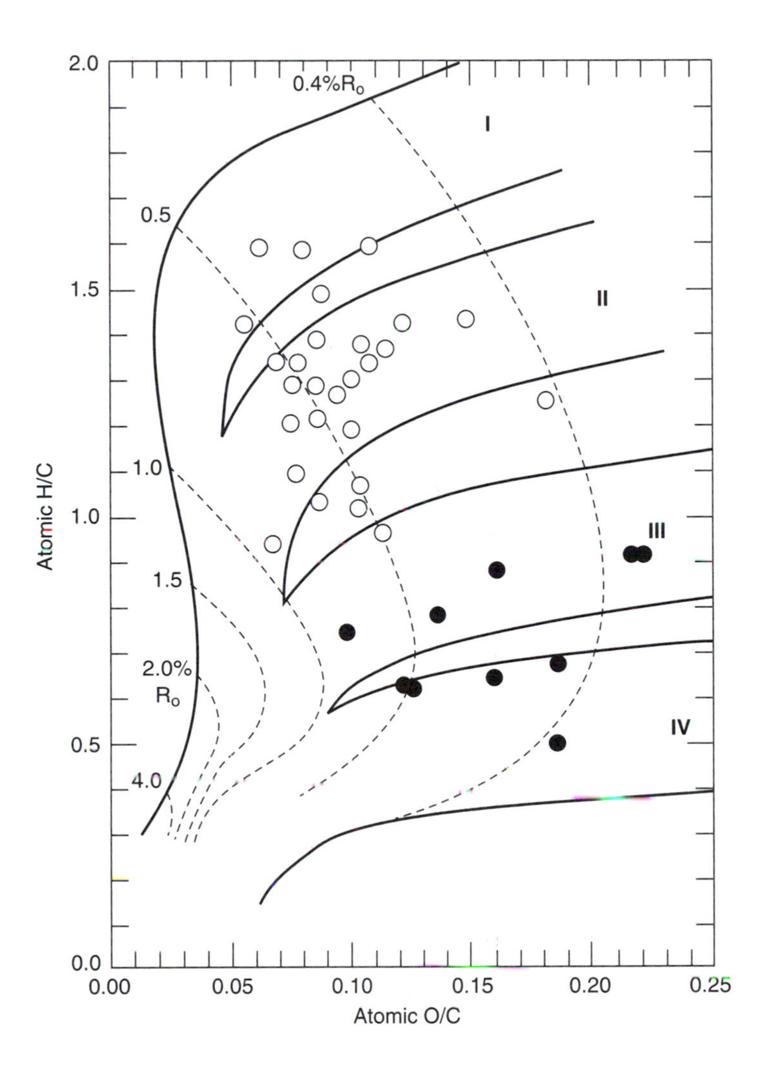

Figure 10-6

Van Krevelen plot showing oil-generating ○ and condensate- and gas-generating ● kerogens of the Cretaceous source rocks of the lacustrine Songliao Basin of China. [Data from Yang et al. 1985]

an atomic H/C ratio above 1.8 combined with an O/C ratio below 0.1. The relative quantities of wax in produced oils might indicate the relative roles of type I compared with those of type II kerogens in forming the oils.

Irrespective of the number and variety of kerogen classification systems, the kerogens with the highest original atomic H/C (hydrogen preservation) and

TABLE 10-4 Hydrogen Content and H/C Ratios of Kerogens

			H/C ratios of kerogens			
			Oil generating		*Gas generating*	
Wt% of H in kerogen		*Corresponding H/C ratio*[a]	*Permian, Texas*	*Viking, Alberta*	*Miocene, Louisiana*	*Atoka, Oklahoma*
9		1.35				
8	Oil	1.2	1.2			
7		1.05				
6		0.9		0.88		
5	Gas	0.75			0.75	
4		0.6				0.6
3		0.45				
2	No gas	< 0.3				

[a]With carbon constant at 80 wt%.

Source: Data from Hunt 1979, p. 343.

the lowest atomic O/C ratios (oxygen depletion) generate the most petroleum. The importance of hydrogen is evident when comparing kerogens that generate oil with those that generate gas. Table 10-4 lists the H/C ratios of kerogens containing varying amounts of hydrogen, along with the ratios for two oil-generating and two gas-generating kerogens (Hunt 1979, p. 243). Oil-generating kerogens usually contain 6% or more hydrogen when immature, whereas gas-generating kerogens have 3 to 5%. In Table 10-4, the Permian and Viking kerogens are generating oil, and the Miocene and Atoka kerogens generate mainly gas with some condensate. The break between oil and gas is at an H/C ratio of about 0.8 on a van Krevelen diagram.

At 2% hydrogen the H/C ratio is less than 0.3, which is in the range where only traces of gas, if any, are formed. The H/C ratio of the kerogen at the bottom of the 9.6 km Bertha Rogers well of Oklahoma was 0.25. The vitrinite reflectance value was > 4%R_o, which is well into the nongenerating range (Price et al. 1981).

Pyrolysis

The van Krevelen atomic H/C–O/C diagram (Figures 10-2 through 10-6) is still the best method for correctly evaluating the quality and maturation state of kerogen and coal in the subsurface. It is not suitable, however, for the rapid screening of well cuttings, such as for a wildcat well that is being drilled. Using HCl and HF to eliminate carbonates and silicates and thereby to isolate the kerogen is a very time-consuming process. Before pyrolysis, some company laboratories analyzed only about 300 cuttings samples a month for elemental carbon, hydrogen, and oxygen in the kerogen. Consequently, a technique was needed that could provide immediate information on source rock potential while wells were being drilled.

Rock-Eval Pyrolysis

In 1977, Espitalié et al. published the first paper on the development and application of the Rock-Eval pyrolyzer. The technique involves passing a stream of helium through 100 mg of pulverized rock heated initially at 300°C. The temperature is then programmed to increase about 25°C/min, up to 550°C (1022°F). The vapors are analyzed with a flame ionization detector (FID), resulting in the peaks shown in Figure 10-7. Peak P_1 (S_1) represents any free hydrocarbons in the rock that either were present at the time of deposition or were generated from the kerogen since deposition. Heating at 300°C simply distills these free hydrocarbons out of the rock. The carboxyl groups in the kerogen break off between 300 and 390°C, yielding CO_2 (P_3 (S_3)), which is trapped and analyzed later during the cooling cycle using a thermal conductivity detector (TCD). Between about 350 and 550°C, hydrocarbons (P_2 (S_2)) are generated by cracking the kerogen until only residual nongenerating carbon remains. In addition, any free high molecular-weight bitumen that was not distilled out in P_1 is cracked into smaller molecules in P_2. Tarafa et al. (1983) noted that hydrocarbons above about C_{24} do not volatilize until cracking temperatures above 350°C are reached.

The areas under P_1, P_2, and P_3 are labeled S_1, S_2, and S_3, respectively. The S_1 and S_2 are proportional to the flame ionization detector (FID) carbon in the vaporized products, which is then calculated as milligrams of hydrocarbon, based on calibrating the detector with standards. The TOC is determined separately. The ratio of mg HC in S_2/g TOC is called the *hydrogen index* (HI) (Espitalié et al. 1977). The ratio of mg CO_2 in S_3/g TOC is called the *oxygen index* (OI). Further studies indicated that the hydrogen index could be roughly correlated with the atomic H/C ratio and the oxygen index with the atomic O/C ratio of the van Krevelen diagram, though with some exceptions. Consequently, by plotting HI versus OI, most pyrolysis data could be interpreted in a manner similar to the elemental analysis data on a van Krevelen diagram.

Figure 10-8 shows an HI/OI plot for the same oil and gas source rocks that were plotted on the van Krevelen diagram in Figure 10-3, except for the Gulf of Alaska samples (Peters 1986). It is incorrect to call the HI/OI plot a modified or pseudo–van Krevelen diagram. Entirely different data are plotted, and the position of the kerogen types is quite different on the van Krevelen diagram compared with the HI/OI plot. The major differences are the large spread between types II and III kerogen and the overlap between types III and IV. The first

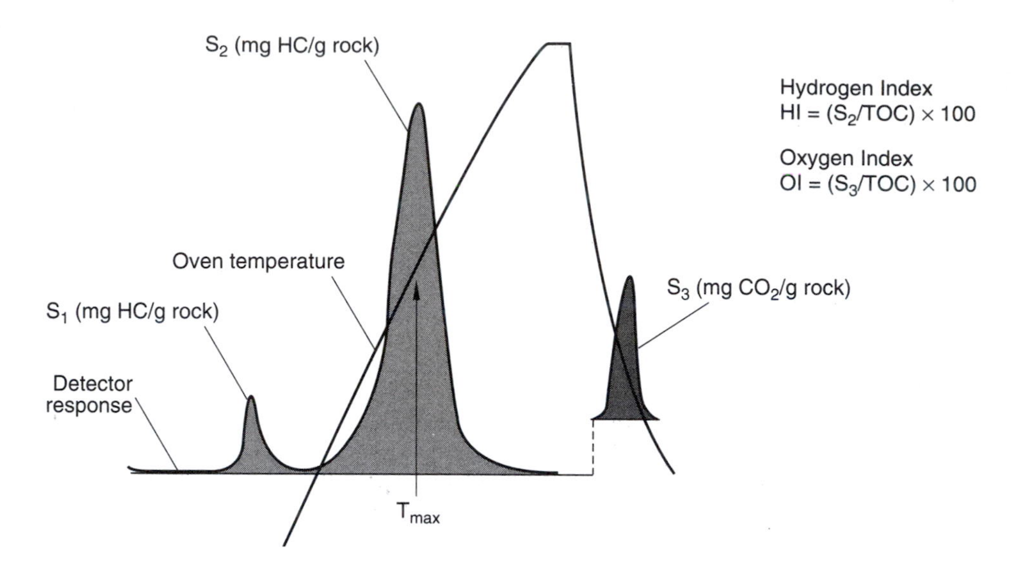

Figure 10-7

Schematic of pyrogram showing the evolution of hydrocarbons and CO_2 from a rock sample during heating (increasing time and temperature from left to right). Important measurements include S_1, S_2, S_3, and T_{max}. Hydrogen and oxygen indices are calculated as shown. [Peters 1986]

difference makes it easy to recognize the best-quality source rocks, but it still is difficult to discriminate among and evaluate the more common source rocks with HI values between 100 and 400. The second difference creates problems in determining whether or not a sample is a potential gas source. These problems are magnified at low TOC values. Nevertheless, pyrolysis has a huge advantage over elemental analysis (van Krevelen) because it is simple, rapid, and inexpensive. Samples can be analyzed in about 20 minutes, making it relatively easy to keep up with a drilling program. Pyrolysis is the best rapid-screening technique for source rock evaluation as a well is being drilled.

The maturation lines in Figure 10-8 represent a median for the kerogen types indicated. The original graph of Espitalié et al. (1977) had two parallel vertical maturation lines in the 100-to-400 HI range, with the midpoint at an OI of about 12. Originally, only a type III line was shown, but this has been modified in Figure 10-8 to show type IV based on data accumulated since 1977 (Peters 1986). The median of the type III line tops out at an HI of about 125, and the median of the type IV tops out at an HI of about 50. The changeover from oil-prone to gas-prone source rocks is usually within the 150-to-200 HI range, although in some areas it is as high as 250.

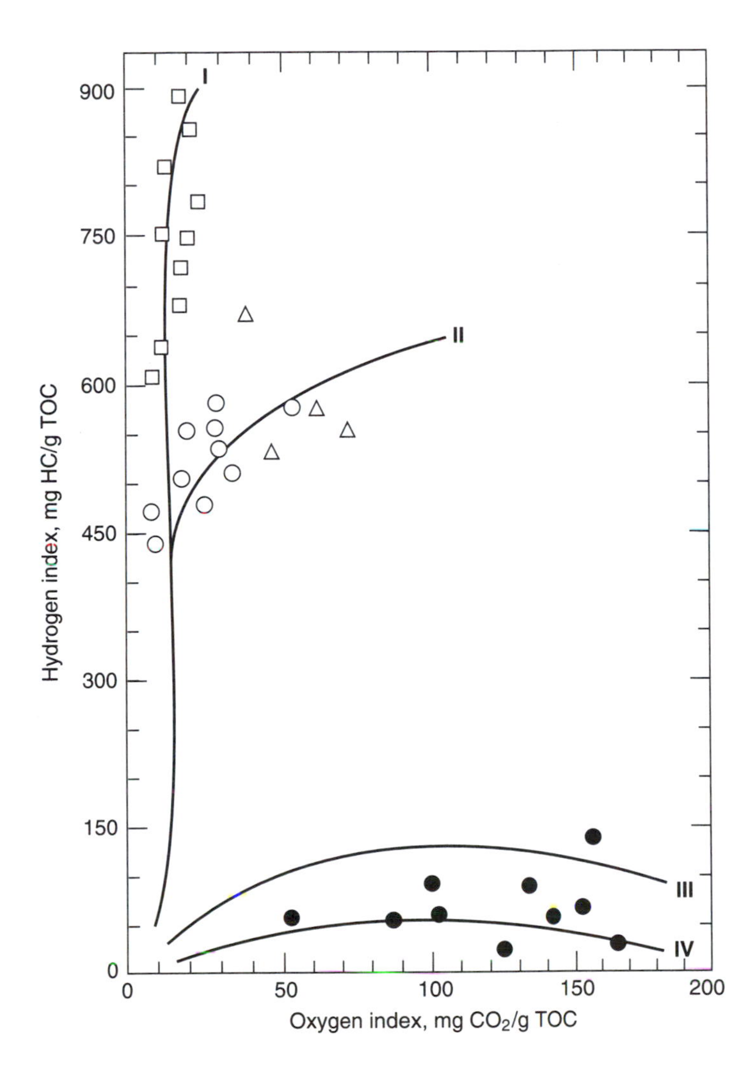

Figure 10-8

Classification of kerogen types on an HI/OI diagram. These are the same kerogens shown in Figure 10-3 except for those from Alaska. Symbols are □ type I Eocene Green River Shale, United States; △ type II Jurassic of Saudi Arabia; ○ Toracian Shale of France; and ● type III Tertiary of Greenland. [Data from Peters 1986]

Figure 10-9 is an HI/OI plot showing the quality and maturation level of Kimmeridgian–Volgian source rocks from the Norwegian and United Kingdom North Sea areas. These samples cover the entire range from rich oil-prone type II kerogens with an HI of 550 to nonsource type IV with an HI of 10. Most of

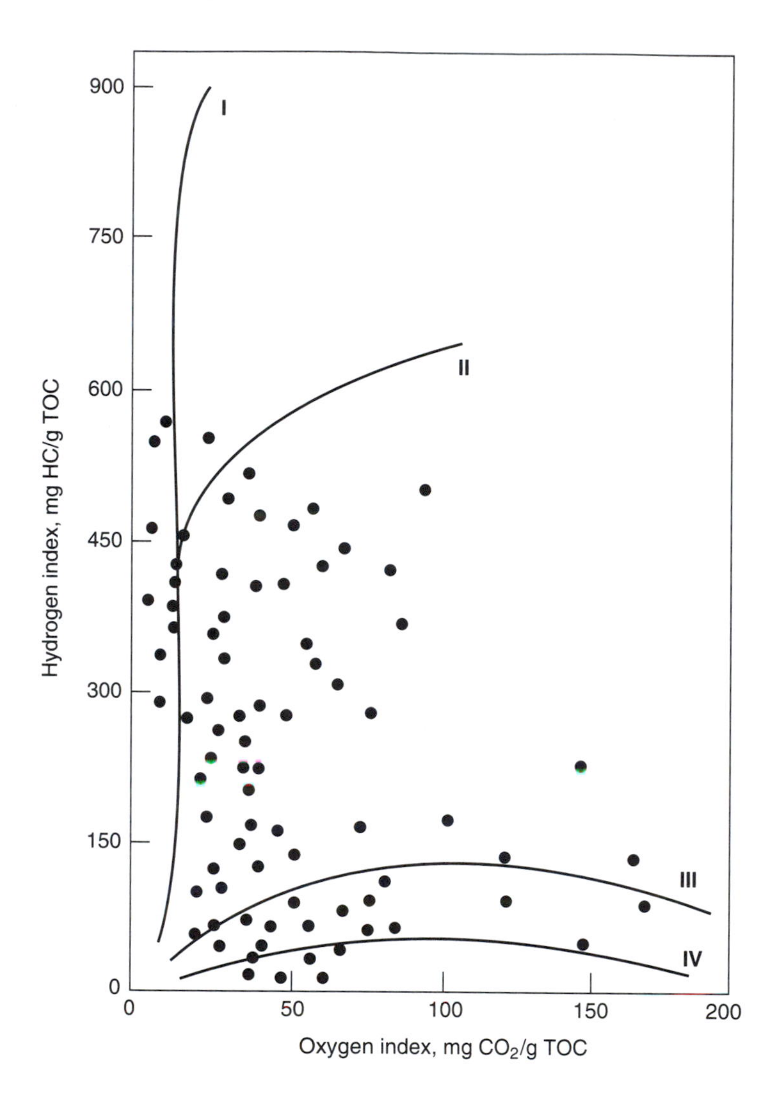

Figure 10-9

An HI/OI plot of kerogen types in Kimmeridgian–Volgian source rocks from the Norwegian and United Kingdom areas of the North Sea. [Data from Peters 1986]

these samples contain more than 60% amorphous OM, which supports previous statements that not all amorphous OM is oil generating. Only the amorphous fluorescent OM in Plate 6D would have a high oil-generating potential.

Moving to the left and down an HI/OI plot, as in Figure 10-8, is presumed to represent an increase in maturation along the evolutionary pathways, as it is in a van Krevelen diagram. In many cases, however, variations in depositional environment and diagenesis can exceed maturation effects. Consequently, other maturation parameters need to be used, and some of these are provided by pyrolysis.

Figure 10-10 shows the free hydrocarbon peak (P_1) and the cracked hydrocarbon peak (P_2) for Tertiary core samples from West Africa analyzed through

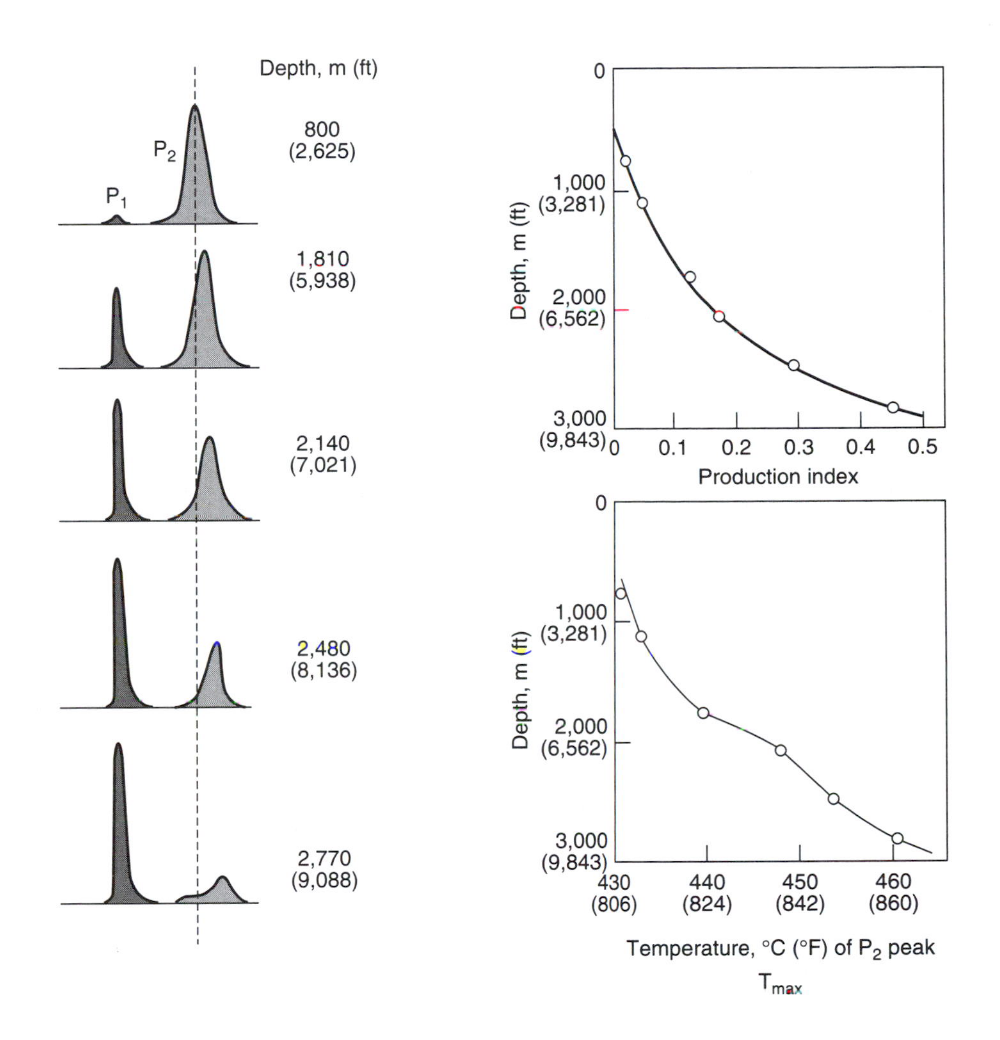

Figure 10-10

Relative hydrocarbon yields from the pyrolysis of cores at different levels of maturation. Free hydrocarbons (P_1) (S_1) increase with depth, and hydrocarbons available from cracking kerogen (P_2) (S_2) decrease with depth. The production index $S_1/(S_1 + S_2)$ and the P_2 peak temperature (T_{max}) increase with maturity. [Espitalié et al. 1977]

the 800-to-2,770 m (2,625-to-9,088 ft) depth (Espitalié et al. 1977). As the kerogen breaks down and loses hydrocarbons with greater burial, P_2 gets smaller and P_1, which represents the released hydrocarbons, gets larger. Dividing the area (S_1) by the combined areas S_1 and S_2 yields the *production index* (*PI*), shown in the upper right of Figure 10-10. Note that this increases steadily during hydrocarbon generation. The beginning of significant oil generation is around a PI of 0.1, and the end is around 0.4.

As generation proceeds, higher temperatures are required to crack the remaining kerogen. This causes the peak of P_2, which is called the T_{max} (the maximum liberation of hydrocarbons), to shift gradually to the right of the dashed line in Figure 10-10. The actual temperature increase is shown in the lower right, with the T_{max} at the beginning of oil generation starting around 430°C and the end being around 460°C. These laboratory temperatures are relative and differ depending on the particular equipment used for pyrolysis.

HI, T_{max}, and the Depositional Environment

Figure 10-11 appears to show a group of types I, II, and III kerogens moving along maturation pathways toward the lower left corner of the HI/OI diagram. But that is not what is happening. Instead, all these data points are from a single 90 m (295 ft) core representing the Cenomanian–Turonian Greenhorn Formation in Colorado (Pratt 1984). The production indices (PIs) are mostly between 0.04 and 0.1, and the T_{max} is between 430 and 440°C. Clearly, the differences in Figure 10-11 cannot be attributed to maturation. Pratt found that they correlated strongly with the extent to which the sediment was bioturbated. Highly macroburrowed strata showing vertical and horizontal burrows greater than 1 mm in diameter indicate well-oxygenated bottom water depositing primarily types III and IV kerogens. The finely layered, laminated strata with no burrows indicate anoxic bottom water. They fall into types I and II. The moderately macroburrowed strata contain types II and III kerogens. This reemphasizes the importance of doing petrographic and microscopic studies on the samples being pyrolyzed. In this example, the PI and T_{max} support the microscopic studies by demonstrating that the differences in HI are not due to maturation. Note also that there are no data points on the type I and type II evolutionary line between about 150 and 300 HI. Maturation of types I and II kerogens might be expected to show some points along this line if thermal effects were occurring.

The Lower Cretaceous Mowry Shale of Wyoming is a well-known, moderately rich source rock whose extracts have been correlated, using biomarkers, with reservoired Cretaceous oils in Wyoming. Figure 10-12 is an HI/OI plot for both immature and mature Mowry shales and the Cretaceous Skull Creek shales in drill cuttings from Wyoming and adjacent states (Burtner and Warner 1984). The data look like a scatter plot with types II, III, and IV kerogens, but this is only part of the story. The depositional environment of the Mowry sea ranged from oxic in the northwest, where TOC values are below 1%, to anoxic in the southeast, where the average TOC exceeds 3%. The T_{max} values for these

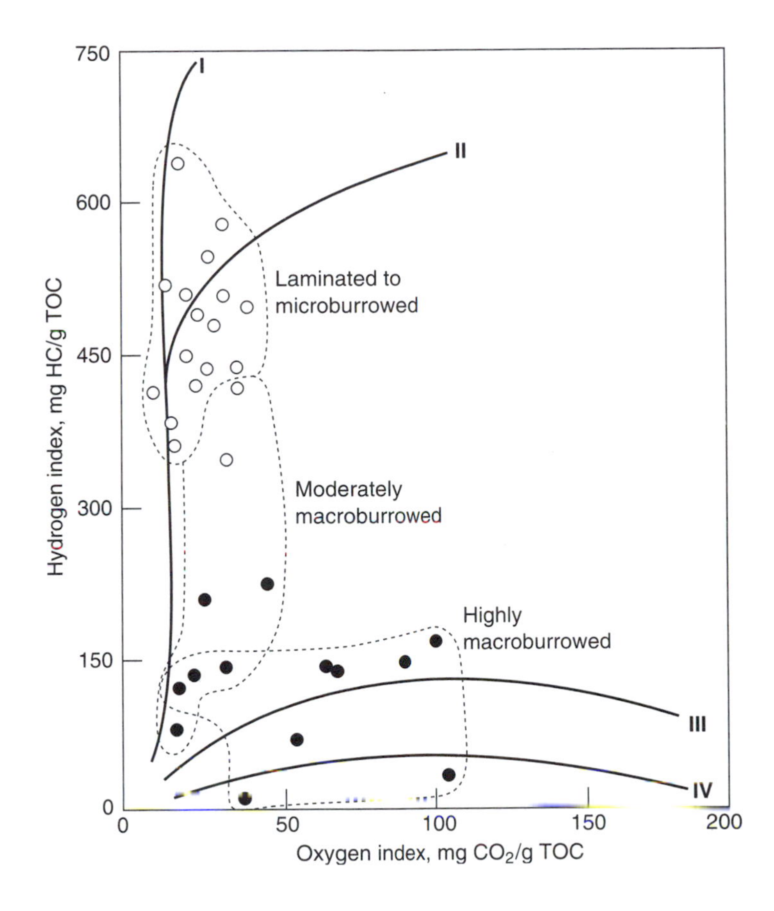

Figure 10-11
The effect of bioturbation on hydrogen and oxygen indices of organic matter in the Greenhorn Formation of Colorado: (○) oil-generating and (●) condensate- and gas-generating kerogen. [Modified from Pratt 1984]

samples ranges from 415 to 450°C (compare these with Figure 10-10). The variations in Figure 10-12 are due to differences in source material, depositional environment, and maturity. All samples between about 150 and 400 HI are type II kerogens, with good oil-generating potential, from southeast Wyoming. Samples from the northwest plot as types III and IV. No type I has been identified in the Mowry Shale.

A later study by Davis et al. (1989) found that the observed trend is due primarily to a difference in the source of OM. Mostly, terrestrially derived OM occurs in the more oxic northwest, whereas abundant marine radiolarians without any recognizable terrestrial OM occur in the less oxic areas of the southeast. They examined only immature Mowry shales, for which nearly all of the hydrogen index values were below 400. This example demonstrates that oil fields can form from source rocks without the high HI values shown in Figure 10-8.

Such wide variations in kerogen type and source potential are characteristic of many source rocks. Thus it is generally not possible to assign a single kerogen type to a particular formation. For example, lacustrine source beds do

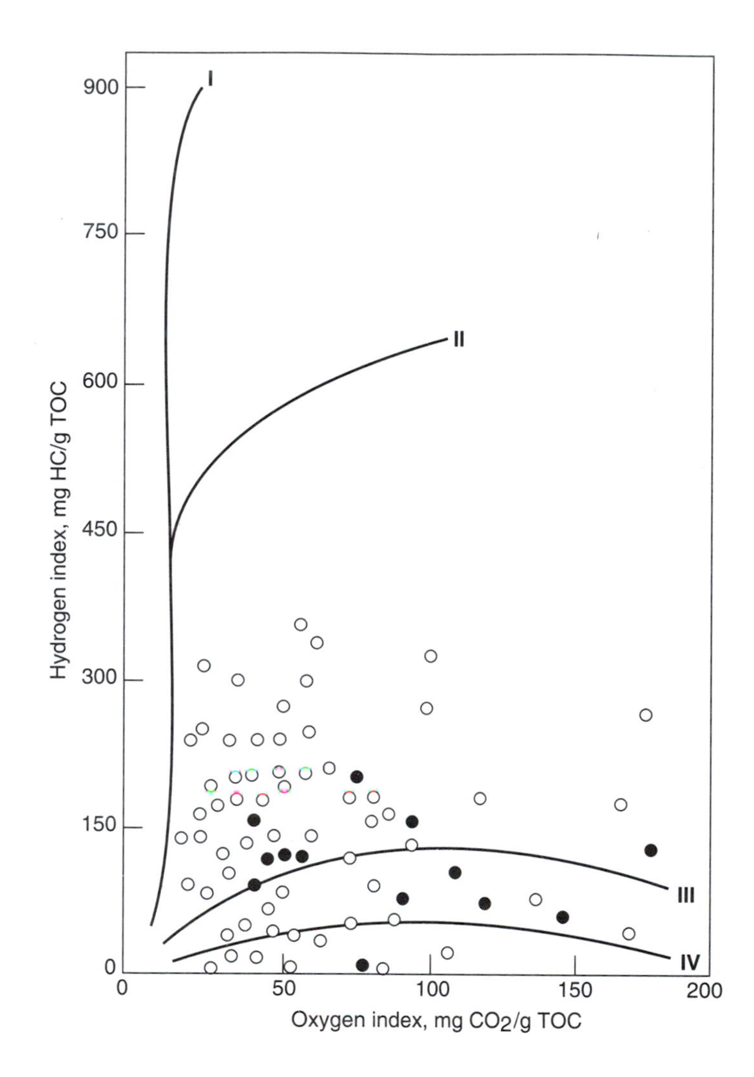

Figure 10-12

An HI/OI plot for the kerogen in Lower Cretaceous Mowry (○) and Skull Creek (●) shales of Wyoming. [Modified from Burtner and Warner 1984]

not contain primarily type I kerogen, as previously shown for the Songliao Basin. Figure 10-13 is an HI/OI plot for kerogens of lacustrine Green River shales and for the laterally equivalent alluvial Wasatch Shales of the Uinta Basin, Utah (Anders and Gerrild 1984). The Green River Formation is a mixture of open lacustrine, marginal lacustrine, and alluvial facies, all of which contribute different kerogen type mixtures. The deeper open lacustrine facies samples are concentrated at the top of Figure 10-13, around the type I line and between types I and II. More mature samples are farther down the graph, going almost as far in maturity as the black shales in Figure 10-5. The most mature samples are buried to depths of 14,050 ft (4,284 m). Five samples of the marginal lacustrine facies are between types I and II, and the rest, including some immature samples, are between types II and III. All the alluvial samples are below an HI of 200, indicating gas-generating capability but little oil.

Figure 10-13 does illustrate that the Green River Formation in general has a much higher potential for generating petroleum than the Mowry Shale in Figure 10-12. Note that none of the Green River Shale samples has an oxygen index of > 60, whereas about half the Mowry Shale samples have OIs above 60. This indicates that most of the Green River lacustrine samples were types I and II when immature, compared with possibly only half the Mowry Shales.

Distinguishing between type III (potential gas) and type IV (nongenerative) kerogen is particularly difficult at low TOC values, because of the mineral matrix effect during whole-rock pyrolysis. Espitalié et al. (1980) found that the pyrolysis of Upper Cretaceous rocks from the Douala Basin containing 0.9 to 1.6% TOC gave lower HI values than the pyrolysis of kerogens isolated from the mineral matrix of the same rocks. For example, shales from a depth of about 800 m in the basin show an HI of around 40 mg/g TOC. Without the mineral matter, the kerogen had a HI of about 170. This shift gradually disappeared with increasing sample depth, until at 3,000 m both kinds of samples had HI values around 15.

A whole-rock pyrolysis of low TOC samples (<1.5 wt%) often results in lower HI and higher OI values than a pyrolysis of pure kerogen. The problem is worst with immature rocks having low TOCs and high contents of smectite and illite, particularly the former. Some of the HI data points for the Mowry Shale in Figure 10-12 may be erroneously low due to this effect. The HI in the whole-rock analyses drops because part of the oil cracked from kerogen (S_2) is adsorbed on clay mineral surfaces. There it undergoes secondary cracking to gas and pyrobitumen at higher temperatures. Consequently, part of the carbon in the S_2 never leaves the rock. In contrast, isolated kerogen releases all its cracked oil, thereby yielding a higher HI than the whole-rock analyses. Overcoming this problem requires dissolving the silicates with HF before pyrolysis (Espitalié et al. 1980; Horsfield and Douglas 1980).

A second problem is that the OI in whole-rock analyses tends to be higher than in isolated kerogen analyses in high-carbonate rocks containing siderite and having TOCs below 2% (Katz 1983; Orr 1983). Siderite, which decomposes to yield CO_2 at lower temperatures than calcite, often occurs in the sedimentary rocks of deltas. An obvious solution to this problem is to remove the carbonates with HCl, but this involves the same time-consuming process as in doing elemental analyses for the van Krevelen diagram. Nevertheless, this should be done if there is a real need to distinguish possible gas sources from nongenerative rocks.

The matrix effect, which is discussed in more detail in Chapter 14, is most noticeable at TOCs of less than around 1.5%. This is not a significant problem if 1% TOC is used as the cutoff for oil source rocks. But it can become important in deltaic areas where large volumes of rock with low TOCs may be generating and expelling considerable quantities of gas and condensate.

The Espirito Santo Basin of Brazil, an offshore oil-producing basin, is a good example of how the HI/OI diagram assisted in changing the direction of

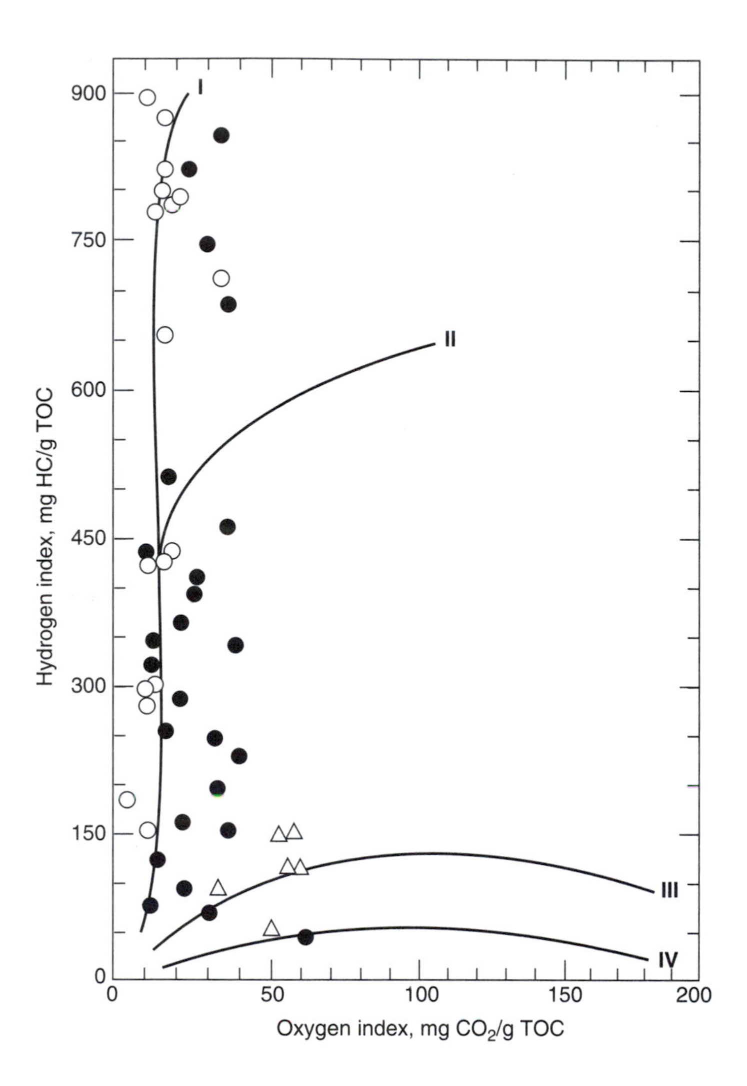

Figure 10-13

An HI/OI plot for kerogens of Green River and Wasatch shales, Uinta Basin, Utah: open lacustrine (○), marginal lacustrine (●), and alluvial facies (△). [Modified from Anders and Gerrild 1984]

the exploration campaign (Estrella et al. 1984). At first the source rocks were characterized according to gas chromatograms of the C_{15+} extracts and microscopic identification of the OM. This indicated that slope shales of Upper Cretaceous to Tertiary age were good marine source rocks. But the HI/OI plot of the well samples revealed that these all were types III and IV kerogens (Figure 10-14). The best source rock was the Neocomian Jiquia stage, and the next best

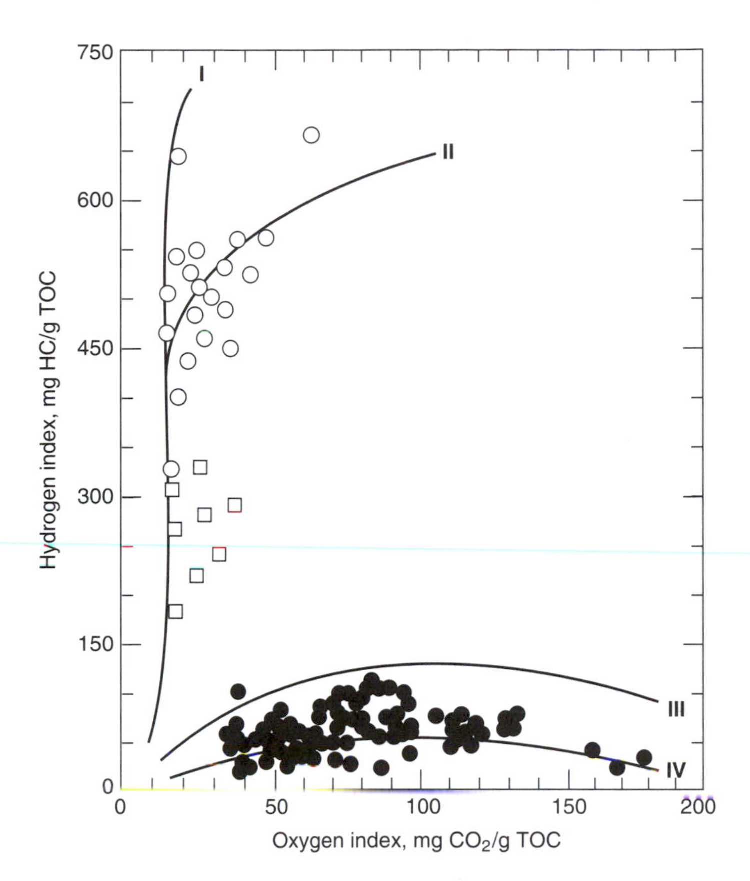

Figure 10-14

An HI/OI plot of the possible source rocks of the Espirito Santo Basin of Brazil: Upper Neocomian Jiquia stage (○), Aptian Alagoas stage (□), Tertiary and Upper Cretaceous (●). [Modified from Estrella et al. 1984]

was the Aptian Alagoas stage. This shifted the drilling to the prospects in the pathway of migrating Jiquia oils. The HI/OI results were confirmed when the biomarker gammacerane was found in the oils and in the Jiquia–Alagoas source rocks, but not in the Tertiary and Cretaceous rocks.

Oil Show Analyzer

Conventional mud logging at a well site usually involves analyzing the drilling mud for C_1-to-C_6 hydrocarbons entrained during drilling. Cuttings also have been analyzed by canning them under water and shipping them back to the laboratory for detailed hydrocarbon GC analyses (see Chapter 14 for details).

Espitalié et al. (1984) invented the Oil Show Analyzer, which measures the free gas in the rock (S_0), the free oil in the rock (S_1), the hydrocarbons from cracking kerogen (S_2), and the CO_2 formed by oxidation of the residual carbon (S_4). The TOC is calculated as the sum of the pyrolyzed and residual organic carbon. In effect, the S_3 for calculating OI is replaced by the S_4 for calculating TOC. Kerogen quality and maturity are determined by plotting HI versus T_{max}

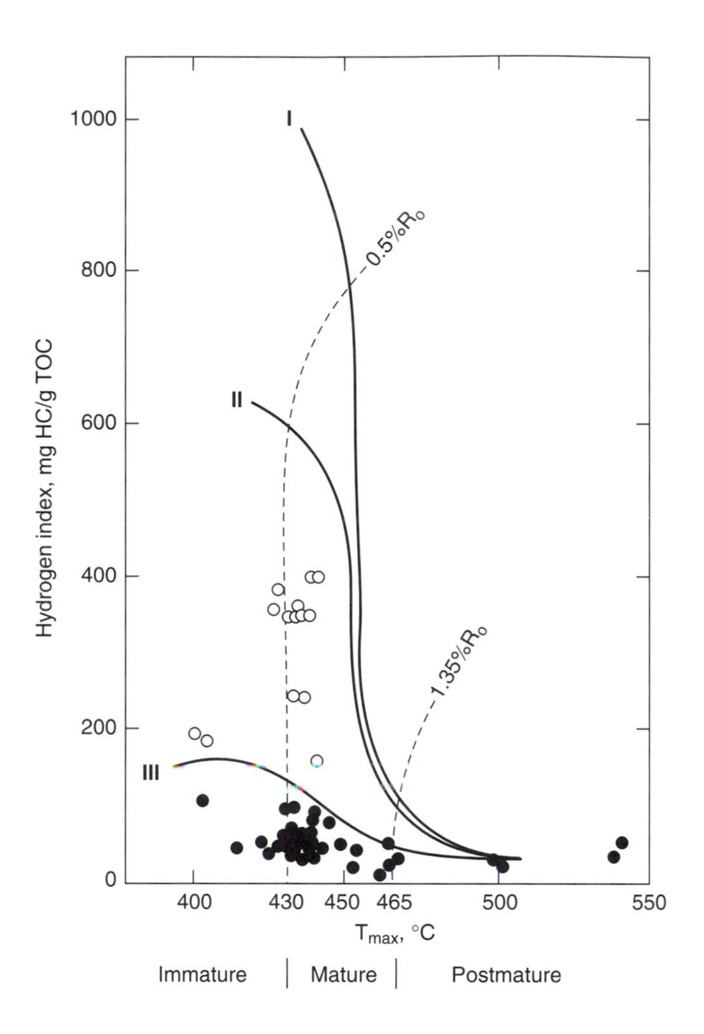

Figure 10-15

A plot of HI versus T_{max} for the Upper Jurassic and Cretaceous source rocks of the Sable Island oil and gas fields off Nova Scotia: (○) types II and III and (●) types III and IV kerogens. [Mukhopadhyay et al. 1995]

rather than HI versus OI. This eliminates the use of OI as a kerogen type indicator (comparable to the O/C in the van Krevelen diagram). The kerogen type designations are based entirely on the HI. The interpretation of maturity, however, is somewhat improved with the HI–T_{max} plot.

Figure 10-15 shows an HI-versus-T_{max} plot for some of the source rocks of the oil and gas fields on the Scotian Shelf off Nova Scotia, Canada (Mukhopadhyay et al. 1995). Most of the samples are in the mature petroleum-generating range (430 to 465°C on a Rock-Eval); several are immature; and a few are overmature. Most kerogens are type III or IV, which fits with the predominance of gas and condensate on the shelf. However, a few oil-prone source rocks in the group were analyzed.

The T_{max} varies with the type of kerogen as well as maturity, particularly in immature samples. Peters (1986) reported varations in the T_{max} of immature samples of up to 20°C due to differences in the type of OM. For example, vitrinite shows a higher T_{max} than sporinite in the same cuttings sample. Such differences tend to disappear at high maturities.

Pyrolysis–Gas Chromatography–Mass Spectrometry (Py–GC–MS)

In 1980, Whelan et al. reported the development and application of a pyrolysis–gas chromatography (PY–GC) technique which became the Chemical Data Systems pyrolyzer. It analyzed for the individual hydrocabons in the P_1 (S_1) and P_2 (S_2) peaks by GC. Figures 10-17, 10-18, and 10-20 are examples of PY–GC analyses. More recently PY–GC and MS have been combined to determine both quantity and identity of individual hydrocarbons obtained by pyrolysis. Figure 10-16 is a simplified picture of a PY–GC–MS instrument. The procedure involves placing about 20 to 30 mg of finely milled rock in a glass tube. The GC column is cryogenically cooled to –5°C and the pyrolysis unit heated to

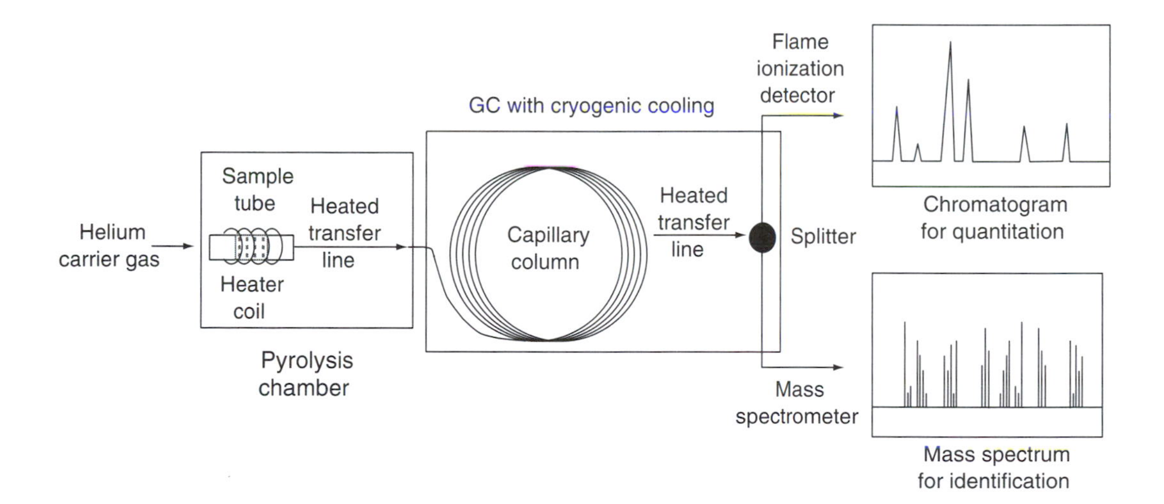

Figure 10-16

Simplified model of pyrolysis–GC–MS analyzer for P_1 (S_1) and P_2 (S_2) fractions. [J. K. Whelan, personal communication]

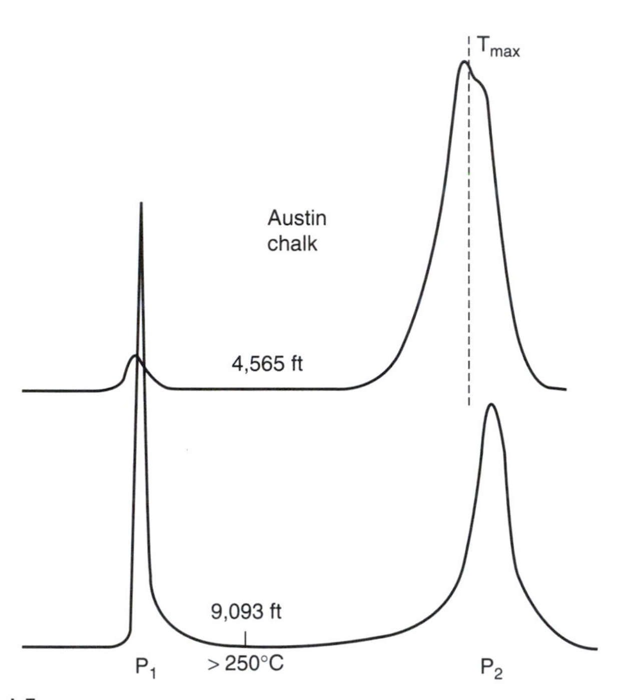

Figure 10-17

Pyrograms of the Austin Chalk showing an increase of P_1 and decrease of P_2 with depth and a shift of maximum temperature (T_{max}) of P_2.

250°C. This drives the P_1 (S_1) free hydrocarbon into the GC capillary column. Subsequent heating of the column transfers the P_1 (S_1) hydrocarbons to the FID and MS (Figure 10-16) thereby providing both a gas chromatogram and mass spectrum of the P_1 (S_1) hydrocarbons. The GC is cooled again to –5°C while the pyrolyzer is heated to 700°C to release the cracked hydrocarbons, P_2 (S_2). These are trapped in the cooled capillary column and subsequently released for the GC–MS analysis.

Figure 10-17 shows the free P_1 and the cracked P_2 hydrocarbons from the Austin Chalk of Texas (Hunt and McNichol 1984). The curve at 4,565 ft (1,390 m) displays a small amount of free hydrocarbon in the chalk (P_1), but the potential to generate hydrocarbons is considerable (P_2). The curve for the sample at 9,093 ft (2,770 m) shows a large P_1, indicating that the chalk already has generated considerable quantities of hydrocarbons but the P_2 is smaller because an appreciable part of the kerogen already has been converted to hydrocarbons. The production index, PI (PI equals the ratio of the area S_1 to the combined areas S_1 plus S_2), is about 0.1 in the shallow sample and 0.7 in the deeper sample, indicating that substantial generation and expulsion from the deeper sample have already taken place.

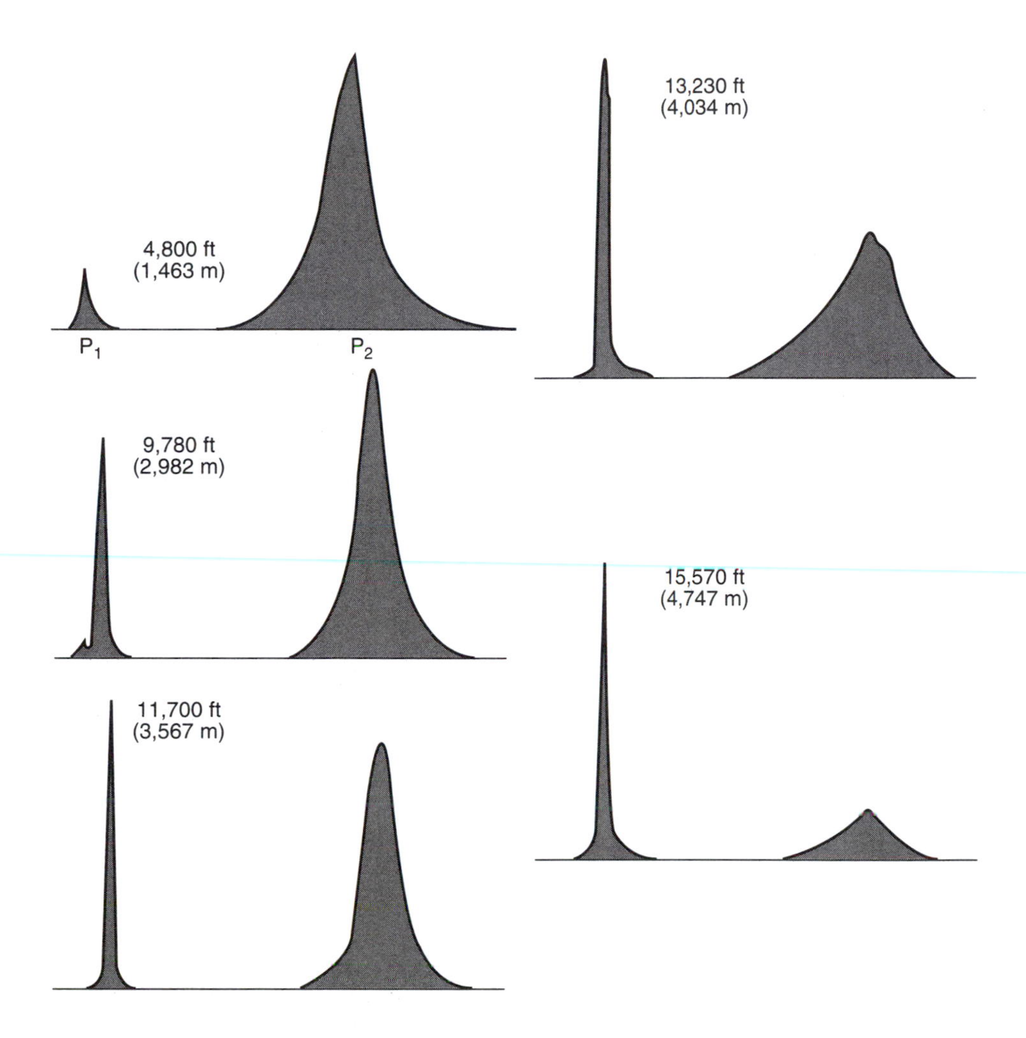

Figure 10-18

Pyrograms for five different depths in a South Padre Island, Texas, well. [Huc et al. 1981]

A Py–GC was used in the analysis of well cuttings from the COST-1 well in the Gulf of Mexico shown in the next series of figures (Whelan et al. 1980). Figure 10-18 shows pyrograms obtained from heating rock cuttings samples through a depth range from 4,800 ft (1,463 m) to 15,570 ft (4,746 m) [Huc et al. 1981]. The free hydrocarbon peaks (P_1) increase steadily in area (S_1) with increasing depth as hydrocarbons are generated and released. As the deeper

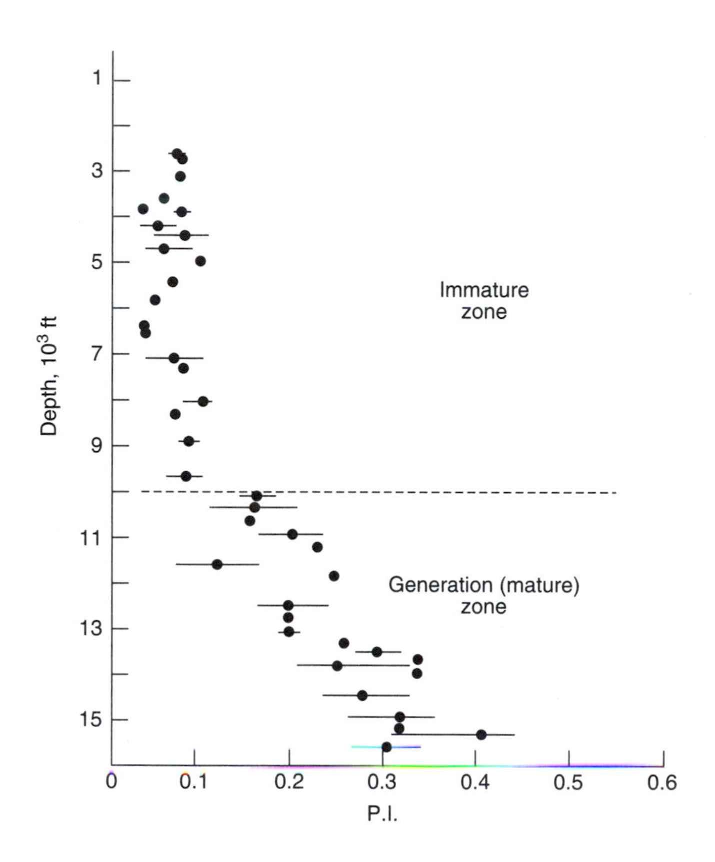

Figure 10-19

Production index versus depth in a South Padre Island, Texas, well. [Huc and Hunt 1980]

samples move beyond the source rock, this peak area diminishes. The area of the P_2 (S_2), representing the potential to generate hydrocarbons, continually decreases in area with depth. The production index, $S_1/(S_1 + S_2)$, remains near 0.1 in the immature zone to a depth of about 10,000 ft (3,280 m), where it begins to increase because of the generation of hydrocarbons (Figure 10-19). The PI maximizes at about 0.35 at 14,000 ft (4,268 m), which is the peak of the oil-generation window (Hunt 1981).

Figure 10-20 shows the distribution of hydrocarbons that make up the P_1 peak at 12,600 ft (3,842 m) and 15,690 ft (4,780 m). Note that the C_{11} *n*-paraffin is present in the highest concentration at the greater depth. At the shallower depth, the earliest eluting C_7 *n*-paraffin is dominant. This shift to larger molecules at greater depths is believed to be due to the preferential migration of the smaller hydrocarbon molecules out of the source rock with increasing matu-

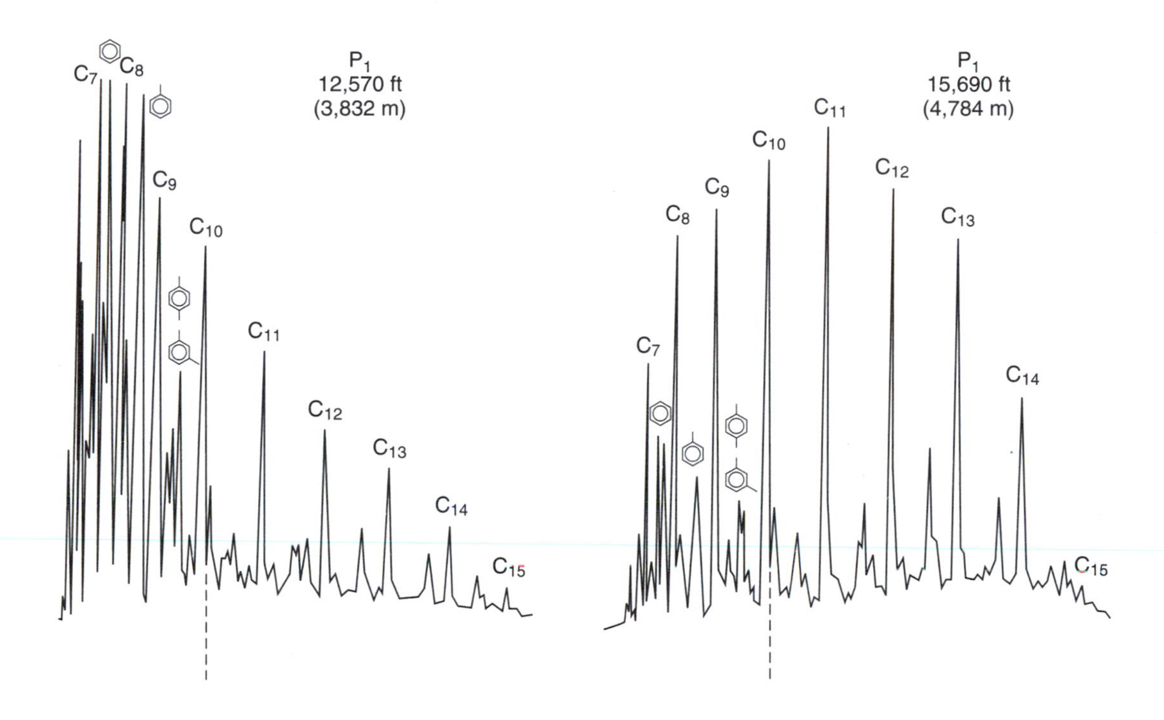

Figure 10-20

Change in distribution of individual hydrocarbons in the P_1 (S_1) peak with depth, South Padre Island, Texas. [Huc et al. 1981]

rity. The quantity of specific hydrocarbon groups such as the *n*-alkanes in the C_7–C_{14} range in mg/g TOC can be calculated with depth, as seen in Figure 10-21 (Hunt 1981). This shows a well-defined oil window peaking at about 14,000 ft (4,268 m).

Applications of Different Pyrolysis Methods

Pyrolysis–GC and GCMS is most useful as a research tool for understanding the processes of individual hydrocarbon formation and migration. Rock-Eval pyrolysis finds its greatest application as a screening tool for quickly defining source rock potential. The concepts that have been discussed are summarized in Table 10-5. In the definitions, P refers to peak number and S to peak area. Rocks are characterized according to the quantity of migrated or generated free hydrocarbons that they contain and their potential to generate hydrocarbons by cracking the OM in the rock. The quality of the OM is defined in terms of

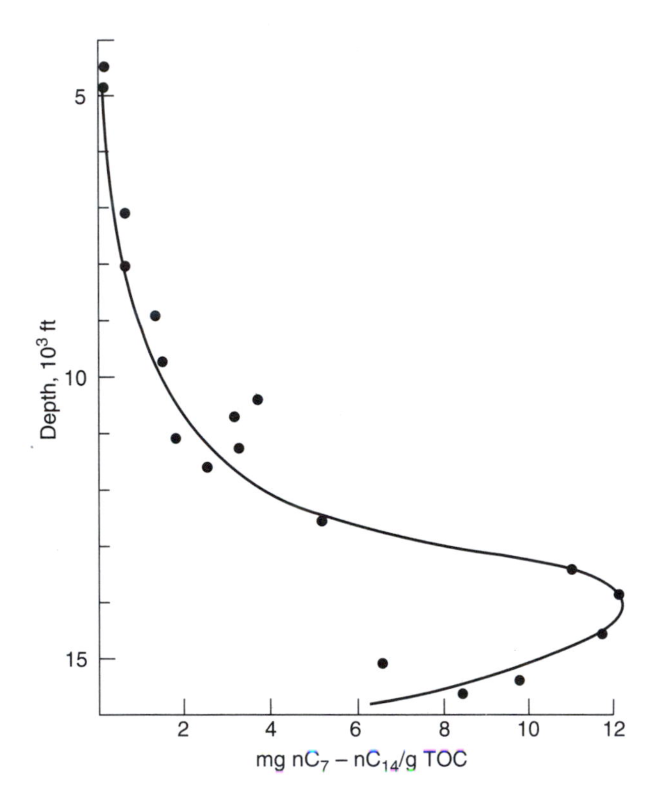

Figure 10-21
Distribution of *n*-alkanes in P_1, South Padre Island, Texas. [Hunt 1981]

kerogen types I through IV from the HI/OI plots and the hydrocarbon type index. The maturation state is determined from the HI/OI plot or the HI–T_{max} plot, the T_{max}, and the production index (PI). Migrated hydrocarbons can be detected at very low levels by anomalies in the PI curve. Potential oil and gas recoveries from possible producing intervals can be monitored by the S_1/S_2, with the best recoveries from those horizons having values greater than 5. Also, S_1/TOC is used in comparing potential recoveries. These techniques are generally used with mud logging and wireline logs.

Clementz et al. (1979) used Rock-Eval pyrolysis at the well site. They found it useful in locating stratigraphic boundaries by differences in organic facies.

Other Maturation Indicators

Other frequently used kerogen maturation parameters are fluorescence, the thermal alteration index (TAI), spore color index (SCI), conodont alteration index (CAI), methylphenanthrene index, vitrinite reflectance, gas chromatogram patterns, and biomarkers. Additional maturation parameters for coals

TABLE 10-5 Source Rock Characterization by Pyrolysis

Definitions

S_0 (P_0)	Free hydrocarbons C_1 to about C_9 thermally liberated at a 90°C isotherm.
S_1 (P_1)	Free hydrocarbons C_1 to about C_{23} thermally liberated at a 300°C isotherm.
S_2 (P_2)	Hydrocarbons cracked from kerogen or from C_{24}+ bitumens by heating to 550°C. Units with TOC go to 600°C.
S_3 (P_3)	Organic carbon dioxide released between 300 and 390°C.
T_{max}	The temperature at the highest yield of S_2 hydrocarbons.
S_4	Total organic carbon in weight percent (TOC). TOC = pyrolysis carbon (PC) + residual carbon (RC). PC = 0.82 ($S_1 + S_2$)/10. RC = S_4/10, where S_4 = mg C/g rock.

Quantity

S_1 = Migrated hydrocarbons if immature sediments (oil and gas shows).

S_1 = Generated or migrated hydrocarbons or contaminants if mature sediments. The migration index, S_1/TOC, is high, and the T_{max} is low if migrated. S_1/TOC = 0.1 to 0.2 for oil expulsion.

S_2 = Potential to generate hydrocarbons if buried deeper. $S_1 + S_2$ is the total genetic potential in mgHC/gm rock. It is < 2 for poor source rocks, 2–5 for fair, 5–10 for good, and > 10 for very good.

Quality (kerogen types I, II, or III)

(S_2/TOC) × 100 = Hydrogen index (mgHC/g TOC).

(S_3/TOC) × 100 = Oxygen index (mgCO_2/g TOC).

S_2/S_3 = Hydrocarbon type index. It is < 2 for gas and > 5 for oil.

Maturation state

$S_1/(S_1 + S_2)$ = Production index. Oil window = 0.08 – 0.4. Higher values are often due to migrating hydrocarbons or contaminants.

T_{max} = S_2 peak temperature. Oil window = ~430 to 470°C.

Reservoir studies

Oil and gas recovery from sands has been monitored by the S_1/S_2, with the best recoveries from horizons with values greater than 5. It is normally used with mud logging and wireline logs. Also, the S_1/TOC can be used.

include coal rank, which is based on oxygen content and volatile and nonvolatile (fixed) carbon. Parameters such as the carbon preference index (CPI and OEP) were discussed in Chapter 4. The change in weight percent C and H and the atomic H/C ratio were discussed in connection with van Krevelen diagrams. The most widely used parameter in this group is vitrinite reflectance.

The methylphenanthrene index (MPI), developed by Radke (1988, and references therein), is one of several aromatic hydrocarbon ratios that have been calibrated against vitrinite reflectance. It is particularly useful in the mature-to-postmature range such as 0.7 to 2.0%R_m, within which many of the previously mentioned indicators are not applicable. It is somewhat less useful in the lower maturity ranges, owing to the effect of different kerogen types on the ratios (Peters and Moldowan 1993, p. 220). Radke et al. (1990) used several aromatic hydrocarbon ratios of crude oils from the Handil field, Indonesia, to estimate the depth of the source rocks.

Fluorescence

Fluorescence microscopy is used to differentiate source rocks from nonsource rocks and to estimate the level of maturation from the fluorescence color. As previously mentioned, amorphous kerogen is potentially oil generating only if it fluoresces.

Fluorescence is the property of matter to emit light under the influence of an exciting light. In the fluorescence microscopy of source rocks, the commonly used exciting light is blue light (477 nm). A xenon lamp equipped with an excitation filter at 495 nm and a barrier filter at 520 nm may be used for qualitative intensity measurements. The fluorescence intensity of most organic substances is much stronger under blue light than under ultraviolet excitation, although the latter would be needed for observing blue or bluish green fluorescence.

In order to fluoresce, organic matter must contain some aromatic structures or chromophore groups, such as pigments with double or triple bonds. Paraffins and cycloparaffins do not fluoresce. Steranes and triterpanes must be aromatized in order to fluoresce. Chlorophyll and carotene fluoresce, but carotane does not.

The first microscopic evidence of oil being formed by coal macerals was made by Teichmüller (1974) using incident blue light on liptinite macerals of coals. She observed a greenish yellow, strongly fluorescent oil being expelled from fine fissures and holes in the macerals. The coals were at a vitrinite reflectance of $R_m = 0.5\%$, equivalent to the beginning of oil generation from kerogens. This expulsion was called the "first coalification jump of liptinites." It consisted of several changes such as a fluorescence maximum of sporinite and the formation of granular micrinite, which appears to be a product of disproportionation reactions of certain liptinites. Micrinite, which has not been identified in lignite and subbituminous coal (< 0.5%R_o), appears to be a secondary maceral formed as a high-carbon residue during oil generation in coals. Teichmüller believes that the hydrogen used to form the liquid oil is released from aromatic structures, which then condense to form the high-carbon micrinite, which has an atomic H/C ratio of 0.5 or less.

Several liptinite macerals were found to expel fluorescent liquids, which were collectively named *exsudatinite* (Teichmüller 1974). Both exsudatinite and micrinite are secondary macerals; that is, they are formed from primary coal macerals or from kerogen during maturation. Since this early work, fluorescence has been widely used to track oil expulsion and migration in source rocks, as shown in Plates 3D, 4A, 4B, and 4C.

The changes in fluorescence colors for several of the components of kerogen and coal when going from immature to mature and postmature stages of oil generation were described by van Gijzel (1979). Fresh pollen grains and spores have fluorescence colors ranging from blue to red, depending on the type or species. Algal material starts out greener than spores. Maturation causes a gradual shift in fluorescence colors from the shorter to the longer wavelengths, that is, blue and green to yellow, orange, and red (van Gijzel 1979).

According to Teichmüller (1982), cutinites and resinites are not well suited for maturation studies because they show varying fluorescent properties at a given rank stage. Sporinite fluorescence is best suited for rank evaluation, even though Carboniferous spores follow a somewhat different maturation track than Mesozoic and Tertiary sporinites, which are mainly pollen.

The most striking changes in fluorescence parameters of liptinites occur at their three coalification jumps, which are at vitrinite reflectance values of 0.5, 0.8 to 0.9, and 1.2 to 1.6%R_m. These three liptinite jumps correlate with the beginning, the maximum, and the end of oil formation from kerogen, as well as from coal (Teichmüller 1982).

Vitrinites and inertinites show very little primary fluorescence at coal ranks > 0.45%R_m. Secondary fluorescence occurs, however, at the first and second coalification jumps. Teichmüller (1982) found that the percentage of fluorescent vitrinites in a series of Ruhr coals increased from 10% at 0.5%R_m to almost 90% at 1.0%R_m, which is through the peak period of oil generation and expulsion. Pyrolysis experiments duplicated those observed in nature. All fluorescence disappears during the third coalification jump, which is in the rank range of medium- to low-volatile bituminous coals (1.2 to 1.6%R_m), equivalent to the end of oil generation.

The change in fluorescence color with maturation can be seen more clearly on the chromaticity diagram shown in Plate 5D. The construction of this diagram, which was developed by the CIE (Commission Internationale de l'Eclairage), is discussed by Thompson-Rizer and Woods (1987) and by Hagemann and Hollerbach (1981). The solid line in Plate 5D shows the maturation change of alginite going from fluorescent colors of dark green (immature) to light green, yellow, and orange (P. van Gijzel, personal communication). Hagemann and Hollerbach (1981) found that the maturity of both coals and sedimentary rocks could be followed by fluorescence color changes in the rock extracts. The fluorescence changed with increasing rank from blue to green, yellow, orange, and finally red.

Subsequently, Thompson-Rizer and Woods (1987) used microspectrofluorescence analysis to predict the maturity of petroleum source rocks and coals to within 0.10%R_o (vitrinite reflectance). They proposed the use of the term R_f to report fluorescence data in equivalent reflectance values. The R_f technique is particularly useful for organic-rich source rocks which frequently have few, if any, vitrinite particles for measuring R_o. They recommended using both R_o and R_f numbers when the data are available, particularly when the R_o numbers are suspiciously low or high. Fluorescence also is useful for evaluating low-maturity samples for which $\%R_o$ values are more difficult to measure.

Some examples of the differences between transmitted white light and reflected blue light are shown in Plates 6A through 6D. Plate 6A shows a Senonian Upper Cretaceous spore and cuticle at 25 times magnification in transmitted light, photographed by H. M. Heck. Plate 6B shows the same spore and cuticle in reflected blue light. Plate 6C is of a Kimmeridge Shale source rock of the North Sea at 40 times magnification in transmitted white light by C. Thompson–Rizer. Plate 6D is the same picture under reflected blue light. Both these examples show a bright yellow fluorescence under blue light, indicating a potential to generate oil.

Henry Hinch, formerly with Amoco, developed a technique for measuring the fluorescence of an entire core in order to evaluate the overall distribution of fluorescent oil-prone OM, bitumen, and inert OM. It also can be used for recognizing intervals saturated with migrated oil. He mounted four 40-watt UV lights, two on each side of the core at about 60 to 70° angles, approximately 3 feet above the core. Pictures were taken with a 35 mm camera with a filter to take out light, which interferes with the fluorescence. Plate 7A shows the heterogeneity in fluorescence of the Antrim Shale source rock of the Michigan Basin. Note that the fluorescent material is concentrated along bedding surfaces and laminae. The nonfluorescent intervals contain varying amounts of inert OM.

At present, sporinite is the most widely used maceral for estimating coal and kerogen rank by fluorescence. It changes color from green through yellow to orange and finally red with increasing rank to the end of the medium-volatile bituminous coals, after which there is no more fluorescence in the visible spectrum. Mao et al. (1994) correlated the vitrinite reflectance ($\%R_o$) scale with a spore fluorescence color (SFC) scale. They used this to estimate maturation for samples with few or no indigenous vitrinite particles. This correlation also was useful for evaluating short-term, high-temperature heating events. They obtained SFC data from samples taken in the hot vent area of the Juan de Fuca Ridge during the Ocean Drilling Program. They found that the majority of the megaspores appeared to have scorched outer walls, in that they fluoresced in medium orange to red colors. A smaller proportion of the spores fluoresced yellow to light orange. The authors concluded that this dichotomy in fluorescence color, which was observed in several holes, was probably due to one or more short but extremely high temperature exposures.

Thermal Alteration Indicators (TAI, SCI)

Palynologists separate hystrichospherids, spores, and pollen grains from sediments by dissolving the mineral matter with hydrochloric and hydrofluoric acids. These microfossils are used for stratigraphic correlations and age dating. In their studies, palynologists noted that the color of spores and pollen change with increasing burial depth from light to dark when viewed under a microscope with transmitted light. Eventually they equipped their microscopes to measure light absorption and used this to follow semiquantitatively the maturation of spores and pollen. By analyzing the spores and pollen in coals of known rank, they were able to correlate the light absorption of spores and pollen with the fixed carbon content of coal analyses.

Staplin (1969) developed the Thermal Alteration Index (TAI) as a relatively simple and rapid technique for evaluating kerogen maturation in well cuttings directly from its change in color. The standard palynological processing technique was modified to eliminate any treatment that might cause a partial loss of color. The color designation was restricted to plant particles that are initially yellow, yellow green, or pale orange. The color designation was made for spores, pollen, plant cuticles, algae, and amorphous organic matter whose colors are in this range. Coaly material was recorded but not used in designating colors. Generally, the lightest-color material was used for color designations, assuming that it was not a contaminant.

The current TAI scale, which is widely used as an exploration tool along with other maturation indicators, is shown in Plate 7B. This chart was made by D. L. Pearson (1990) using Munsell color standards to indicate how the fossil colors are correlated with the TAI scale. The indices from 1 to 5 represent kerogen color changes from light yellow to dark yellow, light brown, brown, dark brown, and black. The yellow is immature; the light brown and brown, mature; and the dark brown and black, postmature. Only dry biogenic gas and, possibly, heavy oil are to be found from the immature facies; oil and wet gas, from the mature facies; and, mainly, dry gas from the postmature facies.

An example of the TAI technique is shown in Figure 10-22. In the 1960s, Imperial Oil of Canada was trying to distinguish, before drilling, between seismic structures that might contain oil and those that might have gas. Imperial was using canned cuttings analysis (see Chapter 14) on all wildcat wells to define the boundary between wet gas and dry gas in the cuttings. But this was not a simple technique and could not be used on outcrops. Evans and Staplin (1971) found that the TAI correlated directly with the composition of the cuttings gas. If the cuttings had wet gas, indicating oil, the TAI would be dark yellow to brown. If the cuttings had dry gas, the TAI would be black. Consequently, it was ultimately used in all wildcat areas with both unweathered outcrops and well cuttings.

Staplin went on to map the boundary between oil and gas occurrences in major sedimentary units from the sixtieth parallel to the Arctic coast. This is the origin of the "hot lines" previously shown in Figure 7-21. Several structures in the Northwest Territories were correctly identified as containing oil instead of gas, based on the combined TAIs and canned cuttings analyses. Meanwhile, Correia (1969) found that oil accumulations in the Jurassic of France's Aquitaine Basin were restricted to sediments containing dark yellow to light brown kerogen. Source beds associated with the dry gas province to the south contained only dark brown and black kerogen.

The TAI is widely used as a rough and inexpensive maturation indicator. Interpretation of the maturation level, however, is difficult in some samples because of the different maturation rates of different species and organisms.

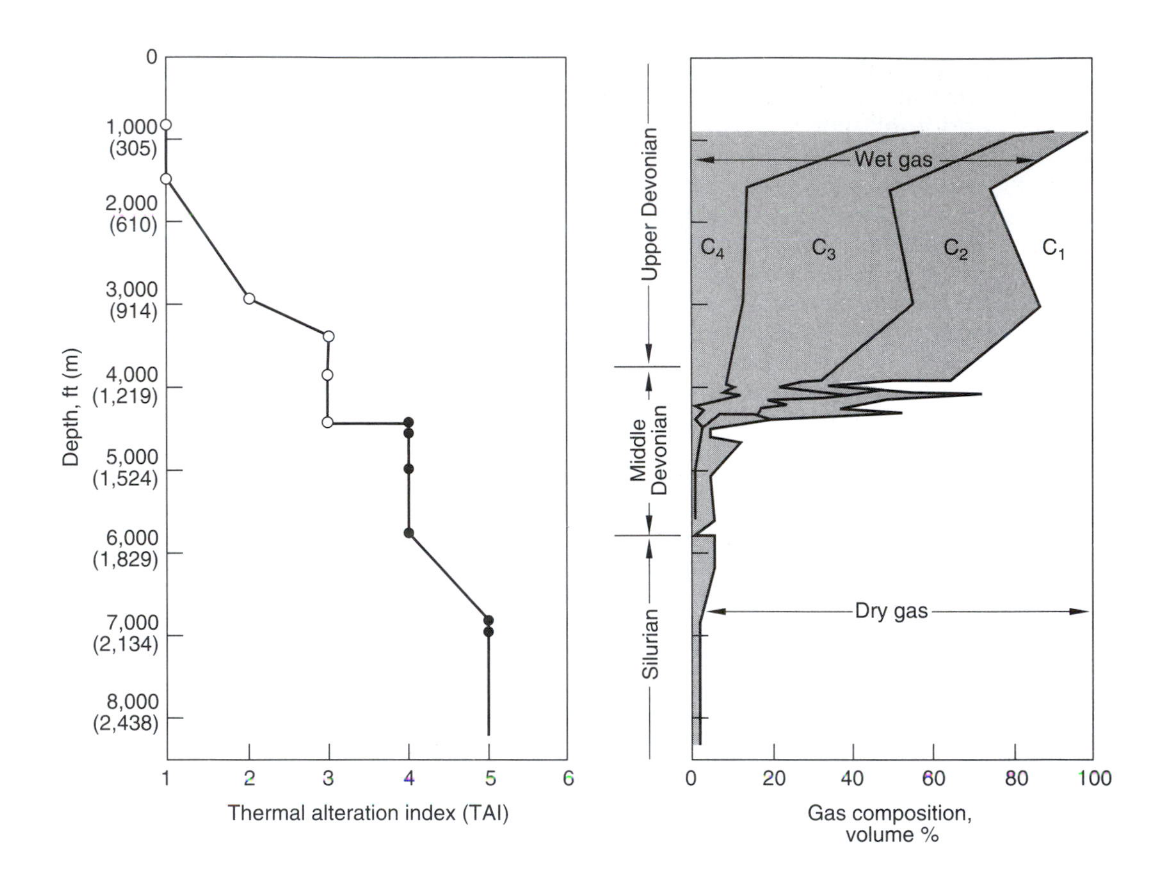

Figure 10-22

Increase in the TAI (kerogen color) with depth and corresponding change in composition of gases from well cuttings of I.O.E. Clare drilled in Northwest Territories, Canada. Index 4 corresponds to the end of wet gas. [Evans and Staplin 1971]

Color interpretations are more accurate when they are limited to individual exinite species. Barnard et al. (1976) developed the spore coloration index (SCI) on a 1-to-10 scale specifically for spores. Both the Barnard and Staplin scales use as standards a set of about twenty color slides.

Experience is needed in all microscopic techniques to understand the problems, which vary in different basins. For example, recrystallization causes a higher level of apparent maturation in the kerogen of carbonates. Such kerogen is frequently black, whereas the stratigraphically equivalent shales may contain dark yellow or brown kerogen. Oxidizing agents and weathering cause substantial changes in translucency. Not everyone recognizes colors in the same way. Smith (1983) correlated the color index scales of TAI and SCI. He then put the spectra of a reference set of color slides for these scales in a computer where they could be matched with new sample data. By using stable illumination, a photomultiplier, and an interference filter of 400 to 700 nm, he was able to minimize some of the subjectivity that occurs with kerogen color analysis.

Maturation techniques involving land-derived plant materials as discussed earlier are not applicable to the Pre-Devonian. However, *conodonts,* which are microscopic marine skeletal (phosphatic) parts of fauna, undergo color changes with maturation. They are widespread from the Cambrian through Pennsylvanian–Permian. They are preserved even in metamorphic facies because they are composed mainly of calcium phosphate. Conodont coloration has been correlated with vitrinite reflectance and appears to be useful for both Pre-Devonian rocks and some organic-poor rocks such as red beds (Epstein et al. 1977). They change from light brown to black during maturation. All these techniques need to be used in combination with other maturation indicators for proper evaluation of source rock maturity.

Vitrinite Reflectance (%R_o)

The use of vitrinite reflectance as a technique for determining the maturation of OM in sedimentary rocks was first described by Marlies Teichmüller in her study of the Wealden Basin (1958). She had observed the same kind of relationship between coal rank and oil described by White (1915) as his carbon ratio theory. In the Wealden Basin, the oil fields are confined to sedimentary units containing low-volatile to medium-volatile bituminous coals. Teichmüller had been using the reflectance of the vitrinite maceral of coal to measure coal rank. It occurred to her that in sediments where coals are not readily available, as in northwest Germany, it should be possible to measure the reflectance of the small vitrinite inclusions that occur in many of the carbonates and shales. She used this technique to classify petroleum regions that did not possess deposits of coal. Later, in a detailed study (1963) of the coalification process in the Munsterland I bore hole—drilled to 5,956 m (19,541 ft), she used the reflectance of vitrinite particles in the shales to extend rank parameters into the deeper horizons where no coal seams were found.

Today, vitrinite reflectance is the most widely used indicator of thermal stress, because it extends over a longer maturity range than any other indicator, and a skilled organic petrologist can make a large number of analyses in a relatively short time.

Vitrinite includes *telinite,* the cell-wall material of land plants, and *collinite,* the organic substance that fills the cell cavities. The reflectance of light on a polished surface of vitrinite increases with greater maturation because of a change in the molecular structure of the maceral. Vitrinite is composed of clusters of condensed aromatic rings linked with chains and stacked on top of one another. With increasing maturity, the clusters fuse into larger, condensed–aromatic ring structures, such as in Figure 7-13 and Plate 3C. Eventually they form sheets of condensed rings that assume an orderly structure. Both the increase in the size of these sheets and their preferred orientation cause greater reflectivity.

Irreversible chemical reactions in which the rate rises exponentially with temperature are responsible for the changes in molecular structure. Consequently, any reflectance measurement of these maturation changes also increases exponentially with a linear rise in temperature.

Ting (1975) demonstrated this relationship in the laboratory when he heated woody lignite particles at a pressure of about 1,000 atmospheres and at a temperature of 100°C for seven days. Vitrinite reflectance was measured, and the experiment was repeated at 200, 300, and 400°C. The %R_o values are plotted as a straight line on a semilog graph with temperature as the ordinate, as shown in Figure 10-23 (solid circles).

Saxby et al. (1986) heated a brown coal for up to six years (see the heating experiments described for Figure 10-2). His R_m measurements also followed a straight line on a semilog plot (open circles in Figure 10-23). The percentage of aromatic carbon atoms in the coal changed from 60 to 93% in this interval, indicating a buildup of the condensed–aromatic ring structures.

Reflectance measurements must be made only on vitrinite group macerals, since the other macerals mature at different rates. Figure 10-24 shows the change in reflectance with the maturation of the three major coal macerals. At a coal carbon content of 88%, the reflectance ranges from about 0.55% for liptinite to 1.7% for inertinite and 1.2% for vitrinite. There also is a maximum and minimum reflectivity of vitrinite, which is measured under polarized light while rotating the stage. It differs considerably in the higher maturity ranges, where anisotropy is greatest. The statistical mean reflectivity (R_m or R_o) of the "representative" population based on a histogram (Figure 10-25) is used in most computations. Data from cavings and recycled OM are excluded wherever possible. If maximum or minimum reflectivities are measured, the symbols R_{max} or R_{min} are used, respectively. The R_o or VR_o involves measuring reflectance with the microscope objective immersed in oil. The R_a is a measurement made in air. (For details on R_a, see Hunt 1979, pp. 328–329). Most source rock maturity results involve the R_o range from 0.2 to 2%.

A problem has arisen with the use of calculated or computer-derived vitrinite reflectance values. Van Gijzel (1990, p. 58) recognized this problem: "Data have been published as if they were actually measured mean vitrinite reflectance values. There is a danger that these R_o data can be misused to make incorrect interpretations and conclusions with respect to the thermal exposure of certain rock formations, disregarding the actual rock data." The term R_o or R_m should be reserved only for real measured values. The R_c is the preferred designation for calculated numbers, as correctly used by Radke et al. (1990), or R_e for equivalent reflectance values, as used by van Gijzel (1990). Kinetic models in themselves can result in geologically unreasonable reflectance values, as observed by Barker (1993). Consequently, they should be clearly distinguished from real measurements in any publications.

There are two main techniques for preparing samples to measure reflectance. The first involves concentrating the organic matter by removing the mineral matter with hydrochloric and hydrofluoric acids. The organic matter is freeze-dried, mounted in epoxy, and polished. The second involves using whole-rock polished pellet mounts or thin sections, rather than the kerogen concen-

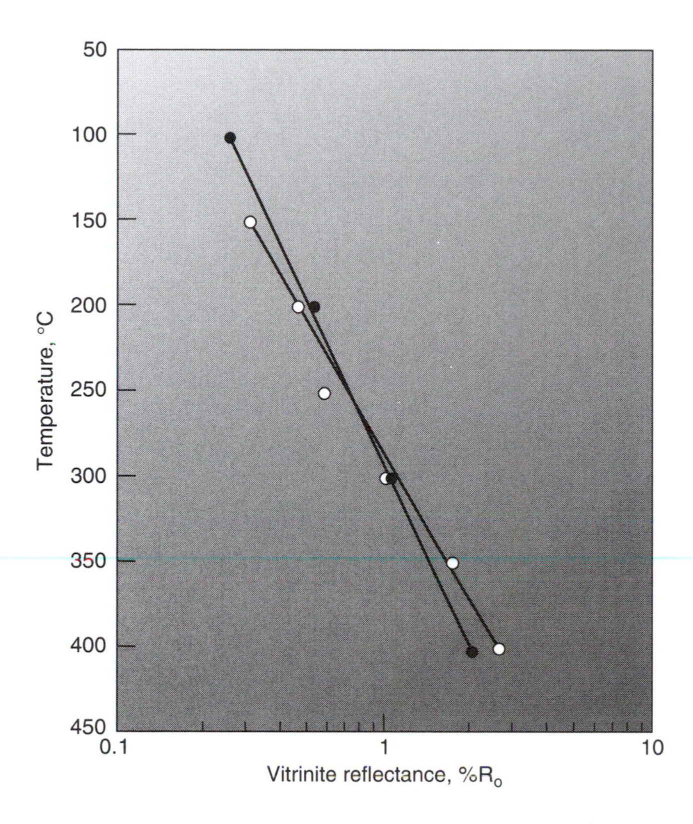

Figure 10-23

Exponential increase in vitrinite reflectance from laboratory heating of wood [solid circles, Ting 1975] and brown coal [open circles, Saxby et al. 1986].

trates. A whole-rock mount has the advantage of showing the entire sample so that the relationship of the vitrinite particles to the rock matrix background can be seen more clearly. Only ten to twenty measurements of the true relevant population need to be made by this technique. Also, the mounts can be reevaluated if anomalous results occur in a succession of samples. However, whole-rock mounts require a very skilled, experienced petrologist, since the determination of the relevant population is quite subjective.

The R_o measurements are made on telocollinite (Table 10-3) with a reflecting microscope using oil-immersion objectives. The term $\%R_o$ refers to the percentage of incident light that is reflected back through the microscope. It goes to 15% with graphite. More than fifty reflectance measurements are made on

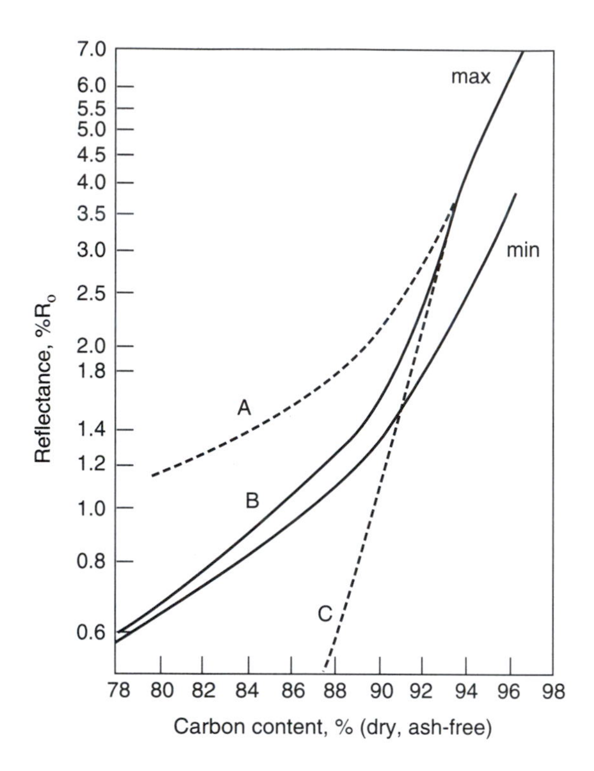

Figure 10-24

Change in reflectance in oil (R_o) for A, inertinite; B, vitrinite; and C, liptinite, with increasing coal rank. [Murchison 1969]

kerogen concentrate pellets and plotted on histograms, as shown in Figure 10-25. This example shows a bimodal distribution with two R_o peaks or modes at 0.8 and 1.6%. The mode with the higher reflectance represents recycled vitrinite, and that with the lower reflectance is primary vitrinite. A more complete description of experimental techniques and data interpretation can be found in Whelan and Thompson-Rizer 1993.

The correlation of reflectance with other maturation indicators and with oil and gas accumulations has resulted in an empirical definition of R_o numbers representing the limits of oil and gas generation. High-sulfur kerogens like some of the Monterey Shales of California may generate and expel heavy oil at R_o = 0.35%. But the lowest value associated with the known generation of conventional oil is about 0.5%, and 0.6% is generally recognized as the beginning of commercial oil accumulations. The peak of oil generation is at an R_o level around 0.8 to 1. At higher R_o levels, the gas/oil ratios increase rapidly with reflectance. The end of oil generation is around R_o = 1.3%, condensate around R_o = 2%, and dry gas around R_o = 3.5% (Dow 1977). To date, no major gas accumulations have been reported in which the vitrinite reflectance of the reservoir rock is higher than 3.5%, except where gas has migrated into more mature reservoirs from less mature source rocks.

Vitrinite reflectance values indicate only the level of maturation of the samples examined. They cannot predict the presence of oil and gas, because these

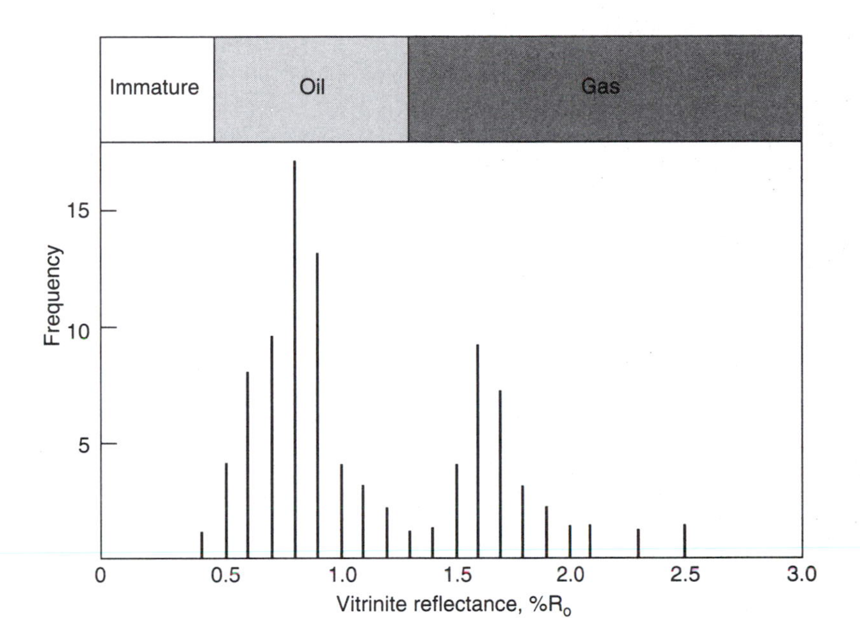

Figure 10-25

Histogram of reflectance of vitrinite particles from a single sediment sample.

frequently migrate updip along permeable beds or through fracture-fault systems to shallower reservoirs at lower levels of maturation and because even nongenerative strata may contain vitrinite. Consequently, oil is found in northwestern Germany, western Siberia, the Volga–Urals districts, and southern Florida where vitrinite reflectivities are only 0.3%R_o, which is too low for a significant generation of conventional oil. The other end of the scale is more clearly defined, because extensive downward migration is less common. Major oil accumulations are not usually associated with reflectivities above about 1.3%. The larger gas fields are found in the R_o range from 1.3 to 3% or lower.

Vitrinite particles are found in about 80% of all well cuttings. Red shales, interbedded red and green shales, and sediments near unconformities that outcrop at the surface contain oxidized vitrinite, which has anomalous reflectance values, frequently in a broad range. Shales and carbonates containing only amorphous kerogen, such as the Green River oil shale and parts of the Bakken Shale from the Williston Basin, have too little vitrinite for accurate evaluation. Pre-Silurian rocks contain no low hydrogen vitrinite because of the absence of vascular plants, although there appears to be a high-hydrogen-suppressed vitrinite in the Cambrian–Ordovician Alum Shale of southern Scandinavia (Buchardt and Lewan 1990).

Vitrinite histograms tend to broaden and flatten as rank rises, owing to increasing vitrinite anisotropy (Figure 14-27). Broadening increases above R_o = 1%

and becomes very noticeable above 2 and 3%. Broadening also occurs in the lower ranks, but it is more typical of high-rank maturation.

Vitrinite maturation is not affected significantly by pressure, only by temperature. If the mean R_o values for samples are plotted against depth on semilog paper, they tend to form a straight-line maturation profile, as on the left of Figure 10-26 (Dow 1977). Such a line for primary vitrinite normally intersects the surface at an R_o of 0.2 to 0.25%. Early vitrinite may be pitted, rough, jellied, or colloidal. The lowest reflectance measurements obtainable are around 0.18 to 0.2%R_o. Good histograms are obtained at depths equivalent to 0.3%R_o and higher. The slope of the primary vitrinite line is determined mainly by the geothermal gradient and the rate of deposition. Any change in these causes a change in the slope. Also, differential heating due to the proximity of salt domes or igneous intrusions alters the slope. Recycled vitrinite tends to mature more slowly than primary vitrinite, so the slope of the line to the right in Figure 10-26 is steeper. Recycled vitrinite is very common, according to Dow (1977). Extrapolation of the recycled vitrinite line to the surface shows the level of maturity that it reached before its second deposition. Thus in Figure 10-26 the recycled vitrinite had a maturity of about 0.5 R_o when deposited. Such information is useful in determining the provenance of the sediments with the recycled vitrinite.

On geological cross sections, the oil and gas source rock intervals can be mapped based on vitrinite reflectance numbers. For example, 0.6 R_o, representing the beginning of major oil generation, is at about 2,650 m (9,694 ft) on Figure 10-26. The 1.35%R_o point, representing the end of oil preservation, is at about 4,200 m (13,780 ft). This defines the major oil-generation interval in this well. It should be recognized, of course, that vitrinite reflectance is only a maturation indicator. Other geochemical data would be needed to determine whether the organic matter in this well is capable of generating oil, gas, or nothing. The primary vitrinite line in Figure 10-26 also indicates that sedimentation has been continuous through this depth interval; otherwise, the line would show a break and offset.

Vitrinite Reflectance and Sedimentary History

Although vitrinite reflectance profiles are used mainly to define zones of oil and gas generation, they also can be used to interpret some of the sedimentary and tectonic history of basins. For example, Unomah and Ekweozor (1993) identified four main unconformities in the Anambra Basin of Nigeria from offsets of the straight-line semilog plots of R_o with depth. The presence of the unconformities also is supported by similar offsets in the depth trends of T_{max} and interval transit times.

Intrusives may cause a large increase in vitrinite reflectance owing to the high temperatures, as depicted in Figure 10-27 (Peters et al. 1983). This shows the change in R_o above and below a diabase sill in Cretaceous black shales of the eastern Atlantic. Those R_o values above the sill increase to higher values than those below the sill, indicating a better transfer of heat upward by advection than downward. Higher temperatures above the sill also are supported

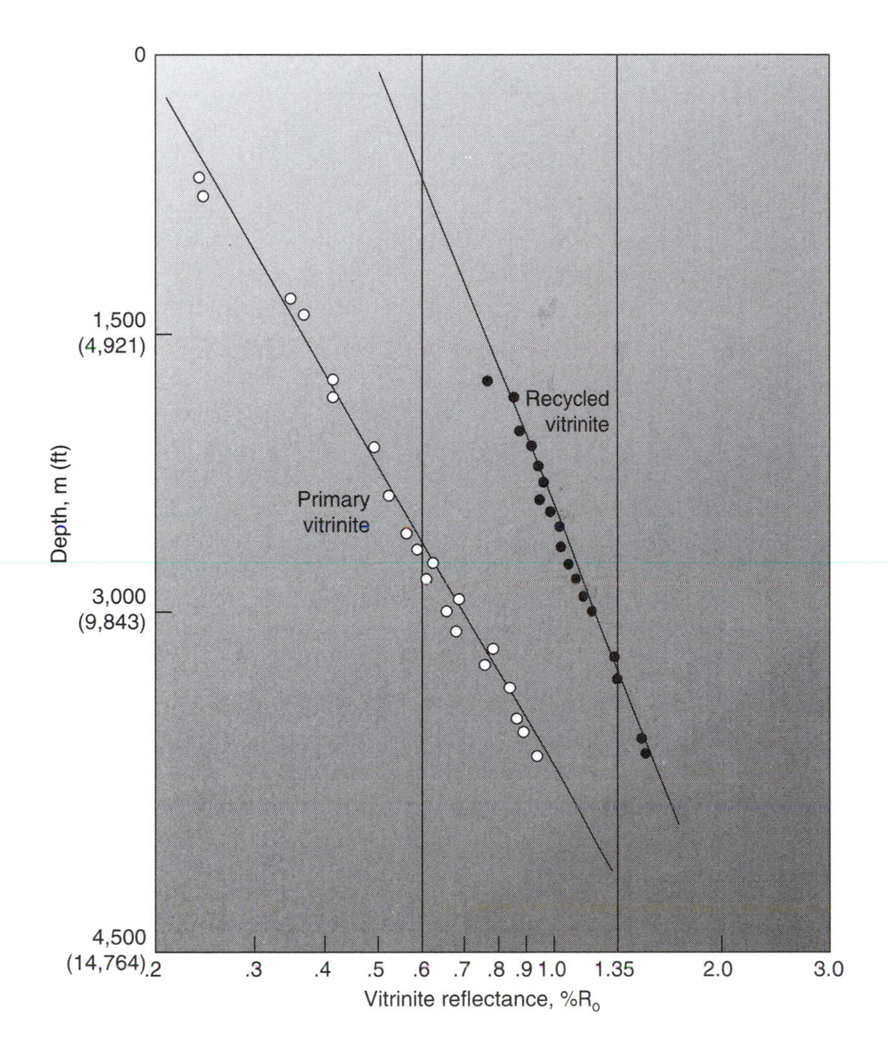

Figure 10-26

Vitrinite reflectance maturation profile of the kerogen in an offshore Texas Gulf Coast Miocene well, comparing primary and recycled vitrinite. [Dow 1977]

by the greater decrease in the atomic H/C ratio of the kerogen above the sill than below it (Figure 10-5).

Pittion and Gouadain (1985) used vitrinite reflectance to follow pre- and postdeformational maturity patterns. An example of predeformational maturity is given in Figure 10-28 for the Tertiary of the Mahakam Delta of Indonesia. In this case, the isoreflectance lines are folded in the same manner as the

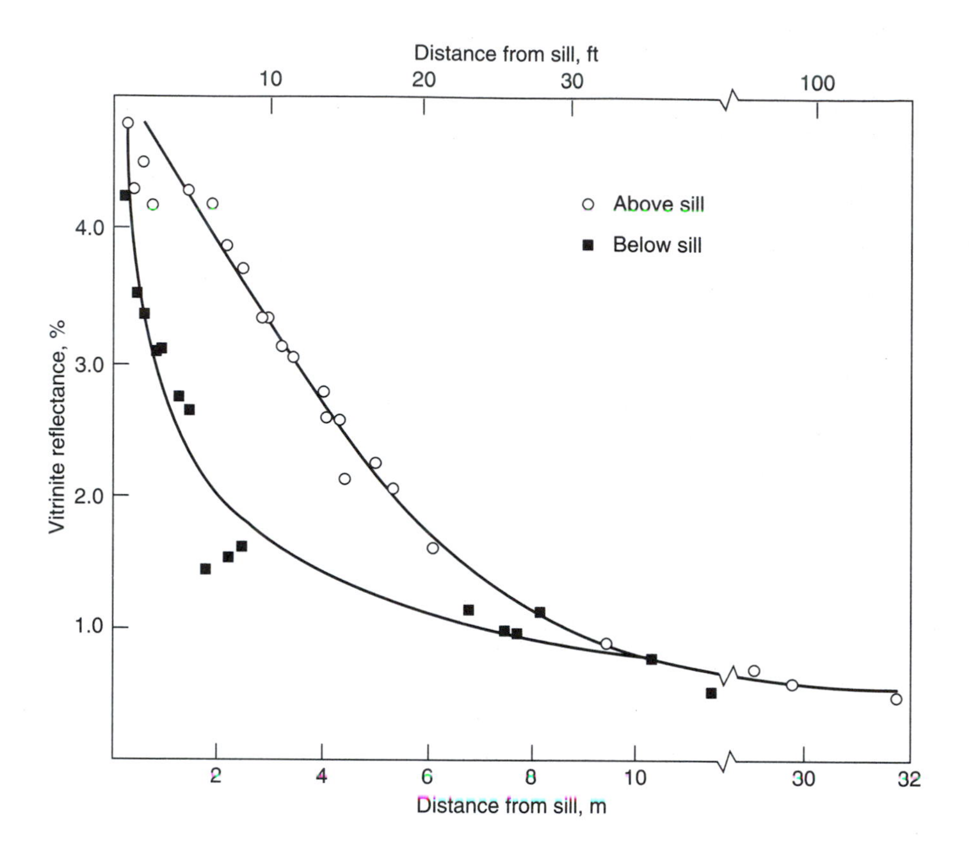

Figure 10-27

Variations in vitrinite reflectance. Values for Cretaceous black shales penetrated by hot diabase sills in the Eastern Atlantic Ocean show that the maximum temperatures reached by shales were higher at a given distance above than below the major sill: (○) above sill; ■ below sill. [Peters et al. 1983]

stratigraphic markers, indicating that the folding occurred after the R_o maturity imprint.

An example of a postdeformational maturity pattern is shown in Figure 10-29 (Pittion and Gouadain (1985). In the Haltenbanken area of the North Sea, the Jurassic coal unit was not buried more than about 1,000 m until down faulting occurred near the end of the Jurassic period. This, plus subsidence during Cretaceous and Tertiary times, caused differential burial of the coal unit with the corresponding differences in maturity. The R_o values in well A are 0.3%, in well B 0.5%, and in well C 1.1%. Only well C has gone through the oil

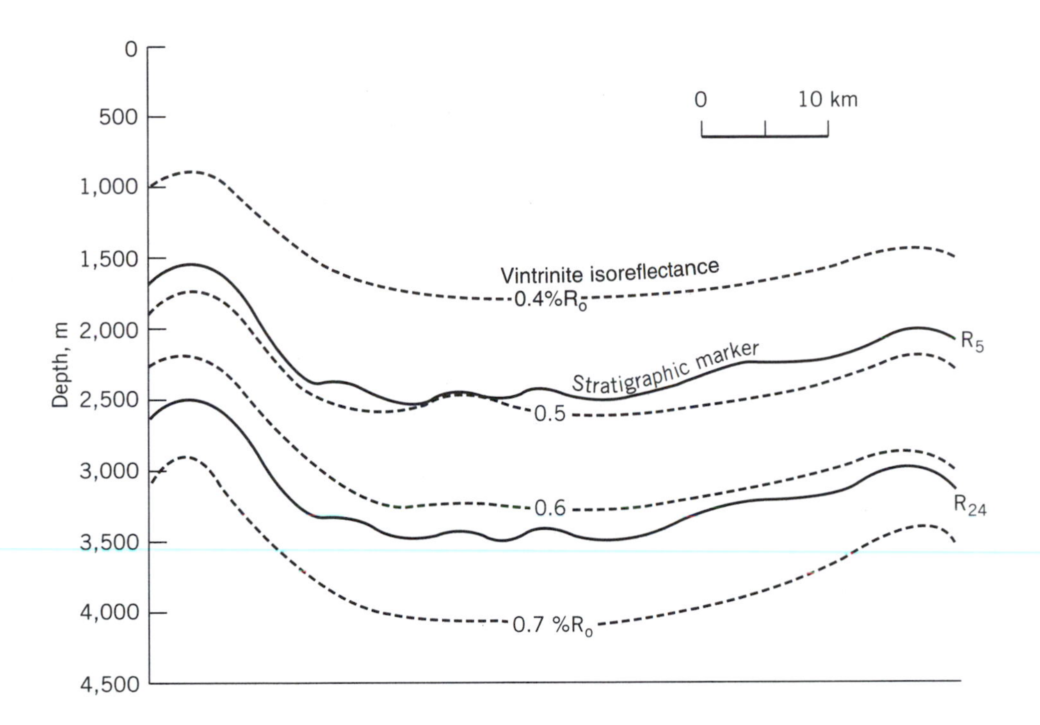

Figure 10-28

Example of a predeformational maturity pattern in the Mahakam Delta, Indonesia. [Pition and Gouadain 1985]

window and is entering gas condensate generation. The isoreflectance lines indicating the oil window transect stratigraphic boundaries.

Vitrinite reflectance differences are common across faults with a vertical displacement. Houseknecht and Matthews (1985) recorded an R_m of 1.69% for the older, more mature hanging wall of a thrust fault in the Ouachitas of western Arkansas, compared with 0.85% for the younger adjacent footwall.

In some areas of uplift, it is possible to estimate the thickness of overburden removed, providing that the primary R_o plots on a straight line, as shown in Figure 10-30. In this example, extrapolation of the line suggests that up to 5,000 ft (1,525 m) of overburden may have been removed. An obvious error in such estimates involves deciding what the surface R_o was before the uplift. Dow's (1977) graphs (like Figure 10-26) start their R_o at 0.2%. Others have used 0.25 or 0.27% in the case of the North Sea. In some parts of the Amadeus Basin of Australia and the North Slope of Alaska, primary vitrinite reflectance is > 1% at the surface, indicating considerable overburden removal.

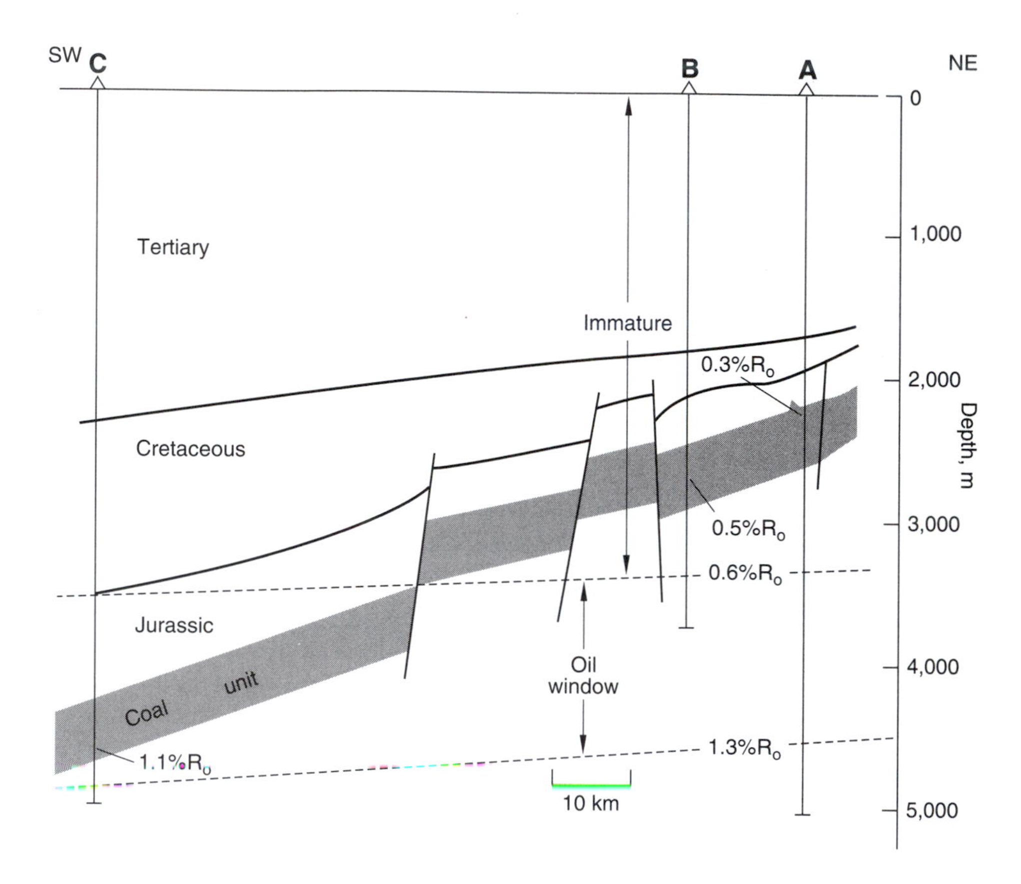

Figure 10-29

Regional variation in the maturity of coal and the location of the oil window in relation to the geological structure, Haltenbanken area, North Sea. The coal unit is immature in wells A and B but mature in well C, where it is nearing the end of the oil window. [Pittion and Gouadain 1985]

Attempts also have been made to estimate erosional removal from offsets in R_o plots occurring at buried unconformities. Such estimates have additional problems such as a reduction in offset with further burial plus complicating factors involving other variations in thermal stress (Katz et al. 1988). Also, thermal perturbations may cause either dogleg or kinky (more than one bend) reflectance profiles.

For example, Figure 10-31 portrays the change in geothermal gradient on entering a high-pressure, high-temperature zone. The change in slope of R_o indicates that oil generation started around 10,000 ft (3,048 m). If the original

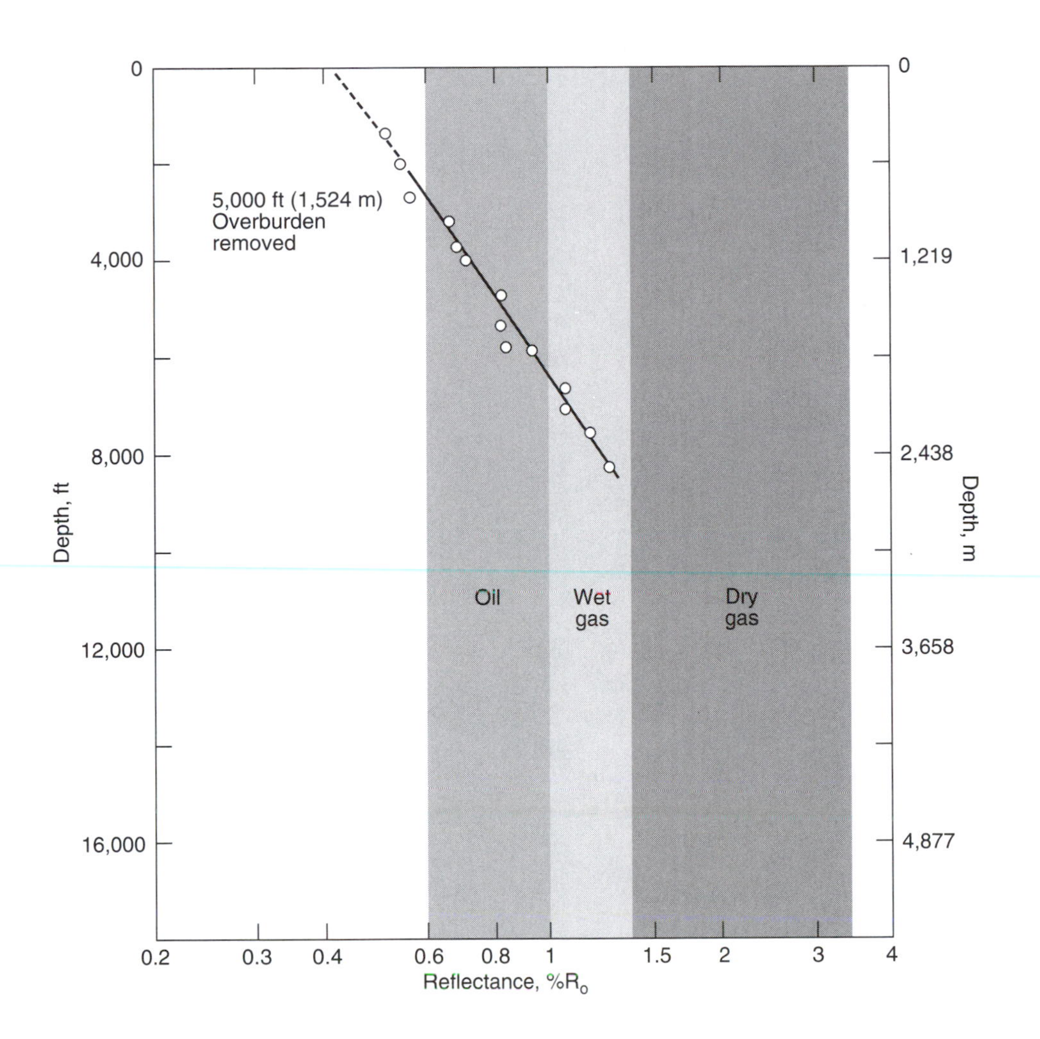

Figure 10-30

Use of a vitrinite reflectance profile for estimating removal of overburden.

gradient of 2.3°C/100 m (1.3°F/100 ft) had been maintained in the deeper high-pressure section, oil generation would not have started until about 14,000 ft (4,267 m). This illustrates the importance of higher geothermal gradients at depth, causing an early generation of oil.

The maturation of vitrinite is an irreversible thermochemical transformation. Consequently, an R_o value of vitrinite cannot fall. It is like a maximum-recording thermometer. Once the dogleg pattern in Figure 10-31 is recorded in the sediments, it becomes a permanent record of the existence of a high-pressure, high-temperature zone. For example, in a shallower uplifted part of

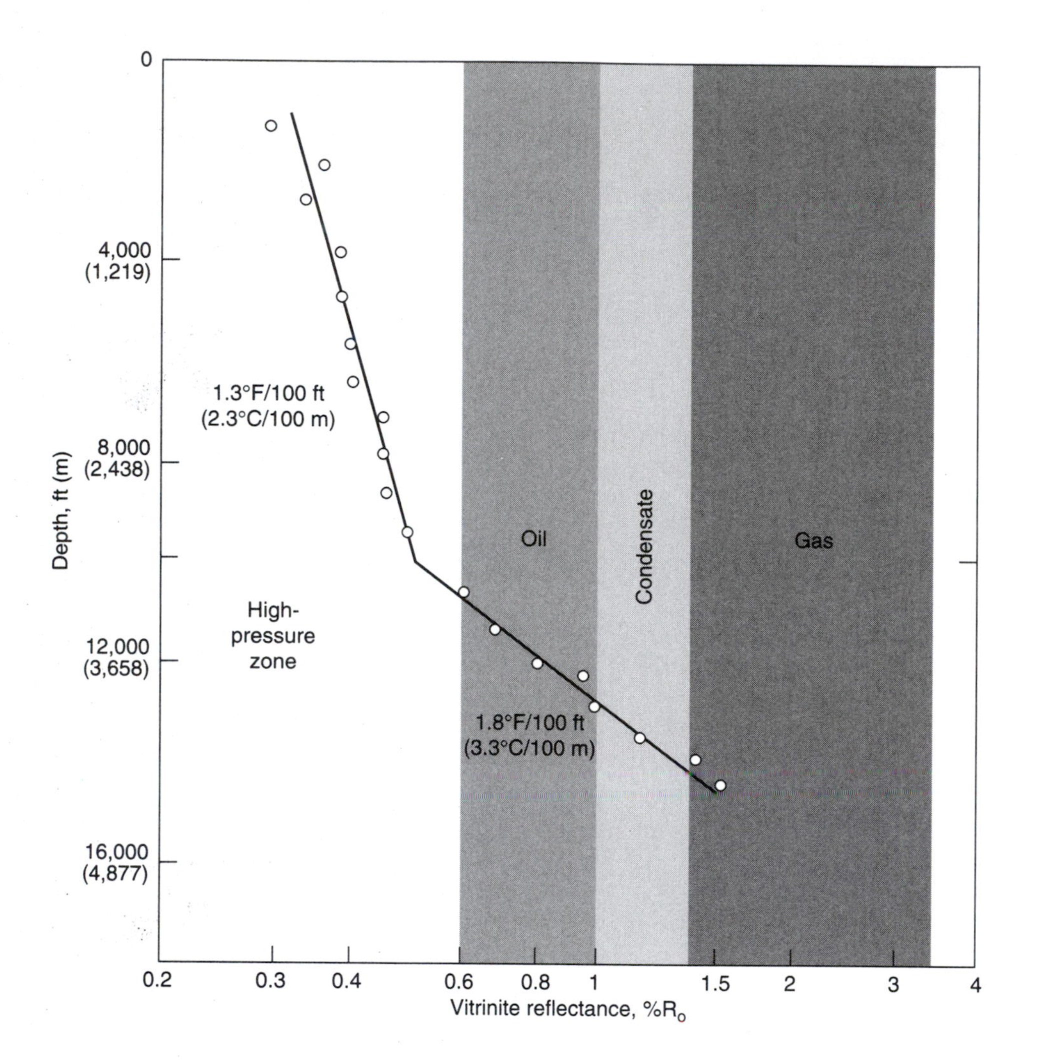

Figure 10-31

Vitrinite reflectance profile showing the effect of a change in the geothermal gradient in a high-pressure zone of a well in the Powder River Basin, Wyoming.

the Powder River Basin of Wyoming there is a dogleg reflectance curve exactly like this, but with the change in slope at about 7,000 ft (2,134 m) and no evidence today of a high-pressure, high-temperature zone. This indicates that a paleo high-pressure zone once existed in this part of the basin before it was uplifted.

Law et al. (1989) observed a large number of kinky vitrinite reflectance (R_m) profiles in the Rocky Mountain basins of the United States and Canada. The kinks in R_m appeared to be restricted to abnormally pressured, low-permeability rock sequences. Law et al. (1989) concluded that the kinks were due to perturbations in the thermal gradient caused by heat transfer processes associated with the development of abnormal pressures.

Figure 10-31 represents a two-segment reflectance profile, whereas Figure 10-32 is a kinky three-segment profile. Law et al. (1989) interpreted the first segment in Figure 10-32 to have originally been in water-bearing, normally pressured rocks, the second in gas- and water-bearing rocks, and the third in overpressured gas-bearing rocks. They believe that the heat transfer through the second segment is by convection caused by a pressure-induced fluid flow. This results in near isothermal conditions and an alteration in the normal geothermal gradient caused by conductive heat transfer. Consequently, the second segment has kinks at both ends. These kinks indicate the tops of the original, not present-day, abnormal pressures.

Vitrinite-1 or A and Vitrinite-2 or B

Two groups of vitrinites occur in coals and shales that Gutjahr (1983) and Buiskool Toxopeus (1983) called *vitrinite-1* (*V-1*) (low hydrogen) and *vitrinite-2* (*V-2*) (high hydrogen). V-1 and V-2 also have been called *A* (high gray) and *B* (low gray), respectively (Whelan and Thompson-Rizer 1993). The V-1 group includes telinite, collinite, telocollinite, and corpocollinite. The principal maceral of the V-1 group is telocollinite, which, by convention, is used for all maturity measurements. It shows no fluorescence during UV radiation. The V-2 group includes desmocollinite, heterocollinite, and degradinite. The principal V-2 member is desmocollinite, which shows weak fluorescence and a lower, more variable reflectance than V-1. The V-1 and V-2 groups are readily distinguished by transmission electron microscopy (Taylor and Teichmüller 1993). The V-2 vitrinites often contain soft inclusions, impregnations, or submicroscopic particles of lipoid-rich OM. There is no well-defined relationship between V-2 reflectance and maturity.

Buiskool Toxopeus (1983) reported a difference of 0.1%R_R between V-1 in a coal and V-2 in a coaly shale at the same coal rank. (R_R [random] involves measuring the reflectance of randomly oriented particles with a polarizer but without rotating the stage.) Measuring V-1 and V-2 in a well, however, showed wider differences with increasing depth (rank). At 6,000 ft (1,829 m), the difference is 0.12%R_R, and at 8,500 ft (2,591 m), it is 0.3%R_R. At the latter depth, V-1 is 0.45%, and V-2 is 0.75%R_R. These differences are significant and emphasize that only telocollinite or another vitrinite-1 maceral should be used for maturity measurements. But even greater differences appear when comparing vitrinites in organic-rich shales with those in coals of the same rank. This is the suppressed vitrinite problem, which is discussed in Chapter 14.

A plot of the atomic H/C ratio of humic coals versus the reflectance (%R_o) of their vitrinites is shown in Figure 10-33. There is a continuous increase in

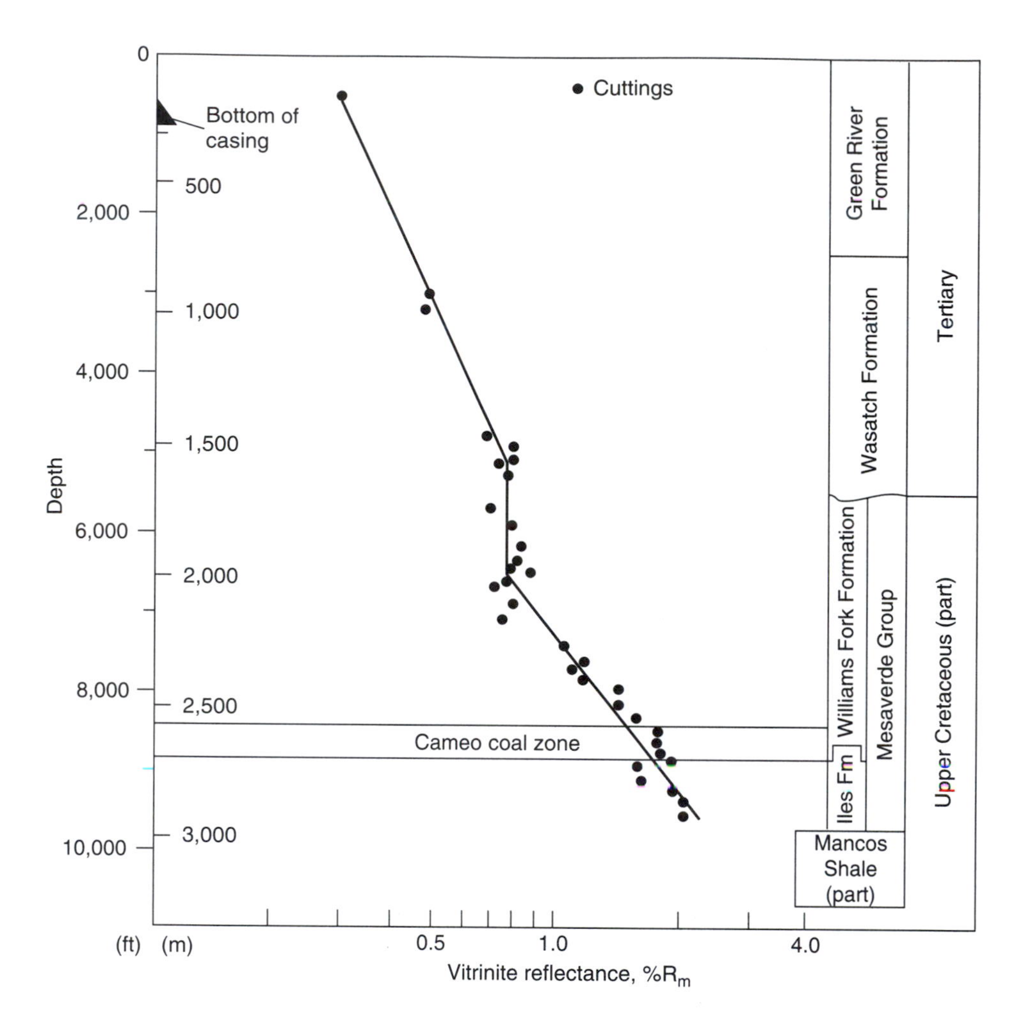

Figure 10-32

Vitrinite reflectance profile of the Crystal Creek A-2 well, Piceance Basin, Colorado. All samples are from drill cuttings. [Law et al. 1989]

reflectivity with a decreasing atomic H/C ratio of the coals, as would be expected. Since the coals are mostly vitrinite, the H/C ratio is decreasing with maturation, because of an increase in the content of the flat aromatic ring structures shown in Plate 3C. This causes high reflection of incident light. Graphite, which is the end product of maturation, has an R_o of about 15% at a zero H/C ratio.

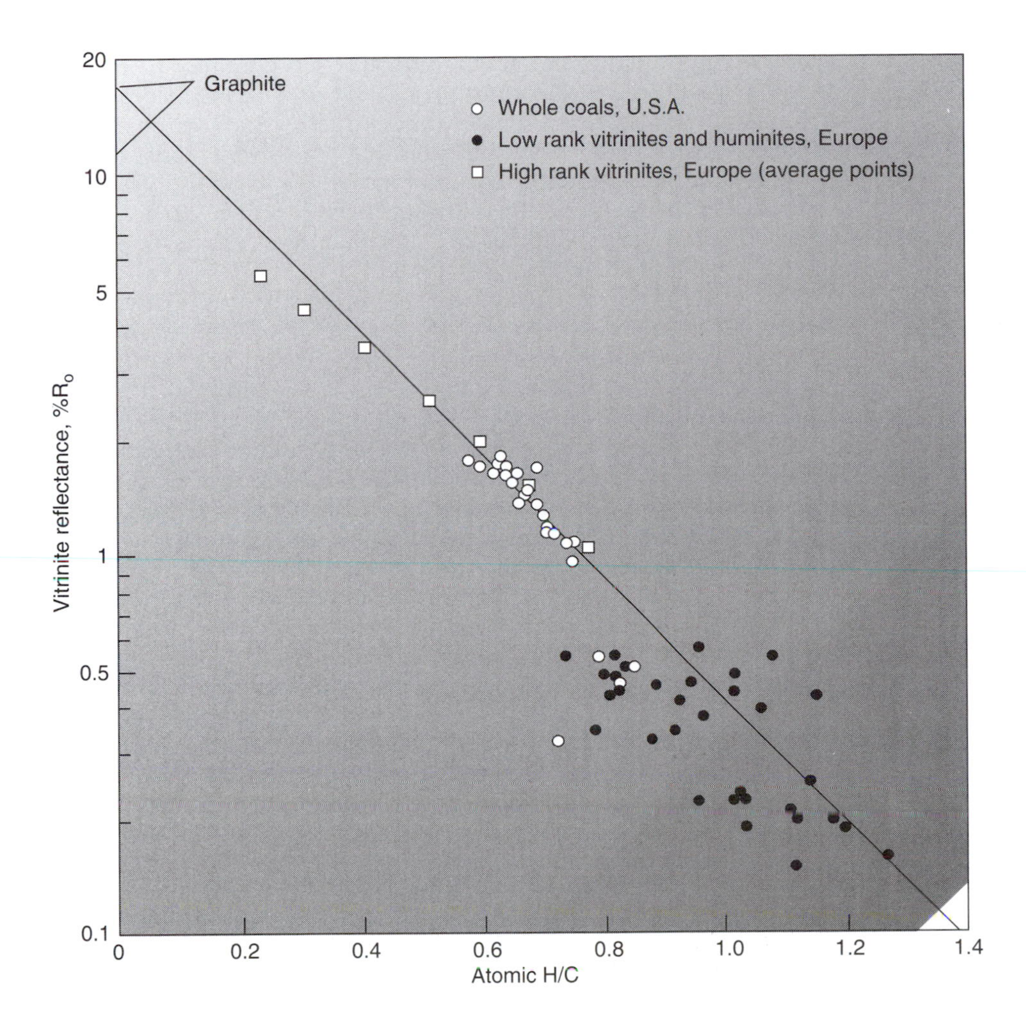

Figure 10-33

Atomic H/C ratios of coals from various areas versus the reflectance of their vitrinites. The line is from McCartney and Ergun 1958, and the vitrinite data are from McCartney and Teichmüller 1972. The European coal analyses are by the Geologisches Landesamt Nordrhein-Westfalen.

Source Rock Extracts

Additional maturation indicators such as the carbon preference index (CPI) and the pristane and phytane indices discussed in Chapter 4, plus gas chromatography patterns and biomarkers, are obtained from the EOM (extractable organic matter) of source rocks. Extraction involves pulverizing about 50 grams of rock to a fine particle size, extracting the hydrocarbons with a solvent such as dichloromethane, evaporating the solvent, separating the residual extract by means of column chromatography, and analyzing the fractions by means of gas chromatography mass spectrometry. (The earlier techniques used are described in Hunt 1979, pp. 262–263, and more recent techniques are listed in Peters and Moldowan 1993, pp. 50–69.)

The patterns of gas chromatograms of rock extracts differ in appearance when going from immature to postmature samples. Typical immature patterns show that isoprenoids such as iso-C_{18}, iso-C_{19} (pristane), and iso-C_{20} (phytane), plus terpane and sterane type naphthenes, are dominant. This is shown in Figure 4-14 for the Irati Shale of Brazil and in Figure 5-10 on the left for the Kimmeridge Shale of the North Sea. These patterns have the characteristic hump of biogenic hydrocarbons, whereas a mature extract shows a smooth *n*-paraffin curve, as for the Kimmeridge on the right of Figure 5-10. This latter pattern shows a type II kerogen extract at full maturity, with the biogenic isoprenoids, terpanes, and steranes buried under the thermally generated *n*-paraffins. For comparison, Figure 10-34 shows a waxy oil generated at an early stage of maturity from a type III kerogen or coal in Indonesia. The original biogenic imprint still is clear in this pattern. Thus, the ratio of pristane to *n*-C_{17} (the peak next to it) is 2.3, compared with less than 1 for the mature curve on the right of Figure 5-10. Also, the CPI of *n*-paraffins in Figure 10-34 is greater than 1 from C_{27} through C_{33}, whereas it is 1 in mature oils and extracts.

Clayton and Bostick (1986) followed the changes in gas chromatograms of the Upper Cretaceous Pierre Shale of Colorado approaching an igneous dike that penetrated the shale. While the vitrinite was changing from an R_o of 0.4 to 2%, the pristane/*n*-C_{17} ratio decreased from 1.3 to 0.3, and the phytane/*n*-C_{18} ratio fell from 0.4 to 0.2. Meanwhile, the *n*-paraffins from about C_{18} to C_{31} gradually disappeared because of cracking. This resulted in the smaller molecules around C_{10} becoming dominant in the GC pattern.

A similar change was noted in gas chromatograms of Duvernay source rocks from east central Alberta covering the vitrinite reflectance range 0.33 to 1.60%R_o (Creaney 1989). Creaney's patterns show a gradual decrease in the hump of biogenic hydrocarbons followed by a postmature shift in the peak of *n*-paraffins from large to small molecules. The mature extract shows a wide range of *n*-paraffins up to C_{30}. The postmature extract shows a narrow range from about C_{12} to C_{17}. Although maturity estimates from gas chromatograms are qualitative, they do provide an additional check on maturity evaluations when other parameters are giving mixed signals.

Biomarker Maturation Indicators

Most of the maturity parameters discussed in this chapter are measured on mixtures of molecules primarily in the disseminated solid OM of sedimentary

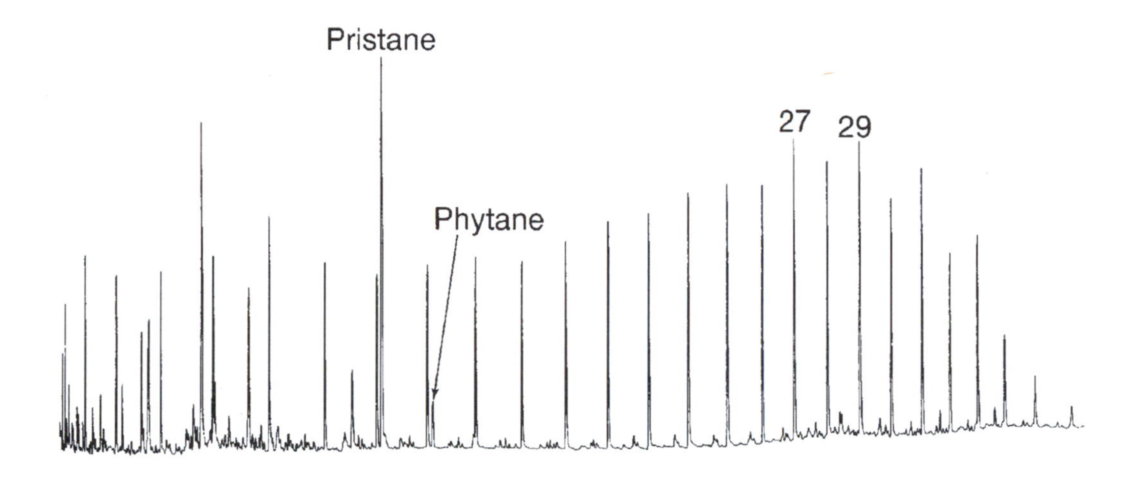

Figure 10-34

Gas chromatogram of a waxy oil (34°API gravity from Indonesia). [Thompson et al. 1984]

rocks. Biomarker indicators differ because they represent single molecules found in both the liquid and solid phases of OM. They are found in the oils, asphalts, asphaltenes, and pyrobitumens as well as in kerogen and coal. Their complex structures provide more information about their source material, depositional environment, thermal maturity, and, in some cases, geological age than do any other geochemical parameter. Biomarker and other maturation parameters can be used to calibrate models of the generation of oil and gas versus the time of trap and fault formation, so that one can predict the volume of petroleum available for migration to reservoir accumulations. Biomarkers can be used to estimate relative contributions of marine, lacustrine, and terrestrial OM as well as to evaluate redox conditions of deposition.

Biomarker maturity changes, however, are limited to the products of immature and mature stages of generation to about an R_o of 1.4% (Peters and Moldowan 1993, p. 226). Also, matrix effects can alter the maturity ratios. Reversals may occur with stereoisomer ratios like 20S and 20R. Recycled OM causes mixing of immature and mature biomarkers. Moreover, mass chromatograms of biomarkers are complex and sometimes have coeluting interfering peaks that alter the apparent maturity ratios. Biomarker separation, analysis, and interpretation require a skilled operator with considerable experience in interpreting the data. They also may be quite expensive.

There are three ways in which the molecular structures called *biological markers* are characterized. These are by (1) the class or family of origin, (2) the size of the molecule, and (3) its shape in space, that is, the three-dimensional side-chain and ring-system configurations. The nomenclature and sterochemistry of biomarkers were described in Boxes 4-1 and 4-2. Biomarker maturities are determined primarily from the molecular shapes, for example, the α, β, R, and S forms. The correlation of oils with other oils, asphalts, and their source rocks is mainly by compound class and molecular size, although the three-dimensional data are also used to determine whether the molecules being compared are of similar maturity.

Compound Class

The compound class is determined in a gas chromatograph mass spectrometer (GCMS) by passing a saturate or aromatic fraction of the oil or extract through a GC column into an MS, which analyzes the individual molecules. Whole oil can be used in some systems. Most biomarkers that are used for maturation and correlation elute between the *n*-paraffins C_{24} and C_{36}. The parent molecular ions initially formed in the MS have a mass-to-charge ratio (m/z), as shown in Figure 10-35. Further bombardment of the parent molecular ion with electrons causes it to crack apart, as indicated along the dashed lines in Figure 10-35. This results in the formation of a major daughter fragment ion plus additional fragment ions. The distribution and relative abundance of fragment ions on the mass spectrum permit identification of the original molecule and its class.

For example, both the parent ions for cholestane and ergostane in Figure 10-35 have a major daughter ion of m/z 217, indicating that they are 14α(H) steranes. All such steranes, regardless of the different side-chain structure, have a major peak at m/z 217, representing the fragmented A, B, and C rings, as shown in the upper right of Figure 10-35. The parent molecular ion also breaks down into many smaller fragment ions, whose distribution identifies the parent. The additional fragments from cholestane show peaks at more than twenty mass-to-charge ratios, including 357, 149, 109, 95, 81, 55, and so forth, whereas ergostane shows peaks at 232, 177, 163, 149, and so forth. Likewise, all hopanes have a dominant daughter fragment ion at m/z 191 in their mass spectra, as shown in Figure 10-35 plus additional fragment ions that identify the individual hopanes.

Molecular Size

Biomarker compound classes contain series of compounds (homologs) that increase or decrease in size by the addition or subtraction of CH_2 or CH_3 groups from the molecule. In the following example, cholestane has 27 carbon atoms. Adding a carbon to the number 24 carbon on the side chain (see Box 4-2 for the numbering system) forms the C_{28} sterane called *ergostane*. Adding another carbon yields the C_{29} sterane called *stigmastane*.

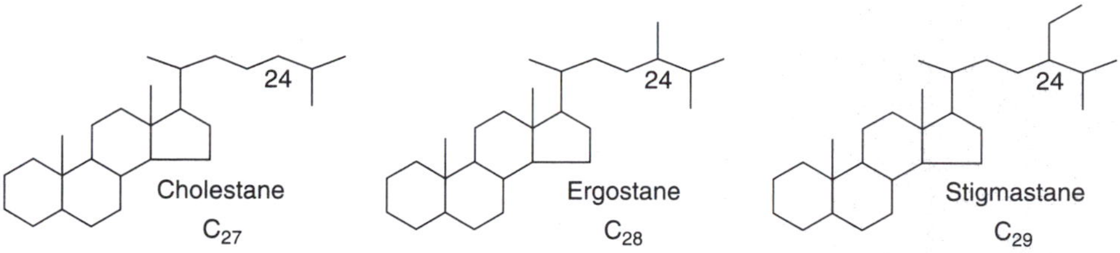

24 24 24

Cholestane C_{27} Ergostane C_{28} Stigmastane C_{29}

This was described in Chapter 3 as a homologous series in which the molecules are identical except for the addition of CH_2. Each additional carbon atom adds four more major peaks on a mass chromatogram: αα20R, ββ20S, ββ20R, αα20S, as shown in Box 4-2. This greatly increases the number of biomarker molecules available for maturation and correlation studies.

Note that the molecular ions in Figure 10-35 also represent a homologous series for each class. Since all these molecular ions have a single positive charge, the m/z is really the mass. Therefore, the sterane masses go from 372 (cholestane) to 386 (ergostane), and the hopanes go from 412 (hopane) to 426 (homohopane). In both cases, this is a difference of 14, which is the mass of the CH_2 being added.

The compound class indicates origin. Steranes are from eukaryotic organisms such as algae and higher plants and animals. Hopanes are from prokaryotes such as bacteria. The molecular size also indicates origin. The C_{27} and C_{28} steranes are mainly from algae; the C_{29} steranes are mostly from higher plants; and the C_{30} steranes are from marine phytoplankton.

The Shape in Space (Stereochemistry)

Thermal stress causes a change in the stereochemistry of the ring system and the side chain of biomarkers. For example, the original biological form of the C_{29} sterane in terrestrial plants is 5α, 14α, 17α, 20R. Thermal stress converts the positions of the hydrogens on the rings and the side chain to 5α, 14β, 17β, 20S. The α-to-β change equilibrates at 70% conversion, and the R-to-S change ends at 50% conversion in the oil window, as shown in Figure 10-36. These maturity parameters are generally expressed as ratios such as 20S/(20S + 20R). In the ideal system there is no 20S at the start and no 20S or R added to or subtracted from extraneous sources. But since conditions rarely are ideal, it is customary to use several parameters for relative maturity assessments.

In addition to isomerizations, some ratios of different biomarkers have been developed empirically based on differences in the thermal stability of individual molecules. Applications of these over the years have resulted in some ratios being superior to others based simply on experience. (An excellent discussion of the positive and negative aspects of most biomarker maturity parameters used in petroleum geochemistry can be found in Peters and Moldowan 1993, pp. 220–251.)

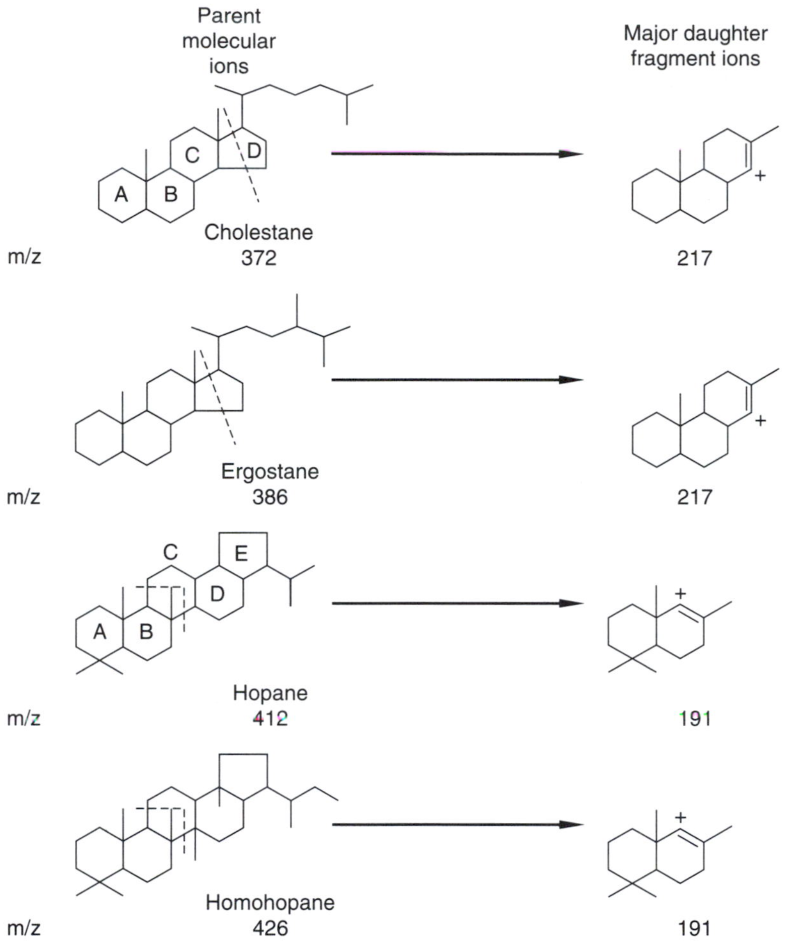

Figure 10-35

Identification of biomarker compound class by gas chromatography mass spectometry (GCMS).

Frequently Applied Biomarker Maturity Parameters

Nine of the twenty ratios discussed in Peters and Moldowan's book are listed in Figure 10-36, along with their maturity ranges compared with vitrinite reflectance. The various ratios reach constant equilibrium values, indicated by the

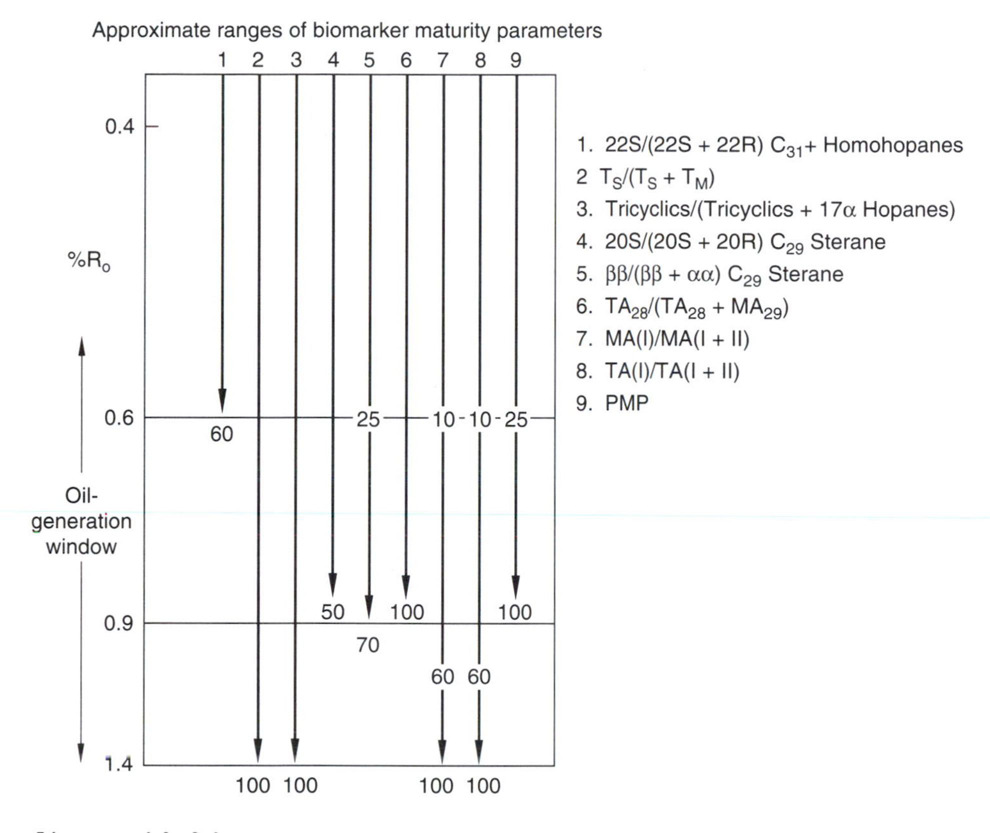

Figure 10-36

Biomarker scales compared to vitrinite reflectance. All biomarkers reach maximum maturity within or by the end of the oil window. [Data from Peters and Moldowan 1993]

numbers at the end of the arrows. Only four ratios continue through the oil window. The change in maturation is not linear, as can be seen from the intermediate values listed on the vertical maturity lines. These parameters are best applied on a relative rather than an absolute basis, owing to the variations in heating rates, lithofacies, and organic facies of the source rock for the oil being evaluated. Also, there can be interference from coeluting biomarker peaks. (All these problems are discussed in more detail in Peters and Moldowan 1993.) In the following brief discussion, the numbers refer to the numbered ratios in Figure 10-36.

1. Biological hopane precursors have a 22R configuration, which gradually converts to a mixture of 22S and R under thermal stress. Isomerization at C-22 in the C_{31}-to-C_{35} 17α(H)-homohopanes reaches equilibrium earlier than any other reaction in this group. Each homohopane forms a doublet of R and S peaks, as shown in Figure 15-25 for the Mowry Shale source rock and its corresponding oil. Ratios of 22S/22R for each homolog, especially C_{31} and C_{32} are useful for comparing low-maturity oils and source rock extracts.

2. The T_s, which is 18α(H)-22,29,30-trisnorneohopane is more stable to thermal maturation than T_m, which is 17α(H)-22,29, 30-trisnorhopane. This ratio is best used as a maturity indicator when evaluating oils from a common source of the same organic facies. There is a source effect on the T_s/T_m ratio. Oils from carbonate source rocks have anomalously low ratios compared with those from shales. Also, bitumens from hypersaline source rocks show high ratios. Spurious ratios sometimes result because T_m and T_s commonly coelute with tricyclic or tetracyclic terpanes on the m/z 191 mass chromatogram.

3. The tricyclic/17α-hopane ratio increases because more tricyclic terpanes than 17α-hopanes are released from kerogen at higher levels of maturity. Laboratory experiments suggest that the tricyclics either migrate faster or are released more easily from the rock matrix than the hopanes. This ratio is source dependent because of the different biological precursors for tricyclic terpanes compared with those for hopanes. This is why these compounds are more useful as correlation tools (see Chapter 15).

4. This increase in the 20S, relative to 20R, epimer in C_{29} steranes may be due to isomerization or to the greater stability of the 20S epimer compared with the 20R. This ratio can be affected by facies, weathering of source rocks and partial biodegradation of oils. Anomalous results are common in samples from hypersaline or evaporitic environments.

5. The ββ and αα refer to isomerization at the C-14 and C-17 positions in the 20S and 20R-C_{29} regular steranes. This ratio reaches equilibrium at 70% conversion, at about the peak of oil generation (Figure 10-36). Peters and Moldowan (1993) recommend plotting ββ/(ββ + αα) versus 20S/(20S + 20R) for the C_{29} steranes, as shown in Figure 10-37 for defining the thermal maturity of source rocks or oils. This involves plotting ratio 4 against ratio 5, in Figure 10-36. Such plots enable the two maturity parameters to be cross-checked against each other. The data in Figure 10-37 are for crude oils in the offshore Santa Maria Basin of California (Peters and Moldowan 1993). Any data off the trend line would suggest checking for analytical problems or differences in heating rates or matrix effects in the subsurface. GCMS improves the accuracy of this and all other C_{29} sterane isomerization ratio measurements.

6. The thermal conversion of a C_{29}-monoaromatic (MA) to C_{28}-triaromatic (TA) steroid is as follows:

A B H C_{29}-MA → C_{28}-TA

Aromatization of the A and B rings results in the loss of a methyl group and seven hydrogen atoms attached to these rings. The ratio increases from 0 to 100% during maturation. Using this ratio as a maturity parameter requires that the same MA and TA components be used for all samples. Peters and

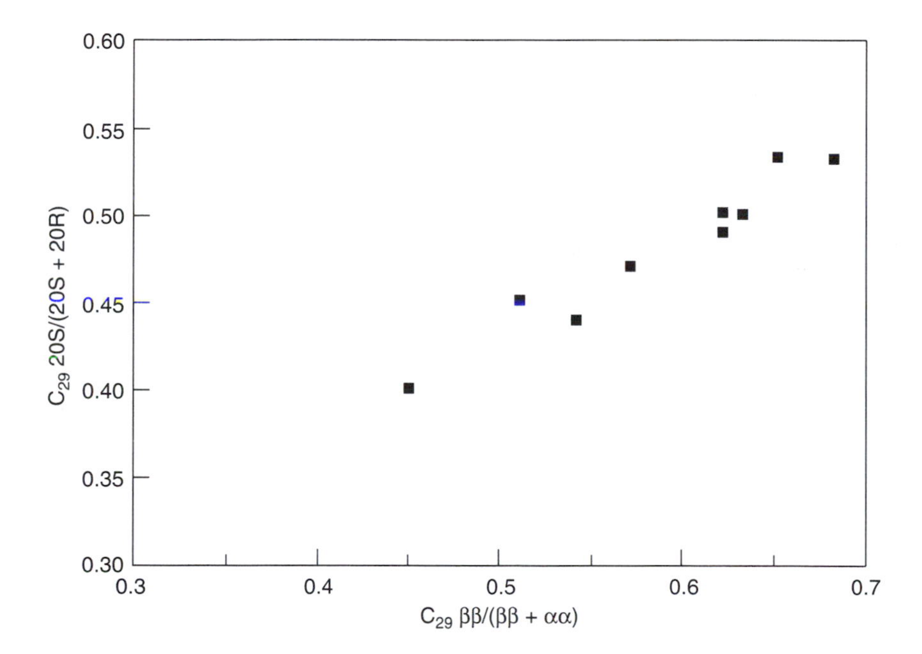

Figure 10-37

Correlation of thermal maturity parameters based on the isomerization of asymmetric centers in the C_{29} steranes for offshore Santa Maria Basin oils, California. [Peters and Moldowan 1993]

Moldowan (1993) use the same twelve MA-steroids and four TA-steroids for all their measurements.

7. Thermal maturation causes an increase in Group I MAs relative to Group II MAs partly because Group I is more thermally stable than Group II. Also, Group II MAs with a side chain are thought to be partly converted to Group I by cleavage of the side chain during maturation, as follows (R = additional methyl groups):

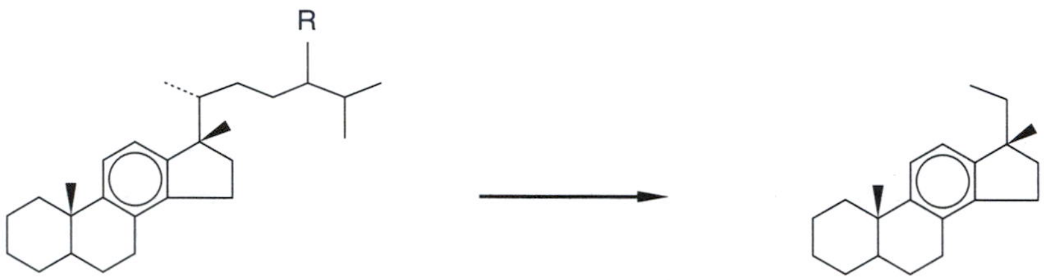

R

MA (II) MA (I)

Changes in the source input such as C_{28} relative to C_{29} and the depositional environment, particularly Eh, also affect the concentrations of these steroids. In order to reduce source input effects, Peters and Moldowan (1993) recommend summing all major C_{27}-to-C_{29} MA steroids as MA II and the C_{21}-plus-C_{22} steroids as MA I.

8. This discussion about MA-steroids also applies to the short-chain (I) and long-chain (II) TA steroids. The change in the TA ratio appears to be due to preferential degradation of the TA II homologues rather than the conversion of TA II to TA I. Peters and Moldowan (1993) prefer to use the sum of C_{26} to C_{28} (20S + 20R) TA steroids as TA II and the C_{20} + C_{21} TA steroids as TA I. This minimizes the effect of the dominance of a single sterol such as C_{29} in the source sediments.

9. PMP is the porphyrin maturity parameter in which E = etioporphyrin and D = DPEP or deoxophylloerythroetioporphyrin, which has an extra ring. The ratio of $C_{28}E/(C_{29}E + C_{32}D)$ vanadyl porphyrins increases with thermal maturity because of either differential generation or desorption from the kerogen. The sharpest increase is at the onset of oil generation. Sundararaman (1992) used laboratory pyrolysis to duplicate the natural maturation. At a Rock-Eval T_{max} of 430°C, the PMP was at about 25% completion (Figure 10-36). It jumped to 100% in the next 15°C increase in temperature.

Correlation of Maturity Parameters

Figure 10-38 shows the approximate correlation of various maturation indicators for OM in coal and kerogen and their relation to the organic maturation

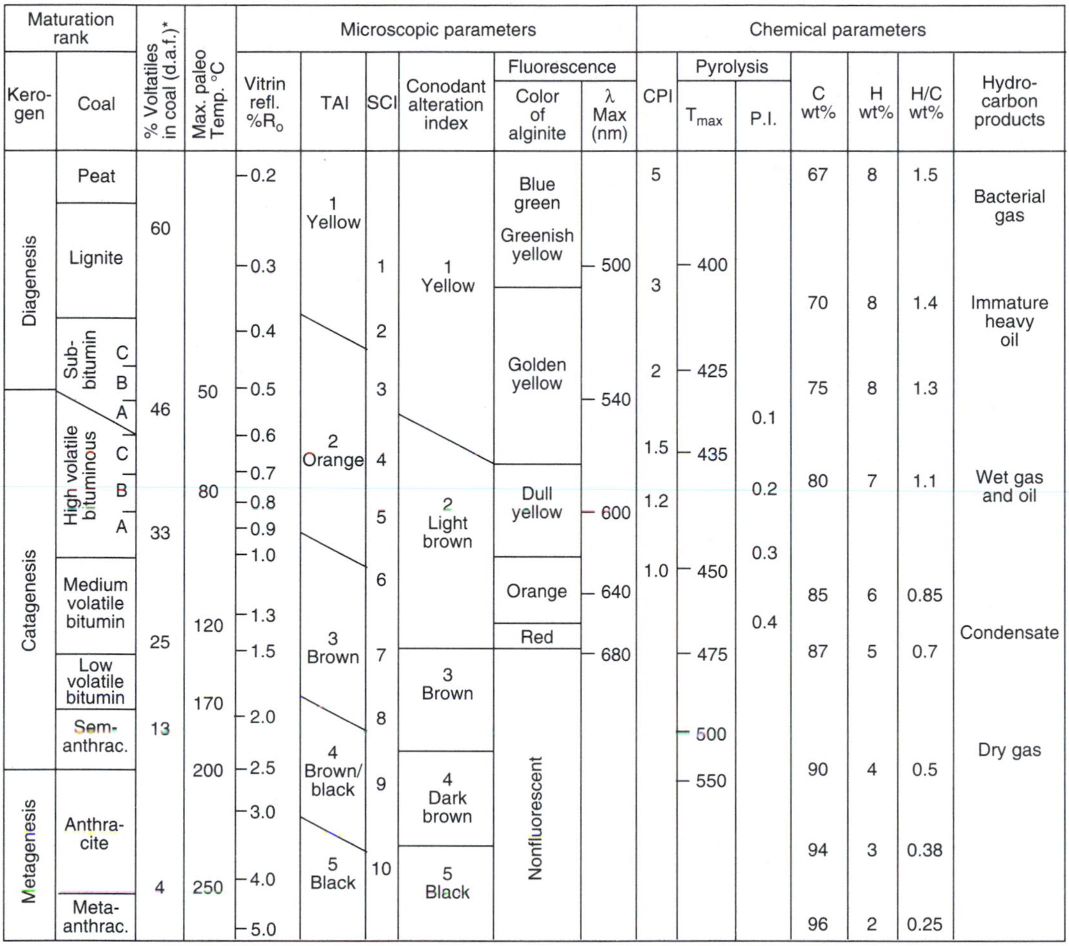

Maturation rank		% Volatiles in coal (d.a.f.)*	Max. paleo Temp. °C	Microscopic parameters						Chemical parameters						
								Fluorescence			Pyrolysis					
Kerogen	Coal			Vitrin refl. $\%R_o$	TAI	SCI	Conodant alteration index	Color of alginite	λ Max (nm)	CPI	T_{max}	P.I.	C wt%	H wt%	H/C wt%	Hydrocarbon products
Diagenesis	Peat			0.2	1 Yellow			Blue green		5			67	8	1.5	Bacterial gas
	Lignite	60		0.3		1	1 Yellow	Greenish yellow	500	3	400					
													70	8	1.4	Immature heavy oil
	Sub-bitumin C			0.4		2										
	Sub-bitumin B		50	0.5		3		Golden yellow	540	2	425		75	8	1.3	
	Sub-bitumin A	46										0.1				
Catagenesis	High volatile bituminous C			0.6	2 Orange	4				1.5	435					
				0.7								0.2	80	7	1.1	Wet gas and oil
	High volatile bituminous B		80	0.8		5	2 Light brown	Dull yellow	600	1.2						
	High volatile bituminous A	33		0.9												
				1.0								0.3				
	Medium volatile bitumin					6		Orange	640	1.0	450		85	6	0.85	
			120	1.3								0.4				
		25						Red								Condensate
	Low volatile bitumin			1.5	3 Brown	7	3 Brown		680		475		87	5	0.7	
			170													
	Sem-anthrac.	13		2.0		8					500					Dry gas
Metagenesis			200	2.5	4 Brown/black	9	4 Dark brown	Nonfluorescent			550		90	4	0.5	
	Anthracite			3.0												
													94	3	0.38	
		4	250	4.0	5 Black	10	5 Black									
	Meta-anthrac.			5.0									96	2	0.25	

*Dry ash free

Figure 10-38

Chart of organic maturation R_o = reflectance with oil immersion objective; CPI = carbon preference index; PI = production index. The chart is a modification of that of Hunt 1979; Mukhopadhyay 1994; Pearson 1990; Staplin 1969; Teichmüller 1974.

stages and the hydrocarbons generated. The temperatures shown are those reached at maximum burial at moderate heating rates and assuming a constant heat flow.

Coal ranks are somewhat arbitrary boundaries. They are very dependent on the kind of OM. A coal filled with wax or spores has a different rank-depth change than one filled with fossil charcoal. Some petrologists use the term *apparent rank* in making comparisons with vitrinite reflectance. Rank is related to the fixed (nonvolatile) carbon and the volatile carbon, which in turn are more related to the coking properties of coal than to reflectance.

Vitrinite reflectance and the atomic H/C ratio are the only maturation parameters that go completely through the oil- and gas-generation windows. All these scales are described in detail in this chapter. They vary somewhat between different laboratories and service companies. Biomarker maturities are listed in detail in Figure 10-37, so they are not included in Figure 10-38.

Other maturation indicators may prove valuable in the future but are not yet widely used. One of these is thermogravimetric analysis combined with Fourier transform infrared analysis (TG-FTIR). It measures the T_{max} on each gas evolved during pyrolysis (Whelan et al. 1990). The T_{max} values shown in Figure 10-39 are for methane generated from kerogens of the Wilcox shales of the U.S. Gulf Coast. The T_{max}s varied from 640 to 910°C over a vitrinite reflectance range from 0.5 to 5.3%R_o. This could be a useful technique for evaluating relative maturities in rocks containing little or no vitrinite. Another example is the transmittance color index (TCI) developed by P. van Gijzel (1990). The TCI analysis replaces subjective visual color estimates of palynomorphs or amorphous OM by objective spectral measurements using a spectral microscope photometer that is properly standardized and calibrated. It is particularly useful in samples containing little or no vitrinite. The OM is analyzed by transmitted light, and the index is related to color changes that occur with progressive thermal maturity. It covers the reflectance range from about 0.2 to 2.5%.

The level of organic maturation (LOM) scale mentioned in Chapter 7 is not included in Figure 10-38 because it is not linear when correlated with a logarithmic R_o scale. It has been well established from the previously mentioned laboratory experiments by Ting and Saxby that vitrinite reflectance (R_o) increases exponentially with a linear increase in temperature. This is also observed in field studies such as an R_o survey of cores in a large number of wells in the Cook Inlet area of Alaska (Castãno, personal communication). This study discovered that R_o plotted on a semilog scale with depth was remarkably linear in all the wells, even through the low-rank zone of 0.3 to 0.5%.

Unfortunately, LOM was developed before vitrinite reflectance was available as a maturation tool in organic petrology. Consequently, LOM was made linear to a coal-rank scale that in later years was recognized as partially incorrect. When attempts were made to correlate LOM with a logarithmic R_o, it was found that the LOM was more accordionlike than linear. Consequently, the use of LOM has been declining.

Geochemical service laboratories now routinely do R_o, TAI, SCI, fluorescence, and pyrolysis, which includes T_{max} and PI as maturation indicators. Biomarkers require another level of equipment and technical experience. Bio-

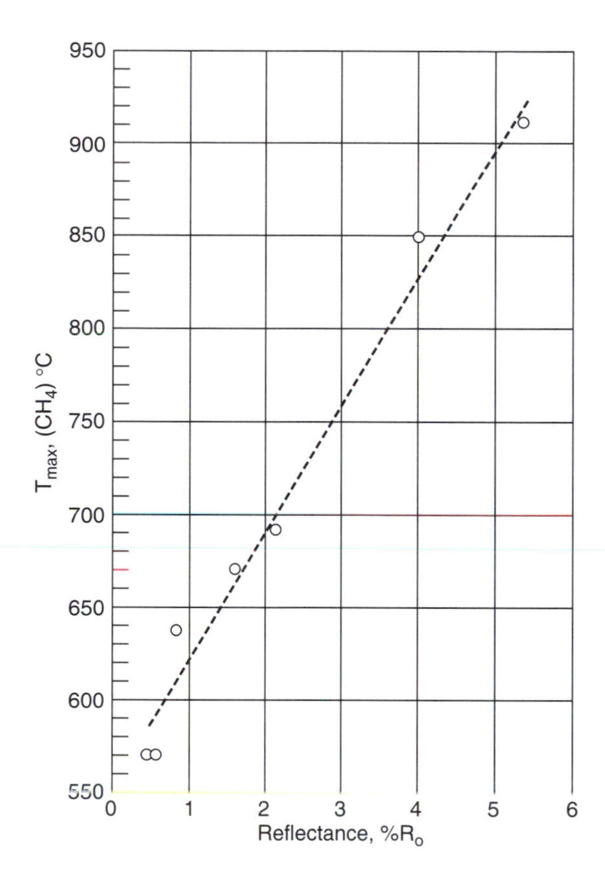

Figure 10-39

Plot of the pyrolysis T_{max} of methane generated by the Wilcox Shale of the U.S. Gulf Coast versus the vitrinite reflectance of the Wilcox kerogen that generated the methane. The correlation coefficient is 0.98. [Whelan et al. 1990]

marker maturities are more accurate than any other indicator, including R_o up to the onset of petroleum generation. Beyond this, vitrinite reflectance is the most widely used maturity parameter through the gas-generation window. All maturity indicators, however, including biomarkers and R_o, have problems. Fortunately, enough indicators are available for cross-checking to solve even the most complex maturity problems.

SUMMARY

1. A petroleum source rock is any rock that is capable of generating and releasing enough hydrocarbons to form an accumulation of oil or gas.

2. The relative ability of source rocks to generate petroleum is defined by the kerogen's quantity (TOC), quality (high or low in hydrogen), and state of maturation (immature, mature, or postmature with respect to oil).

3. Most world-class source rocks with high TOCs are calcareous, dolomitic, siliceous, or phosphatic shales or argillaceous limestones containing clay-size quartz and minor amounts of clay minerals. Those rocks with the finest grain size have the highest TOCs.

4. The minimum TOC required to generate and expel commercial quantities of oil from an immature source rock is around 1.5%. Gas and condensate appear to be expelled down to at least 0.5% TOC.

5. The ratio of reservoired oil to generated petroleum in place in a typical petroliferous basin is usually less than 15%, indicating that the overall process of generation, migration, and trapping is very inefficient.

6. The most important factor controlling the generation of oil and gas is the hydrogen content of the organic matter. All OM may be classified as either sapropelic or humic; the former is high in hydrogen, and the latter is low in hydrogen, so the former generates most of the world's petroleum.

7. The three major maceral groups in coals are liptinite, vitrinite, and inertinite. Liptinite forms sapropelic boghead and cannel coals, and vitrinite and inertinite form the more common humic coals.

8. The kerogen in typical source rocks contains mostly amorphous OM mixed with disseminated liptinite, vitrinite, and inertinite. Kerogens are classified as types I, II, III, and IV, based on their maturation pathways on an atomic H/C–O/C van Krevelen diagram or an HI/OI pyrolysis diagram. Types I and II generate oil accompanied by and followed by gas; type III generates gas, condensate, and some waxy oil; and type IV generates some CO_2 and traces of methane. A continuum exists among these four types.

9. The maturation of both coal macerals and kerogen types due to increasing thermal stress can be followed on an H/C–O/C van Krevelen diagram or on an HI/OI pyrolysis diagram. The floor for all hydrocarbon generation occurs where the atomic H/C ratio of the kerogen reaches about 0.25.

10. Dry, open nonisothermal pyrolysis is the most rapid screening technique available for source rock evaluation. The OM's quantity, quality, and maturation can be estimated from the S_1, S_2, S_3 data, hydrogen and oxygen indices, PI, and T_{max}. It is less suitable than closed isothermal hydrous pyrolysis for determining kinetics and hydrocarbon yields.

11. Kerogen that fluoresces strongly under incident blue light is capable of generating oil. Weakly or nonfluorescing kerogen generates primarily condensate and gas.

12. Currently used maturation indicators include vitrinite reflectance, TAI, SCI, fluorescence, T_{max}, production index, atomic H/C ratios, CPI, and biomarkers. Biomarkers provide the most accurate maturity assessments up to the onset of petroleum generation, but vitrinite reflectance provides the greatest range in maturity assessment of all the indicators.

13. Laboratory experiments indicate that oil generation peaks where the carbon content of kerogen is between 77 and 87% and that gas generation peaks where it is between 85 and 92%. Most oil-generating kerogens have H/C ratios above 0.75 and a hydrogen content of 6% or more. The H/C ratios of gas-generating kerogens are below 0.8, and their hydrogen contents range from 3 to 5%.

14. Semilog profiles of depth versus R_o can be used in some instances to interpret sedimentary events such as the quantity of overburden removed, changes in geothermal gradients or depositional rates, and paleo high-pressure, high-temperature zones.

15. Biological markers are characterized by compound class, the size of the molecules, and their shapes in space. Class and molecular size are used mainly in crude oil and source rock correlations, whereas changes in the stereochemistry (shape in space) are used for maturity assessments. Maturity also is determined from aromatization of the naphthene rings and cleavage of the side chains on the rings.

SUPPLEMENTARY READING

Brooks, J. (ed.). 1981. *Organic maturation studies and fossil fuel exploration.* London: Academic Press, 441 p.

Brooks, J., and A. J. Fleet. 1987. *Marine petroleum source rocks.* Geological Society Special Publication 26. Oxford: Blackwell Scientific Publications for the Geological Society, 444 p.

Fleet, A. J., K. Kelts, and M. R. Talbot (eds.). 1988. *Lacustrine petroleum source rocks.* Geological Society Special Publication 40. Oxford: Blackwell Scientific Publications for the Geological Society, 391 p.

Mukhopadhyay, P. K., and W. G. Dow (eds.). 1994. *Vitrinite reflectance as a maturity parameter: Applications and limitations.* ACS Symposium Series 570. Washington, DC: American Chemical Society, 294 p.

Palacas, J. G. (ed.) 1984. *Petroleum geochemistry and source rock potential of carbonate rocks.* AAPG Studies in Geology 18. Tulsa: American Association of Petroleum Geologists, 208 p.

Thomas, B. M., et al. (eds.). 1985. *Petroleum geochemistry in exploration of the Norwegian shelf.* London: Graham and Trotman for the Norwegian Petroleum Society, 337 p.

Woodward, J., F. F. Meissner, and J. L. Clayton (eds.). 1984. *Hydrocarbon source rocks of the greater Rocky Mountain region.* Denver: Rocky Mountain Association of Geologists, 557 p.

Chapter

11

Coals, Oil Shales, and Other Terrestrial Source Rocks

Terrestrial organic matter (OM) includes the land-derived waxes, spores, and pollen of higher plants, whereas aquatic OM (marine or lacustrine) is mainly plankton and bacteria living underwater. High-wax oils result primarily from the mixing of the terrestrial waxes, discussed in Chapter 4, with the aquatic OM. The terrestrial component is highest in coal-forming swamps, next highest in deltas, lower in lacustrine shales, and lowest in open marine environments. Terrestrial OM contains both the highest and lowest hydrogen contents of all OM, as exemplified by wax and pyrofusinite. This is one reason that it varies so much in its petroleum-generating capability.

In 1968, Hedberg wrote his classic paper on the significance of high-wax oils with respect to the genesis of petroleum. He reviewed forty regional stratigraphic sequences throughout the world that contained such oils, and he reached the following conclusions:

1. High-wax oils are found predominantly in shale–sandstone lithologies deposited under nonmarine or brackish-water conditions.

2. Most of the sequences contain coal beds, oil shales, or other sediments with a high content of organic matter.

3. The deposits are in continental, paralic, or nearshore marine environments such as lakes, bays, gulfs, and deltas, ranging in age from Devonian to Pliocene, inclusive.

4. The oils produced in these sequences are high wax and low sulfur, in contrast to the low-wax, high-sulfur crudes formed in marine environments.

Hedberg's paper was written at a time when petroleum geologists had a strong bias in favor of marine source beds. Geologists would seek a marine source whenever oil was discovered in nonmarine beds, even though such a source would be a long distance from the reservoir in both time and space. Since then it has been recognized that lacustrine and fluvial sediments have sourced large quantities of waxy oils in basins around the world and that many of these terrestrial sediments include interbedded carbonaceous (coaly) shales and coals as sources. Nevertheless, some bias still exists in the belief that coals and coaly shales are limited to being gas sources.

This bias is understandable when the ultimate reserves of coal and conventional oil are compared through time. The largest reserves of coal were deposited in the Late Carboniferous and Permian, with lesser amounts in the Jurassic through the Tertiary. In contrast, the largest reserves of oil are found in the Jurassic through the Early Tertiary. Klemme and Ulmishek (1991) estimated that about 50% of the world's oil formed from Late Jurassic and Cretaceous source rocks. Only 20% of the world's coal is in these rocks. Also, 75% of Hedberg's forty worldwide high-wax oils are in Cretaceous and Tertiary reservoirs, and only 5% are in reservoirs of the coal-bearing Carboniferous and Permian periods. Finally, large accumulations of oil are not normally associated with coals, although they are associated with coaly shales containing predominantly type III kerogens, in several areas like Indonesia and the Gippsland Basin of Australia.

Gas from Coal

Coals are a major source of gas, mainly methane and carbon dioxide (Chapter 7). Meissner (1984) estimated that the Fruitland coals of the San Juan Basin of New Mexico and Colorado generated 55 TCF of methane, of which 26 TCF is adsorbed by the coal. Figure 7-14(A) indicates that between 150 and 200 liters of methane are generated from a kilogram of coal matured through the meta-anthracite stage. Much less gas than this is expelled, because of its adsorption on coal surfaces and in micropores of the coal. Consequently, coal acts as both a source and a reservoir. The adsorbed gas contents of 30 ft^3/ft^3 of coal in the San Juan Basin are three times the volume of gas in a cubic foot of an adjacent sand reservoir. This adsorption capacity of coal increases with increasing pressure and coal rank. When the capacity is exceeded, the excess gas is expelled, creating giant fields such as Groningen of north Holland, which originated from the underlying Carboniferous coal measures. Other examples include the Cooper Basin gas fields of Australia (Plate 1C) formed from Permian coals and the Deep Western Canada Basin fields formed from Lower Cretaceous–Jurassic coals. During its lifetime, coal may generate 3,000 ft^3 of methane per ton, although only 10 to 20% of this is adsorbed at any one time, owing to variations in the properties of the coal and its rank and pressure (Wyman 1984). In the Deep Western Canada Basin there is about 50 TCF of gas in the coal beds in addition to free gas produced from the adjacent sands. As the pressures are reduced, some of this gas will be desorbed and made available for recovery.

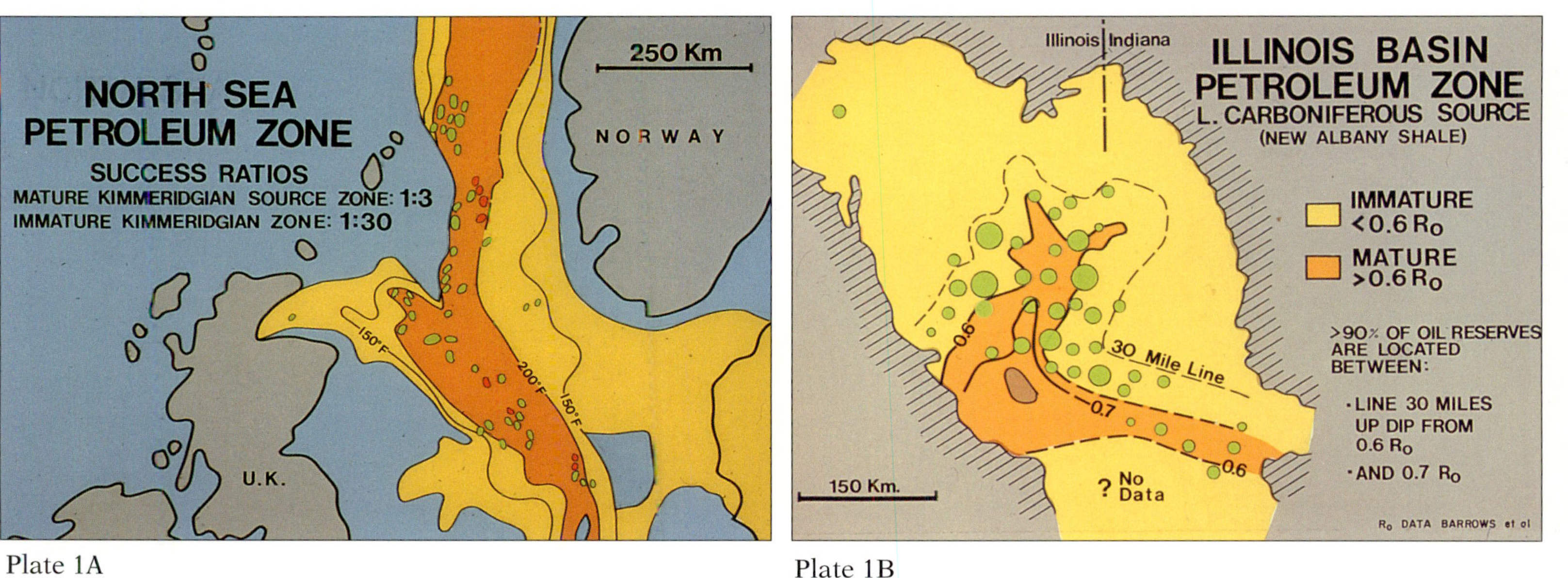

Plate 1A

Plate 1B

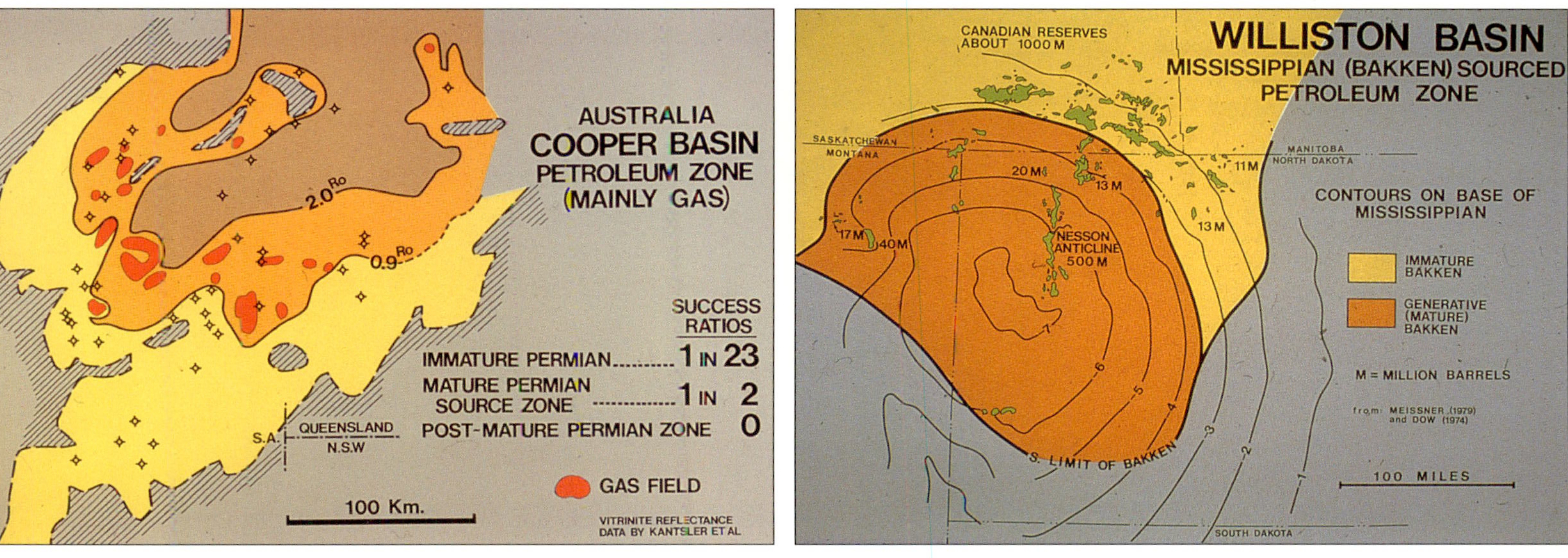

Plate 1C

Plate 1D

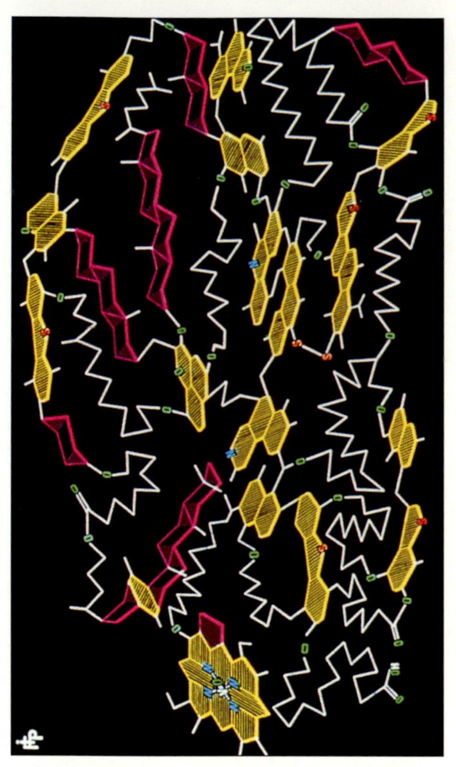
Plate 2A

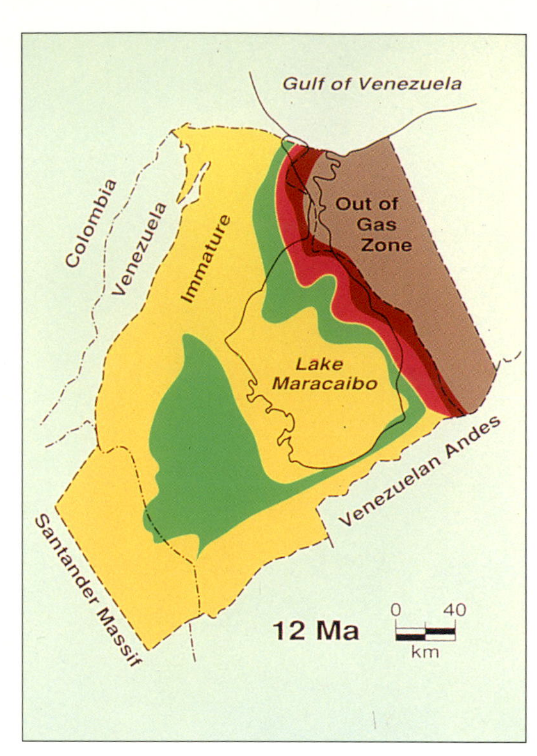

Plate 2B

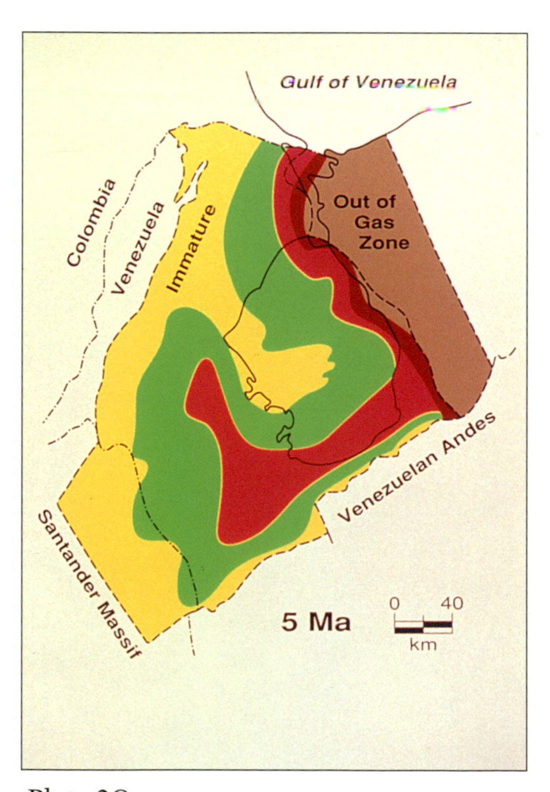

Plate 2C

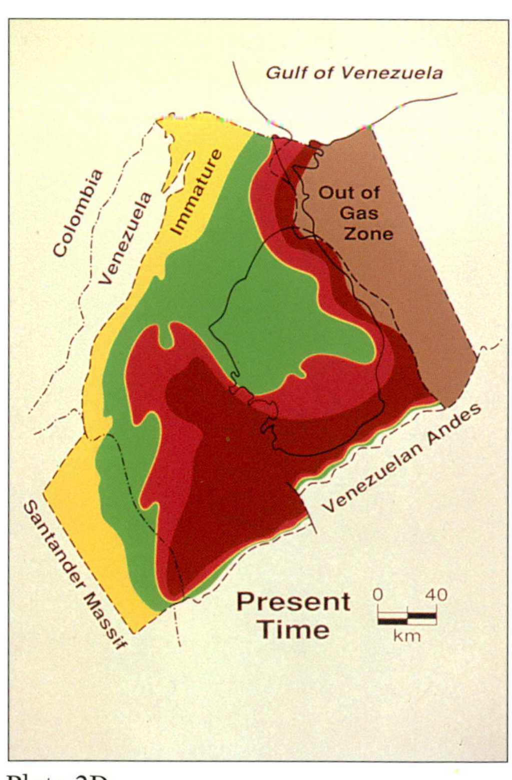

Plate 2D

Plate 3A

Plate 3B

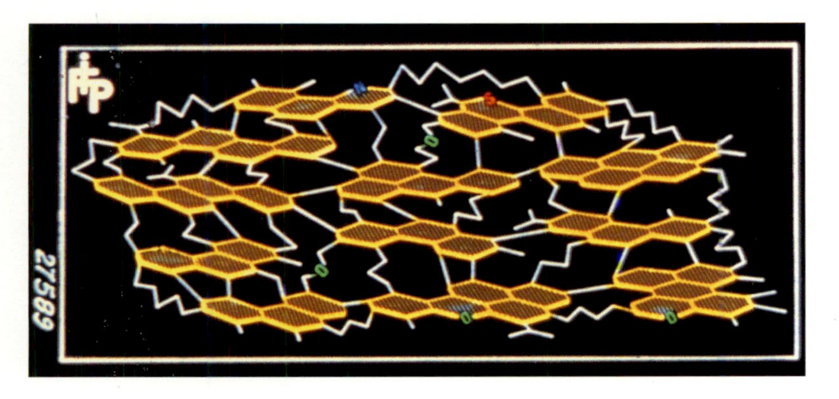

Plate 3C

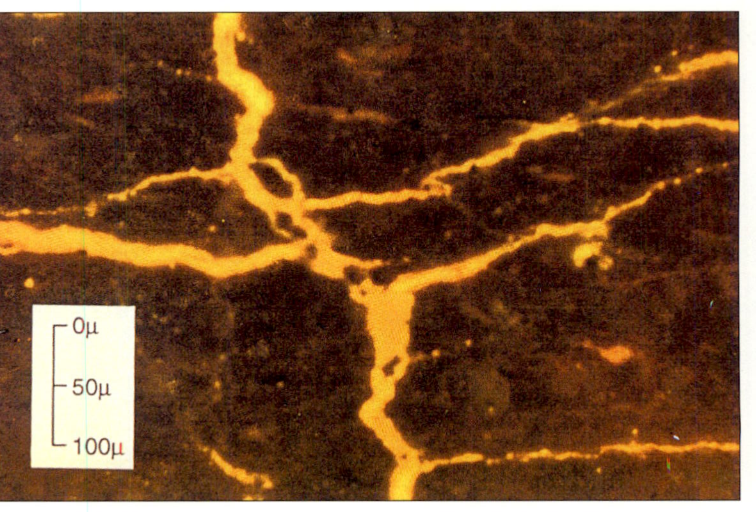

Plate 3D

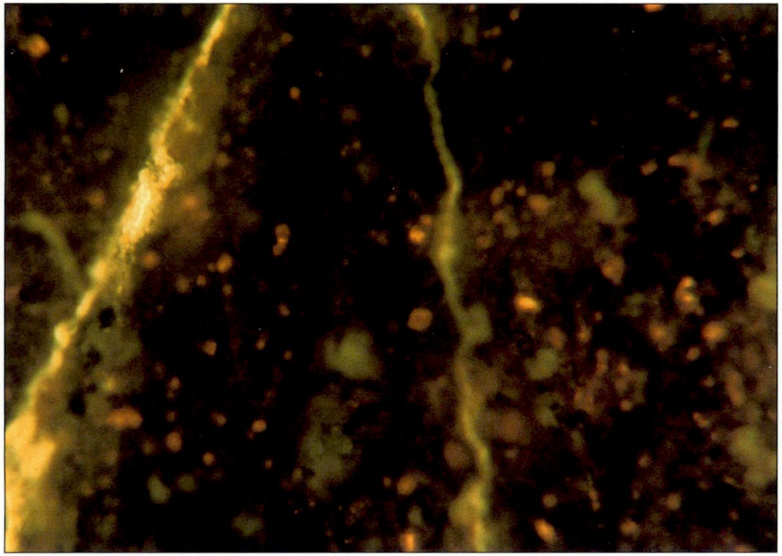

Plate 4A

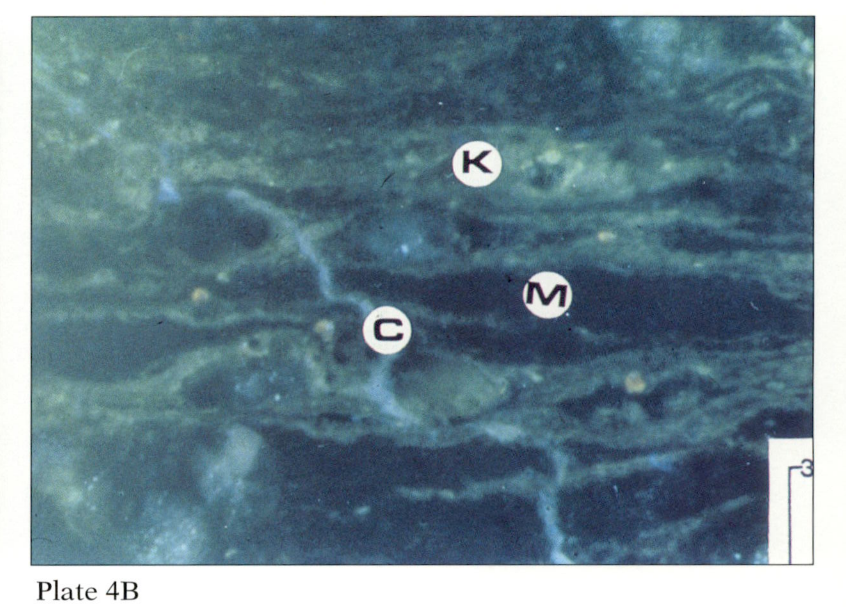

Plate 4B

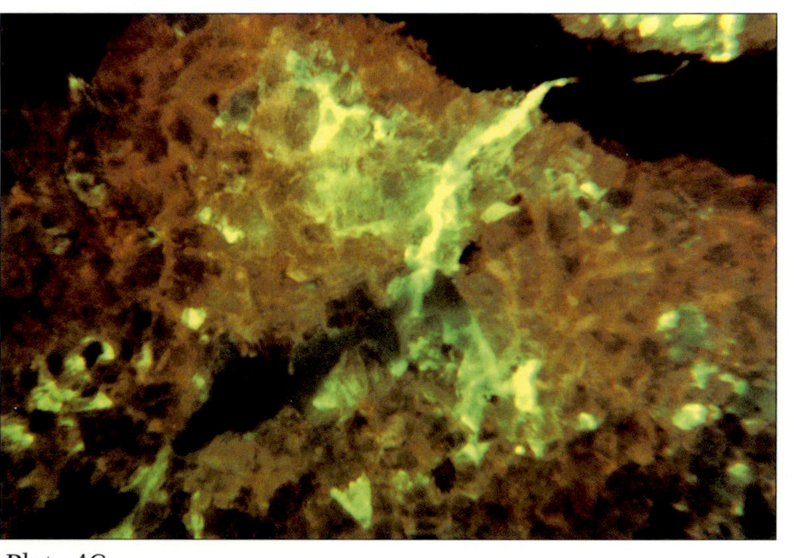
Plate 4C

Plate 4D

Plate 5A

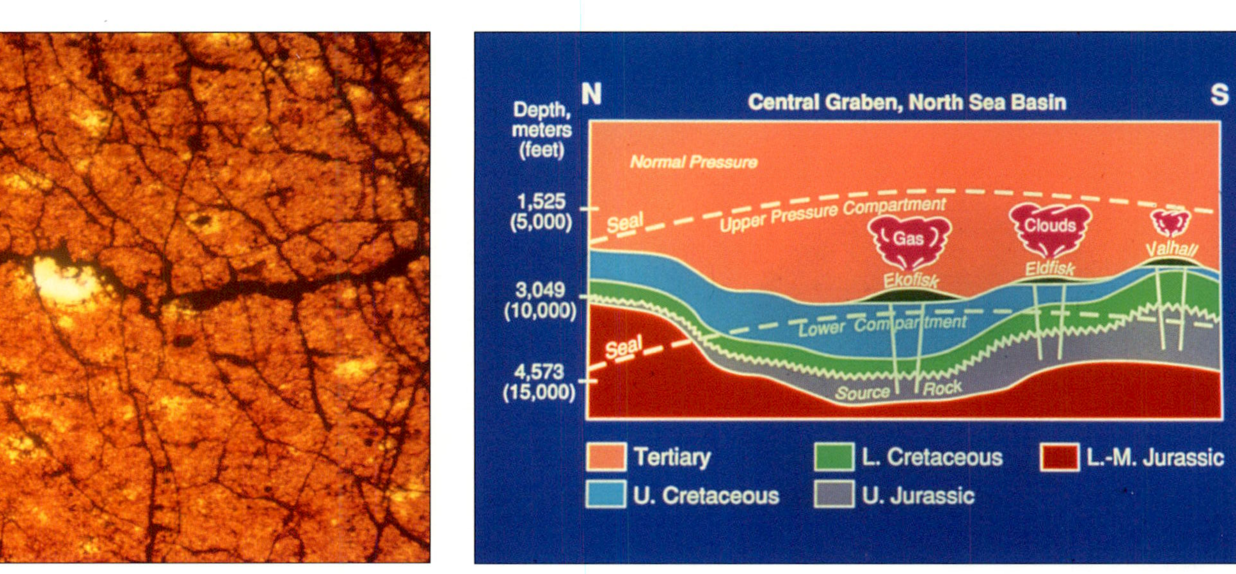

Plate 5B

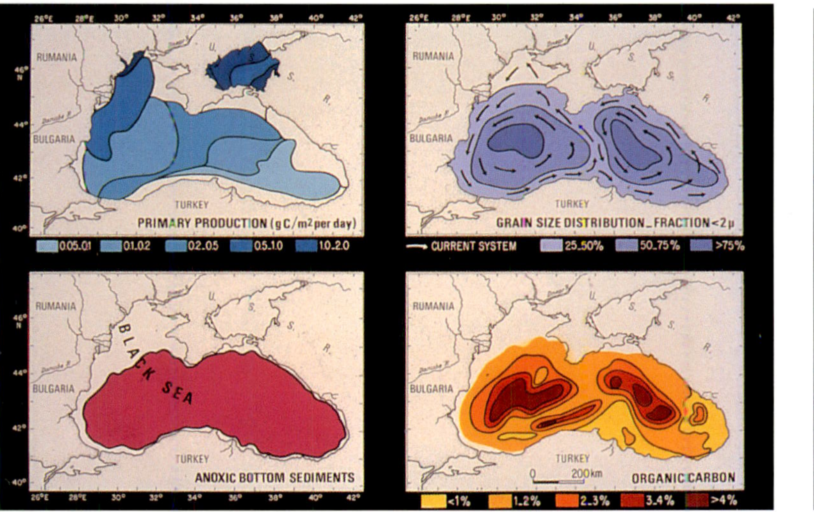

Plate 5C

Plate 5D

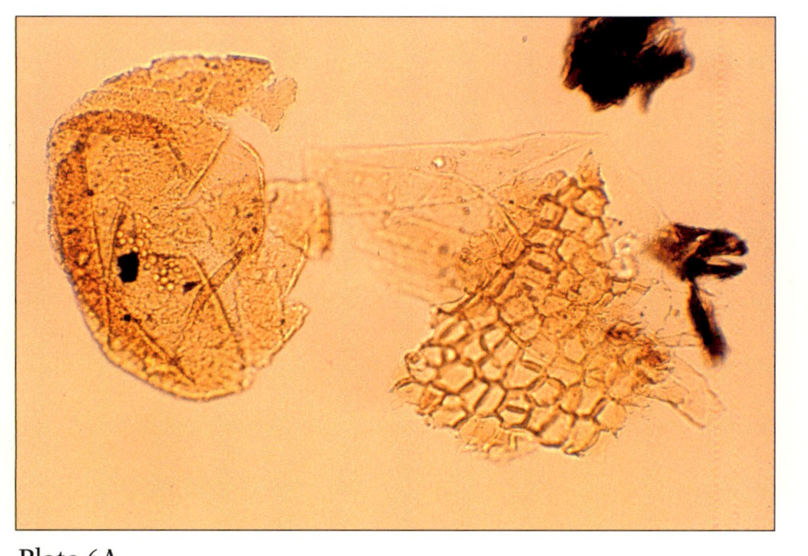

Plate 6A

Plate 6B

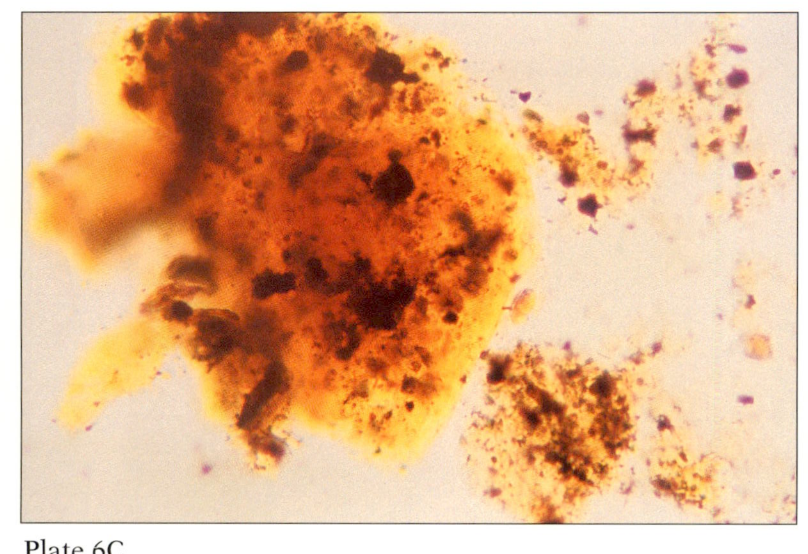

Plate 6C

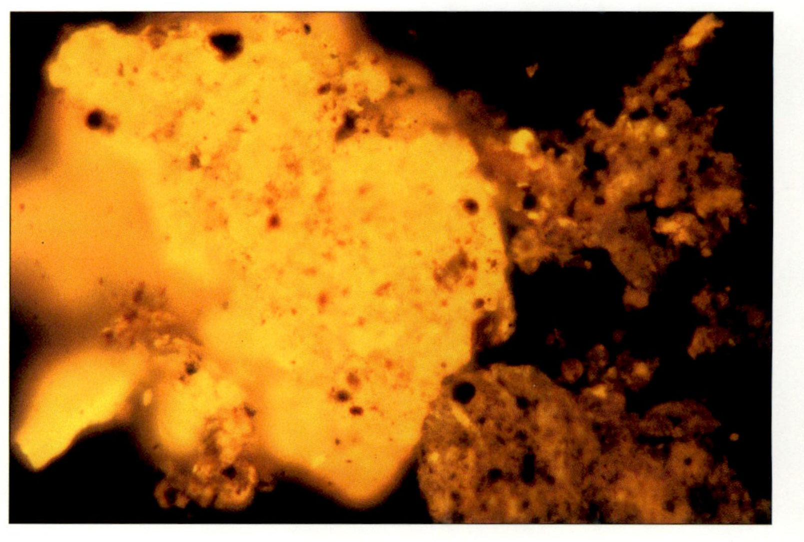

Plate 6D

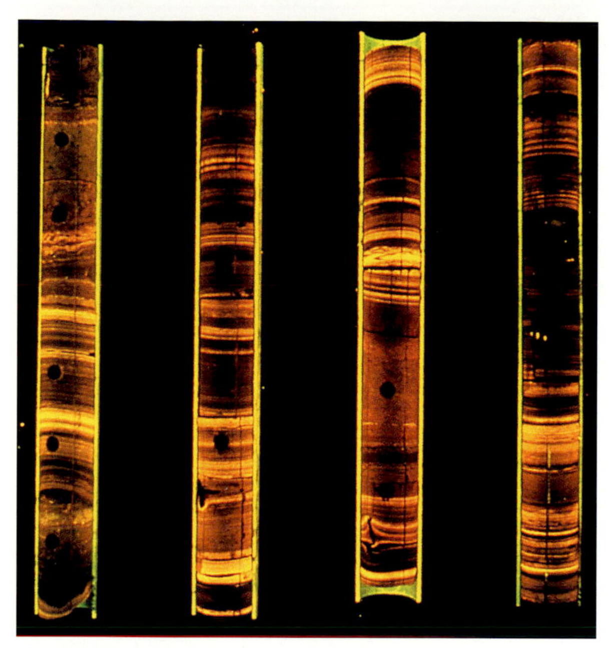
Plate 7A

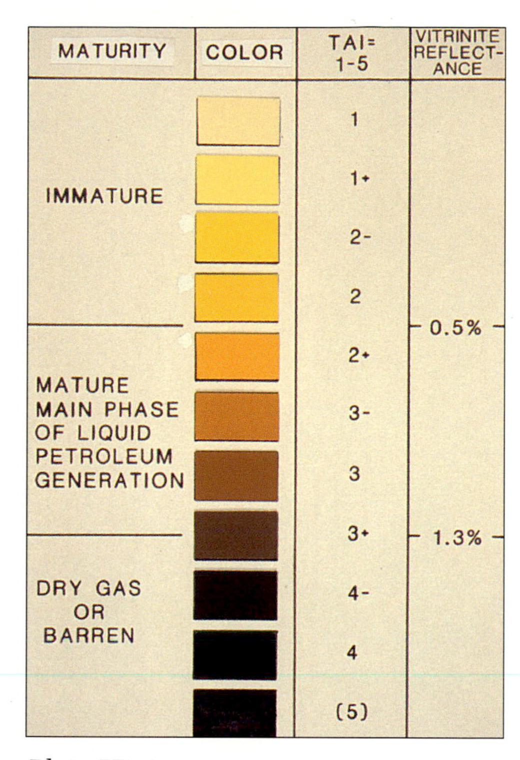

Plate 7B

Plate 7C

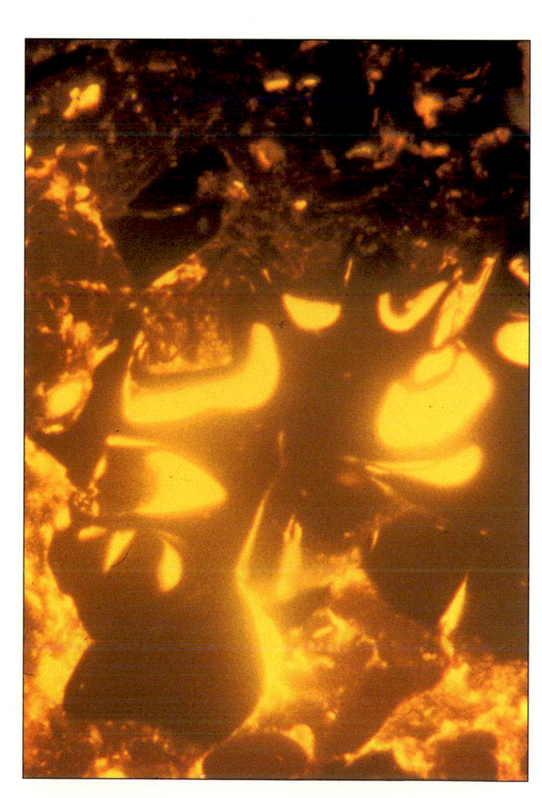
Plate 7D

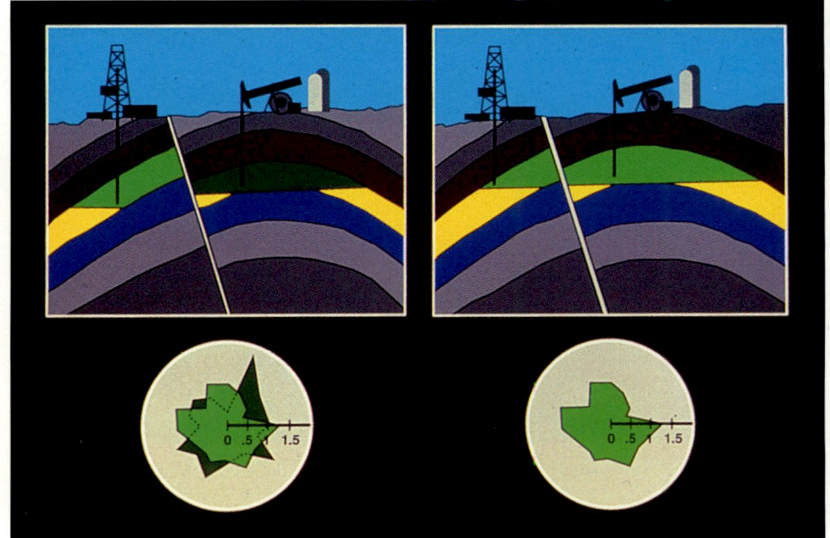

Plate 8A

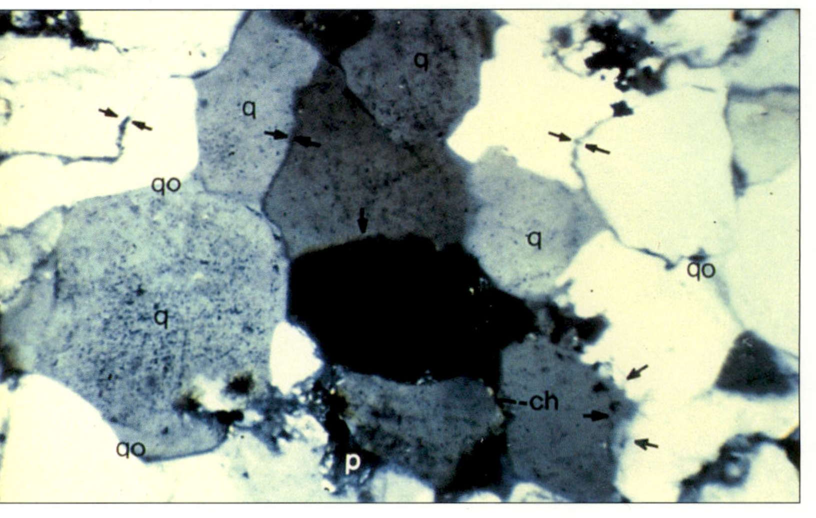

Plate 8B

Plate 8C

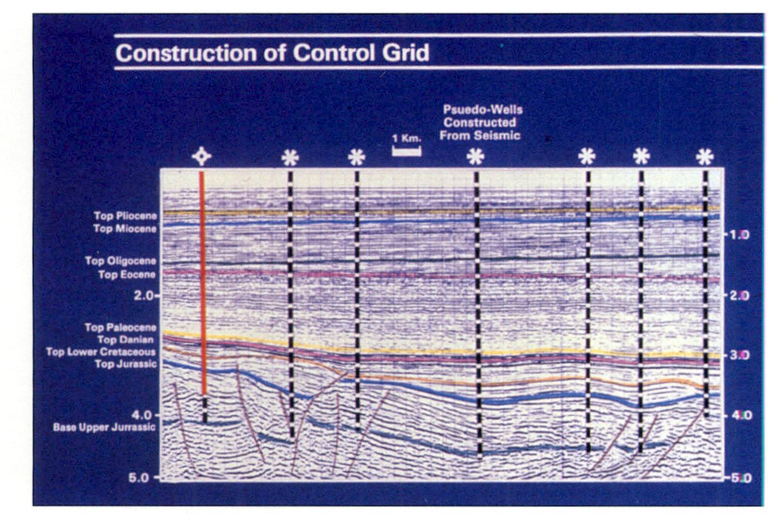

Plate 8D

Gas from coal ranges from dry to wet (high in C_2+), with CO_2 and N_2 as the main nonhydrocarbon gases. The C_2+ fraction is as high as 70% in some Late Carboniferous coal mines of western Germany, 49% in the Late Carboniferous lower Silesian coal basin of Poland, and 34% in coal mines of eastern China. Coal-bed production yields methane with up to 18% C_2+ in the Piceance Basin and 14% in the San Juan Basin of Colorado (Rice 1993). These are mainly humic coals that clearly have the ability to generate hydrocarbons heavier than methane. The generation of gas causes overpressures in many coal beds, pressures greater than those of adjacent sand–shale sequences. Some of the gases from the Piceance and San Juan basins also are associated with waxy oils.

Producing adsorbed gas directly from a coal seam (coal-bed gas) involves removing the water and lowering the hydrostatic pressure on the coal seam. This dehydrates the coal so that the desorbed gas can migrate through the coal micropores and then along fractures leading to the well bore. Usually there are three to four months of initial water production followed by gas, with peak flows of 150 to 200 Mcfd (Willis 1991). By 1991 there were more than 6,000 coal-bed gas wells in western basins of the United States.

Oil from Coals and Coaly Shales

The hydrogen content of OM exerts great control over its physical properties. As the hydrogen content increases, the OM changes from solid to liquid to gas. For example, Table 11-1 shows that the carbon contents of gas, oil, and coal are about the same, around 80%. But there is almost three times as much hydrogen in oil and five times as much in gas, compared with that of a typical humic coal. It follows that the higher the hydrogen content of coal is, the greater will be the ability of coal to generate petroleum.

Table 11-2 shows the hydrogen-to-carbon atomic ratio of the three major kerogen and coal types. The coal types have the same range of hydrogen content as the kerogens. Kerogen and coal have about the same carbon content at the beginning of catagenesis, so the difference in ratios by types is a difference in hydrogen content. Theoretically, a boghead coal should be able to generate

TABLE 11-1 Elemental Composition of Fossil Fuels (wt%)

	Gas	*Oil*	*Coal*
Carbon	76	84.5	81
Hydrogen	24	13	5
Sulfur	0	1.5	3
Nitrogen	0	0.5	1
Oxygen	0	0.5	10

TABLE 11-2 Approximate Hydrogen-to-Carbon Atomic Ratios of Major Coal and Kerogen Types[a]

Kerogens	*Ratio*	*Coals*	*Ratio*
Type I	1.45	Boghead	1.5
Type II	1.25	Cannel	1.2
Type III	0.8	Humic	0.8

[a]All samples are at the beginning of catagenesis.

as much oil as the Green River oil shale, the standard type I kerogen. Both plot in the same position on a van Krevelen HC/OC diagram when immature, as shown in Figures 10-2 and 10-3.

An indirect measure of hydrogen content is the previously discussed hydrogen index determined by pyrolysis. In Figure 11-1, petrographically identified

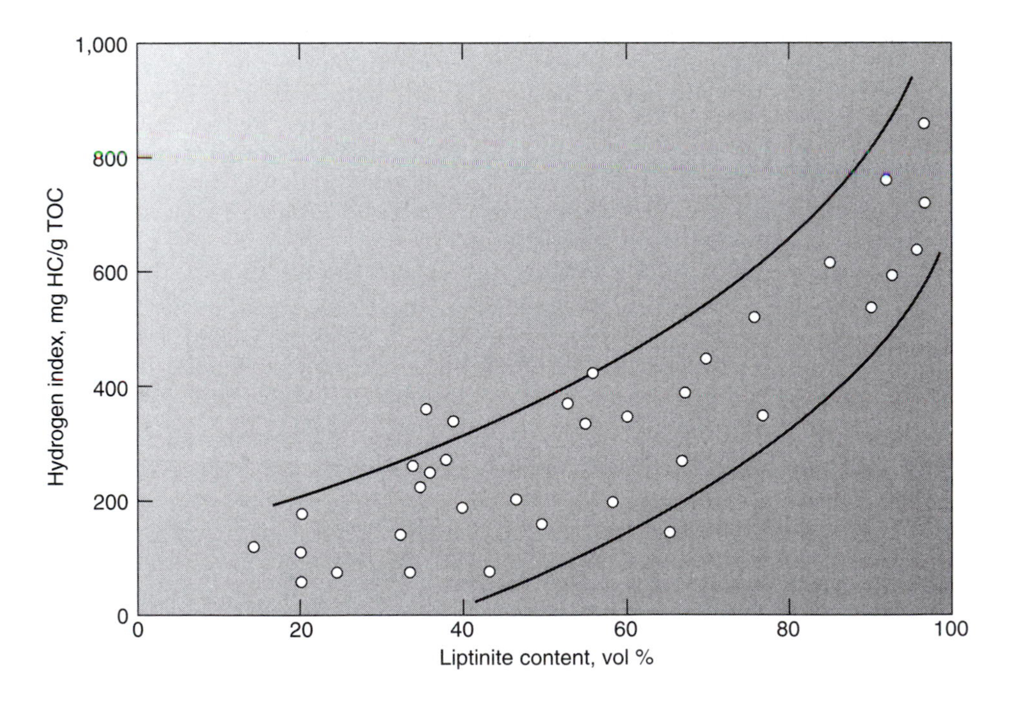

Figure 11-1

Trend of increasing hydrogen index with liptinite content of coals. [Mukhopadhyay and Hatcher 1993]

liptinite is plotted against the hydrogen index for thirty-five coals from around the world (Mukhopadhyay and Hatcher 1993). The broad area between the lines encompasses seventy kerogens and coals reported by Mukhopadhyay et al. (1985a). The general trend is clear. As the liptinite content of the OM increases, so does the hydrogen content and the potential to generate oil, as indicated by the hydrogen index.

The effect of the increasing hydrogen content on oil generation from coal has been observed in the laboratory. Lewan (1990) subjected thirteen humic coal samples from the King Coal Mine, Utah, to hydrous pyrolysis and obtained expelled oil yields ranging from 5.5 to 16 wt% on a mineral-free coal basis. The yield of high-wax oils rose with an increase in the hydrogen content of the coal. The hydrogen content was the only factor measured that showed a good correlation with the oil yield. Those coals for which the mol% ratios of H/(H + C) were around 0.45 yielded 8% of the coal as expelled oil, whereas the coals with ratios of about 0.49 converted 16% of the coal to expelled oil (Figure 11-2).

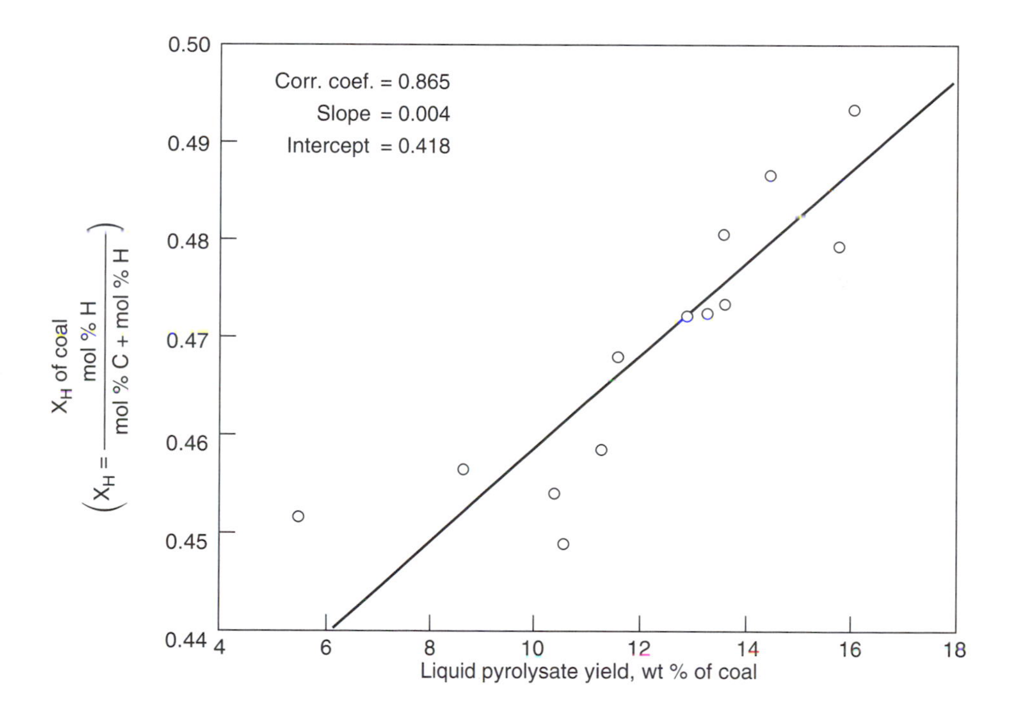

Figure 11-2

Yield of expelled liquid pyrolysate (oil) from the hydrous pyrolysis of humic coals versus the ratio of mol% hydrogen to mol% carbon plus hydrogen in the coals. [Lewan 1990]

Sapropelic Versus Humic Coals

Although sapropelic boghead and cannel coals have the potential to be major oil sources, they represent less than 10% of the world's coal. So sapropelic coals can only be a minor contributor to the world's oil.

Over 80% of all coals are humic, and over 70% of most humic coals consist of the maceral vitrinite. But as explained in Chapter 10, there are two groups of vitrinites. The high hydrogen vitrinite-2 group contains oil-generating liptinitic material, even though it is classified in the gas-prone vitrinite group. The problem with any classification system such as that for coal types and macerals is that the pure individual components are rare in nature. Mixed boghead–cannel coals are more common than either individual type. Likewise, some humic coals contain appreciable quantities of liptinitic materials as solid fillings, coatings, or invisible impregnations. Many humic coals that appear to be made up primarily of vitrinite and inertinite under ordinary reflected light show laminations or pockets of fluorescence under reflected blue light. This is characteristic of oil-generating sapropelic OM. For example, Plate 7C is a polished pellet mount of an Indonesian coal sample under reflected white light. The vitrinite groundmass is medium gray, and the inertinite is a whitish gray (Carolyn Thompson-Rizer 1992, personal communication). Cell fillings appear black. Plate 7D is the same field under reflected blue light. All the fluorescence is coming from the liptinite cell fillings in the inertinite and small lipitinite fragments in the groundmass of the vitrinite. Such data indicate that some humic coals and associated coaly shales may contain more oil-generating OM than do good source rocks of equivalent volume.

In Figure 11-3, the percentage of exinite and resinite in coals is plotted against the H/C atomic ratio of the coals, which is a rough indicator of their hydrocarbon-generating capacity (Jones 1987). As discussed in Chapter 10, OM with an H/C ratio larger than about 0.8 definitely has some liquid-generating ability. Figure 11-3 suggests that coals with more than about 10 to 15% exinite plus resinite are capable of generating oil. This would include the Gippsland Basin coals of Australia and some of the coals of Indonesia, the Philippines, and Alaska. All the samples in Figure 11-3 are immature, as indicated by the $\%R_o$.

In the Mahakam Delta of Indonesia, Huc et al. (1986) found no significant difference in the maturation pathways of natural coals and kerogens isolated from coaly shales interbedded with the coals. The shale and coal samples were taken from well depths of 1 to more than 4 km. The maturation pathways were determined by van Krevelen's original technique, namely, doing elemental analysis on the kerogens and coals isolated from their mineral matrix. In some wells, the coals were richer in hydrogen than the kerogens, but all tended to follow the type III kerogen track in Figure 10-3. Huc et al. concluded that there is no reason to favor shales over coals as possible source rocks in the Mahakam Delta, based on their hydrocarbon potential.

Huc et al. noted that the Miocene Kerbau coals at Mahakam do not generate an extractable bitumen intermediate when pyrolyzed, whereas the shales do generate bitumen, which later cracks to light hydrocarbons. Also, the shales in

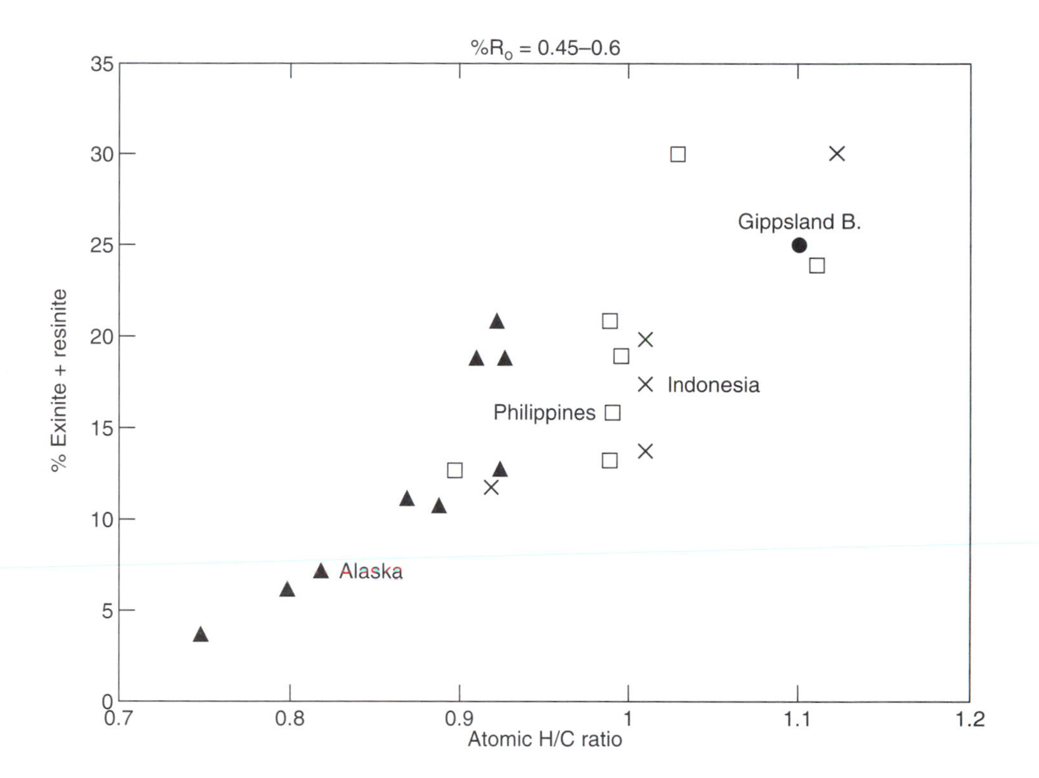

Figure 11-3

Volume percent exinite plus resinite versus atomic H/C ratio of coal: ▲, Alaska; □, Philippines; ×, Indonesia; ●, Gippsland Basin. [Jones 1987]

the Paris and Douala Basins produce bitumen intermediates, but the coals do not. Sterane and triterpane biomarker patterns of oil extracts from thin coal seams and clays at Mahakam were found to be identical to the patterns for oils in reservoirs above (Schoell et al. 1983).

Thompson et al. (1985) examined hydrogen-rich coals in three Indonesian basins: Kutai, Northwest Java, and Sunda. Coal samples from these basins gave hydrogen index values in the range of 250 to 450 mg HC/gTOC. Organic matter with HI values above 200 is usually considered capable of generating some liquid hydrocarbons. Microscopic analysis of the Indonesian coals indicates that they contain desmocollinite, the hydrogen-rich vitrinite-2. Layers of liptinitic material are interspersed in vitrinite along with impregnations of bitumen and other fluorescent materials. The coals were estimated to contain 15 to 65% liptinite (Thompson et al. 1985).

Gas chromatograms of crude oils for these Indonesian basins have the characteristic odd–even predominance in the C_{27}-to-C_{33} *n*-alkane range, along with a pristane-to-phytane ratio greater than 5, both of which are typical of an oxic depositional environment (Figure 10-34). Mass fragmentograms of extracts from the coals show oleanane and C_{30}–C_{31} hopane patterns similar to those of the crude oils believed to be generated by the coals. Oleanane comes from flowering plants that were widespread in Indonesia during Tertiary times. Thompson et al. (1985) concluded that a process of liptinite enrichment like that currently occurring in Indonesian deltas was responsible for the formation of coals as oil source rocks.

Qin et al. (1994) investigated the petroleum-generating potential of the Huanxian brown coal from the Bohai Basin of east China. An immature sample of this coal has an atomic H/C ratio of 1 and a liptinite content of 15 to 25%. It yields 10% oil by the Fischer assay, which goes to 570°C. Artificial maturation of the coal by hydrous pyrolysis gave very different products and yields than by anhydrous pyrolysis. Maturation by hydrous pyrolysis followed the type III evolution pathway, whereas the anhydrous pyrolysis followed the type IV pathway (Figure 11-4). Also, the hydrous pyrolysis yielded the most liquid hydrocarbons, whereas the anhydrous pyrolysis formed the most resins, asphaltenes, and gas. Qin et al. (1994) concluded that hydrous pyrolysis is more analogous to the natural geological conditions than the anhydrous procedure. Their work and the previously described studies by Lewan (1990) demonstrate that humic coals with liptinite contents >15% can yield oil in quantities comparable to shale source rocks under laboratory conditions.

Noble et al. (1991) used hydrous pyrolysis to generate oil from coals and coaly shales of the Oligocene Talang Akar Formation of northwest Java, Indonesia. Gas chromatograms of the oil expelled from the coals more closely match the crude oils in the basin than the oil expelled from the shales. Also, the gas/oil ratios of the expelled hydrocarbons were higher from the shales than from the coals because the coals yielded more oil.

High-hydrogen vitrinites are common in some areas. For example, Newman and Newman (1982) noted a difference in vitrinite reflectance of 0.28% between two New Zealand coal seams separated by only 100 meters. The lower-reflecting coal was formed in swamps with high bacterial activity and little drainage, which produced a hydrogen-rich vitrinite (V-2). The vitrinite fluoresced in reflected blue light from dull yellow to orange brown in color, indicating that it had some oil-generating capability. The higher-reflecting coal was from a well-drained swamp facies and did not exhibit fluorescence. Consequently, humic coals that are mostly vitrinite may generate oil from their hydrogen-rich liptinite cell fillings and laminations or from an increase in hydrogen of the vitrinite itself due to more bacterial activity during burial.

In the intermontane basins of Thailand, such as the Mae Tip Basin, oil shales up to 1 m thick are interbedded with subbituminous coals (Gibling et al. 1985). The oil shales were formed when peat swamps close to a steep basin margin were flooded by shallow lakes, permitting algae to dominate the organic facies. During the more oxic, regressive periods, peat swamps formed humic coals. The formation of oil shales and coals alternates with transgressive

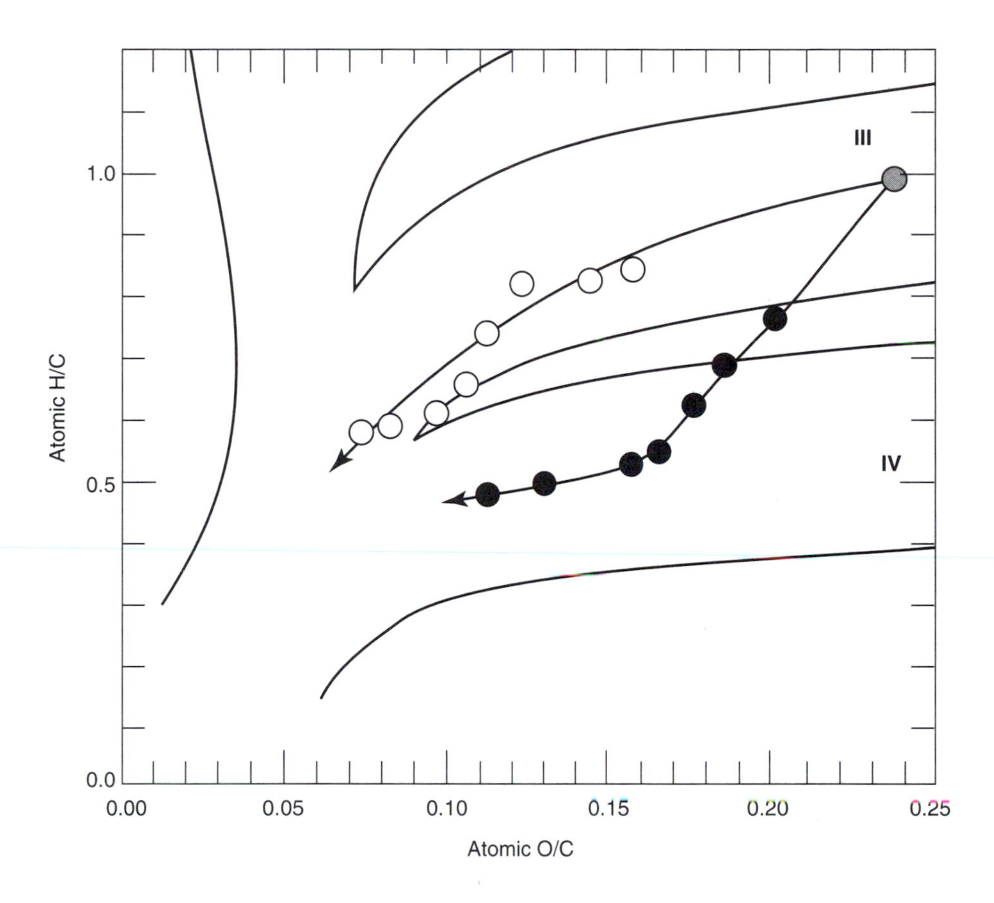

Figure 11-4

Maturation (coalification) pathway of the humic Huanxian (L. Tertiary) brown coal from the Bohai Basin of east China by laboratory pyrolysis: ○ hydrous; ● anhydrous. [Qin et al. 1994]

and regressive cycles. Such coals tend to be high in hydrogen-rich vitrinites. The Upper Triassic in the Shanxi-Gansu-Ningxia Basin of China contains abundant coal in the swamp facies along the basin margin. This grades into oil shales in the anoxic lacustrine facies toward the middle part of the basin (Luo Binjie et al. 1988).

The Gippsland Basin of southeastern Australia had initial reserves of about 3 billion barrels of oil, as well as 8 trillion cubic feet of gas and 800 million barrels of condensate. The oils are high wax and low sulfur, with C_{29} steranes dominant and pristane/phytane ratios usually above 5 (Shanmugam 1985). These all are characteristics of oils originating from OM deposited with terristrial sediments. Sesqui- and diterpanes and other land-derived biomarkers also have been identified in the oils (Philp and Gilbert 1986).

The source of these oils is believed to be in the Lower Tertiary Latrobe coals and coaly shales, which contain oil-generating wax, resinite, suberinite, and other liptinites. Smith and Cook (1984) divided the coal and carbonaceous shales of the Latrobe group into the Latrobe Valley facies and the Upper and Lower Eastern View facies. The liptinite contents of the last two, which are the most likely source beds, range up to 25%, with an average around 15%. The vitrinite averages about 80%, and the inertinite, 5%. Thick coal seams grade into carbonaceous shales and sands. The shales contain disseminated coaly matter of the same composition as the coals. Although the vitrinite content is high, much of it consists of low-reflecting, hydrogen-rich vitrinite-2 that fluoresces under reflected blue light. Consequently, much of the oil may be coming from both the liptinite and the vitrinite-2 (Smith and Cook 1984). Shanmugam (1985) reported hydrogen-index values ranging from 201 to 312 for coal samples from the Halibut field offshore and 391 to 422 for Latrobe coal seams onshore. These values are within the oil-generating range of known shale source rocks, as explained in Chapter 10.

Although the Latrobe Valley group contains some of the world's thickest coals seams (up to 165 m according to Shanmugam), the proportion of coal in the Latrobe group decreases from about 50% onshore to about 5% offshore. The dispersed OM in shales and sands between the coal seams offshore is nearly all type III, with maceral compositions similar to the coals. Shanmugam (1985) also carried out hydrous pyrolysis of coals and resin bodies from the Latrobe group. He concluded that the more paraffinic oils in the Gippsland Basin are derived from the waxy coals and coaly shales, and the naphthenic oils are formed largely from the resin bodies.

The Permian–Triassic rocks of the Cooper–Eromanga Basin in South Australia comprise several cycles of fluvial sandstones, coal measures, and lacustrine shales. All hydrocarbons in the basin are believed to have originated from these terriestrial sediments. About 100 gas fields and 10 oil fields have been discovered in the basin, with recoverable reserves of 5 trillion cubic feet of gas and 300 million barrels of condensate and oil. The Permian coal measures have long been considered the major source of gas, but not necessarily of oil (Vincent et al. 1985).

The dispersed OM of the shales and the macerals of the coals are principally vitrinite and inertinite, with liptinite (mainly sporinite and cutinite) less than 10%. This is probably below the threshold for forming and expelling oil. The Permian Toolachee Formation has hydrogen indices as high as 320, which are in the range of oil generation. However, the crude oils are found in the overlying fluvial–lacustrine Jurassic reservoirs where there are much better source rocks, such as the Jurassic Birkhead and the Basal Jurassic Formations, with dispersed liptinite giving HI values up to 450. These formations contain sporinite, cutinite, resinite, and suberinite, all of which have oil source potential (Vincent et al. 1985). All but one of the crude oils analyzed are paraffinic with low-sulfur contents. The biomarker patterns, such as a high C_{29} sterane content, are characteristic of oils derived from terrestrial source rocks.

The San Juan Basin of Colorado and New Mexico has more than 2,000 wells producing dry coal-bed methane. A few wells, however, have produced noncommercial quantities of wet gas and oil from the coal seams, in volumes totaling 10 to 100 bbl/well. The oil has a CPI greater than 1, a pristane/phytane ratio around 7, a saturate/aromatic ratio >10, large amounts of C_{20+} waxy *n*-alkanes, and a strong predominance of C_{29} steranes (Clayton et al. 1991). Gas chromatograms of the C_{10+} alkanes are similar to that in Figure 11-5 (B), described in the next section. Whole-oil GC analysis showed that all oils also contain a predominance of C_5–C_{12} alkanes, with methylcyclohexane being the most abundant. Since the indigenous oil and wet gas were encountered in only some of the wells, they may be related to the distribution of high-hydrogen vitrinites (V-2) in the Fruitland coals. One coal sample contained a significant amount of desmocollinite, which is the main maceral of the V-2 group (Clayton et al. 1991).

Oil Shales and Other Terrestrial Source Rocks

In the years since Hedberg's paper, kerogen disseminated in lacustrine and fluvial shales has become widely recognized as a source of both waxy oil and gas. Source rocks vary from the type III kerogens found in Tertiary deltas such as the Mississippi, Mahakam, Niger, and Mackenzie to the types I and II kerogens found in the oil shales deposited by saline, anoxic lakes, mostly of Tertiary age. The lacustrine oil accumulations in the Songliao and Bohai Basins of China were unknown at the time of Hedberg's paper. Today they produce over 80% of China's oil, most of which is high in wax.

Powell (1986) wrote a comprehensive review of the petroleum geochemistry and depositional environment of lacustrine source rocks. A key point of his paper is that there is a wide variation in the TOC, from <1% to >20%; and in kerogen types, from I to III; and in the source of organic matter (land plant, algal, or bacterial) in lacustrine sequences. Most crude oils are paraffinic and waxy due to OM sources such as alginite, cutinite, sporinite, and various plant waxes. More naphthenic oils tend to come from resinite, as mentioned earlier. It also was stated in Chapter 10 that lacustrine oil shales, such as the Green River Formation of Utah, contain a mixture of kerogen types instead of mainly type I (Figure 10-13).

According to Powell (1986) the Songlia and Uinta Basins and other lacustrine areas are similar in that type I kerogen is restricted to part of the deepest lake facies, with types II and III increasing from the basin's center to the edge of the lake deposits. These variations suggest that the time–temperature requirements for oil generation in the deepest lake facies are higher than in sediments more toward the margins. The activation energies required to crack type I kerogen are significantly higher than those required for types II and III, as explained in Chapter 6.

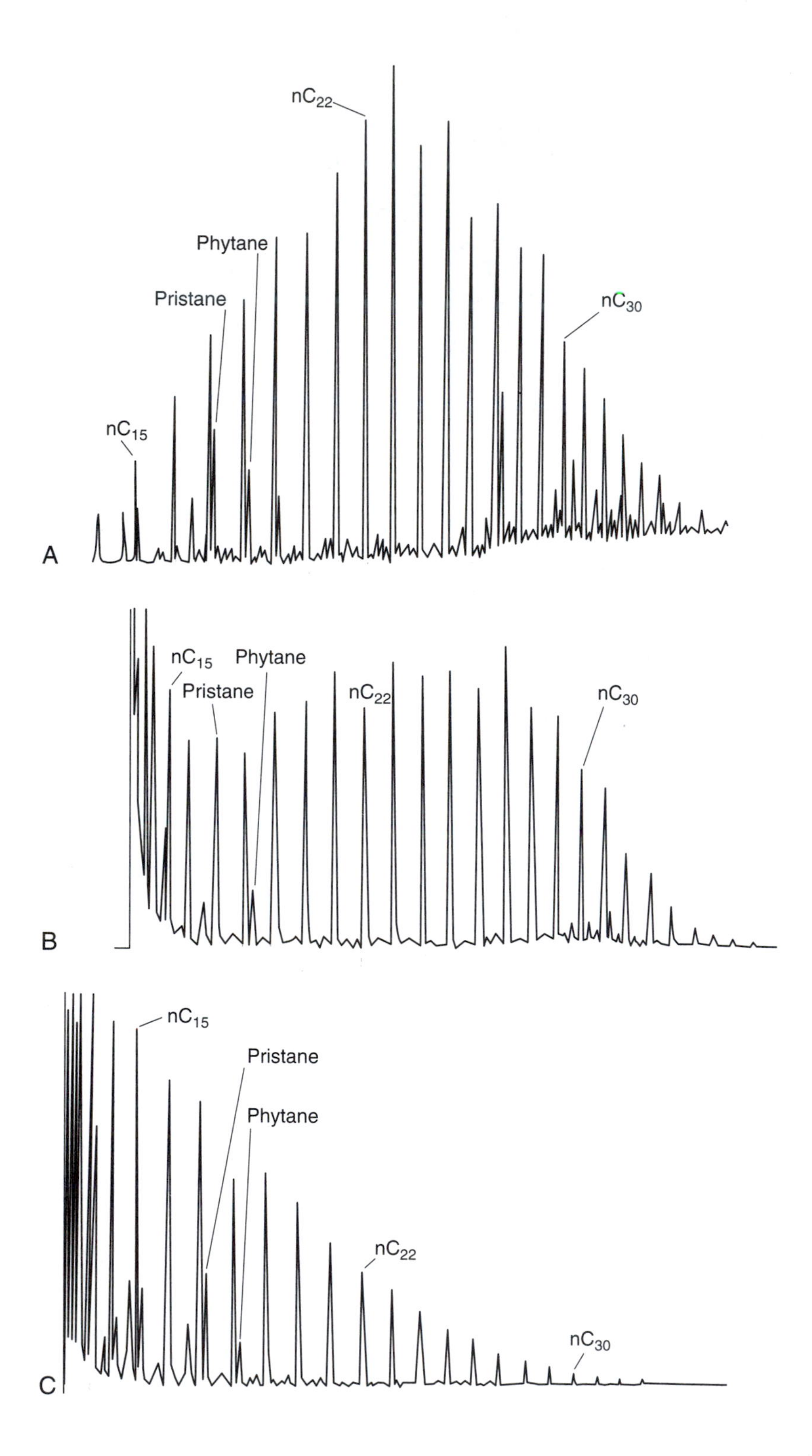

A
nC15
Pristane
Phytane
nC22
nC30
B
nC15
Pristane
Phytane
nC22
nC30
C
nC15
Pristane
Phytane
nC22
nC30

Figure 11-5 (left)

GC profiles of saturate fractions of crude oils: (A) lacustrine-source oil from China's Songliao Basin [Fu and Sheng 1992]; (B) lacustrine-source oil from the South China Sea; and (C) marine-source oil from the Gulf of Mexico. [Katz 1990]

Lacustrine crude oils differ from marine oils in having a high concentration of C_{21}-to-C_{35} n-paraffins, a CPI >1, a pristane/phytane ratio > 1, a low sulfur content, and a high ratio of hopanes to steranes. The first three of these characteristics are apparent in the gas chromatograms of lacustrine oils from China (Figures 11-5 [A] and [B]). The n-paraffins of the Songliao Basin oil peak at C_{23} and show an odd–even predominance through C_{29}. Also, the Pr/Ph ratio exceeds 1 (Fu and Sheng 1992). The South China Sea oil (Figure 11-5 [B]) has similar characteristics, except that the chromatogram is more typical of high-wax crudes in general. Both of these plots are distinctly different from the typical marine-source oil in Figure 11-5 (C) (Katz 1990).

Waxy oils and gas occur in Eocene to Miocene sediments in offshore basins west of Bombay, India. Mukhopadhyay et al. (1985b) found a high content of desmocollinite and bituminite in immature kerogens of these sediments, some containing up to 70% desmocollinite. Half the samples they studied had TOC values from 2 to 14%, with the other half below 2%. The H/C ratios were high, ranging from 0.9 to 1.55. Pyrolysis of the kerogens yielded waxy oils. The oils probably came from desmocollinite, the high-hydrogen vitrinite-2 maceral, which yields oil on maturation (Mukhopadhyay et al. 1985b). The desmocollinite had a weak brown fluorescence under reflected blue light, and the bituminite fluoresced reddish orange in color. Resinite with a yellow fluorescence also was present. This is a good example of a hydrogen-rich disseminated terrestrial kerogen generating a waxy oil.

In deltaic sequences such as the Niger and Mahakam, both the Tertiary sediments and the crude oils have been found to contain oleanane which, as previously mentioned, is considered to be a product of flowering plants. In the Mackenzie Delta, the OM of the sediments has the characteristics of terrestrial OM, such as the presence of oleanane, a high C_{29} sterane content, and a high carbon preference index (CPI) in the wax range. In addition, bisnorlupanes were found in relatively high concentrations in the crude oils (Brooks 1986). Analyses of the organic matter in the sediments showed that only one formation, the Eocene Richards, contains bisnorlupanes. Consequently, the Richards Formation is considered an important contributor to the oils of the Mackenzie Basin. It contains mainly type III kerogens, typical of most deltas (Brooks 1986).

Deltaic sedimentary rocks tend to have low TOCs (less than 5%) and mostly type III kerogen with some type II, but they still generate a lot of oil, condensate, and gas. Oil shales, in contrast, have very high TOCs and mostly oil-prone type I and II kerogens, yet they are rarely associated with significant oil accumulations. There are probably several reasons for this. As previously mentioned, type I kerogen requires a higher activation energy for cracking.

Consequently, by the time that oil shales are buried deeply enough to generate oil, they are no longer recognized as oil shales. Most oil shales that have been analyzed are immature. They have not been buried long enough at sufficiently high temperatures to generate substantial amounts of oil. Second, the oils that are generated tend to be high-wax oils that need further cracking for ease of migration. Third, there is frequently very little water in fractured oil shale reservoirs, such as the Green River Formation in the Altamont field of the Uinta Basin, Utah. In the absence of water, there is no driving mechanism except gas pressure to assist the migration of a thick, waxy oil. Also, some formations like the Green River are overpressured by the formation of a hydraulic seal that prevents the free flow of water and the migration of generated hydrocarbons (Hunt 1990).

Kerogen types I and II are the dominant constitutents of oil shales. Most such kerogens have atomic H/C ratios from 1.3 to 1.8, indicating a high potential for generating oil (Cook et al. 1981). Some oil shales grade into the sapropelic boghead and cannel coals. Much of the OM of immature oil shales fluoresces under incident blue light. Nearly all of them form from algal oozes deposited under suboxic to anoxic conditions. The total world reserves of shale oil (proved and possibly recoverable) are about 3×10^{12} barrels (Cook et al. 1981). However, the recovery costs and disposal of the by-products are such that the economic use of oil shale is still far in the future.

Adsorption–Migration of Hydrocarbons from Coal and Terrestrial Kerogen

There is no doubt that hydrogen-rich coal and terrestrial kerogen can generate economic quantities of liquid petroleum. This has been recognized in the field and in the laboratory experiments discussed in this chapter. Why, then, do we not find more oil accumulations generated from coal beds? One problem is that most thick coal beds are humic coals relatively low in hydrogen, whereas thinner beds interbedded with shales and sometimes oil shales often contain hydrogen-rich vitrinite or sapropelic coals rich in liptinites. In addition, problems related to adsorption and migration phenomena are characteristic of coal and seem not to be critical to the disseminated kerogen or coaly particles of shales.

Charcoal is one of the most widely used adsorbents for taking bitumens and other organic contaminants out of a flowing water system. Laboratory experiments have shown that between high-volatile bituminous coal and anthracite, the adsorption of hydrocarbons more than doubles (Wyman 1984). A greater pressure also increases the adsorption. The larger hydrocarbon molecules are adsorbed more strongly than smaller ones, since adsorption on coal is comparable to that on a packed column in gas chromatography. Wyman (1984) slowly desorbed coal samples obtained at a depth of 2,835 m in the Deep Western Canada Basin. He found that after fifteen days, only the hydrocarbons methane, ethane, and propane were desorbed. By thirty-four days he was also obtaining isobutane, *n*-butane, and isopentane. Extraction of these coals released up to 30 mg HC/g TOC, which apparently was strongly adsorbed on the coal surface.

Barker et al. (1989) studied the adsorption of hydrocarbons and asphaltenes on both wet and dry quartz, clays, and carbonates. They found that an initial monolayer coverage required higher activation energies for removal than did the succeeding multilayers. Pyrolysis of samples with adsorbed hydrocarbons caused their release between pyrolysis peaks 1 and 2, frequently overlapping peak 2, the cracking peak. This suggests that in natural systems, coals with strongly adsorbed hydrocarbons do not release them after generation, and consequently they are cracked to gas and condensate before being expelled.

Barker et al. (1989) also studied the adsorption of bitumens on shales. Figure 11-6 shows bitumen adsorption in arbitrary units plotted against the TOC for a shale with a type II kerogen. They found that above 9% TOC, there was a sudden increase in adsorption efficiency. This suggests that it is difficult for hydrocarbons to migrate out of high-TOC shales or coals before they are converted to gas.

Another property of coals that prevents liquid hydrocarbon migration is the pore diameters. Coal-pore diameters can be divided into macropores >50 nm

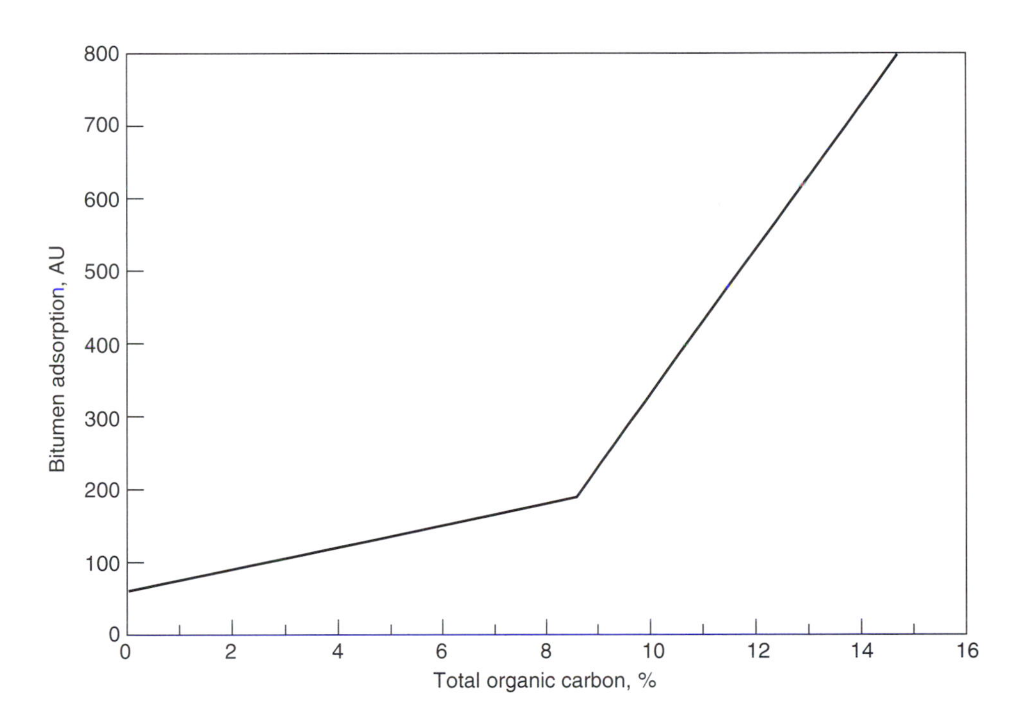

Figure 11-6

Bitumen adsorption on the Excello Shale. AU = arbitrary units. [Barker et al. 1989]

(nanometers), mesopores 2 to 50 nm, micropores 0.8 to 2 nm, and ultramicropores <0.8 nm (Walker 1981). Coals vary widely in both pore volumes and porosities. Parkash and Chakrabartty (1986) found that 30 to 76% of the pore volume in eleven Alberta plains coals is in the micropore range, 19 to 50% is in the macropore range, and the rest is in between. Surface areas of the coals calculated from carbon dioxide adsorption at 25°C ranged between 75 and 500 m^2/g. Since complex hydrocarbon ring structures are about 2 nm in diameter and small asphaltenes are 5 nm, it appears that newly generated liquid bitumens cannot migrate out of the 30 to 76% micropores unless the coal is fractured. *Cleats,* which are natural fractures or joints in coal, may provide a permeability pathway for hydrocarbons. Gas generation in a confined system can open the cleats by raising pressures to lithostatic.

There is other laboratory evidence that coals trap hydrocarbons as they are generated. Landais and Monthioux (1988) pyrolysized coals by different methods. The same coal samples followed different maturation pathways using open, confined, and closed system pyrolysis. After comparing these results with extraction data, they concluded that natural coals retain free hydrocarbons trapped in the pores of their structures.

Youtcheff et al. (1983) found that the quantity of alkanes released from coals by liquefaction in tetralin at 400°C is six to eight times greater than the yield obtained by soxhlet extraction. They concluded that the alkanes obtained by liquifaction had been physically trapped inside the pores of the coals and were released only when these pores were broken open by liquefaction.

When a well is drilled for coal-bed methane, the operators know that it takes many weeks of water production and reduced pressures to dry the coal and initiate gas desorption. Gas production actually increases rather than decreases with time as the pressure declines. This is further evidence that the fracturing of pores and desorption are critical factors in releasing hydrocarbons from coal.

SUMMARY

1. Typical humic coals generate about 150 to 200 cc of CH_4/kg coal during maturation through the meta-anthracite stage. This results in the formation of some large and many smaller gas fields and the adsorption of huge volumes of gas by deeply buried coal beds.

2. Coals and coaly shales containing high-hydrogen vitrinite-2, such as desmocollinite, and with liptinite contents above 15% can generate and release waxy oils as well as gas. Examples are oil fields in Indonesia and Australia.

3. Terrestrial kerogen, deposited in lacustrine, fluvial, deltaic, or brackish-water sediments, ranges from type I to III, with the main oil production coming

from type II in lacustrine rocks, such as the Songlia and Bohai basins of China, and from types II and III in deltaic strata such as Mahakam and Niger.

4. High-wax and low-sulfur oils are formed mainly in sedimentary rocks deposited under lacustrine or brackish-water conditions. Oils formed from coals and coaly shales tend to have high ratios of pristane to phytane and of hopanes to steranes, compared with marine-source oils. They also have relatively high concentrations of (1) C_{21}–C_{35} *n*-alkanes; (2) C_{29} steranes; (3) bicyclic sesquiterpanes; (4) tricyclic diterpanes; and (5) tetracyclic diterpanes and oleananes.

5. Pore-size distributions and adsorption characteristics of bedded humic coals containing principally vitrinite-1 appear to hinder the release of generated petroleum. The eventual conversion of hydrocarbons to gas and condensate is thought to cause the opening of existing cleats and the release of trapped hydrocarbons. Type III kerogens and coaly particles disseminated in shales do not seem to trap and adsorb hydrocarbons as strongly as do bedded coals.

SUPPLEMENTARY READING

Law, B. E., and D. D. Rice (eds.). 1993. *Hydrocarbons from coal.* AAPG Studies in Geology 38. Tulsa: American Association of Petroleum Geologists, 400 p.

Orr, W. L., and C. M. White (eds.). 1990. *Geochemistry of sulfur in fossil fuels.* ACS Symposium Series 429. Washington, DC: American Chemical Society, 708 p.

Scott, A. C., and A. J. Fleet (eds.). 1994. *Coal and coal-bearing strata as oil-prone source rocks?* Geological Society Special Publication 77. London: Geological Society, 213 p.

Stach, E., M.-Th. Mackowsky, M. Teichmüller, G. H. Taylor, D. Chandra, and R. Teichmüller. 1982. *Stach's textbook of coal petrology,* 3rd ed. Berlin: Bebruder Borntraeger, 535 p.

Chapter

12

Petroleum in the Reservoir

The petroleum reservoir is the part of the rock that contains the pool of oil, condensate, or gas. The reservoir rock generally extends beyond the limits of the pool and is considered to be any rock that contains interconnected pores with sufficient permeability to allow oil- or gas-phase production. Petroleum reservoirs occur in nearly every type of rock: sandstones, siltstones, fractured shales, limestones, chalks, dolomites, and fractured or weathered igneous and metamorphic rocks. The fluids migrating from source beds channel through almost any coarse-grained rock in which the permeabilities are the highest. These fluids include oil and gas that are then trapped or sieved out at permeability barriers to hydrocarbons.

The term *petroleum* encompasses gas, condensate, oil, and heavy oil, that is, any bitumen existing in the gaseous or liquid state in its natural reservoir. The variability in the physical and chemical characteristics of petroleum in the reservoir can be due to biodegradation, water washing, thermal alteration, deasphalting, or gravity segregation. The seepage of hydrocarbons out of the reservoir or addition of them from active source rocks can change the original oil. Circulating meteoric waters degrade light oils to heavy oils, and thermal alteration leads to condensates and gas accompanied by natural deasphalting and pyrobitumen formation in the reservoir. All these influence the economics of an oil discovery. Consequently, it is important to understand and predict these processes in order to estimate the value of the petroleum that might occur in a prospect.

Producible gas-free crude oils are those having viscosities less than 10,000 centipoise at original reservoir temperature. Anything more viscous is called a *bitumen,* even though that term correctly includes all natural hydrocarbons. *Extra heavy oils* have densities greater than 1,000 kg/m^3 (< 10°API gravity). Heavy oils have densities from 1,000 to 920 kg/m^3 (10 to 22.3°API). *Medium oils* have densities from 920 to 870 kg/m^3 (22.3 to 31.1°API), and *light oils* have densities less than 870 kg/m^3 (>31.1°API) (Martinez 1984). These densities and API gravities are referenced to 15.6°C (60°F) at atmospheric pressure. The 1,000 kg/m^3 (10°API gravity) is the density of water at 4°C.

Gas associated with the oil may be in a free-gas phase above the oil accumulation (gas cap) or entirely dissolved in the oil. A condensate pool is a hydrocarbon accumulation that exists in a gas phase in the reservoir but changes to a liquid having a gravity > 50°API at 15.6°C and atmospheric pressure. Also, the gas/oil ratio (GOR) should exceed 5,000 standard cubic feet per barrel of oil (SCF/bbl). If the pressure of a condensate pool is reduced enough during production, the liquid hydrocarbons will condense in the reservoir, resulting in a much lower recovery than through production in the gaseous phase. Repressuring can return part of the condensate to the gaseous phase, but pressure maintenance during production (through water or gas injection) is the preferred conservative production practice.

Biodegradation and Water Washing

Wherever meteoric waters penetrate deeply into a basin, any petroleum accumulations that they contact are altered. Water moving past an oil field preferentially dissolves the most soluble hydrocarbons, such as methane, ethane, benzene, and toluene. Microbes in the water also can consume the small hydrocarbon molecules, producing heavy oil with a low API gravity. Degraded oils associated with meteoric waters have been found as deep as 2,134 m (7,000 ft) in the Niger Delta (Dickey et al. 1987), 1,830 m (6,000 ft) in the Gulf of Mexico (Roadifer 1987), and 3,048 m (10,000 ft) in the Bolivar coastal fields of Venezuela (Bockmeulen et al. 1983). The oxygen content of these meteoric waters generally ranges from 2 to 8 ppm.

Sediments, soils, sedimentary rocks, and interstitial waters contain a wide variety of microorganisms that can utilize hydrocarbons as a sole source of energy in their metabolism. Paraffins, naphthenes, and aromatics, including gases, liquids, and solids, all are susceptible to microbial decomposition. More than 30 genera and 100 species of various bacteria, fungi, and yeast metabolize one or more kinds of hydrocarbons. Also, microbial populations are highly adaptable and can alter their metabolic processes to fit the hydrocarbons available. Such microbes are widely distributed in nature. They also are beneficial in causing the relatively rapid disappearance of petroleum from natural seeps and spills. They are destructive in causing the deterioration of asphalt-base highways, asphalt-coated pipelines, and the contamination of stored gasoline and jet aircraft fuels. Military jet planes have been known to crash because their fuel system clogged when microbially altered fuels were used.

The general mechanism of attack is illustrated in Figure 12-1. Hydrocarbons are oxidized to alcohols, ketones, and acids. For example, *n*-butane is oxidized to methylethylketone and butyric acid. The organism *Pseudomonas methanica,* which grows at the expense of methane, also oxidizes ethane, propane, and butane in a mixture yielding the corresponding acids. Long-chain paraffins are oxidized at the termini of a carbon chain to yield diacids. Naphthenic and aromatic rings are oxidized to dialcohols on adjacent carbon atoms (Figure 12-1). The order in which the hydrocarbons are oxidized depends on a variety of factors, but in general, small molecules up to C_{20} are consumed before large ones. Within the same molecular-weight range, the order is usually *n*-paraffins first, followed by branched isoparaffins, naphthenes, aromatics, and polycyclic aromatics. Single-ring naphthenes and aromatics are attacked before isoprenoids, steranes, and triterpanes. Preferential consumption of the low-molecular-weight components causes the high density (low API gravity) of the unconsumed residue.

Jobson et al. (1972) treated crude oil samples with pure and mixed bacterial cultures and observed considerable degradation during twenty-one days of incubation. A North Cantal (Saskatchewan) oil changed in specific gravity from 0.827 to 1.046 (from 40°API to 5°API). Thirty percent of the paraffin–naphthene fraction was destroyed, along with a small percentage of the aromatic hydrocarbons. Figure 12-1 displays the change in the gas chromatograms of the Saskatchewan oil over the twenty-one-day period. Note that at four days, it has lost a substantial percentage of the *n*-paraffins, with chain lengths shorter than C_{25}. At five days, even these longer chains are being attacked, and at fourteen days, most of the *n*-paraffins have disappeared.

Similar experiments and field observations indicate that after the oxidation of *n*-alkanes, the single-branched alkanes are attacked, followed by the double-branched hydrocarbons. Condensed-ring naphthenes, especially those with six rings, are not readily attacked. Thiophenes and other sulfur compounds are concentrated in the oil because the microbes do not alter them until late stages of biodegradation.

Connan (1984) wrote a good review of biodegradation of crude oils in reservoirs, in which he listed 88°C (190°F) as the maximum temperature of known biodegraded reservoir oil. He found that the effects of bacterial degradation decreased as the temperatures increased. In the Aquitaine Basin of southwest France, the most severe degradation was between 20 and 60°C (68 and 140°F), with only slight alteration occurring up to 77°C (171°F) and none above 80°C (176°F).

Microbial degradation occurs in the entire petroleum range from gas to residuum. James and Burns (1984) observed the preferential destruction of propane, followed by ethane, butanes, and pentanes in gas–condensate reservoirs. This process can form a dry gas containing mostly methane. The dry gases associated with biodegraded heavy oils and the asphalt deposits in many reservoirs may have this origin. Residual propane in biodegraded gases has anomalously heavy $\delta^{13}C$ values because the aerobes prefer to consume propane containing ^{12}C. Gas chromatograms of the condensates associated with degraded gas display varying degrees of degradation. Figure 12-2 (A) (from James

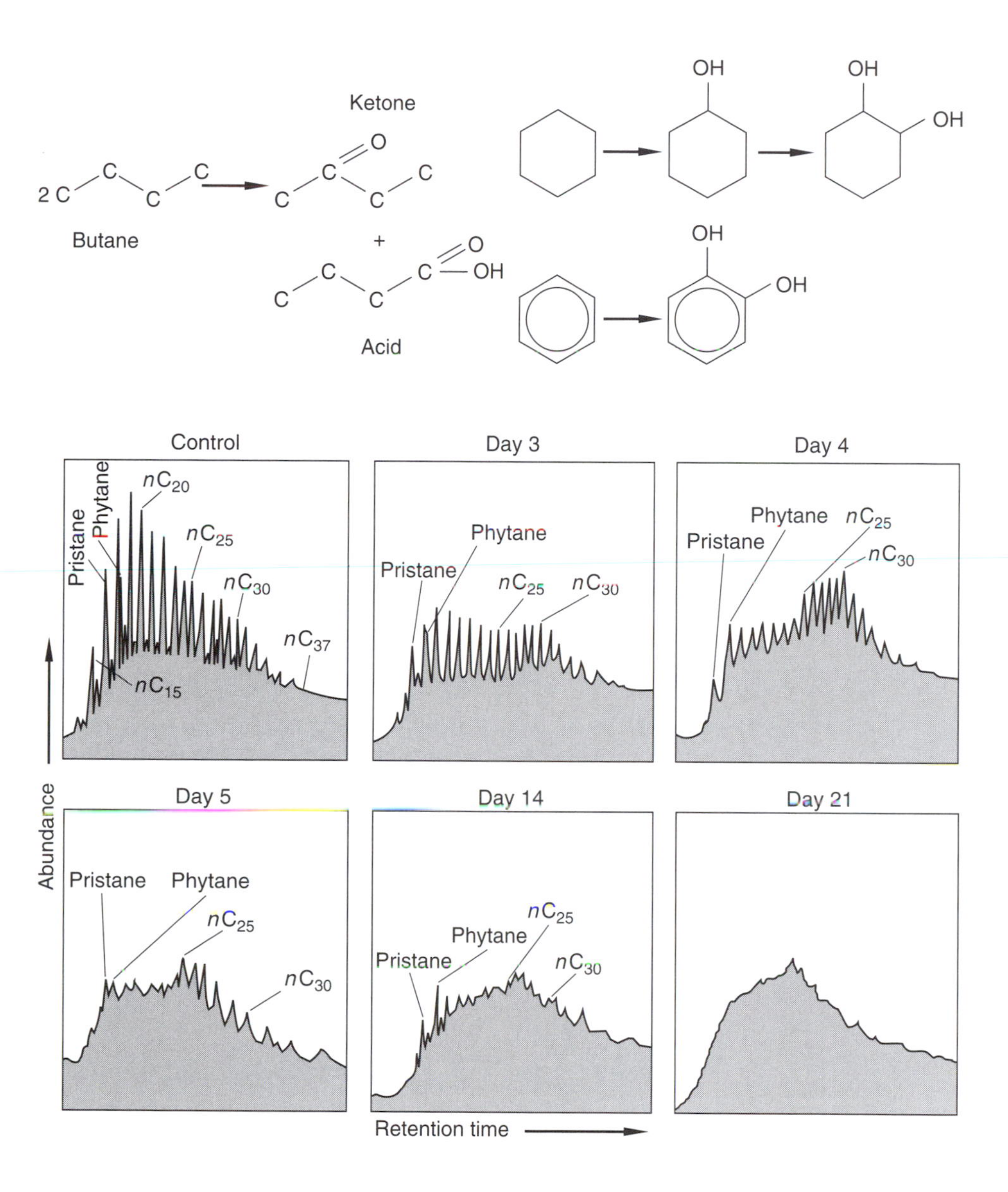

Figure 12-1

Microbiological oxidation of hydrocarbons showing the conversion of *n*-butane to a ketone and acid and the conversion of cyclohexane and benzene to alcohols. Gas chromatograms show the disappearance of *n*-paraffin peaks, first in the C_{16}–C_{25} range and later in the entire range during the incubation of Saskatchewan crude oil with a mixed microbe population at 30°C. The original oil is shown along with changes over a twenty-one-day period. [Jobson et al. 1972]

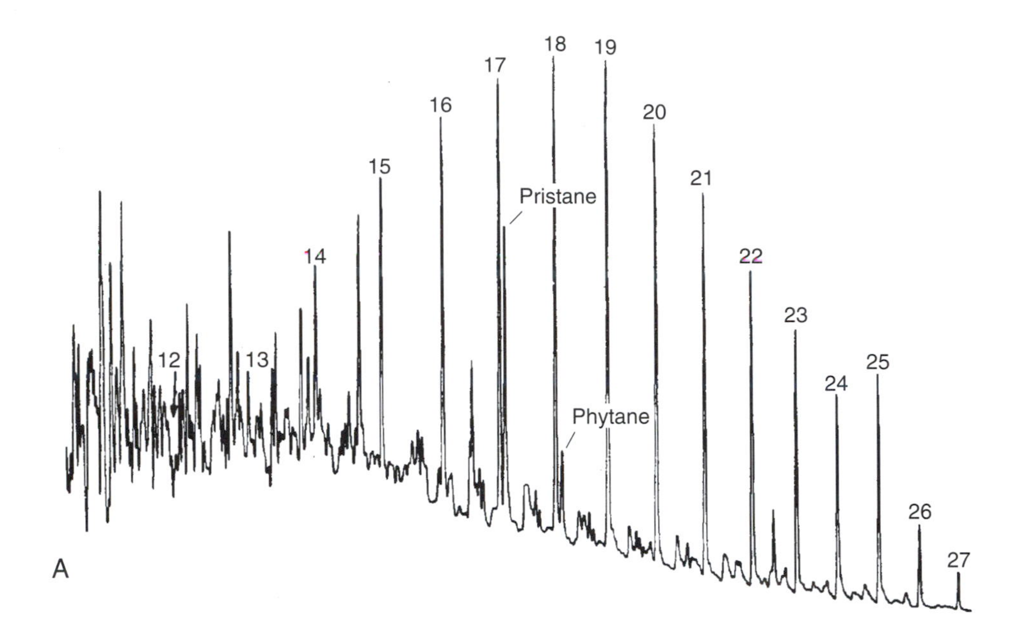
12
13
14
15
16
17
18
19
20
21
22
23
24
25
26
27
Pristane
Phytane
A

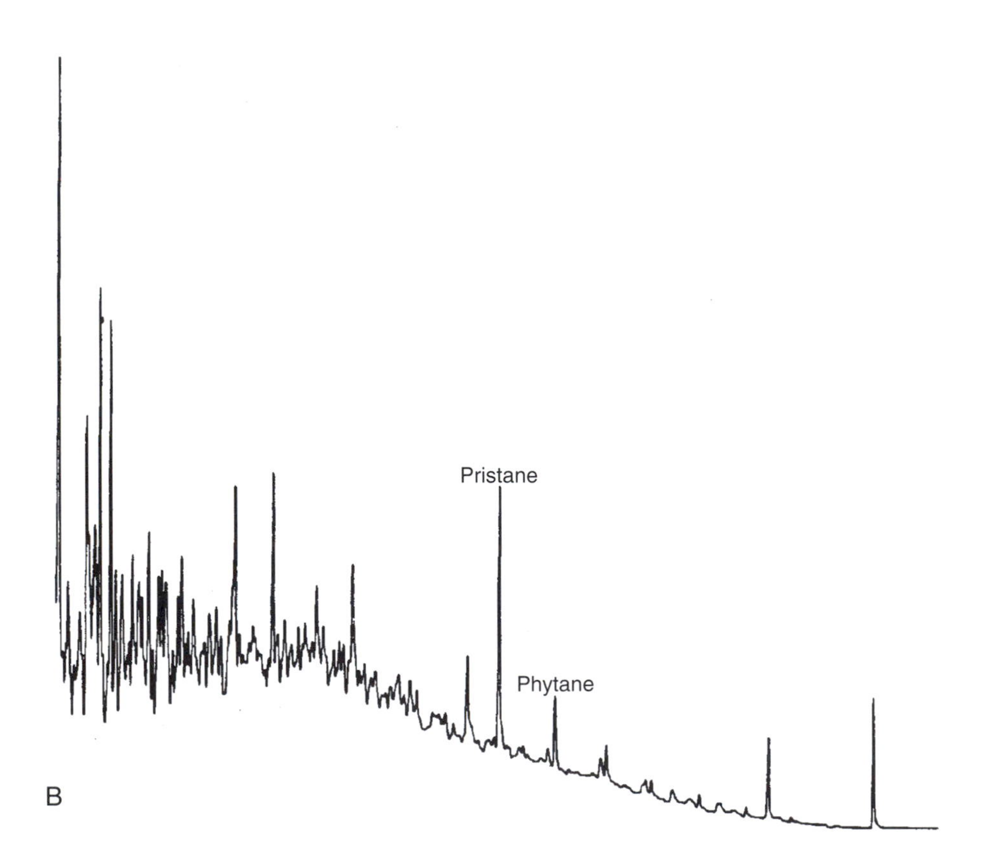
Pristane
Phytane
B

Figure 12-2 (left)

Whole-oil capillary gas chromatograms of the C_{12}–C_{27} boiling range. Normal paraffins are identified by their carbon numbers. (A) Partial loss of *n*-paraffins. (B) Nearly complete loss of *n*-paraffins. [James and Burns 1984]

and Burns 1984) indicates nearly complete removal of the *n*-paraffins through C_{12} in a condensate at 3,019 m (9,902 ft) from the Exmouth Plateau, offshore Australia. There is decreasing loss of *n*-paraffins through about C_{18}, which classifies this as minor biodegradation. Figure 12-2 (B) represents an oil from 2,651 m (8,695 ft) in the same area. It has lost most of its *n*-paraffins, leaving primarily isoprenoids such as pristane and phytane (light degradation). The $\delta^{13}C$ of the propane associated with the oils in Figure 12-2 was –20.98‰ for (A) and –3.67‰ for (B). This compares with –28.95‰ for the ethane in both wells. The $\delta^{13}C$ of the propane would normally be about –27‰, but biodegradation has left behind increasingly heavy propane.

Biodegraded oils are generally associated with meteoric waters low in dissolved solids, including sodium sulfate or bicarbonate. For example, the northeast edge of the Williston Basin in Saskatchewan contains Mississippian oil fields that were not biodegraded until uplift and erosion introduced meteoric sodium sulfate waters across the northern edge of the basin (Bailey et al. 1973). These waters caused various levels of biodegradation, depending on the extent of water invasion. In the area of saline brines to the southeast, the oils are about 36°API gravity and 1% sulfur. Moving north the waters become increasingly fresh, and the oil changes from 35 to 31, 27, 20, and less than 15°API gravity. The sulfur increases proportionately from 1% to more than 3% because of the loss of nonsulfur compounds rather than the introduction of additional sulfur. Table 12-1 compares oils in the Stoughton and High Prairie fields where the difference in the salinity of the pore waters is 100,000 ppm. These two fields are less than 25 mi (40 km) apart. Yet there is a large decrease in the API

TABLE 12-1 Oil and Water Analyses in Two Nearby Mississippian Oil Fields of the Williston Basin, Saskatchewan

Field	*Stoughton*	*High Prairie*
Water salinity, ppm	135,000	35,000
API gravity	29	15
Wt % sulfur	2.1	3.0
GOR	250	50
Percent gasoline	11	0.8

gravity, percentage of gasoline, and gas/oil ratio, along with a corresponding increase in sulfur content due to meteoric water invasion at High Prairie.

Figure 12-3 compares the gas chromatograms of the saturated hydrocarbons of the Stoughton and High Prairie oils. The *n*-paraffins in the High Prairie oil have been biodegraded, leaving only the isoprenoids, pristane and phytane. The ratio of pristane to n-C_{17} and phytane to n-C_{18} increases steadily from the unaltered crudes in the southeast to the altered crudes in the northwest. The proportion of saturated hydrocarbons in the whole oils decreases from 47 to 19%; the nitrogen, sulfur, and oxygen compounds in the crude increase 37%; and the asphaltenes increase 100%. One- and two-ring hydrocarbons were attacked more readily than the larger polycyclic hydrocarbons (Bailey et al. 1973).

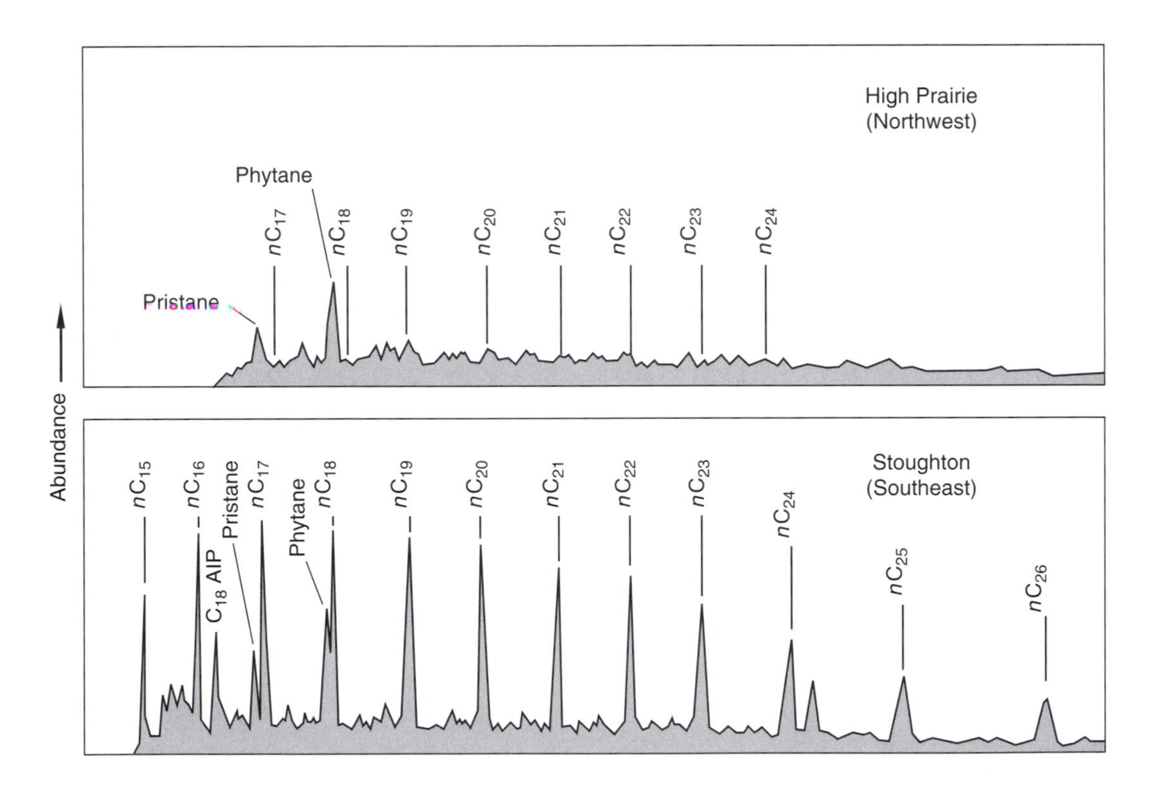

Figure 12-3

Gas chromatograms of a saturated fraction of Mission Canyon oils of the Williston Basin, comparing biodegraded High Prairie oil with unaltered Stoughton oil. The C_{18} AIP is an 18-carbon anteisoprenoid alkane. [Bailey et al. 1973]

The decreases and increases in various gross properties of degraded oils are listed in Table 12-2. As previously mentioned, propane is attacked first in the wet-gas range. The oils are made heavier by the loss of gasoline and the increase in aromatics and asphaltics, which have higher densities than paraffins. The increase in asphaltics, sulfur, nitrogen, and heavy metals is by concentration rather than addition. Although water washing causes some of these changes, it is not as destructive a force as microbial degradation. Connan (1984) pointed out that toluene and xylenes should decrease with respect to *n*-alkanes because of their much greater solubility in water. He found that they were concentrated relative to paraffins, however, because the biodegradation of the C_7 and C_8 *n*-alkanes is more efficient than the water washing of toluene and xylenes.

Williams et al. (1986) made a detailed study of the effects of biodegradation on aromatics, biomarkers, and paraffins. They discovered that biodegradation, particularly for aromatics, follows different routes in different oil fields. Their paper, incidentally, contains several references to case studies of biodegradation. All these studies have gradually created a level or ranking of the severity of biodegradation, as summarized in Table 12-3. It shows changes in hydrocarbon and biomarker molecules with degradation.

The compilation of degradation ranks is from the publications by Alexander et al. (1983), Volkman et al. (1984), Connan (1984), and Peters and Moldowan (1993). Biodegradation levels from 1 to 10 are described as minor to extreme. Peters and Moldowan (1993) emphasized that levels or rankings in tables such as Table 12-3 are quasi-sequential in that a higher-ranked compound class can be attacked before a lower-ranked class is completely destroyed. Also, variations

TABLE 12-2 Changes in Gross Properties of Biodegraded Petroleum

Decrease in

1. Wet gas (C_2–C_6) with propane selectively attacked
2. Gasoline–kerosene range (C_6–C_{15})
3. API gravity (increase in density)
4. All *n*-paraffins, wax content, and pour point
5. Gas/oil ratio GOR

Increase in

1. Asphaltic compounds (NSOs and asphaltenes)
2. Sulfur and nitrogen content
3. Viscosity
4. Vanadium and nickel content

TABLE 12-3 Changes in Molecular Composition of Oil with Increasing Biodegradation

Level or rank	*Compositional changes*	*Extent of biodegradation*
1.	*n*-Alkanes C_1 to ~C_{15} depleted.	Minor
2.	Over 90% C_1-to-C_{35} *n*-alkanes gone.	Light
3.	Isoalkanes, including isoprenoids, attacked; alkylcyclohexanes and alkylbenzenes removed.	Moderate
4.	Isoprenoid alkanes and methylnaphthalenes removed.	Moderate
5.	C_{14}–C_{16} bicyclic alkanes removed.	Extensive
6.	25-Norhopanes may be formed; steranes attacked, with the smaller molecules first.	Heavy
7.	Steranes gone; diasteranes unaffected.	Heavy
8.	Hopanes attacked.	Very heavy
9.	Hopanes gone; diasteranes attacked. Oleanane, tricyclic terpanes, and aromatic steroids survive.	Severe
10.	Diasteranes and tricyclic terpanes destroyed; aromatic steroids attacked; vanadyl porphryins survive.	Extreme

in the different microbes causing degradation can change the rankings of some of the compounds listed. The levels shown are only guidelines indicating the direction of increasing degradation rather than fixed intervals for individual molecules. It is clear, however, that very few molecules (such as vanadyl porphyrins) survive extreme cases of biodegradation.

Asphalt Seals and Tar Mats

Water washing, microbial alteration, oxidation, inspissation, natural deasphalting, and gravity segregation are mechanisms by which reservoir oils are thought to form asphalt seals at the outcrop and asphalt (tar) mats at the oil–water interface of pools in contact with meteoric water.

Asphalt seals are more important than most geologists realize. In the San Joaquin Basin of California—the giant Coalinga field, with more than 600 million barrels of recoverable oil—and Kern River field—with more than 700 million barrels—are both sealed in by thick asphalt covers. The giant Lagunillas field of Venezuela, with reserves measured in the billions of barrels, is sealed by an asphalt outcrop. The combination of microbial and other degradation processes has caused a large decrease in API gravity (Figure 12-4). Fresh water covers a belt 5 to 10 km (3 to 6 mi) wide from the outcrop to the shallowest oil

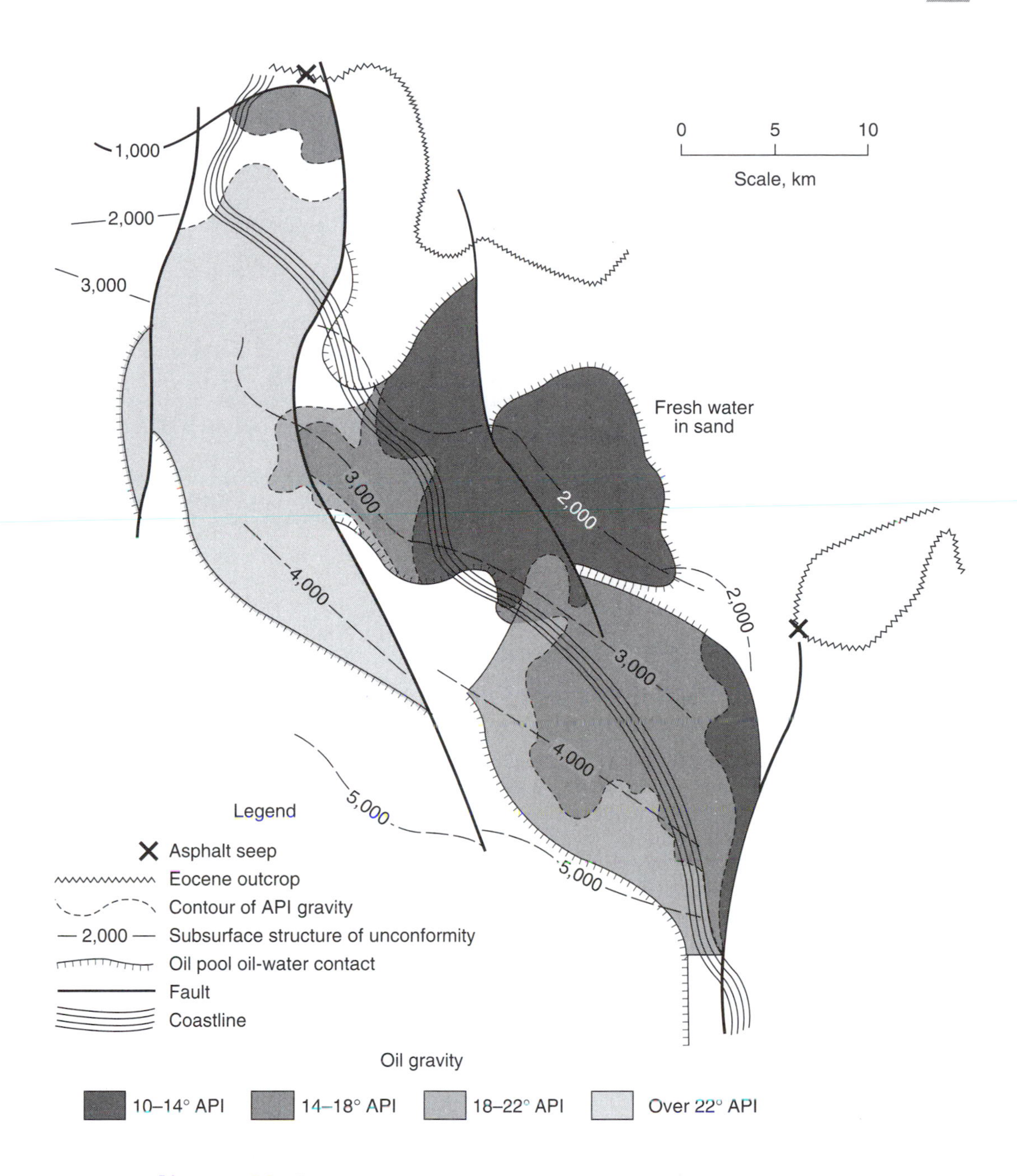

Figure 12-4

Asphalt seal at Lagunillas field, Venezuela. Fresh water overlies heavy oil, which increases in °API gravity downdip. The oil–water contact on southwest edge drops more than 300 m in 20 km because of post-Miocene tilting. [Dickey and Hunt 1972]

with 12°API gravity. The oil becomes lighter downdip, being about 20°API at 1,500 m (4,921 ft) below the outcrop. In the deepest sections, gravities rise to 36°API. In both Eocene and Oligo-Miocene formations in the Bolivar coastal fields, heavy- and medium-gravity oils alternate in successive sandstones. Heavy oils are found near bottom or edge waters and near overlying waters whose freshness indicates meteoric origin. The formation waters in parts of the Bolivar coastal fields are so fresh that those from depleted oil reservoirs are used for drinking.

In the northern part of the South Sumatra Basin, the Palembang Formation contains an asphalt seal at the outcrop. Downdip at the subsurface depth of 100 m (328 ft) is a heavy asphaltic oil with no gasoline and 70% residuum. At a 250 m (820 ft) depth, the oils contain 17% gasoline and 49% residuum, and at 500 m (1,640 ft) depth, 63% gasoline and 14% residuum. The deepest oils run from 40 to 50°API gravity, compared with 8 to 10°API near the surface. Clearly, this is a case of water washing, biodegradation, and loss of gas, and light ends near the outcrop (Dufour 1957).

The Seria field of Borneo, with more than a billion barrels of recoverable oil, contains an asphaltic, biodegraded, nonwaxy crude at about 300 m (984 ft); an intermediate, nonwaxy crude at about 600 m (1,969 ft); and a light, waxy, unaltered oil from 2,000 to 3,000 m (6,562 to 9,843 ft). The oil gravities are 19, 26, and 37°API, respectively. In the Balikpapan trend of the East Borneo oil fields, heavy oil without gasoline is found at a shallow depth, with a lighter paraffinic oil rich in gasoline in the deeper horizons. In the La Brea–Parinas field of Peru, a well may penetrate a low-pour-point, nonwaxy crude in one sandstone followed by a high-pour-point, waxy crude in another, followed by another nonwaxy crude. In such cases, in which each sandstone unit acts as an individual reservoir, biodegradation can influence all the oils in which there has been meteoric water invasion and leave the other oils untouched. Wherever high-pour-point and low-pour-point crude oils are in juxtaposition, biodegradation is the probable explanation.

Formation waters flowing past an oil accumulation remove hydrocarbons by solution up through C_{15} and probably higher, considering the geologic time periods involved. When oxidation by a sulfate ion or dissolved oxygen is added to this, an asphalt mat is created at the oil–water interface. Amosov and Kozina (1966) observed that the Tertiary oils in some Sakhalin reservoirs of Russia become heavier, approaching the oil–water contact. They found a direct correlation between the specific gravity of the oil and the biocarbonate content of the formation water. The oils became lighter when going from the bottom to the crest of the pool, but only about one-fourth of this distribution was attributed to gravity segregation. Most of the difference was caused by the degrading effect of the meteoric water at the oil–water contact. In the Permian oil fields of the Urals, Yarullin (1961) noted asphalt mats of 30 to 80 m (98 to 262 ft) in thickness at the oil–water contact of many pools. They attributed these to oxidation of the oils by the formation waters that were high in sulfate ion. Tar mats at the oil–water interface also may be caused by natural deasphalting, which is discussed under phase changes. The Hawkins field of east Texas is a good example.

Thermal Alteration

One characteristic of reservoir oil that is observed worldwide is the decrease in specific gravity (increase in °API) with maturation that usually, but not always, follows depth. The combination of lighter oils generated from kerogen with depth and of reservoir oils maturing to lighter oils with depth causes an inevitable progression toward higher °API gravity (lower-density) oils. Figure 12-5

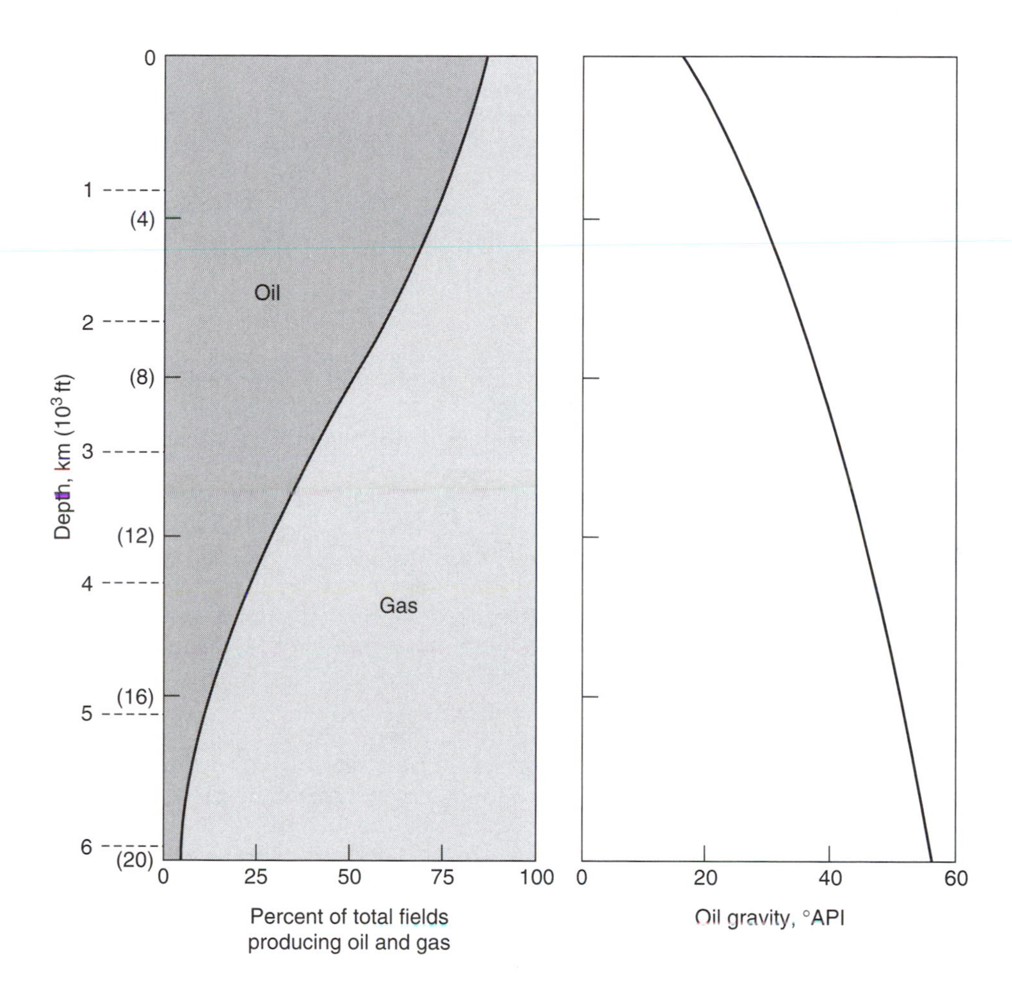

Figure 12-5

Decrease in the percentage of commercial oil fields versus gas fields with depth and the corresponding increase in °API gravity.

shows the change in the percentage of oil fields with depth and the gravities of the oil from data in the International Petroleum Encyclopedias. There is considerable variation in any one producing area, so this cannot be used for predicting gravity changes with depth except very roughly. For example, in the 2,000 to 3,000 ft (610 to 914 m) depth range worldwide, the °API gravities range between 12 and 43. In the 11,000 to 12,000 ft (3,353 to 3,658 m) depth range, they extend from 20 to 53. Some extreme anomalies include the Eocene Barracouta field of Australia with a 63°API gravity condensate at 4,700 ft (1,433 m), the Boscan field in Venezuela with 10°API gravity crude at 7,500 ft (2,286 m), the Ragusa field of Italy with 20°API gravity oil at 12,460 ft (3,798 m), and the Kaplan field in Louisiana with 48°API gravity oil at 21,200 ft (6,463 m) and a reservoir temperature of 378°F (192°C).

There are many explanations for such anomalies. For example, biodegradation and washing by meteoric waters can make an oil heavy at depth. Eroded reservoirs contain a heavy asphalt residue from weathering. They can be reburied, and the asphalt will not have enough hydrogen to change. Oils may be lightened by gas deasphalting, and condensates may escape through vertical permeability to shallow reservoirs, causing wide variations in depth–gravity patterns. Despite these anomalies, all crude oils move relentlessly toward condensate, gas, and pyrobitumen as they are exposed to higher temperatures at depth.

The continual maturation of oil in both the source and reservoir rock results in the number of oil and oil-plus-gas fields gradually diminishing with depth. In the 6,000 to 10,000 ft (1,820 to 3,048 m) depth range of most basins, there is a shift from an oil to a gas majority in the reservoirs. Beyond the 12,000 to 14,000 ft (3,658 to 4,267 m) range, less than a fourth of the reservoirs contain oil, and below 20,000 ft (6,096 m) it is only a small percentage.

Petroleum accumulations, like everything else, continuously move toward a state of lower free energy. The most stable hydrocarbon molecules are those with the lowest free energy. Different hydrocarbons have large differences in free energy. Figure 12-6 illustrates the thermal stability of aromatic, naphthene, and paraffin hydrocarbons in terms of free energy of formation in kilocalories per carbon atom. The zero line corresponds to the free energy of the elements carbon and hydrogen. This figure is valid only at the unit activity of the compounds and at a pressure of 1 atmosphere. At higher pressures, the relative differences are essentially the same, but the absolute values change somewhat. The figure shows that at the low temperatures of shallow drilling, the aromatic hydrocarbons are the least stable and the low-molecular-weight paraffin hydrocarbons are the most stable. At the high temperatures encountered in deep drilling, only the hydrocarbon gases are stable, and at very high rock temperatures, only methane is stable. This is usually the only hydrocarbon found in deep, high-temperature reservoirs.

The following conclusions result from Figure 12-6: First, paraffins are the most stable hydrocarbons at the lower temperatures comparable to sedimentary rocks, whereas aromatics are the most stable at very high (>1,200°C) temperatures. Second, the stability of paraffins increases (free energy decreases) as the number of carbon atoms in the molecule drops. Methane is

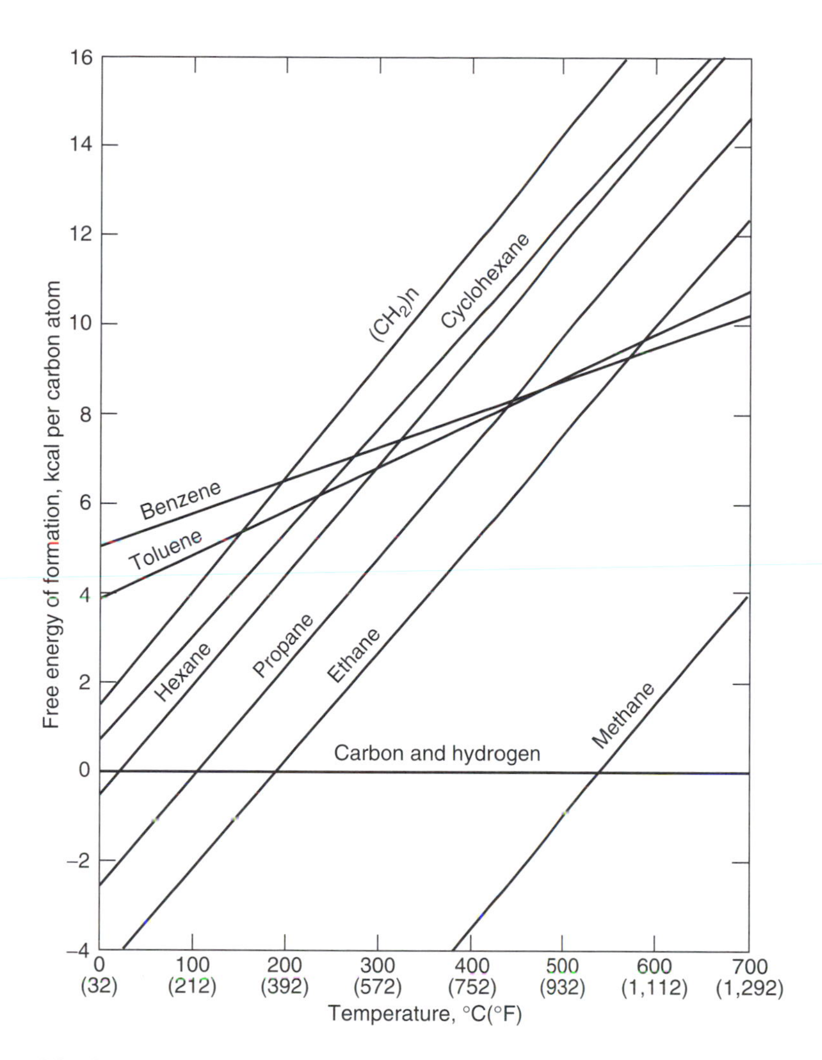

Figure 12-6

Thermal stability of hydrocarbons with increasing temperature. [Hunt 1975; data from National Bureau of Standards, C-461, 1947]

stable at temperatures up to 550°C (1,022°F), considerably higher than that of any sedimentary rocks. Third, the naphthenes are intermediate in stability between the aromatics and paraffins.

The thermal alteration of reservoired oil to more and smaller paraffin molecules requires a source of hydrogen. There is very little free hydrogen in most reservoirs, but fortunately, there are thermodynamically favored reactions such as the condensation of aromatics that can make hydrogen available. The free energy of formation of benzene at 25°C (77°F) is 4,930 cal per carbon atom. In going from benzene with one ring to naphthalene with two, phenanthrene with three, and pyrene with four, there is a steady decrease in free energy to 4,015 cal. This decrease continues with further ring condensation all the way to graphite. Consequently, the conversion of polycondensed aromatic hydrocarbons to an asphaltite or pyrobitumen in the reservoir is favored.

This disproportionation of hydrogen scheme is schematically illustrated in Figure 12-7 by Connan et al. (1975). The phenanthrene shown at the top of this figure is a katacondensed polycyclic aromatic hydrocarbon. Its rings are fused in a linear series. Katacondensation releases four hydrogen atoms with each new added ring. The simple linking of rings releases only two hydrogen atoms, as in the example shown in Figure 12-7 of two phenanthrene molecules being linked instead of condensed. The hydrogen can be used to form the C_5 and C_{12} *n*-paraffins, as indicated. The further rupture of the long hydrocarbon chains to small C_3, C_5, and C_6 molecules utilizes hydrogen from pericondensation of the aromatics. Pericondensed polycyclic aromatics are formed by the fusing of rings on all sides, which can release more than four hydrogen atoms. Eventually the end products in Figure 12-7 are methane, ethane, and propane, plus a large pericondensed, polycyclic aromatic molecule that is beginning to approach the structure of graphite. The combination of kata- and pericondensation plus the aromatization of naphthenes and the linking of rings provides enough hydrogen to form the hydrocarbon gases.

A process like the one just described has been occurring in reservoirs throughout geological time since the first oil began accumulating, as evidenced by the precipitation in reservoirs of black asphaltenes and pyrobitumens, along with gases. These bitumens have been identified in many reservoir cores. But they are not always present, because the gases may migrate from their area of origin in the reservoir. Also, many gases form directly from kerogen in source rocks.

The following equations give a simplified picture of the overall reactions leading to gas and pyrobitumen:

In source rocks:

$$C_5H_5 \text{ (kerogen)} \rightarrow CH_4 \text{ (gas)} + C_4H \text{ (pyrobitumen)}$$

In source and reservoir rocks:

$$C_5H_9 \text{ (oil)} \rightarrow 2CH_4 \text{ (gas)} + C_3H \text{ (pyrobitumen)}$$

About one-fifth of the carbon in gas-generating kerogen ends up as methane, with the rest remaining in the rock as pyrobitumen if the reaction goes to completion. About two-fifths of the carbon in oil forms methane. Theoretically, these reactions should provide a lot of methane in deep reservoirs, but much of it is lost by seepage, solubility in pore waters, and chemical destruction, as explained later.

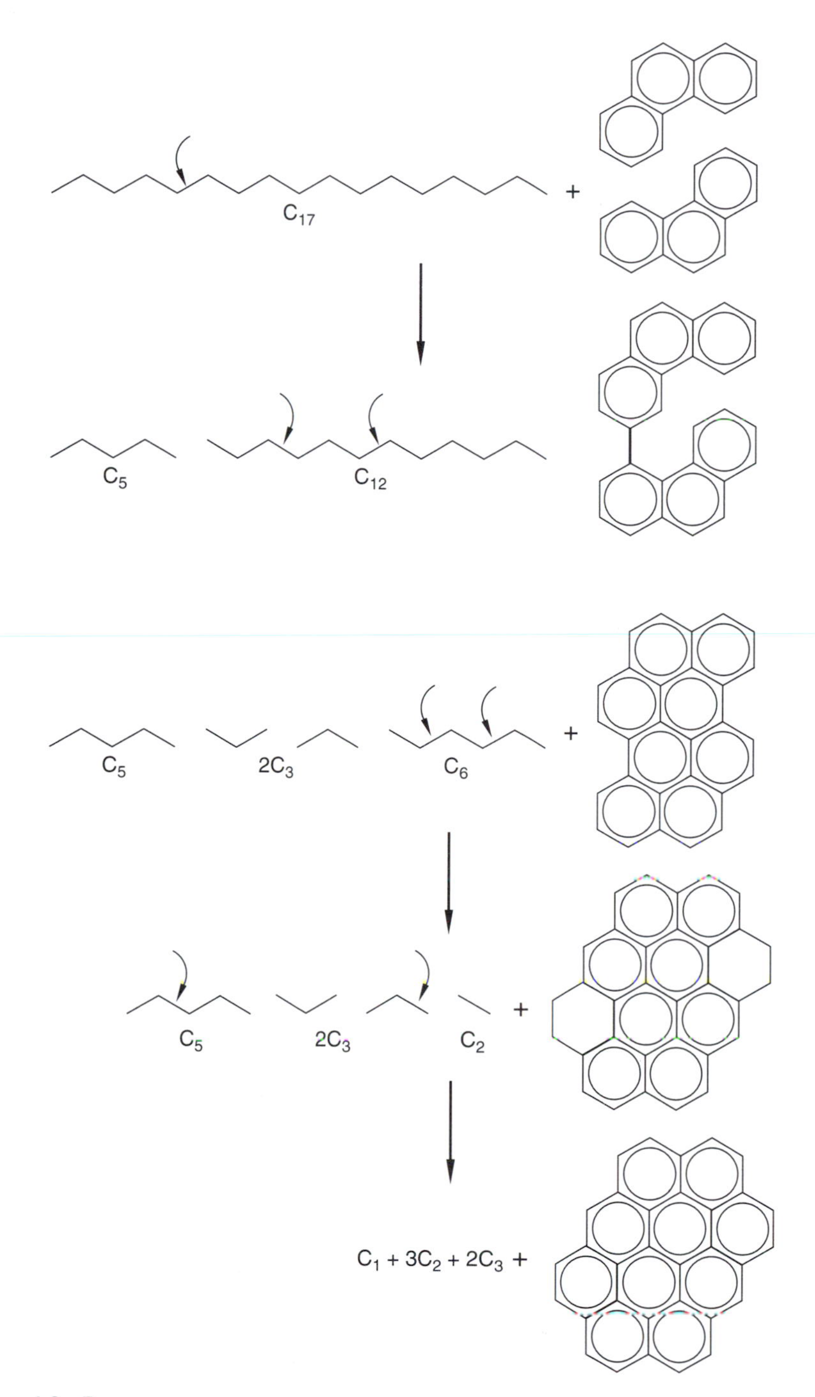

Figure 12-7

The formation of hydrocarbon gases and a pyrobitumen or graphite residue from hydrogen disproportionation reactions of oil molecules during the thermal alteration of petroleum in a reservoir. [Connan et al. 1975]

Connan et al. (1975) observed that the proportion of saturated hydrocarbons (paraffins plus naphthenes) in pooled oils of the Upper Jurassic–Lower Cretaceous fields in the Aquitaine Basin of France rises with increasing depth, being about 8% at 1,400 m (4,593 ft) and 40% at 3,400 m (11,155 ft). There also is a shift from large molecules to small molecules, with the *n*-alkanes peaking at C_{19} around 2,500 m (8,202 ft) and at C_{17} around 3,400 m (11,155 ft). The hydrogen for this increasing saturation probably came from the asphaltene fraction of the oil. Connan simulated this change in the laboratory by heating at 300°C (572°F) for several months a shallow, immature crude oil from the Aquitaine Basin. The aromatic/saturate hydrocarbon ratio in the crude oil changed from about 3.6 to 1 over a twelve-month period, and the *n*-alkane peaking decreased from C_{18} to C_{16}. The solid bitumen content of the crude oil increased from near zero to about 40%.

Change in Oil Gravity with Maturity

The maturation of oil in a sequence of stacked reservoirs usually shows up as an increase in °API gravity (decrease in specific gravity) with reservoir depth. An example is illustrated in Figure 12-8, in which the data for four oil fields in the Volgograd region of Russia are plotted. Gabrielyan (1962) also noted that the paraffin content increased with reservoir depth and the sulfur content was highest in the shallowest samples.

Makarenko and Sergiyenko (1970) observed the same kind of change in the eastern Ciscaucasia, where the specific gravity decreases from 0.89 to 0.81 (equivalent to the change from 26 to 42°API) in the subsurface temperature range from about 30 to 130°C (86 to 266°F). In this same interval, the paraffin hydrocarbons in the oil increase relatively from 35 to 56%; the naphthenes decrease relatively from 47 to 31%; and the aromatics decrease from 18 to 13%. The naphthene/paraffin ratio therefore changes from 1.34 to 0.55.

Claypool and Mancini (1989) analyzed fifty-five oil and condensate samples from the Smackover and Norphlet formations of southwestern Alabama. Figure 12-9 shows the trend of increasing API gravity with depth. This is accompanied by a decrease in sulfur and increases in C_1–C_6 hydrocarbons and the ratio of saturated to aromatic C_{15+} hydrocarbons with depth. Liquid hydrocarbons are converted to gas at depths greater than 14,000 ft (4,270 m). Only gas is found at depths greater than about 20,000 ft (6,100 m). Thermochemical sulfate reduction (TSR) causes increased H_2S and CO_2 in the gas and the depletion of saturated hydrocarbons in some condensate liquids. The gravity of the Smackover oils increases about 1°API for 377 ft (115 m) of burial. This compares with 1°API for 215 ft (66 m) of burial in the previously shown Volgograd oils, for 333 ft (102 m) for Tensleep oils in Wyoming, and for 486 ft (148 m) for Ellenberger oils in the Delaware–Val Verde Basin of Texas. Most reservoir oils increase 1°API (decrease 0.005 in specific gravity) for every 200 to 400 ft (61 m to 122 m) of increasing burial, the differences being mainly

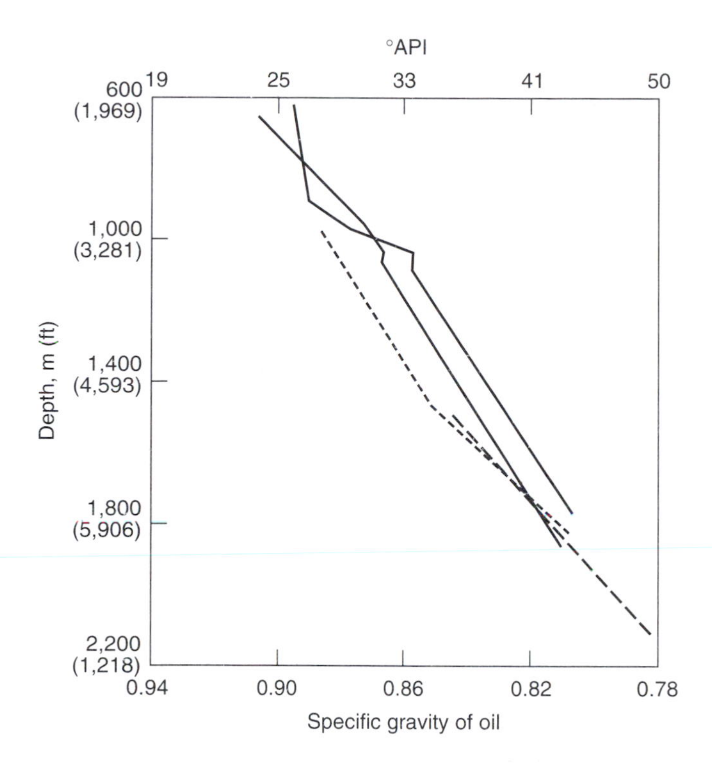

Figure 12-8

The decrease in specific gravity of oil with depth from pools of the Bakhmet'ev, Zhirnov, Archedin, and Klenov fields of the Volgograd region, Russia. [Gabrielyan 1962]

attributable to differences in geothermal gradients (and therefore reservoir temperatures) and slightly to differences in geologic age (duration of heating).

Two of the Tensleep oil reservoirs in the Wind River Basin of Wyoming, Riverton Dome, and Pilot Butte, are only 32 mi (52 km) apart but have a depth difference of 5,400 ft (1,646 m). Table 12-4 shows the compositional differences between these two nearby oil reservoirs. All fractions lighter than gas oil with fewer than twenty carbon atoms per molecule have gained relatively in volume, whereas the heavier oil fractions have lost volume. The reservoir maturation of the Tensleep oils appears to have cracked into smaller molecules those hydrocarbons containing more than twenty carbon atoms, which become part of the gasoline, kerosine, and light gas–oil range. The largest contribution of cracked products comes from the residuum.

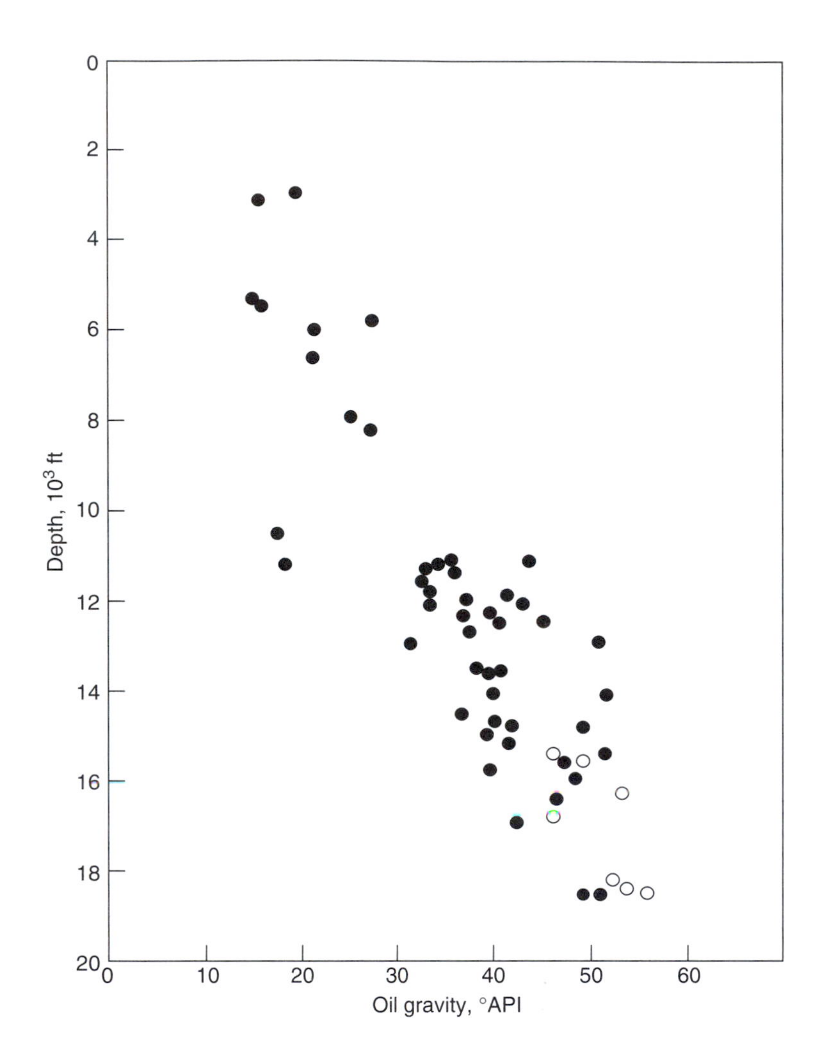

Figure 12-9

API gravity versus depth for southwestern Alabama oil (solid circles) and gas condensate (open circles). [Data from Claypool and Mancini 1989]

Tensleep oils at depths of less than 3,000 ft (914 m) have gravities below 25°API. Many of these are in contact with meteoric waters, which cause degradation through oxidation, microbiological decomposition, and aqueous solution of the lighter components. This has not affected the deeper Tensleep oils.

The importance of reservoir maturation to exploration is apparent when one realizes that many deep horizons were not tested in the past because they

TABLE 12-4 Compositional Differences Between Deep and Shallow Tensleep Oils in Wyoming

Constituent	*Volume percent difference between Riverton Dome (11,700 ft) and Pilot Butte (6,300 ft)*
Light gasoline, C_4–C_7	+16.3
Heavy gasoline, C_8–C_{10}	+20.0
Kerosine, C_{11}–C_{12}	+3.5
Gas oil, C_{13}–C_{20}	+2.6
Nonviscous lube oil, C_{21}–C_{30}	–3.1
Medium lube oil, C_{31}–C_{35}	–5.0
Viscous lube oil, C_{36}–C_{40}	–5.3
Residuum, $>C_{40}$	–29.0

Source: Hunt 1979, p. 370.

were believed to contain uneconomic heavy oil comparable to that in shallow reservoirs. The Tensleep is an example of a formation where deep tests were not planned because the Tensleep oil was generally known to be heavy at the time when the first pools were discovered at depths less than 5,000 ft (1,525 m). Today, many Tensleep oils are produced with gravities higher than 40°API from fields deeper than 10,000 ft (3,048 m).

Phase Changes in Oil and Associated Gas

When oil flows out of a reservoir sandstone into the well bore, there is a drop in pressure, but the temperature stays constant. The *bubble point* is the pressure at which gas starts to come out of solution, or exsolves. When the oil–gas mixture arrives at the surface, it is channeled into a separator that directs the oil into a stock tank. The separated gas may contain considerable amounts of heavier hydrocarbon that normally are liquid. Consequently, there is an appreciable shrinkage of the crude oil. The decrease in volume of the oil from the reservoir to the stock tank is called the *formation volume factor* (FVF). Ordinary oils are characterized by GORs up to approximately 2,000 ft^3/bbl (360 m^3/m^3), oil gravities up to 45°API (0.8 g/cm^3), and FVFs of <2 bbl/bbl (<2 m^3/m^3).

Near-critical oils are those existing in a reservoir at a temperature near the critical point. This is the point beyond which a single phase of dense gas exists. Near-critical oils have FVFs greater than 2 and compositions characterized by 12.5 to 20 mol% C_{7+} and more than 35% C_1 through C_6. Reservoir fluids that contain more than 12.5 mol% C_{7+} are almost always in the liquid phase in the reservoir. Those with less than that are usually in the gas phase. There are exceptions. Oils have been observed with C_{7+} concentrations as low as 10 mol%. A few condensates that are in a gas phase in the reservoir have as high as 15.5 mol% C_{7+}, but they are rare (Moses 1986).

Figure 12-10 is an outline of the various phase changes that may occur in reservoirs. Vaporization–condensation phenomena include retrograde condensation and evaporative fractionation. The former exists as a single-phase dense-gas system. When the pressure is reduced, it converts to a two-phase system of

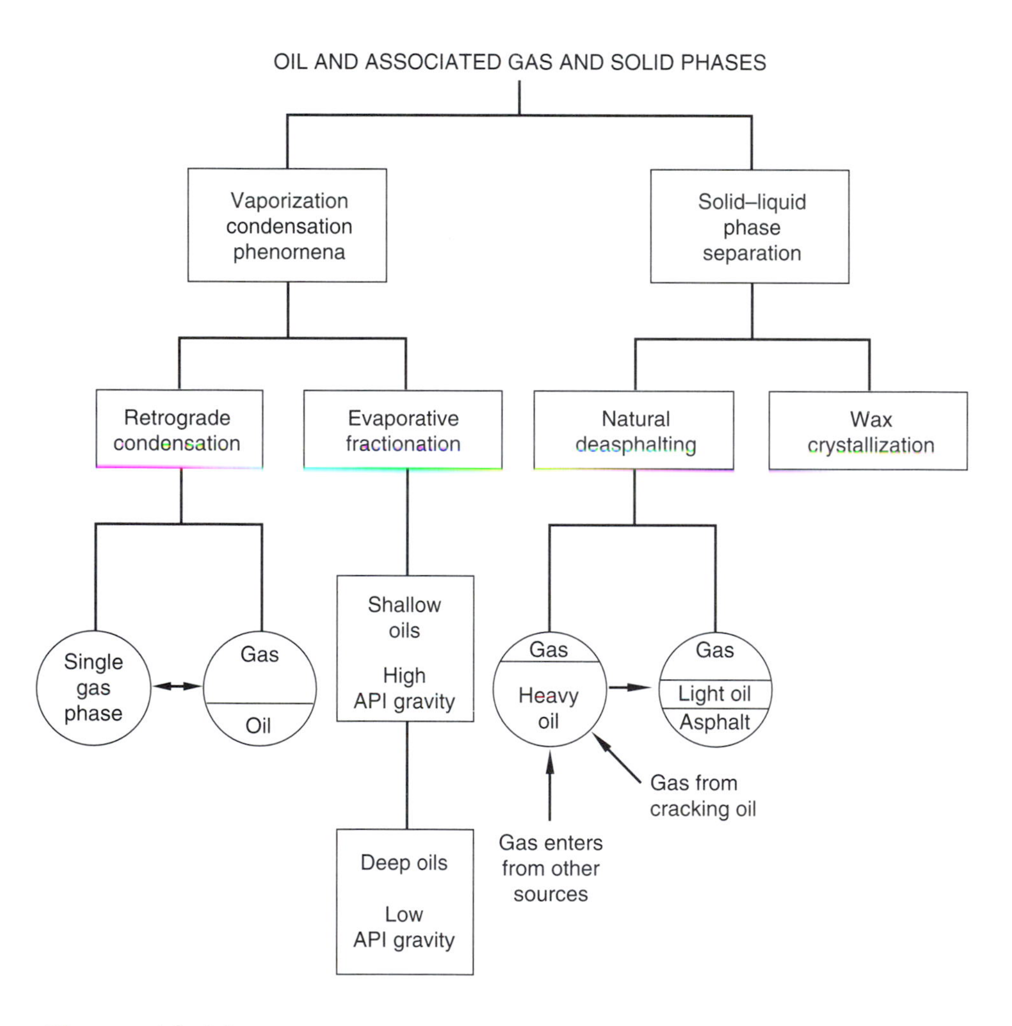

Figure 12-10

Flow diagram showing the various phase changes that may occur in reservoirs.

free gas over oil. Such reservoirs have GORs from 3,000 to 150,000 ft^3/bbl (540 to 27,000 m^3/m^3). The gravities of the separated liquids usually range from about 40 to 60°API (0.83 to 0.74 g/cm^3), although condensate gravities as low as 29°API (0.88 g/cm^3) have been reported (Moses 1986). Retrograde behavior is generally observed at pressures above 2,500 psi (17.2 MPa).

The evaporative fractionation described in Chapter 7 is a process that can form aromatic condensates (Thompson 1987). As an oil matures, it generates increasing amounts of gas. In addition, gas from deeper maturing oils and source rocks can pass through a carrier bed or a reservoir, picking up light hydrocarbons (gas stripping). As reservoir pressures build up toward lithostatic, as explained in Chapter 7, the trapped gas escapes, carrying with it some dissolved oil. The oil condenses out at successively shallower levels up the section as the pressures and temperatures decrease. This results in the nonbiodegraded, shallower oils having higher API gravities than deeper oils (Thompson 1987). This is in contrast to thermal condensates that usually show the general trend of increasing API gravities at greater depths.

In addition, the aromaticity of oils and condensates left behind increases as they lose the lighter hydrocarbons, because paraffins are lighter than aromatics and so volatilize more easily. Thompson's B value is the ratio of the aromatic toluene (sp. gr. = 0.866) to the paraffin *n*-heptane (sp. gr. = 0.684). The ratio is high in the deeper oils that have lost light ends and low in the shallower oils and condensates.

Talukdar et al. (1990) observed this process in the Columbus Basin offshore southeast Trinidad. Oil and gas occur there in thirty-two stratigraphic units over a depth range of about 2,440 m (8,000 ft). A heavier 28.2°API oil is at 3,305 m (10,842 ft), and a lighter oil with 40.5°API is at 1,893 m (6,209 ft), with oils in between showing increasing API going upward. This is the opposite of what occurs by thermal maturation (Figure 12-9), but it is exactly what happens in a refinery. Nature is acting like a distillation tower (Figure 3-6) in moving light hydrocarbons upward and leaving heavier ones behind. The ratio of toluene (heavy) to *n*-heptane (light) in the Columbus Basin oils and condensates is 2.55 at 10,000 ft (3,200 m), 1.84 at 5,000 ft (1,525 m), 1.40 at 4,000 ft (1,220 m), and 0.67 at 1,200 ft (366 m). The ratio is not uniform with depth, however, because of variabilities in restrictions of vertical fluid flow.

Dzou and Hughes (1993) made a detailed study of twenty-one condensates and light oils showing evidence of evaporative fractionation at the K field offshore Taiwan. All samples were generated from the same deltaic source rock that is in the maturity range of 0.4 to 0.8% R_o. This indicates that they were not formed by thermal cracking. Also, the composition of condensates showed that they were not derived from resinite-rich source rocks. The authors used bulk properties, GC patterns, and a fractionation index plus the B value to establish that the oils had undergone evaporative fractionation. Figure 12-11 gives their plot of the fractionation index and Thompson's B value (toluene/n-C_7) versus depth in a single well. The index measures the distribution of light (n-C_{10}) versus heavy (n-C_{16+}–n-C_{25}) hydrocarbons. Note that the hydrocarbons are getting lighter and decreasing in aromaticity going up section. This is a clear indication of evaporative fractionation.

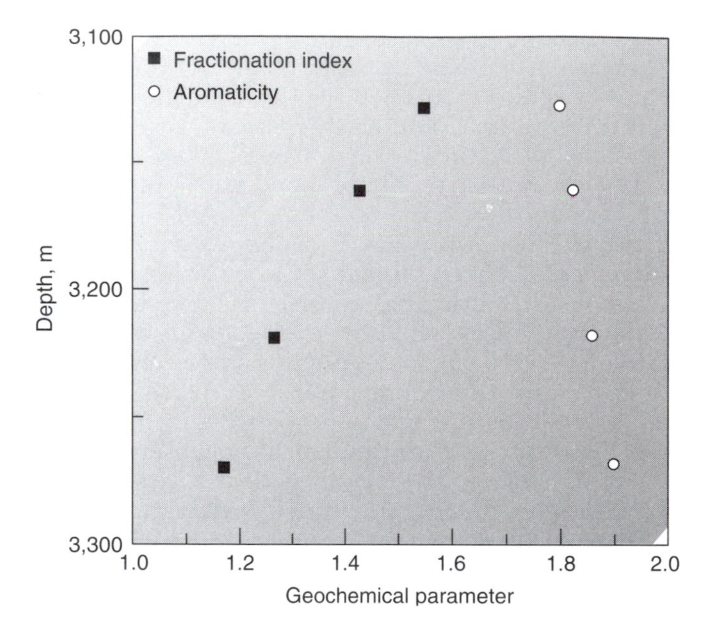

Figure 12-11

Depth versus fractionation index (FI = n-C_{10}/(n-C_{16} + n-C_{25}) and aromaticity (toluene/n-C_7) for K field, Taiwan, oils and condensates. [Dzou and Hughes 1993]

Evaporative fractionation has been reported in other fields such as Handil in Indonesia (Hunt 1990). It probably occurs to some extent in most deltas.

Natural Deasphalting

The wet gas and light oil formed during the maturation of reservoired oil causes a natural deasphalting, with the precipitation of asphaltenes in reservoir pore spaces. Propane deasphalting, which is a widely used refinery process for removing asphaltenes from residuum, was discussed in Chapter 3. Its counterpart in nature is reservoir deasphalting, which can occur either from the C_1–C_6 fraction being generated during maturation or from C_1–C_6 hydrocarbons entering the reservoir from other sources. These low-molecular-weight hydrocarbons dissolve in the reservoir oil, expanding it and precipitating out the asphaltenes. This causes a high API gravity oil to coexist with asphalt.

Natural deasphalting was considered by Dahl and Speers (1986) to be the major cause of the thick asphalt mat at the base of the Oseberg field in the Norwegian sector of the North Sea. The asphalt mat that sits below the present oil–water contact contains steranes, triterpanes, and triaromatic steroids identical to those in the overlying oil column. The asphalt also contains a distribution of C_{15+} *n*-paraffins similar to that of the oil. This indicates that the asphalt mat was not formed because of biodegradation. The main difference is that the asphalt has about 50 wt% asphaltenes, whereas the overlying oil has less than 5%. Asphalt mats are more common in sandstone reservoirs than in carbonates.

Based on their studies of reservoirs in the Western Canada Basin, Rogers et al. (1974) decided that reservoir bitumens have two sources. These are (1) the thermal alteration of oils to dry gas and solid bitumen and (2) the deasphalting of heavy oils after solution with large amounts of wet gas. They found that the thermal process changes the $\delta^{13}C$ from –30.5‰ in the oil to –28.5‰ in the bitumen, whereas deasphalting showed no significant change in $\delta^{13}C$ between the oil and bitumen. This is because thermal cracking forms methane with a more negative $\delta^{13}C$ value than the starting product, whereas deasphalting is a physical change that has no effect on carbon isotopes. Dahl and Speers (1986) found the oils and asphalt mat at Oseberg to have the same carbon isotope values, indicating that natural deasphalting probably formed the mat.

Wilhelms and Larter (1994) conducted a comprehensive study of asphalt mats at the Oseberg and Ula fields of the North Sea and a North American field. Data from GC, pyrolysis–GC, oil–asphalt and biomarker analyses clearly established that the asphalt mats were derived from the overlying oil columns in all three cases. The oil–asphalt mat was recognized in all three by a sharp increase in the asphaltene content, 20–60 wt% in the mat, compared with 1–5 wt% in the oil leg. The authors compared the relative volumes of the oil and asphalt mats and concluded that it would take six volumes of the current Oseberg oil leg and four volumes of the current Ula oil leg to create their existing mats. This suggests either a continuous process of migration and asphalt precipitation or that the volume of oil originally reservoired could have been much larger than today.

Concerning mechanisms of asphalt mat formation, Wilhelms and Larter (1994) eliminated gravity segregation, biodegradation, water washing, and adsorption on clays as feasible mechanisms in their examples. Early thermal alteration (Ula field), gas injection, and some migration-related asphaltene precipitation mechanisms were considered the most likely processes. Also, the influx of an asphaltic-rich oil may contribute to formation of the mats.

Wax Crystallization

The *cloud point* is the temperature at which solid hydrocarbons begin to separate from solution when an oil is chilled. If a high-wax oil in a reservoir migrates to a temperature below its cloud point, the wax will be partly crystallized. The oil solidifies at the pour point. The cloud point of the Altamont field oil in the Uinta Basin of Utah is 170°F (77°C). The pour point (Tuttle 1983) ranges up to 125°F (52°C). When the Duchesne field was discovered at 7,596 ft (2,315 m) in the Uinta Basin of Utah, the oil was found to have essentially the same composition as the solid-wax ozocerite found at the surface. It is 82% paraffin and naphthene hydrocarbons. The oil was liquid at a reservoir temperature of about 90°C (194°F), but it solidified to a wax at the surface and had to be trucked out. It had the highest pour point, 130°F (54°C), for an oil ever recorded by the U.S. Bureau of Mines. Ultimately, the well was hooked up with the nearby Roosevelt field, which contained enough aromatics in its oil to keep the Duchesne oil from solidifying.

A high-temperature gas chromatogram (HTGC) of ozocerite wax from the Duchesne area is shown in Figure 12-12. It has a high *n*-paraffin content in the C_{30}–C_{45} and C_{57}–C_{67} ranges. The H/C ratio of ozocerites ranges between about 1.7 and 1.99, revealing them to be almost a pure mixture of saturated hydrocar-

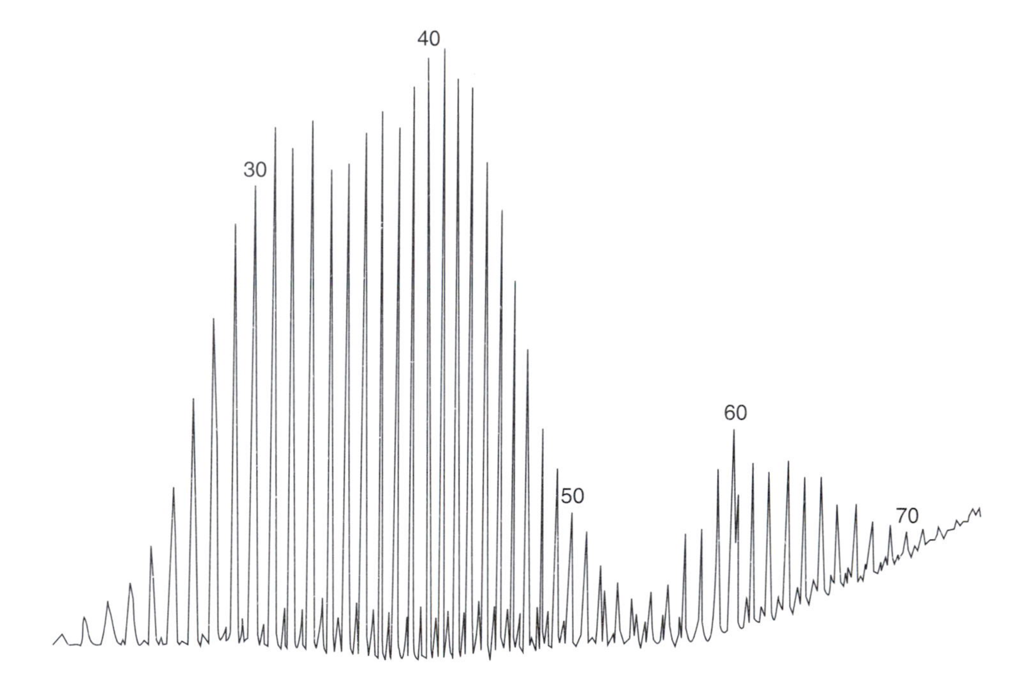

Figure 12-12

High-temperature gas chromatogram (HTGC) of ozocerite, Uinta Basin, Utah. Hydrocarbon chain lengths are noted above the corresponding peaks; *n*-hexacontane (*n*-C_{60}) was added as a sample spike. The *n*-paraffins begin to die out above ~C_{50} and are replaced by another series of normal alkanes extending past C_{70}. [Courtesy of R. M. K. Carlson]

bons. Most such waxes found at the surface are derived from the inspissation of high-wax oils. Wells producing high-wax oils must be treated with hot oil or reamed to prevent the wax from clogging the casing and tubing. The crudes are either dewaxed near the well sites or mixed with low-pour-point aromatic oils to prevent them from solidifying in the pipelines.

Oil to Gas and Pyrobitumen

The terms *bitumen, tar,* and *pyrobitumen* are defined according to their physical properties and not their source or maturity. In fact, tar is a misnomer because it is not a natural product according to the original definition. Bitumens, by definition, are fusible, are soluble in carbon disulfide, and have viscosities greater than 10,000 centipoise (mPa-seconds), whereas pyrobitumens are infusible, are only slightly soluble or insoluble in CS_2, and are solid at reservoir temperatures. Source and relative maturities of bitumens can be determined from biomarkers, as explained in Chapters 10 and 15.

Asphalt (tar) mats are composed of bitumens, whereas the pore walls and throats of reservoirs where the original oil has been converted to dry gas are coated with both bitumen and pyrobitumen. These have been reported in the Ellenberger Formation of the Delaware–Val Verde Basin (Holmquest 1965), the Western Canada Basin (Rogers et al. 1974), and many other areas (Hunt 1979, pp. 366–368). Hydrocarbons of the Smackover and Norphlet Formations of the Mississippi Salt Basin and the Tuscaloosa Formation of the Louisiana Gulf Coast are two good examples of oil destruction forming gas, condensate, and pyrobitumen. Geothermal gradients deeper than 10,000 ft (3,049 m) in these areas are approximately 1.6°F/100 ft (3.0°C/100 m).

Smackover–Norphlet Formations

The Jurassic Smackover Formation of Mississippi, Alabama, and Florida is a good petroleum system in which to follow the changes from oil to condensate and gas, because the Smackover source and reservoir rock plus the underlying Norphlet sandstone are sealed in by the Buckner anhydrite above and the Pine Hill anhydrite and Louann salt below. It is like a pressure cooker generating oil and gas with limited pathways for migration. Pressures in parts of the Mobile area, offshore Mississippi, are greater than 18,000 psi (124 MPa), which is close to lithostatic pressure at 20,000 ft (6,098 m). Updip where the Jurassic is at 10,000 ft (3,049 m), there is an oil trend at normal hydrostatic pressure.

The general boundaries of these hydrocarbon trends are shown in Figure 12-13 by Mancini et al. (1986). The oil trend at depths approximately from 10,000 to 14,000 ft (3,049 to 4,268 m) changes to a gas-condensate trend extending to 18,000 ft (5,488 m) and ultimately to a deep dry gas trend extending below 20,000 ft (6,098 m). These trend lines are generalized, since there is considerable overlap in the oil, gas-condensate, and dry gas fields. All the shallow

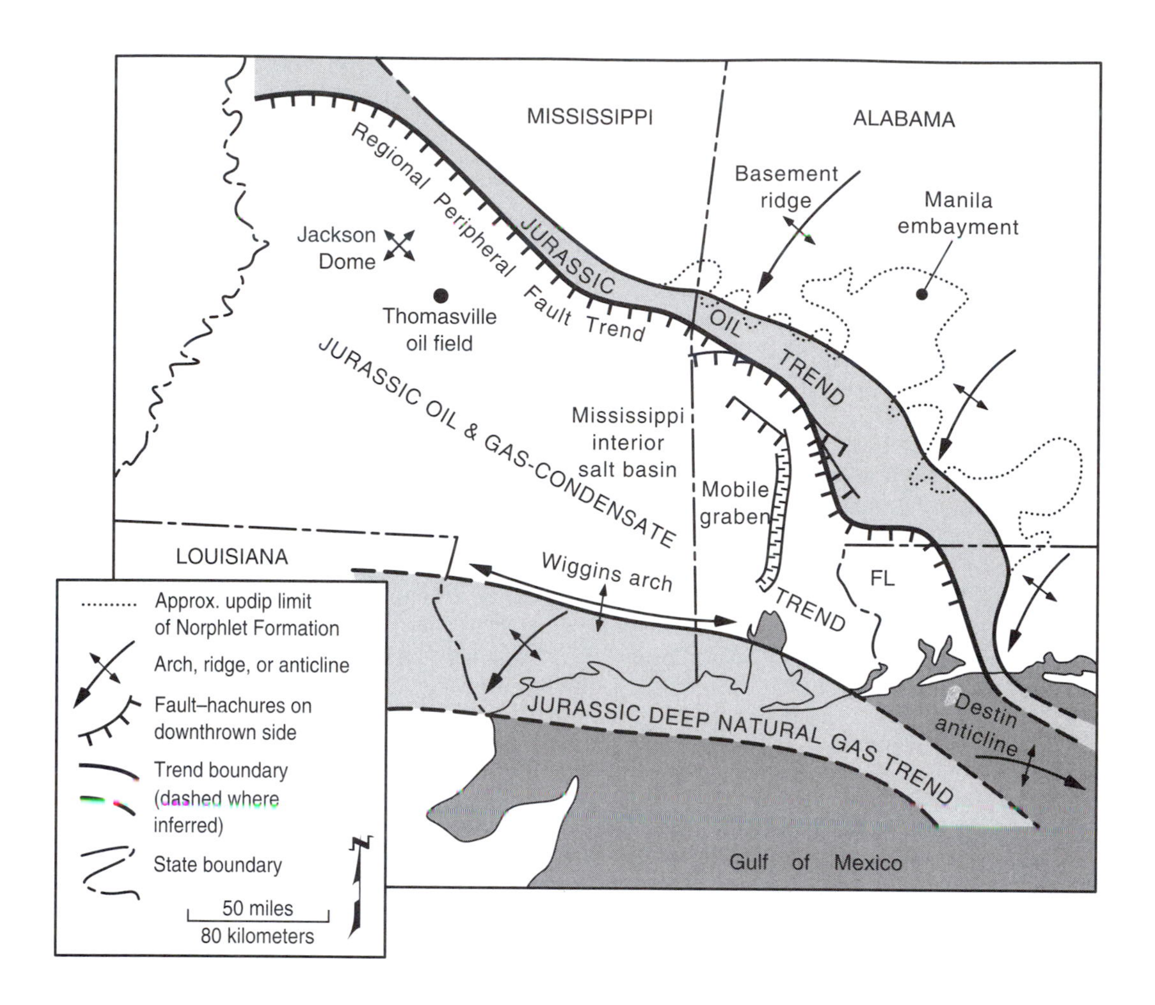

Figure 12-13

Upper Jurassic oil, condensate, and deep natural gas trends for the onshore tristate area of Mississippi, Alabama, and Florida, eastern Gulf of Mexico. [Mancini et al. 1986]

oils contain some gas, and even the deepest dry gas fields contain small amounts of condensate.

Table 12-5 shows the change from oil to condensate and gas in Smackover and Norphlet reservoirs extending over a depth range of almost 10,000 ft (3,049 m). The level of maturation is indicated by vitrinite reflectance measurements from reservoir cores (W. G. Dow 1988, personal communication). Starting with the low API gravity Quitman field, which is in the oil-generation win-

TABLE 12-5 Thermal Alteration of Oil in Jurassic Reservoirs of the U.S. Gulf Coast

Field	*Pay depth (ft)*	*(m)*	*Temp. °F*	*°C*	*Vit. ref., R_o at pay*	*API gravity*	*Mol% H_2S*	*GOR*
Quitman, Mississippi	11,382	3,470	200	93	0.80	22°	0.03	100/1
Pachuta Creek, Mississippi	12,500	3,811	223	106	0.95	37°	0.02	500/1
Goodwater, Mississippi	14,563	4,440	243	117	1.05	39°	0.08	800/1
Jay, Florida	15,550	4,726	296	147	1.20	51°	9	1,200/1
Chatom, Alabama	16,000	4,878	305	152	1.40	55°	17	3,000/1
Hatter's Pond, Alabama	17,712	5,400	322	161	1.45	61°	0.6	4,000/1
South Stage Line, Mississippi	18,100	5,518	333	167	1.50	67°	27	10,000/1
Thomasville, Mississippi	19,730	6,015	350	177	1.80	Gas	35	Gas
Black Creek, Mississippi	19,900	6,067	360	182	2.05	Gas	78	Gas
Mary Ann, Alabama	21,100	6,433	415	213		Gas	9	Gas

Source: Data courtesy of W. G. Dow.

dow, there is a gradational increase in the oil gravity of reservoired oil toward a condensate at an R_o of about 1.4%. This is accompanied by a jump in the GOR as reservoir temperatures exceed 305°F (152°C). Between 1.5 and 1.8% R_o, the accumulations become predominately gas.

Olsen (1980) made a detailed petrographic study of the Thomasville reservoir listed in Table 12-5 and shown in Figure 12-13. The discovery well produced 9.8 million ft^3 gas/day (277,505 m^3/day) and 1,300 bbl of hydrogen sulfide-rich condensate. The H_2S yielded 100 tons of sulfur per day. The gas was 55% CH_4, 35% H_2S, and 9% CO_2. The pressure gradient was 0.88 psi/ft (20 kPa/m), which is close to lithostatic. Pyrobitumen was 2 to 3% of the total rock. Olson concluded that this once had been a huge oil reservoir that was converted to gas and condensate by exposure to the present high temperatures.

The H_2S data in Table 12-5 are mainly from Wade et al. (1989). The H_2S comes from thermochemical sulfate reduction (TSR), in which hydrocarbon accumulations contact anhydrites at high temperatures, as explained in Chapter 7. The rate of the TSR reaction increases with H_2S partial pressure, and the rate of CH_4 destruction increases with the concentration of H_2S and polysulfides in the reservoir. Wade et al. (1989) found that the highest H_2S contents were in the thickest, low-porosity, low-permeability Smackover carbonates. Lower H_2S concentrations occur where the Smackover is thin and porous, such as over basement uplifts. Lower concentrations also were found in the Norphlet, possibly because its higher content of metal ions converts any H_2S

in pore waters to pyrite and other sulfides. Even so, the Mary Ann field in the Norphlet formation is large enough to ultimately produce 225 tons of sulfur/day from the H_2S in the gas stream. Pyrobitumens recovered from about 6,000 m (19,680 ft) had TAIs of 4 and hydrogen indices of 0, indicating a very advanced state of maturation (Sassen 1988). Electron micrographs by Sassen show the pyrobitumen occurring as hollow globules, mostly less than 5 μm in size on calcite crystals. Similar individual bitumen globules also occur in the Ellenburger Formation of the deep Delaware–Val Verde Basin and the Hunton Limestone in the deep Anadarko Basin.

Plate 8A shows a typical Norphlet sandstone in thin section with a pyrobitumen coating on grains of chlorite (Chevron, U.S.A. Inc.). These are from about 20,000 ft (6,098 m) in offshore Mississippi. The top of the Norphlet is frequently dark with bitumen at the Smackover contact. Some cores gradually lighten with increasing depth below the Smackover, whereas others remain dark for 200 ft (61 m). The lower 150 ft (45 m) of the Smackover has organic laminations mixed with pyrite. Smackover carbonates with porosity tend to be overpressured due to the generation of gas.

Diamondoids

Some very deep (> 20,000 ft, 6,098 m) gas condensate reservoirs in the U.S. Gulf Coast contain small amounts of diamondoids (Lin and Wilk 1994). These are rigid fused-ring cycloalkanes with a diamondlike structure. They also are called *polymantanes*. The simplest member of this hydrocarbon group is adamantane (Figure 12-14). Molecules like these with as many six fused-ring structures (hexamantane) have been identified in deep Gulf Coast reservoirs such as the Norphlet and Smackover formations (Wingert 1992). The GC analyses of

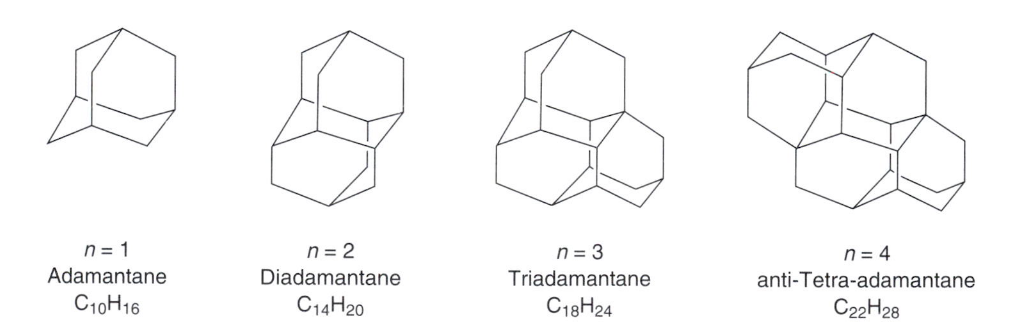

Figure 12-14

Diamondoid structures of polycyclic hydrocarbons with the homologous series molecular formula $C_{4n+6}H_{4n+12}$. They occur in condensates at temperatures >200°C (392°F).

some Norphlet gases have revealed crystalline hydrocarbons in the C_{10}–C_{18} range, which are adamantane and di- and triadamantanes (Figure 12-14). Because of their condensed-cage structure, diamondoids have an API gravity below 10 and density of 1.07. They are found most often in high-temperature gas reservoirs. As pressures are decreased in surface production equipment, the diamondoids sublime and condense out as solids like talcum powder. Their high energy of crystallization causes them to change directly from the gas to the solid phase. They plug up surface production equipment but are readily dissolved out with solvents.

Concentrations in gas rise to 100 ppm, which represents a lot of diamondoids when millions of cubic feet of gas are being produced per day. The ratio of ethane to diamondoids is about 5 to 1 in dry gases. A small amount of condensate recovered from the Mary Ann field had an API gravity of 30 because it was mostly benzene, toluene, xylenes, naphthalenes, and 10 to 20% dissolved diamondoids. Most normal alkanes had disappeared, owing to H_2S destruction. Diamondoids also have been reported in dry gases and condensates in deep reservoirs in Canada and Russia. They are found most often in reservoirs at temperatures >200°C (392°F).

Tuscaloosa Formation

The Upper Cretaceous Eagle Ford Shale is a major source rock in the U.S. Gulf Coast, having generated the oil in the Woodbine Formation of the giant East Texas field. The eastern extension of the Woodbine is the Tuscaloosa Formation, which includes oil-generating shale source rocks interbedded with reservoir sands. Oil has been produced for many years from a large number of structural and stratigraphic fields in Mississippi and north Louisiana at depths between 10,000 ft (3,049 m) and 14,500 ft (4,421 m). This oil trend grades into gas condensate going southwest and eventually into gas at depths from 16,000 ft (4,878 m) to 22,000 ft (6,707 m).

The approximate location of these trends is shown in Figure 12-15. Two fields that show the beginning of the shift from oil to gas are Baywood and Beaver Dam Creek, both about forty miles south of the Mississippi–Louisiana border, between the oil and gas trends, as shown in Figure 12-15. Beaver Dam Creek contains 8×10^6 bbl of light oil, whereas Baywood, only five miles away, contains 50×10^9 ft^3 gas plus 5×10^6 bbl condensate (Harrison and Parrish 1990). Both fields are at 14,500 ft (4,421 m). West of this, the Freeland field at 15,400 ft (4,695 m) produces a light 53°API oil plus wet gas. About nine miles farther south, the Port Hudson field at 16,000 ft (4,878 m) produces wet gas with a high condensate content (Figure 12-15). Going west from there at 20,000 ft (6,098 m), the False River and Digby fields are in the transition from wet to dry gas, with TAIs of 3.5 to 3.7, H/C ratios of 0.5 to 0.55, and a vitrinite reflectance of 2.0% R_o (Funkhouser et al. 1980).

The deep fields are highly overpressured with average pressure–depth gradients of 0.85 psi/ft (19 kPa/m). The pressure–depth plot for the Morganza field

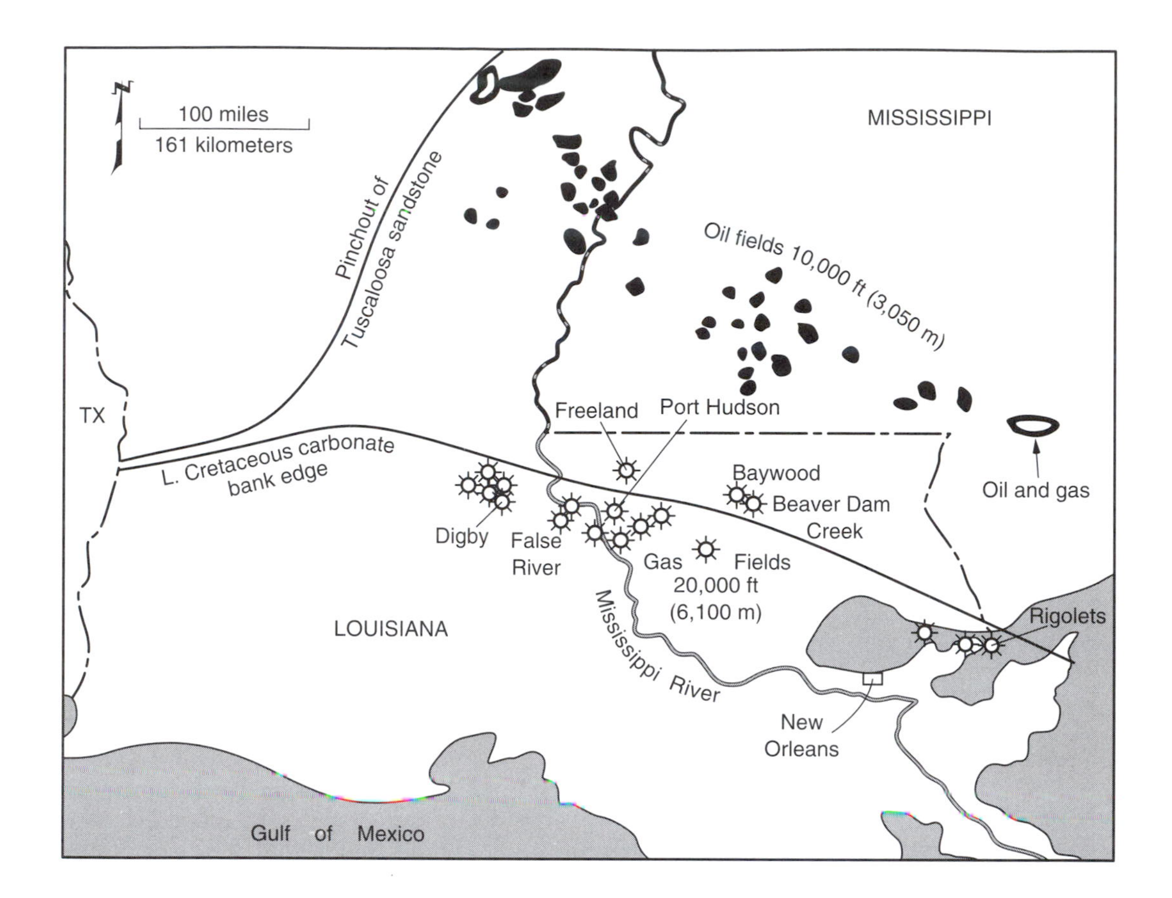

Figure 12-15

Oil and gas pools in the Lower Tuscaloosa Formation showing oil trend to the north and deep gas trend to the south with gas condensates in between. [Funkhouser et al. 1980]

at the western end of this trend was previously shown in Figure 9-3. It has a seal at 18,500 ft (5,640 m), across which is one of the highest pressure steps ever recorded, 24.2 psi/ft (544 kPa/m). The GORs in these high-pressure fields exceed 150,000/1. The estimated reserves for the deep-gas trend exceed 5 TCF of gas. The reservoir temperatures of these gas fields are around 360°F (182°C). Interbedded Tuscaloosa shale source rocks contain from 0.5 to 2.3 wt% TOC. The mean TOC is 0.9 wt%, and the original mean TOC is estimated to be around 1.4 wt% (Funkhouser et al. 1980). Some solid bitumen has been observed in the Rigolets fields at 17,000 ft (5,183 m). The bitumen occurs as individual particles on quartz overgrowth.

Oil and Gas Phaseout Zones

The depth at which oil no longer exists in commercial quantities can be estimated or predicted from thermal history models, such as that described in Chapter 6. The application of Arrhenius kinetics to Canadian Hunter's Brassey field (Table 6-6) fits the observed occurrence of condensate and gas. The various maturity indicators discussed in Chapter 10 also are useful in estimating the oil floor. So many factors are involved, however, that any modeling program can give only estimates at best. Oil and gas-condensate fields can occur at the same maturation level as shown in the previously discussed Beaver Dam Creek oil field. It is at the same depth and only five miles away from the Baywood gas-condensate field.

The gas floor is even more difficult to identify. The maximum vitrinite reflectance value of an economic gas accumulation occurs in the Delaware–Val Verde Basin at 3.2%R_o. Gas has been found at higher R_os, but either it is not in commercial quantities or the thermal history of the reservoir is unrelated to that of the gas. Methane generally tends to rise through the sedimentary column, owing to its low density and small molecular size. Consequently, the floor would be expected to occur where the presumed source rock is no longer generating any gas. Where the H/C ratio for kerogen in the source rock reaches 0.25 (or the HI is 0), it is difficult to see how any more methane could be generated. This is a reasonable floor for gas. The source rocks in the deep gas Tuscaloosa trend where the R_o equals 2.0% and the H/C equals 0.55 would still have some gas-generating capability.

Because methane is thermally indestructible at sedimentary rock temperatures, some geologists maintain that the only limiting control on deep methane production is reservoir porosity. Methane, however, is soluble in pore fluids, as shown in Table 8-4, and it also can be destroyed chemically by TSR, as previously discussed. Consequently, deep prospects in the proximity of anhydrite beds would be risky.

SUMMARY

1. The physical and chemical characteristics of petroleum in its natural reservoir vary because of biodegradation, water washing, thermal alteration, gravity segregation, vaporization–condensation phenomena, deasphalting, and dewaxing, as well as the loss of hydrocarbons by seepage or their addition from other sources.

2. Meteoric waters carrying bacteria degrade petroleum by oxidizing the hydrocarbons at temperatures below 80°C (176°F). Biodegradation causes decreases in wet gas, gasoline, kerosene, all *n*-paraffins, API gravity, wax content, pour point, and gas/oil ratio. It causes increases in density and viscosity and

relative concentrations of asphaltic compounds, sulfur and nitrogen content, and vanadium and nickel content.

3. Biodegradation rankings from minor to extreme are based on the selective removal of biomarkers. The overlapping order of removal starts with *n*-alkanes, followed by branched alkanes, alkylcycloalkanes, alkylbenzenes, bicyclic alkanes, steranes, hopanes, diasteranes, tricyclic terpanes, aromatic steroids, and finally porphyrins.

4. Reservoir oils can form updip asphalt seals at the outcrop and downdip asphalt mats at the oil–water interface of pools in contact with meteoric water through water washing, microbial alteration, oxidation, vaporization, natural deasphalting, and gravity segregation. The total heavy and extra heavy oil in the world trapped by asphalt seals is believed to exceed 2×10^{12} bbl.

5. The thermal maturation of reservoir oil means that the number of oil fields gradually decreases and the number of gas fields gradually increases at depth. If reservoirs are deeper than 12,000 to 14,000 ft (3,658 to 4,267 m), fewer than one-fourth of them will contain oil, and of those deeper than 20,000 ft (6,096 m), all but a small percentage will contain gas.

6. Thermodynamic calculations lead one to predict that *n*-paraffins are the most stable type of hydrocarbons in the gas phase at sedimentary basin temperatures. Their stability increases as the number of carbon atoms in the molecule decreases, with methane being the most stable.

7. The maturation of crude oil in the reservoir involves several hydrogen disproportionation and cracking reactions, with the large molecules giving up hydrogen to permit increased formation of low-molecular-weight paraffins. The large molecules condense to polycyclic aromatic hydrocarbons, which eventually form bitumens or pyrobitumens in the reservoir.

8. Thermal maturation in the reservoir causes the gravity of most crude oils to increase one degree API (decrease 0.005 in specific gravity) for every 200 to 400 ft (61 to 122 m) of increasing depth. Reservoir maturation appears to initially crack the hydrocarbons containing more than twenty carbon atoms into smaller molecules that become part of the gasoline range.

9. Vaporization–condensation phenomena include retrograde condensation and evaporative fractionation. The former exists as a single-phase dense-gas system that, on pressure reduction, converts to a two-phase free gas over oil. Evaporative fractionation involves gas with dissolved oil moving vertically upward from a deep reservoir and condensing out oil accumulations at successive levels as pressures and temperatures decrease. This causes the shallow oils to have higher API gravities and lower aromaticities than the deeper oils.

10. The formation of wet gas from the cracking of heavy oil or from the addition of gas from deeper sources causes natural deasphalting with the precipitation of asphalt in the reservoir along with a lighter oil and gas.

11. Bitumens by definition are fusible, are soluble in carbon disulfide, and have viscosities greater than 10,000 MPa-seconds (centipoise), whereas pyrobitumens are infusible, are only slightly soluble or insoluble in CS_2, and become solid at reservoir temperatures. Both occur at 1 to 3% concentrations in many deep gas reservoirs as a by-product of the thermal destruction of oil.

12. The Smackover–Norphlet Formations of the Mississippi salt basin and the Tuscaloosa Formation of the U.S. Gulf Coast are good examples of the change in the reservoir from oil to condensate to gas and pyrobitumen.

13. The floor for most conventional < 50°API gravity oil is approximately 15,000 ft (4,573 m) in areas where the deep geothermal gradient is 1.6°F/100 ft (3.0°C/100 m). The R_o is ~1.2% at this depth, and the temperatures are ~290°F (143°C).

14. Light oil (>50°API), condensate, and wet gas in such areas are found mainly between 15,000 and 19,000 ft (4,573 and 5,793 m) where the R_o = 1.2 to 1.8%.

15. Dry gas becomes dominant at depths > 19,000 ft (4,573 m) where the R_o exceeds 2.0%. The gas floor is expected to occur where the H/C ratio of the kerogen in the presumed gas source rock reaches 0.25 (R_o > 3.5%).

16. All hydrocarbons, including methane, can be oxidized to CO_2 by H_2S, elemental sulfur, and polysulfides in high-temperature reservoirs that are contacting anhydrites ($CaSO_4$).

17. Diamondoids are crystalline C_{10}–C_{22} hydrocarbons with a condensed-cage structure that occur in gas accumulations in concentrations up to about 100 ppm. They are restricted to high-temperature (>200°C, 392°F), high-pressure reservoirs generally containing some H_2S. They condense as solids in gas-processing equipment at the surface.

SUPPLEMENTARY READING

Meyer, R. E. (ed.). 1987. *Exploration for heavy crude oil and natural bitumen.* AAPG Studies in Geology 25. Tulsa: American Association of Petroleum Geologists, 731 p.

PART FOUR

APPLICATIONS

Chapter

13

Seeps and Surface Geochemical Exploration

The four principal requirements for petroleum accumulations are (1) mature petroleum source beds with a permeable pathway to reservoir beds, (2) adequate reservoir rock porosity and permeability, (3) structural or stratigraphic traps, and (4) a relatively impermeable cover during and since accumulation. Visible oil and gas seeps are important to an exploration program because their presence implies that the first requirement has been satisfied and possibly others if it is a major seep. Many large seeps represent tertiary migration, that is, migration from an accumulation that has been disturbed by tilting of strata, changes in depth of burial, and erosion or development of new avenues of escape to the surface, such as fractures and faults.

A *petroleum seep* is defined as visible evidence at the earth's surface of the present or past leakage of oil, gas, or bitumens from the subsurface. This definition does not include microseeps or invisible seeps, which are discussed in the section entitled "Surface Geochemical Exploration."

Seeps: Onshore

Oil seeps were reported in the earliest recorded history. The use of asphalt as a building material from seepages in the Middle East dates back to 3,000 B.C. Burning gas seeps have existed in the Baku area since several centuries before Christ. Egyptian mummies dating from 1,295 B.C. to A.D. 300 have been found to contain bitumens in their embalming material (Connan and Dessort 1991). It is thought that the bitumens came from the Dead Sea and the Hit-Abu Jir asphalt deposits in Iraq, based on a comparison of biomarker patterns of the balms and the asphalt deposits.

Seeps are particularly important to exploring new basins or areas. Only about one explored basin in three in the world is petroliferous enough to contain producible oil or gas, and only one in six contains even one very large oil field (North 1985, p. 498). The importance of seeps tends to be minimized in this era of increased use of highly sophisticated instrumentation and decreased use of ground surveys. Nevertheless, nearly all the important oil-producing regions of the world were first discovered by surface oil and gas seeps. The first oil wells of Canada, Pennsylvania, Oklahoma, California, and Texas were drilled near oil seeps. The giant Masjid-i-Suliman field in Iran was the first big oil discovery in the entire Middle East. The discovery well was drilled near an oil and gas seep that had leaked through the caprock (Figure 13-1).

When L. G. Weeks, one of the outstanding petroleum geologists of this century, retired from Exxon in 1958, he went to Australia to explore for the Broken Hill Proprietary Company (BHP). While there he remembered that a

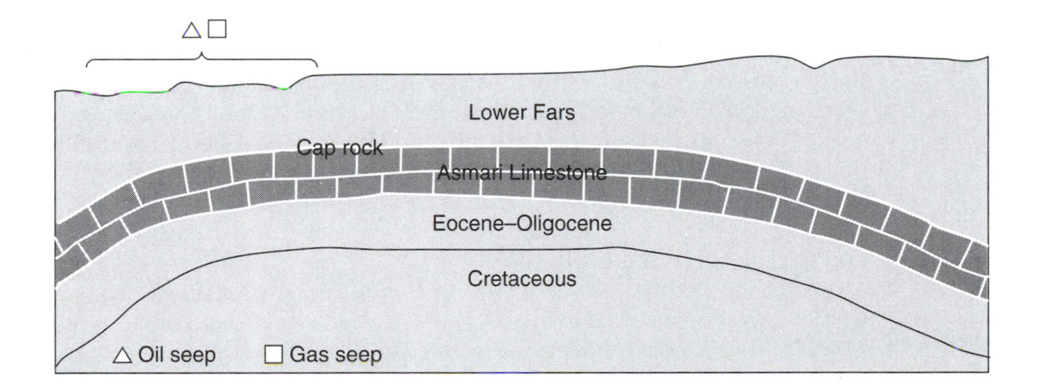

Figure 13-1

Seepage of oil and gas from the Asmari Limestone at Masjid-i-Sulaiman field in Iran. [Link 1952]

friend had told him that back in the 1930s he had seen bitumen washing up on the Gippsland Coast, presumably from a seep somewhere out in the Bass Strait (Weeks 1978). Weeks followed up on this and convinced BHP to initiate the first major offshore exploration in Australia. Today the Gippsland Basin is Australia's most prolific petroleum province, with reserves approaching 4×10^9 bbl of oil and condensate and 9 TCF of gas.

Geology of Seeps

Chapter 10 stated that less than 15% of the hydrocarbons generated by source rocks becomes recoverable oil and gas in reservoirs. The rest is partly disseminated in rocks throughout the subsurface and partly lost at the surface (see Figure 16-1). Seeps occur wherever a permeable pathway leads to the surface from mature source strata or from leaking petroleum reservoirs. They are common at the outcrops of unconformities and permeable homoclinal beds such as in the Western Canada and Eastern Venezuala Basins. They occur along normal faults, as at Gebel Zeit in Egypt, and thrust faults, as at the Infantas field in Colombia (Link 1952). Seeps also are commonly associated with intrusions such as the mud volcanoes in the Caspian Sea off Baku and the piercement salt domes in the U.S. Gulf Coast. The asphalt lake of western Trinidad is a giant seep, as are the saber-toothed tiger-bearing La Brea Tar Pits near Los Angeles, California.

Young sedimentary rocks in tectonically active areas host the most seeps. The small Tertiary basins of California contain hundreds of oil seeps. Seeps also are numerous in small intermontane basins such as in Indonesia and the Magdalena Valley of Colombia. There are many seeps on the mobile sides of basins, such as in the Mesopotamian geosyncline, the Monagas Basin of eastern Venezuela, and the eastern foothills of the Andes mountain chain from Colombia to Cape Horn. Seeps are particularly common along the margins of basins where unconformities and oil-producing formations come to the surface. A typical example is northeastern Oklahoma, where the oil-bearing rocks come to the surface on the flanks of the Ozark uplift, and in southern Oklahoma, where the producing sands are sharply upturned along the Arbuckle and Wichita uplifts. Because these are Paleozoic formations, there has been sufficient time for part of the asphalt to be converted to the asphaltite grahamite and the pyrobitumen impsonite.

Another example of seeps along the edges of a basin or where structural uplifts have exposed the oil-bearing stratigraphic sequences is in the Lake Maracaibo area of Venezuela (Hunt 1979, pp. 418–419; Link 1952). There are more than 200 oil and gas seepages in western Venezuela, many of which occur along the flanks of the Venezuelan Andes and the Perija Mountains. The seepages range from a few barrels of oil to asphalt lakes that cover several square kilometers. Along the northwest edge of the Venezuelan Andes, small seeps issue from fractured igneous rocks that have been thrust basinward over Tertiary or Cretaceous sediments. Seeps are associated with Cretaceous, Miocene, and Eocene sediments. The Mene Grande field, El Mene, La Paz, and the Bolivar coastal fields all were drilled because of seeps.

Link (1952) recorded the worldwide occurrence of seeps and classified them into five groups, depending on their origin:

1. Seeps emerging from homoclinal beds, the ends of which are exposed where these beds reach the surface.
2. Seeps associated with beds and formations in which the oil was formed.
3. Seeps from large petroleum accumulations that have been bared by erosion or reservoirs that have been ruptured by faulting and folding.
4. Seeps at the outcrops of unconformities.
5. Seeps associated with intrusions such as mud volcanoes, igneous intrusions, and piercement salt domes.

Link (1952) did his study before the concept of plate tectonics became accepted. Since then it has been recognized that the rupturing of caprocks and seepage of oil through small fractures and faults is common in earthquake-prone areas along the edges of crustal plates where the continents collide. For example, in the Middle East (Figure 13-2) the collision between the Arabian plate and the Eurasian plate regularly jolts Iran and Iraq, resulting in several seepages near the plate boundary. Associated fracturing and faulting have permitted the upward escape of oil and gas from Cretaceous reservoirs. Near the center of this earthquake belt, the supergiant Burgan field of Kuwait has a large heavy-oil seep directly over its structural crest. Farther northwest, the giant Kirkuk field of Iraq has oil and gas seeps all over its structure, including "the Eternal Fire" of Baba Gurgur (Link 1952). In the southern and central parts of Saudi Arabia, where there is almost no earthquake activity, there are very few visible seeps, even though this area contains some of the largest oil fields in the Middle East, such as Abqaiq and Ghawar (the largest in the world).

Worldwide, there is a correlation between seeps and earthquake activity, with the majority of the seeps close to plate boundaries where most such activity takes place. For example, the western coast of South America has seeps in Ecuador, Peru, and Chile that follow the earthquake belt caused by the subduction of the Nazca plate beneath South America. Trinidad, southern California, southern Alaska, the Philippines, Indonesia, and Burma are other areas whose numerous seeps are related to the earthquake activity of plate boundaries. The hand-dug wells of the Burmese oil fields, which were yielding oil nearly a century before the Drake well, are located on seeps near the Indo–Australian–Eurasian plate boundary. Exploration for oil and gas in wildcat areas near plate boundaries should emphasize the detection and analysis of seeps.

Well-populated areas of the world generally report many petroleum seeps. Selley (1992) documented 173 seepages and impregnations in Great Britain. The seeps were found primarily around the margins of basins where permeable carrier beds unconformably overlie impermeable "basement" rocks. The main

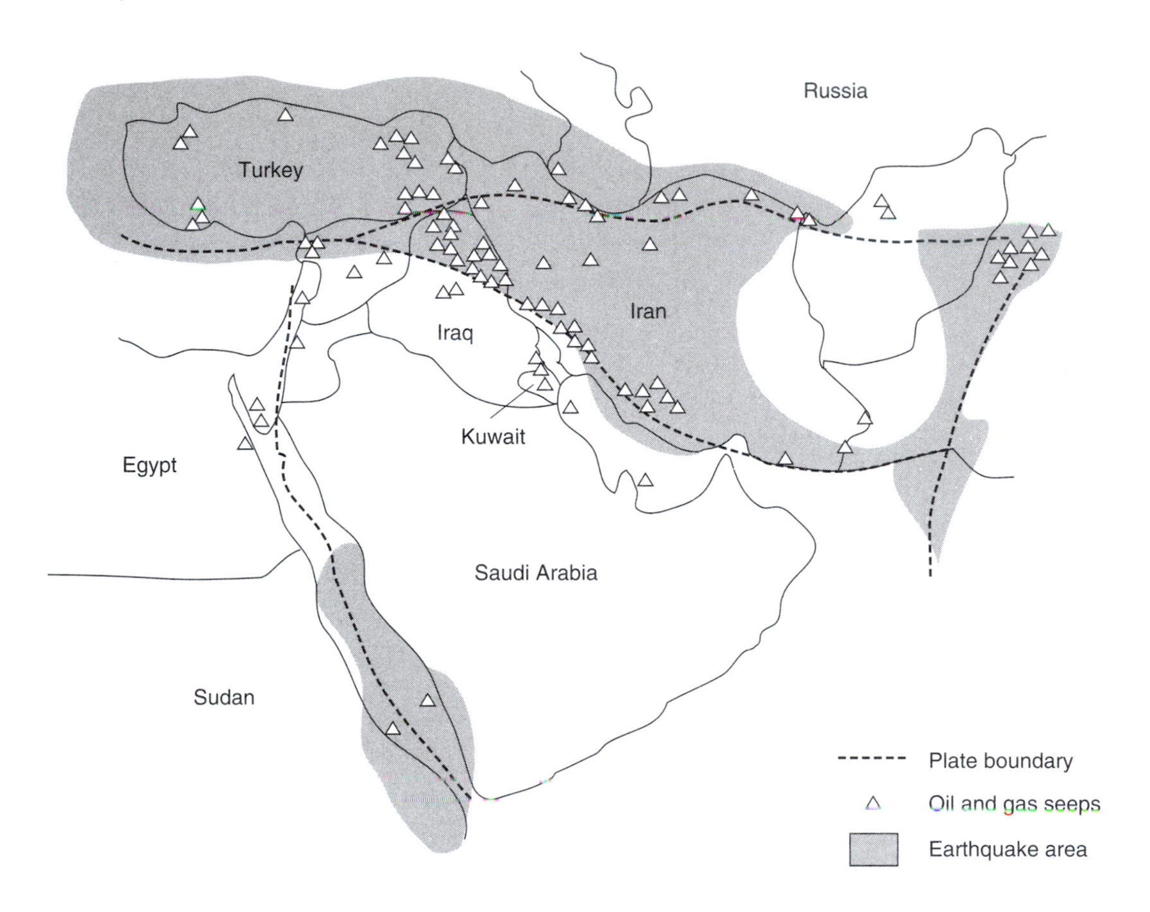

Figure 13-2

The relationship of plate boundaries and earthquake activity to seepage areas of the Middle East. Seeps involving vertical migration through ruptured caprocks are common in petroliferous areas near plate boundaries.

sources of the petroleum seeps were believed to be Devonian oil shales and Carboniferous shales and coals. Seepages were also found associated with faults, particularly in the Wessex and Wealden basins where the Liassic Formation is the probable source (Selley 1992).

Macgregor (1993) compared the global occurrence of seeps with tectonics and subsurface petroleum reserves. In the North and South Sumatra basins he found that most of the visible seeps came from the smaller, shallower traps and those most strongly affected by diapirism or faulting. The larger, deeper fields not connected to the surface by faulting rarely showed visible seeps.

Macgregor's 1993 global study offers the following guidelines for evaluating seeps:

1. Seepage patterns are strongly controlled by regional and local tectonics. Seeps are most common in overpressured diapir-rich basins and in active thrust belts. They are rare in tectonically inactive basins.

2. Most large, deep accumulations do not seep directly to the surface.

3. Intracratonic and foreland basins show a small number of seeps relative to reserves, and thrust belts show an anomalously large number of seeps.

4. The presence of seeps over basins or prospects often considerably reduces the exploration risk. The absence of seeps over a tectonically active basin or shallow-faulted prospect increases the risk.

Macgregor (1993) felt that the prime value of visible seeps in frontier basins is at the regional level, in providing information on the hydrocarbon potential of a basin's source system.

Weathering of Seeps

When oil leaks to the surface, it undergoes a series of changes that considerably alter its physical and chemical composition.

1. *Evaporation of the more volatile hydrocarbons.* In the first two weeks after an oil reaches the surface, it loses its hydrocarbons up through about C_{15}. In subsequent months, additional hydrocarbons are lost, up through about C_{24}.

2. *Leaching of water soluble constituents.* The most soluble nitrogen, sulfur, and oxygen compounds, along with some lighter aromatic hydrocarbons, may be leached out by groundwaters.

3. *Microbial degradation.* As discussed in Chapter 12, hydrocarbons leaking to the surface are subject to microbial attack.

4. *Polymerization.* Polymerization is the combining of some of the intermediate-to-larger molecules to form very large complex structures after the elimination of water, carbon dioxide, and hydrogen.

5. *Auto-oxidation.* Many constituents of petroleum absorb sunlight and oxygen, which convert the oil to an asphalt high in oxygen. With long exposure to air and sunlight, seeps can take up more than 6 wt% oxygen.

6. *Gelation.* The formation of a rigid gel structure may develop over time with some types of petroleum.

All these reactions lead to thickening or solidification of the original oil. The crude is gradually converted from a liquid oil to an asphaltite and eventually to a substance physically close to a pyrobitumen. Consequently, unless a seep is supplied by a continuous flow of fresh oil, it ultimately hardens into a black bitumen or pyrobitumen. The term *pyrobitumen* is most often used to indicate a late stage of thermal maturation. But as explained in Chapter 12, a pyrobitumen is defined through usage as an infusible, CS_2 insoluble solid, regardless of the state of maturation. If degraded seep oil is CS_2 insoluble and infusible, it fits the definition of pyrobitumen.

Studies of surface bitumens reveal that severe weathering can mimic late-stage maturation without any change in maturation indicators such as biomarkers. For example, Table 13-1 lists a series of seep samples collected from a single major seep near Lake Maracaibo. Sample 1 is at the crest of the active flowing Mene Grande oil seep. Samples 2 and 3 represent more weathered bitumens that are successively farther away from the active flow site, and sample 4 is the farthest away (Dickey and Hunt 1972). These all were taken within 100 m of one another across the top of the seep. The decrease in solubility of these samples in organic solvents coincident with an increase in fixed carbon and a decrease in hydrogen content is due to weathering, not maturation. All four samples came from the same seep recently enough to show the same maturity. The change in the H/C ratio is not due to maturation but to oxidation and the other processes just mentioned. Sample 4 would be a pyrobitumen by definition, since it is infusible and insoluble, even though it is no more thermally mature than the fresh oil of sample 1.

Many of the pyrobitumens of the Uinta Basin, Utah, such as wurtzilite, ingramite, and albertite, migrated from the source rocks as liquid bitumens into veins and fissures that opened up when the basin was uplifted. They were probably never buried to high temperatures at great depths, as were the reservoir pyrobitumens discussed in Chapter 12. For example, the H/C atomic ratios of these pyrobitumens vary between about 1.3 to 1.6, whereas the H/C ratios of typical high-temperature reservoir bitumens are generally less than 0.8 (Hunt 1979, p. 401).

TABLE 13-1 Weathering of Seep at Mene Grande, Venezuela

Sample numbers[a]	1	2	3	4
Solubility in *n*-heptane (wt%)	78	43	5	0
Solubility in CS_2 (wt%)	99	78	52	0
Carbon ratio[b]	6	10	28	82
H/C atomic ratio	1.63	1.52	1.36	0.6

[a]Higher numbers are more weathered samples farther from the seep source.

[b]Ratio of nonvolatile organic carbon to total organic carbon.

Both liquid gilsonite and liquid wurtzilite (liverite) were found in the Uinta Basin of Utah during the early years of exploration in that area. These asphalts were in veins that originated directly from the source beds. The liquids that solidified at the outcrop had the properties of asphaltites and pyrobitumens, even though their maximum depth of burial probably never exceeded 1,500 m (4,920 ft). The liquid gilsonite had a H/C ratio of 1.6 and less than 1% oxygen, whereas the solid, weathered gilsonite defined as an asphaltite had a H/C ratio of about 1.35 and an oxygen content of 7 wt% (Hunt 1979).

The immaturity of asphaltites and pyrobitumens such as these was later confirmed by Curiale (1988) in his detailed study of bitumen biomarker patterns. Mass chromatograms of the sterane m/z 217 obtained by Curiale are shown in Figure 13-3. Both these patterns indicate thermal immaturity, with the Utah ingramite being the least mature. The 14β, 17β/(14β, 17β + 14α, 17α) ratio of the ethylcholestanes in the Canadian albertite is 0.39, which represents early maturity, compared with 0.7 for full thermal maturity (see Figure 10-37). The ratio is even less for the Utah ingramite, indicating immaturity. The ingramite also has a very low, immature 20s/(20S + 20R) ethylcholestane ratio of about 0.2. It indicates that the ingramite has never been subjected to the high temperatures of a typical oil-generation window.

Ruble and Philp (1991) evaluated the maturity of Utah wurtzilite using the Ts/(Ts + Tm) hopane ratio plus the 20S/(20S + 20R) and ββ/(ββ + αα) of 24-ethylcholestane as maturity indicators (see Figure 10-37). Wurtzilite had a low Ts/Tm, high concentrations of the 20R cholestane epimer, and low concentrations of the 14β, 17β isomer, all of which indicate a very immature bitumen sample. They also found that tabbyite, an asphalt, and gilsonite, an asphaltite, contain immature biomarker ratios, showing that they were formed by a weathering process rather than a thermal process. (A more detailed discussion of natural asphalts and related substances can be found in Hunt 1979, pp. 398–404.)

Seeps: Offshore

Bitumen washing up on beaches in tectonically active areas may come from underwater seeps, tanker spills, or accidental spills in coastal areas. An interesting example of an accidental spill was reported by Kvenvolden et al. (1993). After the *Exxon Valdez* spill of North Slope oil, carbon isotope analyses were made of a large number of asphalt residues from the islands in Prince William Sound. Two groups of residues were identified. Samples from the North Slope oil spill had $\delta^{13}C$ values averaging –29.4‰. Samples from the second group found in the same areas averaged –23.6‰, which is clearly different. Subsequent studies revealed that the latter group probably came from the destruction of asphalt storage facilities at the town of Valdez during the great Alaska earthquake of 1964. Asphalt from that spill spread into the sound at several similar locations, as did the North Slope oil twenty-five years later. Carbon isotope values for residues from the asphalt storage area and from paving older

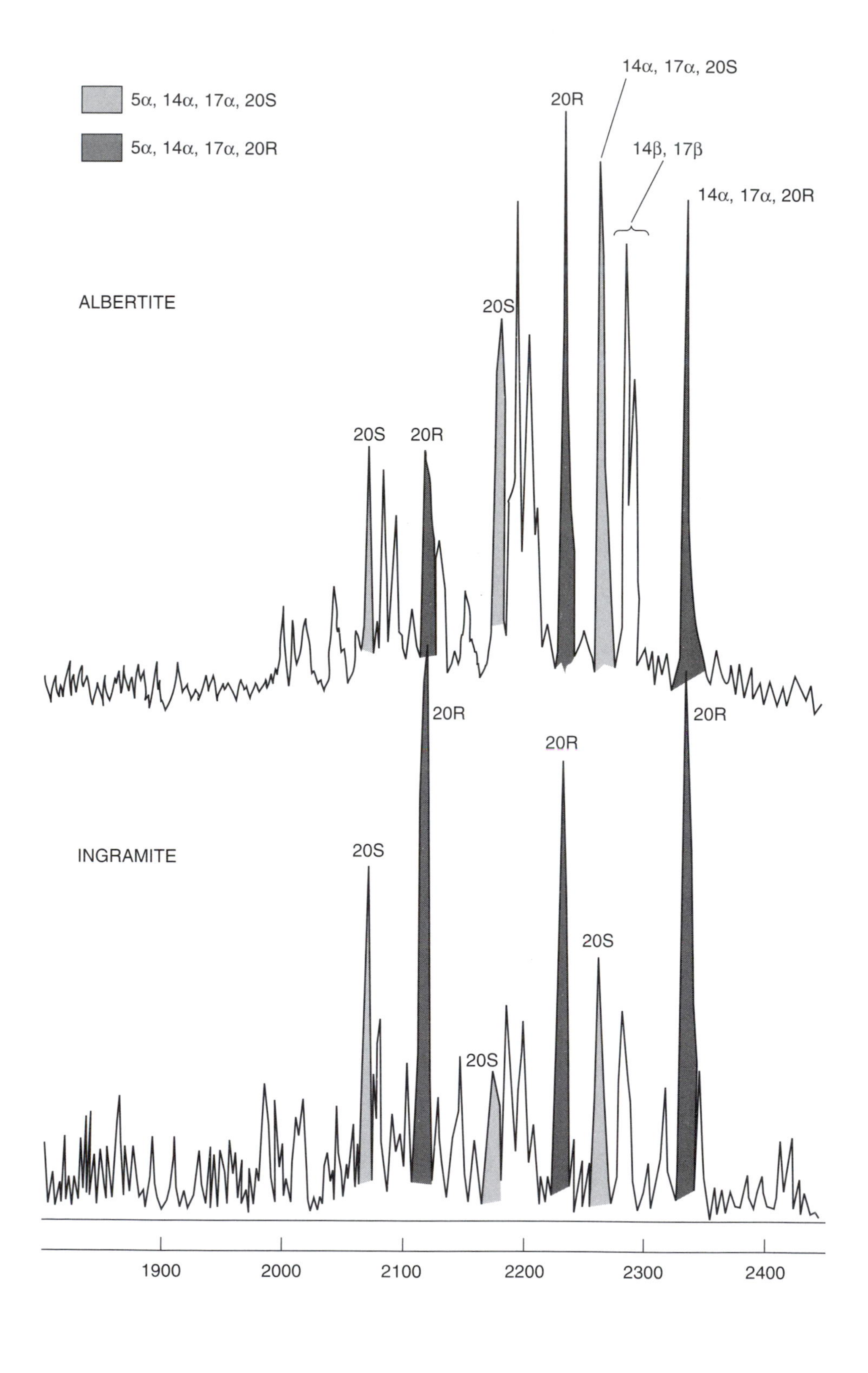

5α, 14α, 17α, 20S
5α, 14α, 17α, 20R
ALBERTITE
20S
20R
20S
20R
14α, 17α, 20S
14β, 17β
14α, 17α, 20R
INGRAMITE
20S
20R
20S
20R
20S
20R
1900
2000
2100
2200
2300
2400

Figure 13-3 (left)

Mass chromatograms of steranes (m/z 217) in two pyrobitumens, Canadian albertite (top) and Utah ingramite (bottom), indicating that both compounds are thermally immature. [Curiale 1988]

than twenty years in the local towns were around –23.8‰. This matched the second group of residues. This second group appeared to be oil and asphalt imported from California before the North Slope discoveries.

Many active seeps are found offshore on the continental shelf, coming from formations that produce oil offshore as well as those that produce oil on the nearby land area. The seaward extension of the oil-rich Ventura Basin in southern California is the source of the prolific Coal Oil Point seep area. It originally released an estimated 50 bbl of oil a day from several seepage vents on the seafloor. These seeps were observed by the early Spanish explorers and have been a chronic source of beach contamination in the southern California area. Vernon and Slater (1963) identified the source of the seeps in the Santa Barbara area by mapping the asphalt mounds on the seafloor, which were particularly abundant near Point Conception. The asphalt mounds range up to 100 ft (31 m) in diameter and 8 ft (2.4 m) in height. They are irregularly distributed along an east–west trend of faulted anticlines. The asphalt becomes denser than seawater through the loss of gas and light hydrocarbons and the accumulation of sediment material. Some of the vented asphalt escapes to the surface, but much of it contributes to the growth of the mounds, which are encrusted by marine organisms. Some of the asphalt mounds, especially those nearest Point Conception, originate from low API gravity crudes that are among the least mature migrant products of type II-S kerogens. Their initial API gravity may be $<10°$.

The Santa Barbara seep area also contains the world's only offshore gas production directly from a seep instead of a well. Twin steel pyramid-shaped tents, each $100 \times 100 \times 20$ ft in size, are on the seafloor, covering a large number of gas vents. The tents collect about 356 million cubic feet (cf) of gas annually. As of early 1993, about 4.54 billion cf of gas and 600 bbl of 24°API gravity crude oil had been produced (Williams 1993).

Landes (1973) summarized the reports of offshore seeps including areas such as the Gaspé Peninsula of Quebec, the U.S. Gulf Coast, the Gulf of Paria near Trinidad, the Gulf of Suez, the Red Sea, the North Slope of Alaska and Canada, and the South China Sea. More recently, Hovland and Judd (1988) described the different styles of underwater petroleum seepages in three areas: (1) the Gulf of Mexico from the Florida escarpment west across the Sigsbee escarpment and south to the Campeche bank off Yucatàn, Mexico; (2) California from the Oregon border to Los Angeles; and (3) offshore Alaska. All these areas contain major oil and gas fields.

The U.S. National Academy of Sciences concluded from its study of petroleum in the marine environment (Wilson et al. 1973) that approximately 1.5

million barrels of oil are being introduced into the world's oceans annually by natural seeps. At least an additional 0.5 million barrels would be seeping on the continents. Since most of this oil was formed in the last 100 million years (Klemme and Ulmishek 1991), it implies that as much as 200×10^{12} bbl of oil could have been lost by seepage, compared with the 10×10^{12} bbl of conventional and heavy oil believed to exist in reservoirs (Roadifer 1987). A mass balance of the global petroleum system in Chapter 16 indicates that this seepage represents possibly 40% of the total bitumen, oil, and gas generated by the source rocks of the world.

Offshore oil seeps usually are first recognized by bitumen floating on the water, but clusters of gas bubbles venting from the seafloor can be detected by the use of 3.5 and 12 kHz acoustical reflection techniques. Bubbles reflect the pulse and show up as a dark vertical line against the reflection-less background response of seawater. Sieck (1973) correlated these bubble clusters with gas-charged sediment cones or chimneys below the seafloor. These cones create volcano-shaped mud lumps on the seafloor, along with circular sea-bottom depressions (pockmarks) where they vent into the seawater. High-resolution data can be obtained by combining several seismic systems such as side-scan sonar, bathometer, tuned transducer, and a sparker or air gun system.

The worldwide occurrences of pockmarks and their relationship to gas seepages were described in considerable detail by Hovland and Judd (1988). Figure 13-4 is a seismic profile from their book showing the relationship between gas-charged sediments and pockmarks on the seafloor of the North Sea (Norwegian block 25/7). The recording shows a pockmark (P.M.) and a mound (M), the latter having an elevation of about 1 m above the seafloor. Pockmarks in this area vary in size from about 5 to 200 m across and 1 to 20 m in depth. Plumes of migrating gas bubbles (gas chimneys) show up as highly reflective areas with an acoustic shadow underneath. Numerous faults show up, some of which seem to be associated with the pockmarks.

The floor and the inside wall of an 80 m wide pockmark in this area contain crusts of sandstone cemented with aragonite and magnesium calcite crystals. The $\delta^{13}C$ of the cement is –59‰, which indicates that the carbon in the cement is coming from methane. Methane escaping from the seafloor was converted by bacteria to CO_2 deficient in C-13. The CO_2 then combined with Ca ions to form C-13-deficient carbonate crystals. Samples of methane collected in the vicinity of these pockmarks had $\delta^{13}C$ values of –39 to –45‰, indicating a thermogenic origin for the gas (Hovland et al. 1985). Most of the gas seeps on the floor of the North Sea have a thermogenic origin, based on Faber and Stahl's (1984) isotope analyses of 350 sediment gases from the central part of the North Sea. The $\delta^{13}C$ values were between – 45 and –20‰, and the $C_1/(C_2 + C_3)$ values were less than 30 (compare this with Figure 7-6(A)).

Many gas chimneys or seismic wipeout zones occur above overpressured reservoirs, as explained in Chapter 9. Buhrig (1989) found such chimneys to be common features in the central part of the Viking and Central Grabens, as well as toward the graben rims. These observations demonstrate that huge volumes of gas are continually being vented through the seafloor. Hovland and Judd

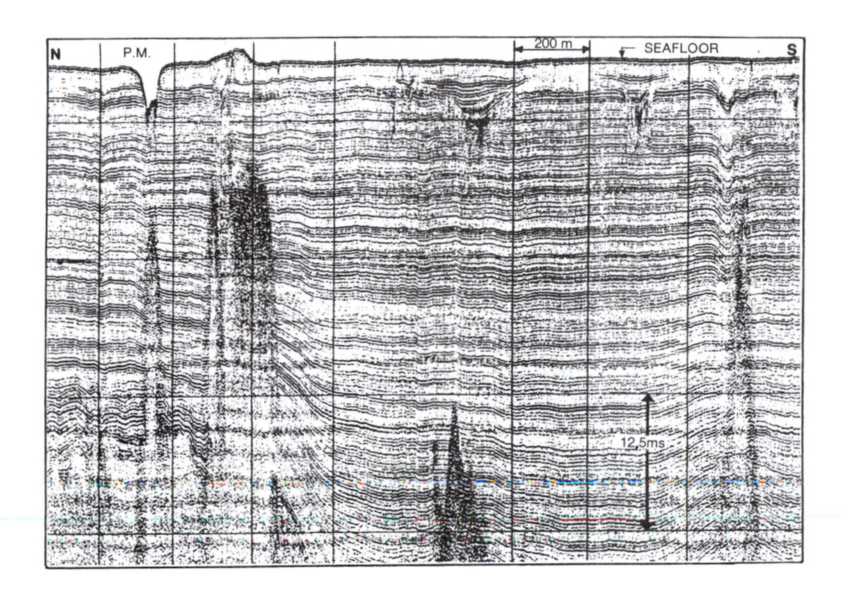

Figure 13-4

Shallow seismic profile showing a pockmark (P.M.) and mound to the right of it due to gas seepage in Norwegian block 25/7, North Sea. [Hovland and Judd 1988]

(1988) estimated that the northern North Sea sediments alone are venting 2.6×10^{12} g of methane per year into the water column.

At what depths do these gas plumes or chimneys originate? In the North Sea they have been linked to petroleum accumulations in the Cretaceous Chalk at 10,000 ft (3,049 m) depth. In the U.S. Gulf Coast some of them appear to be coming from deeper than 15,000 ft (4,573 m). Figure 13-5 is a three-dimensional seismic profile through Plio-Pleistocene sedimentary rocks offshore Louisiana. On the left is the edge of an estimated 7,000 ft (2,134 m) thick Jurassic salt contacting the Plio-Pleistocene sediments. The base of the salt appears pulled up on the time section relative to the surrounding sediments because of the faster seismic velocity in salt.

On the edge of the salt is a gas chimney extending to the surface. These plumes are small (~400 m, 1,312 ft in diameter) and vertically oriented, so they are not usually seen on regional two-dimensional seismic grids. They are not

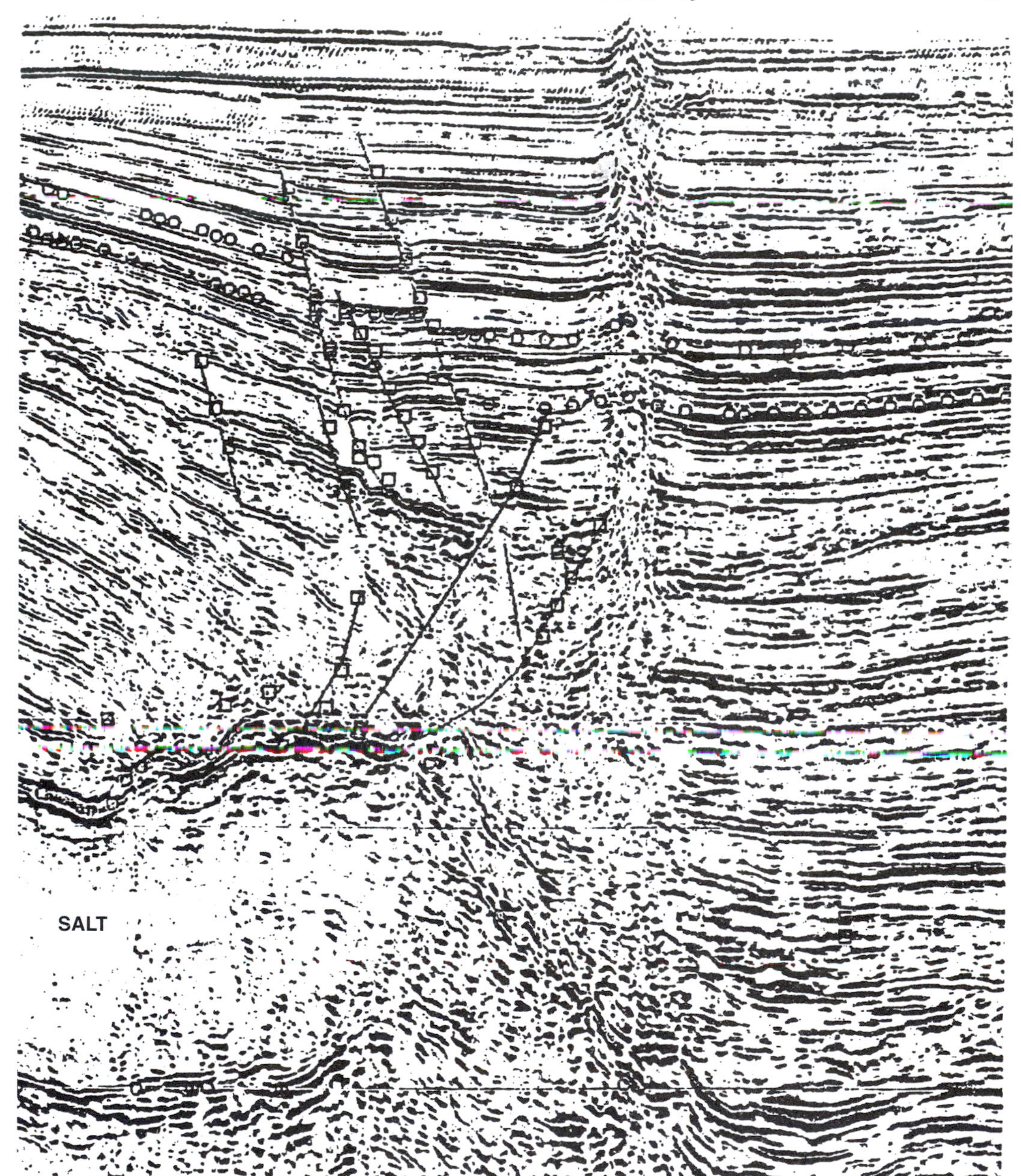

Figure 13-5

A three-dimensional seismic profile of a gas chimney rising from depths greater than 15,000 ft (4,573 m) up through Plio-Pleistocene sediments in the South Marsh Island area, offshore Louisiana. The gas plume is adjacent to a 7,000 ft (2,134 m) thick allocthonous Jurassic salt. The straight lines are faults, and the circles and boxes are cross-line interpretations of horizons and faults.

going up faults because the chimneys are nearly vertical, whereas the faults are at an angle. They are not syndepositional, slow acting, or continuous because there is no evidence of thinning of the sediment layers adjacent to the plumes. They look like high-pressure gas blowouts shooting up like a bullet. Some of them look like wormholes in that they rotate slightly on the way up. The source of the gas in Figure 13-5 is difficult to determine. It may be gas spilling over from an accumulation under the salt or from a deeper reservoir with a fractured seal.

Abrams (1992) found seismic wipeout zones caused by vertical gas migration to be present throughout the Bering Sea Shelf. He took gravity cores and jet cores of sediments over reflection wipeout zones and over near-surface faults related to deeper structures. The jet core recovered samples as deep as 46 m. He found no gasoline-range hydrocarbons at depths shallower than 6 m, because of bacterial action and pore water flushing in this near-surface zone of maximum disturbance. The total gas concentration in the sediments increased by a factor of 17 from the surface to 6 m and a factor of 48 at a depth of 46 m. Also at 6 m, the concentration of C_{2+} hydrocarbons was 162 times greater over a fault, compared with background levels 150 m away. Compositional and isotopic analyses of the gases revealed that the St. George Basin on the shelf contains dry thermogenic gas with no associated liquids and the nearby Navarin Basin contains thermogenic oil and gas. The probable source was believed to be type III Late Cretaceous kerogen.

Jet coring guided by suitably interpreted seismic data is an effective way to trace the origin of gas seeps and to assess the probability of a deep trapping structure containing hydrocarbons. Jet core systems use high-pressure water through a drill pipe and a wireline core to recover 1 m samples from depths as great as 100 m, depending on the nature of the sediment. Some companies have taken jet cores at various depths on the U.S. Gulf Coast to obtain a vertical profile of hydrocarbon distributions in terms of compositional and isotopic variations. Anomalous highs are seen near many faults.

In the 1970s when oil companies were bidding to drill on the giant Destin Dome offshore northern Florida (Figure 12-13), Gulf Oil conducted a detailed geochemical survey over the entire area. It found no hydrocarbon anomalies at all. This discouraged the company from bidding. The winning bidders paid $212 million and drilled fifteen very dry holes, giving it the name "Dusty Dome."

Surface Geochemical Exploration

The term *surface geochemical prospecting* refers to a direct method for finding petroleum that is based on the concept that some hydrocarbons in an oil or gas accumulation migrate vertically to the surface directly over the reservoir. One technique consists of taking samples of soil at a predetermined depth on a grid pattern and analyzing them for hydrocarbons or other materials affected by the presence of hydrocarbons. The method is designed to identify invisible seeps, such as those presumably resulting from hydrocarbon diffusion to the surface. Surface prospecting involves several types of analyses, such as (1) the free hydrocarbon gases in the pore spaces of the soil or in groundwater, (2) hydrocarbon gases adsorbed on soil particles, (3) the fluorescence of soil samples caused by the presence of high-molecular-weight aromatic hydrocarbons, (4) the analysis of soil bacteria that thrive on certain kinds of hydrocarbons, (5) trace metals and radioactive elements in the soil whose adsorption is affected by the presence of hydrocarbons, (6) carbonates formed by the bacterial oxidation of methane, and (7) airborne detection methods.

Unfortunately, when used by itself, surface prospecting has never produced the results claimed for it, despite enormous expenditures by all major oil companies and many independents. Surface prospecting can help an exploration program, but not in the way that it is frequently promoted. The last fiasco occurred in the 1970s when the price of oil was climbing out of sight. A Belgian "count" and an Italian inventor talked the president of France, Giscard d'Estaing, and his prime minister, Raymond Barré, into funding an airborne detection system that was guaranteed to outline oil fields on the ground. After expenditures of some $88 million over nearly two years, during which no qualified geochemist was permitted to see the equipment, the project was quietly shelved (Scott 1984). Although this was obviously a scam, it is still true that even the more professional surface-prospecting methods, when used alone, do not achieve the results expected by the clients, particularly for fields deeper than about 4,000 ft (1,220 m).

The problem is partly a misunderstanding of the processes that cause hydrocarbon anomalies at the surface, partly a failure to realize that microseeps are not very different from visible seeps in locating an oil or gas field, and a failure to utilize all available geophysical and geological data in interpreting the anomalies. Surface anomalies are not caused by diffusion of hydrocarbons from a deeply buried pool (see Hunt 1979, pp. 427–428). Rather, they are caused by migration, mainly in the gas phase along permeable vertical channels such as lineaments, fractures, faults, or permeable beds. They are like visible seeps in that they can reveal that hydrocarbons are present at depth, but except for shallow fields they cannot pinpoint the exact drilling location. Although it is true that the majority of the world's petroleum deposits were discovered by drilling in areas with visible seeps, it is equally true that in many of these cases, a lot of dry holes preceded the first discovery. Oil and gas seeps occur throughout the Western Canada Basin (Link 1952). However, it took 400 dry holes before the Leduc discovery made Canada a major oil producer. The previously mentioned bitumen washing up on the Gippsland coast of Australia caused the drilling of about 100 dry holes onshore before Weeks came along and decided to drill offshore in the Bass Strait. The history of oil exploration is replete with examples of dry holes being drilled in the vicinity of seeps until luck or ingenuity made the big discovery. Neither visible seeps nor surface prospecting can unequivocally outline a petroleum accumulation at depth except where the accumulation is relatively shallow and its caprock is leaking directly upward, as in the Masjid-i-Sulaiman field of Iran, the Burgan field of Kuwait, and the Kirkuk field of Iraq (Figure 13-1). These fields all are in earthquake-prone areas that have caused the caprocks to rupture, and they are shallow enough for direct vertical leakage to be visible.

One of the more competent scientific evaluations of surface prospecting was carried out by GERT (Geochemical Evaluation Research Team). A group of oil companies planned to drill eighteen wildcat wells to depths around 8,000 ft (2,439 m) in the Permian Basin of west Texas. The well locations were based on geological and geophysical data. Before drilling, GERT contracted for the twelve most promising surface-prospecting techniques to be carried out independently at the location of each prospect. The twelve techniques covered a variety of hydrocarbon analyses, trace metal analyses, radiometric measurements, magnetics, and Landsat imagery (Calhoun 1991). The geochemical contracters were asked to predict either a commercial producer or a dry hole. After the surveys were completed, subsequent drilling resulted in three commercial oil discoveries and twelve dry holes, counting marginal wells as dry holes (Towler 1993).

Four of the techniques predicted two of the three discoveries. These were windowed radiometrics measuring gamma radiation in several energy ranges, Landsat imagery using satellite images to define lineaments and tonal anomalies, Mn/K ratio in the soil, and soil fluorescence combined with microbial data. No technique predicted all three discoveries. Also, only two of the twelve techniques correctly predicted the Hunt-2 and Young discoveries. Radiometrics and Landsat imagery predicted the Hunt-2 discovery, and the Mn/K ratio and fluorescence predicted the Young discovery. The Young well was predicted to be a dry hole by radiometrics and Landsat, whereas the Mn/K ratio and fluroescence had null results on the Hunt-2 location. This means that the four apparent best techniques did not support one another as they should have done. Towler (1993) listed the prediction accuracies as ranging between 17 and 83%, counting dry hole predictions, which are made with greater statistical certainty. Some of the techniques had more null results than others. This complicated the comparisons. In order to clarify the benefits of the surveys, Towler (1993) applied risk analysis to the data. He concluded that the overall economics of the exploration program were improved somewhat by using surface prospecting. It would be interesting to repeat this evaluation where drilling is planned in the 1,000 to 9,000 ft (304 to 1,220 m) depth range. That is where surface prospecting has been most successful.

Some of the surface-prospecting projects carried out by major companies in the past are summarized in Hunt 1979, pp. 428–430. Also, Philp (1987) wrote a detailed review of surface-prospecting methods, with more than 200 references. He concluded that the most successful and widely used prospecting methods were those based on direct determination of the hydrocarbons.

The facts that emerged from these and subsequent studies are as follows:

1. Most petroleum reservoirs leak to some extent, particularly those under high gas pressure. Much of the leakage is episodic rather than continuous, due to the periodic opening and closing of migration pathways.

2. Migration to the surface occurs primarily in the gas phase and secondarily in solution, but not by diffusion, which is too slow, even over geologic time. The

migrating hydrocarbons follow the most permeable pathways to the surface faults, fractures, lineaments, fissures, permeable sandstones, piercement salt domes, mud diapirs, and the like.

3. Surface prospecting cannot outline deep subsurface oil fields. It has been most successful in identifying the presence of shallow (4,000 ft; 1,220 m) stratigraphically trapped gas that is leaking. It also can identify low-level hydrocarbon leakage along recognized permeability pathways such as faults.

4. Surface prospecting is a useful auxiliary tool when closely integrated with conventional geological and geophysical exploration methods because it can indicate that hydrocarbons have been generated and are migrating within the area being evaluated.

The most common observation over the years has been the close association of surface hydrocarbon anomalies with faults and lineaments. As pointed out in Chapter 8, many faults periodically open and close because of the episodic release of high-pressured gas. An example of such anomalies is in Figure 13-6. It shows gas anomalies over a fault in a petroleum producing area of northeast Texas (McIver 1985). Gas concentrations here are more than fifty times background levels. Gases migrating up from reservoirs generally spread out in cone-shaped plumes that are not recognized at the surface. Faults, joints, and lineaments, however, tend to focus the gases into narrow channels of migration that are easily identified.

Gas concentrations above 1,000 ppm were observed by McIver (1985) along a major fracture zone in the shallow Scipio–Albion Trend of the Michigan Basin. These analyses were made on the head space gases of canned sediments taken at depths beyond 8 feet (2.4 m) and on free gases taken in vacutainers at the same depths. Aerobic bacteria destroy the gaseous hydrocarbons in the top 6 to 7 ft of most onshore sediments. In the onshore U.S. Gulf Coast, the break between 10 ppm methane and 80 ppm methane in some sediments is at a depth of about 7 ft (2.1 m). Offshore, the oxidation zone is much thinner, but there also is bacterial gas that must be differentiated from the thermogenic gas.

Jones and Drozd (1983) found a clear association of surface hydrocarbon anomalies with fault zones in the Utah–Wyoming overthrust belt, as shown in Figure 13-7. The thrust faults in the Ryckman Creek Anticline area of Wyoming reach the base of the Tertiary, which is between 3,000 and 4,000 ft (900 to 1,220 m) thick. Anomalous seepages of methane, propane, and butanes are associated with the position of the thrust faults at the base of the Tertiary. They indicate that gases from deep reservoirs have migrated updip along the thrust plane and then vertically through the Tertiary section. A similar close association of subsurface faults and high-surface hydrocarbon concentrations was observed at Pineview field, Utah, along the overthrust belt.

Shallow fields may leak without any visible pathways showing up on the seismic records. Jones and Drozd (1983) ran a geochemical survey above a depleted sand that was going to be used for gas storage in the Pleasant Creek area near Sacramento, California. The reservoir is a stratigraphic trap at a depth of 760 m (2,500 ft). After the survey, it was filled to capacity at a pressure

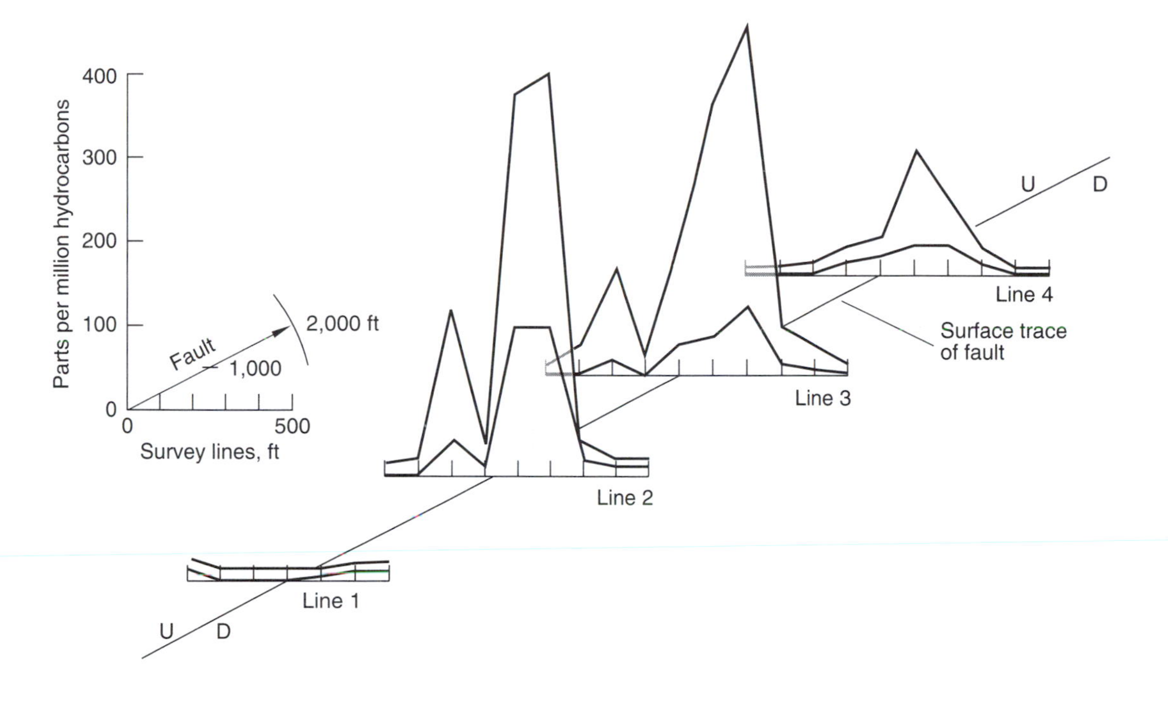

Figure 13-6

Surface geochemical profiles across a major fault in an oil-producing area of northeast Texas. The top curve is total gas, and the bottom curve is ethane plus. Each tick on horizontal lines indicates a sample location. [McIver 1985]

of 10,343 kPa (1,500 psi), and then a second geochemical survey was run. The survey before the gas injection showed no gas, as expected. Depleted oil and gas fields rarely show any hydrocarbons at the surface. Anomalously high C_1, C_2, and C_3 values were noted in the soil gases within a few months after filling the reservoir. This is far too rapid for diffusion, so the gas had to be moving vertically in the gaseous phase, although no obvious pathways showed up in the geophysical data. However, most oil-field geophysical-prospecting methods are insensitive to very small faults or nonoffsetting fractures.

McIver (1985) found methane concentrations in sediments over the Foley gas field of Alabama to be 80 ppm by volume above background levels. The gas had migrated vertically 2,000 ft (610 m) to the surface. McIver had done surface prospecting in many areas in the 1980s and became convinced that the method works not as a direct oil-finding technique but as an indicator of which the best prospects are that are already delineated by seismic data. He also felt that the hydrocarbon signal from a buried field is progressively

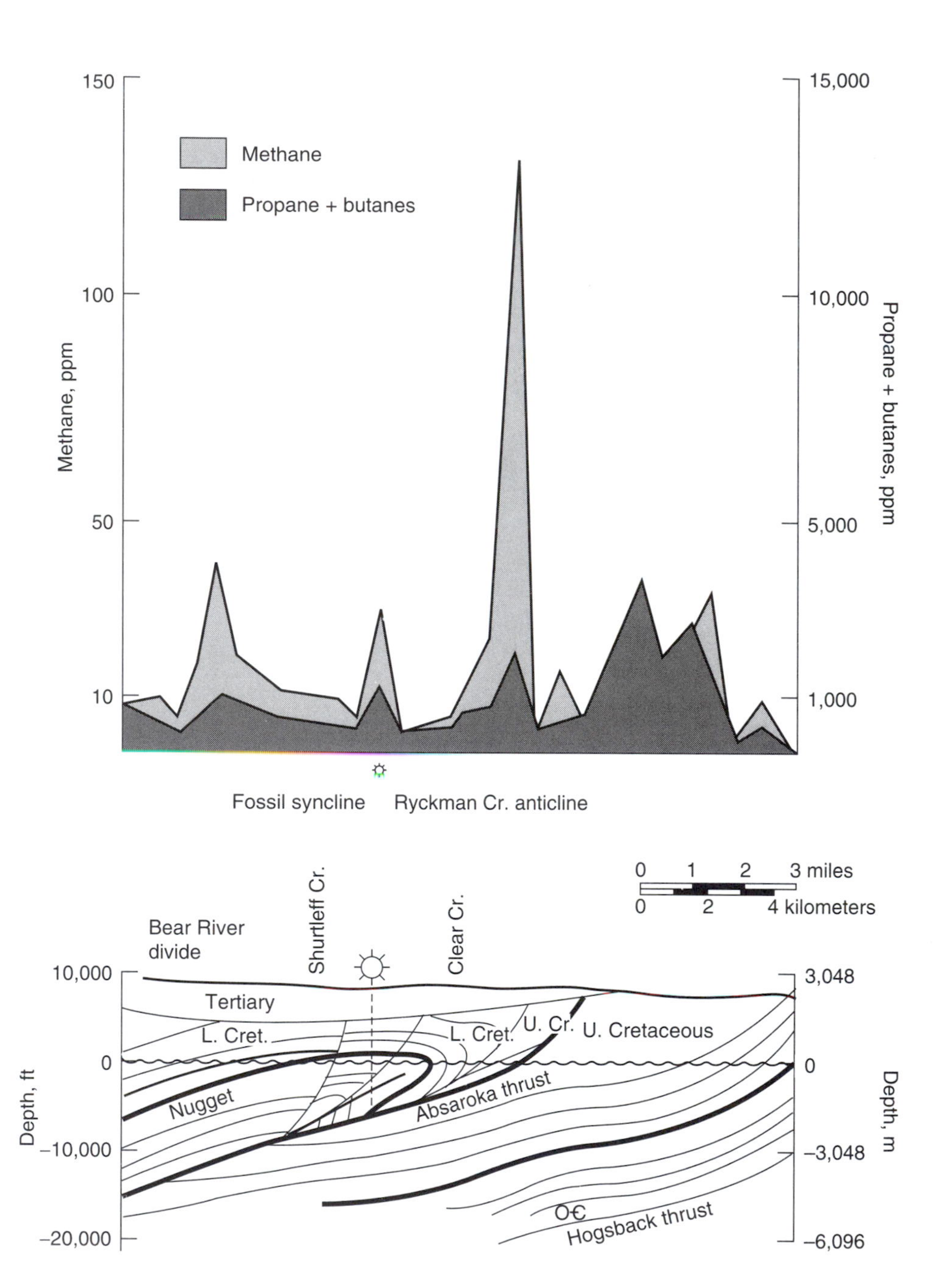

Figure 13-7

Correlation of soil gas with thrust faults reaching the base of the Tertiary at Ryckman Creek, Wyoming Depth is relative to sea level. [Jones and Drozd 1983]

more fragmented as the depth increases. Consequently, surface prospecting is unreliable for fields deeper than about 4,000 ft (1,220 m), except as a regional tool. Also, a tight shale or evaporite as a caprock over a reservoir may prevent leakage, and no signal is detectable. Another problem is source rock gas. In Kentucky the Devonian black shale contains up to 15% TOC and is loaded with gas, thus preventing one from seeing into the reservoirs below the shale. Mapping the lineaments and joints is important because these are the usual migration pathways.

Some prospectors report their data at the ppb (parts per billion) level, which has no significance, according to McIver. In parts of the U.S. Gulf Coast, background levels are 3 or 4 ppm, with anomalies 5 to 100 times that. Working above the Antrim Shale of Michigan or the Ohio black shale gives a background level around 20 ppm.

SUMMARY

1. Visible oil and gas seeps are important to exploration because they indicate that oil or gas has been generated and is migrating in the area. The most important oil-producing regions of the world were discovered through visible oil and gas seeps.

2. When crude oil leaks to the surface, it undergoes evaporation, leaching of water solubles, microbial degradation, polymerization, auto oxidation, and gelation. These reactions lead to thickening or solidification of the oil. A liquid oil is gradually converted to an asphalt, an asphaltite, and eventually a pyrobitumen.

3. Seeps are most numerous on the margins of basins and in sedimentary rocks that have been folded, faulted, and eroded. Seeps originate in five types of geological situations: (1) emerging from monoclinal beds; (2) direct leakage from a source bed; (3) breaching or rupture of a reservoir by erosion, faulting, or folding; (4) migration along unconformities; and (5) migration along mud volcanoes, igneous intrusions, and piercement salt domes.

4. Worldwide, there is a correlation between seeps and earthquake activity, with many seeps in areas of plate boundaries, where such activity is the highest.

5. Seepage patterns are strongly controlled by regional and local tectonics. Seeps are most common in overpressured diapir-rich basins and in active thrust belts. They are rare in tectonically inactive basins.

6. The presence of seeps over basins or prospects often considerably reduces the exploration risk. The absence of seeps over a tectonically active basin or shallow faulted prospect increases the risk.

7. Offshore seeps may be monitored by divers, underwater cameras, gas sniffers, and seismic profiles. Compositional and isotopic analysis of seeping gas is required to distinguish thermogenic from bacterial gas seeps.

8. Offshore gas chimneys or plumes are common in areas with dominantly vertical migration, such as deltas (Mississippi) and rift grabens (North Sea).

9. Surface geochemical prospecting involves analyzing soil samples for hydrocarbons or other materials affected by the presence of hydrocarbons. Gas anomalies measured at sediment depths of 8 ft (2.4 m) or more in the vicinity of joints, faults, fissures, lineaments, or other vertical migration channels frequently are indicators of deeper accumulations.

10. Surface prospecting is most useful as an auxiliary tool to detailed geophysical and geological surveys. It can assist statistically in differentiating structures and areas likely to contain hydrocarbons from those likely to be barren. It is particularly useful in identifying shallow (< 4,000 ft; 1,220 m) fields leaking gas.

11. No surface-prospecting method can outline the dimensions of an oil pool at the surface simply because there is no perfectly vertical mechanism of hydrocarbon migration.

SUPPLEMENTARY READING

Hovland, M., and A. G. Judd. 1988. *Seabed pockmarks and seepages: Impact on geology, biology and the marine environment.* London: Graham & Trotman. 293 p.

Philp, R. P. 1987. Surface prospecting methods for hydrocarbon accumulations. In J. Brooks and D. Welte (eds.), *Advances in petroleum geochemistry,* vol. 2. London: Academic Press, pp. 209–253.

Chapter

14

A Geochemical Program for Petroleum Exploration

The kinds of geochemical analyses that should be used for outcrop and well samples and their value in exploration are discussed in this chapter. These techniques use different approaches to answer the same question, namely, has the sedimentary section of interest generated oil or gas in sufficient quantities to form economic accumulations? And if so, where are the most promising areas? Some of these techniques are so similar that it is not necessary to use all of them on a wildcat well. At the same time, it is equally dangerous to rely on only one methodology, because any geochemical approach is subject to anomalous situations, as are many geological methods. Also, subsurface geochemical data should not be evaluated in the absence of geological information. Geologists and geochemists need to work together closely in interpreting the data if the results are to have any value in exploration. Subsurface techniques have been most valuable when the geologists describe the problems and the questions to be answered with the geochemists, who suggest the best analyses for answering those questions. This chapter also examines the errors and pitfalls in using geochemistry. These are important to recognizing any wrong interpretations in geochemical reports.

A typical geochemical approach is to use the less expensive, more rapid techniques for an initial screening of the hydrocarbon potential of available samples. When these data are combined with data from the hydrocarbon mud logs and wireline logs, one can be more selective in identifying sedimentary intervals for more detailed geochemical analyses.

Surface Reconnaissance

In some areas both the surface sediments and subsurface rocks seem to be alive with hydrocarbons. Seeps are common. Oil and gas shows are encountered at several horizons while drilling. Background gas builds up in the circulating mud. In other areas the rocks seem to be devoid of hydrocarbons. No seeps. The mud logger shows nothing, and there is not the slightest evidence of hydrocarbons from drill-stem tests. This concept of an area's being alive or dead with respect to hydrocarbons can be followed up by mapping both surface and subsurface hydrocarbon shows and using geochemistry to identify their sources.

One factor leading to the formation of giant oil and gas fields is above-normal geothermal gradients, such as those found in basins like Central Sumatra, the Pre-Caucasus, and Los Angeles, California. Klemme (1975) observed that the yield of hydrocarbons per volume of rocks is higher in basins of high heat flow than in those of low heat flow. The effect was greatest for intermediate crustal zone basins that occupied the transition area between continental and oceanic crusts. Typical are the basins associated with the belts along which crustal plates are underthrusting or overriding one another along subduction zones. For example, the west coast of the United States, the basins behind the Indonesian volcanic arc, the Baku Basin, and southeastern Australia are areas of high heat flow associated with large petroleum accumulations. Klemme's worldwide survey showed considerable variation in geothermal gradients, caused by local hot spots or hot belts through sedimentary basins. The hotter areas contained either shallow oil or deep gas.

Heat-flow measurements combined with geophysical data before drilling a well can give a rough estimate of the geothermal gradient in the area. The equation for heat flow is

$$Q = \lambda \frac{\partial T}{\partial Z}$$

where Q = heat flow in cal/cm^2 sec
λ = thermal conductivity of the substance in cal/cm sec °C
$\frac{\partial T}{\partial Z}$ = geothermal gradient in °C/cm, where Z is positive downward

Although heat flow, *Q*, varies considerably in different areas, it is essentially constant at a specific location, assuming no disturbing effects such as a nearby salt dome. Since *Q* is constant, a decrease in thermal conductivity in a sedimentary section results in an increase in the geothermal gradient, and vice versa. Klemme (1975) published heat-flow data for several basins worldwide. By using the lithology of outcrops and extending these into the subsurface with seismic data, it is possible to estimate the thermal conductivity of the rocks at a specific location. Combining this with the heat flow gives an approximate geothermal gradient before drilling. This is needed to predict the probable depth of petroleum generation.

The quality of rock samples for geochemical analyses generally decreases in the following order: conventional whole cores, sidewall cores, drill cuttings, and outcrops. There are exceptions. If the sidewall cores are taken after the well is drilled to its total depth, they may be contaminated with mud filtrate during the long drilling operation.

Outcrops of pre-Tertiary shales and carbonates that are possible source rocks are suitable for many geochemical analyses, assuming that they are taken below the weathered layer. The depth of weathering is quite variable, but in most source rocks it ranges between 0.5 and 5 ft (0.15 and 1.5 m). The exceptions are very young rocks such as the Tertiary of the California coast and the North Sumatra Basin of Indonesia. In some of these areas, weathering extends well beyond the capability of a power auger.

The best visual indicator of weathering is the color change. The greater the discoloration is, the greater the intensity of weathering has been. For example, the effect of weathering was compared with extractable hydrocarbon yields for the Ordovician Viola Limestone and Mississippian Woodford Shale source rocks of southern Oklahoma. The Woodford changes color from a yellow brown to gray black in the first 1 ft (0.3 m) of depth, whereas the Viola limestone does not show the change from yellow brown to gray until 5 ft (1.5 m) of the outcrop is penetrated. The extractable organic matter in the weathered section of both of these rocks is about half that in the unweathered section, with the aromatic hydrocarbons being the most depleted. They are the most soluble in groundwaters.

Clayton and Swetland (1978) found that the weathered surface samples of the Permian Phosphoria Formation of northeastern Utah had 60% less TOC and 53% less C_{15+} extractable hydrocarbons, compared with the unweathered section about 2.8 ft (0.85 m) deeper. The change in extractable aromatic hydrocarbons is shown in Figure 14-1. The aromatic fraction at a depth of 2.8 ft was equivalent to the background level of 780 ppm, compared with only 200 ppm at the weathered surface.

Lewan (1980) documented numerous changes in the organic fraction of black shales from unweathered to severely weathered samples. Outcrop samples were invariably depleted in aromatic compounds. Normal alkanes and isoprenoids were sometimes missing, owing to biodegradation. Many kerogens contained more oxygen and were associated with less pyrite than subsurface samples. Elemental analyses of the Eocene Kreyenhagen Shale of California showed a significant change with weathering. The atomic H/C ratio decreased from about 1.2 to 0.9, and the O/C ratio increased from 0.17 to 0.27 when going from the weathered to the unweathered zone at a depth interval of about 5 ft (1.5 m).

In 1987, Clayton and King reported on biomarker changes in the Phosphoria Formation samples shown in Figure 14-1. They found that there was considerable loss of the 20S diastereoisomer of the C_{29} steranes in the weathered surface sample. Also, the C_{20}-to-C_{26} tricyclic diterpanes were severely reduced, and the C_{19} was eliminated. None of these effects occurred in unweathered samples taken at depths >2.8 ft (0.85 m). In general, outcrop samples taken from below the visible weathered layer (0.5 to 5 ft, 0.15 to 1.5 m) are suitable for most of the geochemical analyses discussed in this chapter.

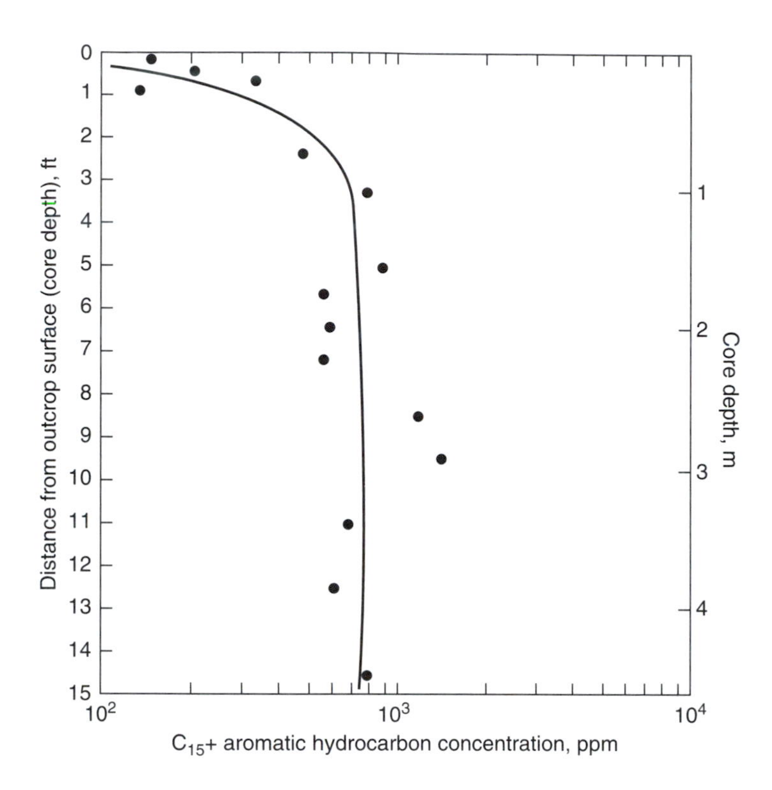

Figure 14-1

Decrease in C_{15+} aromatic hydrocarbons in a Phosphoria source rock outcrop, northeastern Utah. [Clayton and Swetland 1978]

Davis et al. (1989) carried out a successful sedimentological and geochemical study of the Mowry Shale of Wyoming based on 104 outcrop samples. They were able to relate the depositional environment to the petroleum source potential of the Mowry by combining lithofacies studies with Rock-Eval pyrolysis. The Mowry is a siliceous shale that is not deeply weathered.

Hydrocarbon Mud Logging

Hydrocarbon gases in the subsurface exist in the free form, dissolved in formation water, or adsorbed on organic matter and mineral particles. Hydrocarbon mud logging was first used commercially by the oil industry in 1939 to detect pay horizons with rotary drilling. As the drill bit breaks the rock into small fragments or cuttings, the gas present is released into the mud stream. As the mud comes out of the hole, a portion is run into a mud–gas separator. This may be simply a cylinder containing baffles to spread out the mud or a stirring unit to beat up the mud and release the entrained gas. The gas–air mixture is then fed into a gas chromatograph, where the individual hydrocarbons—C_1, C_2, C_3, C_4, and C_{5+}—are analyzed. Some systems using a simple hot-wire detector make only two measurements, the methane and the ethane-plus fraction, which includes all higher gaseous hydrocarbons. Other detectors can be used to analyze nonhydrocarbons such as H_2S. (For details on equipment and interpretations, see Helander 1983, pp. 41–53.)

The technique was designed specifically to test reservoir rocks for hydrocarbons during drilling. In fact, it is still the practice in some areas to use the logger only when penetrating prospective pay horizons, but most companies log continuously in wildcat wells.

In the former Soviet Union, continuous mud logs and intermittent core samples were taken of both reservoir and nonreservoir rock. The cores were hermetically sealed on recovery, and the gas was desorbed in the laboratory by heating to 70°C (158°F). The gas logs they obtained represented a combination of data from both mud and core analyses. This enabled Soviet geochemists to look not only at the reservoir rocks but also at the fine-grained source rocks throughout the entire sedimentary section. In the late 1950s, V. A. Sokolov and B. P. Yasenev noted that sedimentary rocks immediately above productive oil or gas fields frequently contain more gas than do rocks from dry hole areas. For example, in Figure 14-2 the gas logs are shown for two wells drilled in the Kum–Dag region of Russia. Well 41 encountered a commercial petroleum accumulation at a depth of about 780 m (2,256 ft). Well 42, which was located around 1,200 m (3,937 ft) from well 41, was a dry hole. A higher concentration of gases was noted in well 41 throughout the section, but the largest increase occurred below a depth of 580 m (1,900 ft). Here the methane and ethane-plus increased substantially, even though it was 200 m (656 ft) above the oil accumulation (Yasenev 1962). Additional examples of the presence of gas in well cuttings above oil and gas fields and its absence in cuttings above nonproductive structures were cited by Hunt (1979, pp. 436–439). These examples support the concept that sedimentary rocks containing gas are more apt to have underlying structures containing producible hydrocarbons.

In the Western Canada Basin, several hundred mud logs made by commercial logging companies were collected to study gas yields going southeast across Alberta into Saskatchewan and into the Williston Basin. The C_1 and C_{2+} gas yields from the logs were plotted on cross sections, a summary of which is shown in Figure 14-3. Five typical locations in the area are presented. At the first two locations, high gas yields were found on the mud logs from the base of the Devonian up through the Lower Cretaceous Mannville Formation. There are commercial oil and gas fields through this whole section. At location 3, mud gas was observed mainly in the Mississippian and Cretaceous Mannville Formations where the oil was found. There is no oil or mud gas at this location from the Cambrian up through the Lower Devonian. At location 4, immediately west and north of Regina, Saskatchewan, there was no gas evident in the

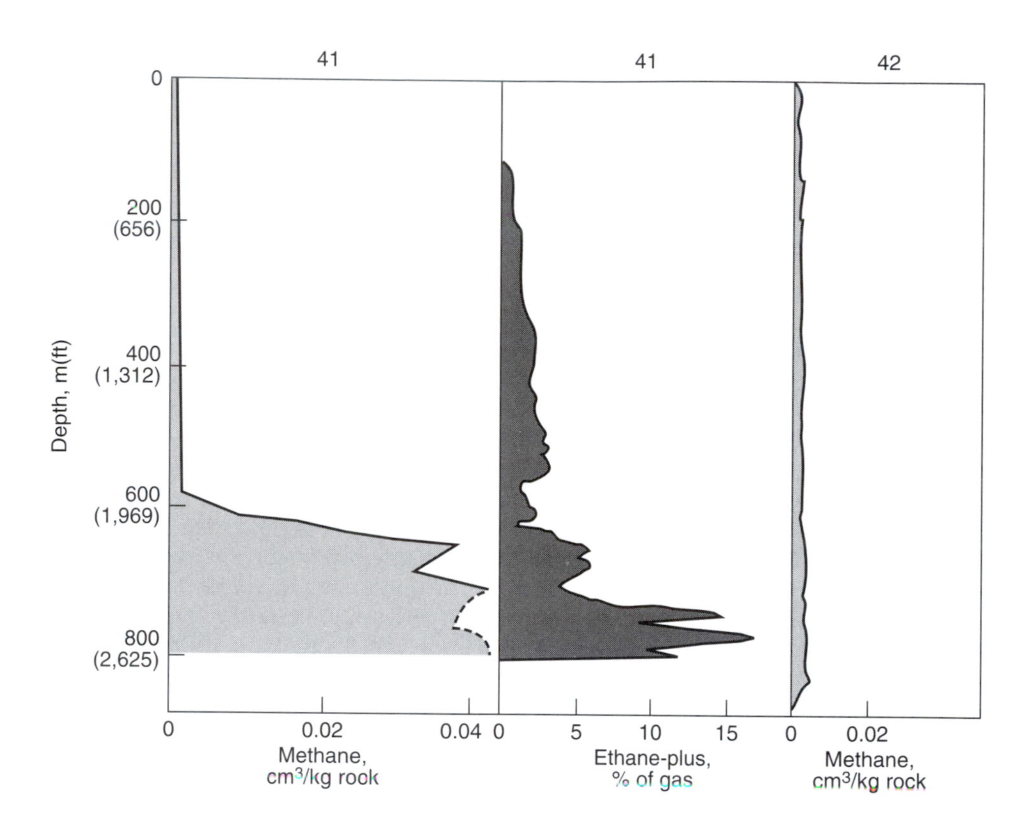

Figure 14-2

Gas logs (mud logs) of two wells in the Kum–Dag region of Russia. Both methane and ethane-plus increase above the pay horizon at 780 m in well 41. Only traces of methane were found in well 42, which was a dry hole. Note that the increase in well 41 does not extend to the surface. [Yasenev 1962]

drilling mud over the entire section from the Cambrian through the Cretaceous Mannville Formation. This is a dead area with little potential for commercial hydrocarbon accumulations, regardless of other geological factors. If no oil or gas is being generated or is migrating anywhere in the section, there can be none in the reservoirs. Some four hundred dry holes were drilled in this area, many of them after the data in Figure 14-3 became available. Southeast of Regina at location 5, the mud logs show a sharp increase in wet gas yields, with particularly high yields in the Mississippian of the Williston Basin where there are oil accumulations. Even here, there is little or no shale gas in the Cretaceous and in the deeper Lower Devonian, Ordovician, and Silurian sections where there also is no production.

From the Devonian through the Cretaceous throughout British Columbia and Alberta, Canada, the shale gas is wet in oil-producing areas and dry in

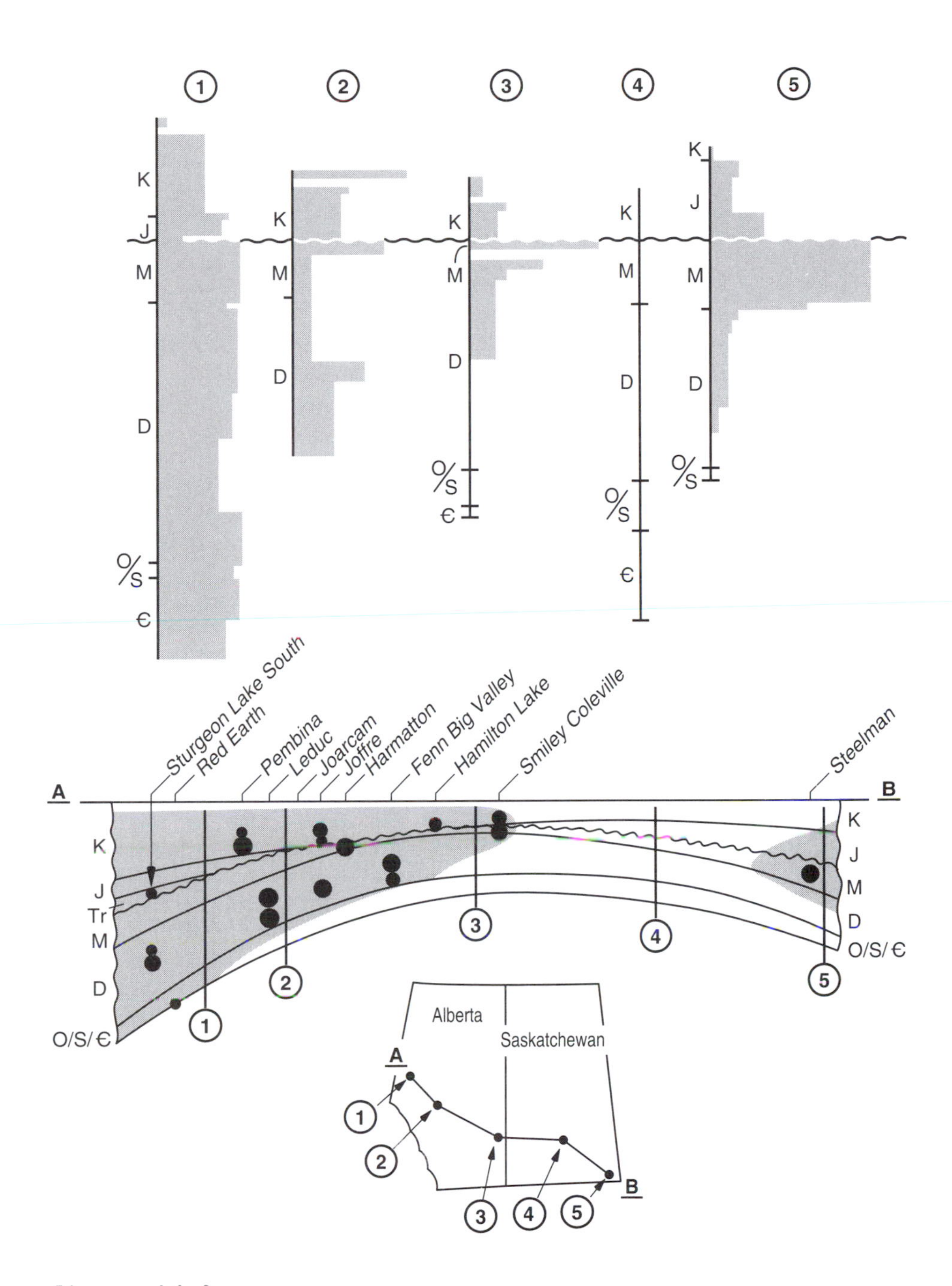

Figure 14-3

Correlation of mud-log gas yields (top) with major oil production (bottom) in the Western Canada Basin. Shaded area (top) represented gas yields. The field names are shown above the fields (solid circles). Letters represent geological periods (see Appendix 2).

gas- and heavy oil-producing areas (Bailey et al. 1974). Pixler (1969) used hydrocarbon ratios in mud logging to determine whether a reservoir is likely to produce oil or gas. The basic concept is that wet gas containing ethane through pentanes indicates oil production, whereas dry gas indicates only gas production or very heavy oil. Pixler's ratios of gases for predicting the probable type of production apply to shales as well as to reservoir rocks.

Correlations of mud-log gas to petroleum production are evident in other areas. Mud logs from the Paradox Basin have higher shale gas yields near production than in the surrounding barren areas. In the Gulf Coast, relatively high shale gas readings were obtained in the Tuscaloosa Formation of Louisiana and Mississippi. Going east, the shale gas yields of the Tuscaloosa decrease to almost zero in Georgia and northern Florida. Drill-stem tests of the Eagle Ford and Tuscaloosa sandstones going from Louisiana to Georgia exhibit the same decrease in the dissolved gas of formation waters (Figure 14-4) as does the shale gas. Gas concentrations as high as 15 cubic feet/barrel of water have been recorded in Louisiana, compared with zero in eastern Georgia (Buckley et al. 1958).

At the time of the study of the formation waters shown in Figure 14-4, more than two hundred dry holes had been drilled through the entire Upper Cretaceous section in the Florida Panhandle and south Georgia, without a single commercial show of oil or gas. The main producing sandstones of the U.S. Gulf Coast from Texas to Alabama contain from 1 to 14 ft^3 gas/bbl water (178 to 2,492 cc gas/liter water) in the oil- and gas-producing regions (Buckley et al. 1958). In the nonproducing areas there is no gas in the formation waters.

Continuous hydrocarbon mud logging or some continuous geochemical analysis of cuttings is important even in developed areas. Too often, production in a specific formation encourages a prejudice against other potentially productive zones. The problem with the rotary drill is that it can seal off producing formations with mud, as the following example demonstrates:

In 1946 the Humble Oil and Refining Company drilled the No. 1 Davis wildcat to test a large structure in the Permian Basin in west Texas. No mud logger was put on the well until about 7,000 ft (2,134 m) before reaching the Ordovician Ellenburger Limestone, because at that time it was the only significant pay horizon in the basin. Meanwhile, the geologist assigned to that area was sitting on another wildcat at the time that the No. 1 Davis went through Pennsylvanian age limestones at 4,000 ft (1,220 m). One of the roughnecks noticed some oil in the mud pit. When the Ellenburger Limestone was tested and found to be dry, the well geologist also wanted to test the Pennsylvanian, but the Houston headquarters did not permit him to do so. They said that "the Pennsylvanian was not recognized as a producing zone at that time." Consequently, the hole was plugged and abandoned. Standard of Texas moved in later and discovered 2.8×10^9 bbl of oil in the Pennsylvanian Scurry Reef. It was the second largest oil field in Texas. Almost any hydrocarbon-monitoring system would have found it.

Another example in the early 1970s was of the Gulf Oil Company wildcat drilled 23 miles west of the Lancashire coast in the United Kingdom's Morecambe Bay. The wireline logs indicated noncommercial gas, even though gas

Figure 14-4

Dissolved gas in the Lower Tuscaloosa Formation waters of the Gulf Coast in cubic feet/barrel of water.

was present on the hydrocarbon mud logger. Gulf Oil dropped the lease. In 1974 British Gas drilled a well on the same structure and discovered 56 TCF (170×10^9 m^3) of gas at 3,600 ft (1,100 m) below the seafloor.

Oil-Rich Muds

Oil-rich and oil-based muds are sometimes used in drilling to eliminate shale hydration and swelling, increase lubrication, provide torque reduction in directional holes, protect against corrosion and stress cracking from H_2S, and ensure formation insolubility when drilling salt (Simpson 1985). Also, oil-rich muds provide mud stability at temperatures above 350°F (177°C). Mineral oils composed of paraffins and naphthenes are gradually replacing the more toxic diesel oil. They are particularly useful for directional drilling in formations where severe hole problems and stuck pipe have been encountered (Johancsik and Grieve 1987).

All oil-rich muds, however, reduce the effectiveness of a mud logger, because the oil in the mud retains much of the gas entering the mud stream from a formation. Also, differential gas absorption fractionates the gases, as illustrated in Figure 14-5. The standard agitation gas separator releases about 65% of the gas in conventional water-based muds for GC analysis. But as oil is added to the mud, the yields drop precipitously, particularly for the higher hydrocarbons. With 3% oil in the mud, only about 6% n-C_4 and iso-C_4, 16% C_3, 35% C_2, and 59% C_1 of the gas entering the mud is released for analyses. The yields continue to drop at higher concentrations of oil in mud, as shown in Figure 14-5. Consequently, it becomes impossible to characterize high-gas shales and to differentiate wet from dry gas sections indicating oil and gas sources.

A conventional mud logger is not very effective with oil-rich muds, even for evaluating reservoirs. This problem can be alleviated by the use of a steam still, which sweeps steam through the mud to release the hydrocarbon gases. Yields of the heavier gases are above 75%, whereas methane yields are close to 100% for water-based muds containing up to 15% oil (Figure 14-5). The steam still is limited, though, in being a batch operation rather than offering the continuous analyses of a mud logger.

Some geochemical analyses can be run on cuttings obtained with oil-based muds. If the cuttings are washed free of the mud and oil with organic solvents, they will be suitable for TOC, kerogen color, vitrinite, and pyrolysis (S_2 only).

Wireline Logs

In 1968, Murray reported that the black petroliferous source rocks of the Mississippian Bakken Formation in the Williston Basin showed very high electrical resistivities, high gamma ray response, and low transit time on the sonic log. The black shales are very radioactive, and their pore spaces are saturated with hydrocarbons. Meissner (1978) expanded on this by showing that the resistivities are very high where the Bakken is mature and has generated oil, whereas they are low where it is immature, with no oil. Many later publications advocate using wireline logs to evaluate source rock potential, including Schmoker (1981), Meyer and Nederlof (1984), Herron (1991), and references therein.

The advantage of wireline logs is that they contain far more data points than are available from conventional geochemical or petrophysical analyses. They provide data where no samples are available, and they can be used to closely follow organic facies variations. In frontier basins, logs may indicate stratigraphic intervals of unidentified source rocks, and in more explored basins they can be used to quantify the vertical and lateral extent of source rock units. On the negative side, logs are only indirect measures of source rock

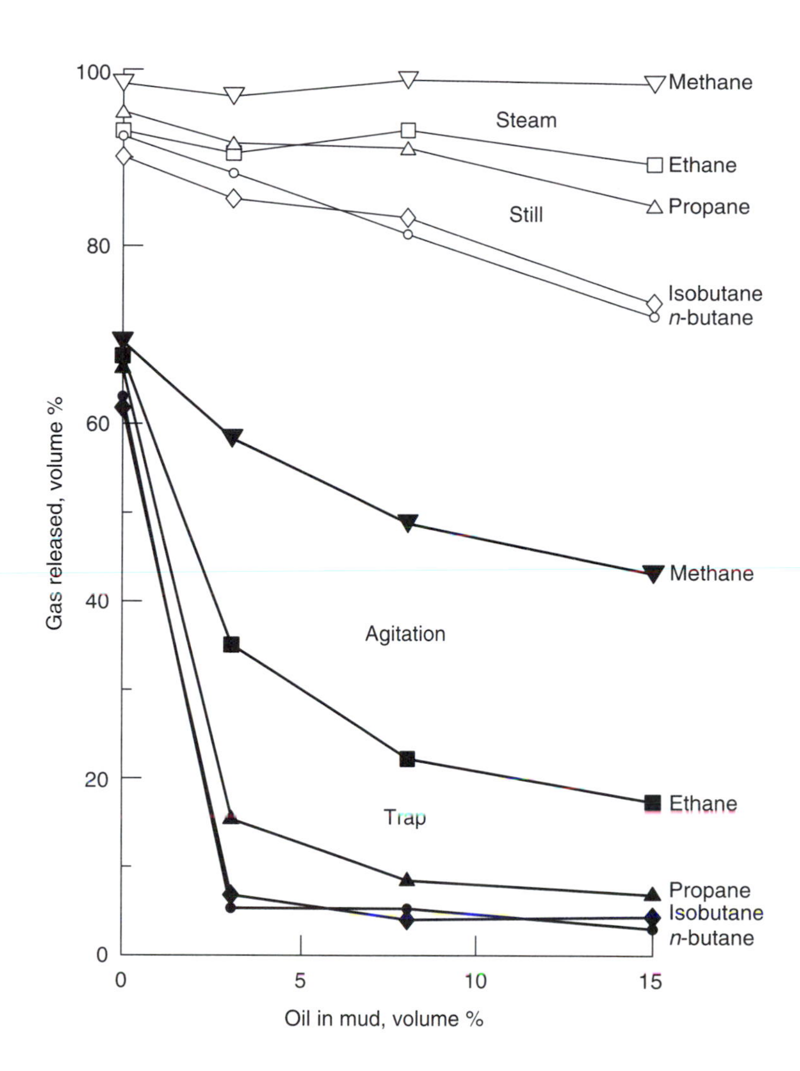

Figure 14-5

The volume percentage of total gas released from drilling mud by an agitation trap (heavy lines) and a steam still (light lines) with increasing amounts of oil in the mud.

characteristics. They are subject to problems related to drilling difficulties, and they must be interpreted by skilled log analysts. The logs most frequently used are gamma ray, resistivity, sonic transit time, and density.

The gamma ray response comes from the three radioactive elements most common in rocks, namely, potassium-40, uranium, and thorium. Background levels of radioactivity that show up on a gamma ray log are primarily from potassium. But when organic matter is deposited in a marine environment, it tends to concentrate the uranium from seawater and pore fluids. Typical gray shales may have around 4 ppm uranium, and black shales may have around 20 ppm. The term *hot shale* describes high TOC anoxic black shales with high gamma ray readings, such as parts of the Kimmeridge Shale in the North Sea. Hot shales may have up to 3,000 ppm uranium.

Gamma ray responses are recorded in API units (API RP 33, 1974). Schmoker (1981) found more than 300 API units in part of the Devonian black shales of the Western Appalachian Basin, compared with background levels of 50 API units in other lithologies above and below the black shales. He used gamma ray logs along with density logs to estimate the TOC values. A similar comparison of gamma ray and TOC by Creaney and Allan (1990) is shown in Figure 14-6. The Jurassic Nordegg is an extremely rich source rock of the Western Canada Basin with TOCs up to 27%. The gamma ray log and TOC measurements both show two cycles of organic enrichment that can be traced across the basin. Goff (1984) also used a gamma ray log interval showing more than 200 API units to correlate a highly radioactive section of the Kimmeridge Shale in the East Shetland Basin with the same stratigraphic level in the Viking Graben. This technique is not applicable, however, to freshwater lacustrine shales, which lack uranium.

Electrical resistivity increases where organic matter or oil replaces water in rock pores, because the former are not electrically conductive. This is characteristic of the Bakken shale in the Williston Basin, as explained in Chapter 9 (Figure 9-9), where the resistivities exceed 100 ohm-meters in the mature section, compared with less than 10 ohm-meters in the immature section of the Bakken. Where this higher resistivity is caused by organic matter, the bulk density is correspondingly lower, as illustrated in Figure 14-7 for the Liassic Posidonien Schiefer source rock in south Germany (Meyer and Nederlof 1984). The maximum TOC in the Posidonien Scheifer is 20%, compared with 0.5% for the overlying shale.

Organic-rich shales also have low sonic velocities, as seen in Figure 14-8 for the Jurassic source rocks of Saudi Arabia (Ayers et al. 1982). Sonic transit times slow from about 60 to nearly 100 μ sec/ft in the most organic-rich zone. Bulk density decreases in the same zone, and the gamma ray log shows an increase. The nature of sonic logs makes for relative changes that are quite sensitive within a given well, but the absolute quantities between wells are not readily compared.

Most evaluations of source rocks use several logs and calibrate them with geochemical data, such as cuttings analyses, in order to estimate quantitatively the TOC and the volume and maturity of the source rocks. Meyer and Nederlof (1984) used cross plots of density versus electrical resistivity and sonic transit time versus electrical resistivity to distinguish source rocks from nonsource rocks. The latter were defined as those having less than 1.5% TOC.

Although wireline logs are a useful supplement to conventional geochemical studies, they must be used with considerable caution, for several reasons.

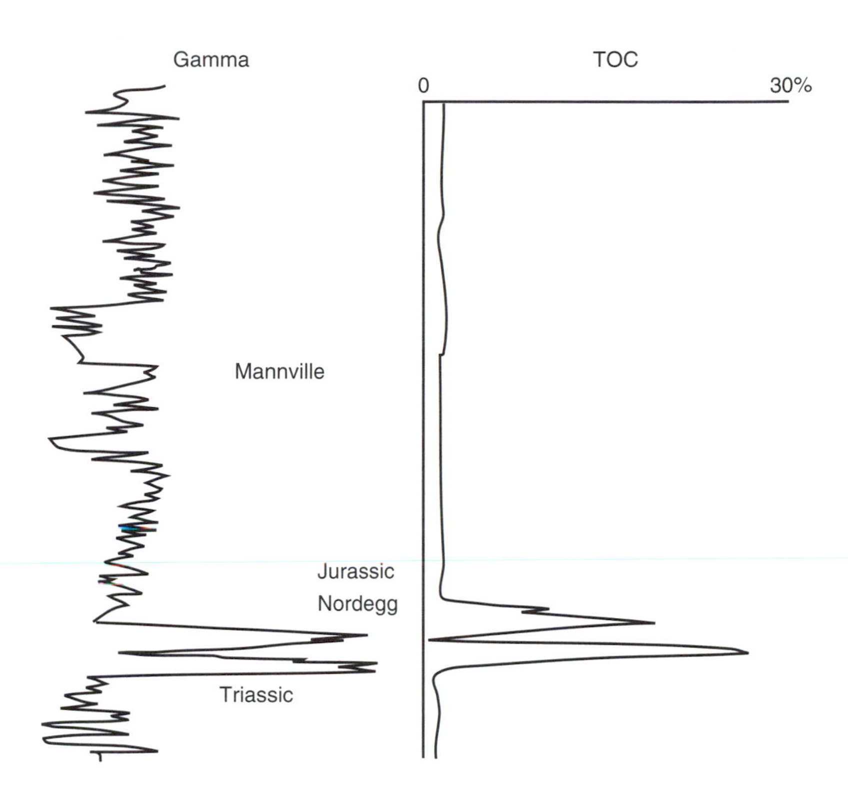

Figure 14-6

Comparison of TOC analyses and the gamma ray logs for the Jurassic Nordegg Shale source rock and adjacent formations in the Western Canada Basin. [Creaney and Allan 1990]

Geochemical data may be calibrated poorly because of (1) pyrite and other minerals producing anomalously high density readings, (2) undercompacted soft formations showing less contrast between organic-rich and lean layers and poor correlations between resistivity and sonic logs, (3) large washouts in boreholes affecting logging tools differently, (4) low-porosity dense rocks such as limestones and dolomites showing anomalously high resistivities because of a lack of electrically conducting fluids, and (5) igneous intrusions causing variable gamma ray intensities. Also, the applicability of logs is limited mostly to rocks with TOCs >1.5%. If these problems are recognized and accounted for, it is possible to use several logs calibrated periodically with geochemical data to increase greatly the available data points for evaluating the source rocks in a basin.

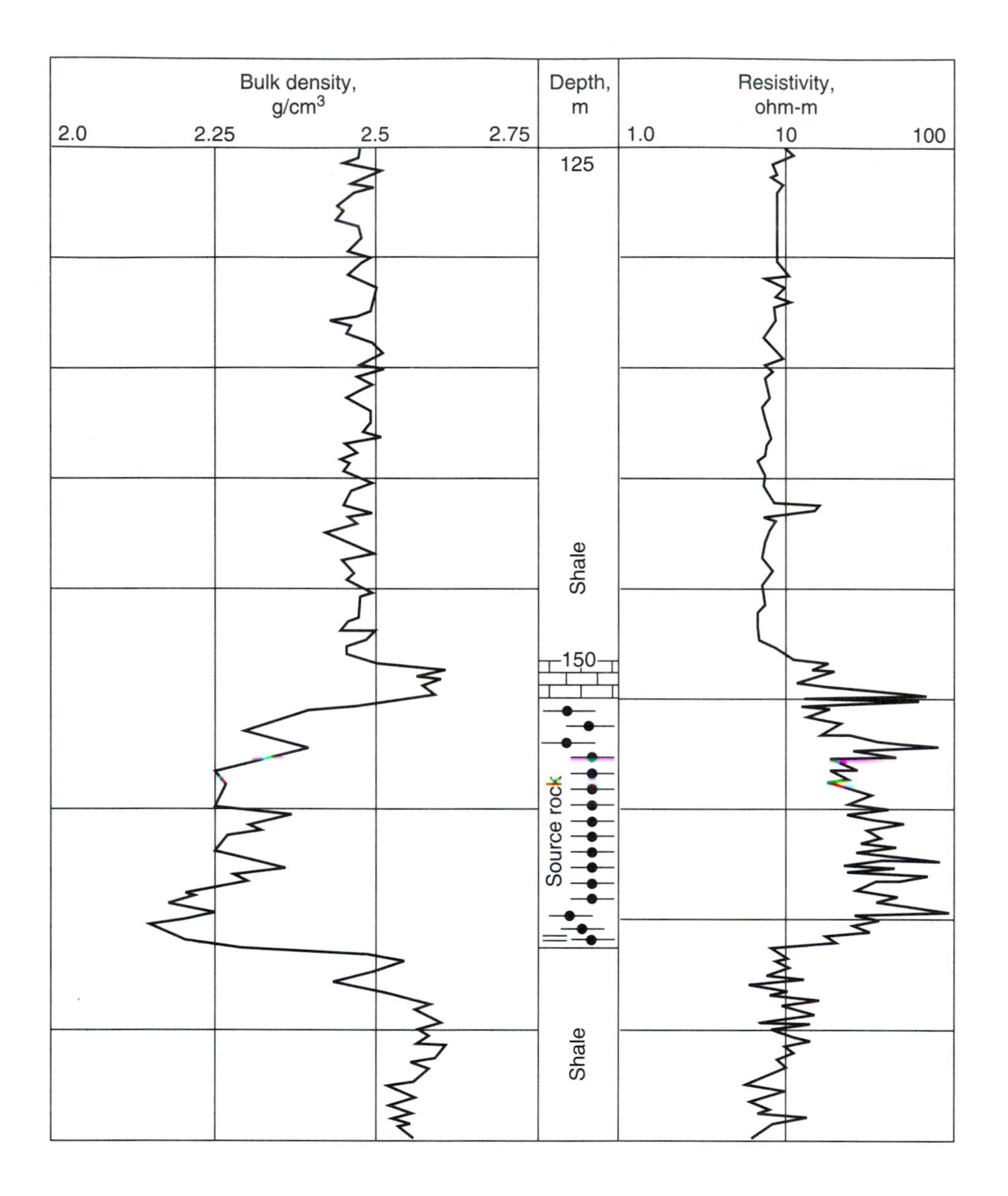

Figure 14-7

Density and resistivity of the Liassic Posidonien Schiefer in a shallow well in south Germany. The resistivity is from Microlaterolog. High resistivity peaks in shale are thin limestone intercalations. [Meyer and Nederlof 1984]

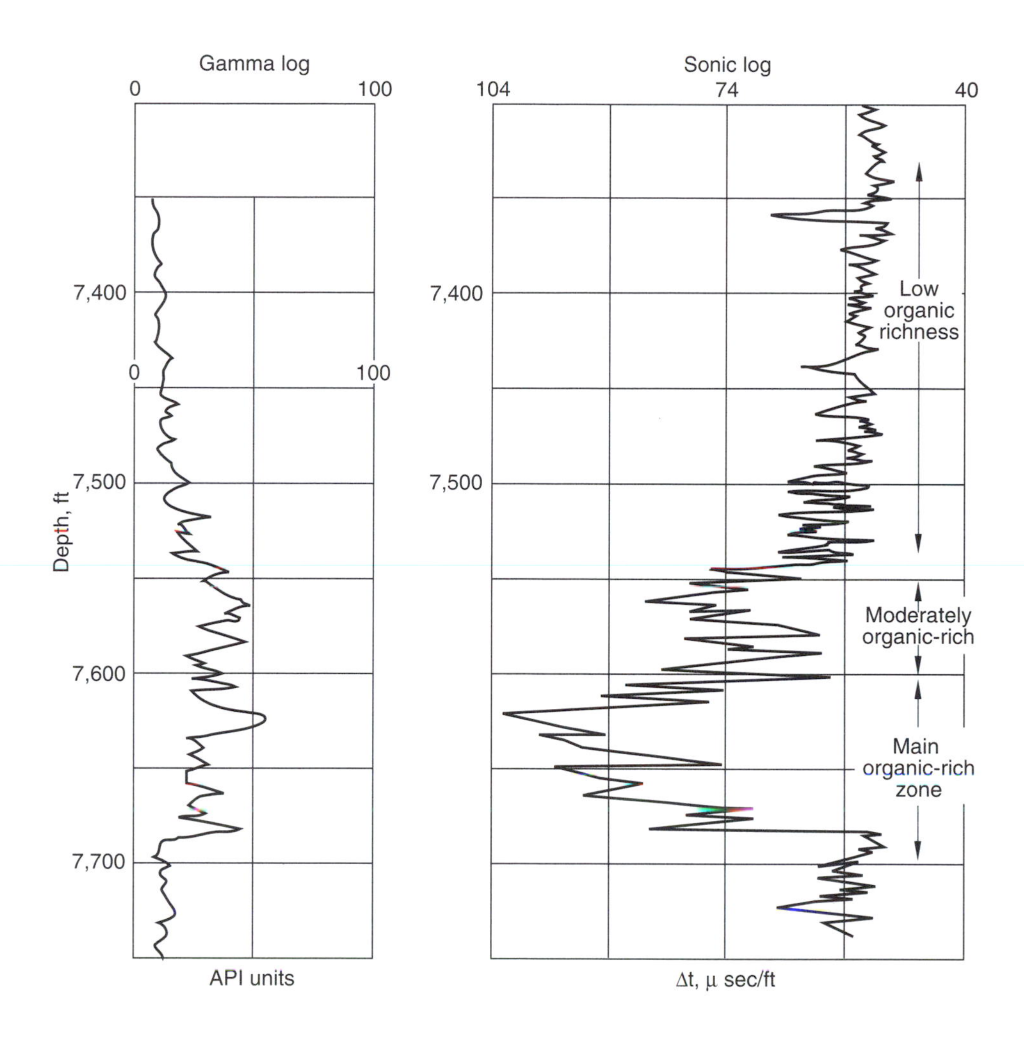

Figure 14-8

Log-character and organic analyses comparing the organic richness of Jurassic (Callovian and Oxfordian) source rocks of Saudi Arabia. [Ayers et al. 1982]

Geochemical Logs

The single most powerful tool for understanding regional geochemistry is the *geochemical log*. This records data such as TOC, HI, OI, S_1 (free hydrocarbons), S_2 (generation potential), S_2/S_3 (gas versus oil), PI (production index) kerogen type, fluorescence, T_{max}, and vitrinite reflectance, along with the depth, age, formation name, lithology, and temperature of the rocks. In addition, mud-log gas, head space, and cuttings gas are sometimes logged to evaluate light hydrocarbons (C_2–C_8) generated in or migrating through the rocks. Some service laboratories divide these into an organic facies log dealing with the quantity and quality (oil versus gas) of the kerogen and a maturity log defining the oil-, condensate-, wet gas-, and dry gas-generation depth intervals.

Figure 14-9 contains an idealized geochemical log based on Rock-Eval pyrolysis (courtesy of K. E. Peters). The kerogen in this well is immature down to about 2,000 m, based on T_{max} and $\%R_o$. In rock interval A, the high OI combined with a very low TOC is typical of organic-lean carbonate rocks, which give off inorganic CO_2, such as in Figure 14-12. Or the high OI could be caused by a drilling-mud contaminant. The high HI and TOC in interval B indicates a good source rock that should be evaluated where it is deeper and more mature in the basin. In interval C the TOC is too low to be a hydrocarbon source. The best mature source rock is in interval D, which appears to have expelled oil into the stratigraphically lower interval E. Interval E has a low TOC, typical of sandstones, but a high S_1, indicating oil saturation. Finally, the high TOC with a very low HI at 4,500 m indicates either a postmature source rock or a rock that probably never had much generation potential.

The kinds of data needed for these geochemical logs were discussed in Chapter 10. This chapter describes some of the problems in getting these data so that any anomalies can be better understood and properly interpreted.

Figure 14-10 shows a core and cuttings analysis chart, with the rapid screening procedures on the left and the more detailed, specialized procedures on the right. Wet well cuttings are required for canned cuttings analysis and also for evaluating the free low-molecular-weight hydrocarbons by means of pyrolysis (S_1). Dried cuttings are suitable for all the other analyses. The pyrolysis procedures were described in Chapter 10. The term *pyrolysis* is applied to both the free (S_1) and the cracked (S_2) hydrocarbons, even though the former is really derived by a distillation process in which the free hydrocarbons are heated until they volatilize. Both analyses are called pyrolysis because they are carried out in a pyrolysis unit such as a Rock-Eval or other type of pyrolyzer. The items listed below the boxes in Figure 14-10 are the various geochemical parameters obtained from the procedures in the boxes above them.

Problems and Pitfalls

The least expensive of the screening processes is the TOC, on the far left of Figure 14-10. The TOC should be run every 60 ft or 18 m from the beginning of lithified rocks to total depth. A TOC can be run at the well site using the Rock-Eval II equipped with a carbon unit. More commonly it is run in the laboratory with the Leco Carbon Analyzer, which was originally designed to analyze carbon in steel. The Leco analyzer usually reports a higher TOC yield than that recorded by the Rock-Eval units, especially at maturities of OM whose R_o is $>1\%$, because it is run at a much higher combustion temperature. It has been argued, however, that most of this additional carbon does not generate oil, although its gas-generating ability has not been ascertained. In practice, if the

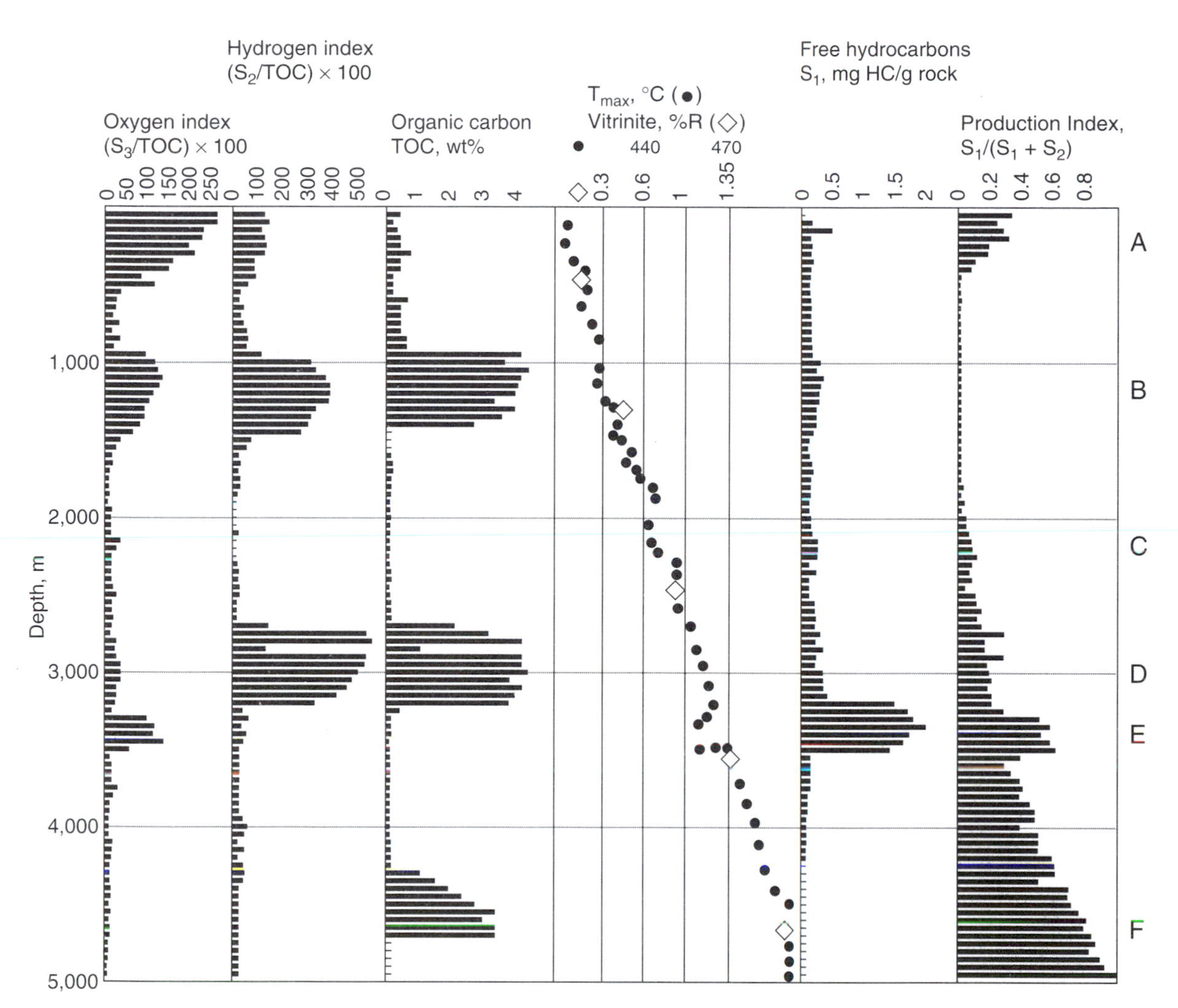

Figure 14-9

Idealized geochemical log based on Rock-Eval pyrolysis, TOC, and vitrinite reflectance of samples from a well showing evidence of (A) oxidized organic matter, which might represent a well additive; (B) thermally immature (potential) source rock containing types II–III kerogen; (C) a nonsource interval; (D) mature source rock containing type II kerogen; (E) reservoir rock containing petroleum apparently derived from and trapped by the overlying effective source rock; and (F) spent (postmature) source rock. [Courtesy of K. E. Peters]

Rock-Eval II plus TOC is available at the well site, it will be used to determine the TOC; otherwise a Leco will be used later in the laboratory. Jarvie (1991) describes the advantages and problems of the two methods.

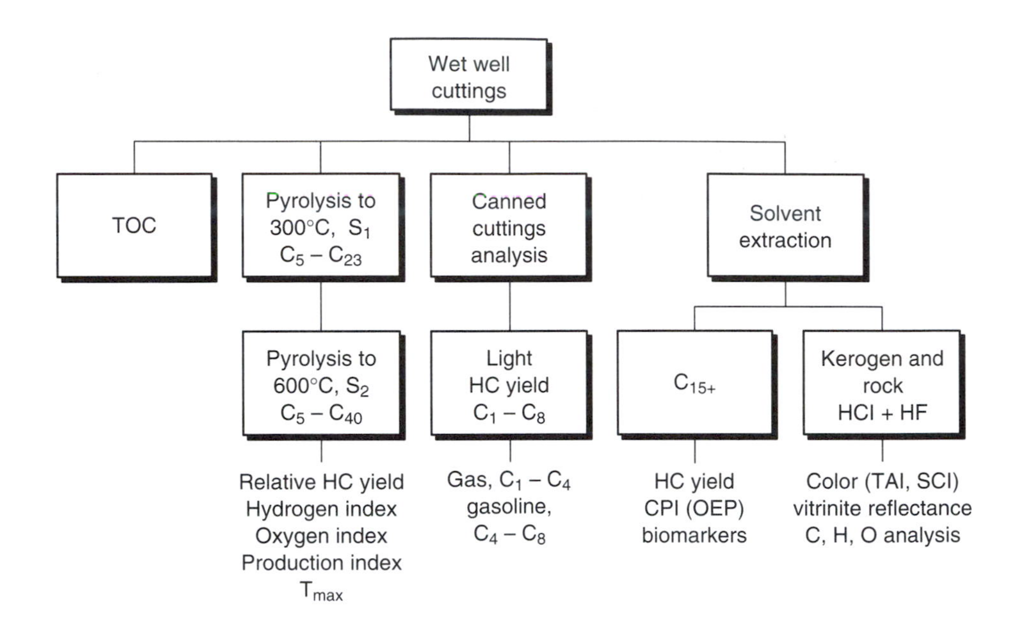

Figure 14-10

Flow chart showing the various geochemical analyses for well samples.

Ratings given to TOC values indicating whether they represent nonsource or fair-to-good source rocks can differ depending on the investigator. These ratings also vary by organic type and basin, as explained in Chapter 10. The important question is what the minimum TOC is that will generate and expel oil and/or gas from the source rock. It is probably different in the Gulf Coast Tertiary from what it is in the West Siberian Basin. Nevertheless, nearly all service laboratories and several oil companies define a minimum for oil expulsion on their geochemical logs. This usually ranges between 0.4 and 1%.

Ronov's (1958) original, detailed study of TOCs on the Russian Platform described in Chapter 10 considered values close to 1.4% to be the minimum for forming commercial oil accumulations. Meyer and Nederlof (1984) defined nonsource rocks as having less than 1.5% TOC. As mentioned earlier, during hydrous pyrolysis, Lewan (1987) found no expulsion of oil from rocks with less than 1.5% TOC. Cooper (1990, p. 20) defined 1% TOC as the minimum for oil generation. Jarvie (1991) considers 0.5 to 1% as marginal, and Peters and Moldowan (1993, p. 51) believe that rocks with less than 1% TOC are unlikely sources for oil. Ronov's (1958) study did indicate that low TOCs around 0.5% could generate and expel gas. The Saratov and Kiev gas fields are believed to have originated from Devonian shales with about 0.5% TOC. This supports the observation that gas and condensate seem to be plentiful in deltaic rocks like the Gulf Coast where many TOCs in the Tertiary are less than 1.5%.

These observations suggest that 1% TOC would be a reasonable cutoff for oil source rocks and 0.5% for gas source rocks. But such cutoffs are only guidelines. Other parameters, like the S_1/TOC discussed later, provide a more quantitative evaluation.

Cuttings sample contaminants may introduce anomalously high TOC values, such as when OM (e.g., walnut hulls, lubrabeads, gilsonite, wood chips) is added to the drilling mud (Peters 1986). Cuttings should be examined with a binocular microscope before they are analyzed so that obvious contaminants can be removed. In addition, the driller's log should be checked to determine what organic materials were added during drilling and at what depths. The addition of diesel oil, crude oil, or other hydrocarbon-bearing additives to the drilling mud may be particularly difficult to treat in some samples. Cuttings samples taken from wells drilled with oil-based muds in onshore north Sumatra proved to be completely intractable for geochemical analyses. Soxhlet extraction failed to erase the diesel imprint, as proved by comparing the results with conventional core samples.

Pyrolysis

Pyrolysis is the second screening process in Figure 14-10. It is generally run on cuttings every 60 ft (18 m), starting with lithified rocks, the same as for TOCs. Samples should be taken every 30 ft (9 m) so that closer intervals may be pyrolyzed in critical sections. Pyrolysis procedures were discussed in Chapter 10 (Figures 10-7 to 10-16). Data from open-system pyrolysis such as Rock-Eval are comparable only on a relative basis, as they commonly overestimate expelled petroleum by a factor of two or more, compared with closed-system pyrolysis (see Chapter 16). Also, different instruments give different results with the same samples. The Rock-Eval pyrolyzer is heated at 300°C for 3 to 4 min to expel the free hydrocarbons in the S_1 peak and then to 550 to 600°C at 25°C/min to form the cracked hydrocarbons in S_2 (Figure 14-10). The corresponding temperatures in the CDS pyrolysis–GC instrument are 250°C and 600°C. Constant sample weights of about 100 mg are recommended for the Rock-Eval because weights below 75 mg tend to reduce the S_1, S_2, and S_3 and increase the T_{max} (Peters 1986). This decrease is attributed to the adsorption of products on the instrument plumbing before reaching the detector.

The Mineral Matrix Effect

A problem with Rock-Eval pyrolysis is the mineral matrix effect, which causes a reduction in HI and a rise in OI for lean rocks. This was briefly mentioned in Chapter 10. Espitalié et al. (1980) noted that the pyrolysis of some rocks with low TOCs gave anomalously low HI values when compared with the pyrolysis of the kerogen separated from the same rock. Their experiments indicated that the difference was due to the retention of the high-molecular-weight hydrocarbons until secondary cracking occurred, which left a carbon residue on the rock. Espitalié et al. also found that surface active minerals such as smectite and illite caused the greatest decrease in HI and that minerals like gypsum and calcite showed the least effect. Moreover, the effect was greatest at low TOCs.

Orr (1983) verified Espitalié et al.'s observations by comparing the pyrolysis of whole rocks with isolated kerogens from the Miocene Monterey Formation of California, the Miocene Puente Formation of California, and the Triassic Shublik Formation of Alaska's North Slope. All three recorded a drop in HI with increasing clay mineral content. Figure 14-11 shows that calcite and kaolinite, which have much smaller surface areas than smectite or illite, have a correspondingly smaller matrix effect (Larter 1984). But there still is a decrease in HI at TOCs <1%.

In natural samples, of course, the smectite and illite would be diluted with quartz and other nonadsorbing minerals, which would reduce the matrix effect. In Chapter 8 it was mentioned that Gulf Coast shales average only 20%

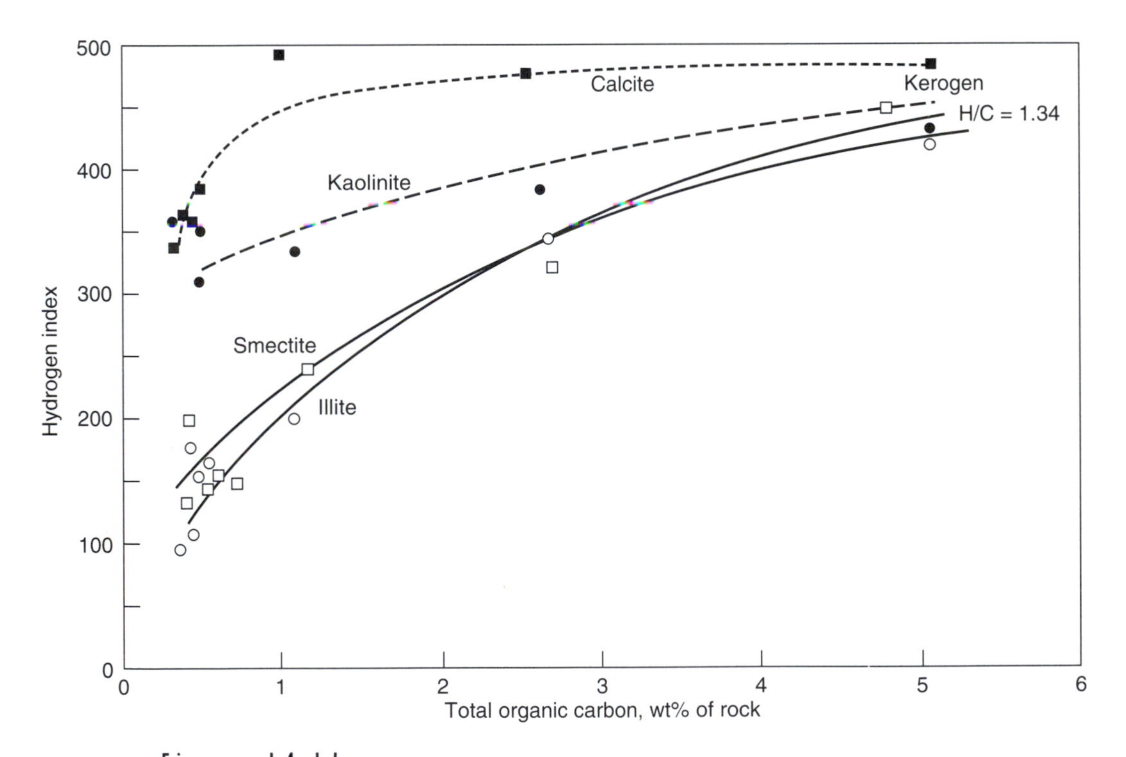

Figure 14-11

Comparison of pyrolysate yields from type II kerogen mixed with pure calcite and clay minerals. [Larter 1985] See Espitalié et al. [1980] and Orr [1983] for similar experiments.

clay minerals, so the retention of hydrocarbons on these shales would be much less than that on pure clay minerals.

Katz (1983) observed anomalously high OI values when pyrolyzing carbonate rocks. The original work of Espitalié et al. (1977) had indicated that CO_2 from the kerogen in rocks came off mostly in the range up to about 400°C (752°F). The CO_2 from pure carbonates evolved mostly at high temperatures: calcite above 600°C (1,110°F), dolomite above 500°C (932°F), and siderite above 400°C (752°F). This indicated that carbonates would not interfere with organic CO_2 evolution. Espitalié et al. (1977) also reported that the yield of organically derived CO_2 was proportional to the O/C ratio of the kerogen in the rock for several areas such as the Paris Basin, Persian Gulf, and Cameroon. This is why the Rock-Eval instrument measures the CO_2 in the temperature range from 300 to 390°C.

In some samples, however, small amounts of carbonates appear to break down at temperatures lower than 400°C, possibly because of the catalyzing effect of aluminum and iron on carbonate decomposition. If a pure carbonate rock with a TOC < 1% is pyrolyzed, the CO_2 released will come mainly from the carbonate instead of the OM, resulting in an anomalously high OI, as seen in Figure 14-12 (Katz 1983). The OM in these Cretaceous carbonates is type II, which would have low OI values, as shown in the samples with more than 4%

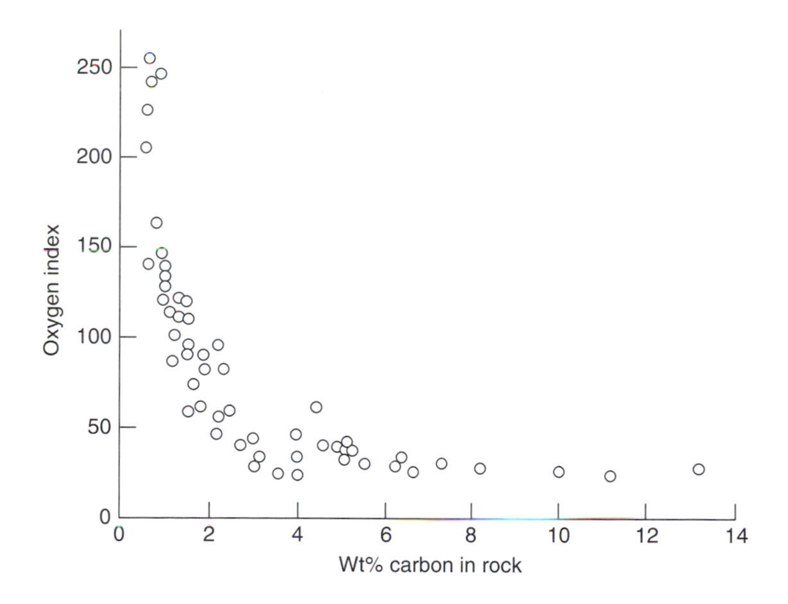

Figure 14-12

The increase in the oxygen index with a decrease in TOC for a suite of carbonate rocks from DSDP Site 535. [Katz 1983]

TOC. But at TOCs less than 2%, the CO_2 from the carbonate overwhelms that from the kerogen.

These problems are not significant when exploring for oil, because rocks with TOCs of < 1% are not worth investigating further. If, however, it is important to distinguish gas-generating type III kerogen from nongenerating type IV in organic-lean rocks, the kerogen should be isolated as shown in Figure 14-13 (Whelan and Thompson-Rizer 1993). This is from the Tertiary section of an East Cameron well offshore Louisiana where the TOCs are generally less than 1.5%. Pyrolysis of the kerogen after removing the mineral matter with HCl and

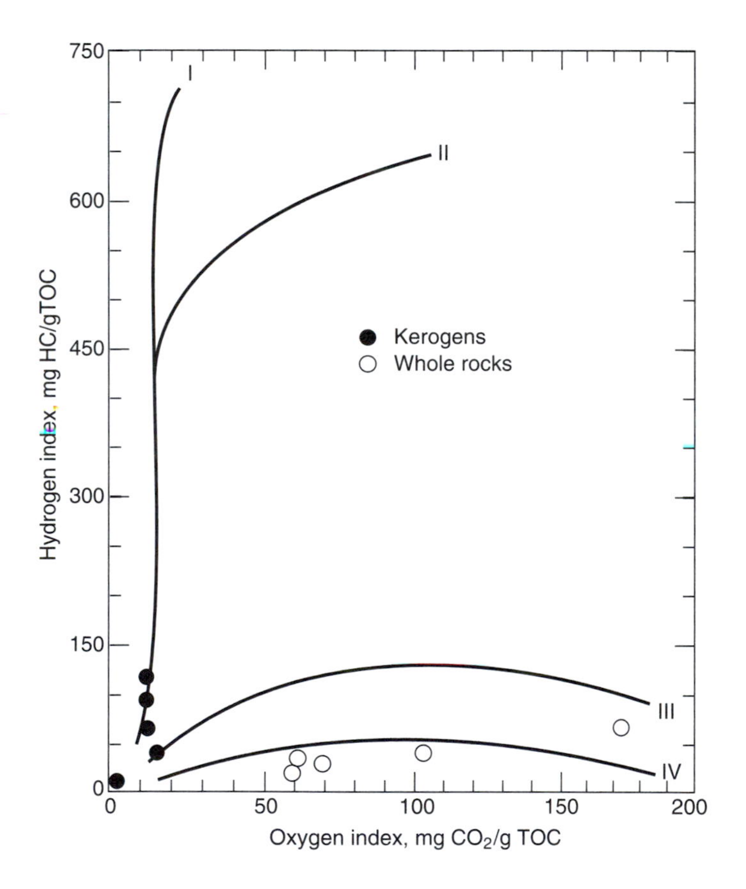

Figure 14-13

HI/OI plot comparing whole-rock pyrolyses with the pyrolyses of the isolated kerogens from the tertiary section of an East Cameron well offshore Louisiana. [Whelan and Thompson-Rizer 1993]

HF acids indicates a possible gas source, whereas pyrolysis of the original rock indicates a nonsource.

Production Index, $S_1/(S_1 + S_2)$

The PI typically climbs from 0.1 to 0.4 from the beginning to the end of the oil-generation window, as illustrated earlier in Figure 10-19. But many PI-versus-depth plots show considerable variation owing to different kerogen types, migrated oil effects, and anomalously high numbers when the S_2 is too low. Because of measurement errors, the PI is meaningless if the S_2 is below 0.2. Many high PI values above 1 mg HC/g TOC indicate migrated oil, especially if the T_{max} decreases and the TOC increases at the same time.

S_1/TOC

Smith (1994) advocates using a plot of S_1/TOC versus depth to determine the depth at which a source rock begins to expel oil. In Chapter 10 it was shown that S_1 increases with depth as oil is being generated. Smith determined from Shell Oil's data base that the ratio of S_1 to TOC should be between 0.1 and 0.2 for oil expulsion to start in the source rock. Figure 14-14 is a depth-versus-S_1/TOC plot for two Norton Basin, Alaska, COST wells. Other studies had indicated that the oil-generation window should be between about 9,000 and 12,000 ft (2,744 and 3,659 m) in these wells (Smith 1994). However, the highest S_1/TOC value is about 0.04, well below the 0.1 that indicates oil expulsion. It was concluded that these shales above the Red Unconformity were incapable of generating oil at any maturation level, although they could generate gas.

When S_1 is high and the TOC is low, migrated hydrocarbons are indicated (see Table 10-5). The Ocean Drilling Program (ODP) uses the graph in Figure 14-15 to distinguish migrated hydrocarbons and contaminants from indigenous hydrocarbons. The slanted line in Figure 14-15 is where S_1/TOC = 1.5. Values above this suggest nonindigenous hydrocarbons, and values below it are indigenous. The graph is based on previously established relationships between C_{15+} extractable hydrocarbons and TOC.

The T_{max}

The T_{max} is supposed to increase steadily with maturation, as shown in Figure 10-10, with the oil-generation window between T_{max} values of about 435°C and 470°C. Usually it does, but sometimes it oscillates back and forth as much as 20°C, because of changes in the organic matter's hydrogen content. A high-hydrogen OM generally has a low T_{max}, and a low-hydrogen OM has a high T_{max}. Consequently, variations in OM caused by the original kerogen type, the addition of recycled OM, or the presence of migrated oil causes sudden increases or decreases in the T_{max} of kerogen with depth.

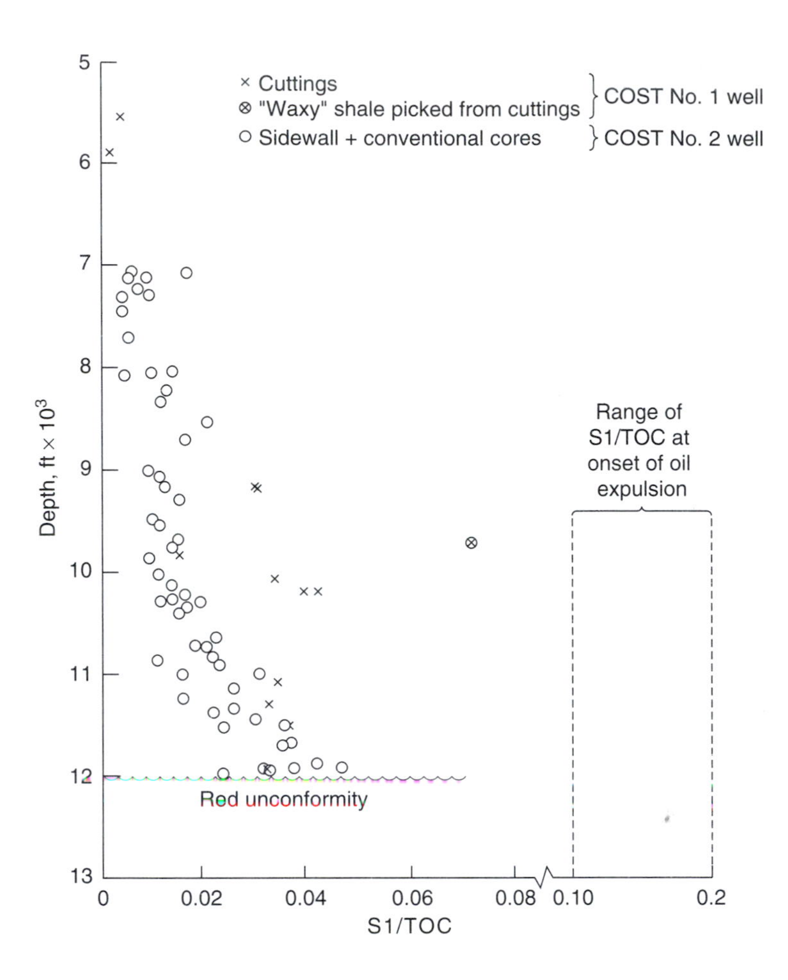

Figure 14-14

Depth versus S_1/TOC plot for two Norton Basin, Alaska, COST wells. [Smith 1994]

In Figure 14-16, the difference in T_{max} between the coal macerals vitrinite and sporinite is about 20°C at the same maturation level based on $\%R_o$ (Dembicki et al. 1983). Pyrolysis of a large number of well cuttings by Peters (1986) revealed that the spread in T_{max} is greatest in the immature OM and least in the postmature OM, as can be seen in Figure 14-17. This would be expected, since with increasing maturity all kerogen types end up with about the same hydrogen content on a van Krevelen diagram, such as in Figure 10-2. Anomalous T_{max} results should always be checked by microscopy to determine any kerogen

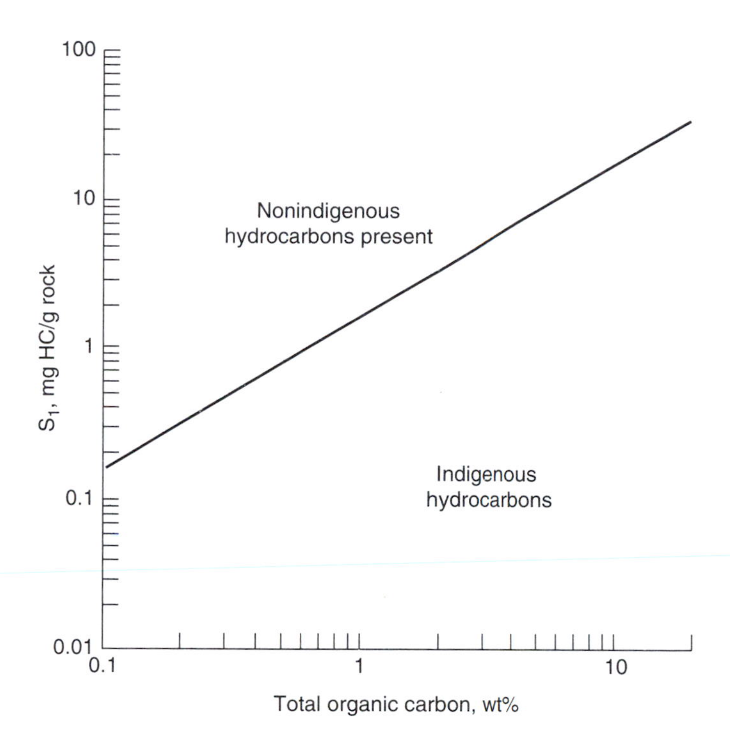

Figure 14-15

S_1 versus TOC for identifying migrating or contaminating hydrocarbons.

type variations. A uniform OM type usually shows the expected increase in T_{max} with maturity, such as for the Australian well in Figure 14-17.

The T_{max} measures the lowest-rank material in a sample that can still generate hydrocarbons. If a sample has 20% cavings of low-rank OM and 80% high-rank OM such as inertinite, the T_{max} will show only the cavings, since the inert material does not generate enough hydrocarbons to form an S_2, even though it is the dominant OM present.

Multiple S_2 Peaks

The reliability of the S_2 peak area that is used to calculate HI depends on the assumption that 300°C will expel all the S_1 free hydrocarbons from the kerogen, leaving only the kerogen decomposition to form S_2. Expulsion, however, depends on the vapor pressure. A $C_{24}H_{50}$ free hydrocarbon has a boiling point of 390°C at atmospheric pressure (Ferris 1955). Consequently, it comes off as

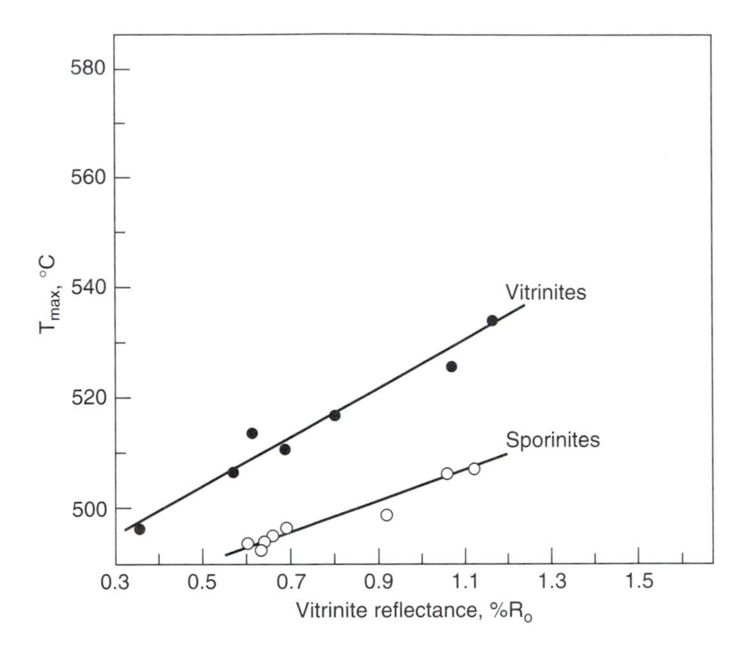

Figure 14-16

Changes in T_{max} with increasing maturity as a function of maceral type. [Dembicki et al. 1983]

part of the S_2 instead of the S_1, as confirmed by laboratory experiments with the CDS pyrolyzer. Three straight-chain paraffins, $C_{16}H_{34}$, $C_{22}H_{46}$, and $C_{24}H_{50}$, were mixed with calcite and pyrolyzed to see if they would come off in the S_1 or the S_2 peak (Tarafa et al. 1983). The C_{16} paraffin did come off in S_1, but the C_{22} was split between the S_1 and S_2 peaks, and most of the C_{24} came off in the S_2 peak. It was concluded that free hydrocarbons containing 24 or more atoms were probably coming off in the S_2 peak and would result in a higher HI than would occur in the absence of these heavy hydrocarbons. Also, since these hydrocarbons contain more hydrogen than kerogen, the S_2 would show a lower T_{max} compared with an S_2 peak free of C_{24+} hydrocarbons.

When such free hydrocarbons occur in well cuttings, they can cause multilobed S_2 peaks in the pyrograms. For example, Figure 14-18 has multilobed $P_2(S_2)$ peaks at two depths in the Kugrua well on the North Slope of Alaska. The different peaks are caused by successive volatilization of hydrocarbons in the high-molecular-weight C_{25}-to-C_{45} range. The T_{max} cannot be used for maturity in such cases because it does not strictly represent the cracking products of kerogen. Nor can the S_2 be used without separating the part of the S_2 due to kerogen from that due to free hydrocarbons larger than C_{23}.

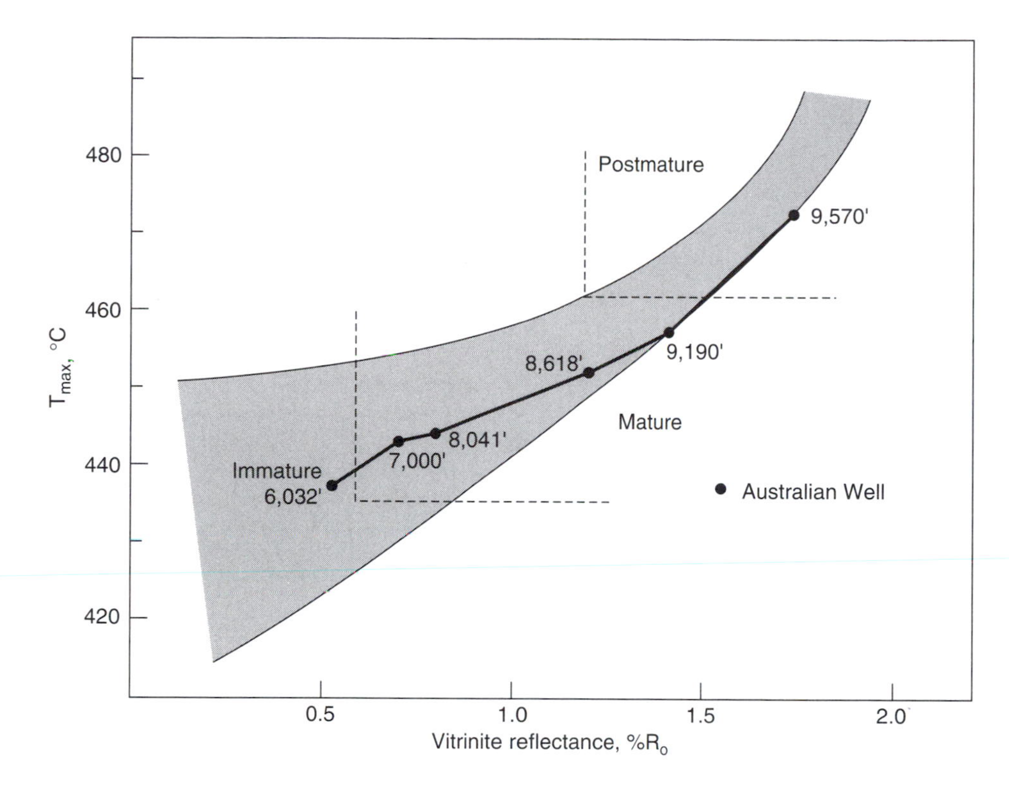

Figure 14-17

T_{max} versus $\%R_o$ based on a large number of analyses made by Chevron. Immature samples show wide variability due to differences in OM type. The OM for the Australian well was uniformly type III. Sample depths are in feet. [Peters 1986]

The heterogeneity of heavy hydrocarbons can cause multiple peaks. A high-temperature gas chromatogram of ozocerite from the Uinta Basin of Utah was presented in Figure 12-12. It contains *n*-paraffins ranging from C_{20} to C_{70}, with peaks around C_{31}, C_{41}, and C_{60}. Interestingly, a CDS pyrolysis of Utah ozocerite (Figure 14-19) also shows three $P_2(S_2)$ peaks, which may correspond to the hydrocarbon distributions shown in Figure 12-12.

Clementz (1979) was able to get a valid S_2 peak by first extracting the free hydrocarbons (bitumens) with organic solvents. However, asphaltites, pyrobitumens, and some resinites are not soluble in conventional organic solvents. Again, microscopic studies can assist in identifying those samples with insoluble bitumen interference. Fortunately, in most samples the quantity of kerogen considerably exceeds the insoluble bitumen. Nevertheless, it can cause anomalous results.

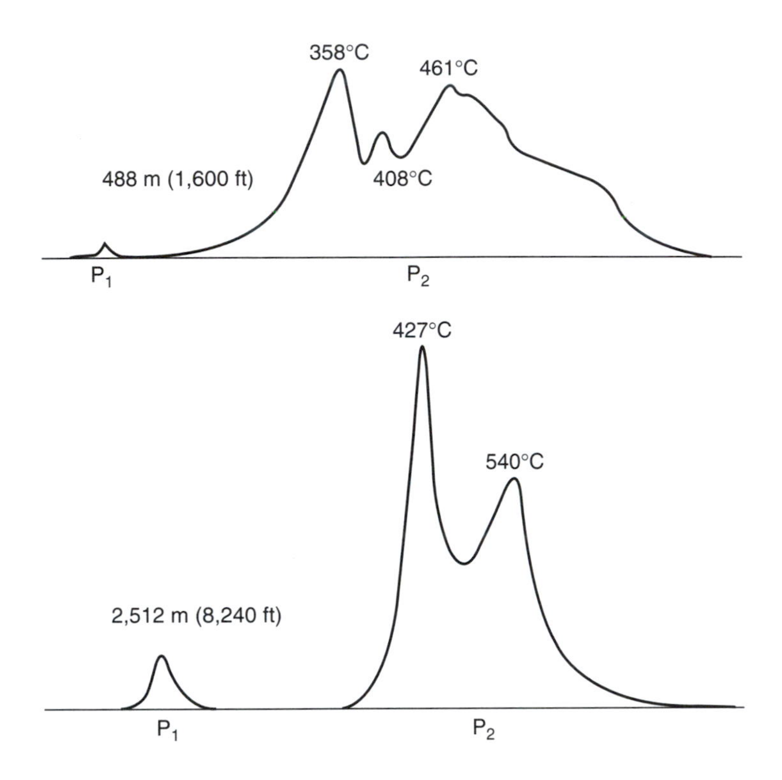

Figure 14-18
Pyrograms from pyrolysis of well cuttings in the Kugrua well, North Slope of Alaska.

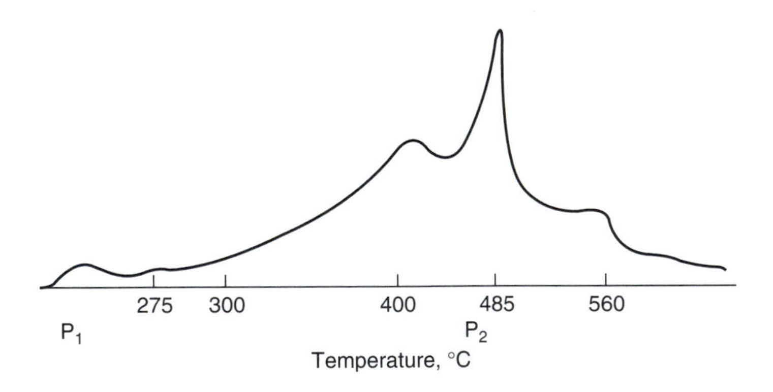

Figure 14-19
Pyrogram of the mineral wax ozocerite from the Uinta Basin, Utah.

Contamination

In addition to heavy hydrocarbons and related natural bitumens causing anomalous pyrolysis results, some contaminants, such as drilling mud additives, can cause misleading interpretations. For example, Figure 14-20 shows the T_{max} for cuttings from a well offshore Louisiana in the Gulf Coast. The T_{max} does not increase in the first 12,000 ft (3,659 m) because it is before oil generation. The oil-generation window begins around 13,000 ft (3,963 m) and continues to the total depth drilled. The T_{max} increases in this interval as expected. However, at around 8,000 ft (2,439 m) there is a sudden drop in the T_{max}, and it does not return to the normal curve until reaching 10,000 ft (3,049 m). Examination of the driller's log showed that at 8,000 ft, diesel oil had been added to the mud. Pyrolysis–GC of the cuttings confirmed that they had been contaminated with cetane ($C_{16}H_{34}$), which is the standard hydrocarbon used in rating diesel fuel (see Chapter 3). Although the cuttings were presumably thoroughly washed at the well site, they were contaminated with the diesel fuel. Fortunately, the

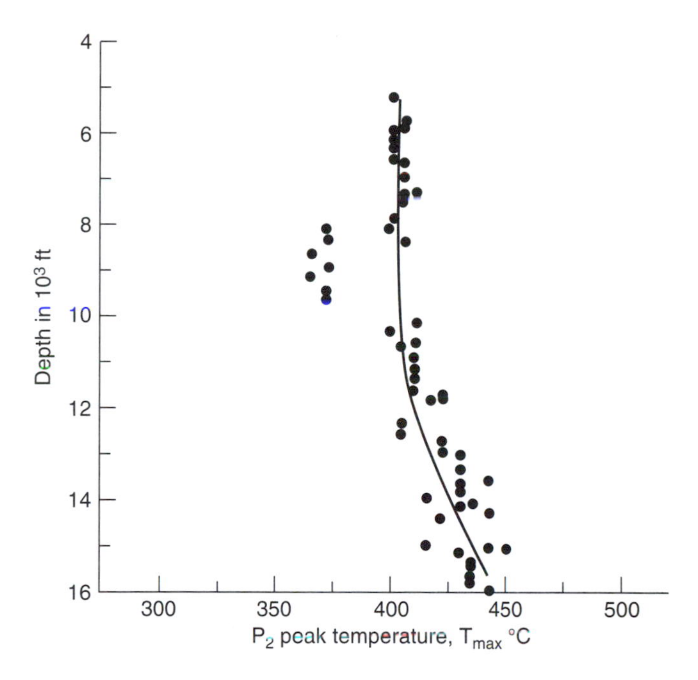

Figure 14-20

A lowering of the T_{max} in the cuttings pyrolyzate from an East Cameron well offshore Louisiana on the Gulf Coast due to diesel oil in the drilling mud.

contamination disappeared in the next 2,000 ft (610 m). Since oil contains more hydrogen than kerogen, any oil additives or any natural oil seeping into the mud from a penetrated reservoir may lower the T_{max}. When the hydrogen in kerogen is increased, both the T_{max} and the vitrinite reflectance go down.

Pyrolysis of various mud additives has indicated that they can either raise or lower the T_{max}, depending on their composition (Peters 1986). Pipe dopes and oils tend to lower the T_{max}, but all organic contaminants increase the TOC, S_2, and HI. Visible particulate contaminants can be removed from the cuttings under a binocular microscope, but some of the additives, like lubrabeads, walnut hulls, and gilsonite, are ground to a fine powder by the drill bit. The powder can form an invisible coating on the cuttings. The T_{max} for some additives and pipe dopes are as follows: lubrabeads (404°C), walnut hulls (425°C), gilsonite (450°C), polyethylene (456°C), pipe dope A (430°C), and pipe dope B (415°C) (Data from Peters 1986). Note that there is a range of 52°C between the lowest and the highest value. This exceeds the 40°C range for the entire oil-generation window, so it does not take much additive to produce anomalous results.

An example of contamination down a well in Alaska is illustrated in Figure 14-21 (Peters 1986). The T_{max} increases gradually until it reaches a depth of about 8,500 ft (2,590 m), at which point it drops abruptly. It returns to normal values around 11,000 ft (3,354 m) and then drops again around 12,000 ft (3,660 m). At both depths where the T_{max} drops, the TOC and HI show a marked increase. Examination of the driller's log showed that a synthetic polymer similar to lubrabeads was added to the mud at the two points where the anomalous T_{max}, TOC, and HI values occur. Subsequent examination of cuttings samples from these intervals using a binocular microscope proved that lubrabeads were present. These intervals might have been interpreted as having good source potential if the mud information had not been available or had been ignored.

Kerogen microscopy before pyrolysis also is advised when drilling through rapidly changing lithologies. For example, Peters (1986) reported a large variation in TOC and all pyrolysis parameters over a 200 ft (61 m) depth interval from a well in Montana. The pyrolysis data are in Table 14-1. The two shallowest samples of calcareous shale have almost the same %TOC, but the S_1, S_2, and HI are entirely different. The shallower sample has a much greater hydrocarbon potential, as indicated by the S_2 and HI. Kerogen microscopy shows it to have 90% amorphous OM, whereas the slightly deeper sample contains mainly structured and inert OM. Separation of the kerogens from the mineral matter, followed by elemental analysis of the kerogen, revealed the shallower sample to be a type II (oil-prone) kerogen with an atomic H/C of 1.15 and the deeper sample to be a type III (gas-prone) kerogen with an atomic H/C of 0.75. This verified the pyrolysis data, in that the rocks had changed from oil prone to gas prone within 25 ft (7 m) of depth.

All the samples in Table 14-1 are immature, based on the T_{max}, and its variation among samples is due to differences in OM type. The sample at 2,007 ft (612 m) has no T_{max} because there is no S_2 peak and no generating capacity. It plots as a type IV kerogen on an HI-versus-OI diagram. Microscopic examination confirms that it is dominantly recycled OM with 60% inertinite.

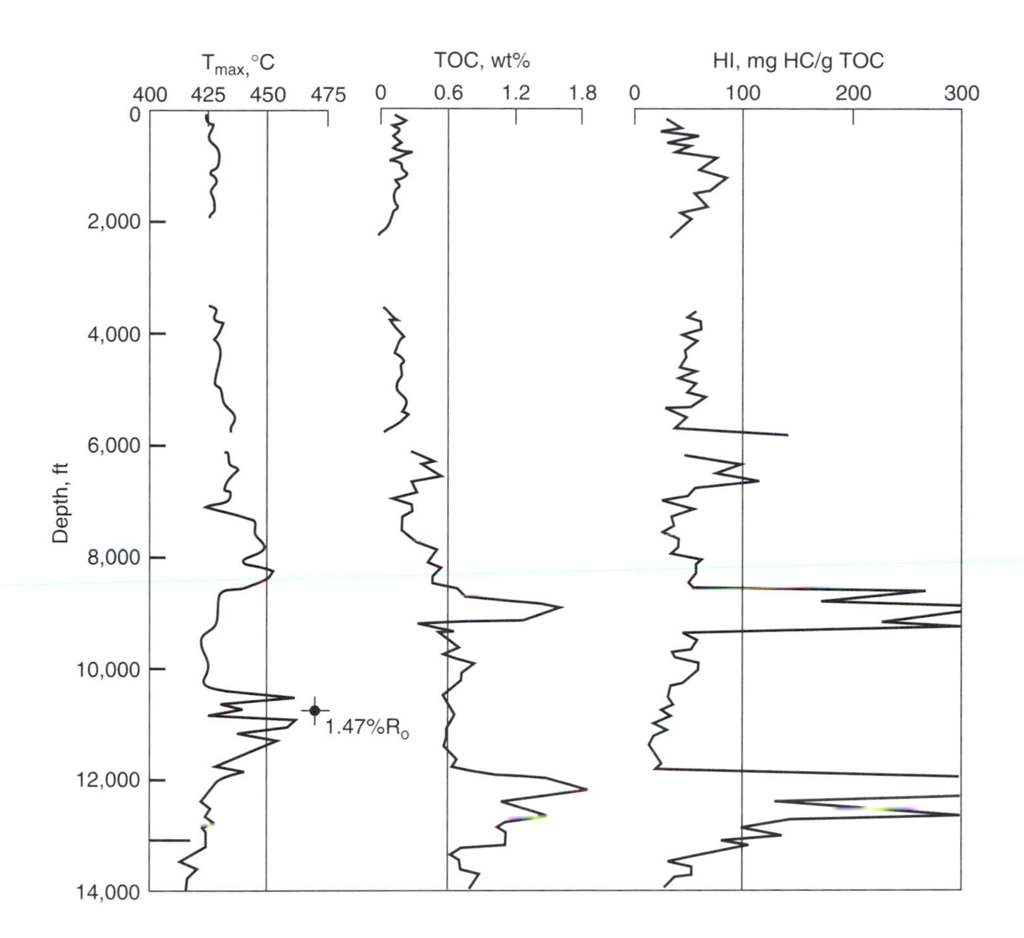

Figure 14-21

Abbreviated geochemical log for ditch cuttings from the Cathedral River well, Alaska. Anomalous results for the T_{max}, TOC, and HI at 8,500 and 12,000 ft (2,590 and 3,660 m) are due to the injection of particulate mud additive. (Data are from Exlog and were plotted by R. G. Huppi in 1982) [Peters 1986]

The 0.61% TOC for the siltstone at 2,090 ft (637 m) is too low to be an oil source rock, so the high HI suggests migrated oil. Samples stained with migrated oil or oil contaminants generally show an S_1 peak greater than 2 mg HC/g rock, a low T_{max}, an anomalously high PI, and a bimodal S_2 peak. This sample had all those characteristics, indicating that it was acting as a micro-reservoir rather than a hydrocarbon source. The well log describes this sample as oil stained. The last sample in Table 14-1 at 2,146 ft (654 m) has an S_2

TABLE 14-1 TOC and Pyrolysis Analyses for Selected Core Samples from a Well in Montana

Depth, ft (m)	*Description*	*TOC (wt%)*	S_1	S_2	S_3	T_{max} (°C)	*PI*	*HI*	*OI*
1,950 (595)	dark gray laminated calcareous shale	3.54	1.77	23.81	1.21	422	0.07	673	34
1,975 (602)	dark gray massive calcareous shale	3.56	0.28	2.96	1.21	427	0.09	83	34
2,007 (612)	medium gray massive shale	1.04	0.04	0	0.55	—	0	0	53
2,073 (632)	black fissile calcareous shale	2.43	0.09	0.56	0.62	432	0.14	23	26
2,076 (633)	medium gray calcareous shale	0.38	0.05	0.25	0.51	432	0.17	66	134
2,090 (637)	brown siltstone (oil stained)	0.61	6.61	4.08	0.12	415	0.47	699	20
2,146 (654)	medium gray massive shale	0.52	0.04	0.14	0.45	422	0.22	27	87

Nomenclature:

S_1 = mg HC/g rock; PI = $S_1/(S_1 + S_2)$;
S_2 = mg HC/g rock; HI = mg HC/g TOC;
S_3 = mg CO_2/g rock; OI = mg CO_2/g TOC

Source: Peters (1986)

<0.2 mg HC/g rock which renders much of the pyrolysis data invalid. As previously mentioned, neither the PI nor the T_{max} is valid when the S_2 is < 0.2.

An interesting point about Table 14-1 is that conventional sample spacing every 60 ft (18 m) would have produced mixtures of lithologies and organic facies, thereby making the interpretations questionable. This is why it is important to sample at close intervals to enable additional analyses to be made if there are any questions about the interpretations. The cost of these extra analyses are minimal compared with the cost of drilling the well.

Many of the problems discussed in this section on pyrolysis can be minimized by using a PC-based computer program designed to interpret large quantities of pyrolysis, TOC, and vitrinite reflectance data from drill cuttings. For example, the Rock-Evaluation Expert System Advisor (REESA) developed by Chevron improves the quality of data by eliminating anomalous inputs based on a set of empirical rules (Peters and Nelson 1992). Thus the T_{max} will be deleted if S_2 is less than 0.2 mgHC/g rock. Most pyrolyses data are rejected if the TOCs are less than 0.4 wt%. Also rejected are T_{max}s below 395°C, PIs if $(S_1 + S_2)$ is less than 2 mgHC/g rock, T_{max} if the HI is less than 50 mg HC/gTOC, and OIs if they are greater than 300 mg CO_2/g rock. This program has more than twenty rules for accepting or rejecting data based on pyrolysis studies from hundreds of exploration wells. After improving the data quality, it can graphically display the results in various geochemical log formats, as requested by the geologist.

Figure 14-22 is an example of a computer-plotted geochemical log with some parameters that differ from those of the log in Figure 14-9 (for additional geochemical logs, see Peters and Cassa 1994, and references therein). This log from the North Sea Basin was obtained by G. J. Demaison and published by Peters (1986). Note the very high HI, TOC, and hydrocarbon source potential (S_2) at the top of the Upper Jurassic Kimmeridgian–Volgian source rocks at 9,500 ft (2,896 m). Also notice how the mud-log gas extends from the bottom of the hole through the mature section and several hundred feet above it to about 8,500 ft (2,591 m). As observed earlier, gas is always seen on the mud log in a section that is alive with hydrocarbons. The overlying Cretaceous and Eocene sections may have some gas potential, but they would require a much deeper burial to reach maturity. By using continuous logs like this, it is possible to identify any anomalous data due to the problems previously discussed and then to investigate the cause of the anomaly with more detailed geochemical tests. Geochemical logs like this can be generated by computer and designed to include any of the data listed in a cuttings analysis program, such as that shown in Figure 14-10.

Canned Cuttings (Headspace) Analysis

Canned cuttings analysis provides valuable information on the generation and migration of light hydrocarbons and gas-plus gasoline (C_1–C_8) through the entire section being drilled. It is particularly useful for a gas province where you cannot do extracts or biomarkers for individual light hydrocarbons. You can analyze for light hydrocarbons by pyrolysis–GC (Figure 10-16), but not by the conventional Rock-Eval pyrolysis. Canned cuttings analysis offers more detail than the conventional mud log in detecting migrated gas and gas and gasoline shows. Migrating gases have been observed in red beds, where they may lead back to the source. Geochemists should not be looking only for source rocks but also for oil and gas shows to determine whether the area is alive with hydrocarbons.

The data obtained represent the hydrocarbons retained by the wet cuttings when they come to the surface. The usual technique is to collect 250 to 500 cc (1/2 to 1 pint) of cuttings from the shale shaker, although smaller samples can be used. The cuttings should not be allowed to dry out. If oil is used in the drilling mud or if the cuttings are heavily coated with mud, they should be washed off. Normally, however, it is preferable not to wash the samples. The wet cuttings are placed in a 1-quart tin can and filled with water to within about 1/2 inch of the top. Service labs can provide cans with small holes on the bottom sealed with a rubber septum. A bactericide such as zephrin chloride is

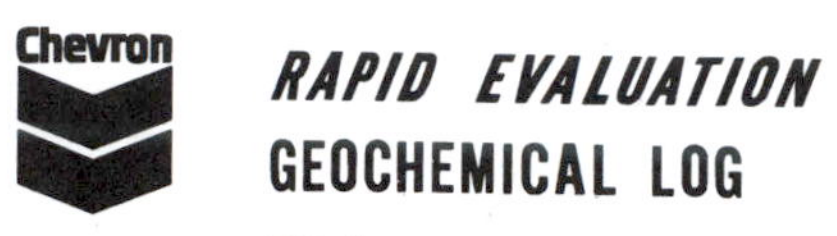

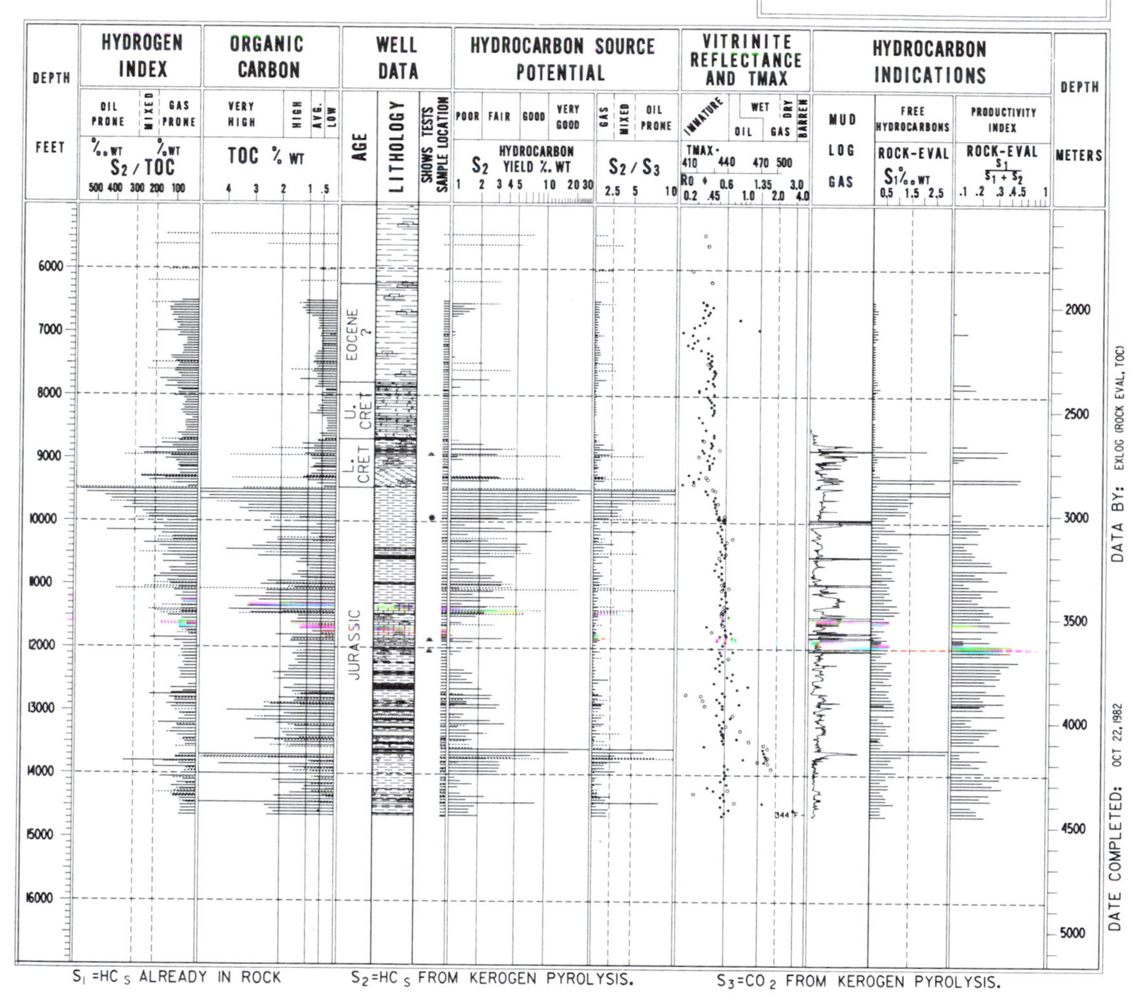

Figure 14-22

Example of a Chevron geochemical log showing Rock-Eval pyrolysis and TOC results from well X in the North Sea. The log includes an option to plot supporting data, such as lithology, vitrinite reflectance, and mud-log gas. Solid T_{max} symbols were determined using cuttings; open T_{max} symbols were determined using cores; and corresponding data in other columns are dashed. [Courtesy of G. J. Demaison]. [Peters 1986]

added to the can to minimize the bacterial formation of methane after the can is sealed. Depending on the type of lid used, the can may be sealed with a crimping device or hammered on with a rubber mallet and secured with clips. After it is labeled, the can is inverted so the rubber septum faces down. The samples are then shipped to a company laboratory or a service lab for detailed analyses of the hydrocarbons.

During shipment, part of the gas in the cuttings gradually comes out and enters the air space above. At the laboratory, a hypodermic needle is inserted through the septum into the air space, and a measured volume of gas is removed for analysis on the GC. That is why it has been called *airspace* or *head space gas*. The method is best suited for fine-grained source rocks, since coarse-grained rocks generally lose much of their gas during the trip to the surface. Gas maturity (iso-C_4/n-C_4 and isotopes), migration, and gas and gasoline concentration relative to TOC can be determined. The data are generally plotted on a geochemical log with depth.

After screening, those stratigraphic intervals showing increased C_{2+} concentrations, because of either generation or migration, can be further analyzed by transferring 5 to 10 g of cuttings into a stainless steel cell fitted with a side arm and septum. As little as 2 g can be analyzed. Two stainless steel balls and water are added in a nitrogen or helium atmosphere to prevent air from entering the cell. Air can oxidize the hydrocarbons released during grinding. The sealed cell is agitated in a paint shaker for 10 minutes, after which is it placed in a 90°C water bath to prevent adsorption of hydrocarbons on the cell walls. The agitation grinds the cuttings to a very fine powder, releasing any remaining gas-plus light gasoline (C_1–C_8). An aliquot of gas is withdrawn through the septum and analyzed by GC. Some companies sum the gas from the ground cuttings and the head space in order to give a restored C_1–C_8 hydrocarbon content of the cuttings, whereas others plot the gas (C_1–C_4) and gasoline (C_4–C_8) data separately.

The original technique by Evans and Staplin (1971) analyzed the ground cuttings and was simply called *cuttings gas*. Examples of cuttings gas logs are in Figure 14-23 for two wildcat wells from the Western Canada Basin. Cuttings from the well on the left recorded only dry gas through the entire section, thus defining it as a gas-prone area. The log on the right shows a high yield of wet gas, particularly through the 3,000-to-4,000 ft (914-to-1,219 m) interval. Suitable traps associated with this interval are more likely to contain oil. Another example was previously shown in Figure 7-10, in which the gasoline-range hydrocarbons (C_4–C_{7+}) were plotted separately from the C_1–C_4 gases.

Cuttings gas data can also be used to plot lateral as well as vertical intervals of oil and gas proneness. For example, Figure 14-24 indicates that all the Middle Devonian shales and carbonates of Alberta are rich in wet gas. The overlying Upper Devonian "shale unit" is barren. Going westward, the gas in the fine-grained rock cuttings changes from wet to dry because the western section has a higher temperature than the east. This is an interesting example because the Imperial Oil geochemists found that the wet to dry conversion occurred just west of the town of Rainbow in northern Alberta. Their data predicted that the Rainbow area to the east would be oil prone, whereas that to the west in British

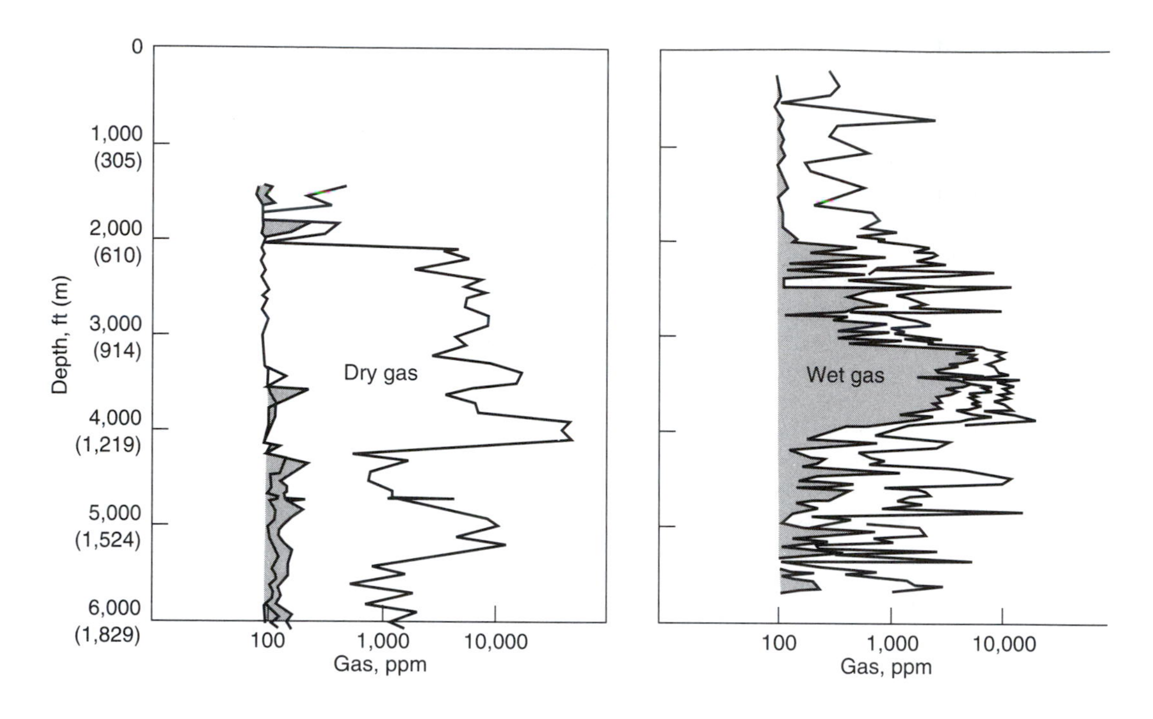

Figure 14-23

Cuttings gas logs from two wildcat wells, Western Canada Basin: 10,000 ppm equals 0.03 ft^3 gas/ft^3 cuttings. [Evans and Staplin 1971]

Columbia would be gas prone. Shortly after this, a major oil field was discovered at Rainbow, confirming their predictions. The transition from wet gas to dry gas in the Middle Devonian of the Western Canada Basin, as defined by extensive cuttings gas surveys, was previously shown as line D in Figure 7-21.

Magoon and Claypool (1983) used cuttings gas to identify C_{2+} generation and migration intervals in the Inigok well on the North Slope of Alaska. This analysis also has been used routinely for the hydrocarbon monitoring of cores in the Ocean Drilling Program. It helps prevent the ODP from drilling into sedimentary rocks that might contain hydrocarbon accumulations that could cause a blowout (*JOIDES Journal* 1992).

Noble (1991) pointed out that the percentage of methane in the total headspace gas is generally higher than that in the gas from the separately ground cuttings. This difference means that gas logs from canned cuttings analyses should clearly state the sampling procedure. Noble (1991) recommends that gas sampled directly from the can be called *airspace gas* and that obtained from grinding the cuttings be called *disaggregate cuttings gas*. Evans and Staplin (Figures 7-10 and 14-23) used the latter procedure, except that they ground the

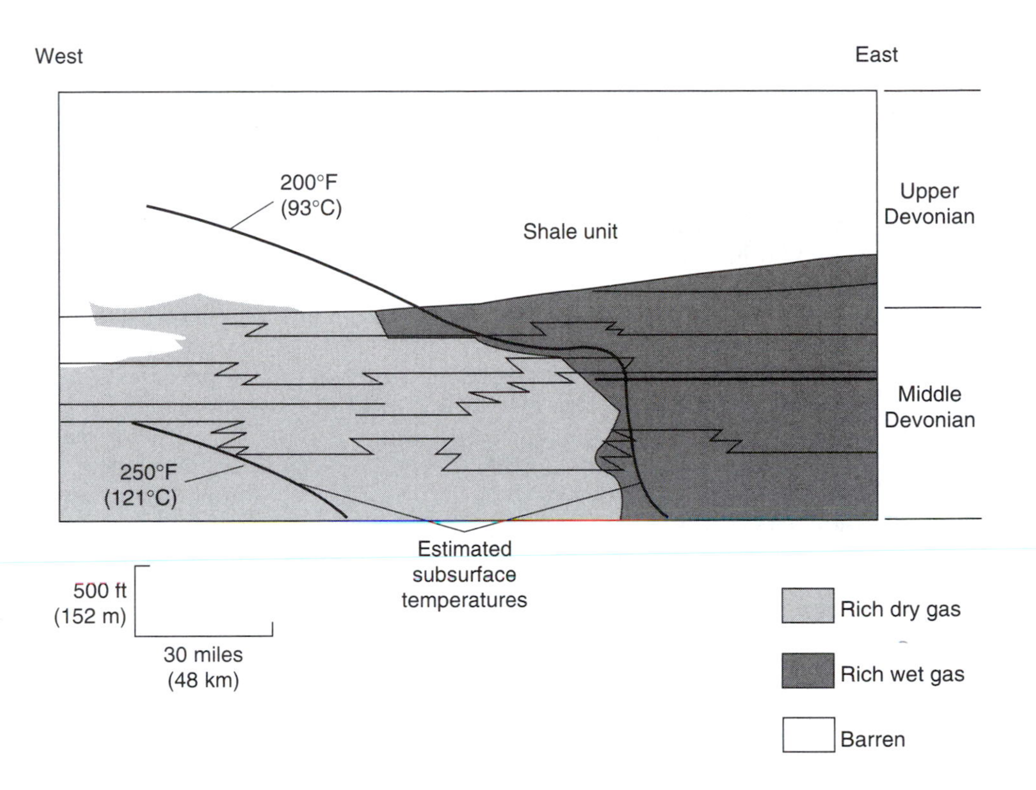

Figure 14-24

Gas facies in fine-grained Devonian shales and carbonates of northeastern British Columbia and northwestern Alberta, Western Canada Basin. [Evans and Staplin 1971]

cuttings in a Waring blender and sent the expelled gas directly to a GC. It is very important when comparing any cuttings gas logs that the same sampling and analysis procedure be used throughout the area under study.

Canned cuttings analysis is the only screening procedure listed in Figure 14-10 that clearly shows individual gas- and gasoline-range hydrocarbons that have already been generated by the rocks or migrated into them. When the rocks are alive with light hydrocarbons, it is a strong indication that petroleum accumulations will be found.

Solvent Extraction

Treating pulverized rocks with organic solvents to isolate the extactable OM (EOM) is required to do CPI or OEP and biomarker analyses (Figure 14-10).

This is a more expensive procedure that is used for those depth intervals determined to be most promising after the previously discussed screening procedures. Conventional large-diameter cores are best for this procedure. Sidewall cores, especially if they are taken after the hole reaches total depth, are not as good, because the walls of the drill hole are flushed with mud filtrate for several weeks or months during the drilling. However, if the sidewall cores are taken episodically while the well is being drilled and if the cores are fine-grained, relatively impermeable shales and carbonates, they are preferable to cuttings. In most wells, however, cuttings are mainly what is available.

The main problems in analyzing extracts are related to biomarkers, which were described in some detail in Chapter 10. It is important to wash the samples thoroughly before extraction and to obtain comparison samples of any organic materials added to the drilling mud. These can be held for later extraction in case any anomalous patterns show up in the GC–MS chromatograms.

Peters and Moldowan (1993) recommend not doing biomarker analyses on the EOM of rocks with less than 1% TOC. The reason is that at low TOCs, the biomarkers of recycled OM may overwhelm the indigenous biomarkers. Thus, Farrimond et al. (1989) found that low-TOC Toarcian Shales had much higher maturities than adjacent high-TOC shales, based on sterane and hopane isomerization ratios. The low TOC shales were recycled.

Kerogen

A rock must be treated with HCl and HF to dissolve the minerals and isolate the kerogen before making an elemental analysis of the kerogen to construct a van Krevelen diagram. This is done on samples from critical depths where the screening techniques indicate good source rocks or where previous analyses do not clearly show the kerogen types present. The van Krevelen diagram still provides the most accurate evaluation of kerogen type and maturity.

Kerogen and spore color (TAI and SCI) and vitrinite reflectance are generally run on the kerogen of cuttings samples taken every 300 ft (91 m), starting with indurated rocks. Since these analyses can be run on dried cuttings, they can be selected after the screening data are evaluated. Vitrinite reflectance also may be run on whole-rock polished pellet mounts, as stated in Chapter 10.

The biggest problem with kerogen and spore color is that it is qualitative and highly subjective. Quantitative light-measuring techniques and sets of reference slides diminish some of this subjectivity, but color is still a very rough maturity indicator. Both the color indices and vitrinite reflectance techniques require skilled organic petrographers familiar with the area being studied to get the best results. The widest applications of color indices have been in mapping oil- and gas-generative areas, such as in the Western Canada Basin by Evans and Staplin (Figure 7-20), the Paris and Aquitaine Basins by Correia (1971), and the Arkoma Basin by Burgess (1974). Vitrinite reflectance is generally more reliable than the color indices, as the next section points out.

Vitrinite Reflectance

All optical techniques are subjective compared with chemical techniques, and both need to be used to verify any conclusions. Vitrinite reflectance is now the most widely used maturity indicator, even though its problems have caused publication of both good and bad data. It has two basic problems, both of which are related to the hydrogen content of the vitrinite: (1) limiting the reflectance readings to the low-hydrogen vitrinite-1 group (telocollinite and telenite) only and eliminating readings of cavings, recycled vitrinite, mud additives, and high-hydrogen vitrinite-2 macerals, as advised in Chapter 10; and (2) limiting readings to coals and coaly particles in coaly shales and silty rocks to avoid high hydrogen–suppressed vitrinite.

The first problem can be minimized by using a reflectogram (Figure 14-25) showing the reflectance of all macerals in a kerogen sample (Peters and Cassa 1994). This is unlike a vitrinite reflectance histogram, which gives the distribution of reflectance values for only those phytoclasts believed by the petrographer to be telocollinite. High-hydrogen macerals like liptinites have low reflectances, and low-hydrogen macerals like fusinites (inertinites) have high reflectances, as Figure 10-24 shows. Also, any fluorescing vitrinites should be eliminated from histograms. Hydrogen-rich vitrinites in low-rank samples fluoresce a dull yellow to an orange brown color under blue light. Hydrogen-poor vitrinites do not fluoresce. The second problem of not reading suppressed vitrinite in organic-rich shales is more difficult, as explained later.

Random reflectance measurements are indicated by R_o or R_r. They plot as a straight line on a semilog plot of depth versus R_o. The R_{max}, which is determined by rotating the stage, is used on pure coal samples and does not make a straight line on a semilog plot. Telocollinite, a high-gray, vitrinite-1, or A, is the submaceral used for maturity measurements rather than the more hydrogen-rich, low-gray, vitrinite-2, or B, including desmocollinite (Figure 14-25), heterocollinite, and degradinite.

Cavings generally have lower R_o values, since the samples are coming from shallower depths than the primary vitrinite. However, cavings under reverse faults or near intrusives may have higher R_o values. Also, recycled cavings have higher R_os. The percentage of cavings in a sample can vary widely, depending on washout zones and casing points, so it is important to refer to the driller's log. Recycled OM generally has higher R_o values, but this can vary with changes in sediment source.

An example of vitrinite histograms with some of these problems is shown in Figure 14-26 (Dow and O'Connor 1982). The histograms are interpreted along with previously obtained data such as TAI, SCI, and pyrolysis. The top histogram in Figure 14-26 was from a sample that was clearly immature based on these latter parameters. Consequently, the first population in it is primary and the second is recycled. This primary population in the top histogram and all later ones follow a least-squares fit line on a semilog plot such as in Figure 10-23 if there are no complications. The maturity at any point in the section would then be taken from this best fit line.

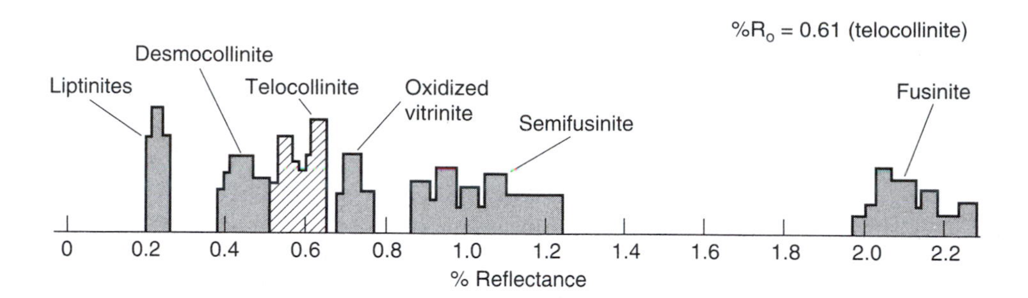

Figure 14-25

A complete reflectogram showing the reflectance of all macerals in a single kerogen sample. In those cases in which the selection of the "true" vitrinite population (telocollinite) is difficult, the trend of R_o versus depth established by many samples is useful for selecting the correct population. Here, telocollinite (hachured) has a mean vitrinite reflectance (%R_o) of 0.61. This sample contains significant amounts of oxidized vitrinite and semifusinite that could be mistaken for vitrinite. [Peters and Cassa 1994]

The most fundamental concept in vitrinite reflectance interpretation is that a linear temperature increase causes an exponential reflectance increase, which plots as a straight line on a semilog graph (Figures 10-23 and 10-26) (Dow 1977). This concept enables the petrographer to extrapolate R_o-depth plots through some stratigraphic intervals that may be devoid of vitrinite. However, geological complications like erosion, faulting, igneous activity, or changes in the geothermal gradient can alter this line. Consequently, for correct interpretation it is critical to understand the geological history of the samples being analyzed.

The standard deviations of the mean reflectivity of the samples in Figure 14-26 increase with depth. The reason is that vitrinite anisotropy increases exponentially with a linear increase in R_o, as shown in Figure 14-27. This broadening increase needs to be considered when interpreting vitrinite reflectance histograms.

Many well samples show groups of reflectance values with multipeaked histograms, as in Figure 14-28 (van Gijzel 1980). This sample contains fragments of fossil wood cells (telinite), for which the histogram showed no

Figure 14-26 (right)

Interpreted primary vitrinite populations and R_o maturity values in a sequence of samples from a well. [Dow and O'Connor 1982]

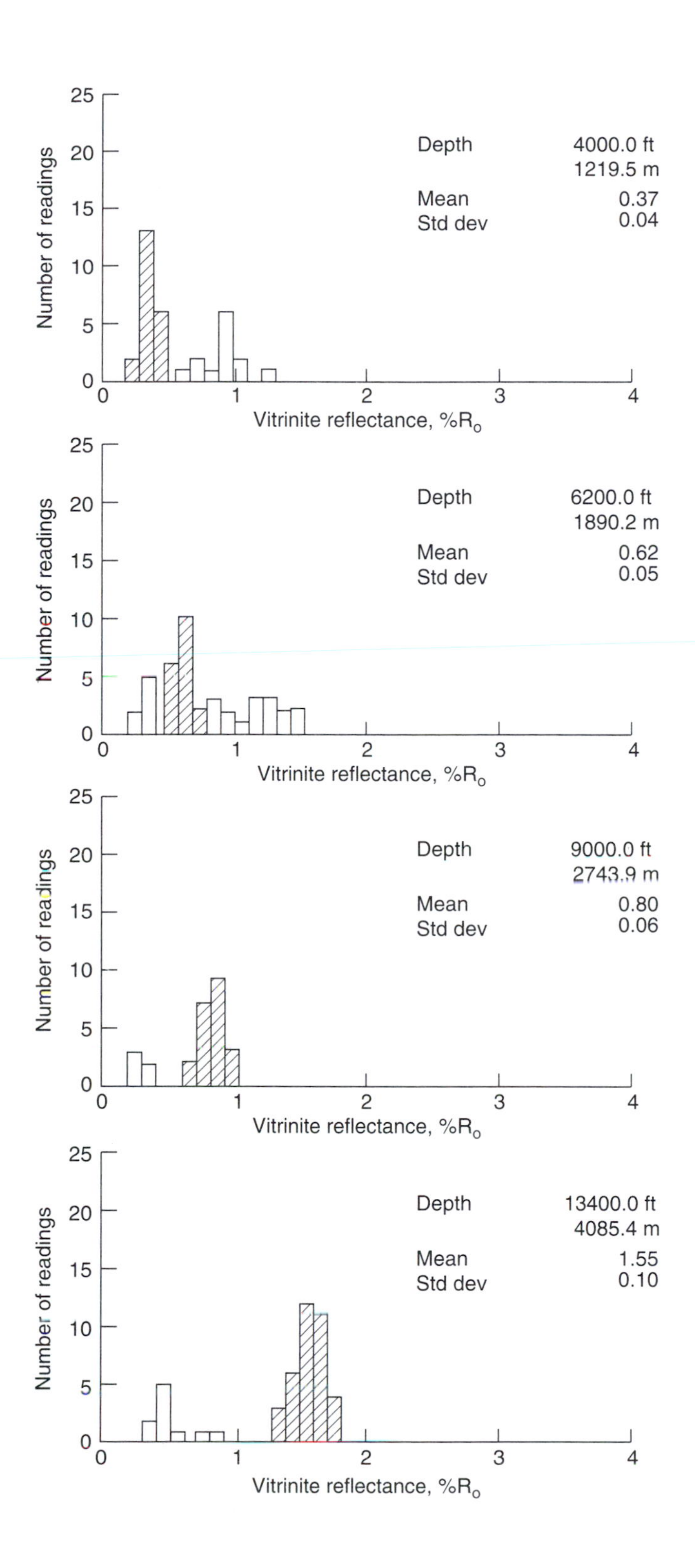
Depth 4000.0 ft
1219.5 m
Mean 0.37
Std dev 0.04
Number of readings
Vitrinite reflectance, %R$_o$
Depth 6200.0 ft
1890.2 m
Mean 0.62
Std dev 0.05
Number of readings
Vitrinite reflectance, %R$_o$
Depth 9000.0 ft
2743.9 m
Mean 0.80
Std dev 0.06
Number of readings
Vitrinite reflectance, %R$_o$
Depth 13400.0 ft
4085.4 m
Mean 1.55
Std dev 0.10
Number of readings
Vitrinite reflectance, %R$_o$

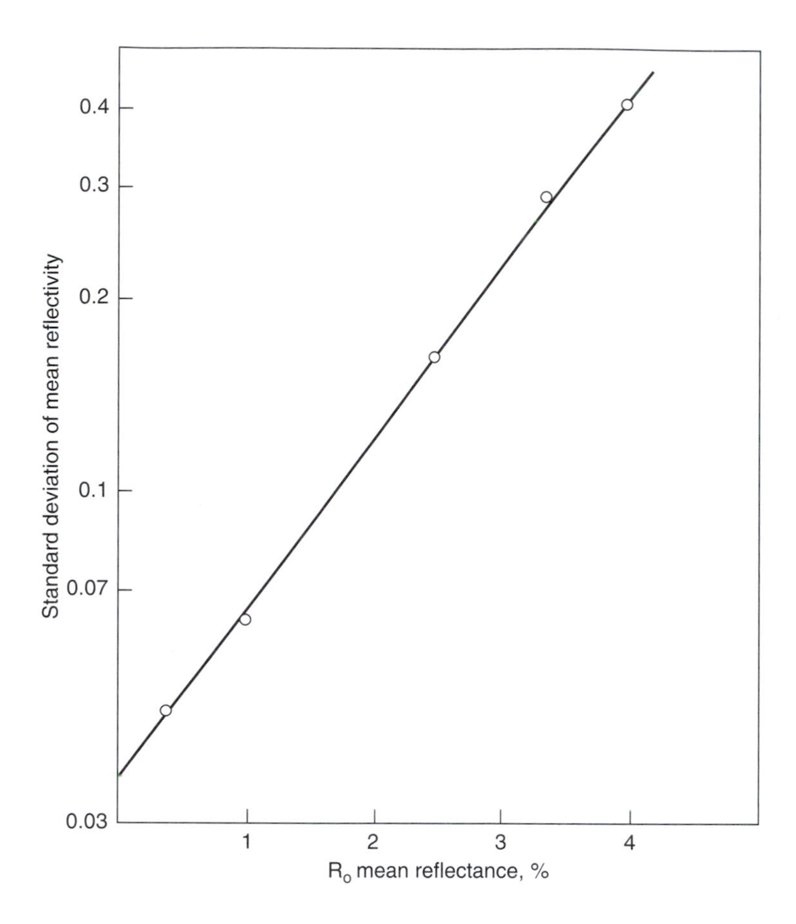

Figure 14-27

Broadening of vitrinite histograms, as indicated by the logarithmic increase in the standard deviation of the mean reflectivity with a linear increase in R_o values. [Hunt 1979]

readings of $R_o < 0.5\%$. This indicated that no lower R_o readings were related to the real vitrinite but belonged to other types of OM with a lower reflectance. Thus, the dark blocks represent fluorescent bitumens or bituminized collinite. In the reflectance range of $R_o > 0.5\%$, the amorphous collinite on the upper line of Figure 14-28 ranges in composition and reflectance more widely than the structured telenite below. Limiting the measurements to telinite leaves two populations at about $R_o = 0.7$ and 1.0%. Different telinite submacerals generally range in R_o by less than 10%, so it was concluded that the higher-reflecting telinite represents recycled wood cells. This leaves the telinite peak at 0.7% as representing the true primary population.

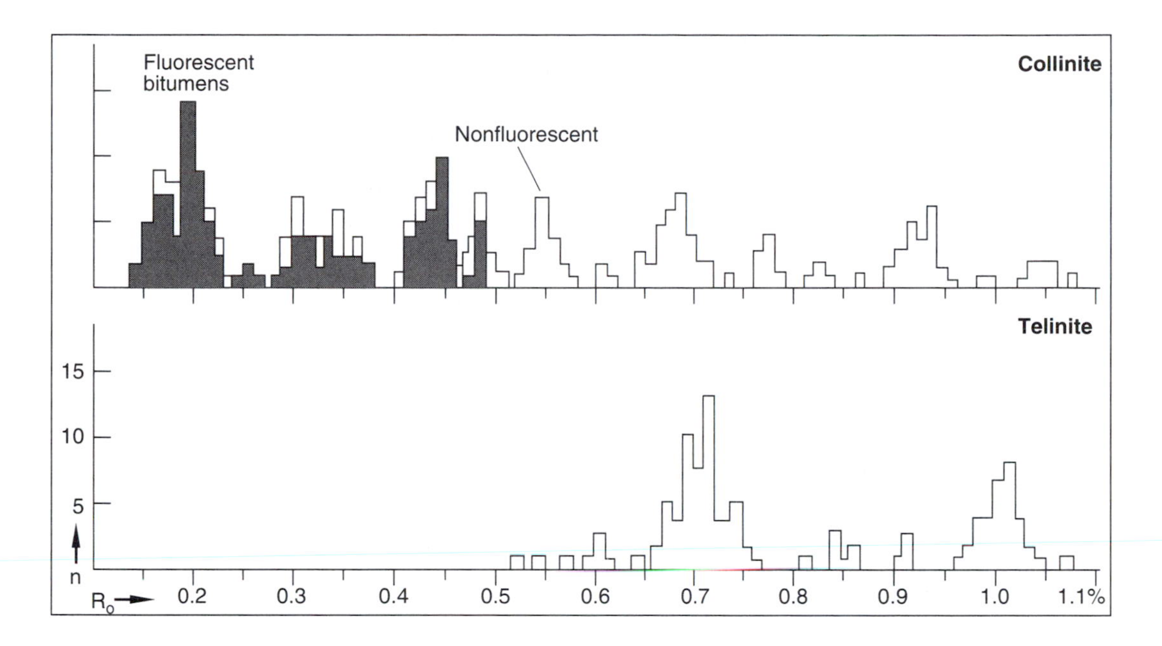

Figure 14-28

Multipeaked histogram from a well sample at a depth of about 4,000 m (13,120 ft). Black blocks are fluorescent bitumens or bituminized collinite whose reflectance values are not used for maturity assessment. Telinite peaks that are higher than 0.95% probably are recycled wood cells. The telinite peak at 0.70% represents the autochthonous population. This sample is in the early mature stage of oil generation. For more details on the interpretation of Figure 14-28, see van Gijzel 1980. [Courtesy of P. van Gijzel]

Suppressed Vitrinite

There is considerable evidence that there is a chemical difference between vitrinites formed in humic coals and coaly shales compared with those formed in shales with large amounts of sapropelic type II kerogen (Price and Barker 1985). Marlies Teichmüller noted that vitrinites of drift woods in oil shales like the Posidonia of Germany have a much lower reflectance (and higher fluorescence) than vitrinites from accompanying normal coals or siltstones (personal communication, 1993). She always avoided measuring vitrinite reflectance in oil source rocks, preferring instead the vitrinite inclusions in the adjacent normal clayey or silty rocks. She felt that the differences were due to variations in source material and early diagenesis.

This lower reflectance in shales compared with coals of equivalent maturity is called *suppressed vitrinite*. It is not the same as the difference observed between telocollinite (V_1) and desmocollinite (V_2). Suppressed vitrinite shows differences in reflectance readings larger than the differences among submacerals.

Seewald and Eglinton (1994) investigated the role of aqueous fluid chemistry, temperature, and time during vitrinite maturation by hydrous pyrolysis. They found that at the same maturity level, pyrolysis using high-pH pore fluids yielded vitrinite with a reflectance of 1.06 to 1.19%R_o, and pyrolysis with a low-pH fluid formed vitrinite with a reflectance of 1.38%R_o. Since marine shales are deposited in a higher pH environment than coals, this could explain part of the suppressed vitrinite difference.

Lewan (1993a) carried out hydrous pyrolysis experiments to compare vitrinite reflectances in coals and shales at various maturation levels. He identified two well-defined maturity trends, a humic coal trend and a suppressed trend, as illustrated in Figure 14-29. The humic coal trend was based on the change in reflectance with laboratory maturation of humic coals and lignite from Utah, Wyoming, and the Gulf Coast. The suppressed trend is from the organic-rich Woodford and Phosphoria Shale source rocks and the Cambrian Alum Shale. Note that the difference in the trends increases with greater thermal stress, reaching a maximum of about 0.7%R_o at 340°C for 72 hours. At this maturity level the reflectance of vitrinite in coals is 1.4%R_o, and it is 0.6%R_o in the Woodford Shale. These two numbers cover the entire range of the oil-generation window, so it is a very significant difference.

Lewan (personal communication, 1995) has never found a humic coal on his suppressed trend in Figure 14-29. Even coals composed of hydrogen-rich vitrinites fall on the humic coal trend. Shales with dispersed vitrinite, however, follow either trend. Organic-rich shales with predominantly type II kerogen, like the Woodford, Phosphoria, and Alum Shales discussed in earlier chapters, all fall on the suppressed trend. The Eocene Kreyenhagen source rock of California also is on the suppressed trend. But the Mowry Shale of Wyoming with low TOCs (1 to 4%), a mixed type II–III kerogen, and no hydrogen indices above 400 (Figure 10-12) falls on the humic coal trend (Figure 14-29). This suggests that dispersed vitrinite in shales with low TOCs and some type III kerogen fit the coal trend and that vitrinites in high TOC shales with predominantly type II kerogen fit the suppressed trend.

Figure 14-29 (right)

Mean random reflectance of vitrinites isolated from aliquots of humic coals and organic-rich shales: (C) Cretaceous coals of Wyoming plus Tertiary coals of Utah and the U.S. Gulf Coast; (L) lignite; (P) Phosphoria Shale of Wyoming; (W) Woodford Shale of Oklahoma; (A) Alum Shale of Scandinavia; (M) Mowry Shale of Wyoming. The samples were heated isothermally in water at temperatures from 300 to 360°C, for 72 hours. [Courtesy of M. D. Lewan]

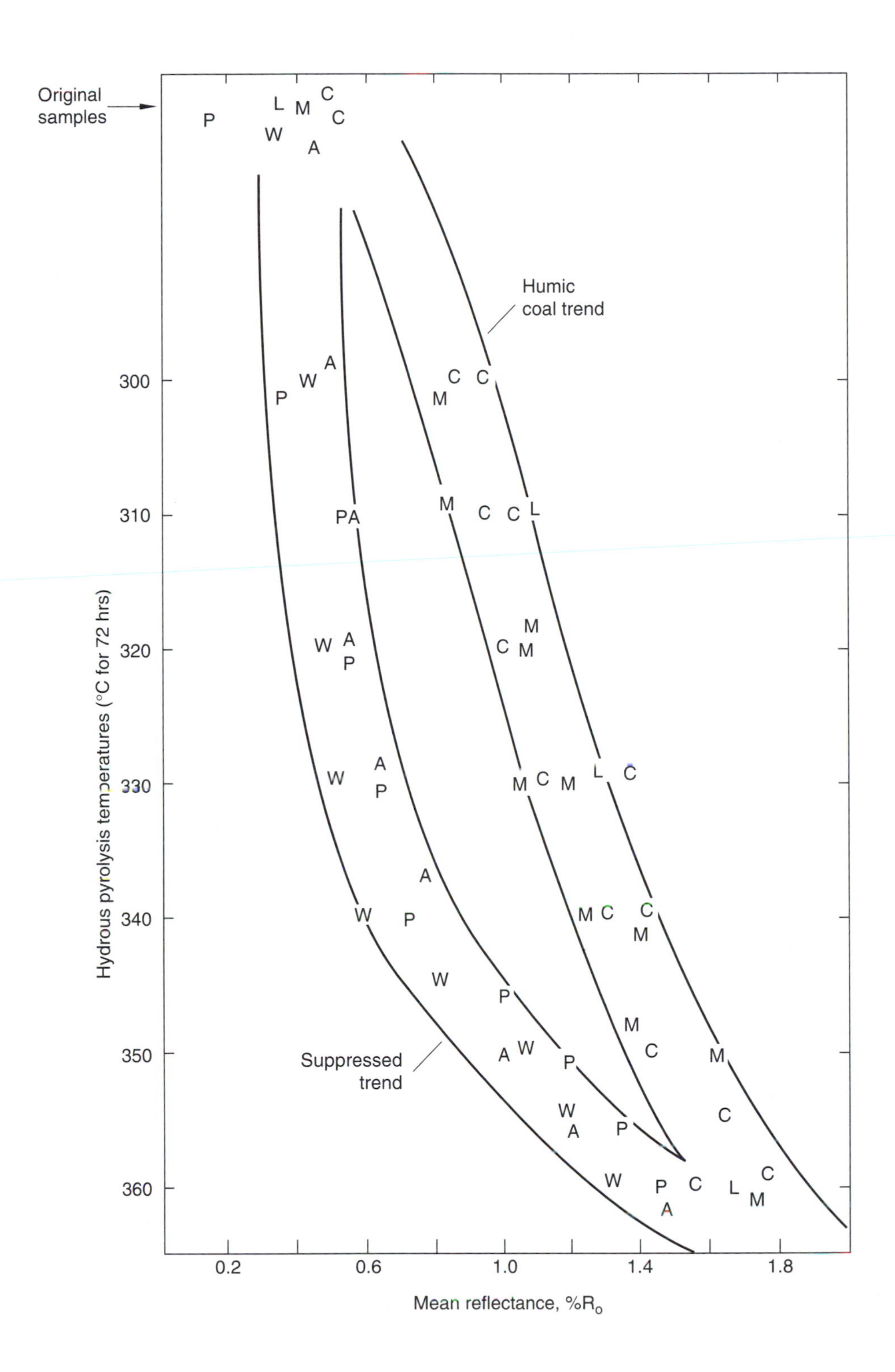

Original samples
Humic coal trend
Suppressed trend
Hydrous pyrolysis temperatures (°C for 72 hrs)
Mean reflectance, %R$_o$
300
310
320
330
340
350
360
0.2
0.6
1.0
1.4
1.8

There is no simple way to correct the suppressed vitrinite values except possibly to use curves like those in Figure 14-29 to obtain rough estimates of the humic coal reflectances. This problem does emphasize that vitrinite reflectance readings need other maturation indicators for confirmation, especially if the data are limited to organic-rich marine shales.

Other Vitrinite Problems

A summary of the kinds of problems that occur in vitrinite reflectance measurements is shown in Table 14-2 from Dow and O' Connor (1982). Only the primary vitrinite gives the correct maturity. Recycled vitrinite from the erosion and redeposition of more mature sedimentary sources gives too high an R_o, and cavings usually give too low a value. Mud additives are more of a problem, since some organic materials cannot be positively distinguished from primary vitrinite in the disseminated form. However, measuring R_o from polished whole-rock pieces minimizes or eliminates this problem, since the mud additives are then more readily distinguished from the indigenous vitrinites (Lo 1992).

Solid bitumens that look like vitrinite can be more easily identified in a polished whole-rock mount because the bitumens usually fill the intergranular spaces (Lo 1992). Bitumens in a polished kerogen concentrate are more difficult to distinguish from vitrinite.

Vitrinite, which has been oxidized in the subsurface, is observed along some major unconformities, in formations in contact with meteoric waters, and in red shales and interbedded red and green shales. Oxidized vitrinite indicates that the sediments have either been eroded or exposed to meteoric waters at some time in their geologic past. Most oxidized vitrinite ranges between 1 and 1.4%R_o. High-rank samples tend to be reduced to this range by oxidation, and low-rank samples are raised to it. Weathering also causes vitrinites to be pitted and corroded with a rough texture. This roughness scatters the incident light, resulting in different R_o numbers. Roughness also can be a polishing problem. Differences that are sometimes observed in R_o values in sandstones compared with adjacent shales may be due to oxidation of the vitrinite in the sandstones by groundwater. The edges of oxidized vitrinite grains are generally more reflective than the center part, causing them to appear to have a white ring.

Interlaboratory comparisons of vitrinite reflectance data on the same samples frequently show wide variations (Dembicki 1984; Lo 1992). These are mainly caused by choosing the wrong indigenous population or measuring the wrong submaceral. Also, the other factors listed in Table 14-2 affect the accuracy of the measurements. Obviously, measuring vitrinite reflectance is not a procedure that can be carried out by someone with only a few weeks of training. As stated earlier, it requires organic petrographers with several years of experience who can recognize all the pitfalls and problems in interpreting R_o profiles. Petrographers do the best job when given the depth and other pertinent information available on the samples. In addition, any other maturity indicators, such as the TAI, SCI, and T_{max}, should be available to assist in the interpretation.

TABLE 14-2 Problems Obtaining True R_o Maturities

A. Properly identified vitrinite

1. Primary
2. Recycled
3. Cavings
4. Mud additives
5. Subtypes with varying R_o (< 0.5)

B. Factors affecting accurate R_o measurements

1. Rough-textured vitrinite
   a. Weathered
   b. Partially dissolved
   c. Fractured
2. Oxidized vitrinite
3. Inclusions
   a. Pyrite
   b. Bitumen
   c. Other macerals
4. Oily vitrinite
5. Natural coking
6. Too few readings

C. Material that may look like vitrinite

1. Solid bitumen (several types)
2. Pseudovitrinite
3. Semifusinite

Problems obtaining maturity data from vitrinite reflectance analysis include the following: Properly identified vitrinite (A) may have several sources, but only the primary population can be used for maturity analysis. A number of factors (B) can affect the accuracy of R_o measurement in each vitrinite population, and some materials look like vitrinite (C) but have different reflectivities at the same maturity level.

Source: Dow and O'Connor 1982.

In analyzing cuttings down a hole, there may be gaps in the occurrence of vitrinite, because organic-rich shales with mostly amorphous kerogen may have little or no primary vitrinite. Also, carbonates that contain mostly marine amorphous kerogen have no vitrinite unless there are some clastic components in the carbonate. Anhydrite and other evaporites rarely have vitrinite. However, if enough data can be obtained on good vitrinite particles at suitable intervals down the hole, the straight line on a semilog plot can be extrapolated through these barren or unusable sections. Since humic vitrinite is derived from vascular land plants, it is not common in rocks older than Devonian, because land plants had not yet evolved. Nevertheless, the reflectance of particles in the Cambrian Alum Shale follow the suppressed vitrinite trend of Figure 14-29 (Lewan, personal communication 1995). This may be vitrinite derived from cellulose.

Vitrinite reflectance has the advantage over other maturation measures of covering the entire temperature range from early diagenesis through catagenesis into metagenesis. It is the most useful subsurface-prospecting tool for determining the present and past stages of maturation.

The importance of having the driller's log was explained in the T_{max} discussion (Figure 14-20). It is also important in vitrinite studies because it indicates where the casing was set in the well. The casing prevents cavings. The log also indicates washout zones, the source of most of the cavings. Recycled material is usually from one or more erosional zones in a well. At $R_o > 2\%$, reflectance values for primary vitrinite increase faster than those for recycled vitrinite. Consequently, profiles of recycled and primary vitrinite eventually merge (Dow and O'Connor 1982).

Coordinating a Geochemical Program

A geochemical program for a wildcat well should be tailored beforehand to fit the questions asked about that drilling area. This is best done if the geochemists who will do the analyses are briefed on the background geology and geophysical interpretations so that they can suggest the best sampling program before drilling the well. If a service laboratory is doing the analyses, their personnel should be brought into the project as early as is feasible so that they can evaluate the needs and risks.

Table 14-3 lists the kinds of analyses suggested for a typical wildcat well. It is a summary of the items covered in this chapter. A considerable amount of geochemical data can be obtained in wildcat areas before the first well is drilled. In addition, if vertical migration pathways, such as faults, are visible on the geophysical profiles, it is justifiable to conduct a surface hydrocarbon survey across the faults, as explained in Chapter 13. Likewise, if gas chimneys are visible in offshore seismic surveys, the surface sediments in the vicinity of the chimneys should be analyzed for gaseous hydrocarbons. Determining molecular-size distributions and analyzing the individual hydrocarbons for carbon isotopes would indicate the maturity of the gas and whether it is thermogenic in origin.

In addition to these laboratory studies, the wireline logs can be checked for intervals of high electrical resistivity, high gamma ray response, exceptionally low density, and low transit time on the sonic log. The combined interpretation

TABLE 14-3 A Suggested Geochemical Program for a Wildcat Well

Before drilling

1. Use heat flow and seismic data plus regional geology to construct burial history curves, and do basin modeling to estimate the oil- and gas-generation windows and possible depths of overpressure.
2. Analyze unweathered outcrops of possible source rocks for TOC. Do pyrolysis on samples with a TOC > 0.5% and vitrinite reflectance on selected samples. Analyze all available seeps for correlation with other oils in the basin, if any.

Screening a wildcat well

1. TOC and pyrolysis every 60 ft or 18 m on well cuttings of lithified rocks (5–10 gms).
2. Light hydrocarbon (headspace) analysis if canned cuttings are available every 90 ft or 27 m (50–100 gms).
3. Vitrinite reflectance every 300 ft or 91 m, with closer spacing in oil and gas windows (10–25 gms).

More detailed studies

1. Kerogen description, fluorescence, and possibly elemental analysis of kerogen on selected samples in critical intervals defined by the screening process.
2. C_{15+} extraction (bitumen) and GC and GCMS (biomarkers) for intervals important to source bed characterization and for correlation with crude oils and sediments in other parts of the basin.

Any conventional cores or sidewall cores of low-permeability shales and carbonates should be analyzed in detail for comparison with the corresponding cuttings.

---

of these logs may point to source rock intervals deserving special analytical or sampling attention. Likewise, mud-log data should be combined with the geochemical screening techniques to help identify the intervals that are alive with hydrocarbons. If the well is a dry hole, it is still important to collect as much geochemical information as possible in order to determine whether further drilling in the area is justified.

SUMMARY

1. Oil and gas accumulations occur in sedimentary sequences that appear "alive" with hydrocarbons in drilled well sections. Prospecting should be directed toward those areas and formations where the shale gas readings on the mud log are high and where geochemical screening analyses show that the fine-grained rocks have a relatively high hydrocarbon potential.

2. Mapping gas yields from mud logs can help identify prospective areas for oil and gas. In the Western Canada Basin, a band of wet shale gas extends from the Peace River area south to the border of Alberta and the United States. All major oil production lies in this wet gas band. West and northwest of this band, the mud logs show dry gas, and the reservoirs contain only gas. Near Regina, Saskatchewan, mud logs show no gas in the entire section from the Cambrian to the Cretaceous. This area lacks commercial production. When oil-rich muds are used in drilling, a special mud–gas separator like a steam still is needed for reliable mud-log readings.

3. Wireline logs may be used to evaluate oil source rock potentials if they have been suitably calibrated with geochemical data and if the TOCs are $>1.5\%$. Good oil source rocks within the oil window show high electrical resistivities, high gamma ray response, low densities, and low transit time on the sonic log.

4. The geochemical screening techniques used with wet well cuttings are TOC, pyrolysis, and canned cuttings analysis. The TOC eliminates samples too low in carbon to qualify as source rocks. Pyrolysis defines maturity, identifies reservoir rocks, and classifies source rocks into three categories: potential (immature), effective (mature), and spent (postmature). Canned cuttings analysis shows relative yields for the gas- and gasoline-range hydrocarbons. Care must be used when interpreting data for cuttings of mixed lithologies, because good, thin source beds may be masked by mixing with cuttings of nonsource rocks.

5. The single most powerful tool for understanding regional geochemistry is the geochemical log. These logs can be generated by computer programs designed to improve the quality of large volumes of data by eliminating or highlighting anomalous inputs. Such logs allow quick recognition of important changes in geochemical parameters with depth. They also provide a basis for identifying the best source intervals, along with any anomalous data.

6. Anomalously low HI values may occur in organic-lean rocks high in smectite and illite, and anomalously high OI values may occur in organic-lean, carbonate-rich rocks, particularly those with siderite. The problem is minimized by eliminating all data for rocks with TOCs $<1\%$ or by pyrolyzing the pure kerogen after isolation from the mineral matrix.

7. The T_{max} can vary by as much as 20°C in immature samples because of variations in kerogen type. This variation diminishes at higher maturities. T_{max} values also are affected by a variety of mud additives, such as walnut hulls,

lubrabeads, gilsonite, and diesel oil. When high-oil or oil-based muds are used, the cuttings need to be extracted with organic solvents before pyrolysis. This prevents the use of the S_1 and biomarkers unless the oil additives are available for comparison.

8. High-molecular-weight bitumens (C_{24+}) generally are recorded in the S_2 peak, resulting in a low T_{max} and a high HI. This is minimized by extracting the bitumen from the sample with organic solvents before analysis.

9. Oil in reservoir rocks, migrating oils, and contaminants in a well can be recognized by an anomalously high S_1, a high PI, and a low T_{max}.

10. Binocular examination of cuttings is strongly advised to eliminate particulate contaminants and also to highlight rapidly changing lithologies for which closer sampling and analysis are required.

11. Vitrinite reflectance is the most widely used maturity indicator. Reflectance should be measured with primary telocollinite or another primary submaceral of the high-gray vitrinite-1 group.

12. When possible, measurements should be made on vitrinite in coals, coaly shales, and siltstones. Organic-rich shales with type II kerogen generally contain suppressed high-hydrogen vitrinite, which gives anomalously low reflectances.

13. The most fundamental concept in vitrinite reflectance is that a linear increase in temperature causes an exponential increase in reflectance, which plots as a straight line on a semilog graph.

SUPPLEMENTARY READING

Bordenave, M. L. 1993. *Applied petroleum geochemistry.* Paris: Éditions Technip.

Cooper, B. S. 1990. *Practical petroleum geochemistry.* London: Robertson Scientific Publications, 174 p.

Helander, D. P. 1983. *Fundamentals of formation evaluation.* Tulsa: OGCI Publications, Oil & Gas Consultants International, 332 p.

Peters, K. E., and M. R. Cassa. 1994. Applied source rock geochemistry. In L. B. Magoon, and W. G. Dow. (eds.), *The petroleum system—From source to trap.* AAPG Memoir 60. Tulsa: American Association of Petroleum Geologists, pp. 93–120.

Chapter

15

Crude Oil Correlation

The correlation of crude oils with one another and with extracts from their source rocks provide valuable tools for helping the exploration geologist answer production and exploration questions and extend existing exploratory trends. When two or more pools are discovered in a relatively unexplored area, the question arises as to both the lateral and vertical extent of the productive trend. Are there one or more families of oils in a particular rock sequence? Are any of these oils related to potential shallower or deeper accumulations? Can known seeps in the basin be related to subsurface pools?

Each family of oils represents one element of a distinct petroleum system. Consequently, identifying the number of oil families is equivalent to defining the number of petroleum systems. By identifying the source rocks of each family, the drilling can focus on prospects within the drainage areas of those sources.

Oil-to-oil correlations also are used in production programs. Is there reservoir continuity between oil and gas accumulations extending across faults? Can the geochemical composition of oils and gases from different production intervals be used to support evidence for their vertical separation by thin shale beds or lateral separation by shale pinchouts? Are tubing or packing leakages the cause of the commingling of oils produced from different reservoirs?

Modern analytical techniques are sufficiently sensitive to establish the similarity or dissimilarity of two oil samples, whether from reservoirs, source rocks, or seeps. More difficult is the problem of interpreting how an oil may change in moving from source to reservoir or how two crude oils of the same origin may undergo different physical and chemical changes after accumulation. A major objective of correlation procedures is to recognize the source fingerprint of the hydrocarbon molecules in oils and to understand how the fingerprint is affected by factors such as migration, water washing, biodegradation, thermal alteration, and evaporative fractionation.

Crude oil samples should be collected in glass bottles or vials with teflon-lined caps. Hydrocarbons can leach contaminants out of some plastic containers, and metal containers can introduce trace elements.

Refinery laboratories estimate the market value of a newly discovered crude oil by determining the API gravity, viscosity, pour point, and sulfur content. The oil is distilled to determine the volume of gasoline, kerosine, gas oil, and residuum. More detailed analyses, including paraffin, naphthene, and aromatic hydrocarbons and nitrogen, vanadium, and nickel content, are made to evaluate performance in various distillation and catalytic cracking units. Although many analytical procedures for petroleum were developed in refinery laboratories, they needed considerable modification to apply them to geological problems. For example, the refinery chemist works with several liters of oil from the pipeline, whereas the exploration geochemist may have a fraction of a milliliter extracted from a source rock.

Correlating Oil Families

In crude oil correlation, genetically related oils are differentiated from unrelated oils on the assumption that the same source material and environment of deposition produce the same oil. A biological marker compound dominant in the source rock would be expected to appear in the oils it generated. A particular ratio of two biomarkers not affected by extraneous factors would be expected to be the same in oils generated from the same source rock. The correlation problem is more complex, however, because the same source rock generates oils of different maturity at different times in its burial history. Also, the accumulated oils can be altered by secondary processes, such as maturation, gravity segregation, water washing, biodegradation, and secondary migration. In practice, a series of parameters are chosen to group oils that are little affected by secondary factors. Depending on the sophistication of the analyses, the oils fall into distinct groups with a few scattered or abnormal samples. The geochemist then correlates the groups with the geological data to arrive at an inferred genetic relationship within a petroleum system.

Bulk Correlation Parameters

Whole-oil properties as determined in the refinery are useful for an initial grouping of the oils and sometimes are sufficient for identifying genetically different oil types. Most oils range in gravity from about 25 to 45°API, as shown in Figure 15-1. The frequency distribution of gravities for 60 oils from the Witch Ground and Viking Grabens of the North Sea are in Figure 15-1 (A) (Fisher and Miles 1983). Two dominant populations are shown with mean gravities of 29 and 37°API. Most heavier, low API gravity oils on such graphs either

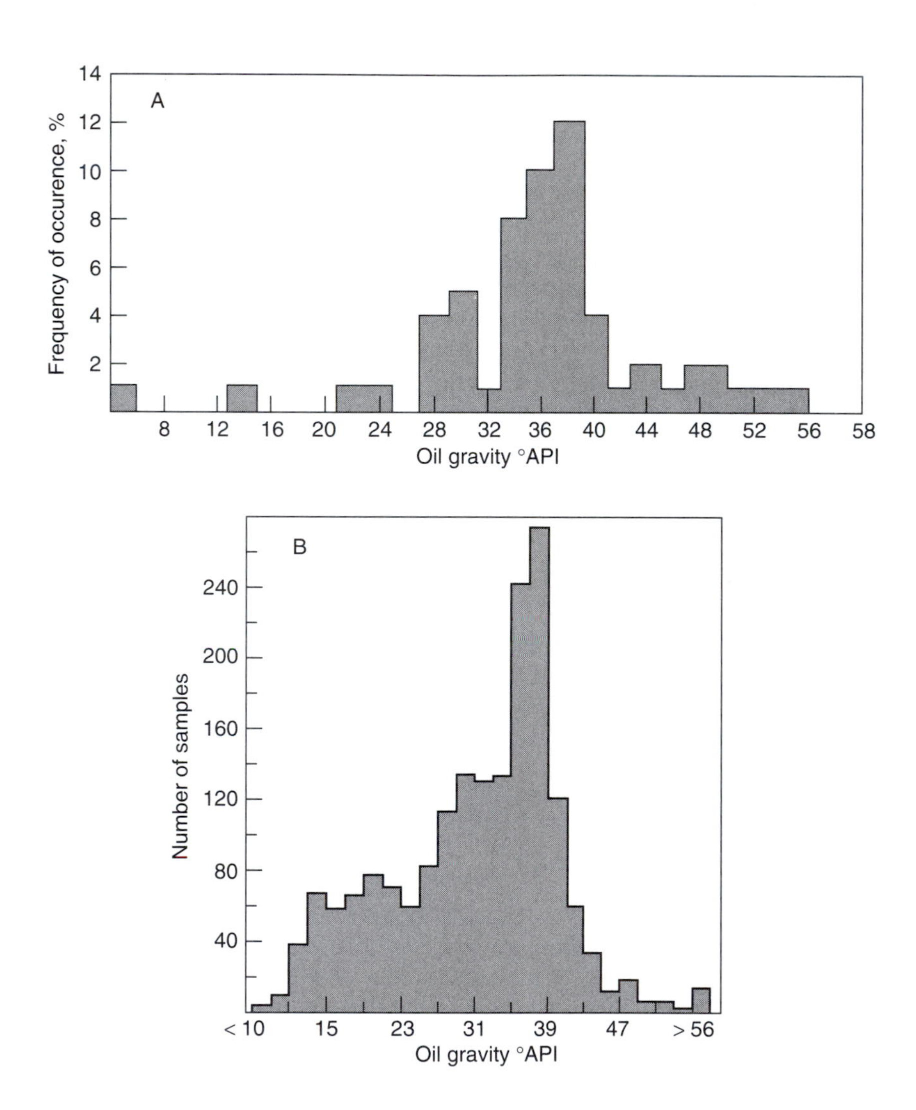

Figure 15-1

Histograms of API gravities for (A) 60 oils and condensates from the Witch Ground and Viking Grabens, North Sea [Fisher and Miles 1983] and (B) 1,800 oils from the United States. [Data from U.S. Bureau of Mines]

are caused by the biodegradation of mature oils or represent the early generation of immature oils. High gravities are characteristic of oils formed from late mature source rocks or of condensates formed by reservoir maturation or evaporative fractionation. A similar histogram in Figure 15-1 (B) represents 1,800 oils from the United States analyzed by the U.S. Bureau of Mines. It shows that most have gravities between 27 and 43°API, with a peak at about 38°API, similar to the second peak in the North Sea oils.

Ternary diagrams are frequently used in oil-to-oil correlations, plotting such things as the relative amounts of paraffins, naphthenes, and aromatics in the oils. Figure 15-2 shows on such a plot the composition of the C_{15+} hydrocarbon fraction of fifty-two oils from the Gulf of Suez (Rohrback 1983). Superimposed on the ternary composition diagram is the API gravity scale for the oils. There is no evidence for the biodegradation of these oils, and they were believed to have a common marine source rock. Rohrback attributed the main differences, based on biomarkers, to maturation. The oils range from 65% aromatics (11°API gravity) to 75% paraffins plus naphthenes (43°API). This

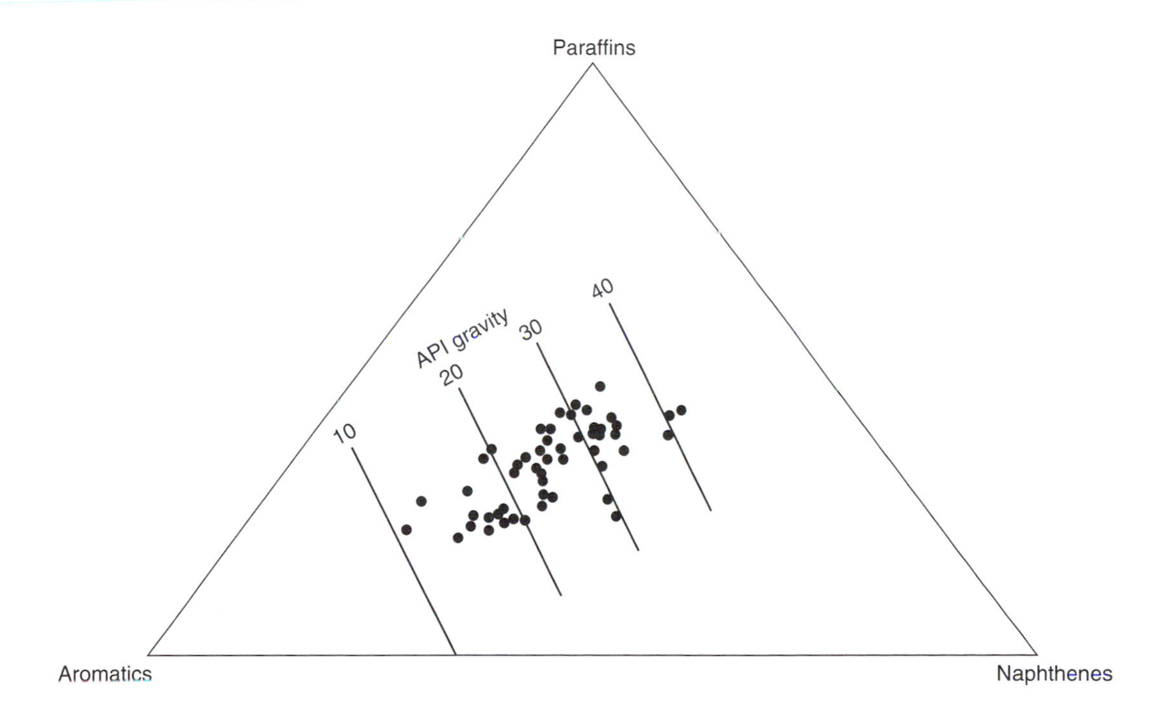

Figure 15-2

Ternary diagram showing the composition and API gravity of C_{15+} fraction of 52 oils from the Gulf of Suez. The increasing API gravity is due to maturation. [Rohrback 1983]

demonstrates that the bulk composition of crude oils can range widely owing to factors other than source. These bulk composition diagrams also are not very useful for correlating biodegraded oils, since biodegradation causes a preferential loss of paraffins.

An example of the effects of gravity segregation on the bulk properties of a reservoired oil is shown in Figure 15-3 and Table 15-1 for the Velma oil pool in Stephens County, Oklahoma. The Pennsylvanian age reservoir has a subsurface dip of 45°, with the downdip limit defined by the porosity pinching out. There is no water drive. Production is by solution gas drive. The first five properties of the oils in Table 15-1, beginning with API gravity, are significantly different for wells A and C, even though these two wells are only 1,320 ft (402 m) apart in the same pool. Thus, the total vanadium plus nickel increases by a factor of six from well A to well C. Wells drilled between A and C, such as B, show a gradationally intermediate change in these properties. However, in terms of hydrocarbon ratios, vanadium/nickel ratios, C_{15+} *n*-paraffins, and carbon isotopes, the oils are similar. This demonstrates that the geochemist must have some knowledge of how well a particular oil sample represents the entire pool. The same problem can occur in a reservoir containing a partially biodegraded

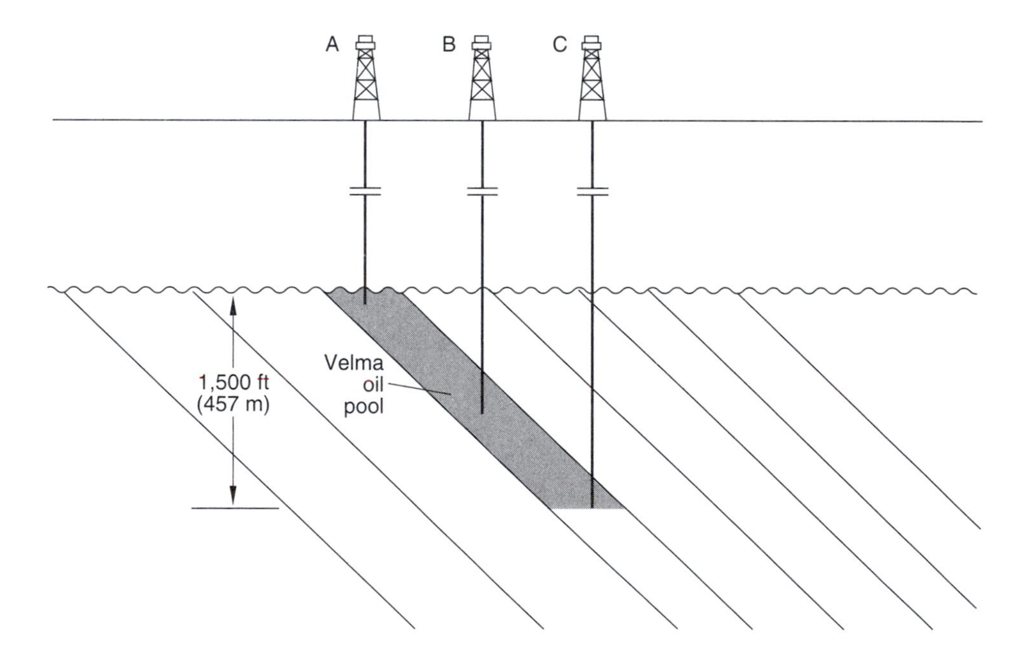

Figure 15-3

Cross section of Velma oil pool, Stephens County, Oklahoma, showing a steeply dipping reservoir.

TABLE 15-1 Properties of Velma Oil Samples

Well	*A*	*B*	*C*
Producing depth, ft	4,350	5,170	5,850
Gravity, °API	37	28	23
Percent boiling off at 200°C	42	19	14
Percent asphaltenes	1.3	3.3	8
Percent sulfur	0.95	1.2	1.4
Vanadium + nickel in ppm	41	106	259
Vanadium/nickel ratio	1.7	1.9	2.1
Percent C_{15}+ *n*-paraffins	6.5	6.1	5.6

oil. The Bell Creek field in Montana has an API gravity of 35° at the updip end in contact with meteoric water and 43° at the downdip end where it contacts connate water (Winters and Williams 1969).

Williston Basin

Somewhat more specific correlation parameters were used by Williams (1974) to separate the Williston Basin crude oils into three types, as in Figure 15-4. This diagram shows the distribution of 11 major saturated hydrocarbons (straight-chain, branched-chain, and cyclic) in the C_4–C_7 fraction of 125 crude oils. Subsequent analysis of the source rock extracts indicated that the type I oil originated from the Ordovician Winnipeg shale, the type II from the Mississippian Bakken shale, and the type III from the Mississippian Tyler formation.

North Slope of Alaska

Another example of using bulk properties to characterize genetically different oil types is the study by Magoon and Claypool (1981) of forty North Slope oils. The locations of the oils and seeps along with their type designations are shown in Figure 15-5. The Simpson–Umiat oils (open circles) and seeps (triangles) occur entirely in reservoirs of Cretaceous age and younger, whereas most of the Barrow–Prudhoe oils (solid circles) are found in Mississippian through Jurassic reservoirs. Some of the results of this study are summarized in Figures 15-6, 15-7, and 15-8 and in Table 15-2. The oil numbers in these figures correspond to those of the map in Figure 15-5. The plot of percent sulfur against API gravity (Figure 15-6) clearly separates the two major oil groups and their biodegraded subgroups. As discussed in Chapter 12, biodegradation lowers the API gravity

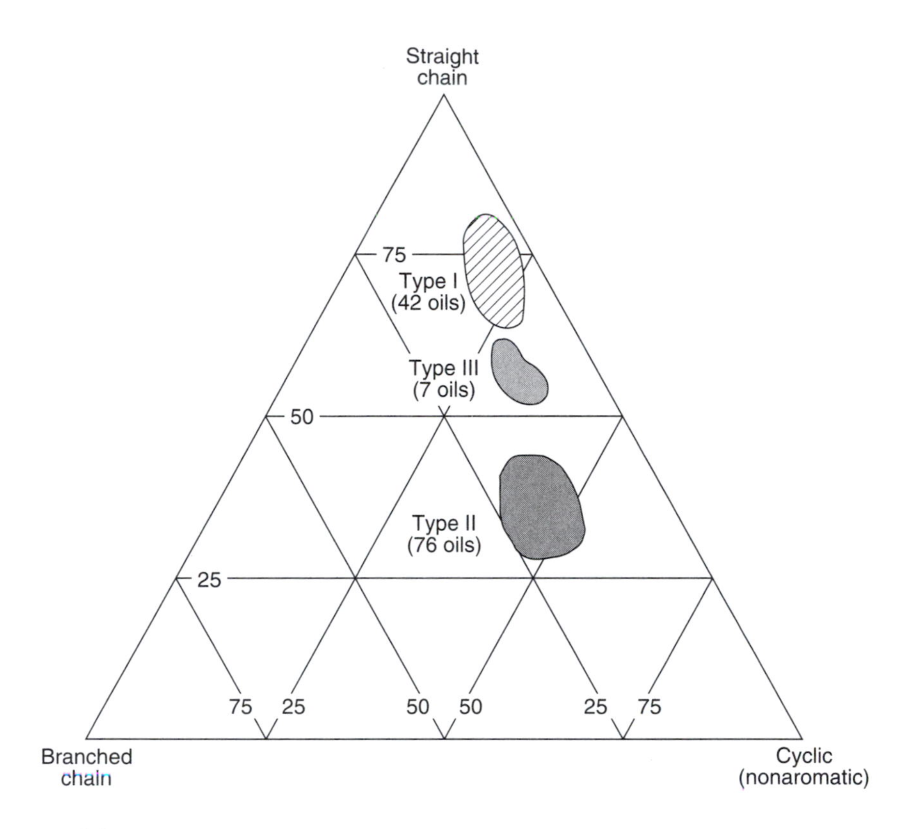

Figure 15-4

Distribution of straight-chain paraffins, branched-chain paraffins, and naphthenes (cycloparaffins) in the C_4–C_7 fraction of Williston Basin oils. [Williams 1974]

(because of the loss of small molecules) and relatively increases the sulfur content by concentration. The biodegradation of these samples was not severe enough, however, to cause an overlap in the sulfur content of the two groups. Biodegradation was recognized by loss of *n*-paraffins from the gas chromatograms of the C_{15+} saturates.

The plot of percent nitrogen versus sulfur (Figure 15-7) also clearly separates the two oil groups. The Barrow–Prudhoe oils show the highest nitrogen and sulfur contents. Sample no. 2, with the anomalously high nitrogen content, is the only solid asphalt in this study. The excess nitrogen probably represents the metabolic by-products of extensive bacterial activity.

The oil groups also are separated based on their stable isotopes. The $\delta^{34}S$ of the no. 2 seep is –10.6‰, and the $\delta^{13}C$ is –28.7‰. This is characteristic of the Simpson–Umiat group (Table 15-2). Figure 15-8 compares $\delta^{13}C$ values for satu-

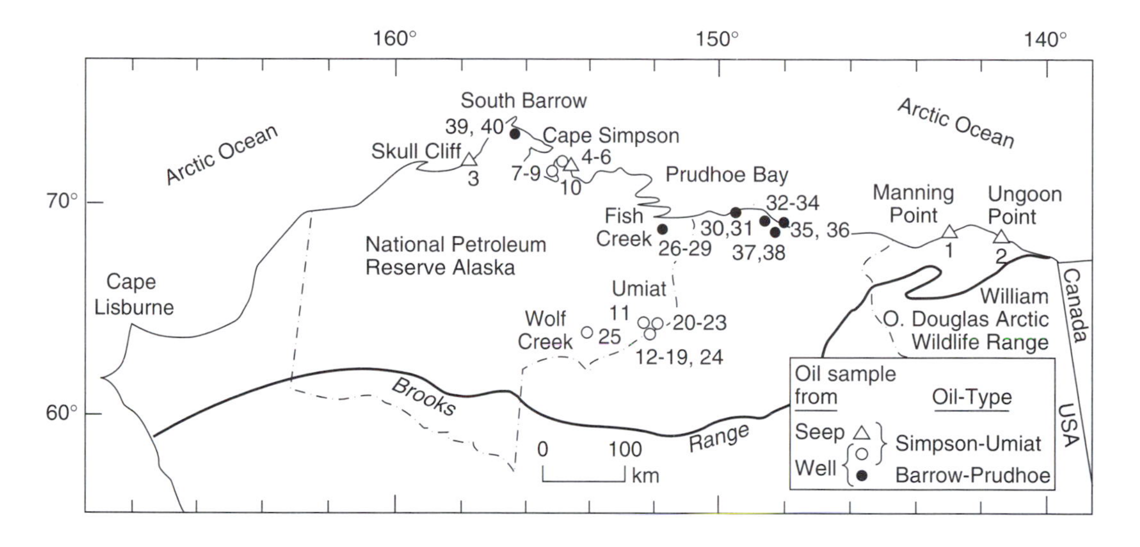

Figure 15-5

Map of the North Slope of Alaska showing the geographic distribution of oil samples (numbered) and two oil types. [Magoon and Claypool 1981]

rated and aromatic hydrocarbons. This plot separates the oils into three groups. The middle group, however, represents biodegraded Simpson–Umiat oils, compared with their normal precursors in the upper part of this figure.

Table 15-2 displays some additional correlation parameters such as the pristane/phytane ratio and the $\delta^{34}S$, which separate the oils into two distinct groups. Subsequent studies by a number of investigators on the North Slope

TABLE 15-2 Properties of North Slope Oil Types

Property	*Barrow–Prudhoe*	*Simpson–Umiat*
Gravity, °API	13 to 28	18 to 37
Percent sulfur	0.79 to 1.92	0.04 to 0.44
Percent nitrogen	0.13 to 0.43	0.01 to 0.05
Pristane/phytane ratio	<1.5	>1.5
$\delta^{34}S$, ppt whole oil	–2.7 to +2.1	–10.6 to –4.6
$\delta^{13}C$, ppt whole oil	–30.3 to –29.8	–29.1 to –27.8

Source: Data from Magoon and Claypool 1981.

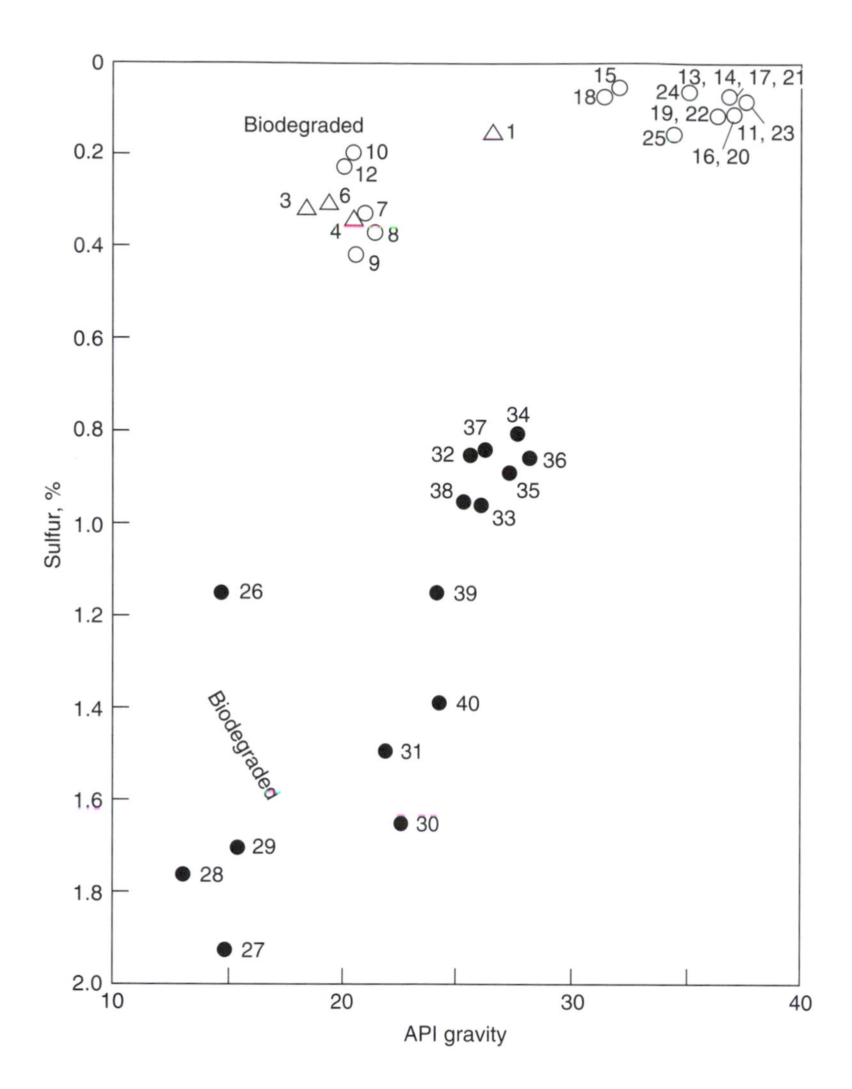

Figure 15-6

API gravity versus percent sulfur for Simpson–Umiat (open symbols) and Barrow–Prudhoe (solid symbols) oil types. [Magoon and Claypool 1981]

support the concept of two families of oils and provide some evidence that the Barrow–Prudhoe oils originated in the Triassic Shublik and Jurassic Kingak shales. The Early Cretaceous Pebble Shale and Torok Formations are believed to be the source of the Simpson–Umiat oils (Claypool and Magoon 1985).

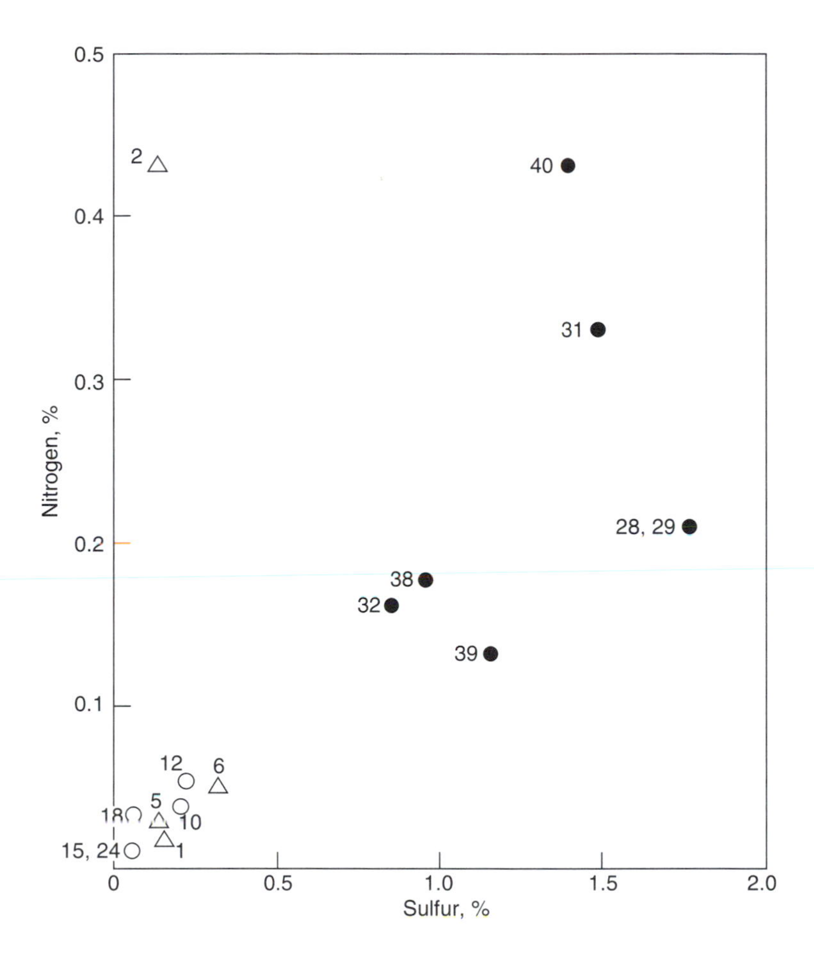

Figure 15-7

Percent sulfur versus nitrogen, showing two distinct populations of Simpson–Umiat (open symbols) and Barrow–Prudhoe (solid symbols) oil types. [Magoon and Claypool 1981]

Trace Elements

The chlorophyll molecule loses its magnesium at the time of deposition. During diagenesis, both vanadium and nickel become complexed to the porphyrin in the place of the magnesium. As porphyrins are introduced into a crude oil from the source rock, they carry the vanadium/nickel distribution with them. Many other trace elements in crude oils are simply a reflection of those picked up during migration or in the reservoir, so they have limited value in correlation.

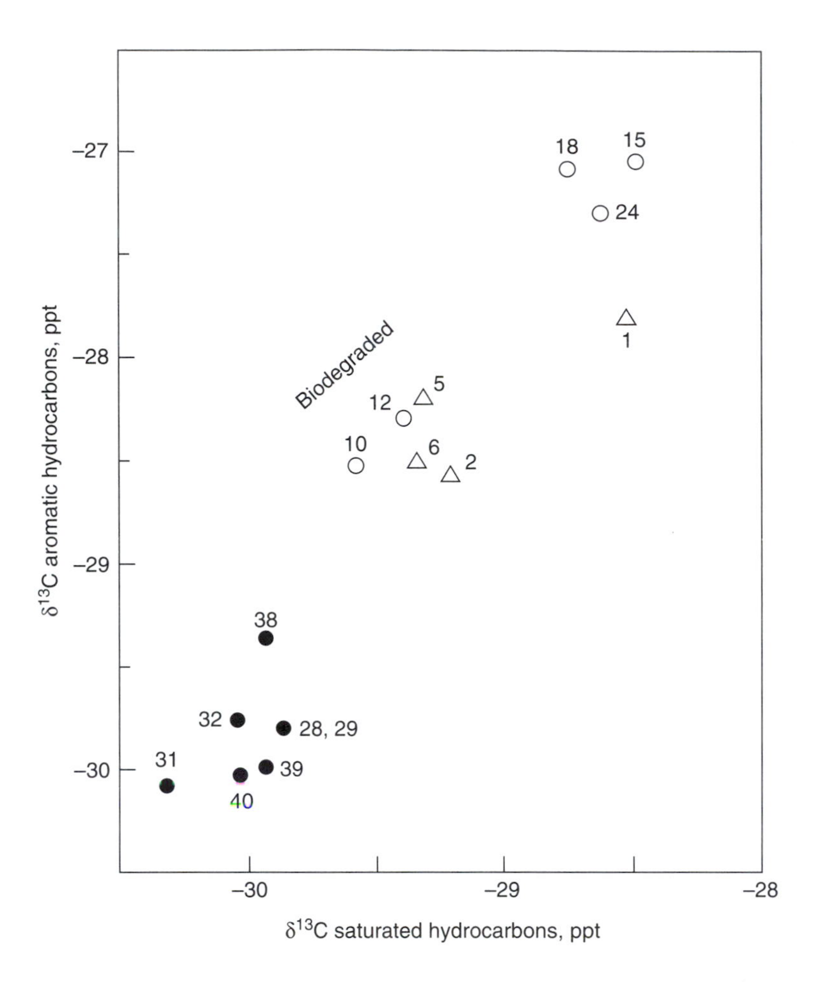

Figure 15-8

Comparison of carbon isotope ratios from saturated and aromatic hydrocarbon fractions for Simpson–Umiat (open symbols) and Barrow–Prudhoe (solid symbols) oil types. [Magoon and Claypool 1981]

Vanadium and nickel differ in having a genetic origin as a metal–organic complex. The concentration of vanadium and nickel in oil is several thousand times what would be expected if these elements were not complexed to the organic structures.

In Figure 15-9 the vanadium and nickel contents of crude oils from four different areas are plotted against API gravity. The best fit lines were calculated

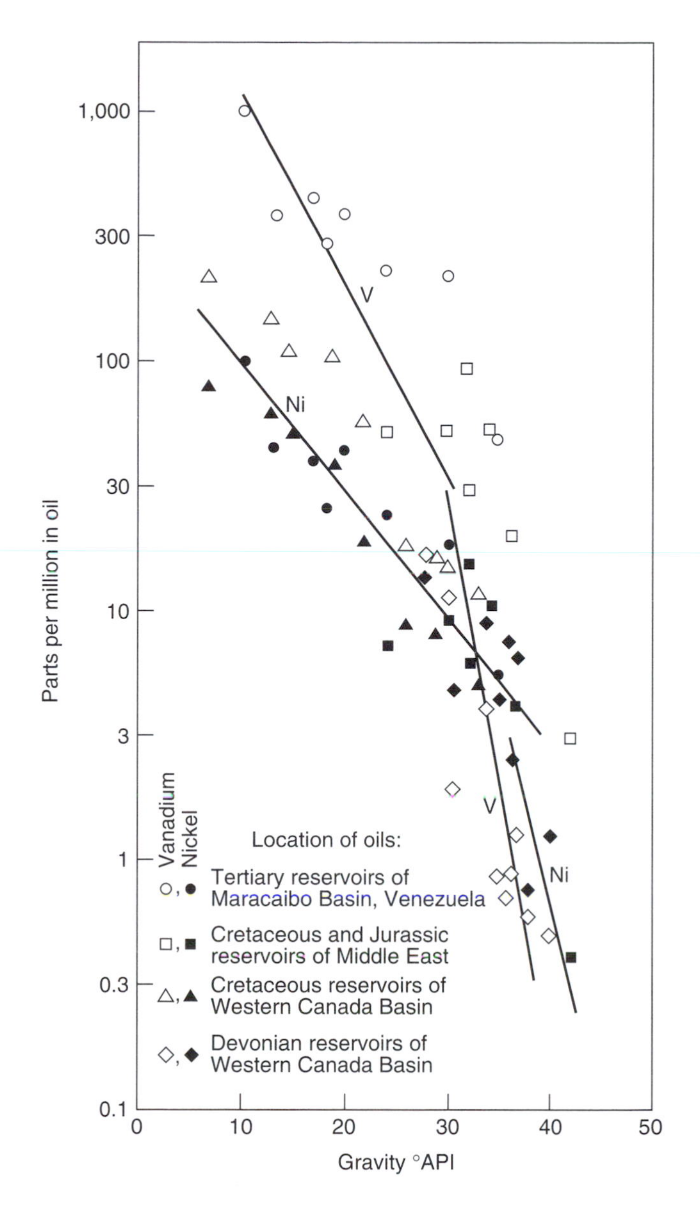

Figure 15-9

Parts per million by weight of vanadium (open symbols) and nickel (solid symbols) in oil versus °API gravity for curves of four areas. [Baker 1964; Hodgson 1954]

by the least-squares method. Two lines were calculated for each metal, the vanadium line splitting at 30 ppm and the nickel at 3 ppm. The correlation coefficients are about 0.9 for the nickel data and 0.7 for the vanadium data. The increasing vanadium and nickel content with decreasing gravity is a reflection of the increase in heavy ends containing the porphyrin structures in the lower-gravity oils. Oils in general contain more vanadium than nickel.

No differences are evident in the crude oils of four oil groups in Figure 15-9 based on their total trace element content. Differences can be recognized by using vanadium/nickel ratios rather than total concentrations, as seen in Figure 15-10. Each of the four groups of oils is characterized by different vanadium/nickel ratios. Hodgson (1954) noted that the vanadium/nickel ratio of crude oils appears to decrease with the age of the reservoir. Rosscup and Bowman (1967) measured the thermal stabilities of vanadium and nickel petroporphyrins and found that the vanadium compound is less stable than the nickel one. In Figure 15-10, the oils in the youngest reservoirs have the highest ratios and those in the oldest reservoirs the lowest, in accordance with the observed stabilities of these metal–porphyrin complexes. The relative maturi-

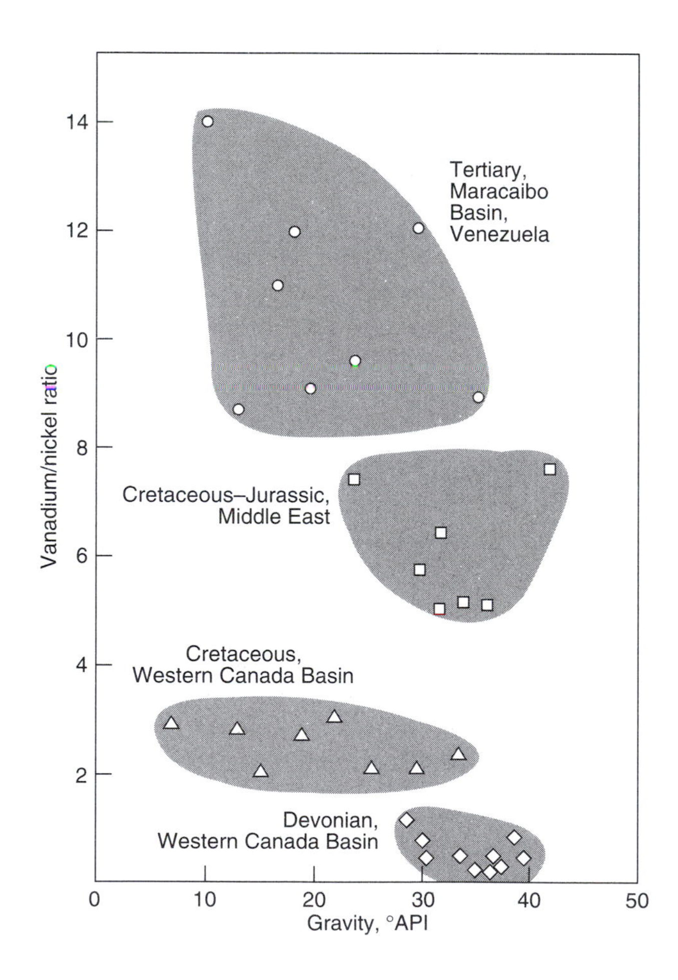

Figure 15-10

Vanadium/nickel ratios versus API gravity for the same oils shown in Figure 15-9.

ties of the oils themselves could not be measured at the time of these studies, so it was assumed that the oldest reservoirs contained the most mature oils.

Vanadium/nickel ratios do not change with the biodegradation or weathering of an oil seep. Consequently, their analysis permits correlation of solid asphalts at the surface with subsurface oil accumulations. Figure 15-11 illustrates the use of these ratios in evaluating seeps in the Wind River Basin of Wyoming. In 1884 a well was drilled near an asphalt seep on the Dallas Dome. It discovered oil in the Triassic at 300 ft (91 m). Subsequent deeper drilling in this century found oil pools in the Permian and Pennsylvanian sections. The question was whether these oils and the seep represented one or several families. All the samples had essentially the same vanadium/nickel (V/Ni) ratios, implying a

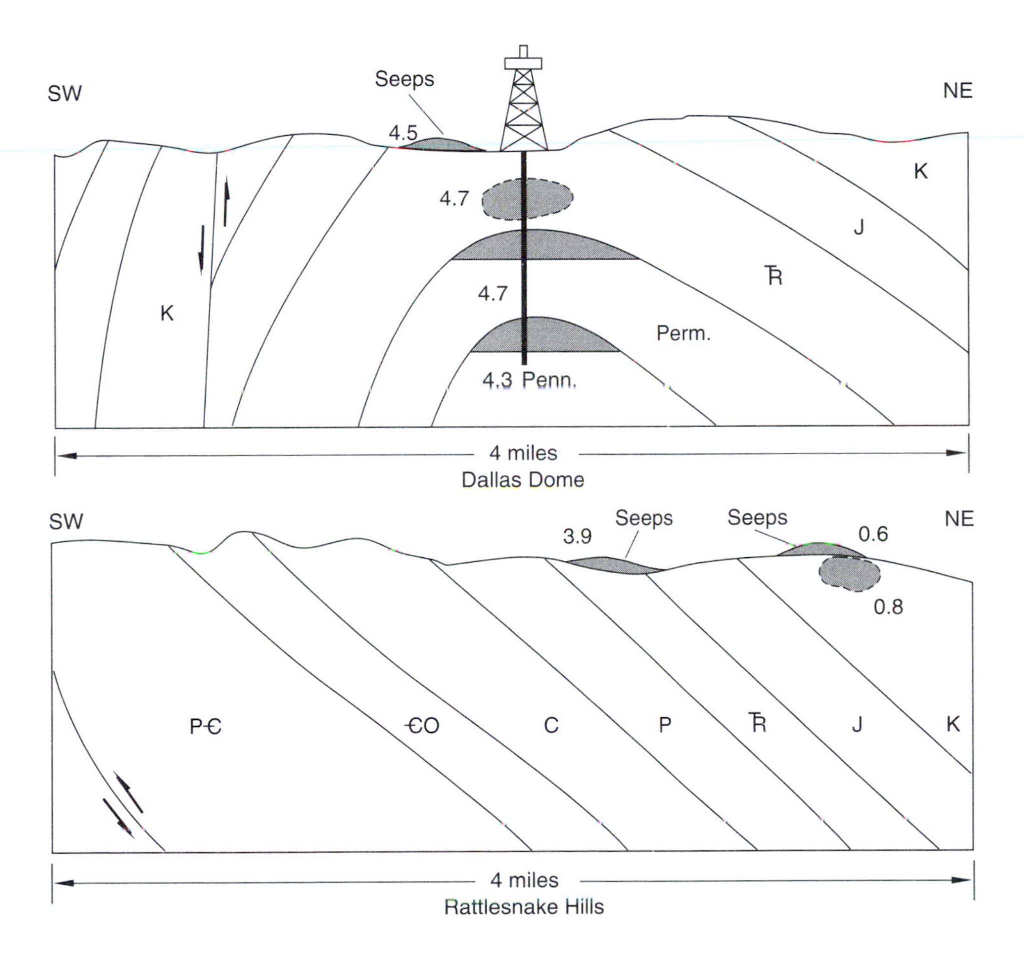

Figure 15-11

Vanadium/nickel ratios in Wind River Basin seeps and oils.

single source for all the oils. In the same basin at Rattlesnake Hills, two seeps from Cretaceous and Triassic outcrops were analyzed. They showed entirely different V/Ni ratios, indicating two oil families with two separate sources for the seeps. Relatively inexpensive surveys such as this can be followed up with more sophisticated biomarker analyses to verify the initial conclusions.

Whole-Oil GC Fingerprints

Obtaining whole-oil GC fingerprints requires analyzing an entire oil for the C_2-to-C_{45} hydrocarbon range on a gas chromatograph with a fused silica capillary column. A software system can be added to provide two gas chromatograms, one showing all hydrocarbons between C_2 and C_{45} and the other a computer-enhanced version showing the hydrocarbons between C_2 and C_{10}. An example of such a plot from DGSI (1993) is shown in Figure 15-12. The upper chromatogram (A) shows the *n*-paraffins numbered from C_4 to C_{45}, whereas the isoprenoids, including pristane and phytane, have letters. The lower pattern (B) shows the C_2-to-C_{10} hydrocarbons that are used in calculating various ratios in order to evaluate such things as maturity, source, biodegradation, and evaporative fractionation. Key hydrocarbons are numbered 1 to 34, with their names listed on the right. For example, Figure 12-11 measures aromaticity by the ratio of toluene to *n*-heptane. These hydrocarbons are shown as numbers 25 and 21, respectively, in pattern B of Figure 15-12.

Light hydrocarbon composition ratios have been used for many years to classify crude oils. Hunt (1979, pp. 494–497) reviewed some of the early oil correlations in the North Sea and Western Canada Basin using hydrocarbon ratios in the C_2-to-C_{10} range. More recently, Thompson (1983) reported a detailed study of hydrocarbon ratios in seventy-six crude oils, representing most of the production in western North America. He used his isoheptane and heptane ratios (I and H in Table 15-3) to define oils as being biodegraded, normal, mature, or supermature. Subsequent papers by Thompson (1987, 1988) described the evaporative fractionation process and listed various hydrocarbon ratios that are useful in characterizing crude oils, as shown in Table 15-3. Thus, an increase in the B ratio of Table 15-3 (toluene/*n*-heptane) reflects an increase in aromaticity and is characteristic of the deeper oils formed by evaporative fractionation. The isoheptane and heptane values show increasing paraffinicity, a characteristic of increasingly mature crude oils that have not undergone phase changes.

The C_7 alkanes are increasingly being used for oil–oil and oil–source rock correlations based on research by Mango (1990, 1992, 1994) indicating that the C_7s are genetically related. Heavy oils, conventional oils, and condensates can be correlated with each other and with their source rocks as will be discussed later.

Figure 15-12 (facing page)

Capillary column gas chromatograms of whole oil: (A) The numbers are *n*-paraffins containing 4 to 45 carbon atoms. The letters are isoprenoids, including pristane (f) and phytane (g); (B) C_1 through C_9 hydrocarbons. The numbers are only for identifying the hydrocarbon peaks, as listed on the right. [Courtesy of DGSI, The Woodlands, Texas]

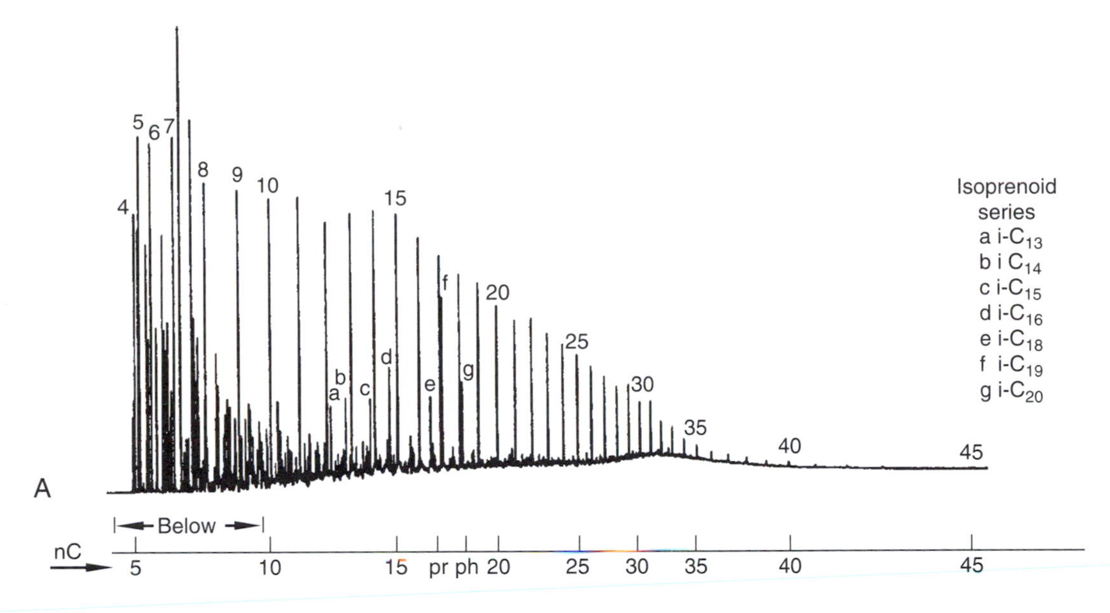

A
Isoprenoid series
a i-C_{13}
b i C_{14}
c i-C_{15}
d i-C_{16}
e i-C_{18}
f i-C_{19}
g i-C_{20}
Below
nC
5 10 15 pr ph 20 25 30 35 40 45

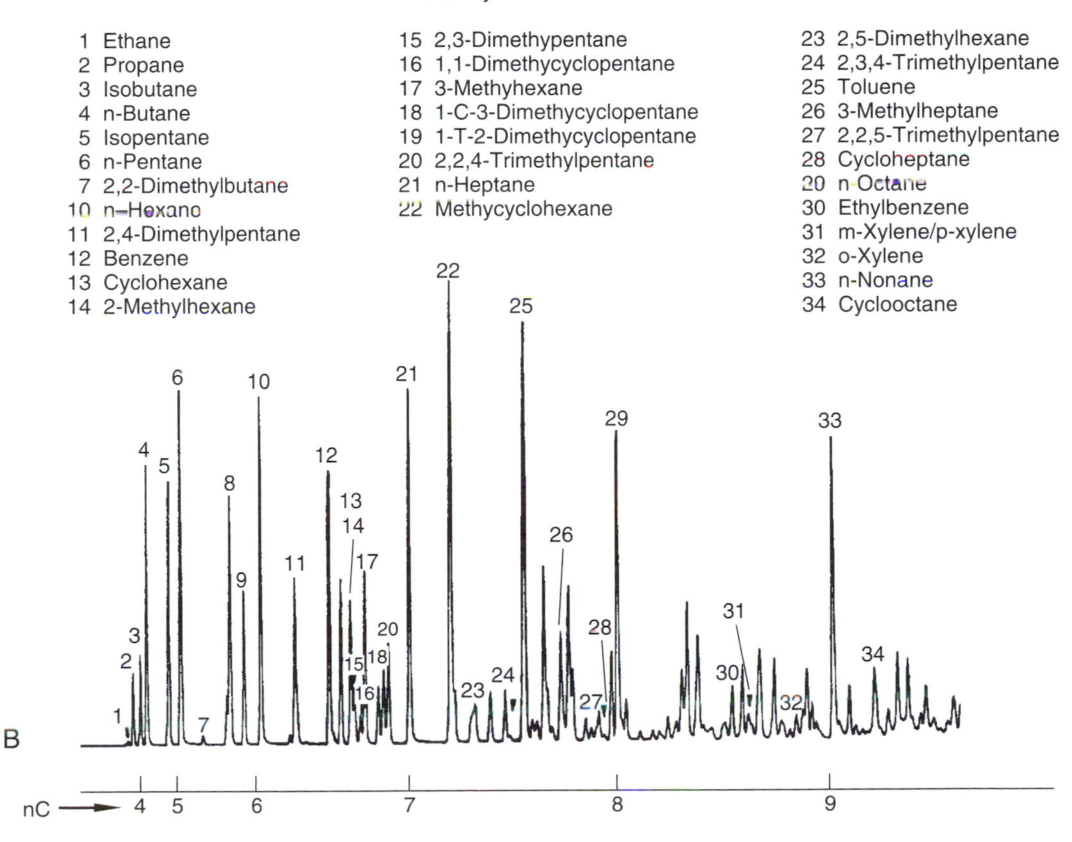

List of Hydrocarbon Peaks
1 Ethane
2 Propane
3 Isobutane
4 n-Butane
5 Isopentane
6 n-Pentane
7 2,2-Dimethylbutane
10 n-Hexane
11 2,4-Dimethylpentane
12 Benzene
13 Cyclohexane
14 2-Methylhexane
15 2,3-Dimethypentane
16 1,1-Dimethycyclopentane
17 3-Methyhexane
18 1-C-3-Dimethycyclopentane
19 1-T-2-Dimethycyclopentane
20 2,2,4-Trimethylpentane
21 n-Heptane
22 Methycyclohexane
23 2,5-Dimethylhexane
24 2,3,4-Trimethylpentane
25 Toluene
26 3-Methylheptane
27 2,2,5-Trimethylpentane
28 Cycloheptane
20 n-Octane
30 Ethylbenzene
31 m-Xylene/p-xylene
32 o-Xylene
33 n-Nonane
34 Cyclooctane
B
nC
4 5 6 7 8 9

TABLE 15-3 Definitions and Significance of Gasoline Compositional Ratios

Designation	*Definition*	*Property assessed*
A	Benzene/*n*-hexane	Aromaticity (fractionation)
B	Toluene/*n*-heptane	Aromaticity (fractionation)
X	Xylene (m & p)/*n*-octane	Aromaticity (fractionation)
C	(*n*-hexane + *n*-heptane)/ (cyclohexane + methylcyclohexane)	Paraffinicity (maturity)
I (isoheptane value)	(Methylhexanes (2- & 3-))/ (dimethylcyclopentanes (1c3-, 1t3-, & 1t2-))	Paraffinicity (maturity)
F	*n*-heptane/methylcyclohexane	Paraffinicity (maturity)
H (heptane value)	100 *n*-heptane/ (Σ cyclohexane through methylcyclohexane)[a]	Paraffinicity (maturity)
R	*n*-heptane/2-methylhexane	Extent of branching
U	Cyclohexane/methylcyclopentane	Extent of branching

[a]Excluding 1,cis-2-dimethylcyclopentane. See hydrocarbon peak numbers 13 through 22 in Figure 15-12B.

Source: Thompson 1987.

Correlating Reservoir Oils

The spacing of wells for primary production and the location of injection wells for secondary and tertiary recovery programs depend on a detailed knowledge of reservoir continuity. This is a problem particularly in deltas where a single field may produce from a series of complex, faulted reservoirs extending vertically over several thousand feet. Conventional pressure tests and wireline logs needed for modeling reservoirs are now being combined with the correlation of GC patterns of the reservoir oils to map some reservoirs. The assumption is that oil in a continuous, high-permeability reservoir will show a similar composition in terms of hydrocarbon ratios throughout the pool. Any faults or thin shale barriers would prevent complete mixing of the oil and could be detrimental to secondary recovery operations. Differences in the hydrocarbon ratios of oils from the same pool are interpreted as indicating such barriers. Consequently, the mapping of reservoirs based on similarities and differences in their hydrocarbon compositions is now being integrated with conventional geological and engineering methods for defining continuous reservoirs.

An example of such a study is in the Columbus basin off Trinidad, which was mentioned earlier in Chapter 12. Rodrigues (1988) concluded from a source rock–crude oil correlation study in the basin that the dominant source of all the oils is the Naparima Hill and Gautier Upper Cretaceous Formations. These are anoxic marine shales with a high source potential. Extracts of the Naparima Hill and Gautier source rocks contain cadalene and other terrestrial biomarkers, which are present in all the oils in varying amounts. The more marine Naparima Hill Marls contain highly fluorescent, amorphous type II kerogens (Heppard et al. 1990). Deep-seated faults connecting these source rocks to the Tertiary reservoirs are considered to be the cross-statal migration pathways that charged the oil fields of the Columbus Basin.

In 1986, Ames and Ross showed that there are both lateral and vertical variations in the composition of oils in numerous pools of the oil fields in this basin. A later study of thirty-four pools in the Samaan field (Ross and Ames 1988) indicated that there is no significant effect of bacterial degradation and no maturity differences based on biomarkers to account for these variations. For example, the C_{20}–C_{21} triaromatic steroids generally increase with advancing maturity. All the Samaan oils analyzed were of low maturity, exhibiting no differences in the triaromatic steroids over a depth range of about 3,700 ft (1,128 m). Consequently, it appeared that major differences in composition are due to migration processes, such as evaporative fractionation. Other differences are due to thin shales and minor faults acting as partial barriers to migration.

Vertical migration up a complex system of growth faults caused oils of increasingly higher API gravities to condense out at shallower depths, as indicated in Figure 15-13. The chromatograms shown here are for the Samaan field, but the classifications are based on more than 100 crude oils from the three fields, Teak, Paui, and Samoan (Ames and Ross 1986). Note that the highest quantity of n-paraffins is in the C_{15}-to-C_{30} hydrocarbon range of the Class III oil. This oil is the deepest stratigraphically and is so waxy that it solidifies at room temperature. Going up Figure 15-13, the n-paraffin concentrations in the C_{15}-to-C_{30} range decrease. The highest peaks are shifting to the lower-molecular-weight ranges, as would be expected with evaporative fractionation. The shallowest oils, Class IV, show the highest hydrocarbon concentrations in the $<C_{10}$ range. Ames and Ross (1986) mapped these oil classes along with several subclasses in order to define the limits of reservoir continuity within the pools of each field. Significant variations in composition based on whole-oil chromatograms were considered to indicate separate reservoirs.

Star Diagrams

Kaufman et al. (1990) developed a sensitive method, called *reservoir oil fingerprinting* (*ROF*), for recognizing differences in the gas chromatograms of oils. It is well known that the n-paraffins, which are the most dominant peaks in whole-oil chromatograms (e.g., see Figure 15-12), are susceptible to secondary processes such as biodegradation (Chapter 12). Also, whole-oil chromatograms sometimes are not specific enough to distinguish among oils of similar composition. Consequently, Kaufman et al. (1990) looked for differences between the smaller naphthenic and aromatic peaks instead of the n-paraffins. Kaufman et al.'s ROF procedure consists of first numbering all small measurable peaks sequentially through n-C_{20}. This might give them 200 peaks through n-C_{15} and 300 through n-C_{20}. An entire chromatogram could be 400. Then they visually select fewer than 25 pairs of peaks (usually 12 or so) and calculate the ratios of their peak heights or areas. Peaks are selected mainly in the C_9–C_{20} range where

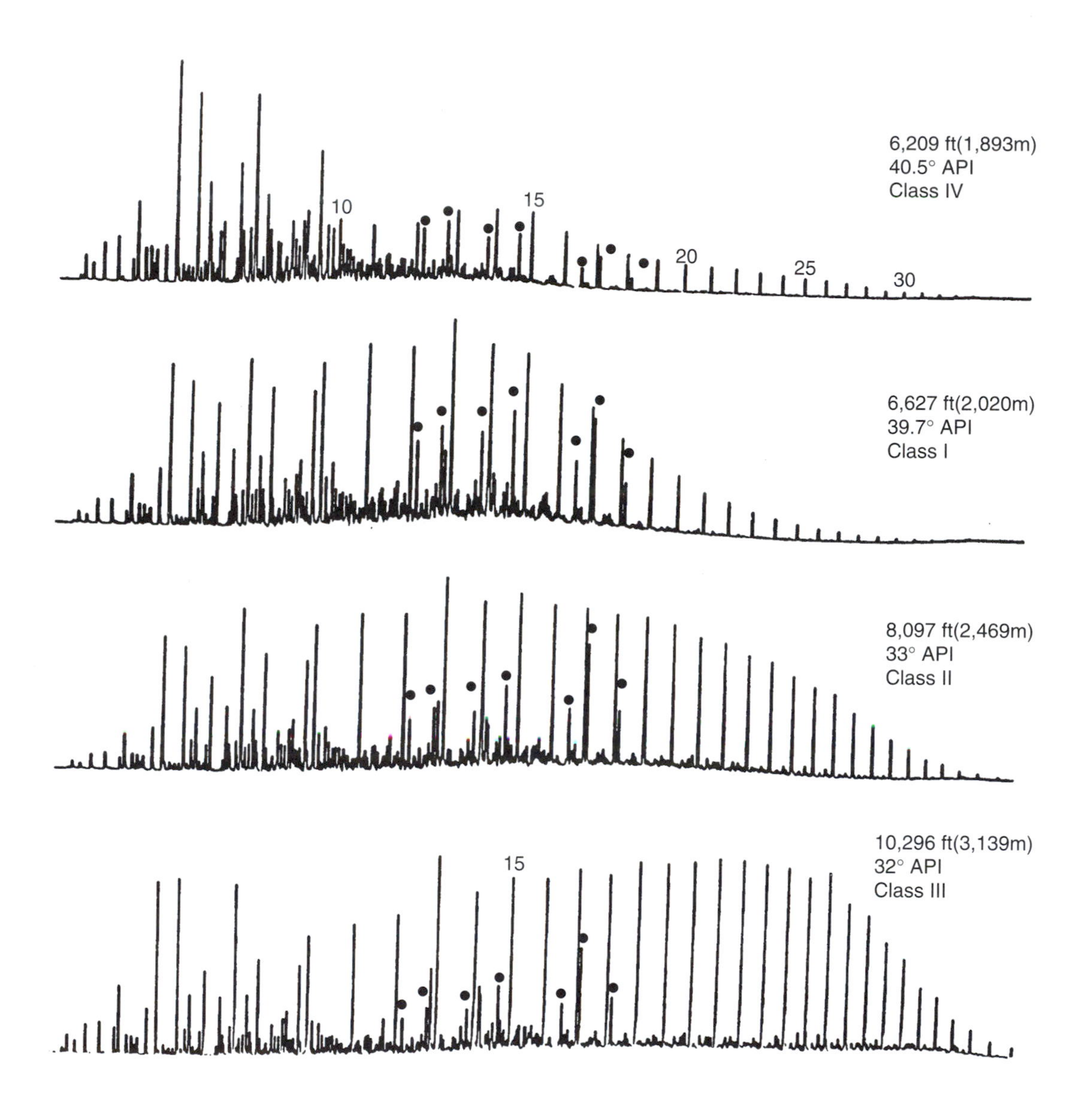

Figure 15-13

Whole-oil chromatograms showing four classes of oils from the Samaan field, Columbus Basin, Trinidad. Isoprenoids are identified by small dots. Pristane and phytane are the last two small dots in each chromatogram. The C_{15} and other carbon numbers in the top chromatogram are at the same location for the lower three profiles. [Ross and Ames 1988]

there is a good distribution of naphthenes and aromatics without too much overlap. Peaks used for comparison must be present in every sample. The peak pairs are then listed by their peak numbers or lettered alphabetically, as in Figure 15-14. These represent expanded chromatograms of the n-C_9 through n-C_{18} range for the Upper and Lower Vermelha Reservoirs in the Takula field offshore Angola.

The next step is to construct a star diagram (polygon plot) by plotting each peak ratio on a different axis of a polar plot, as in Figure 15-15. Each data point is plotted from the center of the concentric circles outward. The points are then connected to create a star-shaped pattern characteristic of each oil. Note that the star pattern for the six Upper Vermelha oils is quite different from the pattern for the two Lower Vermelha oils. For example, the ratio for the H pair in the upper reservoir is about 2.4, compared with a ratio of about 1.8 for the lower reservoir. This plot shows that the ROF is uniform within a continuous reservoir and can be distinctly different from a nearby but vertically separated reservoir.

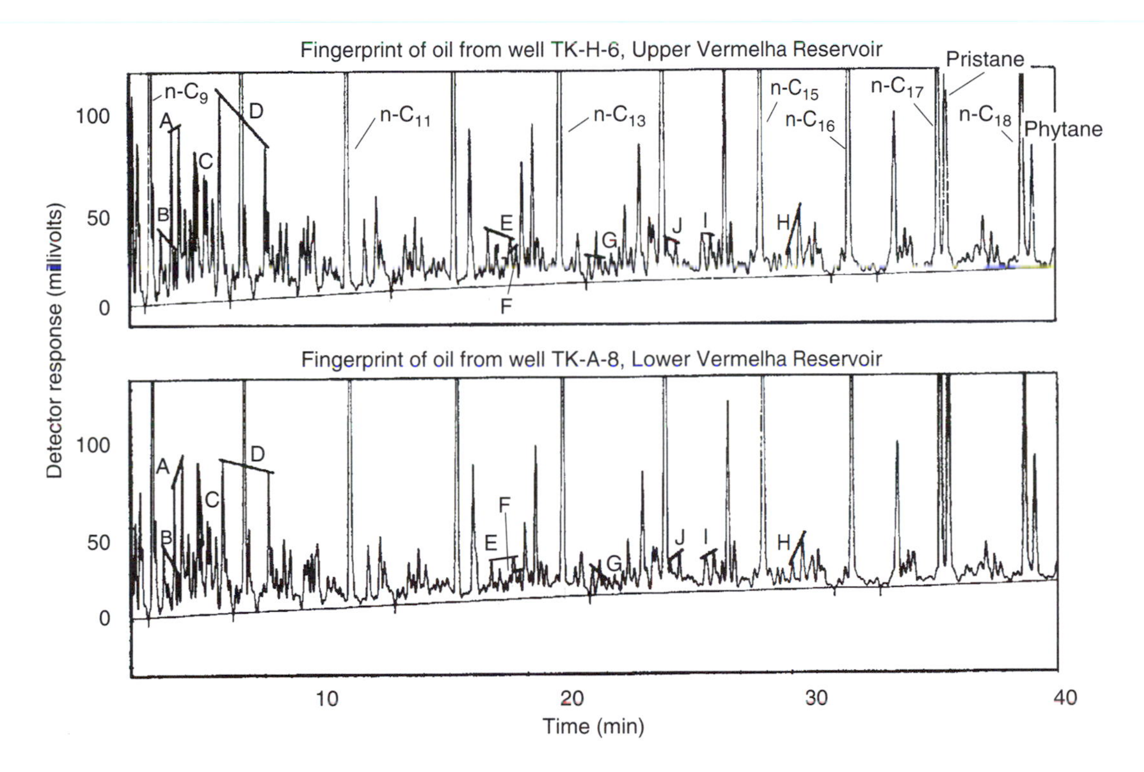

Figure 15-14

Portion of capillary gas chromatograms (n-C_9 through n-C_{18}) of oils from Upper and Lower Vermelha Reservoirs, Takula field, Cabinda, Angola. [Kaufman et al. 1990]

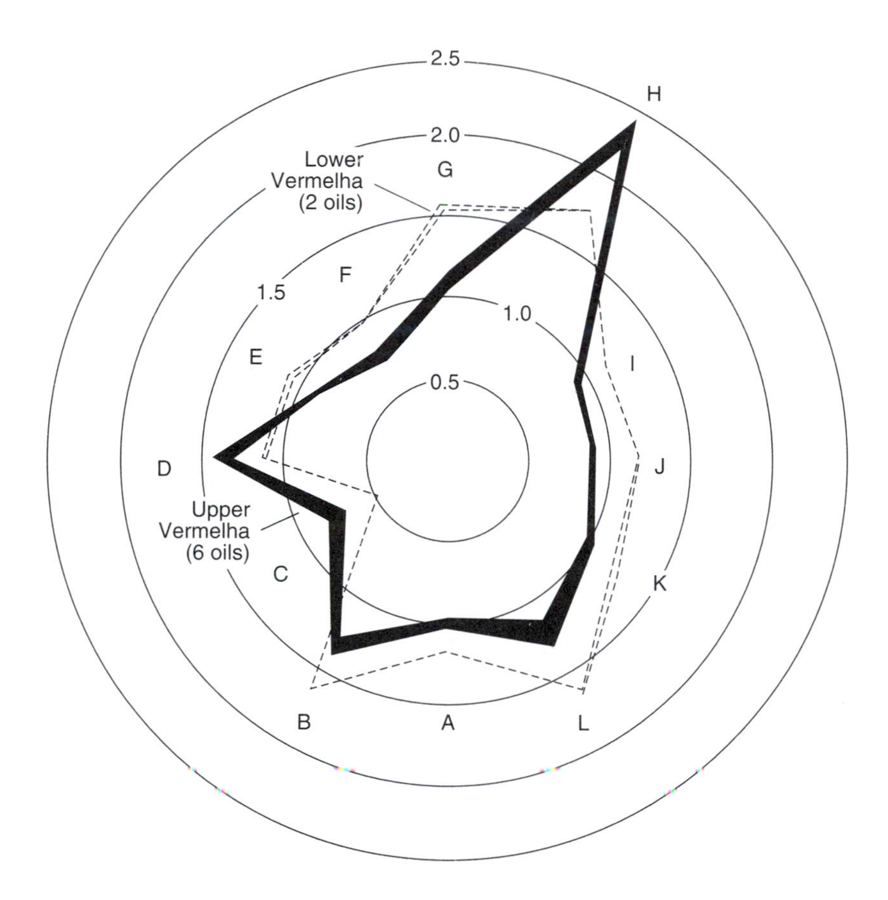

Figure 15-15

Star diagram of oils from the Upper and Lower Vermelha Reservoirs, Takula field, Cabinda, Angola. [Kaufman et al. 1990]

Two hypothetical examples of using star diagrams to evaluate reservoir continuity are shown in Plate 8B (courtesy of DGSI, The Woodlands, Texas). Both these examples show an older producing field on the right and a new discovery well on the left. The question is whether this discovery well represents a separate pool or is simply an extension of the old field. The star diagrams of the two oils on the left define oils of two different compositions indicating two pools, as shown in light and dark green. On the right the diagrams of the two oils are identical, indicating a single oil field extending across the fault.

An alternative way to display fingerprint differences is by constructing a *dendrogram* (*cluster analysis diagram*), using multivariant analysis with peak height or area ratios as the variant. This is particularly useful when comparing a large number of oils or ratios. Peters et al. (1994) used cluster analysis of several source-related parameters to distinguish two major groups of West Siberian oils, those generated by the Upper Jurassic Bazhenov Formation and those presumably coming from the Middle Jurassic Tyumen Formation.

The mapping of reservoir continuity also is being done by carbon isotope analyses in both oil and gas reservoirs, as discussed later in this chapter. Isotope analyses of individual hydrocarbon molecules are called *compound specific isotope analysis* (*CSIA*). According to M. Bjorøy (personal communication), it has been a crucial factor in defining compartmentalization in some oil reservoirs. It also has been useful in establishing continuity in gas reservoirs.

Isoprenoid/*n*-Paraffin Ratios

Source, maturation, migration, and biodegradation are the major factors causing differences in crude oil composition. The ratio of the isoprenoids, pristane to phytane, as a source indicator was discussed in Chapter 4. In crude oil correlation, the ratios of isoprenoids to *n*-paraffins are often used, since they provide information on maturation and biodegradation as well as source (Connan and Cassou 1980; Shanmugam 1985; Talukdar et al. 1993). The *n*-paraffins used in these ratios are those closest to the isoprenoids in a GC chromatogram, as shown in Figure 15-12 (A). Pristane is peak "f" adjacent to n-C_{17}, and phytane, peak "g," is next to n-C_{18}. The ratios of both Pr/n-C_{17} and Ph/n-C_{18} decrease with maturation due to the increasing prevalence of the *n*-paraffins. Both ratios increase with biodegradation due to the loss of these *n*-paraffins. Samples high in pristane indicate an oxidizing source, and those high in phytane indicate reducing conditions, as discussed in Chapter 4. Consequently, a plot of Pr/n-C_{17} versus Ph/n-C_{18} can classify oils and rock extracts in different groups, as shown in Figure 15-16. This shows high-wax crudes from terrestrial source material in the upper part of the diagram. Marine crudes formed under reducing conditions are in the lower part. Crude oils from mixed organic sources are in between.

Correlating Heavy Oils and Asphalts

The use of biomarkers to evaluate the maturity of asphalts and pyrobitumens was examined in Chapter 13 (e.g., see Figure 13-3). Biomarkers also make it possible to correlate heavy oils and asphalts whose gas chromatograms may have no identifiable peaks, such as in Figure 12-1 (Day 21). This "hump" fingerprint is sometimes referred to as the *UCM* (*unresolved complex mixture*). The pattern is characteristic of heavily biodegraded oils like the Athabasca heavy oil of Alberta and the biodegraded La Luna oil from the Lagunillas field in the Maracaibo Basin, Venezuela. The UCM cannot be used for correlation, but

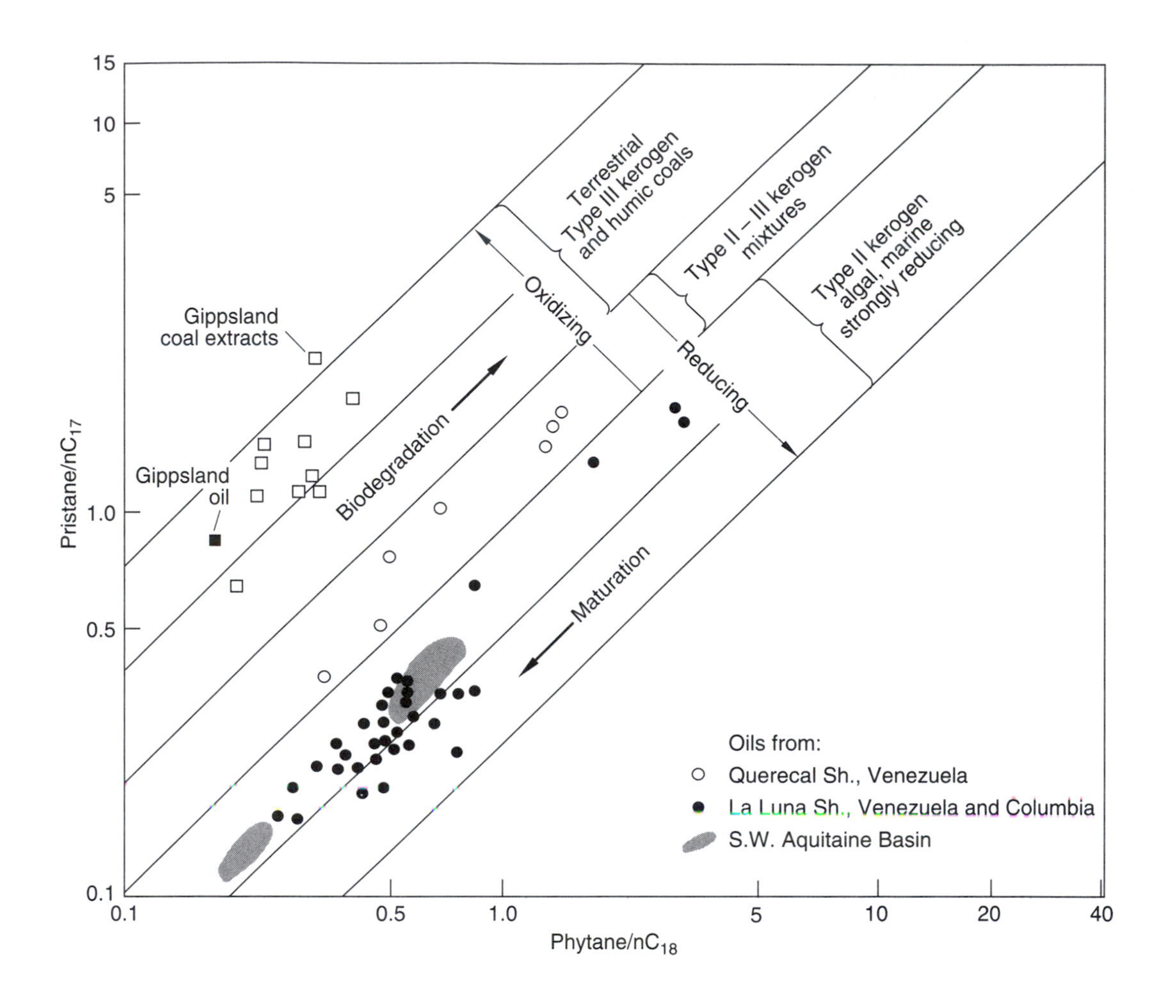

Figure 15-16

Pristane/*n*-C_{17} versus phytane/*n*-C_{18} showing the terrestrial source for Gippsland Basin hydrocarbons [Shanmugam 1985], the marine source for crude oils of the Southwest Aquitaine Basin (shaded area) [Connan and Cassou 1980], and the La Luna—sourced oils of the Maracaibo Basin in Venezuela. The mixed-source oils are from the Barinas–Apure Basin in Venezuela [Talukdar et al. 1993]

biomarkers in the UCM can be used. Thus, the terpanes (m/z 191) in the mass fragmentogram of the biodegraded La Luna oil shows a distribution of tricyclic terpanes and pentacyclic triterpanes that matches that of unaltered La Luna oil from the Motatan field (Talukdar et al. 1986).

It is estimated that there are 1.2×10^{12} bbl of heavy (7°API gravity) oil in the Cretaceous sands of the Western Canada Basin (Roadifer 1987). The source of this giant accumulation is believed to be a group of organic-rich rocks ranging from Late Devonian to Late Cretaceous in age, based on biomarker data (Creaney and Allen 1990). These Athabasca oil sands lie at the base of the Cretaceous on the Cretaceous–Devonian unconformity (Figure 8-18).

Less well known is the 300×10^9 bbl of Upper Devonian Grosmont heavy oil, also 7°API gravity, which is trapped in Paleozoic carbonates beneath and beyond the Athabasca oil sands. Does the Grosmont heavy oil have a different source from the Athabasca? Hoffmann and Strausz (1986) made a detailed study of the biomarkers of both heavy oils. Their fingerprints of the steranes, hopanes, and monoaromatic and triaromatic steroids all were similar, indicating a single source and about the same level of maturity. The main differences appear to be due to the effects of water washing and biodegradation. Also, various ratios such as C_{27}/C_{29} diasteranes are very similar. Figure 15-17 shows that the distribution of C_{21}-to-C_{29} monoaromatic steroids of the Grosmont and Athabasca oils have essentially the same patterns. The C_{21}-to-C_{22} sterane fingerprints also are similar, but the C_{27}-to-C_{29} steranes could not be compared because of biodegradation. The Grosmont sample appears to be more biodegraded and water washed than the Athabasca bitumen. The latter contains phenanthrene and low-molecular-weight triaromatic steroids, which are absent in the Grosmont sample. In addition, bicyclic terpenoid hydrocarbons are common in the Athabasca bitumen but absent in the Grosmont sample.

Curiale (1986) used 58 biological markers to classify 27 solid bitumens from the United States, Spain, Mexico, and Canada. These included grahamites, gilsonites, ingramites, albertites, impsonites, and an ozocerite. Figure 15-18 shows the fingerprint pattern as a histogram for an Oklahoma grahamite. The peaks monitored are listed in the figure caption.

These biomarker fingerprints are derived by area integration of the m/z 191, 217, 231, and 253 mass chromatograms. The relative uncorrected mass spectrometric responses are presented, normalized to the compound having the highest ion fragment area response. Five regular steranes are listed in Figure 15-18 for each carbon number. However, no vertical bar is shown for the 5β steranes of C_{27}, C_{28}, and C_{29} because they are generally absent in mature samples like this grahamite. The 5β steranes are used in biomarker applications involving immature source rocks, particularly oil shales. Also, oils and condensates migrating vertically can extract immature biomarker suites, including the 5β steranes from shallower rocks.

The major control on the fingerprint patterns is source input. Thus the patterns for ingramite and albertite from the Uinta Basin, Utah, are similar, indicating a common source from the Green River Formation. But some bitumen patterns are altered considerably by either thermal processes or biodegradation and weathering. Impsonite, a pyrobitumen, is considered to be a thermally altered grahamite. Curiale analyzed a grahamite and impsonite from the same source and found that the latter had none of the aromatic steroids or terpanes present in the former. Either those compounds were destroyed, or they were incorporated into the insoluble fraction of the impsonite that was not

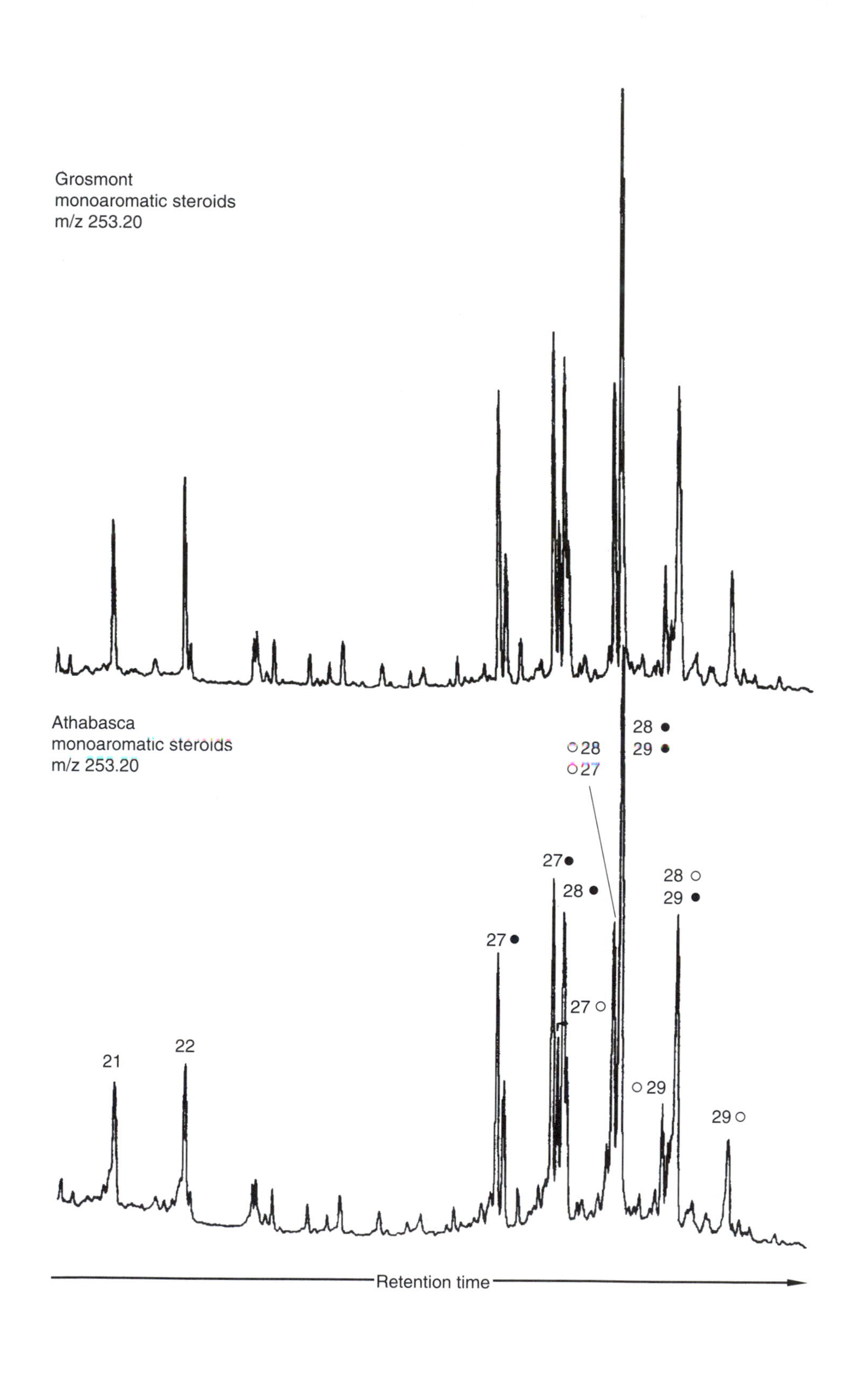

Grosmont
monoaromatic steroids
m/z 253.20
Athabasca
monoaromatic steroids
m/z 253.20
28 ●
29 ●
○28
○27
27●
28 ●
28 ○
29 ●
27 ●
27 ○
21
22
○ 29
29 ○
Retention time

Figure 15-17 (facing page)

Distribution of C_{21}–C_{29} monoaromatic steroid hydrocarbons in Athabaska and Grosmont-O bitumen samples. Carbon numbers and stereochemistry are indicated. Open circles = 5α(H); solid circles = 5 β(H). Most 20S compounds elute before 20R compounds for peaks with identical assignments. [Hoffman and Strausz 1986]

analyzed. Impsonite is much more insoluble than grahamite by conventional extraction techniques.

Curiale (1988) also correlated five Oklahoma grahamites with their probable crude oil source using the C_{27}-to-C_{29} steranes, as will be discussed later.

Oil Gravity and the Aromatization Parameter

Moldowan et al. (1994) found a systematic relationship between the API gravity and the aromatization parameter of crude oils in the Adriatic Basin of Italy. The parameter is C_{28}–TA/(C_{29}–MA + C_{28} – TA) where MA is the monoaromatic steroid (m/z 253) and TA is the triaromatic steroid (m/z 231). This ratio increases from 0.1 to 0.65 as the oil gravities rise from 5°API to 15°API. At a ratio of 0.8 the gravity is around 23°API. The ratio peaks at about 0.9 for oil gravities >40°API. Since this ratio does not change when an oil seeps to the surface, it is possible to predict the gravity of an undiscovered oil from its surface seep. Thus a seep with an aromatization ratio of 0.55 would be expected to originate from a 14°API gravity oil.

Correlation of Source Rocks and Crude Oils

Oil–source rock correlations are more difficult than oil–oil correlations, because many problems are involved in both sampling and interpreting the data. Removing oil from a source rock by conventional extraction results in the loss of hydrocarbons up to the C_{12}-to-C_{15} range. Comparing this extract with a reservoir oil sample requires evaporating the latter to constant weight at a certain temperature, such as 45°C (113°F), in order to remove the same range of volatile hydrocarbons. The C_2-to-C_{15} range can be compared separately using low-temperature heating of the mature source rock and GCMS analyses of the product and the reservoir oil.

These comparisons are analytically straightforward, but the interpretations are difficult. The bulk correlation parameters previously discussed are not as useful for oil–source rock correlations as for oil–oil correlations, for several reasons. First, a few source rock samples cannot adequately represent oil generated from a thick source interval of varying composition. Second, there is considerable evidence that the oil fractionates during the process of leaving the

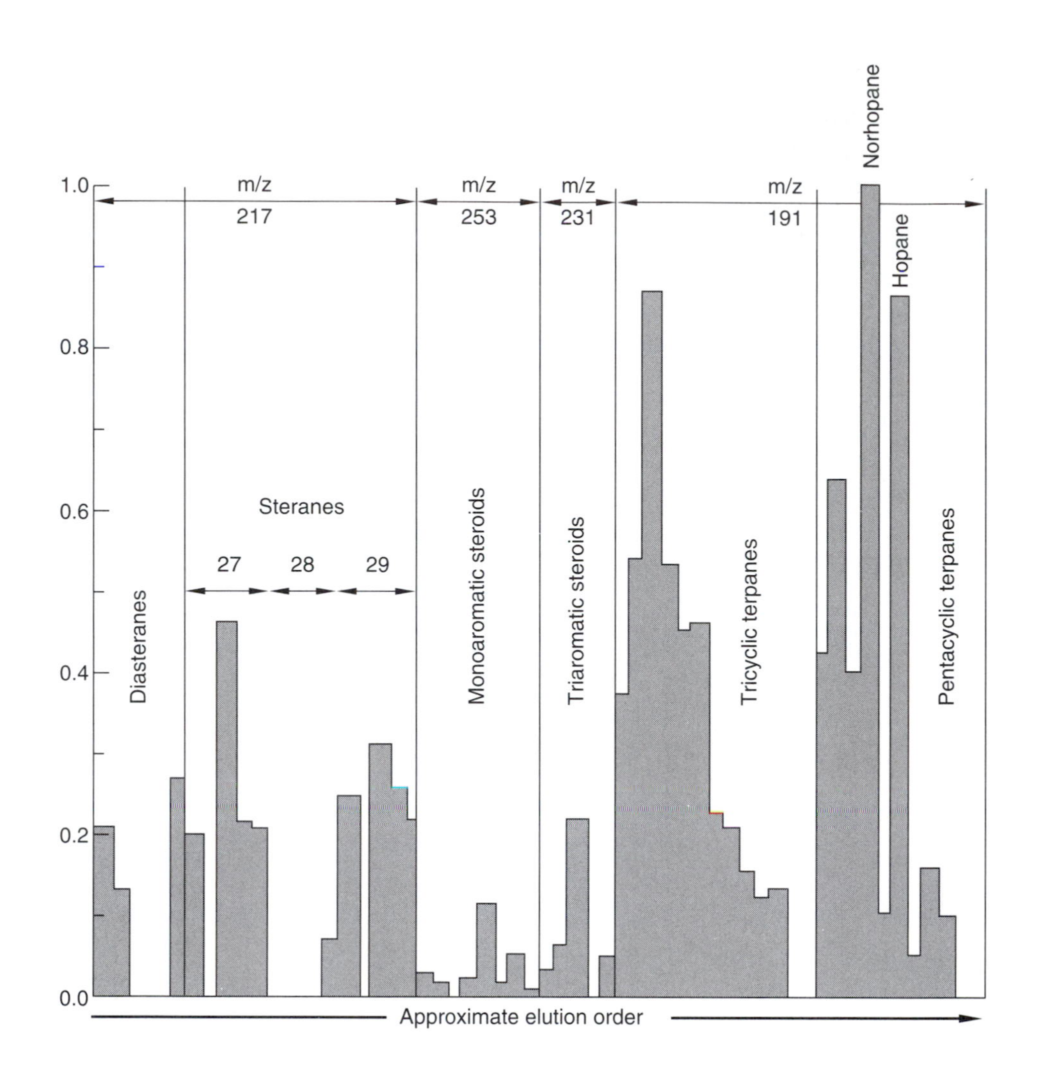

source rock and migrating to the reservoir rock. Third, source rocks do not yield oils of the same composition throughout their generation history. As the rocks are buried to successively greater depths and higher temperatures, the reactions of higher activation energy are initiated, yielding different products. Fourth, the composition of the bitumen extracted from the rock is not necessarily the same in the fine pores of the rock matrix as it is in the microfractures and bedding planes. Consequently, any variation in the extraction procedure can yield extracts of different composition (Price and Clayton 1992, and references therein).

Although these factors affect the bulk properties of extracted oils, they do not cause significant changes in most biomarker compounds. Consequently, the

Figure 15-18 (facing page)

Histogram ("biomarker fingerprint") showing fifty-eight biological markers monitored in an Oklahoma grahamite. [Curiale 1986] Within each class, compounds are shown in order of elution (left to right) from a nonpolar gas chromatographic column (DB-5). A blank space indicates that the biomarker monitored is absent. The term $\alpha\alpha\alpha$ means 5α(H), 14α(H), 17α(H). The five C_{27}, C_{28}, C_{29}-steranes monitored are $\alpha\alpha\alpha$-20S, $\beta\alpha\alpha$-20R (absent), $\alpha\beta\beta$-20R, $\alpha\beta\beta$-20S, and $\alpha\alpha\alpha$-20R. Eight monoaromatic steroid hydrocarbons are shown: 5β,20S-C_{27}; 5β,20R-C_{27}; 5α,20S-C_{27}; 5β,20S-C_{28}; 5α,20R-C_{27} + 5α,20S-C_{28} + 5β,20R-C_{28} + 5β,20S-C_{29}; 5α,20S-C_{29}; 5α,20R-C_{28} + 5β,20R-C_{29}; and 5α,20R-C_{29}. The five triaromatic steroid hydrocarbons are 20S-C_{26}; 20R-C_{26} + 20S-C_{27}; 20S-C_{28}; 20R-C_{27}; and 20R-C_{28}. Thirteen tricyclic terpanes–C_{20}, C_{21}, C_{23}, C_{24}, C_{25} (unresolved)—and both epimers of C_{26}, C_{28}, C_{29}, and C_{30} are shown. Also included are 11 pentacyclic terpanes: 18α(H)-22,29-30-trisnorneohopane; 17α(H)-22,29,30-trisnorhopane; 17α(H), 21β(H)-28,30-bisnorhopane; 17α(H), 21β(H)-30-norhopane; 17β(H), 21β(H)-30-normoretane; 17α(H),21β(H)-hopane, 17β(H), 21α(H)-moretane; and the S & R C-22 epimers of 17α(H),21β(H)-homo- and bishomohopane. This display shows relative distributions only.

comparison of biomarkers is the preferred method of oil–source rock correlation. Biomarkers are affected mainly by the thermal and bacterial alteration processes, as observed in Chapter 12. Thus, the reservoir oil may not have been exposed to the same thermal history as the source rock oil, causing the maturity indicators to differ. Also, bacterial degradation can affect the reservoir oil (Table 12-3) and may affect a source rock extract by uplift and erosion followed by reburial.

In addition, source rock extracts in the C_{15+} range are more subject to sampling contamination than the crude oils obtained for correlation. Sidewall cores and especially cuttings are subject to contamination from both the drilling mud and the handling procedures in the drilling operation. Deroo et al. (1977) observed contamination in the C_{18} range in rock extracts of the Western Canada Basin. GCMS analyses indicated that it was diesel oil. Deroo et al. also observed contamination by iso- and normal alkanes in the C_{20} range in 60 of 110 cores from the central part of the basin. The exterior of the cores showed far more contamination than the interior.

Fractionation of Oil Components Migrating from Shale to Sandstone

The larger and more polar molecules in a generated oil, such as the 3+ fused-ring aromatics, tend to be absorbed by the kerogen of a source rock, whereas the paraffins are released as shown in Table 15-4. This compares the hydrocarbon type analysis of oils from sandstone reservoirs with the oils extracted from their adjacent shale source rocks. The formations shown are in Alberta,

Oklahoma, and Louisiana. Note that the paraffin content of the reservoir oils is always greater than that of the source rock extracts. For Wilcox-sourced oils, the paraffin content is greater than Wilcox source rock extracts by about a factor of six. In contrast, the quantity of 3+ ring aromatics is five times greater in the Wilcox source rock extract than the related crude oil. (Additional examples are described in Hunt 1979, pp. 499–502.) Although ternary diagrams plotting paraffins versus naphthenes versus aromatics plus NSOs are commonly used to classify crude oils, they have little value in comparing crude oils with source rocks because of these fractionation effects.

Figure 15-19 shows these differences more clearly as pie diagrams. The comparison of natural hydrocarbons is only in Figure 15-19 (A). The Woodford Shale source rock oil has more than twice the aromatic hydrocarbons and less than half the paraffin hydrocarbons of the associated Misener reservoir oil (Hunt 1961).

Hydrous pyrolysis products of an immature Monterey siliceous shale are shown in Figure 15-9 (Peters et al. 1990). The oil retained by the shale contains almost twice as many NSOs plus asphaltenes as the expelled oil. The saturates (paraffins plus naphthenes) are five times greater in the expelled oil than in the retained oil. This supports field observations that there is a strong retention of NSOs and asphaltenes in shale source rocks (Hunt 1979, p. 499).

Sandvik et al. (1992) studied the absorption of generated petroleum on the solid organic matter of source rocks and concluded that it is a significant phenomenon that should be considered when modeling expulsion efficiencies from source rocks. They used equilibrium modeling of paraffins, aromatics, and NSOs to estimate the composition of hydrocarbon liquids generated and retained versus those expelled from a coal. Their results, in Figure 15-19 (C), supported the field observations in that the retained oil contained six times as many NSOs as the expelled oil. The numbers in Figure 15-19 (C) are conservative because Sandvik et al. concluded that absorption levels in coals may be

TABLE 15-4 Comparison of Reservoir Oils and Source Rock Extracts

			Aromatics	
Location	*Paraffins*	*Naphthenes*	*1 ring*	*3+ rings*
Viking Shale, Alberta	20	46	6	28
Viking Sandstone, Alberta	42	42	5	11
Cherokee Shale, OK	21	32	2	45
Cherokee Sandstone, OK	27	60	5	8
Wilcox Shale, LA	6	45	9	40
Wilcox Sandstone, LA	34	54	4	8

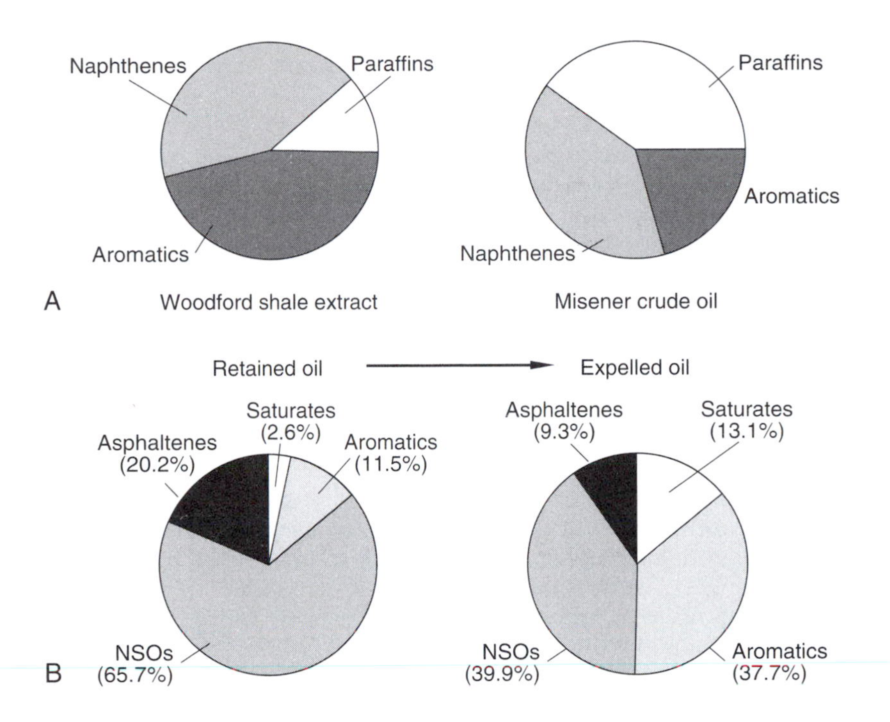

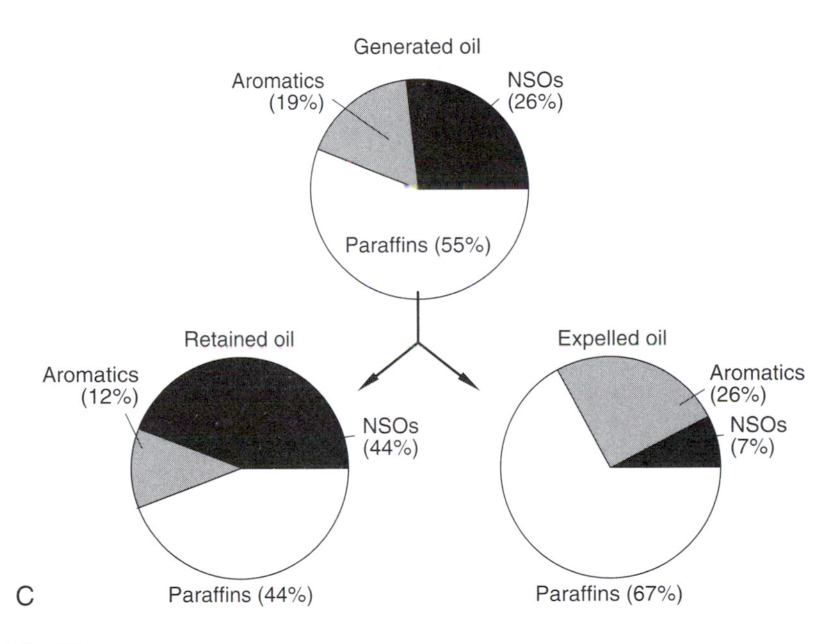

Figure 15-19

(A) Composition of hydrocarbons extracted from the Woodford Shale of Oklahoma compared with crude oil from the adjacent Misener Sandstone. [Hunt 1961] (B) Comparison of bitumen retained by hydrous pyrolysis and oil expelled (floating pyrolyzate) from a Monterey source rock. [Peters et al. 1990] (C) Calculated compositions of retained (absorbed) versus expelled (free liquid) petroleum. [Sandvik et al. 1992]

lower than in dispersed kerogen. They also found that the levels of absorption in solid organic matter followed this order: polar compounds > aromatic hydrocarbons > isoparaffins > normal paraffins.

Biomarker Ratios for Correlation

Some hydrocarbon ratios (Table 15-3), GC patterns (Figure 15-13), and the star diagram of Figure 15-15 are not suitable for oil–source rock correlations because of the problems discussed earlier. Consequently, biomarker ratios and biomarker fingerprint patterns are used mainly in oil–source rock correlation, although some C_7 ratios of paraffins (alkanes) are being used, as is discussed later. Both maturation and biodegradation, however, need to be appraised when comparing biomarkers. Correlations of reservoir and source rock oils tend to concentrate on the biomarker compound class and molecular size, as Chapter 10 pointed out. These are relatively unaffected by maturation compared with the stereoisomers.

One of the strengths of biomarkers is their resistance to biodegradation (see Chapter 12). Nearly all the alkanes used for correlation are long gone before the steranes are attacked (Table 12-3). In severe cases of biodegradation, the aromatic steroids and porphyrins generally survive to be used for correlation. When these various factors are recognized and accounted for, the biomarkers become the most powerful tool available for oil–source rock correlation. Some of the more frequently used ratios are discussed next.

Homohopane Index

Hopane has the formula $C_{30}H_{52}$. The homohopane series (C_{31}–C_{35}) represents hopane, with additional CH_2 groups on the side chain. They are believed to be derived from C_{35} hopanoids in prokaryote microorganisms (Ourisson et al. 1984). The homohopane index is the ratio of $C_{35}/(C_{31}$ to $C_{35})$, with the 17α(H), 21β(H),22S and 22R configurations. The ratio is generally expressed as a percentage. A high ratio indicates strongly reducing conditions, like those found in some marine evaporites and carbonates (Peters and Moldowan 1991). Such environments tend to preserve the C_{35} bacteriohopanetetrol precursors of the homohopanes. A low index, in which C_{31} and C_{32} predominate, is characteristic of a suboxic environment in which oxidation reduces the length of the side chain. Consequently, the variable distribution of the C_{31}–C_{35} homologs is a signature of variations occurring during the deposition of the source rocks. Homohopane distributions were used by Peters and Moldowan (1991) and Moldowan et al. (1992) to characterize seep and oil samples from the Central Adriatic Basin and adjacent areas. They emphasized that a high homohopane index indicates strongly reducing conditions but not necessarily high salinity.

The Monterey oils of California have a high C_{35} index, but there is no evidence of hypersalinity in the Monterey formation. The C_{35} homohopane index did show a decrease with increasing maturity in a suite of oils from the Monterey formation (Peters and Moldowan 1991). Furthermore, extensive biodegradation can alter the index owing to the selective loss of the different homologs, presumably due to variations in the bacterial population of reservoirs.

Oleanane Index

Oleanane was described in Chapter 4 as being derived from angiosperms, higher plants of Cretaceous age and younger. Also, it was pointed out that oleanane has been used for correlation in the Niger, Mahakam, and Mackenzie deltas as well as the Gippsland, Surat/Bowen, and Cooper/Eromanga basins. The oleanane index is $18\alpha(H) + 18\beta(H)$–oleanane/$17\alpha(H)$–hopane. Oils with high index values (>30%) indicate a strong higher plant input, and those with low values (<10%) indicate a marine source with a negligible terrestrial input. Ekweozor and Telnaes (1990) suggest that the oleanane ratio generally increases from low values in immature rocks to a maximum at the top of the oil-generation window. This implies that correlations need to use samples of about the same maturity. Peters and Moldowan (1993, p. 189) recommend cross-plotting the oleanane index with the C_{30} sterane index, which indicates marine inputs. They were able to differentiate oils from the Maracaibo Basin of Venezeula from oils and seeps of Columbia based on a crossplot of these two indices.

Gammacerane Index

Gammacerane is generally associated with environments of increasing salinity, both marine and lacustrine. The index is the ratio of gammacerane/$17\alpha,21\beta(H)$–hopane $\times$ 100. Ruble and Philp (1991) found gammacerane in all bitumens analyzed from the Tertiary lacustrian beds of the Uinta Basin in Utah. The highest gammacerane indices were found in albertite and wurtzilite, 1.6 and 1.76, respectively. These two pyrobitumens were formed from source rocks deposited in high-salinity waters during two periods of maximum lake shrinkage, Middle Green River time and Uinta time. The gammacerane index ranged between 0.27 and 0.60 during more stable periods. Moldowan et al. (1985) reported an index of 8.0 for a Green River oil from the Altamont–Bluebell field deep in the basin center.

Poole and Claypool (1984) identified two families of oils and their source rocks in the Great Basin of Nevada and western Utah, based partly on the gammacerane content. Five oils and source rock extracts from samples of the Tertiary Sheep Pass Formation all had gammacerane contents about five times higher than did a second group of low gammacerane oils. This latter group was correlated with the Mississippian Chainman Shale source rock extracts, which also had a low gammacerane content.

The crude oils in the Espirito Santo Basin offshore Brazil were initially thought to have been generated from Upper Cretaceous/Lower Tertiary slope sediments. But pyrolysis indicated that they have a very poor generation potential (Estrella et al. 1984). Then biomarker analyses showed that all the crude oils contained gammacerane, which was absent in all the possible source rocks except two. These are the Neocomian Jiquia stage and the Aptian Alagoas stage, both of which have high gammacerane concentrations. Other analyses, including pyrolysis (Figure 10-14), confirmed that these high gammacerane facies are the only source rocks for the oils of this basin (Estrella et al. 1984).

The gammacerane indices for four crude oils from the lacustrine Jianghan Basin in eastern China range from 1.20 to 1.34. Their possible source rocks showed a greater range, from 0.35 to 2.16 (Fu Jiamo et al. 1986). This suggested a fluctuating lake level with a corresponding variation in salinity, such as occurred in the previously discussed Uinta Basin.

Tricyclic Terpanes

Among the steranes and terpanes, tricyclic terpanes offer the advantage in correlation of being the least affected by maturity and biodegradation. Seifert and Moldowan (1979) found that heavily biodegraded California and Gulf Coast crude oils still have tricyclic terpanes that can be used for correlation.

Three highly mature Oman oils that contained few if any steranes or hopanes were correlated by comparing the thermally resistant C_{19}–C_{29} tricyclic terpanes (Peters and Moldowan 1993, p. 231).

Sofer (1988) compared twenty-five oils of different maturities from the Alabama and Mississippi–Louisiana Jurassic salt basins and found that they could be grouped according to their tricyclic terpane distributions. The Mississippi and Louisiana oils have large tricyclic C_{23} and tetracyclic C_{24} peaks, whereas the remaining tricyclic peaks from C_{19} to C_{26} are small. In contrast, the Alabama oils have larger C_{19}–C_{21} and C_{24}–C_{26} tricyclic peaks and smaller C_{21}–C_{22} tricyclic and C_{24} tetracyclic peaks. The two tricyclic terpane genetic groups are in agreement with a genetic classification based on carbon isotopes.

Palacas et al. (1984) found the tricyclic terpane distribution in oils and source rocks of the South Florida basin to be the most useful biomarker for correlation. The oils and source rocks fall into two groups: those with a maximum C_{23} tricyclic terpane peak and those with a maximum C_{24} tetracyclic terpane peak, as shown later.

On the North Slope of Alaska, the oil in the Lower Jurassic Sag River sandstone is rich in tricyclic terpanes (Seifert et al. 1980). Among the possible major source rocks, only the Triassic Shublik Formation is rich in tricyclic terpanes in both the bitumen extract and the pyrolysate. This indicates that it was the source of the Sag River oil. In contrast, the bitumen and pyrolysate of the Jurassic Kingak Formation, as well as the associated Kingak crude oils, have low concentrations of tricyclic terpanes, suggesting a Kingak source.

The Permian Phosphoria Formation has long been thought to be the source rock for the Pennsylvanian Tensleep oils of Wyoming (Hunt 1953). But the source of the Cretaceous Mowry oils was not clear. It could have been either the Mowry or the Phosphoria shales. Seifert and Moldowan (1981) noted that the Phosphoria Formation contains relatively high concentrations of tricyclic terpanes. Tensleep oils from the Beaver Creek and Hamilton Dome fields of Wyoming also had high concentrations of the tricyclics. In contrast, all the Cretaceous Mowry oils and Mowry shale bitumen extracts have low concentrations of tricyclic terpanes, indicating a Mowry source.

The ratio of tricyclics/17α(H)–hopanes is basically a source indicator comparing algal lipids (tricyclics) with hopanes coming from different primitive prokaryotes. Peters and Moldowan (1993) used the sum of four tricyclic terpane peaks, 22R and 22S doublets of the C_{28} and C_{29} pseudohomologs of tricyclohexaprenane in the numerator. The denominator contained the sum of the C_{29}–C_{33} 17α(H)–hopanes. The authors used this ratio to show that three Oman-area oils were genetically related, despite the fact that all the oils were highly mature and contained few biomarkers.

C_{30}-Sterane (24-n-propylcholestane) Index

In 1985 Moldowan et al. reported that their analysis of forty-three crude oils for 24-*n*-propylcholestane showed it to be present only in oils from marine source rocks. It was usually absent in oils generated only from nonmarine rocks such as lacustrian formations lacking marine organic input. The presence of these C_{30} steranes is a good indication of a marine contribution, but their absence may not always be source related. That is, the C_{30} steranes can be removed by extreme biodegradation, and they may not survive in highly mature oils, like some condensates. Another problem is that analysis of the 24-*n*-propylcholestanes requires GCMS because of interference with other biomarkers showing similar retention times (Peters and Moldowan 1993). The C_{30} sterane index is the ratio of C_{30}/(C_{27} to C_{30}) steranes. This ratio ranged from 0 to 0.88 for all marine oils analyzed from a variety of environments around the world. None of the nonmarine oils contained these C_{30} steranes (Moldowan et al. 1985).

Subsequently, C_{30} steranes were found in marine oils and bitumens ranging in source rock age from Early Proterozoic (1800 Ma) to Miocene (McCaffrey et al. 1994). This study also found small amounts of 24-isopropylcholestanes to be present along with the dominant 24-*n*-propylcholestane. The ratio of the 24-isopropyl to 24-*n*-propylcholestane varied between 1 and 2.1 for oils and bitumens generated by Vendian and Cambrian marine rocks, whereas it was less than 0.5 for all younger and older bitumens analyzed.

Diasteranes/Regular Steranes

When the Ekofisk oilfield of the North Sea was discovered by Philips Petroleum, it was thought that the Paleocene shales and chalks were the likely source rocks of the oils produced from the Cretaceous Tor and Ekofisk formations. Various biomarker ratios of the crude oils, however, did not match those of the Paleocene formations. Further studies showed that the oil biomarkers matched the extracts of the Upper Jurassic Kimmeridge Shale, which was much deeper than the oil accumulations. For example, the ratio of diasteranes to regular steranes was > 1 in the oils and in the Jurassic shale extracts and <1 in the Paleocene shale extracts. Likewise, the ratio of cholestane to ergostane was high in the oil and Jurassic shale and low in the Paleocene shale (Van den Bark and Thomas 1981).

A plot of pristane/phytane versus diasteranes/steranes was successfully used by Clark and Philp (1989) to differentiate two groups of crude oils and their source rocks in the Middle Devonian Elk Point formations of the Black Creek Basin in northwestern Alberta. The Lower Keg River source rock and oils from the Zamma and Shekilie oil fields had relatively high Pr/Ph and diasterane/sterane ratios. In the second group the Muskeg source rock and crude oils from the Rainbow oil fields had low ratios of these biomarkers. These low ratios plus an abundance of pregnane, squalane, a C_{24} tetracyclic terpane, and aryl isoprenoids (1-alkyl, 2,3,6-trimethylbenzenes) all are characteristic of hypersaline environments. This fits the geological data showing that a thick section of halite was deposited during early Muskeg time in the Rainbow oil-producing area.

Some high diasterane/sterane ratios are due to very high levels of biodegradation or thermal maturation rather than any source effect. Under heavy biodegradation, steranes are selectively destroyed in preference to the diasteranes (Table 12-3).

Sterane Ternary Diagrams

The distribution of C_{27}, C_{28}, and C_{29} homologous sterols on a ternary diagram was first suggested by Huang and Meinshein (1979) as a source indicator. They determined the distribution of these sterols in plankton, sea grass, land plants, and sediments in several areas. From this they defined certain parts of a ternary plot as representing open marine, estuarine, lacustrine, terrestrial, and higher plant ecosystems. Later others found that these areas overlapped too much to be reliable source indicators (see Chapter 4) but that plots of the C_{27}–C_{28}–C_{29} steranes could be used as a correlation tool. Consequently, the ternary diagram and its derivatives became widely used to correlate oils and source rocks. Plots in the literature include (1) C_{27}–C_{28}–C_{29} steranes, (2) C_{27}–C_{28}–C_{29} diasteranes, (3) C_{28}–C_{29}–C_{30} 4-methylsteranes, (4) C_{27}–C_{28}–C_{29} monoaromatic (MA) steroids, and (5) C_{26}–C_{27}–C_{28} triaromatic (TA) steroids.

The distribution of C_{27}–C_{28}–C_{29} sterols in phyto- and zooplankton plus higher plants was discussed in Chapter 4 and also by Huang and Meinschein (1979). Almost all higher plants have C_{29} as the dominant sterol. Since the conversion of sterols to steranes does not change the carbon number, this means that oils and bitumens derived from higher plants usually are dominant in C_{29}. Thus, in Figure 15-20 the two oils from coals and coaly shales of the Gippsland Basin, Australia (Shanmugam 1985), and the lignitic siltstones from the lacustrine Elko Formation of Nevada (Palmer 1984) have a strong C_{29} component.

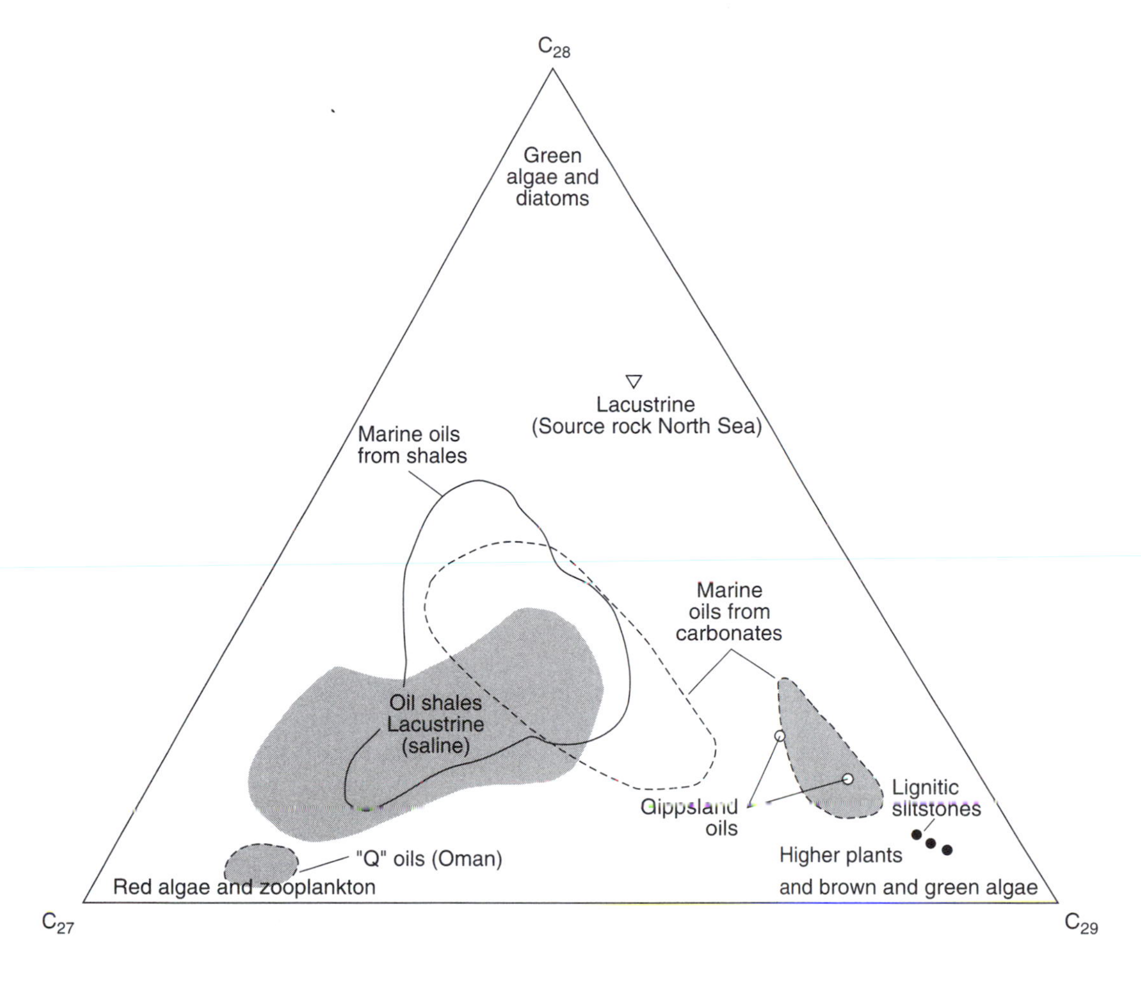

Figure 15-20

Ternary diagram showing the abundances of C_{27}, C_{28}, and C_{29}-regular steranes in marine organisms, higher plant material, crude oils, and shale extracts. See the text for details.

However, C_{29} sterols also are dominant in brown algae and some species of green algae. The dashed outline around the terrestrially sourced Gippsland oils encompasses crude oils from marine carbonate source rocks, namely, the Precambian Huqf Carbonate of Oman and the Permian Mulussa Formations of Syria (Grantham and Wakefield 1988). These marine oils are anomalously high in C_{29} steranes, possibly because of sterol precursors from brown and green algae.

The C_{27} sterols tend to be dominant in most plankton. They are concentrated particularly in red algae and zooplankton (Figure 15-20). The small dashed circle near the C_{27} corner encompasses the Precambrian "Q" oils from Oman (Grantham and Wakefield 1988), which are unusually high in C_{27}. Green algae have a variety of sterols. Ergosterol (C_{28}) is dominant in five species of *Chlorella,* but C_{29} is the major sterol in seven other species (Patterson 1971). Most diatoms contain more than 50% C_{28} sterols. Consequently, the combination of these with other algae results in phytoplankton mixtures having somewhat more C_{28} than zooplankton, although both still have more C_{27} than either C_{28} or C_{29}.

This dominance of C_{27} in plankton in the lower left corner of Figure 15-20 and C_{29} in higher plant material in the lower right corner means that most oils and their source rock extracts fall within a broad arc between these two corners in Figure 15-20. The lignitic siltstones shown have up to 86% C_{29}, and the "Q" oils have 78% C_{27}. The highest C_{28} (60%) reported is for a lacustrine Devonian dolomitic siltstone, shown in Figure 15-20. It contains a type I kerogen that is believed to have cosourced the Beatrice oil of the North Sea (Peters et al. 1989).

In Figure 15-20 the three areas marked marine oils encompass the sterane analyses by Grantham and Wakefield (1988) for more than 400 oils and those by Moldowan et al. (1985) for about 25 oils. The area marked oil shales—lacustrine includes 4 oil shales from the Eocene lacustrine Elko formation of Nevada (Palmer, 1984); 8 lacustrine, brackish/freshwater shales from the Songliao Basin, China (Fu Jiamo et al. 1990); 4 oils and 11 core samples from the hypersaline lacustrine sediments of the Jianghan Basin of eastern China (Fu Jiamo et al. 1986); and about 25 cuttings samples from the Neocomian lacustrine hypersaline Bucomazi shales of the Lower Congo Basin of Angola (Burwood et al. 1992). The Songliao Basin is China's major oil producer. Some Bucomazi samples have TOCs $> 20\%$. It is one of the richest source rocks in the world, with an average source potential index of 46 metric tons HC/m^2 (Demaison and Huizinga 1991).

These examples show that the oil-rich lacustrine shales deposited in brackish-to-hypersaline environments tend to be high in C_{27} steranes, whereas some marine oils from carbonate source rocks are high in C_{29} comparable to the land plants. These results plus the large overlap between marine and lacustrine oils from shales and carbonates in Figure 15-20 make it apparent that sterane distributions cannot be used to unequivocally define depositional environments. In most cases, additional biomarker and petrographic data are needed. Nevertheless, good correlations between oils and source rocks are possible because the unique sterane fingerprint of a specific source rock is preserved in the oils generated from it throughout the oil window.

For example, Figure 15-21 compares crude oils and rock extracts from two areas, the West Siberian Basin of Russia and the Jeanne d'Arc Basin offshore eastern Canada. The sterane distributions for oils from six major fields in West Siberia, including the giant Salym and Federov fields, plot in a small area in Figure 15-21 (Peters et al. 1993). The two most probable source rocks here are the Late Jurassic Bazhenov Shale and the Early to Middle Jurassic Tyumen Shale. The sterane distribution for the Bazhenov bitumen was essentially iden-

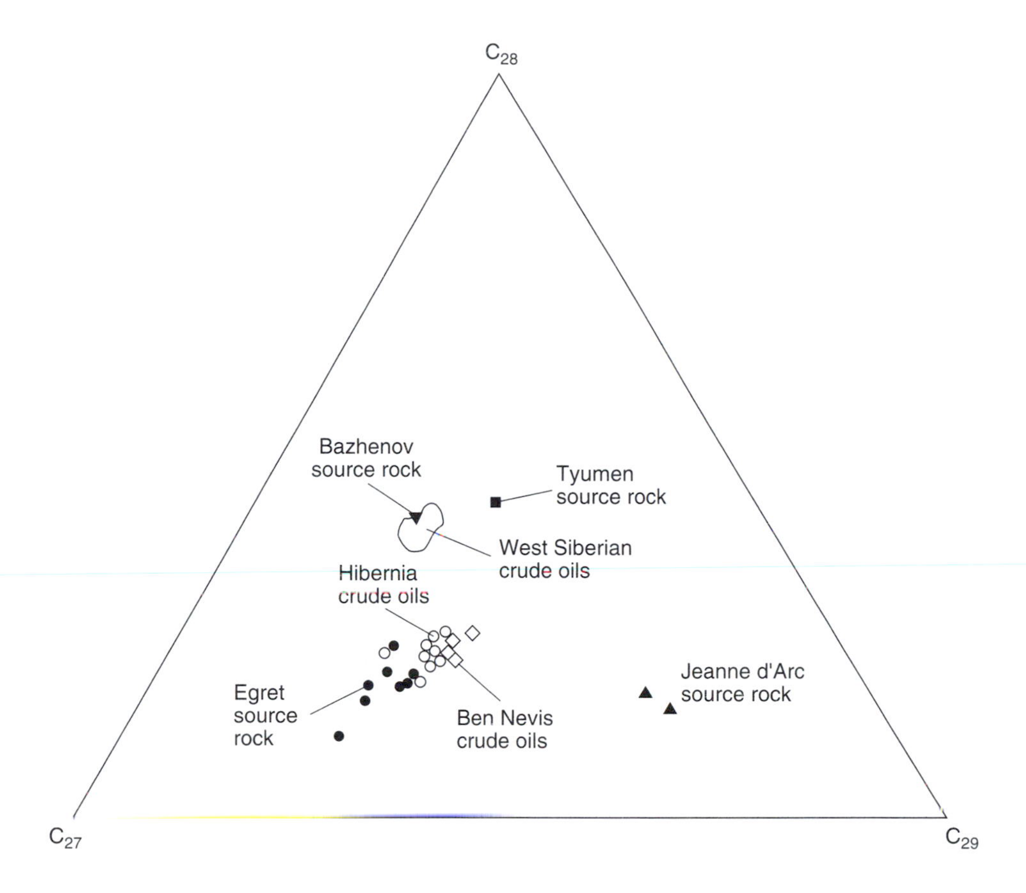

Figure 15-21

The C_{27}, C_{28}, C_{29}-regular sterane distributions in the saturate fractions of crude oils and the extracts of their possible source rocks. [Peters et al. 1993; von der Dick et al. 1989]

tical to the crude oils (Figure 15-21), whereas the Tyumen bitumen contains a slightly higher percentage of the C_{29} sterane, apparently owing to terrestrial influences. The Early Jurassic Tyumen is dominated by a clastic coal-bearing sequence. A ternary diagram of the C_{27}–C_{28}–C_{29} diasteranes showed the same pattern as in Figure 15-21 but with the Bazhenov source rock even more centralized within the crude oil area. The C_{30} sterane indices, the Pr/Ph ratio, and the homohopane distributions also are similar for the Bazhenov bitumen and the oils but differ from the Tyumen bitumen (Peters et al. 1993).

In the Jeanne d'Arc Basin the question was whether the crude oils had been generated by the Tithonian Jeanne d'Arc Shale or the Kimmeridgian Egret

Shale (von der Dick et al. 1989). Whole-oil chromatograms of the oils showed that the Hibernia and Ben Nevis oils are generally similar in molecular distributions except that the latter have more benzene, toluene, xylene, and methylcyclohexane, as well as a higher Pr/Ph ratio. This suggests some terrestrial input to the Ben Nevis oils. The sterane distributions confirm this (Figure 15-21), in showing slightly more C_{29} in the Ben Nevis than in the Hibernia oils. The source rock bitumen comparison shows the Egret Shale to be the primary source rock, since it overlaps some of the Hibernia oils in Figure 15-21. But the Jeanne d'Arc Shale appears to be contributing some terrestrial input to both groups of oils because many of the Hibernia oils are somewhat higher in C_{29} than the Egret source rock bitumen (von der Dick et al. 1989).

An example of the mixing of oils from different source rocks involves the Beatrice crude oil of the North Sea (Peters et al. 1989). The Beatrice oil is unique among North Sea oils in being extremely waxy and in containing β-carotane. These are terrestrial and lacustrine indicators, respectively. They are not found in a Kimmeridge source rock from this area or from the nearby Piper field oil which is more typical of North Sea oils. Epimers of 25,28,30-trisnorhopane and 28,30-bisnorhopane are present in the Piper oil and Kimmeridge Shale bitumen but not in the Beatrice oil or in the Devonian and Middle Jurassic source rock bitumens shown in Figure 15-22.

Vanadyl porphyrin concentrations are high in the Piper and Kimmeridge samples but low in all the others. Other bulk properties like carbon isotopes and sulfur content support the biomarker data in revealing a close relationship between Piper oil and Kimmeridge source rock, but not between the latter and Beatrice oil. Finally, as shown in Figure 15-22 the C_{27}–C_{28}–C_{29} sterane distribution is similar for Piper oil and Kimmeridge bitumen but differs from the Beatrice oil and other source rock bitumens. The same relationship shows up in plots of C_{27}–C_{28}–C_{29} diasteranes and monoaromatic (MA) steroids.

Subsequent studies showed that the Devonian contains β-carotane and a series of terpane biomarkers found also in Beatrice oil, whereas the Middle Jurassic has traces of the C_{30}-steranes (24-*n*-propylcholestanes) and an unusual series of aromatic biomarkers also found in Beatrice oil, but not in the Devonian.

The intermediate position of the Beatrice oil between the Middle Jurassic and the Devonian bitumens in Figure 15-22 also occurs in ternary plots of the C_{27}–C_{28}–C_{29} diasteranes and MA-steroids, suggesting a cosourcing of the Beatrice oil from the Devonian and the Middle Jurassic (Peters et al. 1989). But a different Beatrice oil sample analyzed by Bailey et al. (1990) plotted close to the C_{27}–C_{28}–C_{29} steranes of the hydrous pyrolysate (HP) fractions of Middle Devonian source rocks (Figure 15-22). In addition, Galimov isotope plots of the oil and the HP fractions indicated a predominantly Middle Devonian source, as discussed later (Bailey et al. 1990). Consequently, the extent of Jurassic contributions to the Devonian sourced Beatrice oil is still not clear.

Also shown in Figure 15-22 is the distribution of steranes in five grahamites and a crude oil produced from the Mississippian Stanley Sandstone in the Ouachita Mountain region of Oklahoma (Curiale 1988). Earlier studies of bulk parameters such as carbon and sulfur isotope ratios had indicated that the

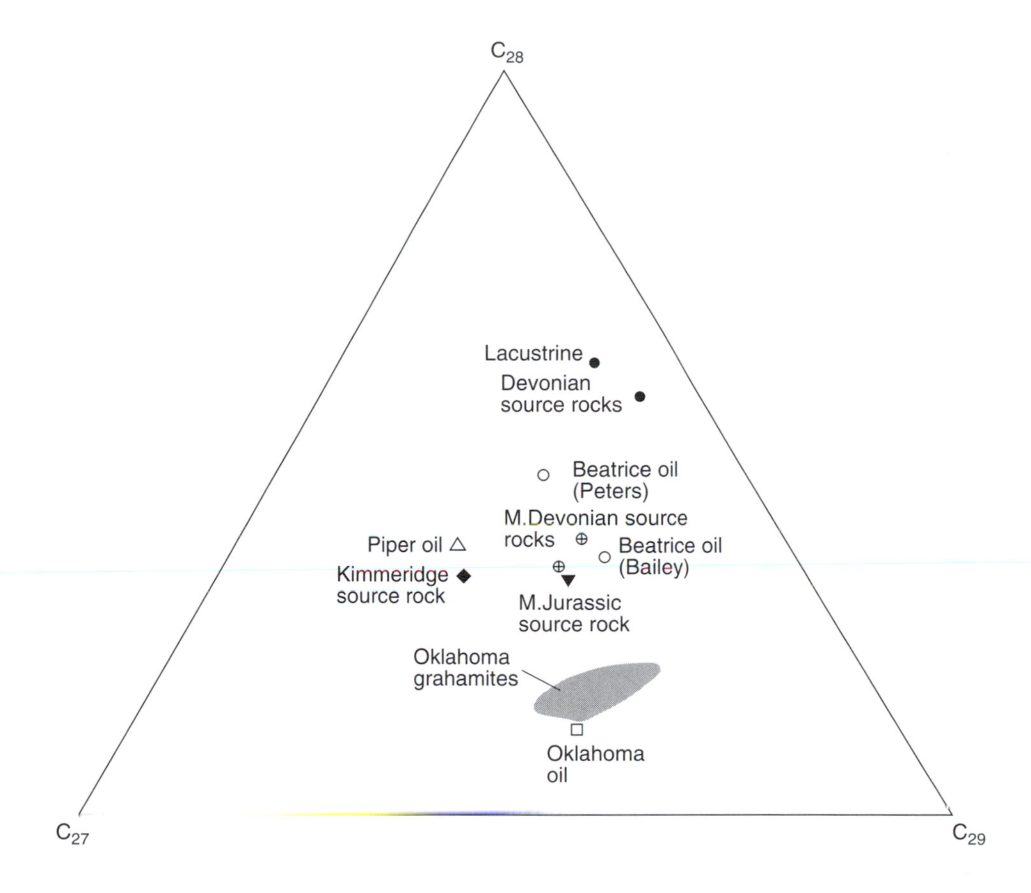

Figure 15-22

Ternary diagram showing the relative abundances of C_{27}, C_{28}, C_{29}-regular steranes in the saturate fractions of North Sea oils and extracts of possible source rocks. [Peters et al. 1989 and Bailey et al. 1990] and steranes in five grahamites and an associated oil from Oklahoma. [Curiale 1988]

grahamite represented a solidified weathered residue of an oil seep in a Stanley Sandstone outcrop. The close correlation of the subsurface Stanley reservoir oil and the grahamites in Figure 15-22 supports the idea that both are from the same source rock.

Riva et al. (1986) identified three main groups of oils and their source rocks in the Po Basin of northern Italy based on a variety of correlation parameters, including the C_{27}–C_{28}–C_{29} steranes (Figure 15-23). The Middle Miocene

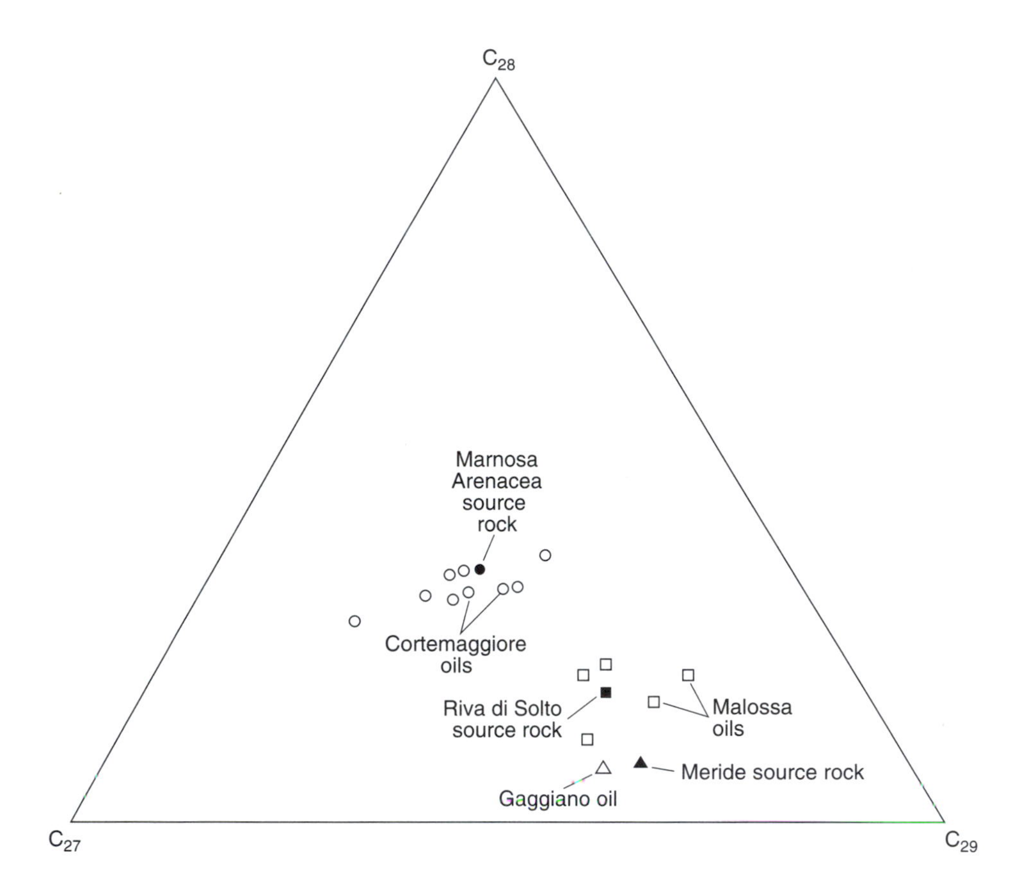

Figure 15-23

Relative abundances of C_{27}, C_{28}, and C_{29}-regular steranes in the saturate fractions of crude oils and extracts of their probable source rocks from the Po Basin of northern Italy. [Riva et al. 1986]

Marnosa Arenacea flysch sequence was considered to be the source of the Cortemaggiore oils. Both the source rock and the oils contain 18α(H)-oleanane, which was not present in the other oils and source rocks. Also, the C_{27}–C_{28}–C_{29} steranes in the Cortemaggiore oils and the Marnosa source rock extracts plotted together, as seen in Figure 15-23.

The Late Triassic Riva di Solto black shale is believed to be the source of the Mesozoic Malossa oils, based on similarities in both the sterane and terpane GCMS fingerprints. Here again, the source rock and oils plot close together in Figure 15-23. The Cortemaggiore and Malossa oils are clearly different from each other based on biomarkers (such as oleanane) and in carbon isotope values and Pr/Ph ratios. The third group, the Gaggiano oil, is similar to the Malossa oils in carbon isotopes and isoprenoid distribution but differs significantly in containing dibenzothiophene and its methyl derivatives plus a V/Ni ratio three times higher than the Malossa oils. The distribution of C_{27}–C_{28}–C_{29} steranes (Figure 15-23) indicates that this oil and its Meride source rock have a smaller percentage of the C_{28} sterane than do the Malossa oils.

Among the problems in using ternary sterane diagrams are differences in the maturity and biodegradation of the samples being compared plus difficulties in separating a sterane peak from interfering peaks, as previously observed for 24-*n*-propylcholestane.

Also, plotting ternary diagram data from different sources has problems such as the following:

1. Integrated peaks may differ depending on which epimers (ααα or αββ, etc.) are used for each homolog.
2. Different GC and MS systems may cause different plot locations for similar samples.

It is not feasible to correlate the extracts of immature source rocks with those of mature rocks or with mature oils. An extreme example of the maturity problem was an attempt to correlate the mature Katia oil of Italy with the extract of its presumed source rock (Monte Prena), which was very immature (Peters and Moldowan 1993, p. 272). The C_{27}–C_{28}–C_{29} sterane distributions in the oil are 35:13:52, compared with 1:17:82 in the extract of the immature source rock. This large difference indicates no relationship. Hydrous pyrolysis of the source rock, however, yielded a bitumen with the sterane distribution of 34:18:48, very close to that of the oil. Conversion of the Monte Prena kerogen from immature to artificially mature caused a large release of the C_{27} sterane relative to the C_{28}–C_{29} steranes. This increase in the C_{27} sterane from the pyrolysis of rocks has been observed before (Mackenzie 1984, p. 189). Apparently, the C_{27} sterane is preferentially formed and expelled in the later stages of the thermal decomposition of kerogen.

Ternary plots of C_{27}–C_{28}–C_{29} diasteranes are used in place of normal steranes when the latter have been altered by extreme biodegradation or high levels of maturity.

A ternary diagram of C_{28}–C_{29}–C_{30}-4-methylsteranes was used by Fu Jiamo et al. (1990) to distinguish the rock extracts from three types of lacustrine environments: freshwater, brackish/freshwater, and saline/hypersaline. The kerogen from the freshwater deposits had the highest concentrations of C_{28}-4-methylsteranes, and that from the saline-water environment had the lowest. The 4-methyl-steranes in petroleum are believed to be derived from dinoflagellates. Since the 4-methylsteranes are common in lacustrine rocks, they may eventually prove to be more useful than the regular steranes in correlating lacustrine oils and their source rocks.

The origin of monoaromatic (MA) steroids is different from that of the regular steranes, so they plot differently on a ternary diagram. Peters and

Moldowan (1993) recommend using both the regular sterane and MA-steroid plots in any correlation problem, since they provide independent results to strengthen the conclusions.

Biomarker Fingerprints

Mass chromatograms, or *fragmentograms,* have been widely used for correlating oils and source rocks since the pioneering work of Seifert (1977). He showed that it was possible to differentiate among crude oils produced from three formations in the San Joaquin Basin of California, on the basis of sterane and terpane fingerprints. Some more recent studies are described in the following section.

Zhanhua Basin, China

Figure 15-24 shows m/z 218 sterane fragmentograms of two oils and two possible source rocks from the Zhanhua Basin in China (Shi Jiyang et al. 1982). Prominent peaks in four molecular sizes are shown: C_{26}, C_{27}, C_{28}, and C_{29}. Four stereoisomers are given for each molecular size, the 5α,14α,17α, and 5α,14β,17β in 20S and 20R configurations. As previously discussed in Chapter 10, the size of the molecules (carbon numbers) are compared for correlation purposes, whereas the stereoisomers are commonly used for comparing levels of maturation.

In Figure 15-24 the dominant peaks for shale LO 14 are the C_{26} and C_{27} steranes, whereas the dominant peaks for oil YI 18 and oil Cheng 15 are C_{27}, C_{28}, and C_{29}. This indicates no relationship between shale LO 14 and the oils. In contrast, shale YI 21 appears to be the source rock for oil YI 18, based on the close similarity of their sterane fingerprints (i.e., C_{27}–C_{29} homolog distribution).

Oil Cheng 15 also appears to show a genetic relationship to source rock YI 21, based on its sterane fingerprint. The fingerprints are very different, however, in stereoisomer distributions, signaling a marked difference in maturity. The steranes of oil Cheng 15 show high proportions of the more mature α,α,α,20S and α,β,β,20S-steranes, which are minor in shale YI 21. If a more mature sample of shale YI 21 were obtained, such as by hydrous pyrolysis of this sample, it

Figure 15-24 (facing page)

Comparison, by the m/z 218 fragmentograms, of the steranes of two oils and two possible source rocks from the Zhanhua Basin of China. [Shi Jiyang et al. 1982]

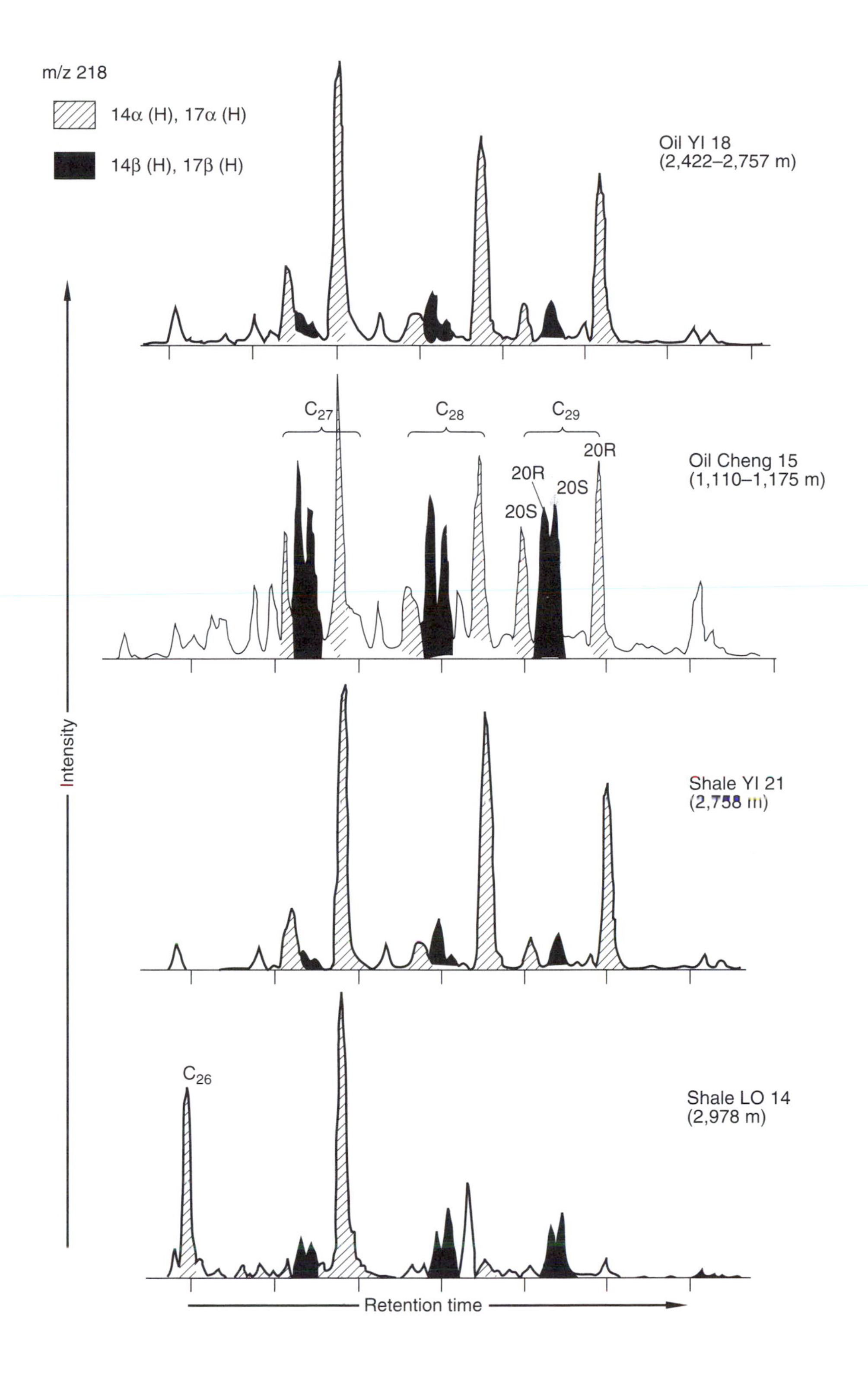

m/z 218
14α (H), 17α (H)
14β (H), 17β (H)
Oil YI 18
(2,422–2,757 m)
C27
C28
C29
20R
20R
20S
20S
Oil Cheng 15
(1,110–1,175 m)
Intensity
Shale YI 21
(2,758 m)
C26
Shale LO 14
(2,978 m)
Retention time

might show the same fingerprint as oil Cheng 15, thereby indicating a genetic relationship. This example demonstrates how differences in maturity (stereoisomers) can interfere with correlations based mainly on the size of the molecules.

Utah–Wyoming Overthrust Belt

It was previously mentioned that tricyclic terpanes are one group of biomarkers used to define the sources of the Pennsylvanian Tensleep and Cretaceous Mowry crude oils of Wyoming (Seifert and Moldowan 1981). The other group used by these authors was the hopanes (Figure 15-25). In this figure, the Ts is C_{27}-18α(H)-trisnorhopane, and the Tm is C_{27}-17α(H)-trisnorhopane. The Ts/Tm ratio is dependent on both maturity and source. Figure 15-25 emphasizes the source difference. Note that in the Tens'eep oil and Phosphoria source rock, Ts is part of a small double peak and Tm is a large peak. Consequently, these samples have low Ts/Tm ratios. In contrast, the Mowry Shales have higher Ts/Tm ratios. Other fingerprint characteristics, such as the pattern of homohopanes, are similar for the Mowry shale and oil but different from the Phosphoria shale and Tensleep oil.

South Florida Basin

The Sunniland Limestone of the South Florida Basin is overlain by the thick Trinity C Anhydrite and underlain by the Punta Gorda Anhydrite. The source of the oils reservoired between these anhydrites is believed to be the Sunniland Limestone itself, since there are no faults or extensive fractures to permit vertical migration through these anhydrites (Palacas et al. 1984). The Sunniland Limestone is divided into upper and lower units, and the question is whether one or both of these units generated the oil. Conventional analyses indicated that the argillaceous carbonates of the Lower Sunniland Formation were the most probable source rock. They had the highest TOC (up to 12%) and the highest content of algal–sapropelic organic matter.

The tricyclic, tetracyclic, and pentacyclic terpane fingerprints confirmed this concept (Figure 15-26). The dominant tricyclic terpane peak in the Sunniland crude oil and in 80% of the extracts of the Lower Sunniland rocks is C_{23}. None of the Upper Sunniland rock extracts showed this pattern. In addition, the relative concentrations of the homohopanes (extended hopanes) decrease

Figure 15-25 (facing page)

Correlation of oils and source rock extracts by the m/z 191 fragmentograms of pentacyclic terpanes for the Cretaceous Mowry Formations (top) and the Permian Phosphoria and Pennsylvanian Tensleep Formation (bottom) in the Rocky Mountain Overthrust Belt, United States. [Seifert and Moldowan 1981]

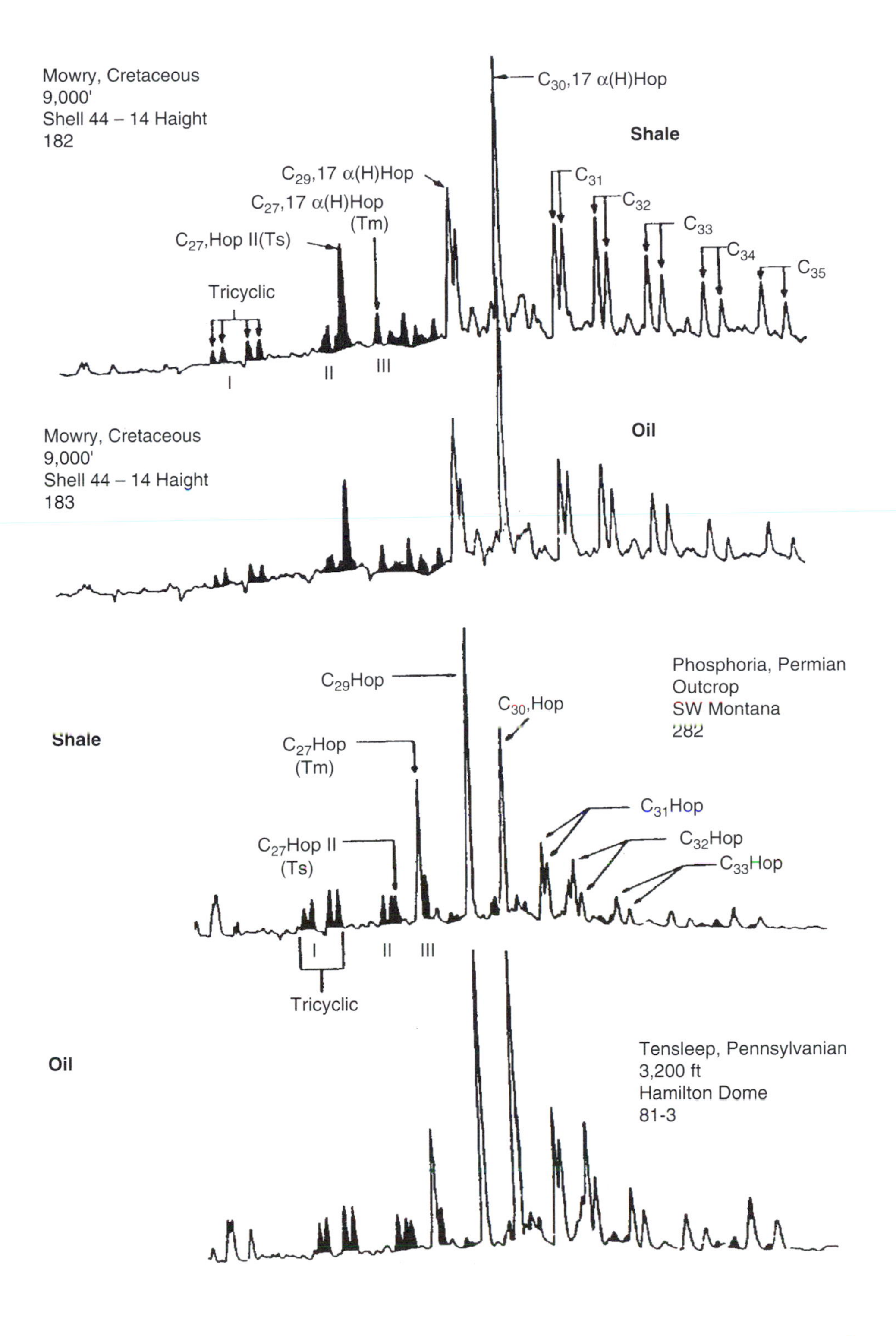

Mowry, Cretaceous
9,000'
Shell 44 – 14 Haight
182
C30,17 α(H)Hop
Shale
C29,17 α(H)Hop
C27,17 α(H)Hop
(Tm)
C27,Hop II(Ts)
C31
C32
C33
C34
C35
Tricyclic
I
II
III
Mowry, Cretaceous
9,000'
Shell 44 – 14 Haight
183
Oil
C29Hop
C30,Hop
Phosphoria, Permian
Outcrop
SW Montana
282
Shale
C27Hop
(Tm)
C31Hop
C32Hop
C33Hop
C27Hop II
(Ts)
I
II
III
Tricyclic
Oil
Tensleep, Pennsylvanian
3,200 ft
Hamilton Dome
81-3

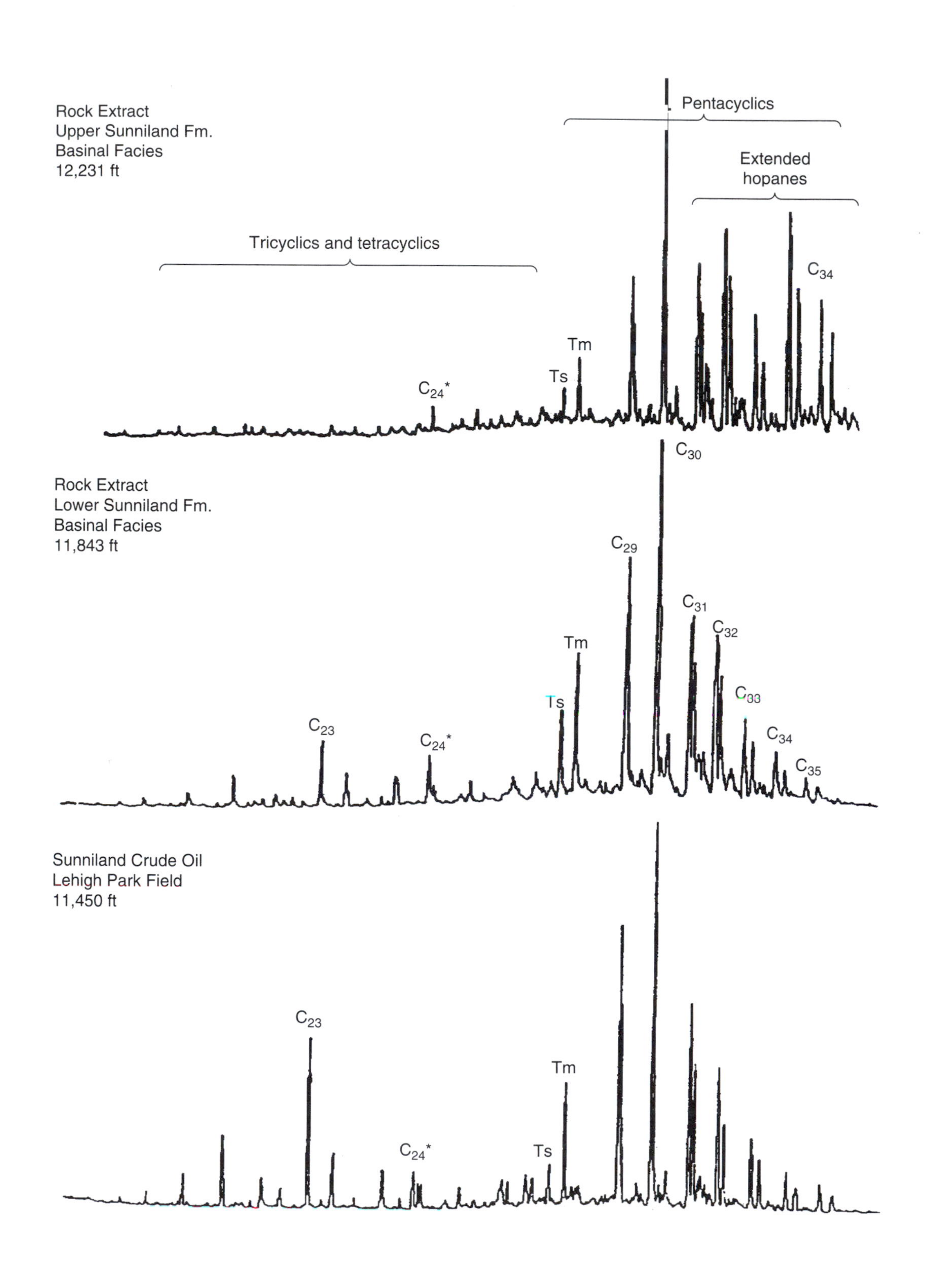

Rock Extract
Upper Sunniland Fm.
Basinal Facies
12,231 ft
Pentacyclics
Extended
hopanes
Tricyclics and tetracyclics
C_{34}
Tm
Ts
C_{24}^*
C_{30}
Rock Extract
Lower Sunniland Fm.
Basinal Facies
11,843 ft
C_{29}
C_{31}
C_{32}
Tm
Ts
C_{33}
C_{23}
C_{24}^*
C_{34}
C_{35}
Sunniland Crude Oil
Lehigh Park Field
11,450 ft
C_{23}
Tm
C_{24}^*
Ts

Figure 15-26 (facing page)

Correlation of Sunniland crude oil with possible source rock extracts using tri-, tetra-, and pentacyclic terpanes (m/z = 191) fragmentograms. The asterisk indicates C_{24} tetracyclic terpane. [Palacas et al. 1984]

continuously from C_{31} to C_{35} in the crude oil and in the Lower Sunniland rocks, but in the Upper Sunniland samples the concentrations oscillate. Generally, the C_{32} or C_{35} peak is the highest. The authors concluded that the Lower Sunniland limestone was the major source of Sunniland oil.

Oils in Italy

Mattavelli and Novelli (1990) identified eleven groups of oils in Italian basins based on detailed geochemical analyses involving both bulk properties and biomarker studies. About 95% of the oils were generated from Middle and Upper Triassic carbonates containing type II-S (IIA and IIB in Chapter 6) kerogen, and the remaining oils were formed by Tertiary shales containing predominantly type III kerogen. Most of the carbonate source oils are heavy and immature because of the kerogen's high sulfur content. Sulfur lowers the activation energy, causing early generation, as pointed out in Chapter 6.

Most heavy oils are more similar to the bitumens extracted from their source rocks than are most light oils. For example, Figures 15-27 and 15-28 show two examples of heavy oil–source rock correlations published by Mattavelli and Novelli (1990). The terpane fragmentogram of the Burano source rock extract is almost identical to that of the 19°API gravity Piropo oil in the Central Adriatic region (Figure 15-27). Both have a similar profile of tricyclic terpanes, Ts/Tm, hopanes, C_{31} homohopane, and gammacerane. High relative amounts of gammacerane indicate that the source rock was deposited under highly reducing, saline conditions. Figure 15-28 compares the bitumen extract from the Late Triassic Noto Formation and its corresponding 15°API gravity Perla oil from southern Sicily. The two fingerprints show a good match, with the gammacerane peak even larger than in the previous example. Note that terpane mass fragmentograms for unrelated samples may look similar if the depositional environments are similar. The advantage of biomarkers is that many different molecules can be compared in order to strengthen the conclusions.

Correlations Using C_7 Hydrocarbons

Mango (1990) found that the ratios of several heptanes, such as 2- and 3-methylhexane to 2,4- and 2,3-dimethylpentanes and of iso- and cyclic C_7 heptanes to dimethylcyclopentanes remain constant during the entire petroleum expulsion and accumulation process. He attributed this to the isocyclic alkanes

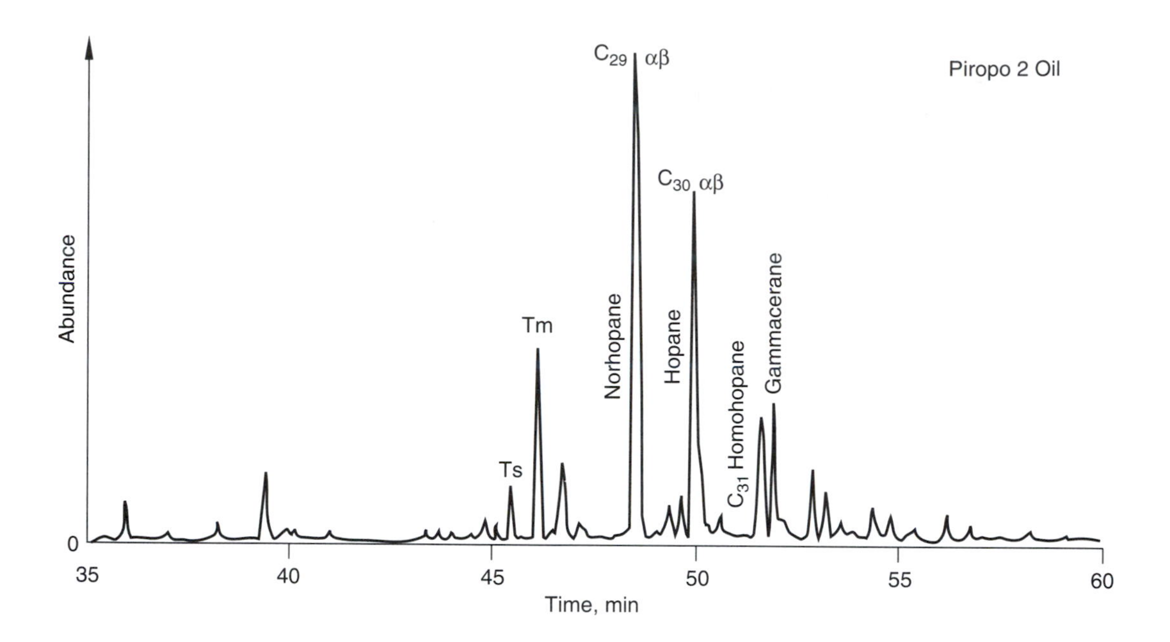

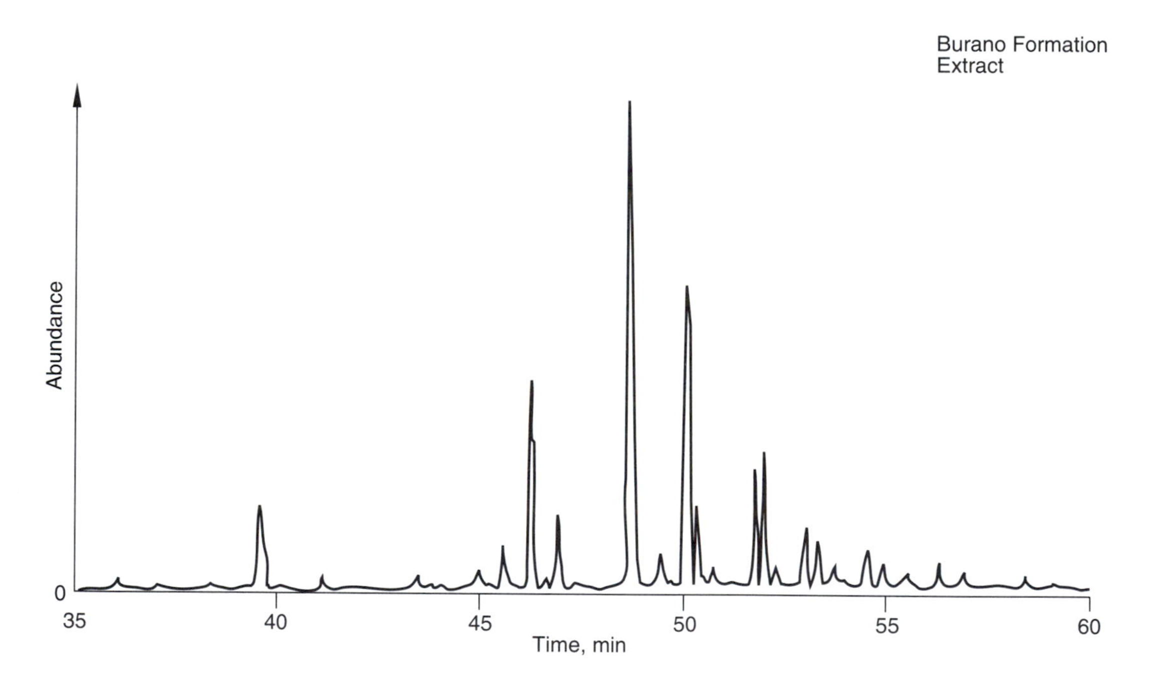

Figure 15-27

Mass fragmentograms (m/z = 191) showing a good correlation between Piropo oil and an extract of the Burano Formation source rock. The peak before the gammacerane peak is 17α,21β(H)22R-C_{31}-homohopane. [Mattavelli and Novelli 1990]

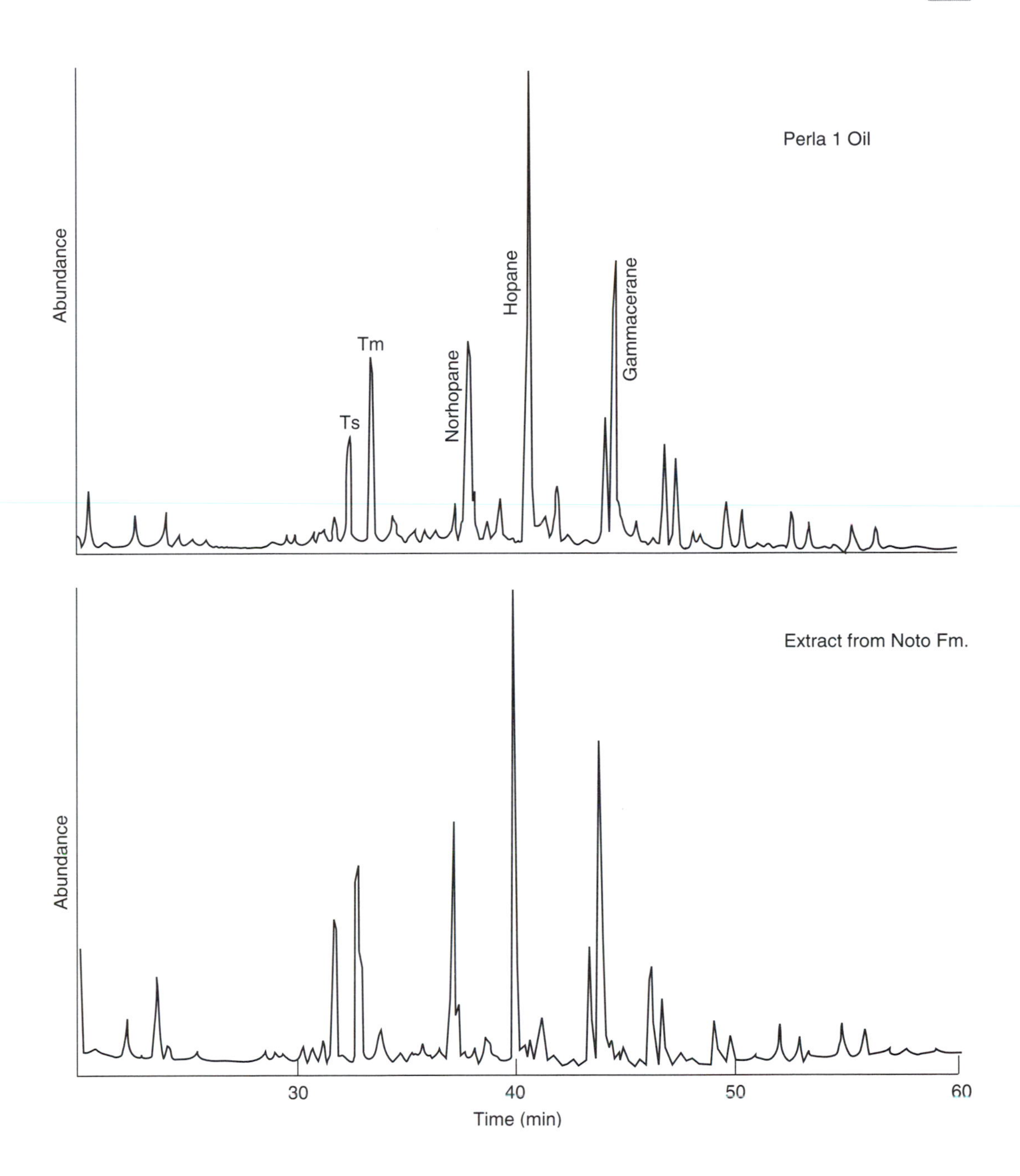

Figure 15-28

Mass fragmentograms (m/z = 191) showing a good correlation between Perla oil and an extract of the Noto Formation source rock. [Mattavelli and Novelli 1990]

originating from *n*-heptane by steady-stage catalysis involving transition metals such as Ni, V, Ti, and Co during burial of the source rocks (Mango 1992). His model indicated that all the C_7 alkanes in petroleum are genetically linked. The distribution of the alkanes is related to the nature of the source rock, catalyst composition, temperature, and pressure. The isomerization of the *n*-heptane to mono- and polybranched isoalkanes plus five- and six-ring cycloalkanes occurs before the expulsion of the oil from the source rock. After expulsion, the C_7 ratios do not change except during biodegradation and phase changes. Consequently, the ratios are valuable source indicators.

Some of the key C_7 parameters in Mango's 1994 kinetic scheme are as follows:

P_1 = *n*-heptane (parent-1)

P_2 = 2-methylhexane + 3-methylhexane (parent-2)

P_3 = 3-ethylpentane + 2,2-dimethylpentane (DMP) + 2,3-DMP + 2,4-DMP + 3,3-DMP (daughters)

N_2 = 1,1-dimethylcyclopentane (DMCP) + cis-1,3-DMCP + trans-1,3-DMCP (daughters)

$X_2 = N_2/P_3$

Maturity parameter = 2,4-DMP/2,3-DMP

For example, a plot of the natural log of 2,4-DMP/2,3-DMP versus maximum temperature of formation indicates that in 95% of 1,600 nonbiodegraded oils, the C_7 alkanes were generated between 95 and 135°C (Bement et al. 1994). At 80°C and 140°C, the ln 2,4 DMP/2,3-DMP values are –4 and 0, respectively. These numbers appear to be independent of source rock age, thermal history, kerogen type, and lithology. However, biodegradation preferentially destroys 2,3-DMP, and gas stripping and evaporative fractionation affect the ratios because of the different solubilities of the isomers.

The X_2 (N_2/P_3) is a selectivity ratio between daughter products of the parent P_2. Kornacki (1993) classified Monterey oils of the Santa Maria Basin, California, into two groups based on a plot of ln X_2 versus P_2. He also correlated the two oil groups with their source rocks, thereby identifying the petroleum systems of the Monterey, as seen in Figure 15-29. The plot of ln X_2 versus P_2 shows that the phosphatic Monterey Shale generates mainly very heavy paraffinic oils, whereas the siliceous Monterey Shale generates lighter naphthenic oils.

Mango (1994) also believes that genetically related ring preferences (RPs) are formed catalytically. These include C_7 cyclohexanes, cyclopentanes, and cyclopropanes, the last of which decomposes to isoalkanes. A ternary plot of these three RPs is useful in correlating crude oils and condensates. The C_7 cyclohexanes tend to be characteristic of a terrestrial-source type III kerogen; the cyclopentanes, of a marine-source type II kerogen; and the isoalkanes of a lacustrine-source type I kerogen. Another plot for separating crude oils into genetic groups is P_3 versus (P_2 + N_2) in wt% of total oil.

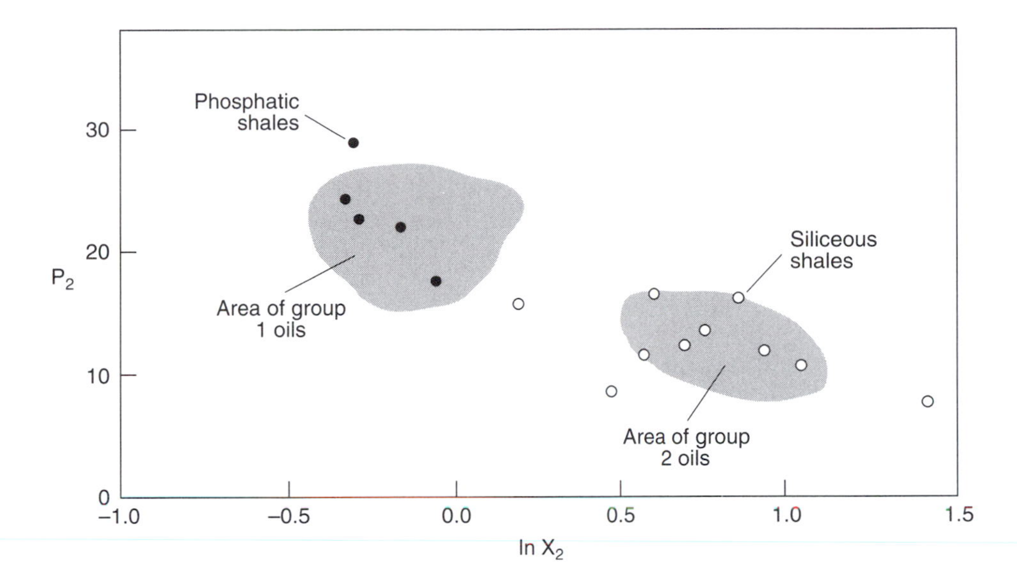

Figure 15-29

Oil to source rock C_7 correlations for the Monterey Shale of the Santa Maria Basin in California. The shaded areas represent the two primary oil groups: 1) heavy, paraffinic oils, and 2) lighter, naphthenic oils. The data points are the values for the C_7 alkanes in samples of phosphatic and siliceous shale source rocks. [Kornacki 1993]

The C_7 analyses can be used in oil–oil and oil–source rock correlations and in maturation studies. They are a less expensive technique than biomarkers. They also are particularly valuable in evaluating high-gravity oils and condensates, which frequently have few or no biomarkers. Condensates contain up to 10% C_7 alkanes. Conventional oils have 1 to 5%, and even very heavy oils contain a small percentage of the C_7 alkanes.

Stable Isotope Ratios for Correlation

Carbon isotope ratios are increasingly being used for both oil–oil and oil–source rock correlations, particularly on oil fractions and on the individual hydrocarbons in oils. Also, isotopes are used to correlate natural gases with one another and their source rocks and to evaluate gas reservoir continuity (Schoell et al. 1993; Whiticar 1994; and references therein).

Carbon Isotope Ratios of Oils and Source Rock Bitumens

Carbon isotope ratios of whole oils by themselves are generally not used for correlation because of the narrow range in $\delta^{13}C$ values for all oils of about 15‰ (–18 to –33‰). In fact, most oils are concentrated between –25 and –31‰, which causes considerable overlap. Consequently, oil-to-oil correlations are usually made using cross plots of carbon and hydrogen isotopes or by comparing oil fractions. Burwood et al. (1990) found that the crude oils of the Lower Congo Coastal Basin of Angola overlapped in $\delta^{13}C$ values. They separated them into two different genetic groups, however, by using a cross plot of $\delta^{13}C$ versus $\delta^{2}H$.

In Magoon and Claypool's 1981 North Slope of Alaska study (Figure 15-8), the oils were nicely separated into two groups by plotting $\delta^{13}C$ of the aromatic-versus-saturated hydrocarbons. Sofer (1984) compared the $\delta^{13}C$ of the C_{15+} aromatic and saturate fractions of 339 oils and concluded that such plots could be used to distinguish oil families and to infer a marine (nonwaxy) versus a terrestrial (waxy) source.

The isotopic relationship developed by Sofer (1984) is as follows:

For marine source: $\delta^{13}C_{ARO} = 1.10\ \delta^{13}C_{SAT} + 3.75$ (nonwaxy oils)

For terrestrial source: $\delta^{13}C_{ARO} = 1.12\ \delta^{13}C_{SAT} + 5.45$ (waxy oils)

Terrestrial waxy oils are considered to originate from land-derived OM deposited in a lacustrine, paralic, or deltaic environment, whereas nonwaxy oils are from OM deposited in open marine environments with minor contributions of terrestrial OM.

Sofer carried out a stepwise discriminant analysis excluding biodegraded oils to determine the best straight line separating the two groups. It is

$$\delta^{13}C_{ARO} = 1.14\ \delta^{13}C_{SAT} = 5.46$$

Figure 15-30 is an example of a Sofer-type plot for bitumens extracted from cores and cuttings in the Cambay Basin of India (Garg and Philp 1994). The $\delta^{13}C$ values for the aromatic and saturate hydrocarbons indicate a dominant terrestrial source of type III kerogen. Pyrolysis–GC analyses of kerogens and their asphaltenes confirmed this observation and also showed that the Cambay Shale had some marine components, as indicated by the Sofer plot.

Figure 15-31 (A) is a cross plot of $\delta^{13}C$ values for aromatic and saturated hydrocarbons in oils, oil seeps, and source rock extracts of the San Joaqin, Salinas, and Cuyama basins of California (Peters et al. 1994) Four oil–source rock groups are indicated. Group I consists of the Eocene Kreyenhagen source rock plus oils and seeps from Eocene and Miocene reservoir rocks in the western San Joaquin Basin. Group II includes the Soda Lake Shale Member of the Lower Miocene Vasqueros Formation in the Cuyama Basin as the source rock. Its extract matched the bitumen extracted from an oil-stained siltstone. Group III represents a family of oils generated by different facies of Miocene Monterey

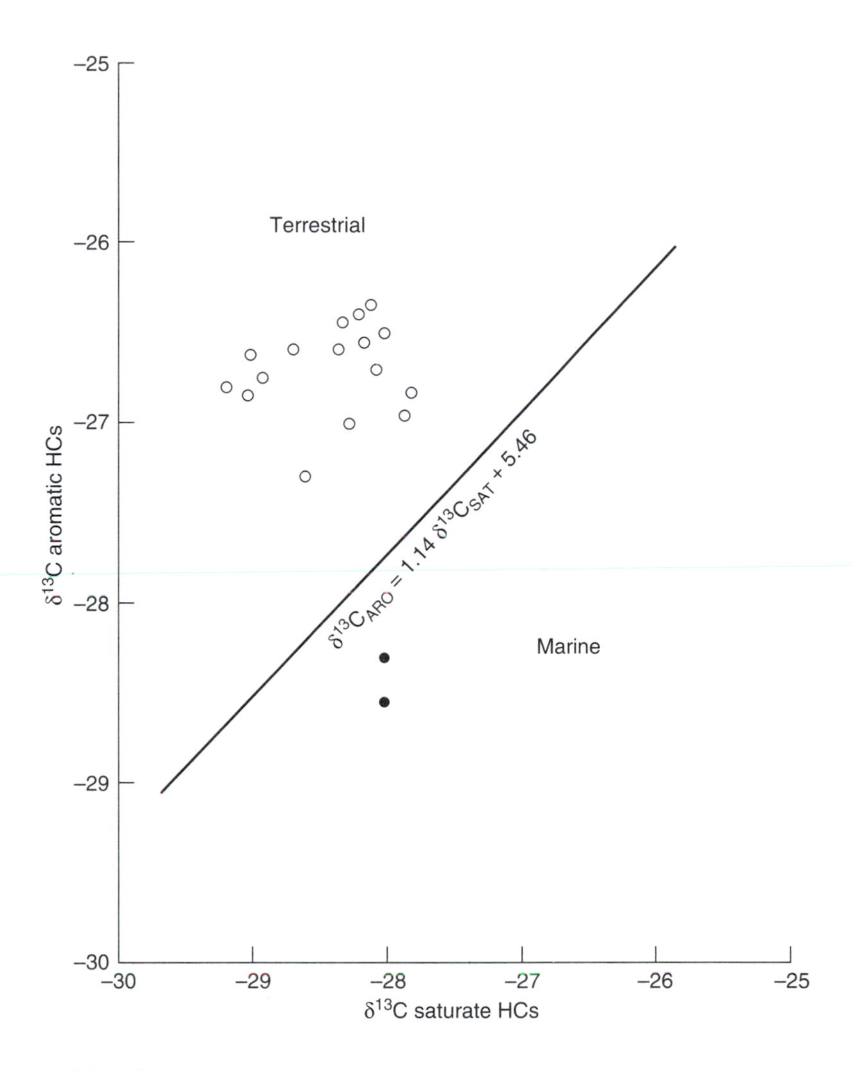

Figure 15-30

A Sofer plot of $\delta^{13}C$ values for the aromatic and saturate hydrocarbons in extracts of Hazad and Cambay shales, Gandhar Field, Cambay Basin, India. All $\delta^{13}C$ values are in ppt(‰) relative to PDB. [Garg and Philp 1995]

source rock in the San Joaquin Basin, as well as several bitumen extracts from the source rock. Oil being produced from the Cretaceous–Paleocene Moreno Formation did not fit any of these categories, so it was designated Group IV. Biomarker differences between the Moreno oil and all other oils in this study verified its uniqueness.

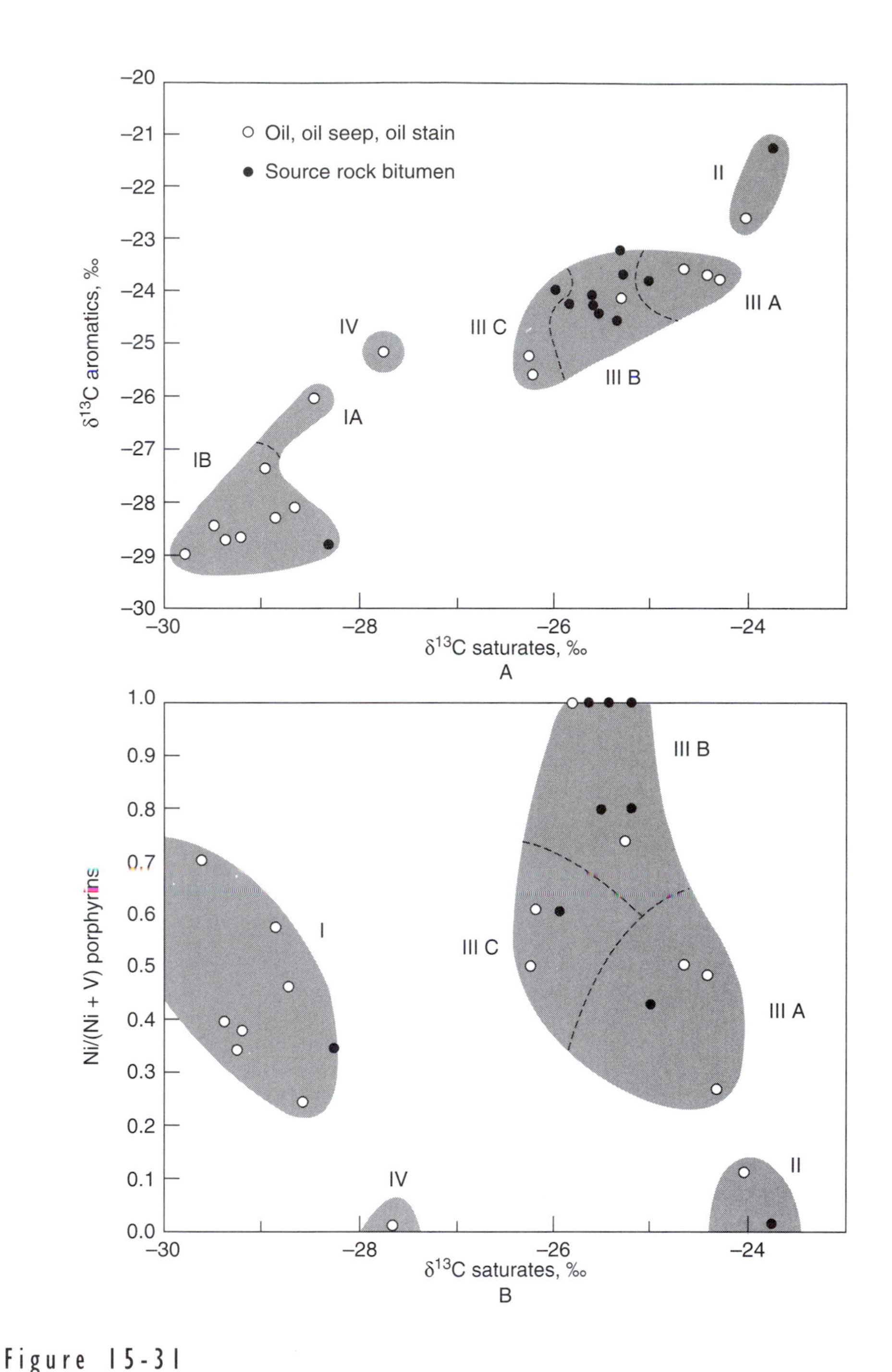

Figure 15-31

(A) The $\delta^{13}C$ of saturate versus the $\delta^{13}C$ of aromatic hydrocarbons showing four oil–source rock groups in the San Joaquin, Cuyama, and Salinas basins of California. The shading outlines group areas, and the dashed lines divide subgroups. (B) The $\delta^{13}C$ of saturate hydrocarbons versus Ni/(Ni + V) porphyrins showing the same four groups as in (A). [Peters et al. 1994]

Cross plotting the $\delta^{13}C$ of the saturated hydrocarbons with the Ni/(Ni + V) porphyrin ratio (Figure 15-31 [B]) divided the preceding four groups even more clearly than in Figure 15-31 (A) (Peters et al. 1994).

Correlations using carbon isotope plots are more successful when combined with GCMS analysis of common biomarker suites. Curiale (1994) cites several such case studies in his concise review of the correlation of oils with their source rocks.

Crude oil–source rock correlations can be made by plotting Galimov curves from $\delta^{13}C$ values of oil and rock extract fractions (Galimov 1973; Stahl 1977). An example of such curves for the Beatrice oil of the North Sea and the hydrous pyrolyzate fractions of its possible source rocks is given in Figure 15-32. In a Galimov curve, the saturates with the least ^{13}C are plotted at the top, and the asphaltenes that are most enriched in ^{13}C are plotted at the bottom. In this correlation the fractions of the three pyrolyzates of Middle Devonian source rocks most closely resemble the Beatrice oil. The Middle Jurassic fractions are isotopically heavier than the Beatrice oil, and the Early Devonian fractions are lighter than the oil. The heavier aromatic and polar fractions of Beatrice oil, however, suggest there may be some contribution from the Middle Jurassic.

Good correlations also can be made by fractionating the oils and making carbon isotope profiles of them, as shown in Figure 15-33 (Northam 1985). This method involves first distilling the oil into eleven fractions between 50 and 300°C. The $\delta^{13}C$ of the fractions and the residuum are then plotted and compared for different oils. The shape of the profile and its differences relative to the $\delta^{13}C$ of the whole oil are used in making the correlation. The profiles are particularly useful for detecting small differences between oils that may have similar biomarker fingerprints because they have a single source.

The isotope profiles of the Statfjord oils from the Viking Graben of the North Sea show considerable variation over the 250°C boiling range (Figure 15-33). In contrast, the Ekofisk oil from the Central Graben has a relatively flat profile and is isotopically heavier than the Statfjord oils. Note also that the two oils produced from adjacent but stratigraphically separate sandstones in the Statfjord field have different profiles, indicating that the reservoirs are not in communication. Northam (1985) used several isotope profiles to distinguish oils produced from different sandstones within the Statfjord field. Consequently, such profiles can be used to evaluate reservoir continuity in the same way as the star diagrams previously discussed.

Isotope profiles should be compared between oils that are at about the same maturity level based on biomarkers or other maturity indicators. This is necessary because thermal maturation causes the $\delta^{13}C$ of the oil to become heavier, owing to the loss of lighter carbon by cracking (Sackett 1978).

Chung et al. (1991) used isotope profiles to differentiate Smackover light oils and condensates that formed during thermochemical sulfate reduction (TSR) from those that formed only by thermal maturation (TM). Whereas the TM causes a shift in $\delta^{13}C$ of the original oil of about 2‰, the TSR causes an additional shift of as much as 4‰ toward the heavier carbon. Interpreting isotope profiles requires an understanding of not only thermal maturation but also other factors, such as biodegradation and water washing. Both processes can alter the curves.

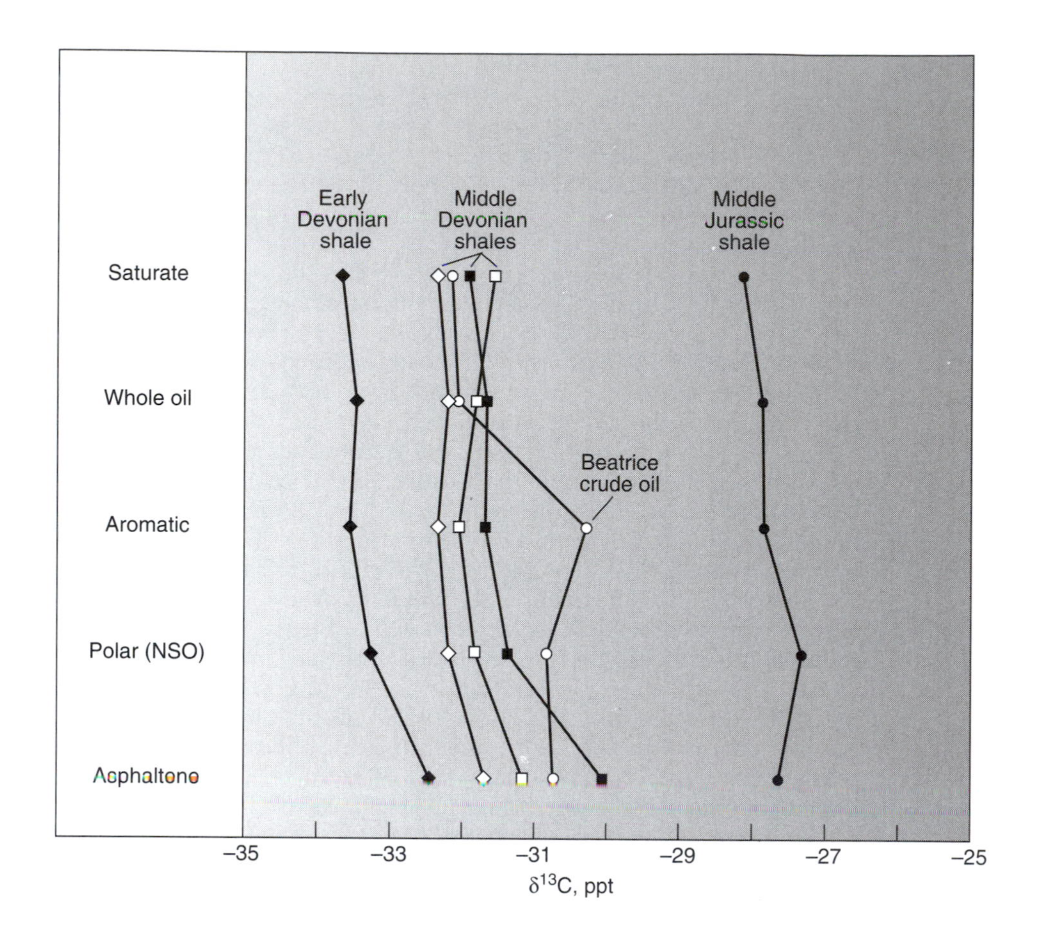

Figure 15-32

Galimov curves of Beatrice crude oil and hydrous pyrolyzate fractions of its potential source rocks. [Bailey et al. 1990]

Chung et al. (1994) published isotope profile data for sixty-nine oils derived from post-Silurian marine source rocks in sixteen petroleum-producing areas worldwide. They initially grouped the oils into marine shale, deltaic, and carbonate-sourced oils. Representatives of these three groups are plotted in Figure 15-34 along with a Stratfjord oil analyzed by Chung et al. The authors found that their profiles of oils from marine shales (e.g., Los Angeles Basin) tended to have gentle negative slopes, starting with Fraction 3 (100°C). Deltaic oils (off-

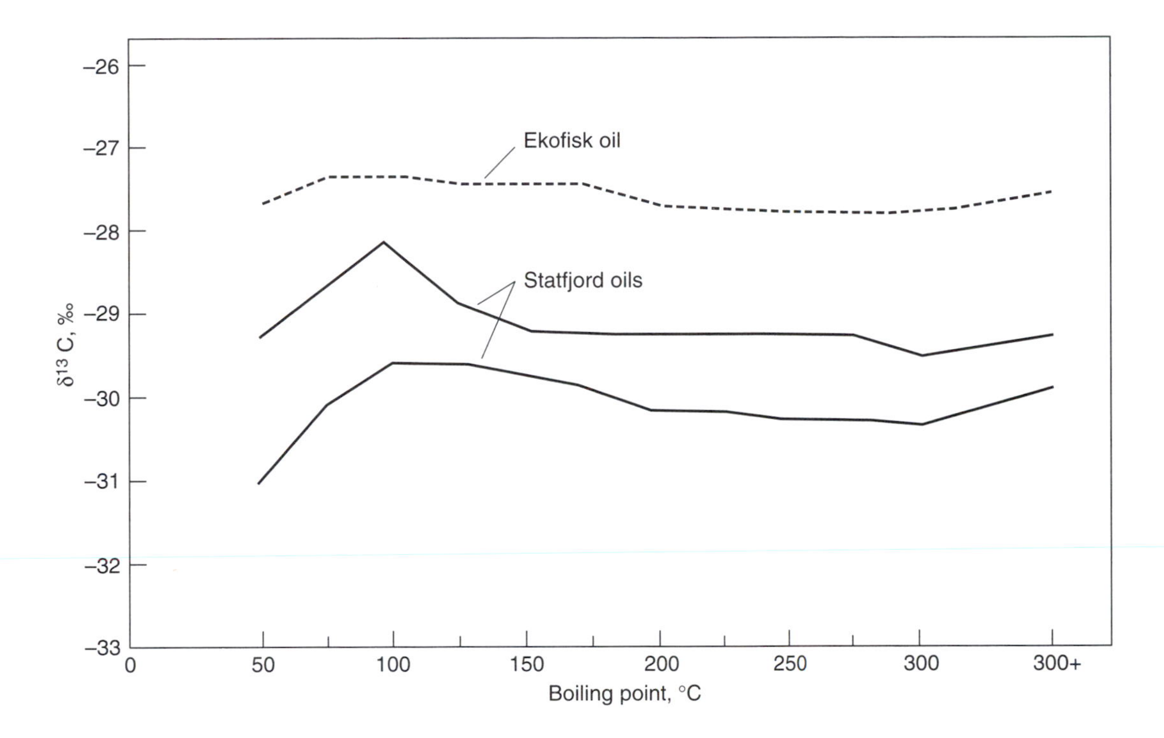

Figure 15-33

Carbon isotope profiles of fractions of oils from Ekofisk and Statfjord fields, North Sea. [Northam 1985]

shore Louisiana) had somewhat steeper negative slopes, and carbonate oils (Saudi Arabia) had gentle positive slopes. Some oils such as the Statfjord oil analyzed by Chung et al. (1994) showed no significant change in slope (Figure 15-34).

Oil–source rock correlations with carbon isotopes are more difficult than oil–oil correlations because differences in maturity may cause isotope variations of more than 3‰. Also, the carbon isotope value of kerogen includes the non-oil-generating kerogen components, whose isotope values may be different from those of the generated oil. Finally, migrated oil or contaminants may alter the isotopic composition of the bitumen extract. The two approaches, oil–oil and oil–source rock correlation, are best viewed as complementary and so both should be used.

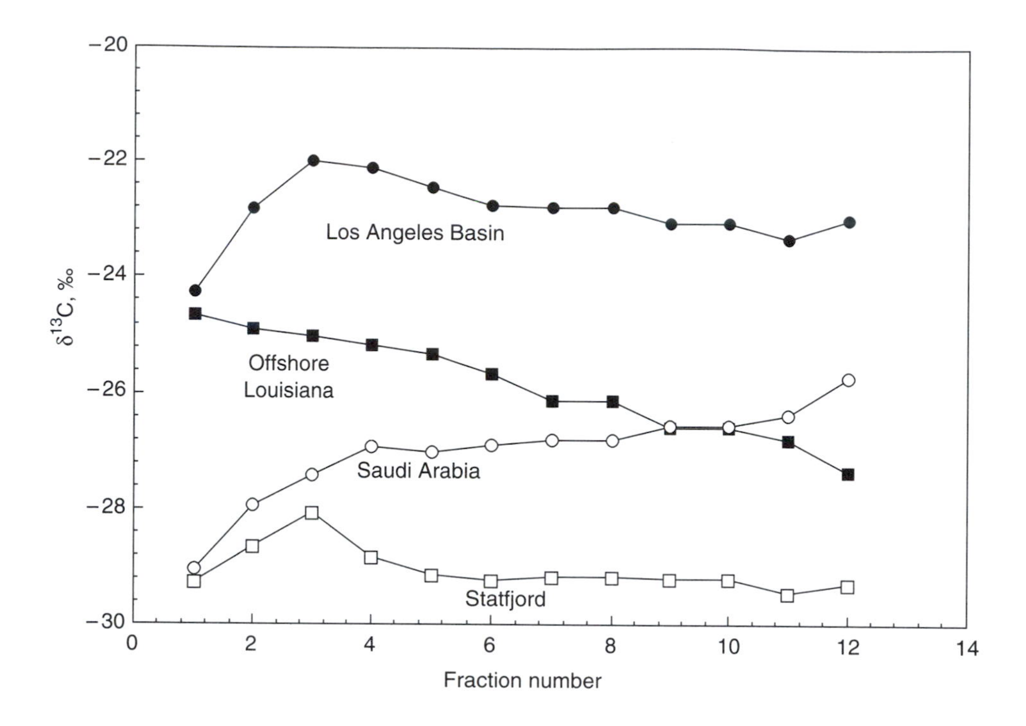

Figure 15-34

Isotope profiles of fractions of oils generated by marine shales of Los Angeles, deltaic shales offshore Louisiana, and carbonate source rocks of Saudi Arabia. A Statfjord oil (North Sea) is shown for comparison. [Chung et al. 1994]

Compound Specific Isotope Analysis (CSIA)

The development of *gas chromatography isotope ratio mass spectrometry* (GC-IRMS) has enabled correlations to be made between the $\delta^{13}C$ values of individual hydrocarbons in condensates, oils, and asphaltene and kerogen pyrolysates. Examples of this technique for distinguishing oils and condensates in the North Sea are described in Bjorøy et al. (1994). These authors compared carbon isotope ratios for selected *n*-alkane, isoalkane, and cycloalkane hydrocarbons. The differences displayed in their curves are due mainly to a combination of source and maturity. Maturity causes an overall enrichment in ^{13}C, as previously stated, but the larger *n*-alkanes ($> C_{10}$), tend to disproportionally lose more ^{12}C than the smaller *n*-alkanes (Northam 1985). This is because the larger *n*-alkanes crack more easily than the smaller ones (see Table 12-4). This effect is much less apparent with the cycloalkanes and aromatics. Consequently, their differences are mainly due to source. Another possible factor is evaporative fractionation, but Bjorøy et al. (1994) found that it did not cause significant isotope changes, at least in the C_4–C_{20} range of oils and condensates.

The $\delta^{13}C$ values for cycloalkanes in three oils and a condensate from the Central Graben of the Southern Norwegian Shelf are shown in Figure 15-35 (Bjorøy et al. 1994). The similar Ekofisk and Albuskjell patterns in this figure are characteristic of oils sourced by the Upper Jurassic Kimmeridge Shale. The Cod condensate with a different pattern is located north of the Greater Ekofisk oil group and is thought to have a different source. The Edda oil pattern is clearly anomalous, indicating a very different kerogen source compared with

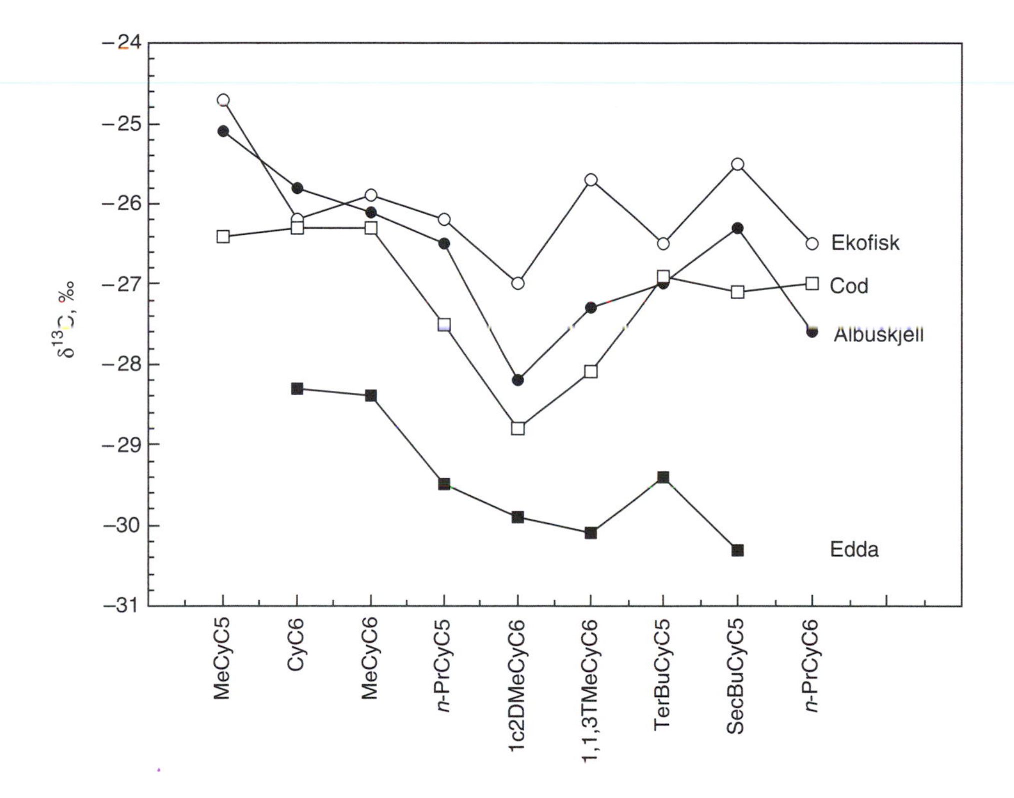

Figure 15-35

Variation in isotopic composition of cycloalkanes in oils from the Ekofisk, Albuskjell, and Edda fields, plus the Cod condensate of the Central Graben, Southern Norwegian Shelf. CyC5 = cyclopentane; CyC6 = cyclohexane, [Bjorøy et al. 1994]

the other Greater Ekofisk oils. Similar conclusions were made from the *n*-alkane and branched-alkane isotope patterns.

McCaffrey et al. (1994) used biomarkers and CSIA to classify 26 Beaufort Sea oils into three genetic groups. Group 1 (21 oils) contains abundant oleanane and 17α- and 17β-24,28-bisnorlupanes, and Group 2 (4 oils) and Group 3 (1 oil) lacked bisnorlupanes and had little or no oleanane. The Group 3 oil differs from Group 2 oils in its sterane and diasterane distributions and in having an *n*-alkane isotope distribution overlapping the Group 1 oils. The *n*-alkanes of Group 1 and 2 oils, however, are clearly separated by CSIA (Figure 15-36).

Variations in the $\delta^{13}C$ of hydrocarbons generated by kerogen pyrolysis provide insight into the components causing $\delta^{13}C$ source differences. Eglinton (1994) flash-pyrolyzed the isolated kerogens of a variety of shales and measured the isotopic composition of their products. Figure 15-37 shows his $\delta^{13}C$

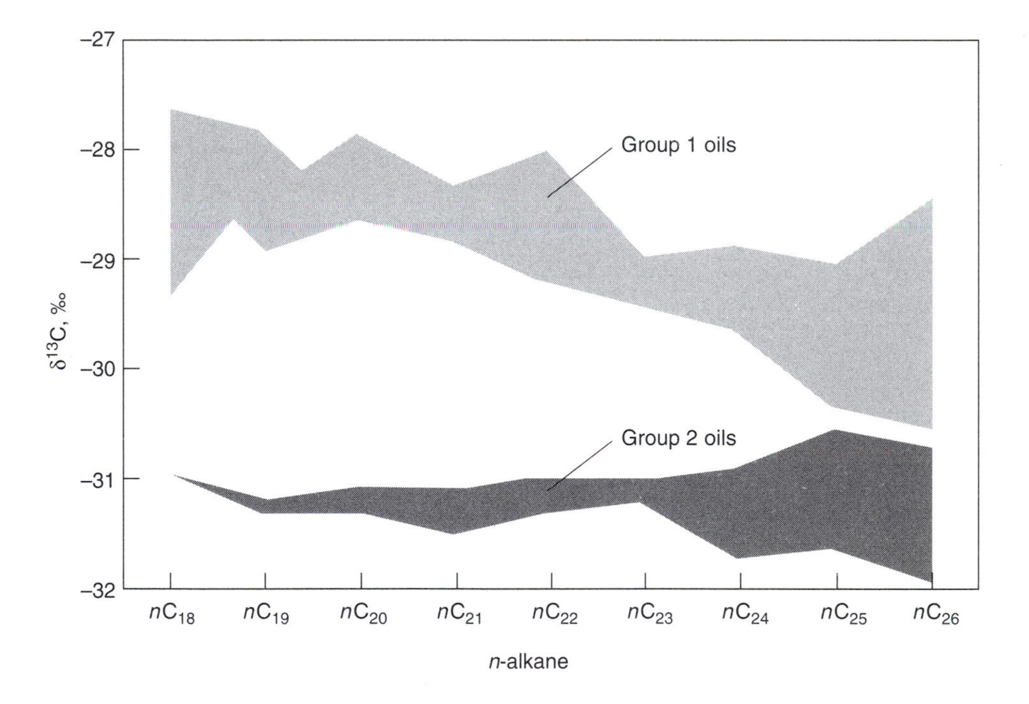

Figure 15-36

CSIA *n*-alkane patterns for two groups of oils from the Mackenzie Delta of Canada. The lower group was deposited under more reducing conditions with an increased marine OM input compared to the upper group. [McCaffrey et al. 1994]

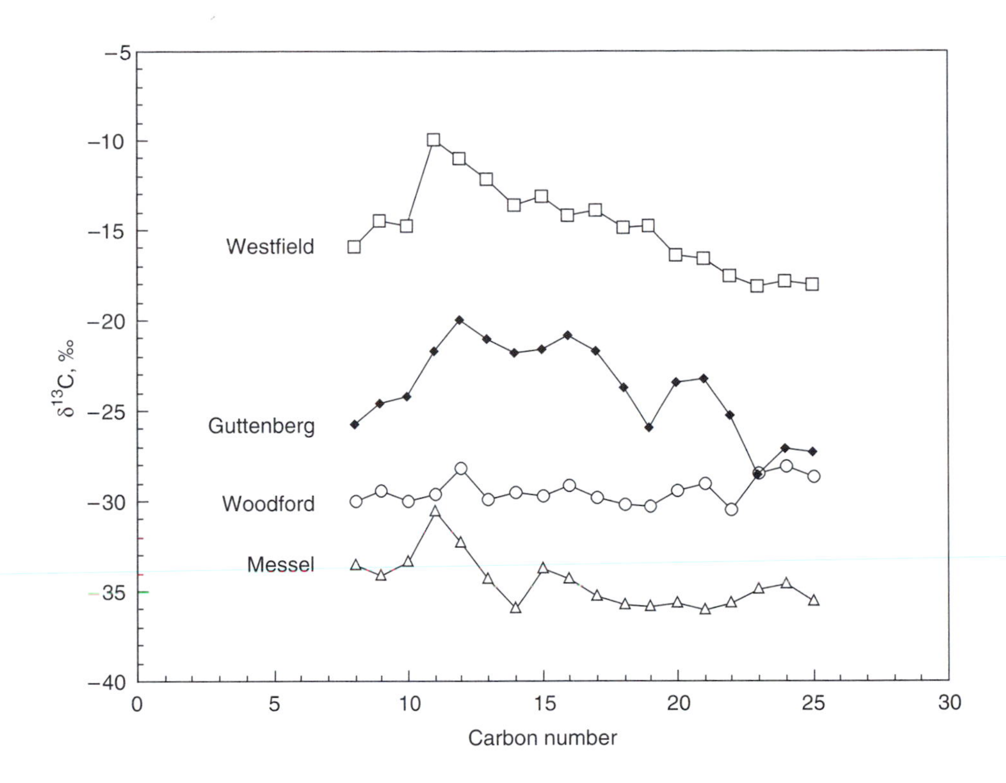

Figure 15 37

Cross plot of $\delta^{13}C$ versus carbon number for *n*-alkane pyrolysis products of kerogens isolated from the Westfield and Messel lacustrine oil shales of Scotland and Germany, respectively, plus the marine source rocks; Woodford Shale of Oklahoma; and Guttenberg Shale of Wisconsin. [Eglinton 1994]

pyrolysis profiles for the *n*-alkanes from two lacustrine oil shales (Westfield and Messel) and two marine source rocks (Guttenberg and Woodford). The $\delta^{13}C$‰ for the bulk TOCs is Westfield –14.3, Guttenberg –24.4, Woodford –29.9, and Messel –27.6. The Westfield kerogen contains mostly *B. braunii,* which may be the cause of the ^{13}C enrichment in this sample. Such enrichment has been observed in botryococcane-related compounds in Indonesian oils (Eglinton 1994). The $\delta^{13}C$ values of the Westfield *n*-alkanes are not uniform as a function of carbon number. This indicates that it has a heterogeneous kerogen composition compared with the more homogeneous composition of biopolymer-enriched kerogens like the Woodford.

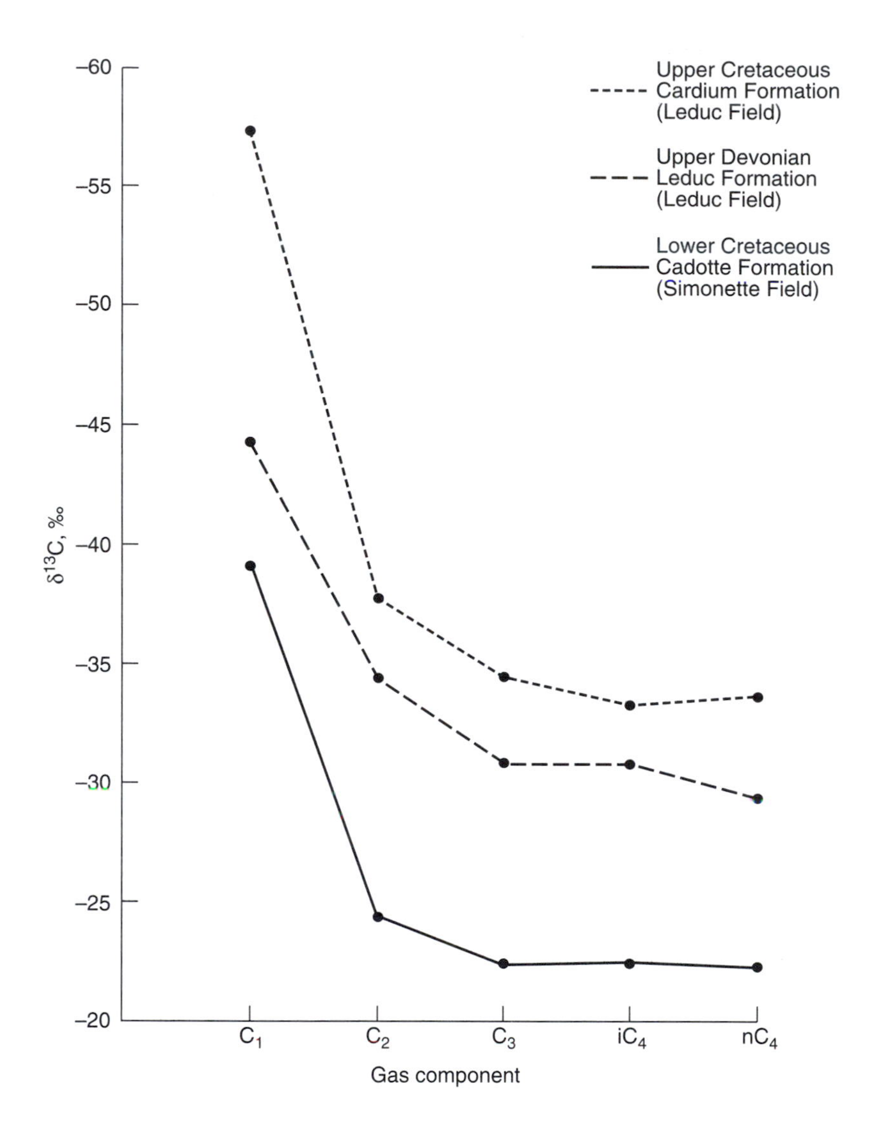

Figure 15-38

Comparison of Western Canada Basin gas families by carbon isotopes. [James 1990]

The relatively uniform $\delta^{13}C$ values of the Messel kerogen *n*-alkanes also indicate a biopolymeric precursor. However, these values average 7‰ lighter than the $\delta^{13}C$ of the TOC, indicating that the biopolymer structure is not a major component of this kerogen (Eglinton 1994).

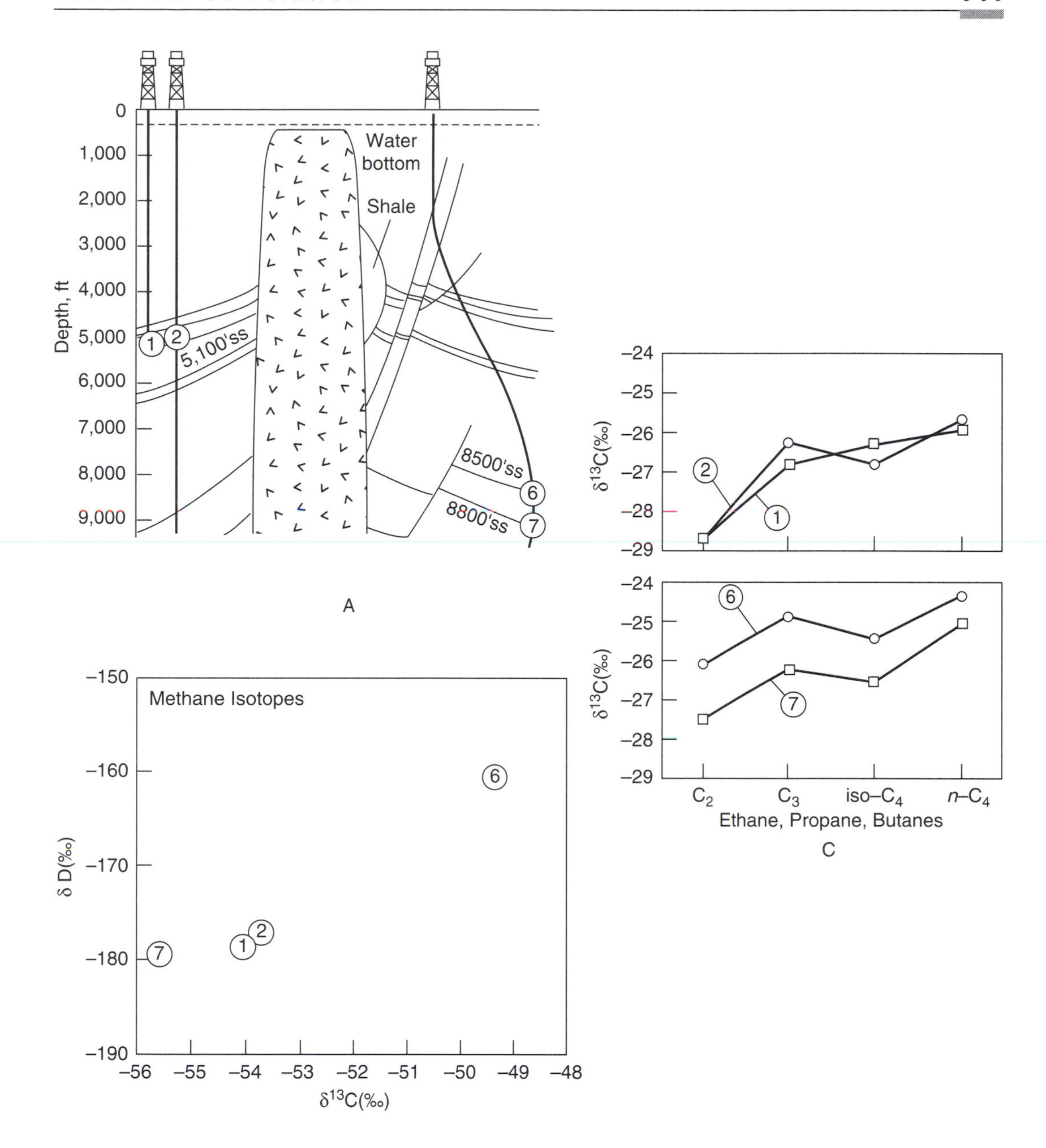

Figure 15-39

Gas isotope variations in reservoirs of a salt dome-related field in the Gulf of Mexico. (A) Location of producing hydrocarbons 1, 2, 6, and 7. (B) Cross plot of carbon and deuterium isotopes of methane in the four gases. (C) Carbon isotopes of individual hydrocarbons in the gases. [Schoell et al. 1993]

Oil–oil correlations with CSIA are becoming routine, but oil–source rock correlations need more research comparing the isotope patterns of kerogen pyrolyzates with those of reservoir oils for known oil–source rock pairs.

Correlating Reservoir Gases

The gas–source rock correlations discussed in Chapter 7 were based on similarities in the maturity of the gas and the rock. Gas–gas correlations can be carried out by plotting $\delta^{13}C$ values for each hydrocarbon component of the gas (James 1990). Figure 15-38 shows three different gas families in the Western Canada Basin based on the $\delta^{13}C$ values of the methane through *n*-butane hydrocarbons. Two of the families produce from different formations in the Leduc field. The differences are attributed to both source and maturity.

Applying Isotope Analyses to Gas Reservoir Problems

Defining reservoir continuity between gas wells is important to gas production operations and to gas storage. In addition, the commingling of gases during production and the possible leakage of gas between closely stacked reservoirs can be monitored with carbon isotopes. Figure 15-39 is an example of how carbon isotopes are used to correlate gases produced from a salt dome–related field in the Gulf of Mexico (Schoell et al. 1993). Figure 15-39 shows (A) the location of the wells, (B) a plot of δD versus $\delta^{13}C$ for the four gases produced. Note that gases 1 and 2 producing from different locations in the same sandstone are identical, whereas gases 6 and 7 producing from two reservoirs 300 ft (91 m) apart are very different. These relationships are further verified by the $\delta^{13}C$ values of the individual gas components in Figure 15-39 (C). Gases 1 and 2 are essentially identical, and gases 6 and 7 are isotopically distinct; that is, there is no communication between gases 6 and 7.

Figure 15-40 illustrates how gas isotopes can assist in mapping reservoir continuity in faulted reservoirs (Schoell et al. 1993). Similar pressure regimes and oil–water contacts are generally used to define the continuity of reservoirs across faults. Small differences, however, are more difficult to map from well data than are isotope analyses of the gases. Figure 15-40 shows a map of three fault blocks (Q, B, C) and five wells from which gas samples were collected. A cross plot of δD versus $\delta^{13}C$ for the five gases is shown below the map. Note that there is a similarity between samples 1 through 4, but sample 5 is isotopically distinct. This means that there is reservoir continuity across faults between the first four samples, but fault block C is probably split into two isolated compartments separating gas samples 1 and 5. It implies a southwestern extension of the fault north of well 5. Combining these results with the engineering data (pressures, water salinities, etc.) is the best way to map fluid-flow patterns in the field.

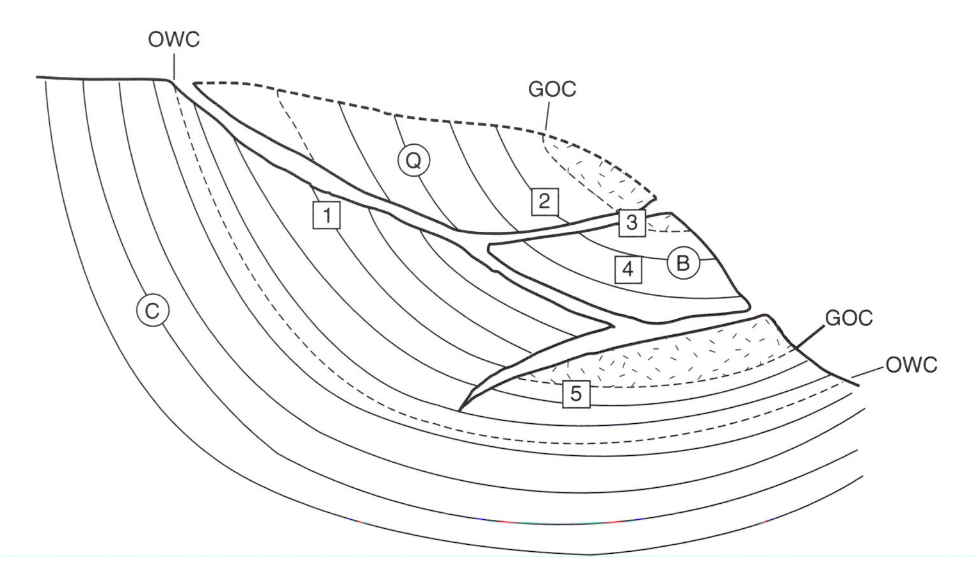

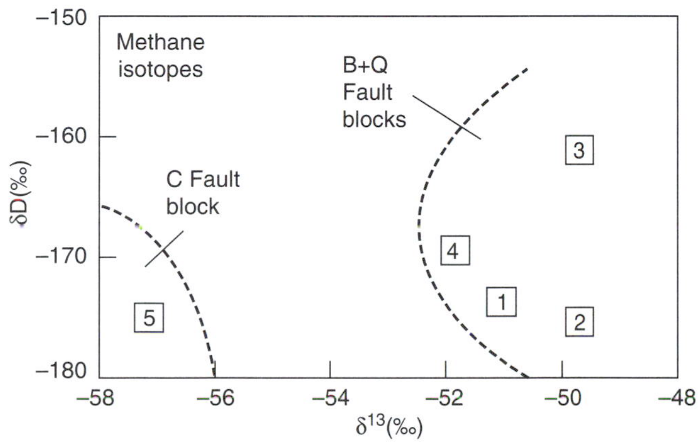

Figure 15-40

Mapping reservoir continuity between fault blocks with gas isotopes. Methane isotope variations in a faulted reservoir sandstone, Gulf of Mexico. GOC = gas–oil contact, OWC = oil–water contact. [Schoell et al. 1993]

Sulfur Isotope Ratios

Sulfur isotope ratios have not been applied to correlation problems as widely as carbon isotopes have, for various reasons, such as the alteration of the original $\delta^{34}S$ of oils by secondary processes (Orr 1974). Nevertheless, sulfur isotope ratios have helped solve some correlation problems. For example, Thode (1981) was able to classify thirty-eight oils from the Williston Basin of North Dakota and Saskatchewan into three major types by their sulfur isotope ratios. The three types correspond to separate source rocks of Ordovician, Mississippian (Bakken), and Pennsylvanian (Tyler) age, as shown in Figure 15-41. The $\delta^{34}S$ of the oils is spread over a range from –5 to +8‰. Thode found no significant change in $\delta^{34}S$ of the oils with migration, although high-temperature maturation and biodegradation can lead to substantial changes in the $\delta^{34}S$.

Bordenave and Burwood (1990) used a cross plot of $\delta^{13}C$ versus $\delta^{34}S$ to separate two groups of Iranian oils. One group generated by the Albian Kazhdumi source rock has $\delta^{34}S$ values of about –9 to +2‰ and $\delta^{13}C$ values more negative than –26‰. The second group formed by the Paleogene Pabdeh source rock has $\delta^{34}S$ values from +3 to +12‰. Its $\delta^{13}C$s are more positive than –26‰.

Figure 15-42 compares the $\delta^{34}S$ in crude oils of the Santa Maria Basin of California with the $\delta^{34}S$ of organic sulfur from potential source rocks of the area (Orr 1986). Also shown is the weight percent organic sulfur in the kerogen of the rocks. Note that the $\delta^{34}S$ of this sulfur increases with depth in the Pliocene, to a relatively constant value of about 0‰ through the upper part of the Miocene Monterey Formation. The organic sulfur of the kerogen in this

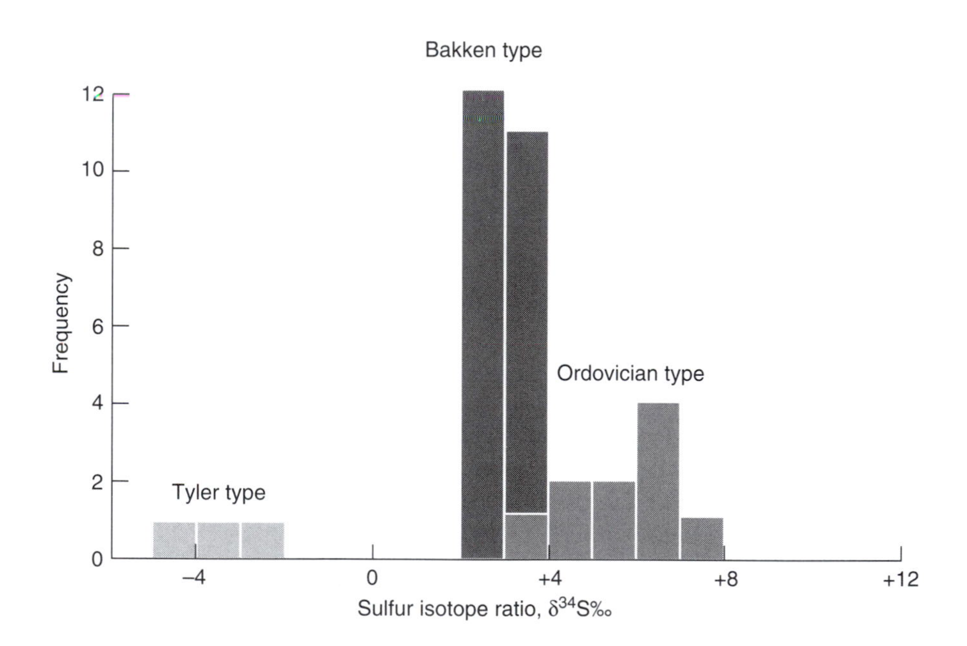

Figure 15-41

A $\delta^{34}S$ histogram defines three major oil types in the Williston Basin. [Thode 1981]

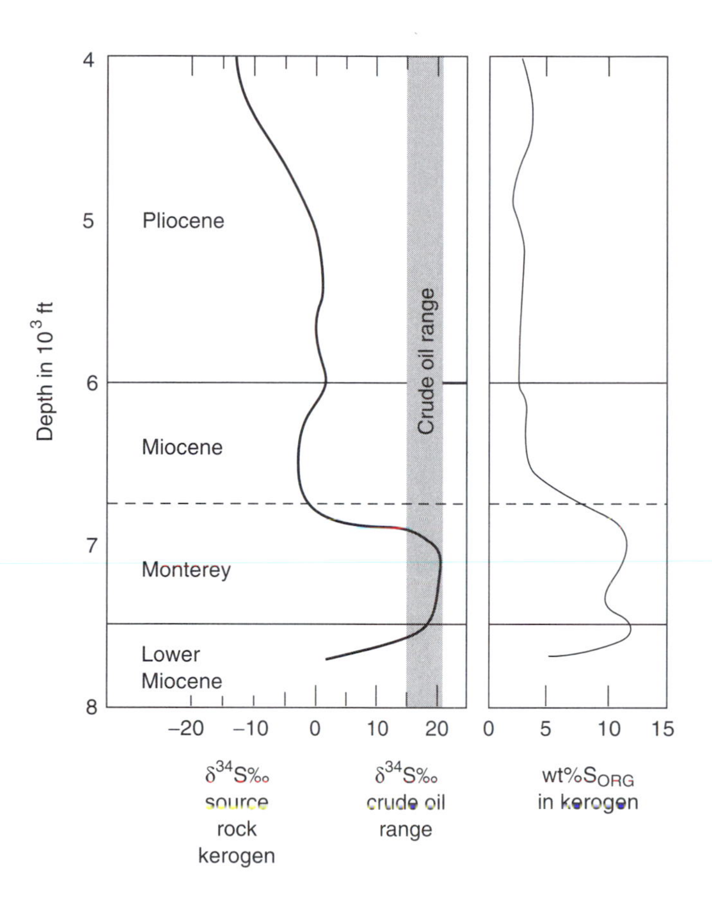

Figure 15-42

Comparison of $\delta^{34}S$ in crude oils and potential source rocks with the weight percent organic sulfur in the kerogen of the rocks of the Santa Maria Basin of California. [Orr 1986]

interval is about 4%. In the lower part of the Monterey this increases to around 12%, and the $\delta^{34}S$ of the kerogen sulfur goes to +20‰, which is equivalent to that of the crude oils. This means that the heavy immature Monterey oils of this area are being generated primarily by the Lower Monterey Shale section, in which the organic sulfur of the kerogen is >10%. This high sulfur content is characteristic of type IIA (II-S) kerogen. This is further evidence that the high-sulfur kerogens are the prime generators of immature, heavy oils, as noted in Chapter 6.

SUMMARY

1. A major objective of correlation procedures is to recognize the source fingerprint of the hydrocarbon molecules in oils, condensates, gases, and asphalts and also to understand how the source imprint is affected by factors such as migration, water washing, biodegradation, and thermal alteration and phase changes like evaporative fractionation.

2. Bulk parameters used in oil-to-oil correlation include API gravity; pour point; percent sulfur, nitrogen, vanadium, and nickel; stable isotopes of carbon, sulfur, and hydrogen; and whole-oil chromatograms.

3. More specific correlation parameters include separation of oils into saturate and aromatic hydrocarbons and NSO compounds; subdivision of saturates into straight-chain, branched-chain, and cyclic hydrocarbons in narrow boiling ranges such as C_4–C_7; stable isotopes of oil fractions; and ratios of individual hydrocarbons such as Pr/C_{17}, Ph/C_{18}, toluene/heptane, and genetically related C_7 branched and cyclic alkanes.

4. Correlations using multiple parameters are enhanced by the use of cross plots and ternary diagrams. Oil–oil correlations designed to interpret reservoir continuity can be made by matching whole-oil GC fingerprints. Subtle differences can be quantified by constructing star diagrams and dendrograms based on a series of hydrocarbon peak ratios in specific ranges such as C_9 to C_{20}. This is a good approach when comparing a group of oils that have the same biomarkers because of similar histories of origin and migration.

5. The most widely used correlation parameters for comparing oils of different source, maturation, migration, and biodegradation, for both oil–oil and oil–source rock correlations are the biomarkers. They also are the most useful method for comparing conventional oils with very heavy oils, asphalts, and pyrobitumens. The more successful correlations involve a multiparameter approach using biomarkers; isotopes; gas chromatograms; and the content of heavy metals, nitrogen, sulfur, and the like.

6. Cross plots and ternary diagrams of genetically related C_7 alkanes are useful in both oil–oil and oil–source rock correlations. They are particularly valuable in evaluating high-gravity oils and condensates that may lack biomarkers.

7. Oil–source rock correlations are generally more difficult than oil–oil correlations because rock extracts are not necessarily comparable to the expelled oils in molecular-weight distribution and in the level of thermal maturation and biodegradation. Oil fractionates during migration from the source to the reservoir rock. The larger, more polar molecules are preferentially left behind in the source rock.

8. Good correlations of oils with their source rocks involve comparing mass fragmentograms of steranes and terpanes and using as correlation parameters

individual biomarker ratios such as the homohopane, oleanane, and gammacerane indices and the tricyclic terpanes. Ternary plots of the normal steranes, diasteranes, methylsteranes, C_{30}-steranes, MA steroids, and TA steroids also are useful for correlation. Carbon numbers of the biomarker molecules are used in correlation, and the stereoisomers and ratios of compounds showing different stabilities are used for comparing levels of maturation.

9. Good oil-to-oil correlations can be made using cross plots of carbon and hydrogen isotopes and by constructing carbon isotope profiles of crude oil distillation fractions. The latter are particularly useful in defining small differences in oils, such as in reservoir continuity studies. Such profiles also can distinguish between influences from thermal maturation and those from thermochemical sulfate reduction.

10. A $\delta^{13}C$ analysis of individual alkane molecules (CSIA) is being used in oil–oil and oil–condensate correlations. Similar analyses of source rock kerogen and bitumen pyrolyzates show promise for oil–source rock correlation by CSIA.

11. Gas-to-gas correlations are made by using cross plots of δD versus $\delta^{13}C$ and plots of $\delta^{13}C$ for individual gases C_2 through n-C_4. Such plots are used for evaluating reservoir continuity, fault block mapping, and intervals of commingled production.

12. Cross plots of $\delta^{34}S$ versus $\delta^{13}C$ and the direct comparison of $\delta^{34}S$ in crude oils and in the organic sulfur of presumed source rocks represent additional correlation parameters that are particularly useful with high-sulfur oils.

SUPPLEMENTARY READING

Moldowan, J. M., P. Albrecht, and R. P. Philp (eds.). 1992. *Biological markers in sediments and petroleum.* Englewood Cliffs, NJ: Prentice-Hall, 411 p.

Peters, K. E., and J. M. Moldowan. 1993. *The biomarker guide: Interpreting molecular fossils in petroleum and ancient sediments.* Englewood Cliffs, NJ: Prentice-Hall, 363 p.

Yen, T. F., and J. M. Moldowan (eds.). 1988. *Geochemical biomarkers.* New York: Harwood Academic Press, 438 p.

Chapter

16

Prospect Evaluation

In 1974 the Lone Star Producing Company's Bertha Rogers well in the Anadarko Basin of Oklahoma was plugged and abandoned at a depth of 31,343 ft (9,558 m), at a cost of about $7 million. A high-pressure pool of liquid sulfur was discovered in the Arbuckle Formation, and the last samples to reach the surface were bright yellow sulfur crystals (Wroblewski 1975). Ten years later, the most expensive dry hole in the history of petroleum, Sohio's Mukluk 1, was drilled in the Beaufort Sea on a gravel island built for $100 million. Sohio and eleven partners spent a total of $140 million to test the Sadlerochit Sandstone and Lisburne Limestone, the main producers in the Prudhoe Bay field. Extensive oil stains were found in the Sadlerochit and a minor amount of 12°API oil was produced on a test, but it appeared that any trapped oil had escaped long ago through a breach in the Pebble Shale seal (Boyd and Hiles 1985).

Today only 1 in about 50 onshore wildcats in the United States finds a field larger than 1 million bbl of producible oil. More than 400 dry holes were drilled in the Western Canada Basin before the first commercial field, Leduc, was discovered. In the North Sea, more than 100 dry holes were drilled before Ekofisk was discovered, and in the western overthrust belt of the United States, more than 500 were drilled before production was found. Obviously, any geochemical application that can reduce this dry hole/discovery ratio will pay for itself many times over. Using established geochemical methods in prospect evaluation should therefore be part of every prospecting procedure.

There are some 900 sedimentary basins around the world, about 600 of which may contain petroleum. Of these, 160 are commercially productive; 240 have been partially explored; and 200 are still unexplored (Halbouty 1986). Most of the unexplored basins are offshore (Hedberg et al. 1979), where larger fields and highly productive reservoirs and wells are required to justify the costs of development.

The major objective of exploration geochemistry is to help reduce the risk of drilling unnecessary dry holes. In a well-run exploration division, every person proposing a wildcat drill site is asked the pertinent geochemical questions, along with those concerning the geology and geophysics of the prospect. Where are the source rocks? Have outcrops or well samples of the source rocks been analyzed? What is their maturity? Have they generated enough hydrocarbons to fill the trap being considered? Where are the migration pathways to the structure? What are odds that the structure contains oil or condensate or dry gas or CO_2 or water? How good is the seal? Are hydrocarbons currently escaping to the surface? Offshore, can you see gas chimneys over the structure on the seismic profiles? Are there any seeps in the area?

Such questions can be only partly answered by putting numbers into a basin-modeling program. There also is a judgment factor here that requires a reasonable knowledge of petroleum geochemistry and some operating experience in using it. One of the world's most eminent petroleum geologists, Wallace E. Pratt, had remarkable insight concerning the principles of petroleum geochemistry, such as the changes in oil composition with maturation and the natural cracking of oil to gas in the earth. He used these concepts in his exploration thinking (Pratt 1943, pp. 9, 10, 17, 18). His University of Kansas lecture, "Where Oil Is," ended with the statement that "oil must be sought first of all in our minds. Where oil really is, then, in the final analysis, is in our own heads" (Pratt 1943, p. 52). He meant that the decision of where to drill still involves someone whose mind can and must intuitively put together all the data available from whatever source—geological, geophysical, or geochemical—to reach a sound decision.

If a company has no one with a strong interest and training in geochemistry, it should use one of the more reputable service laboratories to work closely with its operating personnel. Geochemists and geologists should always work together. There are numerous examples of such cooperation reducing risk. In Chapter 1 the experience of Shell International was cited, in which the forecasting efficiency that used geochemistry and geology to rank 165 prospects was 63%, compared with only 18% if the prospects had been ranked by trap size only (Murris 1984). Chapter 14 used the example of Imperial Oil's geochemists and palynologists predicting that the Rainbow area of northern Alberta would be oil prone and to the west in British Columbia it would be gas prone. The discovery of a major oil field confirmed this prediction. In Chapter 15 it was mentioned that the source rocks of the Esperito Santo Basin offshore Brazil were initially thought to be the Tertiary slope sediments. Subsequent pyrolysis and biomarker analysis caused Estrella et al. to state, "These early conclusions had to be radically revised and furthermore, previously unsuspected source systems became clearly identified" (1984, p. 253). This changed the direction of the exploration targets and the campaign.

A serious problem in geochemistry that Demaison (1986) recognized is the tendency of some geochemists to write up overly optimistic reports in order to support proposals for certain drilling sites. This goes back to the days when the field geologist might send in only the blackest outcrop sample of a presumed source rock section for analysis. Today, geochemists must objectively evaluate the entire sedimentary section under consideration and report to the best of their ability exactly what the data indicate. Otherwise, in the long run, they will lose credibility with management. Some examples of recognizing the probability of dry holes are mentioned here.

The importance of prospecting in areas alive with hydrocarbons was emphasized in Chapter 14. An example was cited in which shale gas yields from mud logs in the Tuscaloosa Formation were found to be high in Louisiana and to decrease going east to almost zero in Georgia. Dissolved gas in formation waters (Figure 14-4) also decreased going east. This indicated that the Tuscaloosa Formation in the east is dead with respect to hydrocarbons. Subsequently, three large structures were tested in Georgia and were abandoned as dry holes, at a cost of about $3 million. There was no gas on any of the mud logs.

W. G. Dow of DGSI once mentioned a client who had drilled three dry holes in a shallow basin where there were large structures and good porosity but no shows. The client wanted to know what was wrong. Geochemistry showed the client that there were no source beds and a low thermal maturity and that migration was unlikely from adjacent basins, where prospects were more promising. The client was committed to drill three more wells, so Dow recommended farming these out to other parties, as discoveries were extremely unlikely.

In Chapter 13 it was stated that Gulf Oil was discouraged from making a high bid on the Destin Dome, offshore northwest Florida, partly because geochemical studies of the entire area had shown no evidence of hydrocarbons. Ultimately, other companies drilled fifteen dry holes without any significant hydrocarbon shows appearing on the mud logs. All these examples show that prospect evaluation is not just drilling the biggest structures but, instead, making a rational decision after analyzing the risks for each structure based on all the available geochemical, geological, and geophysical data. A way to bring together such data involves the petroleum systems concept.

The Petroleum System

A petroleum system includes all those geological elements and processes that are essential for an oil and gas deposit to exist in nature (Magoon 1988). These basic elements are a petroleum source rock, migration paths, reservoir rocks, seals, traps, and the geologic processes that created each of them. Such a system implies a genetic relationship between the source rock and the petroleum accumulations, but proof of that relation requires a geochemical correlation. One sedimentary basin may have several petroleum systems.

The concept of a petroleum system was first used in describing three major source–reservoir oil systems in the Williston Basin of the United States and Canada (Dow 1974). Over the next twenty years the petroleum system was described in different ways, finally culminating in a specific set of definitions and characteristics (for historical details, see Magoon and Dow 1994).

The term *petroleum* in the context of a system includes high concentrations of either thermal or biogenic gas, condensates, crude oils, natural waxes, or asphalts found in both conventional and tight reservoirs, fractured shales, gas hydrates, coal beds, and bituminous sandstones.

The name of a petroleum system is a combination of the source and reservoir rock names followed by a symbol indicating the level of certainty of the source–reservoir rock association. The symbols are (!) for known, (.) for hypothetical, and (?) for speculative systems (Magoon and Dow 1994). Examples of known systems are the Woodford–Misener (!) of Oklahoma (Figure 15-19), the Phosphoria–Tensleep (!) of the Rocky Mountain Overthrust Belt, United States (Figure 15-25), and the La Luna–Misoa (!) of the Maracaibo Basin, Venezuela (Plate 2D).

Petroleum systems are best described by (1) a table listing all the field names, year of discovery, name of producing unit, depth, reservoir seal lithology, trap type, cumulative production, and reserves; (2) a burial history chart at one or more locations; (3) a map showing the geographic extent of the system; (4) a geologic cross section drawn at the critical moment illustrating the spatial relationships of the essential elements; and (5) an events chart indicating the time intervals of the essential elements and processes, including the preservation time and the critical moment (Magoon and Dow 1994). The *critical moment* is the time when most of the petroleum in the system is forming and accumulating in its primary trap.

For example, Talukdar and Marcano (1994) prepared a table listing thirty-five oil fields with estimated reserves plus cumulative oil production totaling 51 billion barrels from the La Luna–Misoa (!) petroleum system of the Maracaibo Basin, Venezuela. Along with the table they published a burial history chart for a field in the center of Lake Maracaibo, a map, a cross section, and an events chart. Maps for this basin are shown in Plates 2B, 2C, and 2D. Two phases of oil generation occurred at different times in different parts of the basin. During the Late Eocene (Phase 1, 40 Ma), oil generation from the La Luna source rock was restricted to the elongate area now shown in red in Plate 2B. Much of this early oil is thought to have been destroyed on uplift. Any remaining oil became heavily biodegraded in the last 35 m.y. Consequently, most of the conventional petroleum in the basin today is considered to have formed and accumulated during Phase 2 in the last 12 m.y. Figure 16-1 (A) shows the events chart for Phase 2 (Talukdar and Marcano 1994), and (B) gives the burial history curve. The critical moment for the La Luna is today. The La Luna is actively generating oil and gas over most of the Maracaibo Basin.

The *generation-accumulation efficiency* (*GAE*) of a petroleum system is the quantity of in-place hydrocarbons plus accumulated production expressed as a percentage of the total petroleum generated. Talukdar and Marcano (1994) calculated a GAE of 14% for the La Luna–Misoa petroleum system. This is at

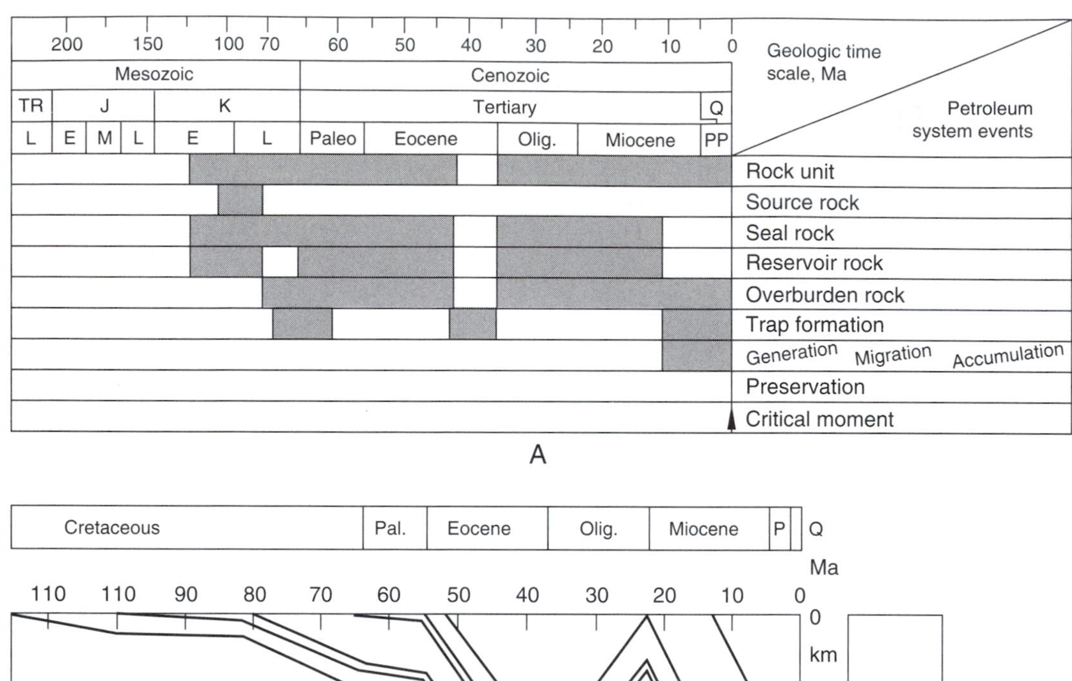

Figure 16-1

(A) Events chart for Phase 2 of the La Luna–Misoa (!) petroleum system; (B) burial history curve for the La Luna source rock in the Centro oil field in the Maracaibo Basin, Venezuela. [Talukdar and Marcano 1994]

the high end of the range of published values (Table 16-1) for an entire system, although it is in the center of the range for individual accumulations on the Southern Norwegian Shelf, as explained later.

Mass Balance

How does one determine whether there are enough mature source rocks and a suitable migration pathway to fill a particular structure with petroleum? Back in 1936, Trask estimated that the oil in the Santa Fe Springs field in the Los Angeles Basin originated from 4% of the organic matter (OM) in its source rock. Subsequent studies indicated that the percentage of reservoired oil in place relative to the oil generated in many basins ranges from a few percentage points to about 15% (Hunt 1979, p. 263).

Mass-balance assessments are made on a play, basin, or global scale. Table 16-1 lists several of these assessments that have been published or quoted since 1965. Good examples of mass-balance assessments are as follows: for a play, Leonard 1988; for a basin, Lewan et al., 1995; and for a global analysis for an entire period (Upper Jurassic), Klemme 1994. Klemme published a value of 0.86 for the ratio of recoverable oil to generated oil in the Hanifa–Arab system, so this was tripled to 2.6 to obtain the ratio of oil in place to generated oil for

TABLE 16-1 Estimated Percentages of Original Reservoired Oil to Generated Oil or to Expelled Oil in Various Basins Worldwide

Area	*Percent reservoired oil in place/ Generated oil*	*In place/ Expelled oil*
Arabian/Iranian Basin[a]	10	
Southern Norwegian Shelf[b]		8
Illinois Basin[c]		6
Surat Basin, Australia[d]	5	
Phosphoria Formation, western United States[e]	4	
Cretaceous, Powder River Basin, Wyoming[f]	4	
Williston Basin, North Dakota[g]	3	
Permian Basin of west Texas and New Mexico[a]	3	
Hanifa–Arab System, Arabian/Iranian Basin[h]	2.6	
West Siberian Basin[a]	2	

Sources: [a]McDowell 1975, [b]Leonard 1988, [c]Lewan et al., 1995, [d]Coneybeare 1965, [e]Claypool et al. 1978, [f]Table 10-2, [g]Webster 1984, [h]Klemme 1994.

Table 16-1. This table reaffirms that the overall process from generation to accumulation is highly inefficient. Consequently, there is a minimum amount of oil that must be generated in order to allow for all these losses. This minimum decreases in the more efficient systems.

The percentage of reservoir oil in place relative to the oil expelled from the source rock is always higher than the percentage in place relative to the oil generated, since some oil never leaves the source rock. Leonard (1988) calculated the first percentage in a localized area of the Southern Norwegian shelf. He showed that the percentage of oil in place to oil expelled for eleven fields in the area ranged between 1 and 31%, with an average of 8%, as shown in Table 16-1.

Many factors contribute to the differences listed in Table 16-1. The main one is how the volume of source rock is calculated. For example, in the Hanifa–Arab system in Table 16-1, if the mass balance is limited to the area from Ghawar to Qatar, the percentage in Table 16-1 will rise from 2.6 to more than 15. Large drainage areas with lateral long-distance migration pathways have lower overall efficiencies than small localized drainage areas expelling hydrocarbons vertically into overlying structures or laterally into interbedded sandstones. Mass-balance calculations require more assumptions in large drainage areas because of the multiplicity and nonuniformity of source rocks. Lateral variations in organic facies, TOC, heating rates, mineralogy, maturity, and permeability result in rough estimates, at best, for average generation and expulsion efficiencies. The problem is augmented by the fact that the geometry of the petroleum system that exists today is not necessarily what existed when the oil migrated.

Some people might say that the numbers in Table 16-1 indicate that there is plenty of oil to fill all structures, so why study source rocks? If so, why are so many structures dry? The problem is not just the amount of oil generated but also the amount expelled and channeled into a narrow pathway leading to the structure. Most petroleum tends to move in different directions after expulsion, with only a small percentage of it reaching a specific trap. Losses to the surface and migration into nonreservoir rocks appear to far exceed the quantity reaching producible reservoirs. The quantity generated and expelled, the migration pathway, and the effectiveness of the trap, seal, and preservation all must be favorable to fill a structure. This is why the close integration of geology and geochemistry is essential to any exploration effort.

Petroleum Generated, Expelled, Trapped, and Lost

The formation of oil and gas in the natural environment was described in Chap ters 4, 5, and 10. Figure 16-2 is a flow diagram on a global scale showing what happens to these products as source rocks mature with deeper burial. The numbers in these boxes are speculative but the ratios of the numbers to each other are reasonable based on what is known concerning the generation, retention, expulsion, migration, and accumulation of oil and gas. The term *oil* in the

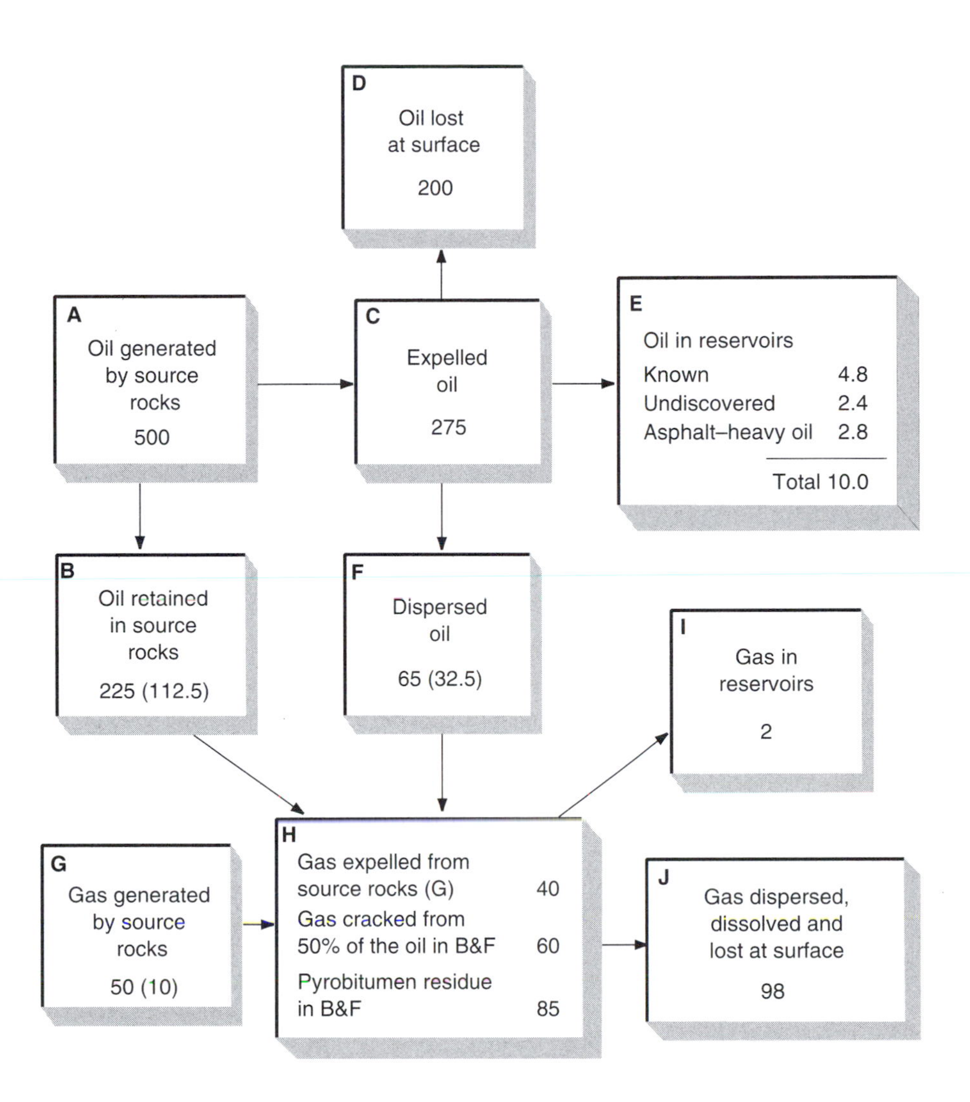

Figure 16-2

Flow chart showing the ultimate sinks for an estimated 550×10^{12} BOE of oil and gas generated in the last 100 million years (Boxes A plus G). The numbers shown are the original BOE for each box. The numbers in parentheses in Boxes B and F represent the residual oil after 50% of the oil cracks to gas and the gas moves to Box H. The number in parenthesis in Box G is the residual gas left adsorbed on the source rocks after primary migration. The 550×10^{12} BOE generated is estimated to be distributed at present as follows: B(112.5), D(200), E(10), F(32.5), G(10), H(85), I(2), and J(98).

figure includes all liquids, light oil, condensate, and heavy oil. The boxes are lettered alphabetically to identify the various steps in the process. The numbers represent the estimated distribution of oil, gas, and pyrobitumen in 10^{12} (trillion) barrels of oil equivalent (BOE) for the last 100 million years (m.y.). About 80% of the oil currently accounted for was generated during this period (Klemme and Ulmishek 1991). The source of these numbers is discussed in the next three sections.

Note particularly that this analysis separates the oil generated (Box A) from the oil expelled from the source rock (Box B). This is an important distinction that is sometimes overlooked in published mass-balance estimates. In going from the source rock to the reservoir (Box A to Box E), there is a loss of 98% of the generated products, although not quite in the way that is sometimes depicted. This loss is distributed over Boxes B, D, F, H, and J in Figure 16-2. About 58% of this loss involves oil retained in both the generative and nongenerative portions of the source rock as well as dispersed in nonreservoir rocks along migration pathways and in tight nonproductive reservoir rocks (Boxes B and F).

Petrographic examinations of source rocks during and after oil generation have shown refractory bitumen and pyrobitumen to be common constituents of good source rocks (Comer and Hinch 1987; Lewan 1987; Littke et al. 1988; Talukdar et al. 1986). Polygonal fracture networks containing bitumen in fractures with widths $<10\ \mu$ remain in rocks that have already expelled considerable oil. The gradual disappearance of the oil-generation window based on extraction data (Figures 5-10 to 5-15) does not mean that all the oil has been expelled but, rather, that part of it has been converted to an insoluble pyrobitumen in the source rock.

In Chapter 8 it was mentioned that a significant amount of the total porosity of shale source rocks in the oil window is in micropores and ultramicropores less than 2 nm in width (Hall et al. 1986). Any generated bitumen filling such pores would have great difficulty migrating out unless gas generation fractured the rock.

The widespread occurrence of bitumen-filled joints, faults, fractures, lineaments, and bedding planes was described in the section "Migration Pathways" in Chapter 8. Olson (1982) observed bitumen in the nongenerating portions of the Woodford Shale source rock of Oklahoma. It had migrated from the generative portion (see Plate 4D). The importance of exploring in an area alive with hydrocarbons, such as shale gas on mud logs and oil shows on DSTs was emphasized in Chapter 14. This vast amount of dispersed hydrocarbons in a petroleum-generating basin reemphasizes that much of the generated oil and gas never makes it into a migration channel leading to a structural or stratigraphic trap large enough to be a commercial reservoir.

Using Pyrolysis for Mass Balance

There is no laboratory process that can measure quantitatively the amount of oil and gas that moves in nature from one part of a rock to another, as depicted in Figure 16-2. Reasonable mass-balance numbers can be obtained for the hydrocarbons in reservoirs (Box E). Demaison and Huizinga's (1991) source potential index (SPI) gives the estimated total quantity of oil and gas generated by the source rock (Boxes A and G), but not all of this is expelled. The expelled oil and gas (Boxes C and H) can be determined by hydrous pyrolysis of the source rocks. The numbers for the other boxes in this figure are best estimates based on published reports and field observations.

The two most commonly used laboratory methods for determining the quantity of hydrocarbons generated by the source rock are open, anhydrous pyrolysis (Rock-Eval) and closed, hydrous pyrolysis. In Chapter 10 it was stated that Rock-Eval pyrolysis provides the best method for comparing relative hydrocarbon yields of source rocks. But these yields cannot be used in mass-balance calculations as expelled oil because they include all the oil and bitumen that are not expelled in nature (Box B). This makes it appear as though generation and expulsion are occurring at high efficiencies.

The main reason for this apparent increase in yield is that the $S_1 + S_2$ that represents the generated hydrocarbon includes all the bitumen, oil, and gas in Boxes B, C, and G of Figure 16-2. At a pyrolysis temperature ranging up to 600°C and with the carrier gas rapidly removing all the pyrolysates, there is essentially no pyrolyzable organic matter left in the source rocks. Only inert material unreactive at 600°C is left. This is in marked contrast to the situation in nature in which bitumens and adsorbed oil remain in postmature source rocks, as explained earlier. Also, removing the oil and bitumen in Rock-Eval before secondary cracking is contrary to the reactions in nature, in which such cracking begins in the oil window. In addition, pulverizing the rock for dry pyrolysis causes the expulsion of bitumen from micropores (<2 nm) that under natural conditions might remain closed in the source rock. Finally, high-temperature open pyrolysis generates and expels hydrocarbons from rocks with 0.5 to 1.5% TOC, whereas closed hydrous pyrolysis expels generated oil only from rocks containing more than 1.5% TOC (see "Quantity of Organic Matter," Chapter 10). Consequently, adding $S_1 + S_2$ yields for the leaner parts of a source rock would increase the apparent yield, even though the rock may be incapable of expelling oil under natural conditions.

Another problem is deciding what volume of source rock has generated and expelled oil. If a source rock retains oil until it cracks to gas, that oil cannot be counted as available for oil accumulations. For example, in the Illinois Basin, Lewan et al. (1995) found that 40% of the volume of the New Albany Shale source rock was between vitrinite reflectance R_o values of 0.6 and 0.7%, at which oil expulsion has not yet occurred but oil has been generated. Consequently, using the rock volume, starting at 0.6%R_o, would give an erroneously high expelled oil value and a correspondingly low ratio of oil in place to expelled oil. The Rock-Eval pyrolysis method considers all generated oil to be expelled, so it would include all New Albany source rock oil between an R_o of 0.6 and 0.7% in its mass balance, whereas hydrous pyrolysis would include only oil from the rock with an $R_o > 0.7\%$ that has actually expelled oil in nature.

Hydrous pyrolysis is a natural process that generates petroleum in hydrothermal vent systems such as in the Guaymas Basin of California

(Simoneit 1992). The hydrothermal system heats the rocks at temperatures approaching 400°C, which covers the range of hydrous pyrolysis experiments in the laboratory. Fragmentograms showing hydrocarbon distributions from hydrous pyrolysis in the labratory are similar to those obtained from the natural hydrothermal generation of petroleum in the Guaymas Basin (Simoneit 1992).

Table 16-2 compares the yield of oil and gas by open anhydrous Rock-Eval pyrolysis and closed hydrous pyrolysis for two well-known source rocks (Lewan 1993b). Note that the total yield of pyrolysate for the Phosphoria Retort Shale is essentially the same using both methods. But the Rock-Eval pyrolysis counts all generated oil as expelled, that is, 13.45 wt%. Hydrous pyrolysis measures the expelled oil directly. It shows the yield available in nature to be 6.52 wt%. This is less than half the Rock-Eval yield. Basically, hydrous pyrolysis is making the division between the yields of Boxes B and C in Figure 16-2, whereas Rock-Eval pyrolysis is lumping these together in Box C, thereby assuming bitumen-free source rocks in Box B that do not exist in nature even at the end of the oil window. Hydrous pyrolysis yields show that if anhydrous open pyrolysis (Rock-Eval) analyses are used for mass balance, the S_2 yields should be divided by 2 or more to obtain the numbers for Box C in Figure 16-2.

TABLE 16-2 Comparison of Rock-Eval and Hydrous Pyrolysis Yields from the Woodford Shale and Phosphoria Retort Shale[a]

	Yield (wt% of rock)	
Type of pyrolysis	*Woodford Shale WD-5*	*Phosphoria Retort Shale P-64*
Rock-Eval pyrolysis[b] of original sample		
Volatile HC (S_1)	0.25	0.50
Generated HC (S_2)	7.11	13.45
Total pyrolysate	7.36	13.95
Hydrous pyrolysis[c] at the end of primary-oil generation		
Retained heavy oil	2.56	6.52
Expelled oil	3.14	6.52
Generated HC gases	0.63	0.94
Total pyrolysate	6.33	13.98

[a]Lewan 1985

[b]Rock-Eval II, cycle 1.

[c]Experimental conditions: 72 hr at 355°C for WD-5 and at 350°C for P-64.

In Table 16-2, about 19.3% of the Woodford kerogen and 20.6% of the Phosphoria kerogen are converted to expelled oil by hydrous pyrolysis. Another Woodford Shale sample yielded 4.15 wt% of the rock as oil. About 13% of the kerogen was converted to oil. According to Lewan (1993b), the difference between hydrous pyrolysis and Rock-Eval can range between a factor of two and four.

Comments on Mass-Balance Estimates

A properly run basin-modeling program generally requires a large number of hydrous pyrolyses analyses to arrive at representative numbers for the barrels of oil equivalent (BOE) expelled (Box C, Figure 16-2). One approach is to run about twenty-five standard source rocks, including the three major oil-generating kerogen types, by both hydrous and Rock-Eval pyrolysis, and then to make cross plots using the hydrous pyrolysis yields to correct the yields for the Rock-Eval.

Table 16-3 contains typical yields of expelled oil and gas for kerogen types I, II, and III, by means of hydrous pyrolysis (Lewan, personal communication, 1995). Type III is represented by humic and sapropelic coals. As expected, type I yields the most oil and gas, and type III, the least. Type III yields a waxy oil. The thermally immature kerogens were subjected to hydrous pyrolysis at 350°C for 72 hours to reach a maturity of 1.5%R_o, representing the end of primary oil generation. Heating at 400°C for the same period of time raised the thermal stress into the gas-generation stage at 1.8%R_0. Some gas is generated in the oil window, but yields greatly increase in the gas window, partly because of the conversion of oil to gas. The total pyrolysate (S_2) yields by conventional Rock-Eval analysis are 70 to 80% for type I, 45 to 50% for type II, and 10 to 25% for type III (Bordenave et al. 1993, p. 242). Dividing these numbers by 2 gives roughly the same numbers for expelled oil by hydrous pyrolysis as those shown in Table 16-3.

Another factor indicating the retention of pyrobitumen in the source rock is the change in TOC during maturation (Boxes B and H in Figure 16-2). Lewan

TABLE 16-3 Hydrous Pyrolysis Yields as Weight Percent of Immature (R_o = <0.5) Kerogen

Kerogen type	*Expelled oil to 1.5%R_o*	*Gas to 1.5%R_o*	*Gas to 1.8%R_o*
I	40–60	3.2	13.4
II	15–30	3.1	10.6
III	0–12	0.7	2.0

Source: Data from M. D. Lewan, personal communication, 1995.

observed from numerous hydrous pyrolysis analyses of type II kerogens that the TOC generally decreases by no more than about one-third to one-half in going from an R_o of 0.5 to 1.8%. Since some of the carbon in kerogen is converted to CO_2, this indicates that less than half the carbon in kerogen is expelled as petroleum from most kerogens reaching this maturity interval. Any postmature type II kerogen with a TOC of 1% today probably did not have more than 1.5 or 2% when it was immature. Based on Lewan's data, from 15 to 45% of a type I and II kerogen mixture could be expelled as oil plus gas, with varying amounts of bitumen being retained in the source rocks.

Mass Balance of the Global Petroleum System: Estimates and Constraints

The mass-balance numbers in Figure 16-2 were obtained in the following manner: The total worldwide volume of known, in-place, conventional, heavy to extra heavy oil, plus reasonably considered undiscovered resources is estimated to be 10×10^{12} barrels (Box E, Figure 16-2) (Roadifer 1987). This number is probably low because part of the 2.8×10^{12} barrels of extra heavy oil could have had a volume 50 to 75% larger at the time of migration. The total volume of gas in place plus undiscovered resources has been estimated at 2×10^{12} BOE (Box I) (Masters 1993; Wyman 1985). The recoverable conventional oil in Box E would be $(7.2)\,(0.28) = 2 \times 10^{12}$ BOE assuming 28% recovery. Consequently, the ratio of oil to gas would ultimately be 1. At present it is 1.5, but more gas than oil is being discovered so it is approaching 1.

The U.S. National Academy of Sciences concluded from its study of petroleum in the marine environment that approximately 1.5 million barrels of oil are seeping into the world's oceans annually from natural sources (*Oil in the Sea*, 1985, p. 49; Wilson et al. 1973). At least an additional 0.5 million barrels would be seeping out on the continents. Assuming that seepage variations balance out over time, the seepage in the past 100 m.y. would be 200×10^{12} BOE (Box D, Figure 16-2).

Estimates of the total amount of oil generated in 100 m.y. (Box A) have ranged widely. If the numbers in Boxes D and E are reasonable, then the minimum amount of oil expelled (Box C) would have to be 210×10^{12} BOE. But this does not allow for petroleum dispersed in nonsource, nonreservoir rocks (Box F).

Large accumulations of unrecoverable oil are common in some formations. For example, numerous dry holes have been drilled in the huge Sprayberry trend, which covers six counties of the Midland Basin in West Texas. It is a fine-grained, oil-saturated siltstone that has been called "the world's largest reserve of unrecoverable oil" (Landes 1970, p. 337). A major cause of dry holes is not the absence of oil but the absence of good porosity and permeability. Rocks having permeabilities of $< 10^{-4}$ darcies are called *tight sands.* They typically produce less than 5 barrels of oil per day. As yet, no one has tried to quantify the volume of oil worldwide dispersed in nonproductive rocks such as tight sands or silty shales. The estimated 65×10^{12} barrels in Box F of Figure 16-2 is probably a minimum, considering the volume of rocks worldwide containing oil shows but no recoverable oil. Adding the amounts in D, E, and F equals C (275×10^{12} BOE).

Based on Lewan's ratios of expelled to retained oil in Table 16-2, it would follow that the 275×10^{12} BOE in Box C is about 55% of the generated oil in Box A, so the latter is estimated at 500×10^{12} BOE. Miller (1992, p. 500) estimated that the global quantity of oil generated and expelled (equivalent to Box C in Figure 16-2) is 2.7×10^{6} bbl/yr. This would be 270×10^{12} bbl in 100 m.y., which is essentially the same as the 275×10^{12} bbl in Box C.

The 45% of the 500 TBOE (trillion bbl of oil equivalent) that is not expelled from Box B (225 TBOE) plus the 65 TBOE in Box F would eventually become pyrobitumen and gas in Box H if it goes to the end of the gas window ($\sim R_o = 3.5\%$). Many source rocks, however, are still in the mature oil stage, so it is estimated that only 50% of the oil in Boxes B and F have cracked to gas. The quantity of gas and pyrobitumen in Box H was calculated from the following equation in which an average H/C ratio of 1.6 is used for the oils in Boxes B and F. The H/C ratio of the pyrobitumen residue at the end of cracking is 0.3.

$$10\ C_{10}H_{16}\ (\text{oil}) = C_{65}H_{20}\ (\text{pyrobitumen}) + 35\ CH_4\ (\text{gas})$$

Using this equation the 145 TBOE (50% of Boxes B + F) equates to 60 TBOE of gas and 85 TBOE of pyrobitumen (Box H). The 50 TBOE of gas in Box G represents the oil to gas ratio of 10 to 1 (Box A and G) based on Table 16-3 at $1.5\%R_o$ prior to extensive cracking of oil. The 100 TBOE of gas moving from Box H to Boxes I and J would be divided 2 to 98 based on the previous statement of gas reserves and resources being 2 TBOE.

Note that the quantity of oil and gas that ends up in the final sinks (Boxes B, D, E, F, G, I, and J plus the pyrobitumen in H) adds up to the 550 TBOE (Boxes A + G) originally generated. A good seal such as the evaporite seals in Saudi Arabia or the permafrost in West Siberia would alter the quantities in Boxes D and E, with much more petroleum being reservoired and less being lost.

A good seal could easily increase the reserves by a factor of five. Also, some of the dispersed gas dissolved in formation waters (Box J) would come out of solution as the fluids migrated vertically up through faults, eventually being trapped under evaporite beds. Most of the world's major gas fields underlie either permafrost or evaporite seals. So the big variations in the final eight sinks are mainly among Boxes D, E, I, and J. In a confined petroleum system with vertical migration, much more of the generated hydrocarbons end up in reservoirs. For example, in the global mass balance of Figure 16-2, only 2.2% of the expelled oil and gas is accumulated in reservoirs. On the Southern Norwegian Shelf at Ekofisk (discussed next), this number is 8%.

The weakest part of this exercise is the mass balance for gas. Laboratory experiments indicate that the generation of oil versus gas from kerogen is in the range 5 or 10 to 1 on a BOE basis. But adding bacterial gas, coal gas, and gas from oil with deeper burial greatly complicates the estimate of available gas.

Meanwhile, field observations suggest gas losses by dispersion, solution, and volitalization at the surface are greater than oil. The previously mentioned figures of Zor'kin and Stadnik (1975) (see Chapter 7) showed that the formation waters of five petroliferous basins in Russia contain more than fifty times the amount of gas reservoired in those basins. The ratio of Box J to I in Figure 16-2 is 49, but if J is all dissolved gas it leaves nothing for surface losses.

Unfortunately, Wilson et al. (1973) who documented the seepage of oil offshore did not attempt any estimate of gas losses. However, the gas losses at the sea bottom described by Hovland and Judd (1988) plus others documented in Chapter 13 make it very likely that the number in Box J is too low.

Hydrocarbon Accumulations on the Southern Norwegian Shelf: A Geochemical Model Study

The following geological and geochemical basin study was carried out by Ray Leonard of Amoco (1988). The area covered by the study is approximately 40,000 square kilometers, as shown on the tectonic map of Figure 16-3. It is an interesting area to demonstrate the use of geochemical modeling for ranking drilling prospects in an exploration program. Structures and traps are similar throughout the area, but there are marked contrasts in discovered reserves. At the time of the study, about 1×10^9 m^3 (6.3×10^9 bbl) of recoverable oil equivalent had been discovered in the Central Graben portion in the southeast, but only 12×10^6 m^3 (76×10^6 bbl) had been found in the equally large area of the northeast. In addition, the exploration drilling success ratio even in the prolific Central Graben is only 1 in 4. Some structures contain no hydrocarbons, and others are only partially filled. The information obtained from the study enabled Amoco to be much more selective in deciding on future areas and prospects for drilling, thereby reducing its risk of drilling dry holes.

Figure 16-3 is a tectonic map of the study area. There are a number of highs, such as the Utsira High and the Jaeren High, situated in the northern portion of the area. There also are several major grabens, the largest being the Central Graben. Far to the south is the Mandal Subbasin, and to the north are the Viking Graben and the Egersund Subbasin. Most of these features were formed by a rifting event from Triassic through Jurassic time, with gradual subsidence taking place from the Cretaceous to the Recent.

The general stratigraphy of the area is presented in Plate 5B. The section consists of about 3 kilometers of shale accumulated from Paleocene to Recent times underlain by several hundred meters of Upper Cretaceous Chalk (blue). Below this is the Lower Cretaceous Shale (green) underlain in turn by the major source rock in the area, the Upper Jurassic Kimmeridge Shale (gray). The Lower to Middle Jurassic (red) is shallow marine to continental. Below this is the Permian Zechstein Salt.

The geochemical basin analysis has four steps: (1) constructing a three-dimensional model, (2) mapping the subsurface temperatures, (3) evaluating the thickness and richness of the source rocks in each drainage area, and

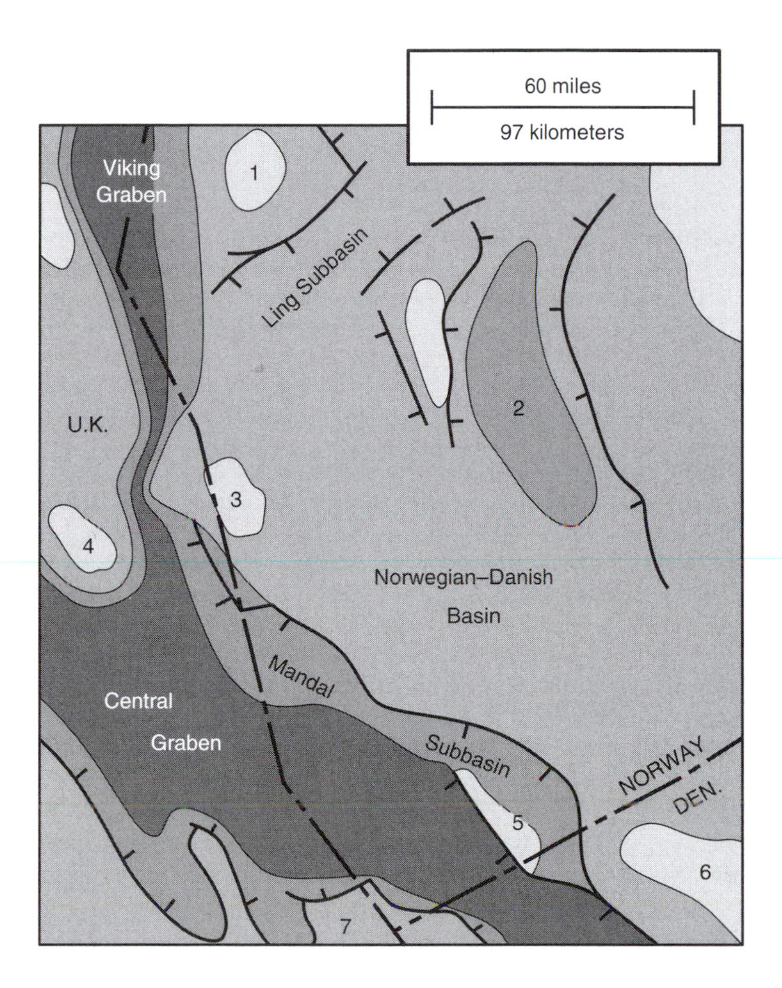

Figure 16-3

Tectonic map of the southernmost Norwegian shelf. (1) Utsira High, (2) Egersund Subbasin, (3) Jaeren High, (4) Montrose Forties High, (5) Mandal High, (6) Ringkøbing-Fyn High, and (7) Mid North Sea High. [Courtesy of R. C. Leonard]

(4) calculating the quantity of hydrocarbons generated for each structure. Twelve thousand kilometers of seismic data were available, along with stratigraphic data from 63 wells. An additional 605 pseudowells were constructed from these data, as shown in Plate 8C. Regional maps were made of the nine horizons listed to the left of Plate 8C, with depths shown as seismic time on the right. The 3.0 seconds equals 3 km, and 4.0 seconds is about 4.5 km. The well control is at the far left.

Table 16-4 shows the burial through time of a point at the base of the Upper Jurassic. The data in Table 16-4 were used to construct the burial history curve in Figure 6-18. Such curves were constructed for all the control wells and pseudowells.

The regional geothermal gradient map in Figure 16-4 was derived from bottom-hole temperatures recorded for the 63 control wells. The map shows an increase in the gradient from east to west of about 28°C/km (15°/1,000 ft) in the Egersund Subbasin to about 40°C/km (22°F/1,000 ft) in the Central Graben. This means that the source rocks will have to be buried to greater depths in the east than in the west to generate equivalent amounts of oil, all other factors being the same.

The next step is identifying the source rock, determining its yields, and mapping variations in its thickness. Two source rocks were identified. The upper part of the Upper Jurassic, the Kimmeridge Shale, with a TOC of 5 to 12% and a high potential yield, is the main source rock, and the lower part of the Upper Jurassic, the Heather Shale, with 2 to 4% TOC and moderate yields, also contributes petroleum.

The generation potential of the Kimmeridge Shale, however, decreases considerably going from southwest to northeast. This was reported earlier by Thomas et al. (1985) and Baird (1986). They found type II kerogen concentrated along the Southern Central Graben, but the content of type III kerogen increased going northeast. Paleogeographic maps have shown that this is due to a change from a marine to a more terrestrial environment going from south-

TABLE 16-4 Burial History of Data Points, Southern Norwegian Shelf

		Depth	
Data point	*Time (Ma)*	*Meters*	*Feet*
Recent	0	5,767	18,916
Top of Pliocene	2.0	5,111	16,764
Top of Miocene	5.2	4,585	15,039
Top of Oligocene	24.0	3,862	12,667
Top of Eocene	37.0	3,383	11,096
Top of Paleocene	53.5	2,791	9,154
Top of Danian Chalk	60.0	2,669	8,754
Top of Lower Cretaceous	100.0	1,039	3,408
Top of Upper Jurassic	135.0	457	1,499
Base of Upper Jurassic	143.0	0	0

Source: Data from R. C. Leonard 1988.

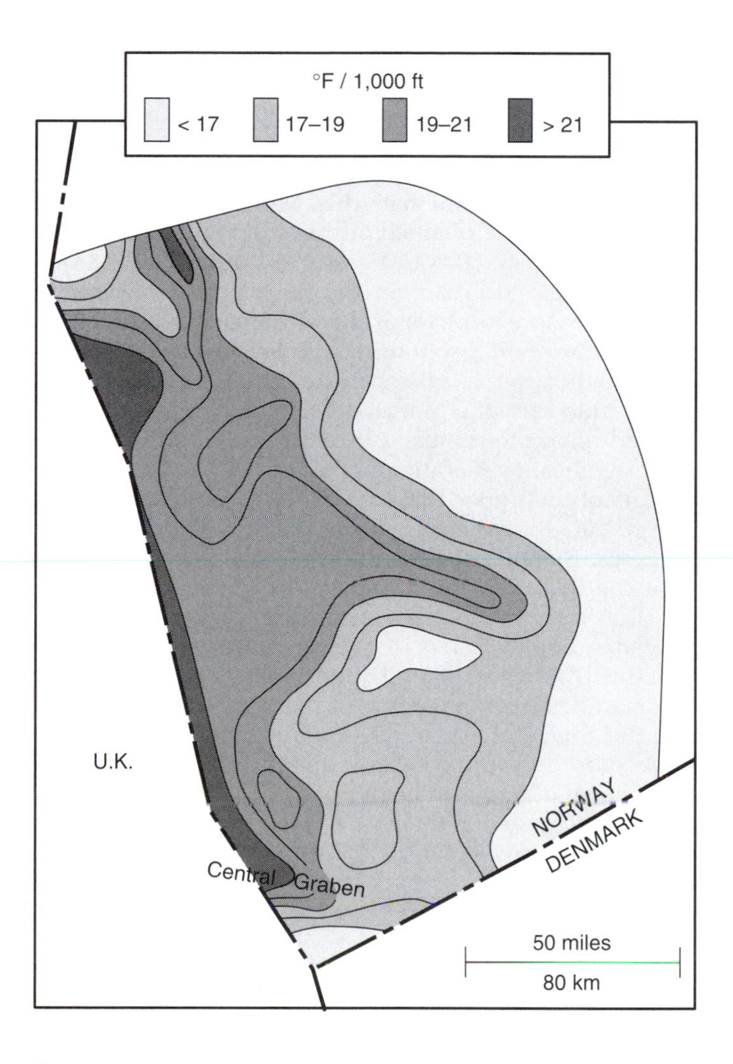

Figure 16-4

Geothermal gradient map of the Southern Norwegian Shelf showing a general increase of the gradient from east to west; 28°C/km in the Egersund Subbasin to about 40°C/km in the Central Graben. [Courtesy of R. C. Leonard]

west to northeast. The Upper Jurassic source rocks also become thinner going northeast. In the Egersund Subbasin where there are three small oil accumulations, the Upper Jurassic is about half as thick as in the Central Graben (Thomas et al. 1985).

Baird (1986) used Rock-Eval pyrolysis to determine hydrocarbon yields from the Kimmeridge Shale in this area, and Leonard (1988) used hydrous pyrolysis supplemented with Rock-Eval pyrolysis which, as pointed out earlier, was corrected with HP data. The maximum yields obtained in the most southwestern part of this area by Baird were 625 bbl/acre-ft, whereas Leonard's maximum yields were about half this, which, again, would be expected. Baird's analysis includes the retained oil (Box B, Figure 16-2), but Leonard is counting only the expelled oil (Box C, Figure 16-2). Baird's (1986) map of generative potential showed a decrease in the petroleum yield of the Kimmeridge Shale going northeast, except for isolated anoxic areas such as the Egersund Subbasin. Leonard (1988) obtained similar results. Application of the Arrhenius equation indicates that oil generation started around 35 Ma, peaked around 15 Ma, and ended about 5 Ma, as described in Chapter 6. At 10 Ma, oil was being generated throughout much of the southern part of the Central Graben. There was no significant generation at 65 Ma.

Many of the Upper Jurassic source rocks in the Central Graben of the area in Figure 16-3 have moved through the oil window and are now generating gas (Table 6-5). Significant generation also has occurred in the Mandal Subbasin east of the Central Graben and in the deepest portion of the Egersund Subbasin. But in the rest of the entire area the source rocks are in the early stages of oil generation, with less than 20% of the reactive kerogen being converted to petroleum (Leonard 1988). This explains why reserves are far greater in the Central Graben than in the rest of the area.

In this area of the North Sea, the migration pathways from source to reservoir are short and vertical. Consequently, there is a closer relationship between the volume of oil expelled from the source rocks in their drainage areas and the volume accumulated in their reservoirs. Calculation of the former was from the following equation:

$$V = (A)(H)(R)(KC)$$

where A is the drainage area of the structure, H is the thickness of the source rock, R is the richness of the source rock, and KC is the extent of kerogen conversion. Figure 16-5 illustrates, based on seismic maps, the drainage areas for each structure in this southernmost area of the Norwegian shelf. Each of these drainage areas leads to a single accumulation on structure. The chalk fields are numbered. The efficiency of trapping and migration was calculated as follows:

$$\%E = \frac{\text{oil in place on closure} \times 100}{\text{oil expelled within drainage area}}$$

Table 16-5 lists the migration and trapping efficiencies for six of the chalk fields in Figure 16-5. Although these are small areas with direct vertical migration, it is apparent that the overall efficiencies can range widely. The range from 1 to 31% covers the range recorded for all the petroleum systems listed in

TABLE 16-5 Migration and Trapping Efficiencies for Oil Fields in the Central Graben, Southern Norwegian Shelf

Field	*Hydrocarbons generated and expelled in 10^9 barrels*	*Barrels in place in reservoir in 10^9 barrels*	*Efficiency (%)*
Ekofisk	20.4	6.34	31
Eldfisk	18.1	2.24	12
Tor	9.8	1.05	11
Valhall	36.9	2.87	8
Albuskjell	9.3	0.44	5
Southeast Tor	8.9	0.13	1

Source: Data from R. C. Leonard 1988.

Table 16-1. Particularly noteworthy is the difference between the 11 and 1% efficiency for Tor and Southeast Tor, which are adjacent in Figure 16-5. This was explained in Chapter 8 (Figure 8-17) as a migration effect. The faults that are migration pathways under the Tor field lead directly from the Upper Jurassic source shales to the chalk reservoir, whereas at Southeast Tor, a salt dome blocks the passage of oil from the source rock to the faults leading to the reservoir. Some oil has migrated to Southeast Tor, probably through small faults or microfractures, but such migration is more difficult and much more inefficient than the pathway at Tor.

The migration and trapping efficiencies in Tables 16-5 through 16-8 are percentages of generated and expelled oil and gas. They do not include oil and bitumen retained in the source rock. Thus, the 20.4×10^9 barrels of oil for Ekofisk in Table 16-5 would be in Box C of the flow chart in Figure 16-2. The 6.34×10^9 barrels of oil in place in the reservoir would be in Box E. Box A would contain 37×10^9 barrels of oil generated in the source rock, and Box B would show the 17×10^9 barrels not expelled based on ratios of the numbers in Figure 16-2. The oil lost at the surface and dispersed in nonreservoir rocks would be $20 \times 10^9 - 6 \times 10^9 = 14 \times 10^9$ barrels.

The source–reservoir relationships in Leonard's (1988) study led to his concept that these fields could be divided into three categories depending on migration pathways and kerogen conversion. The best situation, called type A, is illustrated in Figure 16-6. In this type, 100% of the reactive kerogen has been converted to petroleum, and the Kimmeridge Shale source rock is directly under the crest of the field, with faults present to act as migration conduits. The fields classified as type A are listed in Table 16-6 with their migration and trapping efficiencies. Although these numbers cover the same range as in Table 16-5, they do so for an entirely different reason. All these structures are filled to

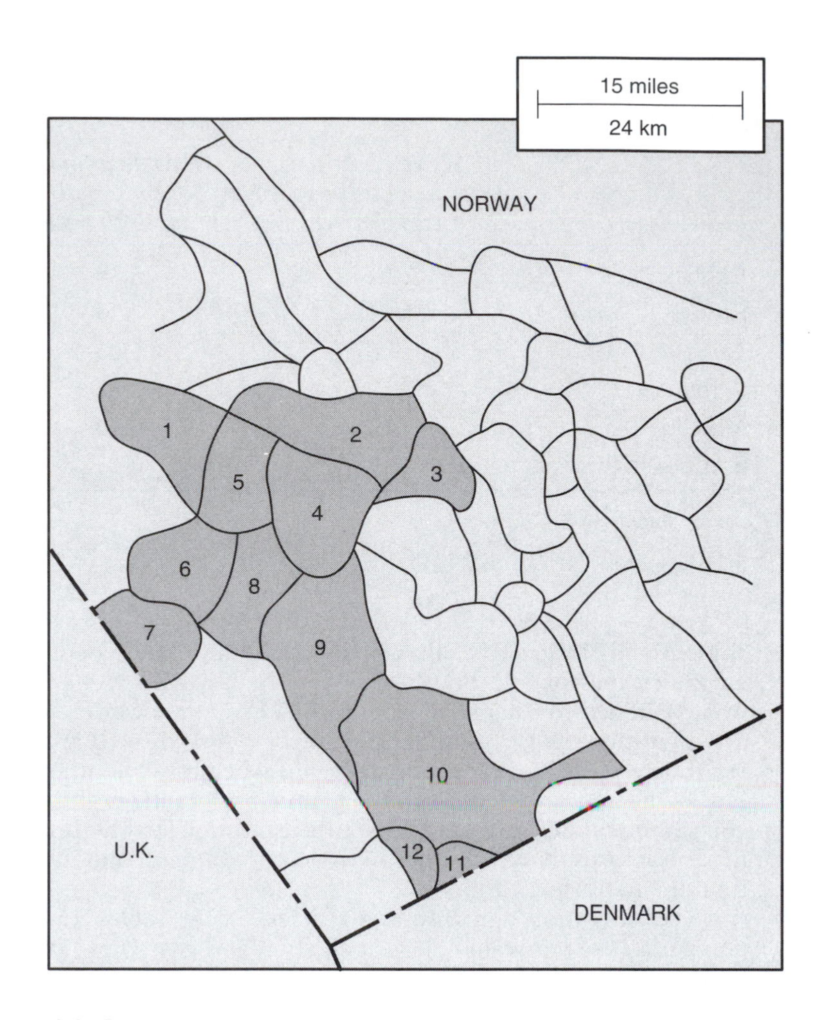

Figure 16-5

Drainage areas for each of the chalk structures in the Central Graben. The following structures are identified with numbers: (1) Albuskjell, (2) Tor, (3) Southeast Tor, (4) Ekofisk, (5) West Ekofisk, (6) Tommeliten North, (7) Tommeliten South, (8) Edda, (9) Eldfisk, (10) Valhall, (11) East Hod, and (12) West Hod. [Courtesy of R. C. Leonard]

the spill point. All could have had much higher efficiencies if larger structures and reservoir volumes had been present. The conditions of type A unquestionably provide the highest efficiencies possible.

Type B fields are illustrated in Figure 16-7. These are examples in which the source rock is under the crest of the field and faults are present to act as migra-

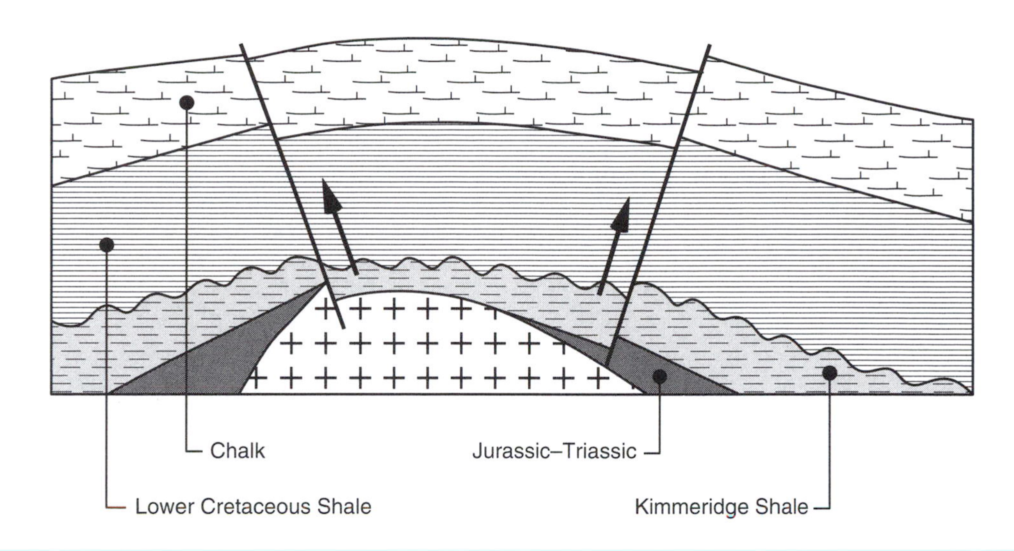

Figure 16-6

Type A source–reservoir relationship: closures with 100% kerogen conversions, Kimmeridge Shale under the crest of the field, and faults to act as migration conduits. All type A structures are filled to the spill point. [Courtesy of R. C. Leonard]

tion pathways but the reactive kerogen has only partially converted to petroleum. Table 16-7 lists the migration and trapping efficiency for the three fields that fall into this category. The low number for east Hod is due to its being a very small structure with a small reservoir volume under closure. The higher

TABLE 16-6 Migration and Trapping Efficiency for Type A System

Field[a]	*Migration and trapping efficiency (%)*
Ekofisk	31.1
Eldfisk	12.4
West Ekofisk	4.3
Edda	1.8
Tommeliten	1.2

[a]Structure filled to spill point.

Source: Data from R. C. Leonard 1988.

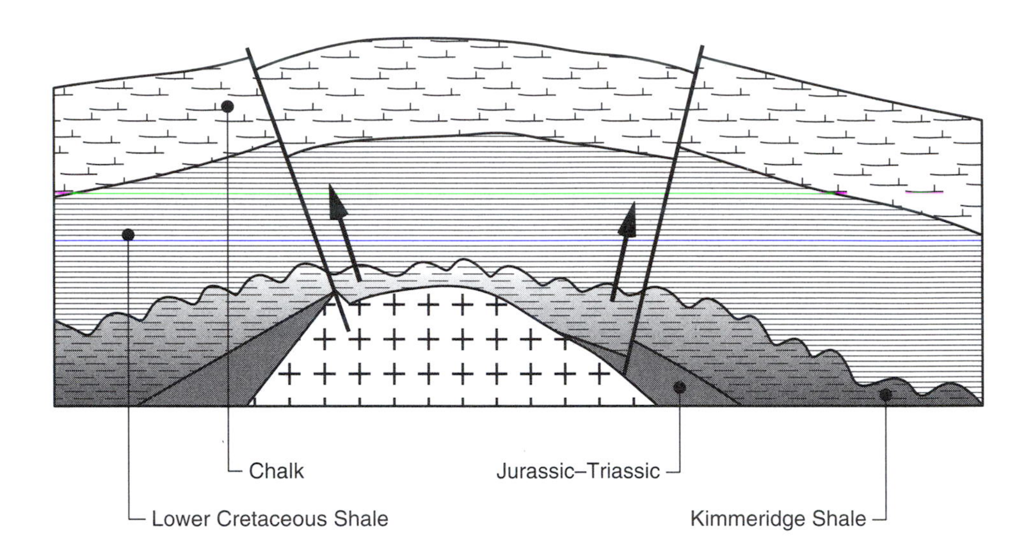

Figure 16-7

Type B source–reservoir relationship: closures with partial kerogen conversion, Kimmeridge Shale under the crest of the field, and faults to act as migration conduits. Oil generation is occurring today and has not been completed. [Courtesy of R. C. Leonard]

numbers for Tor and Valhall are more typical of this category, so this is still a reasonably efficient overall system for accumulating the generated oil.

The type C source–reservoir relationship does not have the source rock under the crest of the structure, although kerogen has been converted to oil on the flanks of the structure, and faults are present to act as migration conduits

TABLE 16-7 Migration and Trapping Efficiency for Type B System

Field	*Migration and trapping efficiency (%)*
Tor	10.7
Valhall	7.8
East Hod[a]	1.7

[a]Structure filled to spill point.

Source: Data from R. C. Leonard 1988.

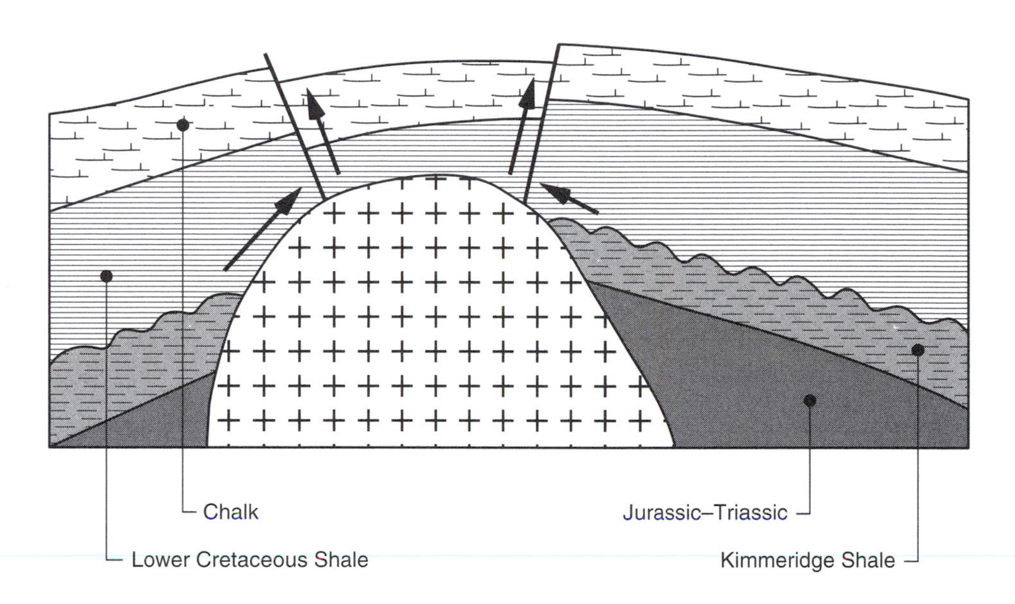

Figure 16-8

Type C source–reservoir relationship: Kimmeridge Shale is not present under the crest, although kerogen has been converted to oil on the flanks of the structure. Faults are present to act as migration conduits. [Courtesy of R. C. Leonard]

(Figure 16-8). Table 16-8 lists the four fields in this category. Interestingly, the only field with an efficiency above 1.5% is Albuskjell. The reason for this is that this structure has two lobes, one of which lacks reservoir rock (Lower Cretaceous shale directly overlies salt), and the other has some of the Kimmeridge Shale source rock under the chalk crest. This second lobe raised the efficiency.

TABLE 16-8 Migration and Trapping Efficiency for Type C System

Field	*Migration and trapping efficiency (%)*
Albuskjell	5.2
West Hod	1.5
Southeast Tor	1.5
Cod	1.2

Source: Data from R. C. Leonard 1988.

There is another source–reservoir relationship that could be called type D, in which there are no faults acting as migration pathways between the source rock and the chalk. Drilling on such structures has invariably resulted in dry holes in this area. Faults are necessary here as migration conduits, as they are in many settings.

In summary, Leonard's 1988 study was the first to demonstrate that the three critical factors leading to these Central Graben accumulations are (1) the size of the trap, (2) the maturity of the source rock under the crest of the structure, and (3) the presence of faults to act as migration pathways from source to reservoir. His study emphasized that the best structural closures still require mature source rocks in the drainage area of the structure in order to contain any petroleum.

Leonard's study also showed that there are at least four reasons that more than eighty times as much oil is trapped in the southwest than in the northeast: (1) The kerogen is marine type II in the southwest, compared with more terrestrial type III in the northeast; (2) the geothermal gradient decreases substantially from the southwest to the northeast; (3) the thickness of the source rocks decreases going northeast; and (4) the source rock kerogen is now in the gas window in the southeast, but only 20% is through the oil window in the northeast. The model dramatically reduced the number of chalk prospects remaining. Amoco was able to concentrate on the most promising types A and B source–reservoir relationships in those areas where the source rocks were thickest with mature type II kerogen.

Genetic Classification of Petroleum Systems

Demaison and Huizinga (1991) developed a genetic classification of petroleum systems to help reduce risk, particularly when exploring in new areas.

The objective of their classification is to estimate the relative quantity of petroleum available and describe the probable location of zones of petroleum occurrence or plays for each petroleum system in a basin. As such, it is more a regional tool than a method for evaluating individual prospects, such as in Leonard's 1988 studies of the Southern Norwegian Shelf. The genetic classification of petroleum systems depends on three factors: (1) the petroleum charge available for trapping, (2) the migration–drainage style, and (3) the entrapment style, as shown in Figure 16-9 (Demaison and Huizinga 1994).

The Petroleum Charge

Demaison and Huizinga (1991) used the *source potential index* (*SPI*) to evaluate the relative charge potential of various source rocks. The SPI is the maximum quantity of bitumen, oil, and gas that can be generated in a column of source rock under 1 square meter of surface area. It is calculated from the following equation:

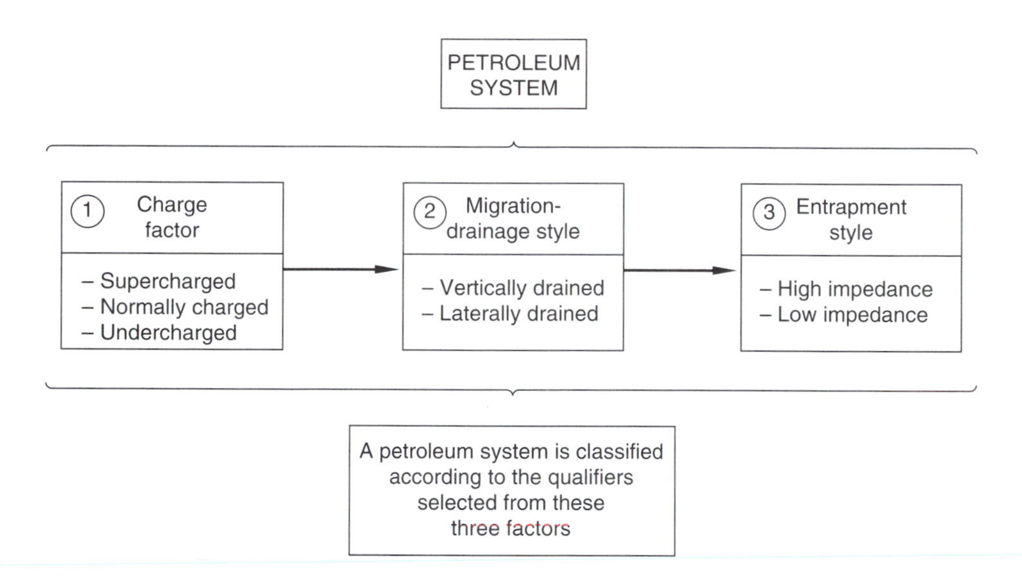

Figure 16-9

Flow diagram for the genetic classification of a petroleum system. [Demaison and Huizinga 1994]

$$\text{SPI} = \frac{h\,(S_1 + S_2)\,\rho}{1{,}000}$$

where the SPI is in metric tons of bitumen, oil, and gas per square meter.

The h equals source rock thickness in meters, excluding intervals that lack significant source potential. In practice the authors included only fine-grained rocks showing $(S_1 + S_2) > 2$ kg HC/metric tons rock unless biomarker analyses correlate oils with leaner source rocks.

The $\overline{(S_1 + S_2)}$ equals the average genetic potential in kilograms of bitumen, oil, and gas that would be obtained by the pyrolysis of a metric ton of the most immature to early mature sample of the source rock available. This is usually determined by Rock-Eval pyrolysis. As stated, Rock-Eval provides the total quantity of hydrocarbon generated (Box A in Figure 16-2), which is more than the quantity actually expelled under natural conditions. If expelled hydrocarbon values are needed (Box C, Figure 16-2), a modified $S_1 + S_2$ would have to be determined by hydrous pyrolysis or by using cross correlations of hydrous pyrolysis with Rock-Eval.

The ρ equals source rock density in metric tons per cubic meter. The authors used a density of 2.5 metric tons/m^3 for the source rocks discussed in their paper. But SPI determinations in specific areas should use measured rock densities when available.

Some SPI values are listed in Table 16-9 for various source rock sequences from Demaison and Huizinga (1991). One metric ton/m^2 equals 30,190 BOE/acre for a 36° gravity oil. These SPI values should be used primarily as a guide to show the relative differences in the charging capacity of various petroleum systems or specific sectors in a basin. Generally, SPI maps are prepared and overlain by maturity maps to highlight regional trends in which geophysical structures are most likely to be charged with petroleum. The SPI values should not be used alone but instead in combination with the drainage and entrapment styles.

For example, simply comparing the numbers in Table 16-9 shows that the Upper Jurassic of Saudi Arabia has the same SPI (14) as the Niger Delta, even though the oil potential of Saudi Arabia is known to be considerably greater. The difference is that the Central Arabian sequence is a laterally drained system that accumulates oil from a huge drainage area, compared with the narrowly focused vertical drainage in the Niger Delta. Consequently, an SPI of 14 would be very high for a lateral drainage system, as in Table 16-10, but only moderate for a vertical drainage system like the Niger Delta. Also, the SPI of 15 for the vertically drained North Sea (Table 16-9) barely qualifies it as high, whereas the value of 14 for Saudi Arabia is twice the minumum high SPI for lateral drainage systems. To be classified as low, moderate, or high, as in Table 16-10, all the vertical drainage systems require higher SPI values than do the lateral drainage systems. In addition, the continuous multiple evaporite seals of central Arabia provide a much higher level of seal integrity than either the North Sea or the Niger Delta.

A high SPI does not necessarily reduce the risk of dry holes unless other factors such as maturation, drainage, and entrapment styles also are highly favorable. The lesson learned from Leonard's study of the Southern Norwegian Shelf was that even though the Upper Jurassic source rock of the North Sea had an SPI of 15, there was no oil in those structures where there were no faults acting as migration pathways from source to reservoir.

A low SPI does not necessarily condemn a petroleum system if all other factors are highly favorable. For example, the Bakken source rock of the Williston Basin has an SPI of 3, which rates it as a moderate lateral drainage system in Table 16-10. Nevertheless, Webster (1984) estimated the quantity of bitumen and oil generated by the Bakken to be 92 billion barrels (Box A in Figure 16-2). According to Dow (1974), at least 3 billion barrels of oil from the Bakken are in place in the Williston Basin reservoirs of the United States and Canada.

Caution also should be used in evaluating SPI numbers that are high owing to a great thickness of a rich source rock like an oil shale. Such rocks may generate oil but may not have suitable migration pathways, like interbedded sands, from the source rock to a reservoir.

For example, the mature oil-generation zone for the Black Shale facies of the Green River lacustrine rocks in the Uinta Basin of Utah with types I and II kerogen is 4,000 ft (1,220 m) thick. The immature lacustrine facies of the Green River has an average $S_1 + S_2$ of about 30 mg HC/g rock (Anders and

TABLE 16-9 Examples of Average Source Potential Indices for Individual Source Rocks from Various Basins

Basin (country)	*Source rock sequence*	*Kerogen type*	*Average SPI (metric tons HC/m²)*
Junggar (China)	Upper Permian	I	65
Lower Congo (Cabinda)	Lower Cretaceous	I	46
Santa Barbara Channel (U.S.)	Miocene	II	39
San Joaquin (U.S.)	Miocene	II	38
East Venezuela FTB[a] (Venezuela)	Middle to Upper Cretaceous	II	27
Offshore Santa Maria (U.S.)	Miocene	II	21
Middle Magdalena (Colombia)	Middle to Upper Cretaceous	II	16
North Sea (U.K.)	Upper Jurassic	II	15
Central Arabia (Saudi Arabia)	Upper Jurassic	II	14
Niger Delta (Nigeria)	Tertiary	III	14
Gulf of Suez (Egypt)	Upper Cretaceous–Eocene	II	14
San Joaquin (U.S.)	Eocene–Oligocene	II to II–III	14
Maracaibo (Venezuela)	Middle to Upper Cretaceous	II	10[b]
West Siberia (Russia)	Upper Jurassic	II	8[b]
Paris (France)	Lower Jurassic	II	7
Barrow–Dampier (Australia)	Middle to Upper Jurassic	II–III to III	6
Illinois (U.S.)	Lower Carboniferous	II	6[b]
Northwest Arabian (Syria)	Triassic and Upper Cretaceous	II	5
Plato (Colombia)	Oligocene–Miocene	II–III to III	5
Northwest Arabian (Turkey)	Upper Silurian–Lower Devonian	II	4
Celtic Sea (Ireland)	Lower Jurassic	II to II–III	4
Malvinas (Argentina)	Lower Cretaceous	II to II–III	3
Williston (U.S.)	Upper Devonian–Lower Carboniferous	II	3
Senegal (Senegal)	Paleocene–Eocene	II–III	2
Cantabrian (Spain)	Lower Jurassic	II	2
Metan (Argentina)	Upper Cretaceous	II	1
Parana (Brazil)	Lower Permian	II	1

[a]FTB = Fold and thrust belt.

[b]Estimated.

Source: Data from Demaison and Huizinga 1991.

TABLE 16-10 Relation of SPI to Migration–Drainage Style

	Vertical	*Lateral*
High SPI	≥15	≥7
Moderate SPI	5 to < 15	2 to < 7
Low SPI	< 5	< 2

Source: Data from Demaison and Huizinga 1991.

Gerrild 1984). This calculates to an SPI of 92, thereby placing it at the top of Table 16-9. However, the only big discovery to date is the marginally economic 2-billion-barrel Altamont–Bluebell field producing from low-porosity fractured sandstones with secondary quartz overgrowths. There are abundant noncommercial oil shows and a few small fields scattered throughout the Uinta Basin, but no large commercial production. Anders and Gerrild (1984) calculated that at least 99×10^9 barrels of oil have been generated, with 19×10^9 barrels expelled (Boxes A and C in Figure 16-2), from Green River lacustrine rocks.

Migration–Drainage Style

Migration–drainage styles are classified as either vertical or lateral, as shown in Tables 16-10 and 16-11. Table 16-11 contains some typical examples of provinces containing petroleum systems classified by Demaison and Huizinga (1991) according to charge factor, drainage style, and entrapment style. The terms *super, normal,* and *undercharged* are assigned according to the magnitude of SPI as being high, moderate, or low in Table 16-10.

It was stated in Chapter 8 that although the long-distance lateral migration typical of foreland basins is less efficient than vertical migration, it results in the capture of more oil because of the larger drainage areas. The five supercharged, laterally drained systems on the right of Table 16-11 contain more than half the world's conventional oil (Klemme 1988) and possibly three-fourths of the world's heavy to very heavy oils (Roadifer 1987). The supercharged, vertically drained systems on the left of Table 16-11 focus the oil better through fault and fracture systems, but their drainage areas are small, as shown in Figure 16-5. Focused vertical migration is typical of deltas, but only about 5% of the world's conventional oil is reservoired in those systems (Klemme 1988).

Entrapment Style

The oil and gas expelled from a source rock have many more pathways to follow than those leading to structural–stratigraphic traps. This is why 96% of

TABLE 16-11 Oil and Gas Provinces Classified Genetically by Their Petroleum Systems

Drainage style	*Vertical*		*Lateral*	
Entrapment (impedance)	*High*	*Low*	*High*	*Low*
Supercharged	Los Angeles (U.S.) San Joaquin (U.S.) Lower Congo (Cabinda) Sirte (Libya) Campeche–Reforma (Mexico) Tampico (Mexico) Permian (U.S.) Central Sumatra (Indonesia) Zagros FTB[a] (Iran, Iraq) Campos (Brazil) North Sea (U.K., Norway)	Sectors of Iraq FTB[a] Southeast Turkey San Joaquin (U.S.) Eastern Venezuela FTB[a]	Central Arabia (Saudi Arabia) West Siberia (Russia) North Slope (U.S.)	Eastern Venezuela Foreland Western Canada Foreland
Normally charged	Gulf of Suez (Egypt) Maracaibo (Venezuela) Gulf Coast (U.S.) Niger Delta (Nigeria) Grand Banks (Canada) Cook Inlet (U.S.) Barrow–Dampier (Australia)	Muglad (Sudan) Ceara–Potiguar (Brazil)	Volga–Urals (Russia) Ghadames (Algeria) Illinois (U.S.) Appalachian (U.S.) Illizi (Algeria)	Williston (U.S.) Oriente (Ecuador) Paris (France)
Undercharged	Rhine Valley Graben (France, Germany) Indus (Pakistan) South Florida (U.S.) Vienna (Austria) Pannonian (Hungary, Yugoslavia) Rharb (Morocco) Pelagian (Tunisia)	Central Adriatic (Yugoslavia) Essaouira (Morocco) Perth (Australia) Barreirinhas (Brazil) Takatu (Guyana)	Cooper (Australia): Mesozoic Canning (Australia)	Denver (U.S.) Molasse (Austria)

[a]Fold and thrust belt.

Source: Data from Demaison and Huizinga 1991.

the expelled oil in Box C, Figure 16-2, is either dispersed in nonsource, nonreservoir rocks or lost at the surface (Boxes D and F, Figure 16-2). Demaison and Huizinga (1991) termed the degree of physical resistance working against this dispersion of petroleum as *impedance*. The two factors controlling the degree of impedance are the tightness of the seal and the structural deformation. Thus the permafrost and methane hydrates over the West Siberian gas fields and the evaporite beds over the Hugoton field (United States) are high-impedance systems. Both these systems have laterally continuous regional seals combined with a moderate-to-high degree of structural deformation. Low-impedance systems have either a low degree of structural deformation or a low degree of regional seal effectiveness.

Actually, if the seal is deficient, the extent of structural deformation is not relevant. The supercharged Eastern Venezuela, Western Canada, and Central Arabian petroleum systems all accumulated giant quantities of oil through lateral drainage, but only the high impedance of the Arabian system retained the world's largest reserves of conventional oil. The trillion barrels of 6-to-10° API oil in each of the other two systems resulted from seal deficiency.

The SPI is still in the development stage, but the concept of combining charge factors with drainage and entrapment styles has promise for the regional evaluation of petroleum systems. The authors found that in most extensively explored areas there is a positive correlation between high SPIs, as defined for vertical and lateral drainage systems, and petroleum reserves.

Effective Petroleum Source Rocks of the World

Klemme and Ulmishek (1991) compared stratigraphic and depositional factors pertaining to the world's source rocks with the quantity of conventional recoverable oil and gas reserves originating from those rocks. They defined source rock effectiveness in terms of the overall process of generation, expulsion, migration, and trapping. Giant heavy oil and bitumen accumulations such as in the Eastern Venezuela Basin were excluded, since they probably never were pools of conventional oil, owing to biodegradation and water washing on approaching the surface.

Klemme and Ulmishek's 1991 study shows clearly that there is an uneven areal and stratigraphic distribution of the world's petroleum reserves. About 91.5% of the world's discovered original reserves of oil and gas came from source rocks of six stratigraphic intervals, representing only 35% of Phanerozoic time. The six intervals shown in Figure 16-10 are (1) the Silurian, which generated 9% of the world's recoverable reserves; (2) the Upper Devonian–Tournaisian (8%); (3) the Pennsylvanian–Lower Permian (8%); (4) the Upper Jurassic (25%); (5) the Middle Cretaceous (29%); and (6) the Oligocene–Miocene (12.5%). The original recoverable reserves used were 2.2×10^{12} barrels of oil equivalent (BOE) (from Masters et al. 1987). Klemme and Ulmishek (1991) published global maps of lithofacies and structural form plus oil reserves generated by source rocks, along with source rock maps showing the distribution and maturity of kerogen types I, II, and III for these six intervals.

During the Silurian period, organic-rich graptolitic shales with type II kerogen were deposited on extensive platforms along continental margins. About two-thirds of the original Silurian source rocks have been deformed and metamorphosed, and most of the remaining third are overmature. Consequently, 85% of the Silurian petroleum reserves are gas (Figure 16-10).

About two-thirds of the Upper Devonian–Tournaisian source rocks with predominantly type II kerogen have been preserved. These are the black shale facies of organic-rich siliceous shales, marls, and limestones deposited on platforms and in intracratonic circular sags. About 80% of the reserves are oil.

The Pennsylvanian–Lower Permian sedimentary rocks contain more coal and type III kerogen than the older Paleozoic rocks because of a global marine regression during this time. This caused almost two-thirds of the petroleum reserves to be gas. The authors found that the preservation of this gas is probably due to the wide distribution of evaporites in the Permian. Evaporites sealed 60% of the petroleum reserves during this interval.

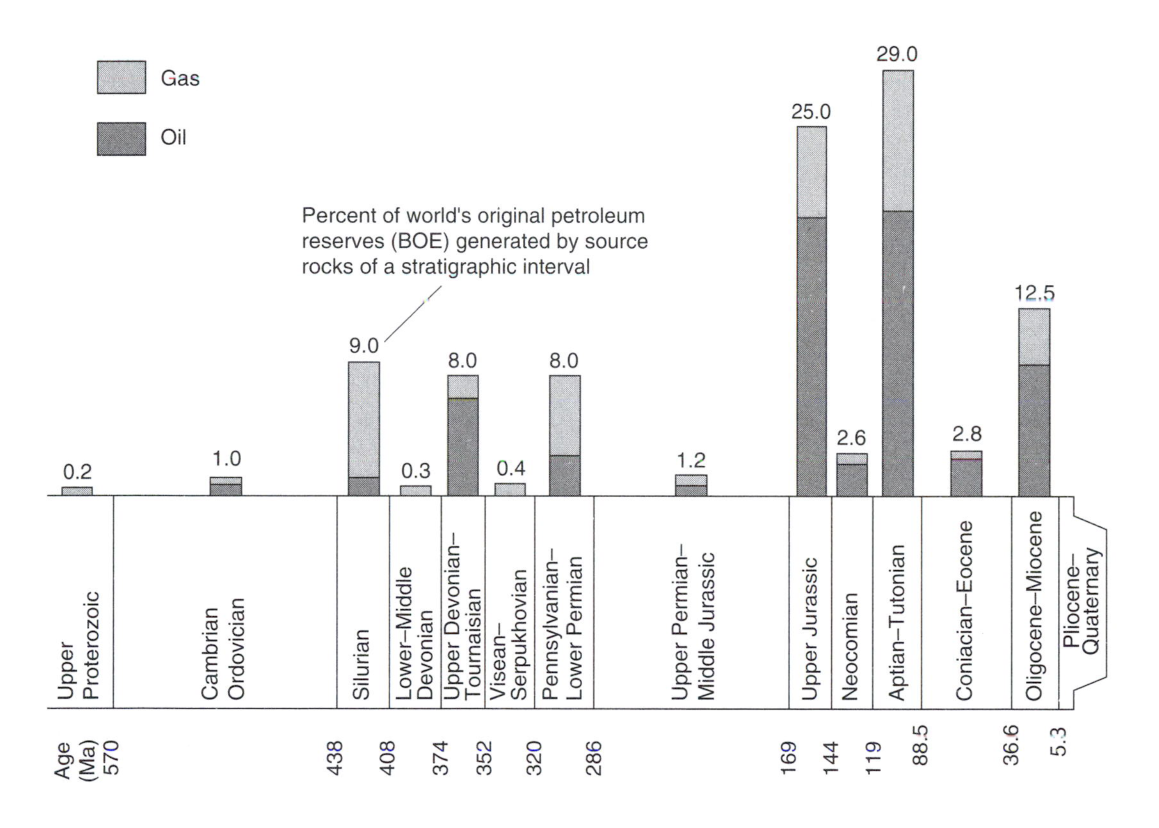

Figure 16-10

Stratigraphic distribution of effective source rocks given as a percentage of the world's original petroleum reserves generated by these rocks. [Klemme and Ulmishek 1991]

A worldwide marine transgression that started in the Jurassic deposited type II kerogen in the silled marine basins of large linear and circular sags. This stratigraphic interval generated 25% of the world's original recoverable reserves. About 74% of it is oil. This includes the Arabian–Iranian, West Siberian, North Sea, Gulf of Mexico, and North Caucasus Basins. Source rocks with type III kerogen and coal are rare in the Upper Jurassic.

The marine transgression of the Middle Cretaceous period reached its maximum in the Turonian, with the sea penetrating deep into most continents. More petroleum reserves originated in Aptian–Turonian source rocks than those of any other stratigraphic interval, as shown in Figure 16-10. These also were deposited mainly in widespread, silled linear and circular sags. About 65% of the reserves are oil. The Middle Cretaceous also saw an increasing abundance of type III kerogen and coal, particularly in high latitudes. The giant reserves of dry gas in the Cenomanian reservoirs of Northern West Siberia are believed to be bacterial, originating from the type III organic matter of the Albian–Cenomanian Pokur Formation.

A Late Tertiary regression resulted in type III kerogens and coals occupying 84% of the total source rock area during the Oligocene–Miocene. The type I and II kerogens are limited to rifted basins of eastern China, back-arc basins of Indonesia, and rift settings on continental margins. The 12.5% of the world's reserves from this stratigraphic interval are about two-thirds oil, dispersed through a large number of basins. Most of the major deltas are in this interval.

Geologic Factors Controlling Source Rock Effectiveness

Klemme and Ulmishek's (1991) source rock maps illustrate that there is no systematic increase in source rock deposition through time, despite the apparent increase in organic matter. A comparison of the source rock areas of the six principal stratigraphic intervals of Figure 16-10 shows a slight areal decrease from Silurian through Lower Permian, followed by an increase in the Upper Jurassic. The source rock area from Silurian through Lower Permian is about 45%, and the Upper Jurassic through Oligocene–Miocene is 54% of the total source rock area for the six provincial intervals in Figure 16-10.

In regard to paleolatitudes, Klemme and Ulmishek found that two-thirds of the source rocks of these six principal intervals were deposited between the paleoequator and 45° paleolatitudes. Much of this occurred in the Tethyan realm, a Silurian–Holocene latitudinal seaway between Gondwana and the Hercynian collision zone of the northern group of continents. This realm contains both clastic rocks and widespread carbonate reservoirs with evaporite seals. The Tethyan basins cover less than one-fifth of the world's land area and continental shelves, yet they contain more than two-thirds of the original petroleum reserves.

Three structural forms—platforms, circular sags, and linear sags—contain the source rocks for about 78% of the original reserves of the six principal intervals in Figure 16-10. The rift–sag structural forms favorable for the formation of anoxic silled basins were developed particularly during Tethyan tectonics, that is, during the opening and closing of the Laurisian and Gondwanan continental plates.

Worldwide marine transgressions favorable for the deposition of black shale source rocks were dominant during the Silurian, Upper Devonian–Tournaisian, Upper Jurassic, and Aptian–Turonian (Figure 5-6). Klemme and Ulmishek also concluded that about 60% of the world's conventional oil migrated mainly vertically upward or stratigraphically downward and that 40% migrated mainly laterally from the source rock to the reservoir. Again, this does not include the heavy oil like that in the Western Canada and Eastern Venezuela Basins.

Geochemical Factors Controlling Source Rock Effectiveness

Interestingly, Klemme and Ulmishek (1991) calculated that only 2.7% of the original reserves of world petroleum came from source rocks with type I kerogen. This validates the discussion in Chapter 10 that concluded that most lacustrine basins, such as the Songliao Basin of China, contain more type II than type I kerogen.

Another important observation was that source rocks with type II kerogen were dominant from the Early Paleozoic through the Upper Jurassic, during which the ratio of type II to III was 3 to 1. By the Middle Cretaceous, the ratio was 1 to 1, and in the Oligocene–Miocene, there was five times as much type III kerogen and coal than type II. The worldwide marine regressions of the Pennsylvanian–Lower Permian and Oligocene–Miocene were the only stratigraphic intervals of the six principal intervals in which there was more type III kerogen and coal formed than types I and II.

A particularly significant observation by Klemme and Ulmishek is that about 80% of the world's recoverable reserves were generated and trapped in the last 100 m.y., as shown in Figure 16-11. The time of maturation of many of the source rocks of Figure 16-10 was long after their times of deposition. For example, the 400-million-year-old Silurian source rocks of the northern Sahara of Algeria were deposited on a platform that was not buried deep enough to generate oil until the end of Albian time. This was 300 million years after their deposition (Poulet and Roucache 1969). All the oil and gas in the Hassi Massaoud field of Algeria was formed from the Silurian source rocks in the last 100 million years. According to Figure 16-11, only 2% of the world's present original recoverable petroleum reserves were formed by the end of the Lower Permian, about 250 million years ago.

Klemme and Ulmishek estimated that about 6% of the world's petroleum reserves are bacterial gas from still-immature source rocks, mostly in northern west Siberia.

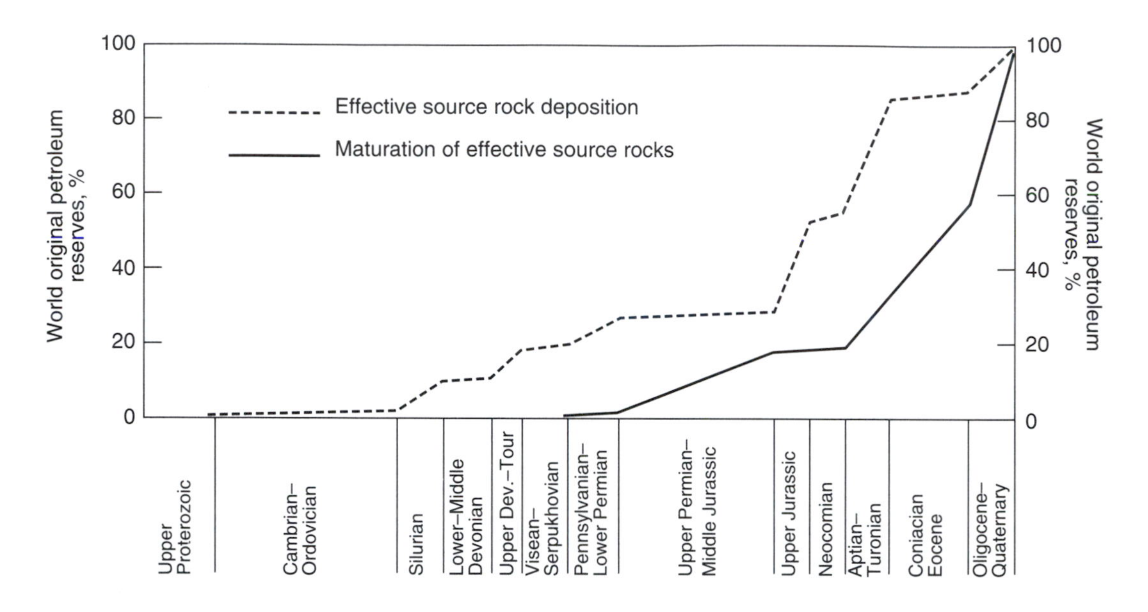

Figure 16-11

Cumulative chart of effective source rock deposition and source rock maturation in the stratigraphic succession, given as a percentage of world's original petroleum reserves. [Klemme and Ulmishek 1991]

SUMMARY

1. The major objective of exploration geochemistry is to reduce the risk of drilling dry holes, by applying geochemical concepts in the exploration programs. Case studies have shown that forecasting efficiencies are considerably improved by using geochemistry combined with geology to rank future prospects.

2. Risk reduction involves identifying and selectively mapping the quantity, quality, and maturation of petroleum source rocks, the possible migration pathways from the mature sources to the prospective traps, and the effectiveness of regional or local seals.

3. A petroleum system includes a source rock, migration paths, reservoir rocks, seals, traps, and the geologic processes that created each of them. One sedimentary basin may have several petroleum systems.

4. The generation-accumulation efficiency (GAE) of petroleum systems ranges from 1 to 15%. The expulsion-accumulation efficiency ranges from 1 to more

than 30%. In small, focused, vertical migration systems such as Ekofisk, the latter efficiency is 31%.

5. A basin model study of the Southern Norwegian shelf found three critical factors leading to petroleum accumulations in this part of the Central Graben of the North Sea: (1) the size of the trap, (2) the maturity of the source rock under the crest of the structure, and (3) the presence of faults acting as migration pathways from source to reservoir. In order to contain petroleum, the most attractive structural closures still require mature source rocks within the drainage area of the structure.

6. Hydrous pyrolysis data indicate that about 16 to 33 wt% of a type II kerogen is converted to expelled oil and gas, up to a maturity of $R_o = 1.5\%$. For type I kerogens, the yields range between 43 and 63%, and for humic type III kerogens, between 0.7 and 13%.

7. The genetic classification of petroleum systems uses three factors: (1) the petroleum charge available for trapping, (2) the migration–drainage style, and (3) the entrapment style.

8. Source potential index (SPI) maps combined with maturity maps can be used to highlight regional trends where geophysically identified structures or traps are most likely to be charged with petroleum.

9. To be classified as supercharged, normally charged, or undercharged, all vertical drainage systems require higher source potential index values than do lateral drainage systems.

10. The entrapment style is defined by the degree of impedance. High-impedance systems have laterally continuous regional seals combined with a moderate-to-high degree of structural deformation. Low-impedance systems have either a low degree of structural deformation or a low degree of regional seal effectiveness.

11. About 91.5% of the world's discovered original reserves of oil and gas came from source rocks of six stratigraphic intervals, representing only 35% of Phanerozoic time. The six intervals are (1) the Silurian, which generated 9% of the world's recoverable reserves; (2) the Upper Devonian (Tournaisian) (8%); (3) the Pennsylvanian–Lower Permian (8%); (4) the Upper Jurassic (25%); (5) the Middle Cretaceous (29%); and (6) the Oligocene–Miocene (12.5%).

12. Four of these intervals occurred during the worldwide marine transgressions that deposited type II kerogen in silled marine basins. The remaining two, Pennsylvanian–Lower Permian and Oligocene–Miocene, deposited mainly type III kerogens and coals during extensive regressions.

13. Two-thirds of the world's original recoverable petroleum reserves were generated in Tethyan basins, which cover less than one-fifth of the world's land area and continental shelves. About 78% of the original reserves in the aforementioned six stratigraphic intervals occur in platforms, circular sags, and linear sags.

14. Rift–sag structural forms favorable for developing anoxic silled basins were formed mainly during Tethyan tectonics.

15. About 60% of the world's conventional recoverable petroleum migrated vertically upward or stratigraphically downward from the source rock to the reservoir.

16. About 80% of the world's recoverable petroleum was generated and trapped in the last 100 million years.

17. A mass balance of the Global Petroleum System indicates that 40% of the generated oil is lost at the surface, 58% is dispersed as oil, gas, and pyrobitumen between source and reservoir type rocks in petroliferous basins and 2% is in oil accumulations.

18. The estimated reserves plus ultimate undiscovered resources for recoverable conventional oil and for gas are the same. Each is about 2×10^{12} barrels of oil equivalent (BOE).

SUPPLEMENTARY READING

Lewan, M. D., J. B. Comer, T. Hamilton-Smith, N. R. Hasenmueller, J. M. Guthrie, J. R. Hatch, D. L. Gautier, and W. T. Frankie. 1995. *Feasibility study on material-balance assessment of petroleum from the New Albany Shale in the Illinois Basin.* U. S. Geological Survey Bulletin 2137.

Magoon L. B., and W. G. Dow (eds.). 1994. *The petroleum system—From source to trap.* AAPG Memoir 60. Tulsa: American Association of Petroleum Geologists, 655 p.

Appendix 1

Units of Measurement

The units of measurement used in this book include both those customarily used in the petroleum industry and the international system of units, known as SI. A few examples of SI units and symbols follow. For more details, see the *Metric Practice Guide,* published by the American Society for Testing and Materials (1974).

Quantity	*Unit*	*SI symbol*	*Formula*
area	square meter		m^2
density	kilogram per cubic meter		kg/m^3
force	newton	N	$kg \times m/s^2$
length	meter	m	
mass	kilogram	kg	
pressure	pascal	Pa	N/m^2
time	second	s	
volume	cubic meter		m^3

SI Prefixes

Multiple and submultiple prefixes are used to indicate orders of magnitude, thus eliminating insignificant digits and decimals. It is preferable to use them in steps of 1,000, as shown.

Multiplication factor	*Prefix*	*SI symbol*
10^{12}	tetra	T
10^{9}	giga	G
10^{6}	mega	M
10^{3}	kilo	k
10^{-3}	milli	m
10^{-6}	micro	μ
10^{-9}	nano	n
10^{-12}	pico	p

Note that M is 10^3 and MM is 10^6 when describing cubic feet of gas in the English system (see abbreviations).

Conversions

1,000 ft^3 = 28.3 m^3

1 ft^3 gas/barrel water = 0.1812 m^3/kiloliter

1 metric ton oil = 680 m^3 gas (chemical conversion basis)

1 barrel oil = 3,200 ft^3 gas (chemical conversion basis)

1 metric ton oil = 1,280 m^3 gas (BTU basis)

1 barrel oil = 6,040 ft^3 gas (BTU basis)

1 barrel oil = 42 gallons = 0.159 m^3

1 acre-foot = 1,233.5 m^3

1 atmosphere = 101.3 kilopascals (kPa)

1 kilogram-force/m^2 = 9.807 Pa

1 pound-force/square inch (psi) = 6.895 kPa

1°F/100 ft = 18.2°C/km

1 calorie = 4.18 joules

Density and Specific Volumes of Petroleum at 15.6°C (60°F)

°API gravity	*Specific gravity*	*Barrels per metric ton*
0	1.076	5.86
10	1.000	6.30
15	0.9659	6.53
20	0.9340	6.75
26	0.8984	7.02
30	0.8762	7.19
36	0.8448	7.46
40	0.8251	7.64
46	0.7972	7.91
50	0.7796	8.09
60	0.7389	8.53

Note: Gravity, °API $= \dfrac{141.5}{\text{specific gravity } 60°/60°\text{F}} - 131.5$

Source: Data from Levorsen 1967, p. 687.

Appendix 2

Geologic Time Scale

Era	North America			Europe			Approx. Age
	Period	Epoch	Age	Period	Epoch	Age	10^6 Years
Cenozoic	Quaternary	Recent		Quaternary	Holocene		0.01
Cenozoic	Quaternary	Pleistocene	Wheelerian	Quaternary	Pleistocene	Tyrrhenian Calabrian	2.0
Cenozoic	Tertiary	Pliocene	Venturian	Neogene	Pliocene	Piacenzian	5.1
Cenozoic	Tertiary	Miocene	Mohnian Saucesian	Neogene	Miocene	Messinian Aquitanian	24
Cenozoic	Tertiary	Oligocene	Zemorrian	Paleogene	Oligocene	Chattian	38
Cenozoic	Tertiary	Eocene	Jacksonian Claibornian Wilcoxian	Paleogene	Eocene	Bartonian	55
Cenozoic	Tertiary	Paleocene	Midwayan	Paleogene	Paleocene	Danian	65

Geologic Time Scale *(continued)*

Era	North America			Europe			Approx. Age
	Period	Epoch	Age	Period	Epoch	Age	10^6 Years
Mesozoic	Cretaceous	Upper	Gulfian	Cretaceous	Upper	Maastrichtian Senonian Turonian Cenomanian	100
		Lower	Comanchean		Lower	Albian Aptian Neocomian	144
	Jurassic	Upper Middle Lower	Kimmeridgian Bathonian Toarcian	Jurassic	Upper Middle Lower	Malm Dogger Lias	213
	Triassic	Upper Middle Lower		Triassic	Upper Middle Lower	Rhaetian Anisian Scythian	248
Paleozoic	Permian	Upper Lower	Ochoan Guadalupian Leonardian Wolfcampian	Permian	Upper Lower	Zechstein Rotliegendes	286
	Pennnsylvanian	Upper Middle Lower	Virgilian Missourian Desmoinesian Atokan Morrowan	Carboniferous	Upper	Stephanian Westphalian Namurian	320
	Mississippian	Upper Middle Lower	Chesterian Meramecian Osagean Kinderhookian		Lower	Viséan Tournaisian	360

Geologic Time Scale *(continued)*

	North America			Europe			Approx. Age
Era	Period	Epoch	Age	Period	Epoch	Age	10^6 Years
Paleozoic	Devonian	Upper	Chatauquan Senecan	Devonian	Upper	Famennian Frasnian	360
		Middle	Erian		Middle	Givetian Couvinian	
		Lower	Ulsterian		Lower	Siegenian Gedinnian	
	Silurian	Upper	Cayugan	Silurian	Upper	Ludlovian Wenlockian	408
		Middle	Niagaran				
		Lower	Medinan		Lower	Llandoverian Valentian	
	Ordovician	Upper	Cincinnatian	Ordovician	Upper	Ashgillian	438
		Middle	Champlain-ian		Middle	Caradocian	
		Lower	Canadian		Lower	Arenigian	
	Cambrian	Upper	Croixan	Cambrian	Upper	Tuorian	505
		Middle	Albertan		Middle	Amgan	
		Lower	Waucoban		Lower	Aldanian	
Pre-cambrian	Proterozoic	Upper	Keweenawan	Proterozoic	Upper	Vendian	590
					Middle	Riphean	
		Lower	Huronian		Lower	Huronian	

Glossary*

Abnormal Pressure Any departure from hydrostatic pressure. Overpressures range above 12 kPa/m (0.53 psi/ft), and underpressures range below 9.8 kPa/m (0.43 psi/ft).

Activation Energy (E) The energy that must be absorbed by a molecule, or molecular complex, to break the bonds and form new products. It is expressed in kilocalories or kilojoules per mol.

Angiosperms Flowering land plants that originated in Early Cretaceous time. Oleanane is a biomarker for angiosperms.

API Gravity The density of oil measured with a hydrometer having a scale developed by the American Petroleum Institute. See Appendix 1, "Units of Measurement," for the correlation of API with specific gravity. API gravity is used worldwide to determine the market price of crude oil. *Light oil* is >31.1°API; *medium oil,* 22.3 to 31.1°API; *heavy oil* 10 to 22.3°API; and *extra heavy oil* <10°API.

Aromatic (Arene) (AR) Hydrocarbons containing one or more benzene rings. Monoaromatics have the molecular formula C_nH_{2n-6}. Benzene, toluene, and the xylenes are arenes. Polycyclic aromatic hydrocarbons (PAH) contain several rings with two or more carbon atoms shared between rings.

Asphalt Black to dark brown solid or semisolid bitumens that gradually liquefy when heated. They are composed principally of the elements carbon and hydrogen but also contain appreciable quantities of nitrogen, sulfur, and oxygen. They are usually soluble in carbon disulfide. Petroleum refinery asphalts are either straight-run residues from distilling crude oils or "blown" asphalts produced by the air oxidation of crude

* Miles (1989) has published a detailed 137-page illustrated glossary of petroleum geochemistry.

residues. Asphalts contain heavy oils, resins, asphaltenes, and high-molecular-weight waxes.

Asphaltenes Asphaltic constituents of crude oil that are soluble in carbon disulfide but insoluble in petroleum ether or *n*-pentane. Asphaltenes are agglomerations of molecules with condensed aromatic and naphthenic rings connected by paraffin chains. They have molecular weights in the thousands (see Plate 2A).

Asphaltite Black to dark brown, comparatively hard, solid bitumens that soften at temperatures above 110°C (230°F) and are usually soluble in carbon disulfide. Examples are gilsonite, glance pitch, and grahamite.

Asymmetric Carbon (Center) A carbon atom bonded to four different organic groups, as shown in the mirror images in Box 4-1.

Bacterial Gas Methane formed by bacteria (methanogens) utilizing carbon dioxide and hydrogen.

Base of Crude Oil The "base" of a crude oil describes the chemistry of its main constituents. A paraffin-base oil contains predominantly paraffinic hydrocarbons. An intermediate- or mixed-base crude contains roughly equivalent mixtures of paraffins and naphthenes (cycloparaffins). A naphthenic-base crude contains predominantly naphthene hydrocarbons. An asphalt-base crude is one containing a relatively high proportion of nonhydrocarbon constituents such as nitrogen, sulfur, and oxygen compounds. The term *aromatic base* is not used because there are no oils known to contain predominantly aromatic hydrocarbons.

Biodegradation The destruction of petroleum and related bitumens by bacteria. At temperatures below 88°C, the petroleum in reservoirs, oil seeps, and asphalt paving and the gasoline in storage tanks are susceptible to bacterial degradation, which converts hydrocarbons to alcohols, acids, and other water-soluble products.

Biological Markers (Biomarkers) Organic compounds whose carbon structure, or skeleton, is formed by living organisms and is sufficiently stable to be recognized in crude oil or in the organic matter of ancient sediments. Typical markers are the porphyrins, pristane, phytane, steranes, carotanes, and pentacyclic triterpanes.

Bitumens Native substances of variable color, hardness, and volatility, composed principally of the elements carbon and hydrogen and sometimes associated with mineral matter, the nonmineral constituents being largely soluble in carbon disulfide.

Bubble Point The pressure at which gas starts to come out of solution.

C_{15+} Fraction The fraction of a crude oil or rock extract containing primarily hydrocarbons with 15 or more carbon atoms. Usually it contains some hydrocarbons down to C_{12}. The C_{15+} fraction is used in oil–oil and oil–source rock correlations. It also contains most of the biomarkers.

Carbohydrates Organic compounds with the approximate general formula $(C \cdot H_2O)_n$, where n is equal to or greater than 4. Sucrose (table sugar), glucose, starch, and cellulose are carbohydrates.

Carbon Preference Index (CPI) The ratio of odd to even chain-length normal paraffins in a specific molecular weight range. Originally, Bray and Evans (1961, p. 9) calculated this as the ratio of the sum of the mole percentages of C_{25}–C_{33} odd-carbon *n*-paraffins to the sum of C_{26}–C_{34} even-carbon *n*-paraffins. Slightly modified ratios are used today, such as that in Table 4-6.

Carbon Ratio Theory The theory that the API gravity of oil increases (specific gravity decreases) as the carbon ratio of coals in the same area increases. As coals increase in rank, oils become lighter, eventually changing to gas. The carbon ratio is the ratio of fixed carbon to total carbon (fixed plus volatile) in a coal.

Carotenoids Plant pigments composed of mono- and dicyclic tetraterpenes. Carotenes are the precursors of vitamin A. The yellow, brown, and red colors of leaves in the fall are caused by carotenoids. Carotanes are the alkanes derived from carotenoids.

Casing-Head Gasoline The liquid hydrocarbon recovered from casing-head gas by means of adsorption, compression, or refrigeration. Casing-head gas is the gas recovered at the surface from an oil well.

Catagenesis The process by which organic material in sedimentary rocks is thermally altered by increasing temperature. Catagenesis covers the temperature range between diagenesis and metagenesis, approximately 50–200°C (122–392°F).

Chitin A polysaccharide (carbohydrate polymer) with a structure similar to cellulose, except that two OH groups are replaced by CH_3ONH groups in each $(C \cdot H_2O)_6$ unit. Chitin forms the horny, hard, outer cover of insects and crustaceans and parts of some other invertebrates.

Chromatography A method for separating mixtures of compounds based on the differences in their physical and chemical properties. Gas chromatography (GC) involves passing petroleum in a gas phase mixed with helium or nitrogen through a long column coated with a stationary heavy-liquid phase. The adsorption and desorption of gaseous hydrocarbons on the liquid result in their separation by molecular size and type. Liquid chromatography (LC) involves passing a solution of petroleum in *n*-C_6 or *n*-C_7 through a column of an adsorbent such as alumina or silica gel to separate the saturated hydrocarbons. Following this process, the aromatics, NSOs, and resins are separated by desorption with increasingly polar solvents like toluene and methanol. High-performance liquid chromatography (HPLC) separates high-molecular-weight aromatics and porphyrins.

Coal A readily combustible rock containing more than 50% by weight, and more than 70% by volume, of organic material formed from the compaction or induration of variously altered plant remains. Humic coals are formed from plant cell and wall material deposited under aerobic conditions, whereas sapropelic coals are formed from spores, pollen, and algae deposited under anaerobic conditions.

Coal Maceral Microscopically recognizable constituents of coal that can be differentiated by their morphology. Macerals are analogous to the minerals of inorganic rocks but differ in having less uniform chemical composition and physical properties. The carbon content of macerals increases with increasing temperature. The major maceral groups are vitrinite, liptinite, and inertinite.

Condensate A hydrocarbon mixture that is gaseous in its reservoir but condenses into liquid when produced. Its gravity usually ranges from 55°API upward.

Connate Water Fossil reservoir water that has not been in contact with the atmosphere since its deposition. It is high in chloride and calcium and frequently contains more than 100,000 ppm total dissolved solids (TDS).

Crude Oil A petroleum that is removed from the earth in a liquid state or is capable of being so removed.

Crude Oil Fractions **Gasoline,** the fraction of crude oil boiling between about 15 and 200°C (60 and 392°F). **Kerosine,** the fraction of crude oil boiling between about 200 and 260°C (392 and 500°F). **Gas oil,** the fraction of crude oil having a viscosity less than 50 seconds S.U. (Saybolt Universal) at 38°C (100°F) and a boiling range between about 260 and 332°C (500 and 630°F). **Lubricating oil,** the fraction of crude oil having a viscosity above 50 seconds S.U. at 38°C (100°F) and a boiling range between about 332 and 421°C (630 and 790°F). **Residuum,** the residue obtained from the distillation of crude oil after all fractions, including lubricating oils, have been taken off.

Decarboxylation The loss of CO_2 from an organic compound such as the conversion of acids to alkanes.

Diagenesis The process involving the biological, physical, and chemical alteration of the organic debris in sediments without a pronounced effect from rising temperature. It covers the temperature range up to about 50°C (122°F).

Distillate The hydrocarbon fluid produced from processing natural gas. It is denser than condensate and generally is run into the tanks with the crude oil. Gravities are from 50°API upward.

Drill-Stem Test (DST) A test of the productive capacity of a well while it is still full of drilling mud. The testing tool, attached to the drill pipe, is lowered into the hole and placed opposite the formation to be tested. Packers are set to shut off the weight of the drilling mud, and the tool is opened to permit the flow of any formation fluid into the drill pipe, where the flow is measured.

Dry Gas Natural gas consisting principally of methane and devoid of readily condensable constituents such as gasoline. Dry gas contains less than 0.1 gallon natural gas liquid vapors per 1,000 ft^3 (1.3 liters per 100 m^3).

Euxinic An anoxic restricted depositional environment (barred basin) such as the Black Sea.

Fragmentogram *See* Mass Chromatogram.

Gas Hydrates Crystalline compounds in which the ice lattice of H_2O expands to form cages that contain the gas molecules. Methane hydrates can hold eight methane molecules in each of the 46 H_2O molecules, a formula of $CH_4 \cdot 5.75\ H_2O$.

Hempel Distillation A method of distilling an oil into 15 fractions and the residuum. The first fraction consists of everything distilling up to 50°C (122°F). The next nine fractions are taken at intervals of 25°C (45°F) up to 275°C (527°F) under atmospheric pressure. The last five fractions are distilled under vacuum at 40 mm pressure. The eleventh fraction consists of all materials boiling up to 200°C (392°F), and the succeeding four fractions are taken at intervals of 25 to 300°C (572°F) at 40 mm. The base of the crude and the quantities of gasoline and other constituents in the crude can be calculated from the distillation data.

Heterocyclic Compounds Ring compounds in which one or more of the carbon atoms in the ring is replaced by an atom of another element such as nitrogen, sulfur, or oxygen.

Hilt's Law The volatile carbon in coal decreases proportionally with the depth of the coal in normal stratigraphic sequences. The nonvolatile or fixed carbon of coals increases with increasing depth and temperature.

Homologous Series Compounds having similar structures but differing in the addition or subtraction of a CH_2 group.

Humic Organic Matter The decomposition and polymerization products of the lignin, tannins, and cellulose of plant cell and wall material plus carbonized organic matter deposited in swamps and soils under aerobic conditions with a partial restriction of oxygen. Kerogen types III and IV are humic.

Hydrocarbon A compound composed of only the elements hydrogen and carbon. Bitumens such as petroleum are composed principally, but not exclusively, of hydrocarbons.

Hydrostatic Gradient The pressure increase with depth of a liquid in contact with the surface. The gradient for fresh water is 9.8 kPa/m (0.433 psi/ft).

Inertinite The coal maceral group that shows little or no reaction during the coking process. Inertinite includes fusinite (fossil charcoal), sclerotinite derived from fungal remains, and other high-carbon materials with a low-hydrogen content and a high reflectance.

Isomers Molecules that have the same number and kinds of atoms but are different substances. Structural isomers differ in the way that the atoms are linked together, for example, *n*-butane and isobutane. Stereoisomers differ in the spatial arrangements of groups, for example, cis- and trans- (boat and chair) isomers of 1,2-dimethylcyclopropane. Optical isomers are nonidentical mirror images comparable to right- and left-handed gloves.

Isoprenoid A hydrocarbon whose molecular structure contains the basic unit isoprene, consisting of five carbon atoms, with a branch at the second atom, as shown.

$$\begin{array}{c} \quad C \\ \quad | \\ C=C-C=C \end{array}$$

This is the basic building block of many natural products such as terpenoids, steroids, carotenoids, pristane, and phytane. Many petroleum hydrocarbons are diagenetic derivatives of isoprenoid polymers. The designations ip-18, ip-19, and ip-20, as in Figure 4-14, refer to isoprenoids with 18, 19, and 20 carbon atoms, the latter two being pristane and phytane, respectively.

Isotopes Atoms whose nuclei contain the same number of protons but a different number of neutrons. All carbon atoms have six protons, but three carbon isotopes contain six, seven, and eight neutrons, giving atomic masses of 12, 13, and 14 (written as ^{12}C, ^{13}C, and ^{14}C).

Kerogen The disseminated organic matter of sedimentary rocks that is insoluble in nonoxidizing acids, bases, and organic solvents. The organic matter initially deposited with unconsolidated sediments is not kerogen but a precursor that is converted to kerogen during diagenesis. When heated, sapropelic kerogens yield oil and gas, and humic kerogens yield mainly gas. Kerogen includes both marine- and land-derived organic matter, the latter being identical to the components of coal.

Kitchen The volume of source rock that is generating or has generated and expelled petroleum to a specific structure or closely related structures and traps. Thus the Troll Kitchen and the Oseberg Kitchen of the North Sea are the specific rock volumes expelling oil to those fields.

Lipids A broad term that includes all oil-soluble, water-insoluble substances such as fats, waxes, fatty acids, sterols, pigments, and terpenoids.

Liptinite (Exinite) A coal maceral group that is the dominant organic constituent of boghead coals. Liptinite macerals include sporinite, cutinite, resinite, and alginite, which are derived from spores and pollen, cuticles, resins, and algae, respectively. Bituminite is an amorphous liptinite maceral. Liptinite is widely disseminated in sediments and is an important source of crude oil. The kerogen of oil shales is mostly of liptinite origin.

Lithostatic Gradient The total pressure increase with depth caused by rock grains and water. It averages about 24.4 kPa/m (1.08 psi/ft).

Maceral *See* Coal Maceral.

Mass Chromatogram (Fragmentogram) A recording of the intensity of a specific molecular ion versus the GC retention time, for example, the patterns in Box 4-1 and Figure 13-3. These display the chromatograms for the m/z 218 and 217 sterane ions, respectively. The "fingerprints" show the carbon number and isomer distributions for each ion. Used for correlation. See also Figures 15-24 through 15-28.

Mass Spectrum A plot of the relative intensities of ions formed versus the mass-to-charge (m/z) ratio. Used primarily for identifying compounds.

Mercaptans Compounds containing the sulfhydryl (–SH) group. They are sulfur analogs of the alcohols in that the oxygen in the alcohol (–OH) group is replaced by sulfur.

Metagenesis The high-temperature alteration of organic matter in sedimentary rocks to graphitic residues and methane. The temperature range is from 200°C (392°F) to >250°C (482°F), with vitrinite reflectances changing from 2 to >5%R_o. Metagenesis overlaps with rock metamorphism in which the greenschist facies begin to appear.

Metamorphism The transformation of preexisting rocks into new types by the action of heat, pressure, stress, and chemically active migrating fluids. Metamorphism usually begins at temperatures above 200°C (392°F). At such temperatures, the organic matter is already reduced to a low-hydrogen carbon residue capable of yielding only small amounts of gas.

Meteoric Water Fresh surface water entering a subsurface sedimentary section through permeable outcrops. It is high in sodium, bicarbonate, and sulfate and usually contains less than 10,000 ppm total dissolved solids (TDS).

Mineral Wax A species of bitumen having a characteristic luster and unctuous feel, composed principally of saturated hydrocarbons and containing considerable crystallizable paraffins, the nonmineral constituents being soluble in carbon disulfide. An example is ozocerite.

Naphtha The 60 to 100°C distillation fraction of petroleum (C_6 + C_7). It is used as a solvent and paint thinner.

Naphthene (Cycloalkane, Cycloparaffin) A hydrocarbon ring with the molecular formula C_nH_{2n}. Cyclopentane (C_5) and cyclohexane (C_6) ring structures are the most common in petroleum. Condensed or polycyclic naphthenes contain rings in which two or more carbon atoms are shared. Tetracyclic and pentacyclic naphthenes contain four and five rings, respectively, fused together.

Naphthenic Acids Petroleum acids containing a naphthene or cycloparaffin structure. The most common naphthenic acids contain a cyclopentane (5-carbon atom) ring.

Natural Gas A petroleum consisting of varying proportions of gaseous hydrocarbons such as methane, ethane, propane, and isobutane and occasionally containing liquid hydrocarbons such as pentanes and hexanes and nonhydrocarbon gases such as carbon dioxide, hydrogen sulfide, nitrogen, hydrogen, and helium.

Oil Shale A compact rock of sedimentary origin with an ash content of more than 33% and containing organic matter that yields oil when destructively distilled but not appreciably when extracted with petroleum solvents.

Oil Window The depth–temperature interval in which a petroleum source rock generates and expels most of its oil. Oil windows are formed mainly in the subsurface temperature range of 60°C (140°F) to 160°C (320°F).

Organic Facies Organic facies are mappable subdivisions of stratigraphic units distinguished from the adjacent subdivisions by the character of their organic matter. Different organic facies generate and expel different amounts and types of oil and gas.

Paraffin (Alkane) A hydrocarbon with the molecular formula C_nH_{2n+2}. It includes normal straight-chain paraffins and branched alkanes, such as methane, ethane, propane, and isobutane.

Paraffin Dirt Yellow brown, gummy, soft organic matter composed of nitrogenous–humic–cellulosic remains of plant material heavily impregnated with fungi, yeasts, actinomyces, and bacteria. The word *paraffin* is a misnomer, since this substance contains less than 3% lipid material. The waxy appearance is caused by living and dead microbial cells. Paraffin dirt tends to accumulate in moist soils near hydrocarbon seeps.

Petroleum A species of bitumen composed principally of hydrocarbons and existing in the gaseous or liquid state in its natural reservoir.

Petroleum Ether The 20 to 60°C distillation fraction of petroleum (C_5 + C_6). It is used as a solvent and paint thinner.

Phytane A saturated isoprenoid containing 20 carbon atoms.

Polar Compounds Organic compounds with molecular regions of partial positive and negative charges, such as the double bonds of aromatics and acids. All nitrogen, sulfur, and oxygen compounds and aromatic compounds are polar. Alkanes (paraffins plus naphthenes) are nonpolar. Most polar compounds are in the heavy fractions of petroleum.

Porphyrins Porphyrins are organometallic derivatives of the green chlorophyll in plants and the red hemoglobin in blood. These have a basic structure consisting of four interconnected rings, each ring containing four carbon atoms and one nitrogen atom (see Figure 4-8). In living organisms, hemoglobin is complexed with iron, and chlorophyll is complexed with magnesium. The porphyrins in oil are complexed with vanadium, nickel, iron, copper, and manganese.

Pour Point The temperature at which crude oil will not flow when a tube containing it is first heated in a bath to dissolve all the wax and then cooled slowly. At cooling intervals of 3°C (5°F), the tube is held horizontal until there is no flow for 5 seconds.

Primary Migration The movement of oil and gas within the fine-grained portion of a mature source rock.

Pristane A saturated isoprenoid containing 19 carbon atoms.

Proteins High-molecular-weight polymers of amino acids that constitute more than 50% of the dry weight of animals. The organic nitrogen and sulfur of living organisms are concentrated in the protein fraction. Gelatin, albumin, collagen (connective tissues), keratin (hair, hooves, nails), and serum globulins are proteins.

Pyrobitumen Black to dark brown, hard bitumens that are infusible and relatively insoluble in carbon disulfide. Albertite, wurtzilite, and impsonite are pyrobitumens.

Repeat Formation Tester A wireline device that measures up to 30 formation pressures an hour to within about 6.9 kPa (1 psi). It can also extract formation waters for analysis.

Resin Petroleum resins are the fraction of residuum that is insoluble in liquid propane but soluble in normal pentane. Plant resins are terpenoids ranging in molecular size from sesquiterpenes (C_{15}) to tetraterpenes (C_{40}). They contain the olefinic double bonds of the isoprene building block that, when exposed to air, causes the liquids to polymerize and oxidize to hard resins. Balsam and mastic are plant resins.

Resinite A coal maceral in the liptinite group. Resinites are fossil tree resins derived from balsam, mastic, latex, and some plant gums, waxes, oils, and fats. Amber and copal are resinites.

Retention Time The time required for an individual compound to pass through a chromatographic column. Examples are shown in Figures 15-7, 15-24, 15-27, and 15-28. Compounds are separated according to their different retention times.

Sapropelic Organic Matter The decomposition and polymerization products of high-lipid organic materials, such as spores and planktonic algae deposited in subaquatic muds (marine or lacustrine) under predominantly anaerobic conditions. Kerogen types I and II are sapropelic.

Saturates (Saturated Hydrocarbons) A general term that includes normal and branched alkanes and cycloalkanes (paraffins and naphthenes). Saturates are the nonaromatic hydrocarbon fraction of an oil. They are saturated with hydrogen; that is, they have no double bonds between carbon atoms.

Secondary Migration The movement of petroleum along permeable pathways beyond the mature source rock until it forms a pool in a structural or stratigraphic trap.

Sour Oil An oil containing noticeable quantities of noxious sulfur compounds such as hydrogen sulfide and mercaptans. A sweet oil does not contain these compounds.

Source Rock A fine-grained rock that in its natural setting has generated and released enough hydrocarbons to form a commercial accumulation of oil or gas.

Stereoisomers Molecules having the same molecular formula and the same atomic bonds but a different spatial arrangement of their atoms around one or more asymmetric carbon atoms. Stereoisomers have different retention times on mass chromatograms (for example, see Figure 15-24).

Tar A thick, black, or dark brown viscous liquid obtained by the destructive distillation of coal, wood, or peat. Tar is not a natural product, and it is incorrect to refer to asphalt deposits or seeps as *tars*.

Terpenoid An isoprenoid polymer usually containing 10, 15, 20, 30, or 40 carbon atoms (mono-, sesqui-, di-, tri-, and tetra-, respectively). Terpenoids include hydrocarbons, alcohols, and acids. Among the common terpanes (hydrocarbons) are the hopanes (pentacyclic triterpanes, see Figure 4-11).

Tertiary Migration The movement of an entire oil or gas pool to a different location.

Thermal Alteration Index (TAI) A maturation color index for the particulate organic matter of sedimentary rocks. The index indicates the degree of thermal alteration that the organic matter has undergone. As proposed by Staplin (1969), the index numbers 1 to 5 include color changes from yellow to brown to black, representing immature, mature, and metamorphosed facies of organic matter.

Unsaturates Compounds that have double or triple bonds between carbon atoms, which means that they are unsaturated with respect to hydrogen. Aromatics, olefins, resins, and asphaltenes are unsaturated compounds in petroleum.

Viscosity Index (VI) A series of numbers ranging from 0 to 100 that indicate the rate of change of viscosity with temperature. A VI of 100 indicates a small change in viscosity between the temperatures of 38 and 99°C (100 and 210°F), whereas a VI of 0 indicates a large change.

Vitrinite A coal maceral group that is the dominant organic constituent of humic coals. Vitrinite forms the familiar brilliant black bands of coal. Macerals in the vitrinite group include telinite derived from plant cell walls and collinite from the cell filling. Vitrinite particles are found in about 80% of the shales and sandstones of sedimentary basins.

Wax *See* Mineral Wax.

Wet Gas Natural gas consisting of methane and heavier hydrocarbons. The natural gas liquid vapors amount to 4 or more liters per 100 m^3 (0.3 or more gallons per 1,000 ft^3).

References

Abrams, M. A. 1992. Geophysical and geochemical evidence for subsurface hydrocarbon leakage in the Bering Sea, Alaska. *Marine and Petroleum Geology,* 9, April, 208–221.

Aizenshtat, Z. 1973. Perylene and its geochemical significance. *Geochim. Cosmochim. Acta,* 37, 559–567.

Akhishev, I. M., R. Kh. Muslimov, N. P. Lebedev, and V. I. Troypol'skiy. 1974. Bitumen deposits of the Permian sediments of Tataria and prospects for their exploration. *Geol. Nefti Gaza,* 12 (3), 23–28.

Albaiges, J. 1980. Identification and geochemical significance of long chain acyclic isoprenoid hydrocarbons in crude oils. In A. G. Douglas and J. R. Maxwell (eds.), *Advances in organic geochemistry 1979*. Oxford: Pergamon Press, pp. 19–28.

Albaiges, J., J. Bordon, and W. Walker. 1985. Petroleum isoprenoid hydrocarbons derived from catagenic degradation of archaebacterial lipids. *Org. Geochem.*, 8 (4), 293–297.

Albaiges, J., J. Grimalt, J. M. Bayona, R. Risebrough, B. de Lappe, and W. Walker II. 1984. Dissolved particulate and sedimentary hydrocarbons in a deltaic environment. *Org. Geochem.*, 6 (4), 237–247.

Alexander, R., R. I. Kagi, and R. A. Noble. 1983a. Identification of a bicyclic sesquiterpene drimane and eudesmane in petroleum. *J.C.S. Chem. Comm.*, 226–228.

Alexander, R., R. I. Kagi, G. W. Woodhouse, and J. K. Volkman. 1983b. The geochemistry of some biodegraded Australian oils. *APEA Journal,* 23, 53–63.

Alexander, R., R. A. Noble, and R. I. Kagi. 1987. Fossil resin biomarkers and their application in oil to source-rock correlation, Gippsland Basin, Australia. *APEA Journal,* 27, 63–72.

American Society for Testing and Materials. 1974. *Standard metric practice guide. E*380–74. Philadelphia: American Society for Testing and Materials.

Ames, R. L., and L. M. Ross. 1986. Petroleum geochemistry applied to oilfield development, offshore Trinidad. *Transactions of the First Geological Conference of the Geological Society of Trinidad and Tobago, 1986,* 227–236.

Amosov, G. A., and T. A. Kozina. 1966. Interaction of crude oils and water in Sakhalin oil fields. *Internat. Geol. Rev.,* 9 (7), 883–889.

Anders, D. E., and P. M. Gerrild. 1984. Hydrocarbon generation in lacustrine rocks of Tertiary age, Uinta Basin, Utah–Organic carbon, pyrolysis yield, and light hydrocarbons. In J. Woodward, F. F. Meissner, and J. L. Clayton (eds.), *Hydrocarbon source rocks of the greater Rocky Mountain region*. Denver: Rocky Mountain Association of Geologists, pp. 513–524.

Anders, D. E., and W. E. Robinson. 1973. Geochemical aspects of the saturated hydrocarbon constituents of Green River oil shale–Colorado no. 1 core, U.S. Bureau of Mines, Report of Investigations 7737. Washington, DC: U.S. Bureau of Mines.

Anderson, R. N., L. M. Cathles III, and H. R. Nelson Jr. 1991. "Data cube" depicting fluid flow history in Gulf Coast sediments. Annual geophysical report. *Oil & Gas Journal,* November, 50–55.

Antonov, P. L., G. A. Gladysheva, and V. P. Kozlov. 1958. Diffusion of carbon dioxide across rock salt. *Petrol. Geol.*, 2 (2b), 175–178.

API RP 33. 1974. *Recommended practices for standard calibration and format for nuclear logs*. 3rd ed. Dallas: American Petroleum Institute, Division of Production.

Aquino Neto, F. R., J. N. Cardoso, R. Rodgrigues, and L. A. F. Trindade. 1986. Evolution of tricyclic alkanes in the Espirito Santo Basin, Brazil. *Geochim. Cosmochim. Acta*, 50, 2069–2072.

Aquino Neto, F. R., J. M. Trendel, A. Restle, J. Connan, and P. A. Albrecht. 1983. Occurrence and formation of tricyclic and tetracyclic terpanes in sediments and petroleums. In M. Bjorøy et al. (eds.), *Advances in organic geochemistry 1981.* Chichester: Wiley, pp. 659–667.

Arefev, O. A., M. N. Zabrodina, V. M. Makushina, and A. A. Petrov. 1980. Relic tetra- and pentacyclic hydrocarbons in the old oils of the Siberian Platform. *Izv. Akad. Nauk SSR, Ser. Geol.*, 3, 135–140.

Attaway, D. H., P. L. Parker, and J. A. Mears. 1970. Normal alkanes of five coastal spermatophytes. *Contributions in Marine Science,* 15, 13–19.

Ayers, M. G., M. Bilal, R. W. Jones, L. W. Slentz, M. Tartir, and A. O. Wilson. 1982. Hydrocarbon habitat in main producing areas, Saudi Arabia. *AAPG Bulletin*, 66, 1–9.

Bailey, N. J. L., R. Burwood, and G. E. Harriman. 1990. Application of pyrolysate carbon isotope and biomaker technology to organofacies definition and oil correlation problems in North Sea Basins. *Org. Geochem.*, 16 (4–6), 1157–1172.

Bailey, N. J. L., C. R. Evans, and C. W. D. Milner. 1974. Applying petroleum geochemistry to search for oil: Examples from Western Canada Basin. *AAPG Bulletin*, 58, 2284–2294.

Bailey, N. J. L., A. M. Jobson, and M. A. Rogers. 1973. Bacterial degradation of crude oil: Comparison of field and experimental data. *Chem. Geol.*, 11, 203–221.

Baird, R. A. 1986. Maturation and source rock evaluation of Kimmeridge clay, Norwegian North Sea. *AAPG Bulletin*, 70, 1–11.

Baker, D. R. 1962. Organic geochemistry of Cherokee Group in southeastern Kansas and northeastern Oklahoma. *AAPG Bulletin*, 46, 1621–1642.

Baker, D. R. 1972. Organic geochemistry and geological interpretations. *J. Geol. Ed.*, 21 (5), 221–234.

Baker, E. W. 1964. Vanadium and nickel in crude petroleum of South America and Middle East origin. *J. Chem. and Eng. News*, 42 (15), 307–308.

Baker, E. W., and J. W. Louda. 1986. Porphryins in the geological record. In R. B. Johns (ed.), *Biological markers in the sedimentary record.* Amsterdam: Elsevier Science, pp. 121–225.

Balashova, M. M., A. Z. Koblova, and V. M. Provorov. 1983. Late Precambrian petroleum formation in the northern Ural–Volga region. *Internat. Geol. Rev.*, 25, 1455–1458.

Barker, C. 1972. Aquathermal pressuring–Role of temperature in development of abnormal pressure zones. *AAPG Bulletin*, 56, 2068–2071.

Barker, C. 1987. Development of abnormal and subnormal pressures in reservoirs containing bacterially generated gas. *AAPG Bulletin,* 71, 1404–1413.

Barker, C. 1990. Calculated volume and pressure changes during the thermal cracking of oil to gas in reservoirs. *AAPG Bulletin*, 74, 1254–1261.

Barker, C. E. 1993. Implications for organic maturation studies of evidence for a geologically rapid increase and stabilization of vitrinite reflectance at peak temperature: Cerro Prieto geothermal system, Mexico: Reply. *AAPG Bulletin,* 77, 668–669.

Barker, C., L. Wang, and E. B. Butler. 1989. Distribution of bitumens in shales and its significance for petroleum migration. Paper presented at 14th International Meeting on Organic Geochemistry, Paris, September 18–22.

Barnard, P. C., B. S. Cooper, and M. Fisher. 1976. Organic maturation and hydrocarbon generation in the Mesozoic sediments of the Svedrup Basin,

Arctic Canada. Paper presented at 4th International Palynology Conference, Lucknow.

Barnes, M. A., and W. C. Barnes. 1983. Oxic and anoxic diagenesis of diterpenes in lacustrine sediments. In M. Bjorøy et al. (eds.), *Advances in organic geochemistry 1981*. New York: Wiley, pp. 289–298.

Barrick, R. C., and J. I. Hedges. 1981. Hydrocarbon geochemistry of Puget Sound Region: II. Sedimentary diterpenoid, steroid and triterpenoid hydrocarbons. *Geochim. Cosmochim. Acta*, 45, 381–392.

Baset, Z. H., R. J. Pancirov, and T. R. Ashe. 1980. Organic compounds in coal: Structure and origins. In A. G. Douglas and J. R. Maxwell (eds.), *Advances in organic geochemistry 1979*. Oxford: Pergamon Press, pp. 619–630.

Baskin, D. K., and K. E. Peters. 1992. Early generation characteristics of a sulfur-rich Monterey kerogen. *AAPG Bulletin*, 76, 1–3.

Becker, L. E., and J. B. Patton. 1968. World occurrence of petroleum in Pre-Silurian rocks. *AAPG Bulletin*, 52, 224–245.

Behar, F. H., and P. Albrecht. 1984. Correlations between carboxylic acids and hydrocarbons in several crude oils. Alteration by biodegradation. *Org. Geochem.*, 6, 597–604.

Behar, F. H., and M. Vandenbroucke. 1987. Chemical modelling of kerogens. *Org. Geochem.*, 11, 15–24.

Belayouni, H., and J. Trichet. 1984. Hydrocarbons in phosphatized and non-phosphatized sediments from the phosphate of Gafsa. *Org. Geochem.*, 6, 741–754.

BeMent, W. O. 1994. The temperature of oil generation as defined with a C_7 chemistry maturity parameter (2,4-DMP/2,3-DMP ratio). Paper presented at the First Joint AAPG/AMGP Hedberg Research Conference, Mexico City, October 2–6.

Bendoritis, J. G. 1974. Hydrocarbons of biogenic origin in petroleum–Aromatic triterpenes and bicyclic sesquiterpenes. In B. Tissot and F. Bienner (eds.), *Advances in organic geochemistry 1973*. Paris: Éditions technip, pp. 209–224.

Bergius, F. 1913. Production of hydrogen from water and coal from cellulose at high temperatures and pressures. *Journal Society Chem. Ind.*, 32, 462–467.

Bernard, B. B. 1978. Ph.D. diss., Texas A & M University.

Bernard, B. B., J. M. Books, and W. M. Sackett. 1976. Natural gas seepage in the Gulf of Mexico. *Earth and Planet. Sci. Letters*, 31, 48–54.

Bestougeff, M. A. 1967. Petroleum hydrocarbons. In B. Nagy and U. Columbo (eds.), *Fundamental aspects of petroleum geochemistry*. Amsterdam: Elsevier Science, pp. 73–108.

Bjorøy, M., P. B. Hall, and R. P. Moe. 1994. Variation in the isotopic composition of single components in the C_4–C_{20} fraction of oils and condensates. *Org. Geochem.*, 21 (6–7), 761–776.

Blount, C. W., L. C. Price, L. M. Wenger, and M. Tarullo. 1980. Methane solubility in aqueous NaCl solutions at elevated temperatures and pressures. In M. H. Dorfman and W. L. Fisher (eds.), *Proceedings of Fourth United States Gulf Coast Geopressured–Geothermal Energy Conference: Research and development, October 29–31, 1979.* Austin: University of Texas Press, vol. 3, pp. 1225–1270.

Blumer, M. 1967. Hydrocarbons in digestive tract and liver of a basking shark. *Science*, 156, 390–391.

Blumer, M., R. L. Guillard, and T. Chase. 1971. Hydrocarbons of marine phytoplankton. *Mar. Biol.*, 8 (3), 183–189.

Blumer, M., M. M. Mullin, and R. L. Guillard. 1970. A polyunsaturated hydrocarbon (3,9,12,15,18-heneicosa-hexane) in the marine food web. *Mar. Biol.*, 6 (3), 226–235.

Blumer, M., M. M. Mullin, and D. W. Thomas. 1964. Pristane in marine environments. *Helgol. Wiss. Meeresunters*, 10, 187–201.

Bockmuelen, H., C. Barker, and P. A. Dickey. 1983. Geology and geochemistry of crude oils, Bolivar coastal fields, Venezuela. *AAPG Bulletin*, 67, 242–270.

Bois, C., P. Bouche, and R. Pelet. 1982. Global geologic history and distribution of hydrocarbon reserves. *AAPG Bulletin*, 66, 1248–1270.

Bordenave, M. L. 1993. *Applied petroleum geochemistry*. Paris: Éditions technip.

Bordenave, M. L., and R. Burwood. 1990. Source rock distribution and maturation in the Zagros Orogenic Belt: Provenance of the Asmari and Bangestan Reservoir oil accumulation. *Org. Geochem.*, 16 (1–3), 369–387.

Bordenave, M. L., J. Espitalié, P. Leplat, J. L. Oudin, and M. Vandenbroucke. 1993. Screening techniques for source rock evaluation. In M. L. Bordenave (ed.), *Applied petroleum geochemistry.* Paris: Éditions technip, pp. 219–278.

Bordovskiy, O. K. 1965. Accumulation and transformation of organic substance in marine sediments. *Marine Geol.*, 3, (3–4), 1–114.

Borst, R. L. 1982. Some effects of compaction and geological time on the pore parameters of argillaceous rocks. *Sedimentology*, 29, 291–298.

Borst, R. L., and J. E. Smith. 1982. Pore size distributions of shales. *American Laboratory*, 14 (1), 56–61.

Boudou, J-P. 1984. Chloroform extracts of a series of coals from Mahakam Delta. *Org. Geochem.*, 6, 431–437.

Boyd, K. A., and R. M. Hiles. 1985. Oil and gas developments in Alaska in 1984. *AAPG Bulletin*, 69, 1485–1490.

Bradley, J. S. 1975. Abnormal formation pressure. *AAPG Bulletin*, 59, 957–973.

Bradley, J. S. 1976. Abnormal formation pressure: Reply. *AAPG Bulletin*, 60, 1127–1128.

Bradley, J. S. 1986. Fluid movement in deep sedimentary basins: A review. *Proceedings of 3rd Canadian/American Conference on Hydrogeology, Banff 86*. National Water Well Association, pp. 19–31.

Braduchan, Yu. V. 1990. Bazhenov horizon of west Siberia. *Petroleum Geology*, 24, (1–2), 1–36.

Bralower, T. J., and H. R. Thierstein. 1984. Low productivity and slow deep-water circulation in mid-Cretaceous oceans. *Geology*, 12, 614–618.

Brassell, S. C., and G. Eglinton. 1986. Molecular geochemical indicators in sediments. In M. L. Sohn (ed.), *Organic marine geochemistry*. ACS Symposium series 305, April. Washington, DC: American Chemical Society. pp. 10–31.

Brassell, S. C., G. Eglinton, and Fu Jiamo. 1986a. Biological marker compounds as indicators of the depositional history of Maoming oil shale. *Org. Geochem.*, 10, 927–941.

Brassell, S. C., C. A. Lewis, J. W. de Leeuw, F. de Lange, and J. S. Sinninghe Damsté. 1986b. Isoprenoid thiophenes: Novel products of sediment diagenesis? *Nature*, 320, 160–162.

Brassell, S. C., J. McEvoy, C. F. Hoffmann, N. A. Lamb, T. M. Peakman, and J. R. Maxwell. 1984. Isomerisation, rearrangement and aromatisation of steroids in distinguishing early stages of diagenesis. *Org. Geochem.*, 6, 11–23.

Brassell, S. C., A. M. K. Wardroper, I. D. Thomson, J. R. Maxwell, and G. Eglinton. 1981. Specific acyclic isoprenoids as biological markers of methanogenic bacteria in marine sediments. *Nature*, 290, 693–696.

Bray, E. E., and E. D. Evans. 1961. Distribution of *n*-paraffins as a clue to recognition of source beds. *Geochim. Cosmochim. Acta*, 22, 2–15.

Bray, E. E., and E. D. Evans. 1965. Hydrocarbons in nonreservoir-rock source beds: Part 1. *AAPG Bulletin*, 49, 248–257.

Bromley, B. W., and S. R. Larter. 1986. Biogenic origin of petroleums. *Chem. Eng. News*, August 25, 3, 43.

Brooks, J. (ed.). 1981. *Organic maturation studies and fossil fuel exploration*. London: Academic Press.

Brooks, J., and A. J. Fleet. 1987. *Marine petroleum source rocks*. Geological Society special publication 26. Oxford: Blackwell Scientific Publications for the Geological Society.

Brooks, P. W. 1986. Unusual biological marker geochemistry of oils and possible source rocks, offshore Beaufort–Mackenzie Delta, Canada. *Org. Geochem.*, 10, 401–406.

Brown, D. M., K. D. McAlpine, and R. W. Yole. 1989. Sedimentology and sandstone diagenesis of Hibernia formation in Hibernia oil field, Grand Banks of Newfoundland. *AAPG Bulletin,* 73, 557–575.

Buchardt, B., and M. D. Lewan. 1990. Reflectance of vitrinite-like macerals as a thermal maturity index for Cambrian–Ordovician Alum Shale, southern Scandinavia. *AAPG Bulletin,* 74, 394–406.

Buckley, S. E., C. R. Hocott, and M. S. Taggart Jr. 1958. Distribution of dissolved hydrocarbons in subsurface waters. In L. G. Weeks (ed.), *Habitat of oil: A symposium.* Tulsa: American Association of Petroleum Geologists, pp. 850–882.

Budyko, M. I., A. B. Ronov, and A. L. Yanshin. 1985. *History of the earth's atmosphere.* Berlin: Springer-Verlag.

Buhrig, C. 1989. Geopressured Jurassic reservoirs in the Viking Graben: Modelling and geological significance. *Marine and Petroleum Geology,* 6, 31–48.

Buiskool Toxopeus, J. M. A. 1983. Selection criteria for the use of vitrinite reflectance as a maturity tool. In J. Brooks (ed.), *Petroleum geochemistry and exploration of Europe.* Oxford: Blackwell Scientific Publications, pp. 295–307.

Burgess, J. D. 1974. Microscopic examination of kerogen (dispersed organic matter) in petroleum exploration. Washington, DC: Geological Society of America special publication 153, pp. 19–30.

Burnham, A. K., R. L. Braun, H. R. Gregg, and A. M. Samoun. 1987a. Comparison of methods for measuring kerogen pyrolysis rates and fitting kinetic parameters. *Energy & Fuels*, 1, 452.

Burnham, A. K., R. L. Braun, and A. M. Samoun. 1987b. Further comparison methods for measuring kerogen pyrolysis rates and fitting kinetic parameters. Berkeley, CA: Lawrence Livermore National Laboratory, UCRL-97352.

Burrus, J. (ed.). 1986. *Thermal modeling in sedimentary basins.* Paris: Éditions technip.

Burtner, R. L., and M. A. Warner. 1984. Hydrocarbon generation in Lower Cretaceous Mowry and Skull Creek shales of the northern Rocky Mountain area. In J. Woodward, F. F. Meissner, and J. L. Clayton (eds.), *Hydrocarbon source rocks of the greater Rocky Mountain region.* Denver: Rocky Mountain Association of Geologists, pp. 449–468.

Burwood, R., P. J. Cornet, L. Jacobs, and J. Paulet. 1990. Organofacies variation control on hydrocarbon generation: A Lower Congo Coastal Basin (Angola) case history. *Org. Geochem.*, 16 (1–3), 325–338.

Burwood, R., P. Leplat, B. Mycke, and J. Paulet. 1992. Rifted margin source rock deposition: A carbon isotope and biomarker study of a West African Lower Cretaceous "lacustrine" section. *Org. Geochem.*, 19 (1–3), 41–52.

Calhoun, G. G. 1991. How 12 geochemical methods fared in GERT project in Permian basin. *Oil & Gas Journal*, May 13, 62–68.

Carpenter, A. B. 1978. Origin and chemical evolution of brines in sedimentary basins. *Oklahoma Geological Survey Circular*, 79, 60–77.

Cathles, C. L., and A. T. Smith. 1983. Thermal constraints on the formation of Mississippi Valley–type lead–zinc deposits and their implications for episodic basin dewatering and deposit. *Economic Geology*, 78, 983–1002.

Chandra, K., D. S. N. Raju, and P. K. Mishra. 1993. Sea level changes, anoxic conditions, organic matter enrichment, and petroleum source rock potential of the Cretaceous sequences of the Cauvery Basin, India. In B. J. Katz and L. M. Pratt (eds.), *Source rocks in a sequence stratigraphic framework*. AAPG Studies in Geology 37. Tulsa: American Association of Petroleum Geologists, pp. 131–146.

Chapman, D. J., and J. W. Schopf. 1983. Biological and biochemical effects of the development of an aerobic environment. In J. W. Schopf (ed.), *Earth's earliest biosphere, its origin and evolution*. Princeton, NJ: Princeton University Press, pp. 302–320.

Chappe, B., P. Albrecht, and W. Michaelis. 1982. Polar lipids of archaebacteria in sediments and petroleums. *Science*, 217, 65–66.

Chappe, B., W. Michaelis, and P. Albrecht. 1980. Molecular fossils of archaebacteria as selective degradation products of kerogen. In A. G. Douglas and J. R. Maxwell (eds.), *Advances in organic geochemistry 1979*. Oxford: Pergamon Press, pp. 265–274.

Cherskiy, N. V., V. P. Tsarev, and S. P. Nikitin. 1985. Investigation and prediction of conditions of accumulation of gas resources in gas-hydrate pools. *Petrol. Geol.*, 21, 65–76.

Chibnall, A. C., S. H. Piper, A. Pollard, E. F. Williams, and P. N. Sahai. 1934. Constitution of primary alcohols, fatty acids, and paraffins present in plant and insect waxes. *Biochem. J.*, 28, 2189.

Chilingar, G. V., and L. Knight. 1960. Relationship between pressure and moisture contents of kaolinite, illite, and montmorillonite clays. *AAPG Bulletin*, 44, 89–94.

Chowdhary, L. R., and S. Taha. 1987. Geology and habitat of oil in Ras Budran field, Gulf of Suez, Egypt. *AAPG Bulletin*, 71, 1274–1293.

Chung, H. M., G. E. Claypool, M. A. Rooney, and R. M. Squires. 1994. Source characteristics of marine oils as indicated by carbon isotopic ratios of volatile hydrocarbons. *AAPG Bulletin*, 78, 396–408.

Chung, H. M., M. A. Rooney, and G. E. Claypool. 1991. Thermal maturity of oils. In D. Manning (ed.), *Organic geochemistry: Advances in applications in energy and the natural environment*. Manchester, England: Manchester University Press, pp. 143–146.

Clark, J. P., and R. P. Philp. 1989. Geochemical characterization of evaporite and carbonate depositional environments and correlation of associated crude oils in the Black Creek Basin, Alberta. *Bull. Canadian Petrol. Geol.,* 37 (4), 401–416.

Clark, R. C. Jr. 1966. Saturated hydrocarbons in marine plants and sediments. M.A. thesis, MIT–WHOI Joint Program.

Clark, R. C. Jr., and M. Blumer. 1967. Distribution of *n*-paraffins in marine organisms and sediment. *Limnol. and Oceanogr.,* 12 (1), 79–87.

Claypool, G. E., and K. A. Kvenvolden. 1983. Methane and other hydrocarbon gases in marine sediment. *Ann. Rev. Planet Sci.,* 11, 299–327.

Claypool, G. E., A. H. Love, and E. K. Maughan. 1978. Organic geochemistry, incipient metamorphism, and oil generation in black shale members of Phosphoria Formation, western interior United States. *AAPG Bulletin,* 62, 98–120.

Claypool, G. E., and L. B. Magoon. 1985. Comparison of oil–source rock correlation data for Alaskan North Slope: Techniques, results and conclusions. In L. B. Magoon and G. E. Claypool (eds.), *Alaska North Slope oil/rock correlation study.* AAPG Studies in Geology 20. Tulsa: American Association of Petroleum Geologists, pp. 49–81.

Claypool, G. E., and L. B. Magoon. 1988. Oil and gas source rocks in the National Petroleum Reserve in Alaska. In G. Gryc (ed.), *Geology and exploration of the National Petroleum Reserve in Alaska, 1974 to 1982.* U.S. Geological Survey professional paper 1399. Washington, DC: U.S. Government Printing Office, pp. 451–481.

Claypool, G. E., and E. A. Mancini. 1989. Geochemical relationships of petroleum in Mesozoic reservoirs to carbonate source rocks of Jurassic Smackover Formation, southwestern Alabama. *AAPG Bulletin,* 73, 904–924.

Clayton, J. L., and N. H. Bostick. 1986. Temperature effects on kerogen and on molecular and isotopic composition of organic matter in Pierre Shale near an igneous dike. *Org. Geochem.,* 10, 135–143.

Clayton, J. L., and J. D. King. 1987. Effects of weathering on biological marker and aromatic hydrocarbon composition of organic matter in Phosphoria shale outcrop. *Geochim. Cosmochim. Acta,* 51, 2153–2157.

Clayton, J. L., D. D. Rice, and G. E. Michael. 1991. Oil-generating coals of the San Juan Basin, New Mexico and Colorado, U.S.A. *Org. Geochem.,* 17, 735–742.

Clayton J. L., and P. J. Swetland. 1978. Subaerial weathering of sedimentary organic matter. *Geochim. Cosmochim. Acta,* 42 (3), 305–312.

Clementz, D. M., G. J. Demaison, and A. R. Daly. 1979. Well site geochemistry by programmed pyrolysis. Paper 3410, 11th Offshore Technology Conference, Houston, April 30–May 3.

Cloud, P. 1983. The biosphere. *Scientific American*, 249 (3), 176–189.

Coleman, H. J., J. E. Dooley, D. E. Hirsch, and C. J. Thompson. 1973. Compositional studies of a high-boiling 370–535°C distillate from Prudhoe Bay, Alaska crude oil. *Anal. Chem.*, 45 (9), 1724–1737.

Comer, J. B., and H. H. Hinch. 1987. Recognizing and quantifying expulsion of oil from the Woodford Formation and age-equivalent rocks in Oklahoma and Arkansas. *AAPG Bulletin*, 71, 844–858.

Coneybeare, C. E. B. 1965. Hydrocarbon-generation potential and hydrocarbon-yield capacity of sedimentary basins. *Canadian Petrol. Geol. Bull.*, 13 (4), 509–528.

Connan, J. 1967. Geochemical signficance of the extraction of amino acids from sediments. *Bull. Centre Rech. Pau.–SNPA,* 1 (1), 165–171.

Connan, J. 1974. Time–temperature relation in oil genesis. *AAPG Bulletin,* 58, 2516–2521.

Connan, J. 1984. Biodegradation of crude oils in reservoirs. In J. Brooks and D. Welte (eds.), *Advances in petroleum geochemistry*. London: Academic Press, vol. 1, pp. 229–335.

Connan, J., J. Bouroullec, D. Dessort, and P. Albrecht. 1986. The microbial input in carbonate–anhydrite facies of a sabkha paleoenvironment from Guatemala: A molecular approach. *Org. Geochem.,* 10, 29–50.

Connan, J., and A. M. Cassou. 1980. Properties of gases and petroleum liquids derived from terrestrial kerogen at various maturation levels. *Geochim. Cosmochim. Acta*, 44, 1–23.

Connan, J., and D. Dessort. 1987. Novel family of hexacyclic hopanoid alkanes (C_{32}–C_{35}) occurring in sediments and oils from anoxic paleoenvironments. *Org. Geochem.,* 11 (2), 103–113.

Connan, J., and D. Dessort. 1991. Du bitume dans des baumes de momies égyptiennes (1295 av. J-C.-300 ap. J-C.): Détermination de son origine et évaluation de sa quantité. Paris: C.R. Acad. Sci. 312, vol. 2, pp. 1445–1452.

Connan, J., K. Le Tran, and B. Van Der Weide. 1975. Alteration of petroleum in reservoirs. In *Proceedings of 9th World Petroleum Congress.* London: Applied Science Publishers, vol. 2, pp. 171–178.

Connan, J., A. Restle, and P. Albrecht. 1980. Biodegradation of crude oil in the Aquitaine Basin. In A. Douglas and J. R. Maxwell (eds.), *Advances in organic geochemistry 1979.* Oxford: Pergamon Press, pp. 1–17.

Coogan, A. H. 1970. Measurements of compaction in oolitic grainstone. *J. Sed. Petrology*, 40, 921–929.

Cook, A. C., A. C. Hutton, and N. R. Sherwood. 1981. Classification of oil shales. *Bull. elf-aquitaine*, 5 (2), 353–377.

Cook, E. 1979. The helium question. *Science*, 206 (4423), 1141–1147.

Cooper, B. S. 1990. *Practical petroleum geochemistry*. London: Robertson Scientific Publications.

Correia, M. 1969. Contribution à la recherche de zones favorables à la genese du pétrole par l'observation microscopique de la matière organique figurée. *Rev. l'Inst. français pétrole,* 24, 1417–1454.

Correia, M. 1971. Diagenesis of sporopollenin and other comparable organic substances: Application to hydrocarbon research. In J. Brooks et al. (eds.), *Sporopollenin*. London: Academic Press, pp. 569–620.

Coveney, R. M. Jr., E. D. Goebel, E. J. Zeller, G. A. M. Dreschhoff, and E. E. Angino. 1987. Serpentinization and the origin of hydrogen gas in Kansas. *AAPG Bulletin*, 71, 39–48.

Creaney, S. 1989. Reaction of organic material to progressive geological heating. In N. D. Naeser and T. H. McCulloh (eds.), *Thermal history of sedimentary basins: Methods and case histories*. New York: Springer-Verlag, pp. 37–52.

Creaney, S., and J. Allan. 1990. Hydrocarbon generation and migration in the Western Canada sedimentary basin. In J. Brooks (ed.), *Clastic petroleum provinces*. London: Geological Society special publication 50, pp. 189–202.

Culberson, O. L., and J. J. McKetta Jr. 1951. Phase equilibria in hydrocarbon–water systems: Part 3. The solubility of methane in water at pressures to 10,000 psia. *Petroleum Transactions, AIME,* 192, 223–226.

Curiale, J. A. 1986. Origin of solid bitumens, with emphasis on biological marker results. *Org. Geochem.* 10 (1–3), 559–580.

Curiale, J. A. 1988. Biological markers in grahamites and pyrobitumens. In T. F. Yen and J. M. Moldowan (eds.), *Geochemical biomarkers.* London: Harwood Academic Publishers, pp. 1–24.

Curiale, J. A. 1994. Correlation of oils and source rocks–A conceptual and historical perspective. In L. B. Magoon and W. G. Dow (eds.), *The petroleum system–From source to trap.* AAPG Memoir 60. Tulsa: American Association of Petroleum Geologists, pp. 251–260.

Curiale, J. A., R. D. Cole, and R. J. Witmer. 1992. Application of organic geochemistry to sequence stratigraphic analysis: Four Corners Platform Area, New Mexico, U.S.A. *Org. Geochem.*, 19 (1–3), 53–75.

Dahl, B., and G. C. Speers. 1986. Geochemical characterization of a tar mat in the Oseberg Field Norwegian Sector, North Sea. *Org. Geochem.*, 10, 547–588.

Dahl, B., and M. A. Yukler. 1991. The role of petroleum geochemistry in basin modeling of the Oseberg area, North Sea. In R. K. Merrill (ed.), *Source and migration processes and evaluation techniques*. Tulsa: American Association of Petroleum Geologists, pp. 65–85.

Davis, H. R., C. W. Byers, and L. M. Pratt. 1989. Depositional mechanisms and organic matter in Mowry shale (Cretaceous), Wyoming. *AAPG Bulletin*, 73, 1103–1116.

Degens, E. T. 1969. Biogeochemistry of stable carbon isotopes. In G. Eglinton and M. T. J. Murphy (eds.), *Organic geochemistry: Methods and results.* New York: Springer-Verlag, pp. 304–329.

Degens, E. T., P. A. Meyers, and S. C. Brassell (eds.). 1986. *Biogeochemistry of black shales.* Vol. 60. Hamburg: Geologisch–Pälaontologischen Institutes.

Degens, E. T., R. P. Von Herzen, H. K. Wong, W. G. Deuser, and H. W. Jannasch. 1973. Lake Kivu: Structure, chemistry and biology of an East African rift lake. *Geol. Rundschau,* 62, 245–277.

Demaison, G. J. 1977. Asphalt sands and supergiant oil fields. *AAPG Bulletin,* 61, 1950–1961.

Demaison, G. J. 1984. The generative basin concept. In G. J. Demaison and R. J. Murris (eds.), *Petroleum geochemistry and basin evaluation.* AAPG Memoir 35. Tulsa: American Association of Petroleum Geologists, pp. 1–14.

Demaison, G. J. 1986. Progress and problems in industrial applications of petroleum geochemistry. Paper presented at American Chemical Society Annual Meeting, Anaheim, September 10.

Demaison, G. J., and B. J. Huizinga. 1991. Genetic classification of petroleum systems. *AAPG Bulletin,* 75, 1626–1643.

Demaison, G. J., and B. J. Huizinga. 1994. Genetic classification of petroleum systems using three factors: Charge, migration, and entrapment. In L. B. Magoon and W. G. Dow (eds.), *The petroleum system–From source to trap.* AAPG Memoir 60. Tulsa: American Association of Petroleum Geologists, pp. 73–89.

Demaison, G. J., and G. T. Moore. 1980. Anoxic environments and oil source bed genesis. *Org. Geochem.,* 2 (1), 9–31.

Demaison, G. J., and R. J. Murris (eds.). 1984. *Petroleum geochemistry and basin evaluation.* AAPG Memoir 35. Tulsa: American Association of Petroleum Geologists.

Dembicki, H. Jr. 1984. An interlaboratory comparison of source rock data. *Geochim. Cosmochim. Acta*, 48, 2641–2649.

Dembicki, H. Jr., B. Horsfield, and T. T. Y. Ho. 1983. Source rock evaluation by pyrolysis-gas chromatography. *AAPG Bulletin,* 67, 1094–1103.

Dembicki, H. Jr., W. G. Meinschein, and D. E. Hatton. 1975. Possible ecological and environmental significance of the predominance of even–carbon number C_{20}–C_{30} *n*-alkanes. *Geochim. Cosmochim. Acta,* 39, 203–208.

Deroo, G., T. G. Powell, B. Tissot, and R. G. McCrossan, with contributions by P. A. Hacquebard. 1977. *The origin and migration of petroleum in the western*

Canada sedimentary basin, Alberta: A geochemical and thermal maturation study. Geological Survey of Canada Bull., 262. Ottawa: Geological Survey of Canada.

Deroo, G., B. Tissot, R. G. McCrossan, and F. Der. 1974. *Geochemistry of heavy oils of Alberta*. Canadian Soc. Petrol. Geol. Memoir 3. Calgary: Stacs Data Services, pp. 148–167.

Deuser, W. G. 1974. Evolution of anoxic conditions in Black Sea during Holocene. In E. T. Degens and D. A. Ross (eds.), *The Black Sea–Geology, chemistry and biology*. AAPG Memoir 20, pp. 133–136.

Dickey, P. A. 1969. Increasing concentration of subsurface brines with depth. *Chem. Geol.*, 4, 361–370.

Dickey, P. A. 1972. Migration of interstitial water in sediments and the concentration of petroleum and useful minerals. *Proceedings of 24th International Geologic Congress*. Sec. 5: *Mineral Fuels*, pp. 3–16.

Dickey, P. A., G. O. George, and C. Barker. 1987. Relationships among oils and water compositions in Niger Delta. *AAPG Bulletin,* 71, 1319–1328.

Dickey, P. A., and J. M. Hunt. 1972. Geochemical and hydrogeologic methods of prospecting for stratigraphic traps. In R. E. King (ed.), *Stratigraphic oil and gas fields*. AAPG Memoir 16. Tulsa: American Association of Petroleum Geologists, pp. 136–137.

Dickey, P. A., and C. Soto. 1974. Chemical composition of deep subsurface waters of the western Anadarko Basin. SPE–AIME paper 5178. Dallas: Society of Petroleum Engineers and American Institute of Mining, Metallurgical, and Petroleum Engineers.

Dickinson, G. 1953. Geological aspects of abnormal reservoir pressures in Gulf Coast Louisiana. *AAPG Bulletin,* 37, 410–432.

Dimmler, A., T. D. Cycr, and O. P. Strusz. 1984. Identification of bicyclic terpenoid hydrocarbons in the saturate fraction of athabasca oil sand bitumen. *Org. Geochem.*, 7, 231–238.

Ditkenshteyn, D. Kh., et al. 1986. West Siberian oil–gas province. *Petrol. Geol.,* 22 (3), 96–132.

Djerassi, C. 1981. Recent studies in the marine sterol field. *Pure & Appl. Chem.,* 53, 873–890.

Doligez, B., (ed.). 1987. *Migration of hydrocarbons in sedimentary basins.* 2nd IFP Exploration Research Conference, Carcans, June 15–19. Paris: Éditions technip.

Dorsey, N. E. 1940. *Properties of ordinary water-substance.* New York: Reinhold.

Douglas, A. G., and P. J. Grantham. 1974. Fingerprint gas chromatography in the analysis of some native bitumens, asphalts, and related substances. In

B. Tissot and F. Bienner (eds.), *Advances in organic geochemistry 1973*. Paris: Éditions technip, pp. 261–276.

Dow, W. G. 1974. Application of oil-correlation and source-rock data to exploration in Williston Basin. *AAPG Bulletin*, 58, 1253–1262.

Dow, W. G. 1977. Kerogen studies and geological interpretations. *J. Geochem. Explor.*, 7 (2), 77–79.

Dow, W. G. 1994. An introduction to DGSI & total quality geochemistry. DGSI Report, The Woodlands, TX.

Dow, W. G., and D. I. O'Connor. 1982. Kerogen maturity and type by reflected light microscopy applied to petroleum exploration. In *How to assess organic maturation and paleotemperatures*. Society of Economic Paleontologists and Mineralogists short course.

Dow, W. G., M. A. Yukler, J. T. Senftle, M. C. Kennicutt II, and J. M. Armentrout. 1990. Miocene oil source beds in the East Breaks Basin, Flex-Trend, offshore Texas. In D. Schumacher and B. F. Perkins (eds.), *Gulf Coast oils and gases. Proceedings, GCSSEPM Foundation, Ninth Annual Research Conference, October 1*, pp. 139–150.

Downey, M. W. 1984. Evaluating seals for hydrocarbon accumulations. *AAPG Bulletin*, 68, 1752–1763.

Downey, M. W. 1990. Faulting and hydrocarbon entrapment. *Geophysics: The Leading Edge of Exploration*, January 20–22.

Drummond, K. J. 1992. Geology of Venture, a geopressured gas field, offshore Nova Scotia. In M. T. Halbouty (ed.), *Giant oil and gas fields of the decade (1978–1988)*. AAPG Memoir 54. Tulsa: American Association of Petroleum Geologists, pp. 55–71.

Dufour, J. 1957. On regional migration and alteration of petroleum in south Sumatra. *Geol. en Mijnbouw*, new ser., 19, 172–181.

Dunnington, H. V. 1967a. Aspects of diagenesis and shape change in stylolitic limestone reservoirs. In *Proceedings of Seventh World Petroleum Congress*. London: Elsevier Science, vol. 2, pp. 339–352.

Dunnington, H. V. 1967b. Stratigraphical distribution of oilfields in the Iraq–Iran–Arabian Basin. *Journal of the Institute of Petroleum*, 53 (520), 129–161.

Durmish'yan, A. G. 1973. The compaction of clay rocks. *Izd. Akad. Nauk SSSR*, Ser. Geol. (in Russian), no. 8, pp. 85–89. Trans. in *Internat. Geol. Rev.*, 1973, 16 (6), 650–653.

Dzou, L. I., and W. B. Hughes. 1993. Geochemistry of oils and condensates, K. Field, offshore Taiwan: A case study in migration fractionation. *Org. Geochem.*, 20(4), 437–462.

Egbogah, E., E. Onu, and D. O. Lambert-Aikhionbare. 1980. Possible new oil potentials of the Niger Delta. *Oil & Gas Journal,* April 14, 176–184.

Eglinton, G. 1969. Organic geochemistry: The organic chemist's approach. In G. Eglinton and M. T. J. Murphy (eds.), *Organic geochemistry: Methods and results*. New York: Springer-Verlag, pp. 20–73.

Eglinton, G., and M. Calvin. 1967. Chemical fossils. *Scientific American*, 261, 23–32.

Eglinton, T. 1994. Carbon isotopic evidence for the origin of macromolecular aliphatic structures in kerogen. *Org. Geochem.*, 21 (6–7), 721–735.

Eglinton, T. I., J. S. Sinninghe Damsté, M. E. L. Kohner, and J. W. de Leeuw. 1990. Rapid estimation of organic sulfur content of kerogens, coals and asphaltenes by pyrolysis-gas chromatography. *Fuel*, 69, 1394–1403.

Ekweozor, C. M., J. I. Okogun, D. E. U. Ekong, and J. R. Maxwell. 1979. Preliminary organic geochemical studies of samples from the Niger Delta (Nigeria): 1. Analyses of crude oils from triterpanes. *Chem. Geol.*, 28, 11–28.

Ekweozor, C. M., J. I. Okogun, D. E. U. Ekong, and J. R. Maxwell. 1981. C_{24}–C_{27} degraded triterpanes in Nigerian petroleum: Novel molecular markers of source/input or organic maturation? *J. Geochem. Explor.*, 15, 653–662.

Ekweozor, C. M., and O. P. Strausz. 1983. Tricyclic terpanes in the Athabasca oil sands: Their geochemistry. In M. Bjorøy et al. (eds.), *Advances in organic geochemistry 1981*. Chichester: Wiley, pp. 746–766.

Ekweozor, C. M., and N. Telnaes. 1989. Oleanane parameter: Verification by quantitative study of the biomarker occurrence in sediments of the Niger Delta. *Org. Geochem.*, 16 (1–3), 401–413.

Elvsborg, A., T. Hagevang, and T. Throndsen. 1985. Origin of the gas-condensate of the Midgard Field at Haltenbanken. In B. M. Thomas et al. (eds.), *Petroleum geochemistry in exploration of the Norwegian Shelf*. London: Graham & Trotman for the Norwegian Petroleum Society, pp. 213–219.

Emery, K. O. 1960. *The sea off southern California*. New York: Wiley.

Emery, K. O., and D. G. Aubrey. 1991. *Sea levels, land levels, and tide gauges*. New York: Springer-Verlag.

Emery, K. O., and J. M. Hunt. 1974. Summary of Black Sea investigations. In E. T. Degens and D. A. Ross (eds.), *The Black Sea–Geology, chemistry, and biology*. AAPG Memoir 20. Tulsa: American Association of Petroleum Geologists, pp. 575–590.

Engler, K. O. V. 1913. *Die Chemie und Physik des Erdöls*. Vol. 1. Leipzig: S. Hirzel.

Epstein, A. G., J. B. Epstein, and L. D. Harris. 1977. Conodont color alteration–An index to organic metamorphism. United States Geological Survey professional paper 995.

Erdman, J. G. 1962. Oxygen, nitrogen, and sulfur in asphalts. *API Division of Science and Technology Summer Symposium*, 42 (8), 33–40.

Eremenko, N. A., and I. M. Michailov. 1974. Hydrodynamic pools at faults. *Canadian Petrol. Geol. Bull.*, 22 (2), 106–108.

Espitalié, J., J. L. La Porte, M. Madec, F. Marquis, P. Le Plat, J. Paulet, and A. Boutefeu. 1977. Méthode rapide de caractérisation des roches mères de leur potentiel pétrolier et de leur degré d'évolution. *Rev. l'Inst. français pétrole,* 32 (1), 23–42.

Espitalié, J., J. Madec, and B. Tissot. 1980. Role of mineral matter in kerogen pyrolysis: Influence on petroleum generation and migration. *AAPG Bulletin*, 64, 58–66.

Espitalié, J., F. Marquis, and I. Barsony. 1984. Geochemical logging. In K. J. Voorhees (ed.), *Analytical pyrolysis: Techniques and applications.* London: Butterworth, pp. 276–304.

Estrella, G., M. R. Mello, P. C. Gaglianone, R. L. M. Azevedo, K. Tsubone, E. Rossetti, J. Concha, and I. M. R. A. Bruning. 1984. The Espirito Santo Basin (Brazil) source rock characterization and petroleum habitat. In G. J. Demaison and R. J. Murris (eds.), *Petroleum geochemistry and basin evaluation.* AAPG Memoir 35. Tulsa: American Association of Petroleum Geologists, pp. 253–271.

Evans, C. R., and F. L. Staplin. 1971. Regional facies of organic metamorphism in geochemical exploration. In Canadian Institute of Mining and Metallurgy, *Proceedings, 3rd International Geochemical Exploration Symposium,* special vol., 11. Montreal: Canadian Institute of Mining and Metallurgy, pp. 517–520.

Evans, R. A., and J. H. Campbell. 1979. Oil shale retorting–A correlation of selected infrared absorbance bands with process heating rates and oil yields. *In Situ*, 3, 33–51.

Faber, E., and W. Stahl. 1984. Geochemical surface exploration for hydrocarbons in North Sea. *AAPG Bulletin*, 68, 363–386.

Farrimond, P., G. J. Eglinton, S. C. Brassell, and H. C. Jenkyns. 1989. Toarcian anoxic event in Europe: An organic geochemical study. *Marine and Petroleum Geology*, 6, 136–147.

Federal Power Commission Report. 1973. Report of the Technical Advisory Committee, Data Tape H.

Ferris, S. W. 1955. *Handbook of hydrocarbons.* New York: Academic Press.

Fertl, W. H., R. E. Chapman, and R. F. Hotz (eds.). 1994. *Studies in abnormal pressures*. Developments in Petroleum Science 38. Amsterdam: Elsevier Science.

Fertl, W. H., and W. G. Leach. 1990. Formation temperature and formation pressure affect the oil and gas distribution in Tertiary Gulf Coast sediments. *Transactions–Gulf Coast Association of Geological Societies,* 40, 205–216.

Fisher, M. J., and J. A. Miles. 1983. Kerogen types, organic maturation and hydrocarbon occurrences in the Moray Firth and South Viking Graben, North Sea Basin. In J. Brooks (ed.), *Petroleum geochemistry and exploration of Europe.* Geological Society special publication 12. Oxford: Blackwell Scientific Publications, pp. 195–201.

Fitzgerald, T. A. 1980. Giant field discoveries 1968–1978: An overview. In M. T. Halbouty (ed.), *Giant oil and gas fields of the decade 1968–1978*. AAPG Memoir 30. Tulsa: American Association of Petroleum Geologists, pp. 1–5.

Fleet, A. J., K. Kelts, and M. R. Talbot (eds.). 1988. *Lacustrine petroleum source rocks*. Geological Society special publication 40. Oxford: Blackwell Scientific Publications.

Forgotson, J. M. Sr. 1969. Indication of proximity of high-pressure fluid reservoir, Louisiana and Texas Gulf Coast. *AAPG Bulletin,* 53, 171–173.

Forgotson, J. M. Sr. 1979. Super overpressured gas. *AAPG Bulletin*, 63, 1534–1537.

Fowler, M. G., and A. G. Douglas. 1984. Distribution and structure of hydrocarbons in four organic-rich Ordovician rocks. *Org. Geochem.*, 6, 105–114.

Franks, S. G., and R. W. Forester. 1984. Relationships among secondary porosity, pore-fluid chemistry and carbon dioxide, Texas Gulf Coast. In D. A. McDonald and R. C. Surdam (eds.), *Clastic diagenesis.* AAPG Memoir 37. Tulsa: American Association of Petroleum Geologists, pp. 63–80.

Fu Jiamo and Sheng Guoying. 1992. Source and biomarker composition characteristics of Chinese nonmarine crude oils. In J. K. Whelan and J. W. Farrington (eds.), *Organic matter: Productivity, accumulation, and preservation in recent and ancient sediments.* New York: Columbia University Press, pp. 417–432.

Fu Jiamo, Sheng Guoying, Peng Pingan, S. C. Brassell, G. Eglinton, and Jiang Jigang. 1986. Peculiarities of salt lake sediments as potential source rocks in China. *Org. Geochem.*, 10, 119–126.

Fu Jiamo, Sheng Guoying, Xu Jiayou, G. Eglinton, A. P. Gowar, Jia Rongfen, Fan Shanfa, and Peng Pingan. 1990. Application of biological markers in the assessment of paleoenvironments of Chinese non-marine sediments. *Org. Geochem.*, 16 (4–6), 769–779.

Funkhouser, L. W., F. X. Bland, and C. C. Humphries Jr. 1980. The deep Tuscaloosa gas trend of South Louisiana. Paper presented at AAPG Annual Convention, Denver, June 8–11.

Gabrielsen, R. H., S. Ulvøen, A. Elvsborg, and O. Fredrik. 1985. The geological history and geochemical evaluation of Block 2/2, offshore Norway. In B. M. Thomas et al. (eds.), *Petroleum geochemistry in exploration of the Norwegian Shelf*. London: Graham & Trotman for the Norwegian Petroleum Society, pp. 165–178.

Gabrieylan, A. G. 1962. On the formation of oil and gas pools in the Volgograd–Volga region. *Geologicheskoe stroenie i neftegazonosnost' Volgogradskoi oblasti,* 1, 248–273.

Gagosian, R. B., and E. T. Peltzer. 1986. The importance of atmospheric input of terrestrial organic material to deep sea sediments. *Org. Geochem.*, 10, 661–669.

Galimov, E. M. 1968. Isotopic composition of carbon in gases of the crust. *Internat. Geol. Rev.,* 11 (10), 1092–1104.

Galimov, E. M. 1973. *Carbon isotopes in oil–gas geology*. Moscow: Nedra. Trans. by National Aeronautics and Space Administration, Washington, DC.

Galimov, E. M. 1988. Sources and mechanisms of formation of gaseous hydrocarbons in sedimentary rocks. *Chem. Geol.,* 71, 77–95.

Galimov, E. M., N. G. Kuznetsova, and V. S. Prokhorov. 1968. The composition of the former atmosphere of the earth as indicated by isotopic analysis of Pre-Cambrian carbonates. *Geokhim.,* 11, 1376–1381.

Garg, A. K., and R. P. Philp. 1994. Pyrolysis–gas chromatography of asphaltenes/kerogens from source rocks of the Gandhar Field, Cambay Basin, India. *Org. Geochem.,* 21 (3–4), 383–392.

Gavrilov, A. Ya., and V. S. Dragunskaya. 1963. Aromatic condensate of eastern Turkmen, SSR. *Akad. Nauk Turkm. SSR,* ser. F12. *Khim. Geol. Nauk,* 3, 111–113.

Gehman, H. M. Jr. 1962. Organic matter in limestones. *Geochim. Cosmochim. Acta,* 26, 885–897.

Gelpi, E., H. Schneider, J. Mann, and J. Oro. 1970. Hydrocarbons of geochemical significance in microscopic algae. *Phytochem.,* 9, 603–612.

Gibling, M. R., Y. Ukakimaphan, and S. Srisuk. 1985. Oil shale and coal in intermontane basins of Thailand. *AAPG Bulletin,* 69, 760–766.

Gies, R. M. 1984. Case history for a major Alberta deep basin gas trap: The Cadomin Formation. In J. A. Masters (ed.), *Elmworth: Case study of a deep basin gas field.* AAPG Memoir 38. Tulsa: American Association of Petroleum Geologists, pp. 115–140.

Goff, J. C. 1984. Hydrocarbon generation and migration from Jurassic source rocks in the East Shetland Basin and Viking Graben of the northern North Sea. In G. J. Demaison and R. J. Murris (eds.), *Petroleum geochemistry and basin evaluation.* AAPG Memoir 35. Tulsa: American Association of Petroleum Geologists, pp. 273–301.

Goldberg, I. S. 1973. Solid bitumens in petroleum deposits of the Baltic region as indicators of stages in the migration of petroleum. *Dokl. Akad. Nauk SSSR*, 209 (2), 462–482.

Goosens, J., J. W. de Leeuw, P. A. Schenck, and S. C. Brassell. 1984. Tocopherols as likely precursors of pristane in ancient sediments and crude oils. *Nature*, 312, 440–442.

Gorin, G. E., L. G. Racz, and M. R. Walter. 1982. Late Precambrian–Cambrian sediments of Huqf Group, sultanate of Oman. *AAPG Bulletin*, 66, 2609–2627.

Gorskaya, A. I. 1950. Investigations of the organic matter of Recent sediments. In *Recent analogs of petroliferous facies (symposium).* Moscow: Gostoptekhizdat.

Grabowski, G. J. Jr. 1984. Generation and migration of hydrocarbons in Upper Cretaceous Austin Chalk, south-central Texas. In J. G. Palacas (ed.), *Petroleum geochemistry and source rock potential of carbonate rocks*. AAPG Studies in Geology 18. Tulsa: American Association of Petroleum Geologists, pp. 97–115.

de la Grandville, B. F. 1982. Appraisal and development of a structural and stratigraphic trap oil field with reservoir in glacial to periglacial clastics. In M. T. Halbouty (ed.), *The deliberate search for the subtle trap.* AAPG Memoir 32. Tulsa: American Association of Petroleum Geologists, pp. 267–286.

Grantham, P. J. 1986. The occurrence of unusual C_{27} and C_{29} sterane predominances in two types of Oman crude oil. *Org. Geochem.*, 9, 1–10.

Grantham, P. J., J. Posthuma, and A. Baak. 1983. Triterpanes in a number of Far-Eastern crude oils. In M. Bjorøy et al. (eds.), *Advances in organic geochemistry 1981*. New York: Wiley, pp. 675–683.

Grantham, P. J., and L. L. Wakefield. 1988. Variations in the sterane carbon number distributions of marine source rock derived crude oils through geological time. *Org. Geochem.*, 12 (1), 61–73.

Grenier, A. Ch., C. S. Pyckelle, and P. Albrecht. 1976. Aromatic hydrocarbons from geological sources: 1. New naturally occurring phenanthrene and chrysene derivatives. *Tetrahedron*, 32, 257–260.

Gretener, P. E. 1982. Another look at Alborz nr. 5 in central Iran. *Bull. Ver. schweiz, Petroleum. Geol. u.-Ing.*, 48 (114), 1–8.

Gretener, P. E. 1986. Macrofractures and fluid flow. *Bull. Swiss Assoc. of Petroleum-Geologists and Engineers,* 53 (123), 59–74.

Gretener, P. E., and G. Bloch. 1992. Geopressures: Two distinctly different kinds of conditions. Abstract in program of American Association of Petroleum Geologists Annual Convention, Calgary, pp. 48–49.

Gretener, P. E., and Zengmo Feng. 1985. Three decades of Geopressures—Insights and enigmas. *Bull. Ver. Schweiz. Petroleum-Geol. u.-Ing.*, 71 (120), 1–34.

Grimalt, J., and J. Albaiges. 1987. Sources and occurrence of C_{12}–C_{22} *n*-alkane distributions with even carbon-number preference in sedimentary environments. *Geochim. Cosmochim. Acta*, 51, 1379–1384.

Guseva, A. N., and L. A. Fayngersh. 1973. Conditions of accumulation of nitrogen in natural gases as illustrated by the Central European and Chu-Sarysu oil–gas basins. *Dokl. Akad. Nauk SSSR,* 209 (2), 210–212.

Gutjahr, C. C. M. 1983. Introduction to incident-light microscopy of oil and gas source rocks. In W. van den Bert and R. Felix (eds.), *Geol. Mijnbouw*, special issue in honour of J. D. de Jong, 62, 417–425.

Habicht, J. K. A. 1964. Comment on the history of migration in the Gifhorn Trough. *Proceedings of the Sixth World Petroleum Congress*, paper 19-PD2, sec. 1, p. 480.

Hagemann, H. W., and A. Hollerbach. 1980. Relationship between the macropetrographic and organic geochemical composition of lignites. In A. G. Douglas and J. K. Maxwell (eds.), *Advances in organic geochemistry 1979*. Oxford: Pergamon Press, pp. 631–638.

Hagemann, H. W., and A. Hollerbach. 1981. Spectral fluorometric analysis of extracts: A new method for the determination of the degree of maturity of organic matter in sedimentary rocks. *Bull. Elf-Aquitaine*, 7 (2), 635–650.

Hagen, E. S., and R. C. Surdam. 1984. Maturation history and thermal evolution of Cretaceous source rocks of the Bighorn Basin, Wyoming and Montana. In J. Woodward, F. F. Meissner, and J. L. Clayton (eds.), *Hydrocarbon source rocks of the greater Rocky Mountain region.* Denver: Rocky Mountain Association of Geologists, pp. 321–338.

Halbouty, M. T. 1986. Basins and new frontiers: An overview. In M. T. Halbouty (ed.), *Future petroleum provinces of the world.* Tulsa: American Association of Petroleum Geologists, pp. 1–10.

Halbouty, M. T., A. A. Meyerhoff, R. E. King, R. H. Dott Sr., H. D. Klemme, and T. Shabad. 1970. World's giant oil and gas fields. In M. T. Halbouty (ed.), *Geology of giant petroleum fields.* AAPG Memoir 14. Tulsa: American Association of Petroleum Geologists, pp. 502–528.

Hall, P. L., D. F. R. Mildner, and R. L. Borst. 1986. Small-angle scattering studies of the pore spaces of shaly rocks. *J. Geophys. Res.*, 92 (B2), 2183–2192.

Han, J., and M. Calvin. 1969. Hydrocarbon distribution of algae and bacteria and microbiological activity in sediments. *Proc. Nat. Acad. Sci. U.S.A.*, 64 (2), 436–443.

Hanor, J. S. 1984. Variation in the chemical composition of oil-field brines with depth in northern Louisiana and southern Arkansas: Implications for mechanisms and rates of mass transport and diagenetic reaction. *Transactions—Gulf Coast Association of Geological Societies*, 34, 55–61.

Hanor, J. S., and J. E. Bailey. 1983. Use of hydraulic head and hydraulic gradient to characterize geopressured sediments and the direction of fluid migration in the Louisiana Gulf Coast. *GCAGS Transactions,* 33, 115–122.

Hanor, J. S., and R. Sassen. 1988. Deep basin hydrodynamics of the Louisiana Gulf Coast: Implications for oil and gas migration. In program and abstracts of Ninth Annual Research Conference Gulf Coast Section Society of Economic Paleontologists and Mineralogists Foundation, New Orleans, December 4–7, p. 6.

Hanor, J. S., and R. Sassen. 1990. Evidence for large-scale vertical and lateral migration of formation waters, dissolved salt, and crude oil in the Louisiana Gulf Coast. In D. Schumacher and B. F. Perkins (eds.), *Gulf Coast oils and gases: Their characteristics, origin, distribution and exploration and production significance. Ninth Annual Research Conference Proceedings, GCSSEPM Foundation*. October 1. Austin: Society of Economic Paleontologists and Mineralogists Foundation, pp. 283–296.

Harrison, F. W. Jr., and G. E., Parrish. 1990. The Tuscaloosa rejuvenated: Beaver Dam Creek and Baywood Fields, St. Helena Parish, Louisiana. Paper presented at the AAPG Annual Meeting, San Francisco, June 3–6.

Hayes, J. M., I. R. Kaplan, and K. W. Wedeking. 1983. Precambrian organic geochemistry, preservation of the record. In J. W. Schopf (ed.), *Earth's earliest biosphere, its origin and evolution*. Princeton, NJ: Princeton University Press, pp. 93–134.

Hedberg, H. D. 1926. The effect of gravitational compaction on the structure of sedimentary rocks. *AAPG Bulletin,* 10, 1035–1073.

Hedberg, H. D. 1931. Cretaceous limestone as petroleum source rock in northwestern Venezuela. *AAPG Bulletin*, 15, 229–246.

Hedberg, H. D. 1936. Gravitational compaction of clays and shales. *Am. J. Sci.,* fifth series, 31 (184), 241–287.

Hedberg, H. D. 1968. Significance of high-wax oils with respect to genesis of petroleum. *AAPG Bulletin*, 52, 736–750.

Hedberg, H. D. 1974. Relation of methane generation to undercompacted shales, shale diapirs, and mud volcanoes. *AAPG Bulletin,* 58, 661–673.

Hedberg. H. D. 1980. Methane generation and petroleum migration. In W. H. Roberts III and R. J. Cordell (eds.), *Problems of petroleum migration.* AAPG Studies in Geology 10. Tulsa: American Association of Petroleum Geologists, pp. 179–206.

Hedberg, H. D., J. D. Moody, and R. M. Hedberg. 1979. Petroleum prospects of the deep offshore. *AAPG Bulletin*, 63, 286–300.

Helander, D. P. 1983. *Fundamentals of formation evaluation*. Tulsa: OGCI Publications, Oil & Gas Consultants International.

Henderson, W., V. Wollra, and G. Eglinton. 1969. Identification of steranes and triterpanes from a geological source by capillary gas liquid chromatography and mass spectrometry. In P. A. Schenck and I. Havenaar (eds.), *Advances in organic geochemistry 1968*. Oxford: Pergamon Press, pp. 181–207.

Henrich, W. M., and J. W. Farrington. 1984. Peru upwelling region sediments near 15°S: 1. Remineralization and accumulation of organic matter. *Limnol. Oceanogr.*, 29 (1), 1–19.

Heppard, P. D., R. L. Ames, and L. M. Ross. 1990. Migration of oils into Samaan Field, offshore Trinidad, West Indies. Paper presented at Second Geological Conference of the Geological Society of Trinidad and Tobago, April 3–8.

Herron, S. L. 1991. In situ evaluation of potential source rocks by wireline logs. In R. K. Merrill (ed.), *Source and migration processes and evaluation techniques*. Tulsa: American Association of Petroleum Geologists, pp. 127–134.

Hilt, C. 1873. Die Beziehungen zwischen der Zusammensetzung und den technischen Eigenschaften der Steinkohle. *Sitzber. Aachener Bezirksvereinigung VDI*, 4.

Hinch, H. H. 1980. The nature of shales and the dynamics of hydrocarbon expulsion in the Gulf Coast Tertiary section. In W. H. Roberts III and R. J. Cordell (eds.), *Problems of petroleum migration*. AAPG Studies in Geology 10. Tulsa: American Association of Petroleum Geologists, pp. 1–18.

Hitchon, B. 1963. Geochemical studies of natural gas. *J. Canadian Petrol. Tech.*, 3 (3), 100–116; (4), 165–174.

Hitchon, B. 1974. Occurrence of natural gas hydrates in sedimentary basins. In I. R. Kaplan (ed.), *Natural gases in marine sediments*. New York: Plenum Press, vol. 3, pp. 195–255.

Hitchon, B., and M. Gawlak. 1972. Low molecular weight aromatic hydrocarbons in gas condensates from Alberta, Canada. *Geochim. Cosmochim. Acta*, 36, 1043–1059.

Hite, R. J., and D. E. Anders. 1991. Petroleum and evaporites. In J. L. Melvin (ed.), *Petroleum and mineral resources*. Developments of Sedimentology 50. New York: Elsevier Science, pp. 349–411.

Hobson, G. D. (ed.). 1984. *Modern petroleum technology*. 5th ed., vol. 1. New York: Wiley.

Hodgson, G. W. 1954. Vanadium, nickel and iron trace metals in crude oils of western Canada. *AAPG Bulletin*, 38, 2537–2554.

Hoefs, J. 1969. Carbon abundance in common igneous and metamorphic rock types. In K. H. Wedepohl (ed.), *Handbook of geochemistry.* New York: Springer-Verlag, pp. 6E1-6, 6M-1.

Hoering, T. C. 1976. *Molecular fossils from the Precambrian Nonesuch Shale.* Washington, DC: Carnegie Institute, Yearbook 75, pp. 806–813.

Hoering, T. C., and V. Navale. 1987. A search for molecular fossils in the kerogen of Precambrian sedimentary rocks. *Precambrian Res.*, 34, 247–267.

Hoffman, C. F., and O. P. Strausz. 1986. Bitumen accumulation in Grosmont Platform Complex, Upper Devonian, Alberta, Canada. *AAPG Bulletin*, 70, 1113–1128.

Holder, G. D., D. L. Katz, and J. H. Hand. 1976. Hydrate formation in subsurface environments. *AAPG Bulletin,* 60, 981–988.

Holland, H. D. 1984. *The evaluation of the atmosphere and oceans*. Princeton, NJ: Princeton University Press.

Holland, H. D., B. Lazar, and M. McCaffrey. 1986. Evolution of the atmosphere and oceans. *Nature*, 320, 27–33.

Holmquest, H. J. 1965. Deep pays in Delaware and Val Verde basins. In A. Young and J. E. Galley (eds.), *Fluids in subsurface environments.* AAPG Memoir 4. Tulsa: American Association of Petroleum Geologists, pp. 257–279.

Horsfield, B., and A. G. Douglas. 1980. The influence of minerals on the pyrolysis of kerogens. *Geochim. Cosmochim. Acta*, 44, 1119–1131.

Horstman, E. L. 1988. Source maturity, overpressures and production, North West Shelf, Australia. In P. G. and R. R. Purcell (eds.), *The North West Shelf Australia. Proceedings of North West Shelf Symposium, Perth, W.A.*, pp. 529–537.

Houseknecht, D. W., and S. M. Matthews. 1985. Thermal maturity of Carboniferous strata, Ouachita mountains. *AAPG Bulletin,* 69, 335–345.

Hovland, M. 1983. Gas-induced erosion features in the North Sea. *Earth Surface Processes and Landforms,* 9, 209–228.

Hovland, M., and A. G. Judd. 1988. *Seabed pockmarks and seepages: Impact on geology, biology and the marine environment.* London: Graham & Trotman.

Hovland, M., M. Talbot, S. Olaussen, and L. Aasberg. 1985. Recently formed methane-derived carbonates from the North Sea floor. In B. M. Thomas et al. (eds.), *Petroleum geochemistry in exploration of the Norwegian Shelf*. London: Graham & Trotman for the Norwegian Petroleum Society, pp. 263–266.

Howell, D. G. (ed.). 1993. The future of energy gases. U.S. Geological Survey professional paper 1570. Washington, DC: U.S. Government Printing Office.

Howell, V. J., J. Connan, and A. K. Aldridge. 1984. Tentative identification of demethylated tricyclic terpanes in nonbiodegraded and slightly biodegraded crude oils from the Los Llanos Basin, Colombia. *Org. Geochem.*, 6, 83–92.

Huang, W. Y., and W. G. Meinschein. 1979. Sterols as ecological indicators. *Geochim. Cosmochim. Acta*, 43, 739–745.

Hubbert, M. K. 1953. Entrapment of petroleum under hydrodynamic conditions. *AAPG Bulletin*, 37, 1954–2026.

Huc, A. Y. 1988. Aspects of depositional processes of organic matter in sedimentary basins. *Org. Geochem.*, 13 (1–3), 263–272.

Huc, A. Y., B. Durand, and J. C. Monin. 1978. Humic compounds and kerogens in cores from Black Sea sediments, Leg 42B, Holes 379A, B, and 380A. In D. A. Ross and Y. P. Neprochov (eds.), *Initial reports of the Deep Sea Drilling Project, Leg 42B*. Pt. 2, vol. 42. Washington, DC: U.S. Government Printing Office, pp. 737–748.

Huc, A. Y., B. Durand, J. Roucachet, M. Vandenbroucke, and J. L. Pittion. 1986. Comparison of three series of organic matter of continental origin. *Org. Geochem.*, 10, 65–72.

Huc, A. Y., and J. M. Hunt. 1980. Generation and migration of hydrocarbons in offshore south Texas Gulf Coast sediments. *Geochim. Cosmochim. Acta*, 44, 1081–1089.

Huc, A. Y., J. M. Hunt, and J. K. Whelan. 1981. The organic matter of a Gulf Coast well studied by a thermal analysis–gas chromatography technique. *Journal of Geochemical Exploration*, 15, 671–681.

Huffington, R. M., and H. M. Helmig. 1990. Badak field, Indonesia, Kutai Basin, East Kalimantan, Borneo. In E. A. Beaumont and N. H. Foster (comps.), *Treatise of petroleum geology. AAPG atlas of oil: Oil and gas fields*, A-019, pp. 265–308.

Hunt, J. M. 1953. Composition of crude oil and its relation to stratigraphy in Wyoming. *AAPG Bulletin*, 37, 1837–1872.

Hunt, J. M. 1954. Chemistry applied to exploration. In *Preprints, 10th Southwest Regional ACS Meeting,* Fort Worth, December 2–4. Washington, DC: American Chemical Society.

Hunt, J. M. 1961. Distribution of hydrocarbons in sedimentary rocks. *Geochim. Cosmochim. Acta,* 22, 37–49.

Hunt, J. M. 1963. Geochemical data on organic matter in sediments. In V. Bese (ed.), *Proceedings of 3rd International Scientific Conference on Geochemistry, Microbiology and Petroleum Chemistry, October 8–13*. Budapest: KULTURA, vol. 1, pp. 394–412.

Hunt, J. M. 1968. How gas and oil form and migrate. *World Oil,* 167, 140–150.

Hunt, J. M. 1972. Distribution of carbon in the crust of the earth. *AAPG Bulletin,* 56, 2273–2277.

Hunt, J. M. 1975. Is there a geochemical depth limit for hydrocarbons? *Petrol. Eng.,* 47 (3), 112–127.

Hunt, J. M. 1977. Distribution of carbon as hydrocarbons and asphaltic compounds in sedimentary rocks. *AAPG Bulletin,* 61, 100–104.

Hunt, J. M. 1979. *Petroleum geochemistry and geology.* 1st ed. San Francisco: Freeman.

Hunt, J. M. 1981. Source rock characterization by thermal distillation and pyrolysis. In G. Atkinson and J. J. Zuckerman (eds.), *Origin and chemistry of petroleum.* Oxford: Pergamon Press, pp. 57–65.

Hunt, J. M. 1984. Generation and migration of light hydrocarbons. *Science,* 226, 1265–1270.

Hunt, J. M. 1987. Primary and secondary migration of oil. In R. E. Meyer (ed.), *Exploration for heavy crude oil and natural bitumen.* AAPG Studies in Geology 25. Tulsa: American Association of Petroleum Geologists, pp. 345–349.

Hunt, J. M. 1990. Generation and migration of petroleum from abnormally pressured fluid compartments. *AAPG Bulletin,* 74, 1–12.

Hunt, J. M. 1991a. Generation and migration of petroleum from abnormally pressured fluid compartments: Reply. *AAPG Bulletin,* 75, 328–330, 336–338.

Hunt, J. M. 1991b. Generation of gas and oil from coal and other terrestrial organic matter. *Org. Geochem.,* 17 (6), 673–680.

Hunt, J. M., E. E. Hays, E. T. Degens, and D. A. Ross. 1967. Red Sea: Detailed survey of the hot brine areas. *Science,* 156, 514–516.

Hunt, J. M., and R. J-C. Hennet. 1992. Modeling petroleum generation in sedimentary basins. In J. Whelan and J. Farrington (eds.), *Organic matter: Productivity, accumulation, and preservation in recent and ancient sediments.* New York: Columbia University Press, pp. 20–52.

Hunt, J. M., and G. W. Jamieson. 1956. Oil and organic matter in source rocks of petroleum. *AAPG Bulletin,* 40, 477–488.

Hunt, J. M., and A. P. McNichol. 1984. The Cretaceous Austin chalk of south Texas—A petroleum source rock. In J. G. Palacas (ed.), *Petroleum geochemistry and source rock potential of carbonate rocks.* AAPG Studies in Geology 18. Tulsa: American Association of Petroleum Geologists, pp. 117–125.

Hunt, J. M., M. D. Lewan, and R. J-C. Hennet. 1991. Modeling oil generation with time–temperature index graphs based on the Arrhenius equation. *AAPG Bulletin,* 75, 795–807.

Hunt, J. M., F. Stewart, and P. A. Dickey. 1954. Origin of hydrocarbons of Uinta Basin, Utah. *AAPG Bulletin,* 38, 1671–1698.

Hunt, J. M., J. K. Whelan, L. B. Eglinton, and L. M. Cathles III. 1994. Gas generation—A major cause of deep Gulf Coast overpressures. *Oil & Gas Journal,* July 18, 59–62.

Hunt, J. M., J. K. Whelan, L. B. Eglinton, and L. M. Cathles. Relation of gas generation and shale compaction to deep overpressures in the U.S. Gulf Coast. In B. E. Law, G. F. Ulmishek, and V. I. Slavin (eds.), *Abnormal pressures in hydrocarbon environments.* AAPG Special Publication, in press.

Hunt, T. S. 1863. Report on the geology of Canada. *Canadian Geological Survey Report: Progress to 1863*. Canadian Geological Survey.

Hussler, G., and P. Albrecht. 1983. C_{27}–C_{29} monoaromatic and anthrasteroid hydrocarbons in Cretaceous black shales. *Nature,* 304, 262–263.

Hussler, G., B. Chappe, P. Wehrung, and P. Albrecht. 1981. C_{27}–C_{29} ring A monoaromatic steroids in Cretaceous black shales. *Nature,* 294, 556–558.

Hussler, G., J. Connan, and P. Albrecht. 1984. Novel families of tetra- and hetracyclic aromatic hopanoids predominant in carbonate rocks and crude oils. *Org. Geochem.,* 6, 39–49.

Hutcheon, I., and H. J. Abercrombie. 1989. The role of silicate hydrolysis in the origin of CO_2 in sedimentary basins. *Proceedings of the Sixth International Symposium on Water–Rock Interaction*. Rotterdam: Balkema, pp. 321–324.

Hwang, R. J., and S. C. Teerman. 1988. Hydrocarbon characterization of resinite. *Energy & Fuels,* 2, 170–175.

Illich, H. A. 1983. Pristane, phytane and lower molecular weight isoprenoid distributions in oils. *AAPG Bulletin,* 67, 385–393.

Illich, H. A., and P. L. Grizzle. 1983. Comment on "Comparison of Michigan Basin crude oils" by Volger et al. *Geochim. Cosmochim. Acta,* 47, 1151–1155.

Illing, V. C. 1933. Migration of oil and natural gas. *J. Inst. Petrol.,* 19 (114), 229–274.

Illing, V. C. 1938. The origin of pressure in oil pools. In *The science of petroleum*. Oxford: Oxford University Press, vol. 1, pp. 229–234.

Initial reports of the Deep Sea Drilling Project. 1971–1978. Washington, DC: U.S. Government Printing Office.

Isaacs, C. M. 1987. Source and deposition of organic matter in the Monterey Formation, south-central coastal basins of California. In R. F. Meyer (ed.), *Exploration for heavy crude oil and natural bitumen.* AAPG Studies in Geology 25. Tulsa: American Association of Petroleum Geologists, pp. 93–205.

Issler, R. D., and L. R. Snowdon. 1990. Hydrocarbon generation kinetics and thermal modelling, Beaufort–Mackenzie Basin. *Bull. Can. Petrol. Geol.,* 38, 1–16.

Ivlev, A. A., R. G. Pankina, and G. D. Gal'peri. 1973. Thermodynamics of reactions of sulfurization of oil. *Petroleum Geology*, 11, 70–74.

Jackson, M. J., T. G. Powell, R. E. Summons, and I. P. Sweet. 1986. Hydrocarbon shows and petroleum source rocks in sediments as old as 1.7×10^9 years. *Nature*, 322, 727–729.

James, A. T. 1983. Correlation of natural gas by use of carbon isotopic distribution between hydrocarbon components. *AAPG Bulletin*, 67, 1176–1191.

James, A. T. 1990. Correlation of reservoired gases using the carbon isotopic composition of wet gas components. *AAPG Bulletin*, 74, 1441–1448.

James, A. T., and B. J. Burns. 1984. Microbial alteration of subsurface natural gas accumulations. *AAPG Bulletin*, 68, 957–960.

Jannasch, H. W. 1988. Microbiological studies in the Black Sea. *Woods Hole Oceanographic Institution Annual Report*, pp. 38–40.

Jannasch, H. W., and C. O. Wirsen. 1979. Chemosynthetic primary production at east Pacific sea floor spreading centers. *Bioscience*, 29 (10), 592–598.

Jansa, L. F., and V. H. Noguera Urrea. 1990. Geology and diagenetic history of overpressured sandstone reservoirs, Venture gas field, offshore Nova Scotia, Canada. *AAPG Bulletin*, 74, 1640–1658.

Jarvie, D. M. 1991. Total organic carbon (TOC) analysis. In R. K. Merrill (ed.), *Source and migration processes and evaluation techniques*. Tulsa: American Association of Petroleum Geologists, pp. 113–118.

Jasper, J. P., J. K. Whelan, and J. M. Hunt. 1984. Migration of C_1 to C_8 volatile organic compounds in sediments from the Deep Sea Drilling Project, Leg 75, Hole 530A, Walvis Ridge. In *Initial Reports from the Deep Sea Drilling Project*. Washington, DC: U.S. Government Printing Office, vol. 75, pp. 1001–1008.

Jenden, P. D., K. D. Newell, I. R. Kaplan, and W. L. Watney. 1988. Composition and stable-isotope geochemistry of natural gases from Kansas, Midcontinent, U.S.A. *Chemical Geology*, 71, 117–147.

Jensenius, J., and N. C. Munksgaard. 1989. Large scale hot water migration systems around salt diapirs in the Danish Central Trough and their impact on diagenesis of chalk reservoirs. *Geochim. Cosmochim. Acta*, 53, 79–88.

Jiang, Z. S., and M. G. Fowler. 1986. Carotenoid-derived alkanes in oils from northwestern China. *Org. Geochem.*, 10, 831–839.

Jobson, A., F. D. Cook, and D. W. S. Westlake. 1972. Microbial utilization of crude oil. *Applied Microbiol.*, 23 (6), 1082–1089.

Johancsik, C. A., and W. R. Grieve. 1987. Oil-based mud reduces borehole problems. *Oil & Gas Journal*, April 27, 46–56.

Johns, R. B. (ed.). 1986. *Biological markers in the sedimentary record.* New York: Elsevier Science.

Johnston, D. J. 1990. Geochemical logs thoroughly evaluate coalbeds. *Oil & Gas Journal*, December 24, 45–51.

1992. Ocean drilling program guidelines for pollution prevention and safety. *JOIDES Journal*, 18 (7), 19.

Jones, P. H., and R. H. Wallace Jr. 1974. Hydrogeologic aspects of structural deformation in the northern Gulf of Mexico Basin. *J. Res. Geol. Survey,* 2 (5), 511–517.

Jones, R. W. 1981. Some mass balance and geological constraints on migration mechanisms. *AAPG Bulletin*, 65, 103–122.

Jones, R. W. 1986. Origin, migration, and accumulation of petroleum in Gulf Coast Cenozoic. *AAPG Bulletin*, 70, 65.

Jones, R. W. 1987. Organic facies. In J. Brooks and D. Welte (eds.), *Advances in petroleum geochemistry*. London: Academic Press, vol. 2, pp. 1–90.

Jones, V. T., and R. J. Drozd. 1983. Predictions of oil or gas potential by near-surface geochemistry. *AAPG Bulletin*, 67, 932–952.

Jüntgen, H., and J. Klein. 1975. Origin of natural gas from coaly sediments. *Erdöl und Kohle,* 28 (2), 65–73.

Kalomazov, R. U., and M. A. Vakhitov. 1975. Appearance and nature of anomalously high formation pressures in the Kuyab Mega syncline of the Tadzhik depression. *Nefti gazovaya geolog. geofiz.*, no. 10, pp. 3–6.

Kantsler, A. J., T. J. C. Prudence, A. C. Cook, and M. Zwigulis. 1984. Hydrocarbon habitat of the Cooper/Eromanga Basin, Australia. In G. J. Demaison and R. J. Murris (eds.), *Petroleum geochemistry and basin evaluation.* AAPG Memoir 35. Tulsa: American Association of Petroleum Geologists, pp. 373–390.

Karweil, J. 1955. Die Metamorphose der Kohlen vom Standpunkt der physikalischen Chemie. *Z. deutsch. geol. Ges.*, 107, 132–139.

Karweil, J. 1969. Aktuelle Probleme der Geochemie der Kohle. In P. A. Schenk and I. Havenaar (eds.), *Advances in organic geochemistry 1968*. Oxford: Pergamon Press, pp. 59–84.

Katsube, T. J., B. S. Mudford, and M. E. Best. 1991. Petrophysical characteristics of shales from the Scotian shelf. *Geophysics*, 56 (10), 1681–1689.

Katz, B. J. 1983. Limitation of Rock-Eval pyrolysis for typing organic matter. *Org. Geochem.*, 4 (3–4), 195–199.

Katz, B. J. 1990. Controls on distribution of lacustrine source rocks through time and space. In B. J. Katz (ed.), *Lacustrine basin exploration*. AAPG Memoir 50. Tulsa: American Association of Petroleum Geologists, pp. 61–76.

Katz, B. J., R. N. Pheifer, and D. J. Schunk. 1988. Interpretation of discontinuous vitrinite reflectance profiles. *AAPG Bulletin,* 72, 926–931.

Katz, B. J., and L. M. Pratt (eds.). 1993. *Source rocks in a sequence stratigraphic framework*. AAPG Studies in Geology 37. Tulsa: American Association of Petroleum Geologists.

Katz, D. L. 1971. Depths to which frozen gas fields (gas hydrates) may be expected. *J. Petrol. Tech.,* 23, 419–558.

Katz, D. L. 1983. Overview of phase behavior in oil and gas production. *J. Petrol. Tech.,* 35, 1205–1214.

Katz, D. L., D. Cornell, R. Kobayashi, F. H. Poettmann, J. A. Vary, J. R. Elenbaas, and C. F. Weinaug. 1959. Water–hydrocarbon systems. In D. L. Katz et al. (eds.), *Handbook of natural gas engineering*. New York: McGraw-Hill, pp. 189–221.

Kaufman, R. L., A. S. Ahmed, and R. J. Elsinger. 1990. Gas chromatography as a development and production tool for fingerprinting oils from individual reservoirs: Applications in the Gulf of Mexico. In D. Schumacher and B. F. Perkins (eds.), *Gulf Coast oils and gases: Their characteristics, origin, distribution, and exploration and production significance. Proceedings of the the Ninth Annual Research Conference GCSSEPM, October*. Society of Economic Paleontologists and Mineralogists Foundation, pp. 263–282.

Kelly, W. C., and G. K. Nishioka. 1985. Precambrian oil inclusions in late veins and the role of hydrocarbons in copper mineralization at White Pine, Michigan. *Geology*, 13, 334–337.

Khalimov, E. M. 1980. The principles of classification and oil resources estimation. In *Proceedings of the 10th World Petroleum Congress*. London: Heyden & Son, vol. 2, pp. 263–268.

Kimble, B. J., J. R. Maxwell, R. P. Philp, G. Eglinton, P. Albrecht, A. Ensminger, P. Arpino, and G. Ourisson. 1974. Tri- and tetraterpenoid hydrocarbons in the Messel oil shale. *Geochim. Cosmochim. Acta*, 38, 1165–1181.

Klein, J., and H. Jüntgen. 1972. Studies on the emission of elemental nitrogen from coals of different rank and its release under geochemical conditions. In H. R. von Gaertner and H. Wehner (eds.), *Advances in organic geochemistry 1971*. Oxford: Pergamon Press, pp. 647–656.

Klemme, H. D. 1975. Geothermal gradients, heat flow and hydrocarbon recovery. In A. G. Fischer and S. Judson (eds.), *Petroleum and global tectonics*. Princeton, NJ: Princeton University Press, pp. 251–306.

Klemme, H. D. 1983. The geologic setting of giant gas fields. In C. Delahaye and M. Grenon (eds.), *Conventional and unconventional world natural gas resources*. Laxenburg: International Institute for Applied Systems Analysis, pp. 133–160.

Klemme, H. D. 1988. *Basin classification chart*. Geo Basins Ltd., February.

Klemme, H. D. 1992. *Petroleum advisor package*. Geo Basins Ltd.

Klemme, H. D. 1993. World petroleum systems with Jurassic source rocks. *Oil & Gas Journal*, November 8, 96–99.

Klemme, H. D. 1994. Petroleum systems of the world that involve Upper Jurassic source rocks. In L. B. Magoon and W. G. Dow (eds.), *The petroleum system—From source to trap*. AAPG Memoir 60. Tulsa: American Association of Petroleum Geologists, pp. 51–72.

Klemme, H. D., and G. F. Ulmishek. 1989. Depositional controls, distribution and effectiveness of world's petroleum source rocks. *AAPG Bulletin*, 73, 372–373.

Klemme, H. D., and G. F. Ulmishek. 1991. Effective petroleum source rocks of the world: Stratigraphic distribution and controlling depositional factors. *AAPG Bulletin*, 75, 1809–1851.

Klomp, U. C. 1986. The chemical structure of a pronounced series of iso-alkanes in south Oman crudes. *Org. Geochem.*, 10, 807–814.

Knoche, H., and G. Ourisson. 1967. Organic compounds in fossil plants (*Equisetum*, horsetails). *Angew. chem.*, internat'l ed., 6, 1085.

Koons, C. B., G. W. Jamieson, and L. S. Ciereszko. 1965. Normal alkane distributions in marine organisms: Possible significance to petroleum origin. *AAPG Bulletin*, 49, 301–316.

Kornacki, A. S. 1993. C_7 chemistry and origin of Monterey oils and source rocks from the Santa Maria Basin, California. Abstract in AAPG Annual Convention Program, New Orleans, April 25–28, p. 131.

Korvin, G. 1984. Shale compaction and statistical physics. *Geophysical Journal of the Royal Astronomical Society*, 78, 35–50.

Kozlovsky, U. A. 1984. The world's deepest well. *Scientific American*, 251, 98–104.

Kravets, V. V. 1974. Temperature field of the Dnieper–Donets depression and distribution of oil and gas pools within it. *Geologiya i Geokhimiya Goryuchikh Iskopayemykh*, no. 39, pp. 58–66. English trans., *Petrol. Geol.*, 12 (5), 1975, 275–278.

Krouse, H. R. 1979. Stable isotope geochemistry of non-hydrocarbon constituents of natural gas. In *Source prediction and separation of non-hydrocarbon constituents of natural gas,* preprint of the Tenth World Petroleum Congress, Bucharest, 1979. London: Heyden & Son.

Kruge, M. A. 1986. Biomarker geochemistry of the Miocene Monterey Formation, West San Joaquin Basin, California: Implications for petroleum generation. *Org. Geochem.*, 10, 517–530.

Kudel'skiy, A. V. 1973. Ammonia in subsurface waters of oil–gas regions and its value in oil exploration. *Petroleum Geology*, 8, 49–53.

Kudryatseva, Y. I., A. A. Andreyeva, and O. I. Suprunenko. 1974. Discovery of natural kerosene in southwestern Kamchatka. *Dokl. Akad. Nauk SSSR,* 216 (2), 418–421.

Kushnareva, T. I. 1971. Evidence of hydrothermal activity in oil-bearing carbonate rocks of the Upper Devonian of the Pechora Ridge. *Dokl. Akad. Nauk SSSR,* 198 (1), 175–177.

Kvenvolden, K. A. 1988. Methane hydrate–A major reservoir of carbon in the shallow geosphere? *Chemical Geology*, 71, 41–51.

Kvenvolden, K. A. 1993. A primer on gas hydrates. In D. G. Howell et al. (eds.), *The future of energy gases*. U.S. Geological Survey professional paper 1570. Washington, DC: U.S. Printing Office, pp. 297–298.

Kvenvolden, K. A., and P. R. Carlson. 1993. Possible connection between two Alaskan catastrophes occurring 25 years apart (1964–1989). *Geology*, 21, 813–816.

Lakshmanan, C. C., M. L. Bennett, and N. White. 1991. Implications of multiplicity in kinetic parameters to petroleum exploration: Distributed activation energy models. *Energy & Fuels,* 5, 110–117.

Land, S., and G. L. Macpherson. 1992. Origin of saline formation waters, Cenozoic section, Gulf of Mexico sedimentary basin. *AAPG Bulletin,* 76, 1344–1362.

Landais, P., and M. Monthioux. 1988. Closed system pyrolysis: An efficient technique for simulating natural coal maturation. *Fuel Processing Technology,* 20, 123–132.

Landes, K. K. 1970. *Petroleum geology of the United States*. New York: Wiley-Interscience.

Landes, K. K., 1973. Mother Nature as an oil polluter. *AAPG Bulletin,* 57, 637–641.

Larskaya, Ye. S., and D. V. Zhabrev. 1964. Effects of stratal temperatures and pressures on the composition of dispersed organic matter (From the example of the Mesozoic–Cenozoic deposits of the western Ciscaspian region). *Dokl. Akad. Nauk SSSR,* 157 (4), 135–139.

Larter, S. R. 1985. Integrated kerogen typing in the recognition and quantitative assessment of petroleum source rocks. In B. M. Thomas et al. (eds), *Petroleum geochemistry in exploration of the Norwegian Shelf.* London: Graham & Trotman for the Norwegian Petroleum Society, pp. 269–286.

Law, B. E. 1984. Relationships of source-rock, thermal maturity, and overpressuring to gas generation and occurrence in low-permeability Upper Cretaceous and Lower Tertiary rocks, greater Green River Basin, Wyoming, Colorado, and Utah. In J. Woodward, F. F. Meissner, and J. L. Clayton (eds.), *Hydrocarbon source rocks of the greater Rocky Mountain region.* Denver: Rocky Mountain Association of Geologists, pp. 469–490.

Law, B. E., and W. W. Dickinson. 1985. Conceptual model for origin of abnormally pressured gas accumulations in low-permeability reservoirs. *AAPG Bulletin*, 69, 1295–1304.

Law, B. E., V. E. Nuccio, and C. E. Barker. 1989. Kinky vitrinite reflectance well profiles: Evidence of paleopore pressure in low-permeability, gas-bearing sequences in Rocky Mount foreland basins. *AAPG Bulletin,* 73, 999–1010.

Law, B. E., and Rice, D. D. (eds.). 1993. *Hydrocarbons from coal.* AAPG Studies in Geology 38. Tulsa: American Association of Petroleum Geologists.

Leach, W. G. 1994. Distribution of hydrocarbons in abnormal pressure in south Louisiana, U.S.A. In W. H. Fertl, R. E. Chapman, and R. F. Hotz (eds.), *Studies in abnormal pressures: Developments in petroleum science.* Vol. 83. Amsterdam: Elsevier Science, pp. 391–428.

Leenheer, M. J. 1985. Mississippian Bakken and equivalent formations as source rocks in the western Canadian Basin. *Org. Geochem.*, 6, 521–532.

Leonard, R. C. 1983. Geology and hyrocarbon accumulations, Columbus Basin, offshore Trinidad. *AAPG Bulletin*, 67, 1081–1093.

Leonard, R. C. 1988. Generation, migration, and entrapment of hydrocarbons on southern Norwegian Shelf. AAPG distinguished lecture tour, abstract. *AAPG Bulletin,* 72, 1522.

Lerche, I. 1990. *Basin analysis. Quantitative methods.* Vols. 1 and 2. San Diego: Academic Press.

Le Taiming and Lu Shanfan. 1990. A numerical kerogen type index. *J. Petrol. Geol.*, 13 (1), 87–92.

Le Tran, K. 1972. Geochemical study of hydrogen sulfide absorbed in sediments. In H. R. von Gaertner and H. Wehner (eds.), *Advances in organic geochemistry 1971.* Oxford: Pergamon Press, pp. 717–726.

Le Tran, K., J. Connan, and B. Van Der Weide. 1974. Problemes relatifs à la formation d'hydrocarbures et d'hydrogene sulfure dans le bassin sud-ouest aquitain. In B. Tissot and F. Bienner (eds.), *Advances in organic geochemistry 1973.* Paris: Éditions technip, p. 76.

Levorsen, A. I. 1967. *Geology of petroleum*. 2nd ed. San Francisco: Freeman.

Levy, E. J., R. R. Doyle, R. A. Brown, and F. W. Melpolder. 1961. Identification of components in paraffin wax by high-temperature gas chromatography and mass spectrometry. *Anal. Chem.,* 33 (6), 698–704.

Lewan, M. D. 1978. Laboratory classification of very fine grained sedimentary rocks. *Geology*, 6, 745–748.

Lewan, M. D. 1980. Geochemistry of vanadium and nickel in organic matter of sedimentary rocks. Ph.D. diss., University of Cincinnati.

Lewan, M. D. 1985. Evaluation of petroleum generation by hydrous pyrolysis experimentation. *Philosophical Transactions of the Royal Society, London,* 315, 123–134.

Lewan, M. D. 1986. Organic sulfur in kerogens from different lithofacies of the Monterey Formation. Abstract 94, Division of Geochemistry, 192nd American Chemical Society National Meeting, Anaheim, September 7–12.

Lewan, M. D. 1987. Petrographic study of primary petroleum migration in the Woodford Shale and related rock units. In B. Doligez (ed.), *Migration of hydrocarbons in sedimentary basins:* Paris: Éditions technip, pp. 113–130.

Lewan, M. D. 1989. Hydrous pyrolysis study of oil and tar generation from Monterey Shale containing high sulfur kerogen. Abstract, symposium on geochemistry, American Chemical Society National Meeting, Dallas, April 9–14.

Lewan, M. D. 1990. Variability of oil generation from coals of the Blackhawk Formation as determined by hydrous pyrolysis. Abstract, American Chemical Society National Meeting, Boston, June.

Lewan, M. D. 1993a. Identifying and understanding suppressed vitrinite reflectance through hydrous pyrolysis experiments. *Abstracts and Program, 10th Annual Meeting of the Society for Organic Petrology*, vol. 10, pp. 1–3.

Lewan, M. D. 1993b. Laboratory simulation of petroleum formation: Hydrous pyrolysis. In M. H. Engel and S. A. Macko (eds.), *Organic geochemistry*. New York: Plenum Press, pp. 419–442.

Lewan, M. D. 1993c. Primary oil migration and expulsion as determined by hydrous pyrolysis. *Proceedings of the 13th World Petroleum Congress, Buenos Aires, 1991*. Chichester: Wiley, vol. 2, pp. 215–223.

Lewan, M. D., M. Bjorøy, and D. L. Dolcater. 1986. Effects of thermal maturation on steroid hydrocarbons as determined by hydrous pyrolysis of Phosphoria Retort Shale. *Geochim. Cosmochim. Acta*, 50, 1977–1987.

Lewan, M. D., and B. Buchardt. 1989. Irradiation of organic matter by uranium decay in the Alum Shale, Sweden. *Geochim. Cosmochim. Acta,* 53, 1307–1322.

Lewan, M. D., J. B. Comer, T. Hamilton-Smith, N. R. Hasenmueller, J. M. Guthrie, J. R. Hatch, D. L. Gautier, and W. T. Frankie. 1995. Feasibility study on material-balance assessment of petroleum from the New Albany Shale in the Illinois Basin. U.S. Geological Survey Bulletin 2137.

Lewan, M. D., and J. A. Williams. 1987. Evaluation of petroleum generation from resinites by hydrous pyrolysis. *AAPG Bulletin*, 71, 207–214.

Lewan, M. D., J. C. Winters, and J. H. McDonald. 1979. Generation of oil-like pyrolyzates from organic rich shale. *Science*, 203, 897–899.

Leythaeuser, D, A. Mackenzie, R. G. Schaefer, and M. Bjorøy. 1984. A novel approach for recognition and quantification of hydrocarbon migration effects in shale–sandstone sequences. *AAPG Bulletin*, 68, 196–219.

Leythaeuser, D., R. G. Schaefer, and M. Radke. 1987. On the primary migration of petroleum. *Proceedings of Twelfth World Petroleum Congress, Houston, 1987*. London: Wiley, pp. 227–236.

Lijmbach, G. M. G. 1975. On the origin of petroleum. *Proceedings of Ninth World Petroleum Congress*. London: Applied Science Publishers, vol. 2, pp. 357–369.

Lillack, H., W. Esser, and K. Schwochau. 1991. Evolution of hydrogen and methane by non-isothermal pyrolysis of petroleum source rocks. In *Organic geochemistry: Advances and applications in energy and the natural environment*. Manchester: Manchester University Press, pp. 306–309.

Lindblom, G. P., and M. D. Lupton. 1961. Microbiological aspects of organic geochemistry. *Develop. Ind. Microbiol.*, 2, 9–22.

Link, W. K. 1952. Significance of oil and gas seeps in world oil exploration. *AAPG Bulletin*, 36, 1505–1540.

Littke, R., D. R. Baker, and, D. Leythaeuser. 1988. Microscopic and sedimentologic evidence for the generation and migration of hydrocarbons in Toarcian source rocks of different maturities. *Org. Geochem.*, 13 (1–3), 549–559.

Livingston, H. K. 1951. Knock resistance of pure hydrocarbons in correlations with chemical structure. *I. & E. Chem.*, 43 (12), 2834–2840.

Livsey, A., A. G. Douglas, and J. Connan. 1984. Diterpenoid hydrocarbons in sediments from an offshore (Labrador) well. *Org. Geochem.*, 6, 78–81.

Lo, H. B. 1992. Identification of indigenous vitrinites for improved thermal maturity evaluation. *Org. Geochem.*, 718 (3), 359–364.

Longman, M. W., and S. E. Palmer. 1987. Organic geochemistry of Mid-Continent Middle and Late Ordovician oils. *AAPG Bulletin*, 71, 938–950.

Lopatin, N. V. 1971. Temperature and geological time as factors of carbonifaction. *Akad. Nauk SSSR, Izv. Ser. Geol.*, 3, 95–106.

Lopatin, N. V. 1976. The determination of the influence of temperature and geologic time on the catagenic processes of coalification and oil-gas formation. In *Issledovaniya organicheskogo veshchestva sovremennykh i iskopaemykh osadkov* (Research on organic matter of modern and fossil deposits). Moscow: Akademii Nauk SSSR, Izdatel'stvo, "Nauka," pp. 361–366.

Lopatin, N. V. 1980. Evolution of the biosphere and fossil fuels. *Internat. Geol. Rev.*, 22 (10), 1117–1131. Trans. in *AN SSSR Izv., Ser. Geol.*, 7, 1979, 5–22.

Loucks, R. G., M. M. Dodge, and W. E. Galloway. 1984. Regional controls on diagenesis and reservoir quality in Lower Tertiary sandstones along the Texas Gulf Coast. In D. A. McDonald and R. C. Surdam (eds.), *Clastic diagenesis*. AAPG Memoir 37. Tulsa: American Association of Petroleum Geologists, pp. 5–45.

Louda, J. W., and E. W. Baker. 1986. The biochemistry of chlorophyll. In M. L. Sohn (ed.), *Organic marine geochemistry.* ACS Symposium series 305, April. Washington, DC: American Chemical Society, pp. 107–126.

Lovering, E. G., and K. J. Laidler. 1960. A system of molecular thermochemistry for organic gases and liquids: II. Extension to compounds containing sulfur and oxygen. *Canadian Journal of Chemistry,* 38, 2367.

Luo Binjie, Yang Xinghua, Lin Hejie, and Zheng Guodong. 1988. Characteristics of Mesozoic and Cenozoic non-marine source rocks in north-west China. In A. J. Fleet, K. Kelts, M. R. Talbot (eds.), *Lacustrine petroleum source rocks*. Geological Society Special Publication No. 40. Oxford: Blackwell Scientific Publications, pp. 291–298.

Lutz, M., J. P. H. Kaasschieter, and D. H. Van Vijhe. 1975. Geological factors controlling Rotliegend gas accumulations in the mid-European Basin. *Proceedings of Ninth World Petroleum Congress*. London: Applied Science Publishers, vol. 2, pp. 93–103.

Macgregor, D. S. 1993. Relationships between seepage, tectonics and subsurface petroleum reserves. *Marine and Petroleum Geology,* 10, 606–619.

Machihara, T., and R. Ishiwatari. 1987. Possible carotenoid-derived structures in fossil kerogens. *Geochim. Cosmochim. Acta*, 55, 207–211.

Mackenzie, A. S. 1984. Applications of biological markers in petroleum geochemistry. In J. Brooks and D. Welte (eds.), *Advances in petroleum geochemistry*. London: Academic Press, vol. 1, pp. 115–214.

Mackenzie, A. S., S. C. Brassell, G. J. Eglinton, and J. R. Maxwell. 1982. Chemical fossils: The geological fate of steroids. *Science*, 217 (4559), 491–504.

Mackenzie, A. S., and T. M. Quigley. 1988. Principles of geochemical prospect appraisal. *AAPG Bulletin*, 72, 399–415.

Macko, S. A. 1981. Stable nitrogen istotope ratios as tracers of organic geochemical processes. Ph.D. diss., University of Texas at Austin.

Magara, K. 1978. *Compaction and fluid migration: Practical petroleum geology.* Developments in Petroleum Science 9. New York: Elsevier Science.

Magoon, L. B. 1988. The petroleum system–A classification scheme for research, exploration, and resource assessment. *U.S. Geological Survey Bulletin,* 1870, 2–15.

Magoon, L. B., and G. E. Claypool. 1981. Two oil types on North Slope of Alaska–Implications for exploration. *AAPG Bulletin,* 65, 646–652.

Magoon, L. B., and G. E. Claypool. 1983. Petroleum geochemistry of the North Slope of Alaska: Time and degree of thermal maturity. In M. Bjorøy et al. (eds.), *Advances in organic geochemistry 1981.* New York: Wiley, pp. 28–38.

Magoon, L. B., and W. G. Dow (eds.). 1994. *The petroleum system–From source to trap.* AAPG Memoir 60. Tulsa: American Association of Petroleum Geologists, pp. 3–24.

Magoon, L. B., and Z. C. Valin. 1994. Overview of petroleum system case studies. In L. B. Magoon and W. G. Dow (eds.), *The petroleum system–From source to trap.* AAPG Memoir 60. Tulsa: American Association of Petroleum Geologists, pp. 329–338.

Makarenko, F. A., and S. I. Sergiyenko. 1970. Geothermal zoning of the composition of oil in eastern Ciscaucasia. *Dokl. Akad. Nauk SSSR,* 210, 207–209.

Makogon, Y. F. 1981. *Hydrates of natural gas.* Tulsa: Penn Well Publishing.

Makogon, Y. F., F. A. Trebin, A. A. Trofimuk, V. P. Tsarev, and N. V. Cherskiy. 1971. Detection of a pool of natural gas in a solid (hydrated gas) state. *Dokl. Akad. Nauk SSSR,* 196, 197–200.

Makogon, Y. F., V. P. Tsarev, and N. V. Cherskiy. 1972. Formation of large natural gas fields in zones of permanently low temperatures. *Dokl. Akad. Nauk SSSR,* 205 (3), 215–218.

Mancini, E. A., R. M. Mink, and B. L. Bearden. 1986. Integrated geological, geophysical, and geochemical interpretation of Upper Jurassic petroleum trends in the eastern Gulf of Mexico. *Transactions–Gulf Coast Association of Geological Societies,* vol. 36, pp. 219–226.

Mango, F. C. 1990. The origin of light hydrocarbons in petroleum: A kinetic test of the steady-state catalytic hypothesis. *Geochim. Cosmochim. Acta,* 54, 1315–1323.

Mango, F. C. 1992. Transition metal catalysis in the generation of petroleum and natural gas. *Geochim. Cosmochim. Acta,* 56, 553–555.

Mango, F. C. 1994. The origin of light hydrocarbons in petroleum: Ring preference in the closure of carbocyclic rings. *Geochim. Cosmochim. Acta,* 58 (2), 895–901.

Mao, S., L. B. Eglinton, J. Whelan, and L. Liu. 1994. Thermal evolution of sediments from Leg 139, Middle Valley, Juan de Fuca Ridge: An organic petrological study. In M. J. Mottl et al. (eds.), *Proceedings of the Ocean Drilling Program, scientific results*. Vol. 139.

Martin, R. L., J. C. Winters, and J. A. Williams. 1963. Distributions of *n*-paraffins in crude oils and their implications to origin of petroleum. *Nature,* 199, 110–113.

Martinez, A. R. 1984. Classification and nomenclature systems for petroleum and petroleum reserves. *Proceedings of Eleventh World Petroleum Congress.* Vol. 2: *Geology Exploration Reserves.* Chichester: Wiley, pp. 325–329.

Masters, C. D. 1984. Distribution and quantitative assessment of world crude oil reserves and resources. *Proceedings of Eleventh World Petroleum Congress*. Vol. 2: *Geology Exploration Reserves.* Chichester: Wiley, pp. 229–237.

Masters, C. D. 1993. World resources of natural gas–A discussion. In D. G. Howell (ed.), *The future of energy gases.* U.S. Geological Survey professional paper 1570. Washington, DC: U.S. Government Printing Office, pp. 607–616.

Masters, C. D., E. D. Attanasi, W. D. Dietzman, R. F. Meyer, R. W. Mitchell, and D. H. Root. 1987. World resources of crude oil, natural gas, natural bitumen, and shale oil. Preprint, Twelfth World Petroleum Congress, Houston.

Masters, J. A. (ed.). 1984. *Elmworth: Case study of a deep basin gas field.* AAPG Memoir 38. Tulsa: American Association of Petroleum Geologists.

Mattavelli, L., and L. Novelli. 1990. Geochemistry and habitat of oils in Italy. *AAPG Bulletin*, 74, 1623–1639.

Mattavelli, L., T. Ricchiuto, D. Grignani, and M. Schoell. 1983. Geochemistry and habitat of natural gases in Po Basin, northern Italy. *AAPG Bulletin*, 67, 2239–2254.

Matviyenko, V. N. 1975. Comparative characteristics of geothermal conditions in some fields of west Siberia. *Nefti Gazovaya Geolog. Geofiz.,* 10, 12–14.

Maxwell, J. R., A. G. Douglas, G. J. Eglinton, and A. McCormick. 1968. The botryococcenes—Hydrocarbons of novel structure from alga *Botryococcus braunii Kutzing. Phytochem.,* 7, 2157–2171.

McAuliffe, C. D. 1966. Solubility in water of paraffin, cycloparaffin, olefin, acetylene, cyclo-olefin, and aromatic hydrocarbons. *J. Phys. Chem.,* 70 (4), 1267–1275.

McCaffrey, M. A., J. E. Dahl, P. Sundararaman, J. M. Moldown, and M. Schoell. 1994. Source rock quality determination from oil biomarkers II: A case study using Tertiary-reservoired Beaufort Sea oils. *AAPG Bulletin*, 78, 1527–1540.

McCartney, J. T., and S. Ergun. 1958. Optical properties of graphite and coal. *Fuel*, 37, 272–281.

McCartney, J. T., and M. Teichmüller. 1972. Classification of coals according to degree of coalification by reflectance of the vitrinite component. *Fuel,* 51, 64–68.

McDowell, A. N. 1975. What are the problems in estimating the oil potential of a basin? *Oil & Gas Journal,* June 9, 85–90.

McEvoy, J., and W. Giger. 1986. Origin of hydrocarbons in Triassic Serpiano oil shales: Hopanoids. *Org. Geochem.,* 10, 943–949.

McIver, R. D. 1967. Composition of kerogen–Clue to its role in the origin of petroleum. *Proceedings of Seventh World Petroleum Congress in Mexico City*. London: Elsevier Science, vol. 2, pp. 26–36.

McIver, R. D. 1985. Near-surface hydrocarbon surveys in oil and gas exploration. *Oil & Gas Journal,* 82 (39), 113–117.

McKay, J. F., and D. R. Lantham. 1973. Polyaromatic hydrocarbons in high-boiling petroleum distillates: Isolation by gel permeation chromatography and identification by fluorescence spectrometry. *Anal. Chem.,* 45 (7), 1050–1055.

McKinney, C. M., E. P. Ferrero, and W. J. Wenger. 1966. *Analysis of crude oils from 546 important oilfields in the United States.* U.S. Bureau of Mines Report of Investigations 6819. Washington, DC: U.S. Department of Interior, Bureau of Mines.

McKirdy, D. M., and A. R. Chivas. 1992. Nonbiodegraded aromatic condensate associated with volcanic supercritical carbon dioxide, Otway Basin: Implications for primary migration from terrestrial organic matter. *Org. Geochem.,* 18, 611–627.

McKirdy, D. M., R. E. Cox, J. K. Volkman, and V. J. Howell. 1986. Botryococcane in a new class of Australian crude oils. *Nature,* 320 (6057), 57–59.

McKirdy, D. M., D. J. McHugh, and J. W. Tardif. 1980. Comparative analysis of stromatolitic and other microbial kerogens by pyrolysis–hydrogenation–gas chromatography (PHGC). In P. A. Trudinger, M. R. Walter, and B. J. Ralph (eds.), *Biogeochemistry of ancient and modern environments.* Canberra: Australian Academy of Science, pp. 187–200.

Meissner, F. F. 1978. Petroleum geology of the Bakken Formation, Williston Basin, North Dakata and Montana. *Proceedings of 1978 Williston Basin Symposium, The Economic Geology of the Williston Basin, September 24–27.* Billings: Montana Geological Society, pp. 207–227.

Meissner, F. F. 1984. Cretaceous and Lower Tertiary coals as sources for gas accumulations in the Rocky Mountain area. In J. Woodward, F. F. Meissner, and J. L. Clayton (eds.), *Hydrocarbon source rocks of the greater Rocky Mountain region.* Denver: Rocky Mountain Association of Geologists, pp. 401–431.

Merrill, R. K. (ed.). 1991. *Source and migration processes and evaluation techniques.* Tulsa: American Association of Petroleum Geologists.

Metzger, P., C. Berkaloff, E. Casadevall, and A. Coute. 1985a. Alkadiene- and botryococcene-producing races of wild strains of *Botryococcus braunii. Phytochem.*, 24 (10), 2305–2312.

Metzger, P., E. Casadevall, M. J. Pouet, and Y. Pouet. 1985b. Structures of some Botryococceses: Branched hydrocarbons from B-race of green alga *Botryococcus braunii. Phytochem.*, 24 (12), 2995–3002.

Meyer, B. L., and M. H. Nederlof. 1984. Identification of source rocks on wireline logs by density/resistivity and sonic transit time/resistivity crossplots. *AAPG Bulletin*, 121–129.

Meyer, R. E. (ed.). 1987. *Exploration for heavy crude oil and natural bitumen.* AAPG Studies in Geology 25. Tulsa: American Association of Petroleum Geologists.

Meyerhoff, A. A. 1968. Geology of natural gas in south Louisiana. In B. W. Beebe and B. F. Curtis (eds.), *Natural gases of North America*. AAPG Memoir 9. Tulsa: American Association of Petroleum Geologists, vol. 1, p. 546.

Meyerhoff, A. A. 1980. Geology and petroleum fields in Proterozoic and Lower Cambrian strata, Lena–Tunguska Petroleum province, eastern Siberia, USSR. In M. T. Halbouty (ed.), *Giant oil and gas fields of the decade 1968–1978.* AAPG Memoir 30. Tulsa: American Association of Petroleum Geologists, pp. 225–252.

Michaelis, W., B. Mycke, and H-H. Richnow. 1986. Organic chemical indicators for reconstructions of Angola Basin sedimentation process. In E. T. Degens et al. (eds.), *Biogeochemistry of black shales*. Hamburg: Geologisch–Päleontologischen Instituts der Universität Hamburg, 60, 99–113.

Miknis, F. P., and T. F. Turner, 1988, Thermal decomposition of Tipton Member, Green River Formation of oil shale from Wyoming: Report by Western Research Institute, Laramie, to Department of Energy, September.

Miknis, F. P., T. F. Turner, G. L. Berdan, and P. J. Conn. 1987. Formation of soluble products from thermal decomposition of Colorado and Kentucky oil shales. *Energy & Fuels*, 1, 477–483.

Miles, J. A. 1989. *Illustrated glossary of petroleum geochemistry.* Oxford: Clarendon Press. 137 p.

Miller, R. G.. 1992. The global oil system: The relationship between oil generation, loss, half-life, and the world crude oil resource. *AAPG Bulletin,* 76, 489–500.

Moldowan, J. M., P. Albrecht, and R. P. Philp (eds.). 1992. *Biological markers in sediments and petroleum.* Englewood Cliffs, NJ: Prentice-Hall.

Moldowan, J. M., C. Y. Lee, P. Sundararaman, T. Salvatori, A Alajbeg, B. Gjukic, G. J. Demaison, N. Slougui, and D. S. Watt. 1994. Source correlation and maturity assessment of select oils and rocks from the central Adriatic Basin (Italy and Yugoslavia). In J. M. Moldowan, P. Albrecht, and R. P. Philp (eds.), *Biological markers in sediments and petroleum*. Englewood Cliffs, NJ: Prentice-Hall, pp. 370–396.

Moldowan, J. M., and W. K. Seifert. 1979. Head-to-head linked isoprenoid hydrocarbons in petroleum. *Science*, 204, 169–170.

Moldowan, J. M., and W. K. Seifert. 1980. First discovery of botryococcane in petroleum. *J.S.C. Chem. Comm.*, 912–914.

Moldowan, J. M., W. K. Seifert, and E. J. Gallegos. 1983. Identification of an extended series of tricyclic terpanes in petroleum. *Geochim. Cosmochim. Acta*, 47, 1531–1534.

Moldowan, J. M., W. K. Seifert, and E. J. Gallegos. 1985. Relationship between petroleum composition and depositional environment of petroleum source rocks. *AAPG Bulletin*, 69, 1255–1268.

Momper, J. A. 1978. Oil migration limitations suggested by geological and geochemical considerations. In *Physical and chemical constraints on petroleum migration*. Vol. 1. Notes for AAPG short course, April 9, AAPG National Meeting, Oklahoma City.

Momper, J. A. 1981. The petroleum expulsion mechanism–A consequence of the generation process. Notes for AAPG Geochemistry for Geologists School, Denver, February 23–25.

Momper, J. A., and J. A. Williams. 1984. Geochemical exploration of the Powder River Basin. In G. J. Demaison and R. J. Murris (eds.), *Petroleum geochemistry and basin evaluation*. AAPG Memoir 35. Tulsa: American Association of Petroleum Geologists, pp. 181–191.

Moore, R. E. 1979. Marine aliphatic natural products. In F. D. Gunston et al. (eds.), *Aliphatic and related natural product chemistry*. London: Chemical Society, pp. 55–58.

Morton, R. A., and L. S. Land. 1987. Regional variations in formation water chemistry, Frio Formation (Oligocene), Texas Gulf Coast. *AAPG Bulletin*, 71, 191–206.

Moses, P. L. 1986. Engineering applications of phase behavior of crude oil and condensate systems. *J. Petrol. Tech.*, July, 715–723.

Muir-Wood, R. 1988. Shear waves show the earth is a bit cracked. *New Scientist*, September 21, 44–48.

Mukhopadhyay, P. K. 1994. Vitrinite reflectance as a maturity parameter: Petrographic and molecular characterization and its applications to basin modeling. In P. K. Mukhopadhyay and W. G. Dow (eds.), *Vitrinite reflectance as a maturity parameter: Applications and limitations.* ACS Symposium series 570. Washington, DC: American Chemical Society, pp. 1–24.

Mukhopadhyay, P. K., and W. G. Dow (eds.). 1994. *Vitrinite reflectance as a maturity parameter: Applications and limitations.* ACS Symposium series 570. Washington, DC: American Chemical Society.

Mukhopadhyay, P. K., H. W. Hagemann, and J. R. Gormly. 1985a. Characterization of kerogens as seen under the aspect of maturation and hydrocarbon generation. *Wissenschaft + Technik,* 38 (1), 7–18.

Mukhopadhyay, P. K., and P. G. Hatcher. 1993. Composition of coal. In B. E. Law and D. D. Rice (eds.), *Hydrocarbons from coal.* Studies in Geology series, vol. 38. Tulsa: American Association of Petroleum Geologists, pp. 79–118.

Mukhopadhyay, P. K., U. Samanta, and J. Jassal. 1985b. *Origin of oil in a lagoon environment: Desmocollinite/bituminite source-rock concept*. Vol. 4 of *Compte rendu*. Carbondale: Southern Illinois University Press, pp. 753–763.

Mukhopadhyay, P. K., J. A. Wade, and M. A. Kruge. 1995. Organic facies and maturation of Jurassic/Cretaceous rocks, and possible oil–source rock correlation based on pyrolysis of asphaltenes, Scotian Basin, Canada. *Org. Geochem.,* 22 (1), 85–104.

Muller, J. 1964. Palynological contributions to the history of Tertiary vegetation in N.W. Borneo. In D. Murchison and T. S. Westoll (eds.), *Coal and coal-bearing strata.* New York: Elsevier Science, pp. 39–40.

Munns, J. W. 1985. The Valhall Field: A geological overview. *Marine and Petroleum Geology*, 2, 23–43.

Murchison, D. 1969. Some recent advances in coal petrology. *Congres Internat. Stat. Geol. Carbonif. Compte rendu*. Sheffield, vol. 1. Maastricht: E. van Aalst, pp. 351–368.

Murphy, M. T. J., A. McCormick, and G. J. Eglinton. 1967. Perhydro-ß-carotene in Green River Shale. *Science*, 157, 1040–1042.

Murray, G. H. Jr. 1968. Quantitative fracture study–Sanish Pool, McKenzie County, North Dakota. *AAPG Bulletin*, 52, 57–65.

Murris, R, J. 1980. Middle East: Stratigraphic evolution and oil habitat. *AAPG Bulletin,* 64, 597–618.

Murris, R. J. 1984. Introduction. In G. J. Demaison and R. J. Murris (eds.), *Petroleum geochemistry and basin evaluation.* AAPG Memoir 35. Tulsa: American Association of Petroleum Geologists, pp. x–xii.

Murtada, H., and B. Hofling. 1987. Feasibility of heavy-oil recovery. In R. F. Meyer (ed.), *Exploration for heavy crude oil and natural bitumen*. AAPG Studies in Geology 25. Tulsa: American Association of Petroleum Geologists, pp. 629–643.

Naeser, N. D., and T. H. McCulloh (eds.). 1989. *Thermal history of sedimentary basins*. New York: Springer-Verlag.

National Academy of Sciences. 1975. *Petroleum in the marine environment*. Workshop on Inputs, Fates, and the Effects of Petroleum in the Marine Environment, May 21–25, 1973, Airlie, VA. Washington, DC: National Academy of Sciences, p. 2.

Nechayeva, O. L. 1968. Hydrogen in gases dissolved in water of the western Siberian plain. *Dokl. Akad. Nauk SSSR,* 179 (4), 961–962.

Neglia, S. 1979. Migration of fluids in sedimentary basins. *AAPG Bulletin,* 63, 573–597.

Nelson, R. A. 1985. *Geologic analysis of naturally fractured reservoirs*. Vol. 1 of *Contributions in Petroleum Geology & Engineering*. Houston: Gulf Publishing.

Nelson, W. L. 1972. What's the average sulfur content vs. gravity? *Oil & Gas Journal,* 70 (5), 59.

Nelson, W. L. 1974. What are the amounts of nitrogen and oxygen in U.S. products? *Oil & Gas Journal,* 72 (5), 112–114.

Nelson, W. L. 1978. Where are analyses of 57°–70° API crude oils? *Oil & Gas Journal*, 76, 100.

Neumann, H-J., B. Paczynskaya-Lahme, and D. Severin. 1981. *Composition and properties of petroleum*. New York: Wiley.

Newberry, J. S. 1860. *The rock oils of Ohio*. Agricultural report for 1859.

Newman, J., and N. A. Newman. 1982. Reflectance and anomalies in Pike River coals: Evidence of variability in vitrinite type, with implications for maturation studies and "Suggate rank," New Zealand. *J. Geol. and Geophysics*, 25, 233–243.

Nishimura, M., and E. W. Baker. 1986. Possible origin of *n*-alkanes with a remarkable even-to-odd predominance in recent marine sediments. *Geochim. Cosmochim. Acta*, 50, 299–305.

Nissenbaum, A., M. Goldberg, and Z. Aizenshtat. 1985. Immature condensate from southeastern Mediterranean coastal plain, Israel. *AAPG Bulletin,* 69, 946–949.

Noble, R. A. 1991. Geochemical techniques in relation to organic matter. In R. K. Merrill (ed.), *Source and migration processes and evaluation techniques.* Tulsa: American Association of Petroleum Geologists, pp. 97–102.

Noble, R. A., R. Alexander, and R. I. Kagi. 1986. Identification of some diterpenoid hydrocarbons in petroleum. *Org. Geochem.*, 10, 825–829.

Noble, R. A., C. H. Wu, and C. D. Atkinson. 1991. Petroleum generation and migration from Talang Akar coals and shales offshore N.W. Java, Indonesia. *Org. Geochem.*, 17 (3), 363–374,

North, F. K. 1985. *Petroleum geology*. Boston: Allen & Unwin.

North Sea Letter and European Offshore News. 1981. *Financial Times*, September 23.

Northam, M. A. 1985. Correlation of northern North Sea oils: The different facies of their Jurassic source. In B. M. Thomas et al. (eds.), *Petroleum geochemistry in exploration of the Norwegian Shelf.* London: Graham & Trotman for the Norwegian Petroleum Society, pp. 93–99.

Novelli, L., M. A. Chiaramonte, L. Mattavelli, G. Pizzi, L. Sartori, and P. Scott. 1987. Oil habitat in the northwestern Po Basin. In B. Doligez (ed.), *Migration of hydrocarbons in sedimentary basins.* Paris: Éditions technip, pp. 27–57.

Novokshchenov, A. M. 1982. Formation of intra-salt high-pressure brines in areas of southeast Turkmenia. *Petrol. Geol.*, 18, 190–192.

Oil & Gas Journal. 1987. Eastern Siberia slow to yield reserves. March 9, 49–50.

Oil in the sea: Inputs, fates, and effects. 1985. Washington, DC: National Academy Press, 601 pp.

Olsen, R. S. P. 1980. Depositional environment of Jurassic Smackover sandstones, Thomasville Field, Rankin County, Mississippi. M.A. thesis, Texas A & M University.

Olson, R. K. 1982. Factors controlling uranium distribution in Upper Devonian–Lower Mississippian Black Shales of Oklahoma and Arkansas. Ph.D. diss., University of Tulsa.

Oro, J., T. G. Tornabene, D. W. Nooner, and E. Gelpi. 1967. Aliphatic hydrocarbons and fatty acids of some marine and freshwater microorganisms. *J. Bacteriol.*, 93, 1811–1818.

Orr, W. L. 1974. Changes in sulfur content and isotopic ratios of sulfur during petroleum maturation—Study of Big Horn Basin Paleozoic oils. Part 1. *AAPG Bulletin*, 58, 2295–2318.

Orr, W. L. 1977. Geologic and geochemical controls on the distribution of hydrogen sulfide in natural gas. In R. Campos and J. Goni (eds.), *Advances in organic geochemistry 1975*. Madrid: Empressa nacional adaro de investigaciones mineras, pp. 571–597.

Orr, W. L. 1978. Sulfur in heavy oils, oil sands and oil shales. In O. P. Strausz and E. M. Lown (eds.), *Oil sand & oil shale chemistry*. Weinheim: Verlag Chemie, pp. 223–243.

Orr, W. L. 1983. Comments on pyrolytic hydrocarbon yields in source-rock evaluation. In M. Bjorøy et al. (eds.), *Advances in organic geochemistry 1981*. Chichester: Wiley, pp. 775–787.

Orr, W. L. 1986. Kerogen/asphaltene/sulfur relationships in sulfur-rich Monterey oils. *Org. Geochem.*, 10, 499–516.

Orr, W. L., and T. H. McCulloh. 1993. Guidelines for type II-S kerogens in basin modeling: Organic sulfur content of kerogens and crude oils. Abstract. *AAPG Bulletin*, 77, 1652.

Orr, W. L., and J. S. Sinninghe Damsté. 1990. Geochemistry of sulfur in petroleum systems. In W. L. Orr and C. M. White (eds.), *Geochemistry of sulfur in fossil fuels*. ACS Symposium series 429. Washington, DC.: American Chemical Society, pp. 2–29.

Orr, W. L., and C. M. White (eds.). 1990. *Geochemistry of sulfur in fossil fuels*. ACS Symposium series 429. Washington, DC: American Chemical Society.

Ortoleva, P. (ed.). 1994. *Basin compartments and seals*. AAPG Memoir 61. Tulsa: American Association of Petroleum Geologists.

Ostroukhov, S. B., O. A. Arefyev, V. M. Makushina, M. N. Zabrodina, and A. Petrov. 1982. Monocyclic aromatic hydrocarbons with isoprenoid chains. *Neftekhimiya*, 22, 723–728.

Ourisson, G., P. Albrecht, and M. Rohmer. 1979. The hopanoids, paleochemistry and biochemistry of a group of natural products. *Pure & Appl. Chem.*, 51, 709–729.

Ourisson, G., P. Albrecht, and M. Rohmer. 1982. Predictive microbial biochemistry–From molecular fossils to procaryotic membranes. *Trends Biochem. Sci.*, 7, 236–239.

Ourisson, G., P. Albrecht, and M. Rohmer. 1984. The microbial origin of fossil fuels. *Scientific American*, 251, 44–51.

Owen, E. W. 1975. Trek of the oil finders. In *A history of exploration for petroleum*. AAPG Memoir 6. Tulsa: American Association of Petroleum Geologists, pp. 1–4.

Palacas, J. G. 1984a. Carbonate rocks as sources of petroleum: Geological and chemical characteristics and oil-source correlations. *Proceedings of the Eleventh World Petroleum Congress*. Chichester: Wiley, vol. 2, pp. 31–43.

Palacas, J. G. (ed.). 1984b. *Petroleum geochemistry and source rock potential of carbonate rocks*. AAPG Studies in Geology 18. Tulsa: American Association of Petroleum Geologists.

Palacas, J. G., D. E. Anders, and J. D. King. 1984. South Florida Basin—A prime example of carbonate source rocks of petroleum. In J. G. Palacas (ed.), *Petroleum geochemistry and source rock potential of carbonate rocks*. AAPG Studies in Geology 18. Tulsa: American Association of Petroleum Geologists, pp. 71–96.

Palmer, S. E. 1984. Hydrocarbon source potential of organic facies of the lacustrine Elko Formation (Eocene/Oligocene), northeast Nevada. In J. Woodward, F. F. Meissner, and J. F. Clayton (eds.), *Hydrocarbon source rocks of the greater Rocky Mountain region*. Denver: Rocky Mountain Association of Geologists, pp. 491–511.

P'an Chung-Hsiang. 1982. Petroleum in basement rocks. *AAPG Bulletin,* 66, 1597–1643.

Parkash, S., and S. K. Chakrabartty. 1986. Microporosity in Alberta plains coals. *Internat'l J. Coal Geol.*, 6, 55–70.

Pasley, M., W. Gregory, and G. F. Hart. 1991. Organic matter variations in transgressive and regressive shales. *Org. Geochem.*, 17 (4), 483–509.

Patterson, C. 1956. Age of meteorites and the earth. *Geochim. Cosmochim. Acta,* 10, 230–237.

Patterson, G. W. 1971. The distribution of sterols in algae. *Lipids*, 6, 120–126.

Payzant, J. D., D. S. Montgomery, and O. P. Strausz. 1986. Sulfides in petroleum. *Org. Geochem.*, 9 (6), 357–369.

Pearson, D. L. 1990. Pollen/spore color "standard." Phillips Petroleum Company, Geology Branch. Second printing of version 2, April.

Pelet, R., F. Behar, and J. C. Monin. 1986. Resins and asphaltenes in the generation and migration of petroleum. *Org. Geochem.*, 10 (1–3) 481–498.

Perrodon, A. 1983. Dynamics of oil and gas accumulations. *Bull. centres recherches exploration–production Elf-Aquitaine*. Pau. Memoir 5, 337–339.

Peters, K. E. 1986. Guidelines for evaluating petroleum source rock using programmed pyrolysis. *AAPG Bulletin*, 70, 318–329.

Peters, K. E., and M. R. Cassa. 1994. Applied source rock geochemistry. In L. B. Magoon and W. G. Dow (eds.), *The petroleum system–From source to trap*. AAPG Memoir 60. Tulsa: American Association of Petroleum Geologists, pp. 93–120.

Peters, K. E., T. D. Elam, M. H. Pytte, and P. Sundararaman. 1994. Identification of petroleum systems adjacent to the San Andreas Fault, California, U.S.A. In L. B. Magoon and W. G. Dow (eds.), *The petroleum system—From source to trap.* AAPG Memoir 60. Tulsa: American Association of Petroleum Geologists, pp. 423–436.

Peters, K. E., A. Eh. Kontorovich, J. M. Moldowan, V. E. Andrusevich, B. J. Huizinga, G. J. Demaison, and O. F. Stasova. 1993. Geochemistry of selected oils and rocks from the central portion of the West Siberian Basin, Russia. *AAPG Bulletin*, 77, 863–887.

Peters, K. E., and J. M. Moldowan. 1991. Effects of source, thermal maturity, and biodegradation on the distribution and isomerization of homohopanes in petroleum. *Org. Geochem.*, 17 (1), 47–61.

Peters, K. E., and J. M. Moldowan. 1993. *The biomarker guide: Interpreting molecular fossils in petroleum and ancient sediments.* Englewood Cliffs, NJ: Prentice-Hall.

Peters, K. E., J. M. Moldowan, A. R. Driscole, and G. J. Demaison. 1989. Origin of Beatrice oil by co-sourcing from Devonian and Middle Jurassic source rocks, Inner Moray Firth, United Kingdom. *AAPG Bulletin*, 73, 454–471.

Peters, K. E., J. M. Moldowan, and P. Sundararaman. 1990. Effects of hydrous pyrolysis on biomarker thermal maturity parameters: Monterey phosphatic and siliceous members. *Org. Geochem.*, 15 (3), 249–265.

Peters, K. E., and D. A. Nelson. 1992. REESA—An expert system for geochemical logging of wells. Program abstract, Annual Convention of American Association of Petroleum Geologists, Calgary, p. 103.

Peters, K. E., J. K. Whelan, J. M. Hunt, and M. E. Tarafa. 1983. Programmed pyrolysis of organic matter from thermally altered Cretaceous black shales. *AAPG Bulletin*, 67, 2137–2146.

Petersen, N. F., and P. J. Hickey. 1987 California Plio-Miocene oils: Evidence of early generation. In R. F. Meyer (ed.), *Exploration for heavy crude oil and bitumen*. AAPG Memoir 38. Tulsa: American Association of Petroleum Geologists, pp. 351–359.

Petersil'ye, I. A., Ye. K. Kozlov, K. D. Belyayev, V. V. Sholokhnev, and V. S. Dokuchayeva. 1970. Nitrogen and hydrocarbon gases in ultramafic rocks of the sopcha stock of the Monchegorsk pluton, Kola Peninsula. *Dokl. Akad. Nauk SSSR*, 194, 200–203.

Petrov, A. A. 1984. *Petroleum hydrocarbons*. Berlin: Springer-Verlag.

Philippi, G. T. 1965. On the depth, time and mechanism of petroleum generation. *Geochim. Cosmochim. Acta*, 29, 1021–1049.

Philp, R. P. 1985a. Biological markers in fossil fuel production. *Mass Spectrometry Reviews*, 4, 1–54.

Philp, R. P. 1985b. *Fossil fuel biomarkers, applications and spectra*. Amsterdam: Elsevier Science.

Philp, R. P. 1987. Surface prospecting methods for hydrocarbon accumulations. In *Advances in petroleum geochemistry*. London: Academic Press, vol. 2, pp. 210–253.

Philp, R. P., and Fan Zhaoan. 1987. Geochemical investigation of oils and source rocks from Qianjiang Depression of Jianghan Basin, a terrigenous saline basin, China. *Org. Geochem.*, 11 (6), 549–562.

Philp, R. P., and T. D. Gilbert. 1986. Biomarker distributions in Australian oils predominantly derived from terrigenous source material. *Org. Geochem.*, 10, 73–84.

Philp, R. P., B. R. T. Simoneit, and T. C. Gilbert. 1983. Diterpenoids in crude oils and coals of southeastern Australia. In M. Bjorøy et al. (eds.), *Advances in organic geochemistry 1981*. Chichester: Wiley, pp. 698–704.

Pittion, J. L., and J. Gouadain. 1985. Maturity studies of the Jurassic "coal unit" in three wells from the Haltenbanken area. In B. M. Thomas et al. (eds.), *Petroleum geochemistry in exploration of the Norwegian Shelf.* Proceedings of Norwegian Petroleum Society Conference, Stavanger, pp. 22–24, 1984. London: Graham & Trotman for the Norwegian Petroleum Society, pp. 205–211.

Pixler, B. O. 1969. Formation evaluation by analysis of hydrocarbon ratios. *J. Petrol. Tech.,* 24, 665–670.

Poirer, M. A., and G. T. Smiley. 1984. A novel method for separation and identification of sulphur compounds in naphtha (30–200°C) and middle distillate (200–350°C) fractions of Lloydminster heavy oil by GC/MS. *J. Chron. Sc.*, 22, 304–309.

Polivanova, A. N. 1977. The relationship between the carbon isotopic composition of methane and hydrogen sulphide and saliferous occurrences. In N. B. Vassoevich et al. (eds.), *8th International Congress on Organic Geochemistry, abstracts of reports.* Moscow, May 10–13, vol. 2, pp. 164–166.

Pollard, D. D., and A. Aydin. 1988. Progress in understanding jointing over the past century. *Geological Society of America Bull.*, 100, 1181–1204.

Poole, F. G., and G. E. Claypool. 1984. Petroleum source-rock potential and crude-oil correlation in the Great Basin. In J. Woodward, F. F. Meissner, and J. F. Clayton (eds.), *Hydrocarbon source rocks of the greater Rocky Mountain region.* Denver: Rocky Mountain Association of Geologists, pp. 179–229.

Potonie, H. 1908. Die rezenten Kaustobiolithe und ihre Langerstatten: Die Sapropeliten. *Abh. Kgl. Preuss. Geol. Landesanstalt,* new ser., 1 (55).

Poulet, M., and J. Roucache. 1969. Etude géochemique des gisements du Nord-Sahara (Algerie): *Rev. l'Inst. français du pétrole*, 24, 615–644.

Powell, T. G. 1986. Petroleum geochemistry and depositional setting of lacustrine source rocks. *Marine and Petroleum Geology*, 3, 200–219.

Powell, T. G., and D. M. McKirdy. 1973. The effect of source material, rock type and diagenesis on the *n*-alkane content of sediments. *Geochim. Cosmochim. Acta,* 37, 523–633.

Powley, D. E. 1980. Pressures, normal and abnormal. AAPG Advanced Exploration Schools unpublished lecture notes, 38.

Powley, D. E. 1985. Pressures, normal and abnormal. Lecture notes, techniques of petroleum exploration: II. American Association of Petroleum Geology School, South Padre Island, September 16–19.

Powley, D. E. 1992. Shale porosity–depth relations in normally compacted shale. Second Symposium on Deep Basin Compartments and Seals, Gas Research Institute, Oklahoma State University, Stillwater, September 29–October 1.

Powley, D. E. 1993. Shale compaction and its relationship to fluid seals. Section III: Quarterly report, January–April 1933, Oklahoma State University to the Gas Research Institute, G.R.I. Contract 5092-2443.

Pratt, L. M. 1984. Influence of paleoenvironmental factors on preservation of organic matter in Middle Cretaceous Greenhorn Formation, Pueblo, Colorado. *AAPG Bulletin,* 68, 1146–1159.

Pratt, W. E. 1943. *Oil in the earth.* Lawrence: University of Kansas Press.

Press, F., and R. Siever. 1986. *Earth.* 4th ed. New York: Freeman.

Price, L. C. 1973. The solubility of hydrocarbons and petroleum in water as applied to the primary migration of petroleum. Ph.D. diss., University of California at Riverside.

Price, L. C. 1976. Aqueous solubility of petroleum as applied to its origins and primary migration. *AAPG Bulletin*, 60, 213–244.

Price, L. C., and C. E. Barker. 1985. Suppression of vitrinite reflectance in amorphous rich kerogen–A major unrecognized problem. *Journal of Petroleum Geology*, 8 (1), 59–85.

Price, L. C., and J. L. Clayton. 1992. Extraction of whole versus ground source rocks: Fundamental petroleum geochemical implications including oil-source rock correlation. *Geochim. Cosmochim. Acta*, 56, 1213–1222.

Price, L. C., J. L. Clayton, and L. L. Rumen. 1981. Organic geochemistry of the 9.6 km Bertha Rogers no. 1 well, Oklahoma. *Org. Geochem.*, 3, 59–77.

Proshlyakov, B. K. 1960. Reservoir properties of rocks as a function of their depth and lithology. *Geol. Nefti Gaza,* 4 (12), 24–29. Assoc. Tech. Services Translation RJ 3421.

Purcell, W. R. 1949. Capillary pressure—Their measurements using mercury and the calculation of permeability therefrom. *Petrol. Trans. Am. Inst. Mining and Met. Engr.*, 186, 39–48.

Puttmann, W., M. Wolf, and E. Wolff-Fischer. 1986. Chemical characteristics of liptinite macerals in humic and sapropelic coals. *Org. Geochem.*, 10, 625–632.

Qin, K., D. Chen, and Z. Li. 1991. A new method to estimate the oil and gas potentials of coals and kerogens by solid state ^{13}C NMR spectroscopy. *Org. Geochem.*, 17 (6), 865–872.

Qin, K., Q. Yang, S. Guo, Q. Lu, and W. Shu. 1994. Chemical structure and hydrocarbon formation of the Huanxian brown coal, China. *Org. Geochem.*, 21 (3–4), 333–341.

Quigley, T. M., A. S. Mackenzie, and J. R. Gray. 1987. Kinetic theory of petroleum generation. In B. Doligez (ed.), *Migration of hydrocarbons in sedimentary basins*. Paris: Éditions technip, pp. 649–665.

Radchenko, O. A., I. P. Karpova, and A. S. Chernysheva. 1951. *A geochemical investigation of weathered and highly altered mineral fuels from South Fergana*. Trudy VNIGRI, new ser., no. 5, *Contributions to geochemistry*, no. 2–3, pp. 180–202. Trans. by Israel Program for Scientific Translations, Jerusalem, 1965.

Radke, M. 1988. Application of aromatic compounds as maturity indicators in source rocks and crude oils. *Marine and Petroleum Geology,* 5 (3), 224–236.

Radke, M., P. Garrigues, and H. Willsch. 1990. Methylated dicyclic and tricyclic aromatic hydrocarbons in crude oils from the Handil field, Indonesia. *Org. Geochem.*, 15 (1), 17–34.

Rall, H. T., C. J. Thompson, H. J. Coleman, and R. L. Hopkins. 1972. Sulfur compounds in crude oil. *U.S. Bureau of Mines Bulletin*, no. 659.

Reed, J. C., H. A. Illich, and B. Horsfield. 1986. Biochemical evolutionary significance of Ordovician oils and their sources. *Org. Geochem.*, 10, 347–358.

Reeder, M. L., and I. C. Scotchman. 1985. Hydrocarbon generation–Central and northern North Sea. *Oil & Gas Journal,* 137–144.

Reerink, H., and J. Lijzenga. 1973. Molecular weight distributions of Kuwait asphaltenes as determined by ultracentrifugation: Relation with viscosity of solutions. *J. Inst. Petrol.,* 59 (569), 211–222.

Reitsema, R. H., A. J. Kaltenback, and F. A. Lindberg. 1981. Source and migration of light hydrocarbons indicated by carbon isotopic ratios. *AAPG Bulletin*, 65, 1536–1542.

Rhoads, D. C., and J. W. Morse. 1971. Evolutionary and ecologic significance of oxygen-deficient marine basins. *Lethaia*, 4, 413–428.

Rice, D. D. 1984. Occurrence of indigenous biogenic gas in organic-rich, immature chalks of Late Cretaceous age, eastern Denver Basin. In J. G. Palacas (ed.), *Petroleum geochemistry and source rock potential of carbonate rocks*. AAPG Studies in Geology 18. Tulsa: American Association of Petroleum Geologists, pp. 135–150.

Rice, D. D. 1993a. Biogenic gas: Controls, habitat, and resource potential. In D. G. Howell et al. (eds.), *The future of energy gases.* U.S. Geological Survey professional paper 1570. Washington, DC: U.S. Government Printing Office, pp. 583–606.

Rice, D. D. 1993b. Controls on coal-bed gas composition. Abstract. *AAPG Bulletin*, 77, 1658.

Richardson, J. S., and D. E. Miller. 1983. Biologically-derived compounds of significance in the saturate fraction of a crude oil having predominant terrestrial input. *Fuel*, 62, 524–528.

Rieke, H. H. III, and G. V. Chilingarian. 1974. *Compaction of argillaceous sediments*. New York: Elsevier Science.

Rinaldi, G. G. L. 1985. Presence of monoaromatic secohopanes and benzohopanes in petroleums. Symposium on chemical biomarkers, American Chemical Society Meeting, April 28–May 3, Miami Beach.

Riva, A., T. Salvatori, R. Cavaliere, T. Ricchiuto, and L. Novelli. 1986. Origin of oils in Po Basin in northern Italy. *Org. Geochem.* 10, 391–400.

Roadifer, R. E. 1987. Size distributions of the world's largest known oil and tar accumulations. In R. F. Meyer (ed.), *Exploration for heavy crude oil and natural bitumen*. AAPG Studies in Geology 25. Tulsa: American Association of Petroleum Geologists, pp. 3–23.

Roberts, W. H. III, and R. J. Cordell (eds.). 1980. *Problems of petroleum migration*. AAPG Studies in Geology 10. Tulsa: American Association of Petroleum Geologists.

Robinson, C. J. 1971. Low-resolution mass spectrometric determination of aromatics and saturates in petroleum fractions. *Anal. Chem.*, 43 (11), 1425–1434.

Robinson, J. N., and S. J. Rowland. 1986. Identification of novel widely distributed sedimentary acyclic sesterpenoids. *Nature*, 324, 561–563.

Rodrigues, K. 1988. Oil source bed recognition and crude oil correlation, Trinidad, West Indies. *Org. Geochem.*, 13 (1–3), 365–371.

Rogers, M. A., J. D. McAlary, and N. J. L. Bailey. 1974. Significance of reservoir bitumens to thermal-maturation studies, Western Canada Basin. *AAPG Bulletin*, 58, 1806–1824.

Rogers, M. R., and T. H. Anderson. 1984. Tyrone–Mt. Union cross-strike lineament of Pennsylvania: A major Paleozoic basement fracture and uplift boundary. *AAPG Bulletin*, 68, 92–105.

Rohmer, M., P. Bouvier, and G. Ourisson. 1979. Molecular evolution of biomembranes: Structural equivalents and phylogenetic precursors of sterols. *Proc. Natl. Acad. Sci. USA*, 76 (2), 847–851.

Rohrback, B. G. 1983. Crude oil geochemistry of the Gulf of Suez. In M. Bjorøy et al. (ed.), *Advances in organic geochemistry 1981*. Chichester: Wiley, pp. 39–48.

Romankevich, E. A. 1984. *Geochemistry of organic matter in the ocean*. Berlin, Heidelberg: Springer-Verlag. Trans. Geokhimiia organicheskogo veshchestva v okeane. Vestnic Academii Nauk SSSR, 1978.

Ronov, A. B. 1958. Organic carbon in sedimentary rocks (in relation to the presence of petroleum). *Geochem.,* 5, 497–509.

Ronov, A. B. 1982. The earth's sedimentary shell (quantitative patterns of its structure, composition, and evolution). The 20th V. I. Vernadskiy lecture, March 12, 1978. *Int. Geol. Rev.*, 24 (12), 1365–1388.

Ronov, A. B. 1994. Phanerozoic transgressions and regressions on the continents: A quantitative approach based on areas flooded by the sea and areas of marine and continental deposition. *American Journal of Science,* 294, 777–801.

Ronov, A. B., and A. A. Yaroshevsky. 1969. Chemical composition of the earth's crust. *The earth's crust and upper mantle. American Geological Union Monograph* 13. Washington, DC: American Geological Union, pp. 37–57.

Rosenfeld, W. D., and S. R. Silverman. 1959. Carbon isotopic fractionation in bacterial production of methane. *Science,* 130, 1658–1659.

Ross, L. M., and R. L. Ames. 1988. Stratification of oils in Columbus Basin off Trinidad. *Oil & Gas Journal,* September 26, 72–76.

Rosscup, R. J., and J. Bowman. 1967. Thermal stabilities of vanadium and nickel petroporphyrins. *Preprints of the Division of Petroleum Chemistry, American Chemical Society*. Washington, DC: American Chemical Society, vol. 12, p. 77.

Rossini, F. D. 1960. Hydrocarbons in petroleum. *J. Chem. Ed.,* 37 (11), 554–561.

Rouchet, J. 1981. Stress fields, a key to oil migration. *AAPG Bulletin,* 65, 74–85.

Rowland, S. J., D. A. Yon, C. A. Lewis, and J. R. Maxwell. 1985. Occurrence of 2,6,10-trimethyl-7-(3-methylbutyl) dodecane and related hydrocarbons in the green alga *Enteromorpha prolifera* and sediments. *Org. Geochem.*, 8, 207–213.

Ruble, T. E. 1995. Geochemical investigation of the mechanisms of hydrocarbon generation and accumulation in the Uinta Basin, Utah. Ph.D. dissertations, School of Geology and Geophysics, University of Oklahoma, Norman.

Ruble, T. E., and R. P. Philp. 1991. Geochemical investigation of native bitumens from the Uinta Basin, Utah, U.S.A. *The Compass*, 68 (3), 135–150.

Rulkötter, J., D. Leythaeuser, and E. Wendisch. 1982. Novel 23, 28-bisnorlupanes in Tertiary sediments. Widespread occurrence of nuclear demethylated triterpanes. *Geochim. Cosmochim. Acta*, 46, 2501–2509.

Rumeau, J-L., and C. Sourisse. 1973. Un exemple de migration primaire en phase gazeuse. *Bull. Centre Rech. Pau–SNPA*, 7 (1), 53–67.

Russell, W. L. 1972. Pressure–depth relations in the Appalachian region. *AAPG Bulletin*, 56, 528–536.

Rzasa, M. J., and D. L. Katz. 1950. The coexistence of liquid and vapor phases at pressures above 10,000 psi. *Transactions AIME,* 189, 119.

Sachanen, A. N. 1945. *Chemical constituents of petroleum*. New York: Reinhold.

Sackett, W. M. 1978. Carbon and hydrogen isotope effects during the thermocatalytic production of hydrocarbons in laboratory simulation experiments. *Geochim. Cosmochim. Acta*, 42, 571–580.

Salathiel, R. A. 1973. Oil recovery by surface film drainage in mixed-wettability rocks. *J. of Petrol. Tech.*, 25, 1216–1224.

Sandvik, E. I., W. A. Young, and D. J. Curry. 1992. Expulsion from hydrocarbon sources: The role of organic absorption. *Org. Geochem.* 19 (1–3), 77–87.

Sassen, R. 1988. Geochemical and carbon isotopic studies of crude oil destruction, bitumen precipitation, and sulfate reduction in the deep Smackover Formation. *Org. Geochem.*, 12(6), 351–361.

Saxby, J. D., A. J. R. Bennett, J. F. Corcoran, D. E. Lambert, and K. W. Riley. 1986. Petroleum generation: Simulation over six years of hydrocarbon formation from torbanite and brown coal in a subsiding basin. *Org. Geochem.*, 9 (2), 69–81.

Schaefle, J., B. Ludwig, P. Albrecht, and G. Ourisson. 1977. Hydrocarbures aromatiques d'origine geologique, 11. *Tetrahedron Letters*, 41, 3673–3676.

Scheuer, P. J. 1973. *Chemistry of marine natural products*. New York: Academic Press, pp. 58–87.

Schidlowski, M. 1986. $\delta^{34}S$ evidence for the Precambrian origin of sulfate respiration. *Geochem. Int.*, 17 (1), 107–114.

Schidlowski, M. 1988. A 3,800-million-year isotopic record of life from carbon in sedimentary rocks. *Nature*, 333, 313–318.

Schmidt, V., and D. A. McDonald. 1979. The role of secondary porosity in the course of sandstone diagenesis. SEPM special publication 26, pp. 175–207.

Schmitter, J. M., W. Sucrow, and P. J. Arpino. 1982. Occurrence of novel tetracyclic geochemical markers: 8,14-seco-hopanes in a Nigerian crude oil. *Geochim. Cosmochim. Acta*, 46, 2345–2350.

Schmoker, J. W. 1981. Determination of organic-matter content of Appalachian Devonian shales from gamma-ray logs. *AAPG Bulletin*, 65, 1285–1298.

Schoell, M. 1983. Genetic characterization of natural gases. *AAPG Bulletin*, 67, 2225–2238.

Schoell, M. 1984a. Recent advances in petroleum isotope geochemistry. *Org. Geochem.*, 6, 645–663.

Schoell, M. 1984b. Stable isotopes in petroleum research. In J. B. Brooks and D. Welte (eds.), *Advances in petroleum geochemistry*. London: Academic Press, vol. 1, pp. 215–245.

Schoell, M. 1988. Multiple origins of methane in the earth. *Chemical Geology*, 71, 1–10.

Schoell, M., E. Faber, and M. L. Coleman. 1983. Carbon and hydrogen isotopic compositions of the NBS 22 and NBS stable isotope reference materials: An interlaboratory comparison. *Org. Geochem.*, 5 (1), 3–6.

Schoell, M., P. D. Jenden, M. A. Beeunas, and D. D. Coleman. 1993. Isotope analyses of gases in gas field and gas storage operations. Richardson, TX: Society of Petroleum Engineers, SPE 26171, pp. 334–337.

Schoell, M., M. Teschner, H. Wehner, B. Durand, and J. L. Oudin. 1983. Maturity related biomarker and stable isotope variations and their application to oil source rock correlation in the Mahakam Delta, Kalimantan. In M. Bjorøy et al. (eds.), *Advances in organic geochemistry 1981*. Chichester: Wiley, pp. 156–163.

Scholle, P. A., and R. B. Halley. 1985. Burial diagenesis: Out of sight, out of mind. In N. Schneidermann and P. M. Harris (eds.), *Carbonate cements*. SEPM special publication 36, pp. 309–315.

Schopf, J. W. (ed.). 1983. *Earth's earliest biosphere, its origin and evolution*. Princeton, NJ: Princeton University Press.

Schowalter, T. T. 1979. Mechanics of secondary hydrocarbon migration. *AAPG Bulletin*, 63, 723–760.

Schreiber, B. C. 1988. Introduction. In B. C. Schreiber (ed.), *Evaporites and hydrocarbons*. New York: Columbia University Press, pp. 1–10.

Scott, A. C., and A. J. Fleet (eds.). 1994. *Coal and coal-bearing strata as oil-prone source rocks?* Geological Society special publication 77. London: Geological Society.

Scott, R. W. 1984. Sniffer: L'odor. *World Oil*, March, 5.

Secor, D. T. Jr. 1965. Role of fluid pressure in jointing. *Amer. J. Sci.*, 263, 633–646.

Seewald, J. S., and L. B. Eglinton. 1994. Organic–inorganic interactions during vitrinite maturation: Constraints from hydrous pyrolysis experiments. Abstracts of 11th Annual Meeting of the Society for Organic Petrology, Jackson Hole, September 25–30, vol. 11, pp. 91–93.

Seifert, W. K. 1977. Source rock–oil correlations by C_{27}–C_{30} biological marker hydrocarbons. In R. Campos and J. Goni (eds.), *Advances in organic geochemistry 1975*. Madrid: Empresa nacional adaro de investigaciones mineras, pp. 21–44.

Seifert, W. K., R. M. K. Carlson, and J. M. Moldowan. 1983. Geomimetic synthesis, structure assignment, and geochemical correlation application of monoaromatized petroleum steroids. In M. Bjorøy (ed.), *Advances in organic geochemistry 1981*. Chichester: Wiley, pp. 710–724.

Seifert, W. K., and J. M. Moldowan. 1978. Applications of steranes, terpanes and monoaromatics to the maturation, migration and source of crude oils. *Geochim. Cosmochim. Acta*, 42 (1), 77–95.

Seifert, W. K., and J. M. Moldowan. 1979. The effect of biodegradation on steranes and terpanes in crude oils. *Geochim. Cosmochim. Acta,* 43, 111–126.

Seifert, W. K., and J. M. Moldowan. 1981. Paleoreconstruction by biological markers. *Geochim. Cosmochim. Acta*, 45 (6), 783–794.

Seifert, W. K., J. M. Moldowan, and G. J. Demaison. 1984. Source correlation of biodegraded oils. *Org. Geochem.*, 6, 633–643.

Seifert, W. K., J. M. Moldowan, and R. W. Jones. 1980. Application of biological marker chemistry to petroleum exploration. *Proceedings of 10th World Petroleum Congress, Bucharest, September 1979*, paper SP8, pp. 425–440.

Seifert, W. K., and R. M. Teeter. 1970. Identification of polycyclic aromatic and heterocyclic crude oil carboxylic acids. *Anal. Chem.*, 42 (7), 750–758.

Selley, R. C. 1992. Petroleum seepages and impregnations in Great Britain. *Marine and Petroleum Geology*, 9, June, 226–244.

Senftle, J. T., J. H. Brown, and S. R. Larter. 1987. Refinement of organic petrographic methods for kerogen characterization. *Internat'l J. of Coal Geol.*, 7, 105–117.

Shanmugam, G. 1985. Significance of coniferous rain forests and related organic matter in generating commercial quantities of oil, Gippsland Basin, Australia. *AAPG Bulletin*, 69, 1241–1254.

Shaw, D. B., and C. E. Weaver. 1965. The mineralogical composition of shales. *J. Sed. Petrology*, 35 (1), 213–222.

Shen, M. S., L. S. Fan, and K. H. Castleton. 1984. American Chemical Society, Division of Petroleum. *Chemistry*, 29 (1), 127–134.

Sheng Guoying, F. Shanfa, L. Dehan, S. Nengxian, and Z. Hongming. 1980. The geochemistry on *n*-alkanes with an even–odd predominance in the tertiary Shanhejie Formation of northern China. In A. G. Douglas and J. R. Maxwell (eds.), *Advances in organic geochemistry 1979.* Oxford: Pergamon Press, pp. 115–121.

Shi Jiyang, A. S. Mackenzie, R. Alexander, G. J. Eglinton, A. P. Gowar, G. A. Wolff, and J. R. Maxwell. 1982. A biological marker investigation of petroleums and shales from the Shengli Oilfield of the People's Republic of China. *Chem. Geol.*, 35, 1–31.

Sieck, H. C. 1973. Gas-charged sediment cones pose possible hazard to offshore drilling. *Oil & Gas Journal,* 71, 148, 150, 155, 163.

Silverman, S. R. 1965. Migration and segregation of oil and gas. In A. Young and G. E. Galley (eds.), *Fluids in subsurface environments*. AAPG Memoir 4. Tulsa: American Association of Petroleum Geologists, pp. 54–65.

Simoneit, B. R. T. 1986. Cyclic terpenoids of the geosphere. In R. B. Johns (ed.), *Biological markers in the sedimentary record.* Methods in Geochemistry and Geophysics 24. Amsterdam: Elsevier Science, pp. 43–99.

Simoneit, B. R. T. 1992. Natural hydrous pyrolysis: Petroleum generation in submarine hydrothermal systems. In J. K. Whelan and J. W. Farrington (eds.), *Organic matter: Productivity, accumulation, and preservation in recent and ancient sediments.* New York: Columbia Univeristy Press, pp. 368–402.

Simoneit, B. R. T., P. T. Crisp, B. G. Rohrback, and B. M. Didyk. 1980. Chilean paraffin dirt: II. Natural gas seepage at an active site and its geochemical consequences. In A. G. Douglas and J. R. Maxwell (eds.), *Advances in organic geochemistry 1979.* Oxford: Pergamon Press, pp. 171–176.

Simpson, J. P. 1985. The drilling mud dilemma–Recent examples. *J. Petrol. Tech.*, 201–206.

Sinninghe Damsté, J. S., and J. W. de Leeuw. 1990. Analysis, structure and geochemical significance of organically-bound sulphur in the geosphere: State of the art and future research. *Org. Geochem.*, 16 (4–6), 1077–1101.

Sluijk, D., and M. H. Nederlof. 1984. Worldwide geological experience as a systematic basis for prospect appraisal. In G. J. Demaison and R. J. Murris (eds.), *Petroleum geochemistry and basin evaluation.* AAPG Memoir 35. Tulsa: American Association of Petroleum Geologists, pp. 15–26.

Smith, G. C., and A. C. Cook. 1984. Petroleum occurrence in the Gippsland Basin and its relationship to rank and organic matter type. *APEA Journal*, 24 (1), 196–216.

Smith, J. E., J. G. Erdman, and D. A. Morris. 1971. Migration, accumulation and retention of petroleum in the earth. *Proceedings of Eighth World Petroleum Congress, Moscow*. London: Applied Science Publishers, pp. 13–26.

Smith, J. T. 1994. Petroleum system logic as an exploration tool in a frontier setting. In L. B. Magoon and W. G. Dow (eds.), *The petroleum system–From source to trap*. AAPG Memoir 60. Tulsa: American Association of Petroleum Geologists, pp. 25–49.

Smith, J. W. 1983. The chemistry that formed Green River Formation oil shale. In F. P. Miknis and J. F. McKay (eds.), *Geochemistry and chemistry of oil shales*. Washington, DC: American Chemical Society, pp. 235–248.

Smith, L. W., and N. Hilton. 1980. An occurrence of high gravity oil in an Oligocene Vicksburg age sandstone in Jimhogg County. Texas. *Transactions–Gulf Coast Association of Geological Societies,* 30, 223–227.

Smith, P. M. R. 1983. Spectral correlation of spore coloration standards. In J. Brooks (ed.), *Geological Society special publication 12.* Oxford: Blackwell Scientific Publications. pp. 289–294.

Snarskii, A. N. 1964. Relationship between primary migration and compaction of rocks. *Petrol. Geol.,* 5 (7), 362–364.

Snarskii, A. N. 1970. The nature of primary oil migration. *Izv. Vyssh. Ucheb. Zavedenii, Neft Gaz,* 13 (8), 11–15.

Snider, L. C. 1934. Current ideas regarding source beds for petroleum. In W. E. Rather and F. H. Lahee (eds.), *Problems of petroleum geology.* AAPG Memoir 1. Tulsa: American Association of Petroleum Geologists, pp. 51–66.

Snowdon, L. R., and T. G. Powell. 1982. Immature oil condensate-modification of hydrocarbon generation model for terrestrial organic matter. *AAPG Bulletin*, 66, 775–788.

Sofer, Z. 1984. Stable carbon isotope compositions of crude oils: Application to source depositional environments and petroleum alteration. *AAPG Bulletin*, 68, 31–49.

Sofer, Z. 1988. Biomarkers and carbon isotopes of oils in the Jurassic Smackover of the Gulf Coast states, U.S.A. *Org. Geochem.*, 12 (5), 421–432.

Sokolov, V. A., and S. I. Mironov. 1962. On the primary migration of hydrocarbons and other oil components under the action of compressed gases. In *The chemistry of oil and oil deposits*. Acad. Sci. USSR, Inst. Geol. and Exploit. Min. Fuels, pp. 38–91 (in Russian). Trans. by Israel Program for Scientific Translation, Jerusalem, 1964.

Sokolov, V. A., T. P. Zhure, N. B. Vassoevich, P. L. Antonov, G. G. Grigoryev, and V. P. Kozlov. 1963. Migration processes of gas and oil, their intensity and directionality. Paper presented at 5th World Petroleum Congress, June 19–26, Frankfurt, ME.

Spencer, C. S. 1987. Hydrocarbon generation as a mechanism for overpressuring in Rocky Mountain region. *AAPG Bulletin*, 71, 368–388.

Stach, E., M-Th. Mackowsky, M. Teichmüller, G. H. Taylor, D. Chandra, and R. Teichmüller. 1982. *Stach's textbook of coal petrology.* 3rd ed. Berlin: Bebruder Borntraeger.

Stahl, W. J. 1977. Carbon and nitrogen isotopes in hydrocarbon research and exploration. *Chem. Geol.*, 20, 121–149.

Stainforth, J. G. 1984. Gippsland hydrocarbons–A perspective from the basin edge. *APEA Journal*, 24, 91–99.

Staplin, F. L. 1969. Sedimentary organic matter, organic metamorphism, and oil and gas occurrence. *Canadian Petrol. Geol. Bull.*, 17 (1), 47–66.

Stoll, R. D., J. Ewing, and G. M. Bryan. 1971. Anomalous wave velocities in sediments containing gas hydrates. *J. Geophys. Res.*, 76 (8), 2090–2094.

Storer, A. 1959. Constipamento dei sedimenti argillosi nel Bacino Padano, giacimenti gassiferi dell' Europa Occidentale. Rome: Acad. nazionale dei Lincei, pp. 519–544.

Summons, R. E., and T. G. Powell. 1987. Identification of aryl isoprenoids in source rocks and crude oils: Biological markers for green sulfur bacteria. *Geochim. Cosmochim. Acta*, 51, 557–566.

Sundararaman, P. 1992. Comparison of natural and laboratory simulated maturation of vanadylporphyrins. In J. M. Moldowan, P. Albrecht, and R. P. Philp (eds.), *Biological markers in sediments and petroleum.* Englewood Cliffs, NJ: Prentice-Hall, pp. 313–319.

Sweeney, J. J., A. K. Burnham, and R. L Braun, 1987. A model of hydrocarbon generation from type I kerogen: Application to Uinta Basin, Utah. *AAPG Bulletin,* 71, 967–985.

Talukdar, S. C., B. De Toni, F. Marcano, J. Sweeney, and A. Rangel. 1993. Upper Cretaceous source rocks of northern South America. Abstract. *AAPG Bulletin*, 77, 351.

Talukdar, S. C., W. G. Dow, and K. M. Persad. 1990. Geochemistry of oils provides optimism for deeper exploration in Atlantic off Trinidad. *Oil & Gas Journal,* November 12, 118–121.

Talukdar, S., O. Gallango, and M. Chin-a-lien. 1986. Generation and migration of hydrocarbons in the Maracaibo Basin, Venezuela: An integrated basin study. *Org. Geochem.*, 10, 261–279.

Talukdar, S. C., and F. Marcano. 1994. Petroleum systems of the Maracaibo Basin, Venezuela. In L. B. Magoon and W. G. Dow (eds.), *The petroleum system—From source to trap.* AAPG Memoir 60. Tulsa: American Association of Petroleum Geologists, pp. 463–481.

Tannenbaum, E., and Z. Aizenshtat. 1984. Formation of immature asphalt from organic-rich carbonate rocks: II. Correlation of maturation indicators. *Org. Geochem.*, 503–511.

Tarafa, M. E., J. M. Hunt, and I. Ericsson. 1983. Effect of hydrocarbon volatility and adsorption on source-rock pyrolysis. *J. Geochem Explor.*, 18, 75–85.

Taylor, G. H., and M. Teichmüller. 1993. Observations on fluorinite and fluorescent vitrinite with the transmission electron microscope. *Internat'l J. of Coal Geology*, 22, 61–82.

Teichmüller, M. 1958. Metamorphisme du charbon et prospection du pétrole. *Rev. ind. minerale,* special issue, 1–15.

Teichmüller, M. 1963. Die Kohlenfloze der Bohrung Munsterland, pt. 1. *Fortschr. Geol. Rheinld. 1. Westf.,* 11, 129.

Teichmüller, M. 1974. Generation of petroleum-like substances in coal seams as seen under the microscope. In B. Tissot and F. Bienner (eds.), *Advances in organic geochemistry 1973.* Paris: Éditions technip, pp. 379–407.

Teichmüller, M. 1982. *Fluoreszenzmikroskopische Anderungen von Liptiniten und Vitriniten mit zunehmendem Inkohlungsgrad und ihre Beiziehungen zu Bitumenbildung und Verkokungsverhalten.* Krefeld: Geologisches Landesamt Nordrhein-Westfalen.

ten Haven, H. L., J. W. de Leeuw, and P. A. Schenck. 1985. Organic geochemical studies of Messinian evaporite basin, northern Apennines (Italy): I. Hydrocarbon biological markers for a hypersaline environment. *Geochim. Cosmochim. Acta*, 49, 2181–2191.

Teslenko, P. F., and B. S. Korotkov. 1966. Effect of arenaceous intercalations in clays on their compaction. *Internat'l Geol. Rev.*, 9 (5), 699–701.

Thode, H. G. 1981. Sulfur isotope ratios in petroleum research and exploration: Williston basin. *AAPG Bulletin*, 1527–1535.

Thode, H. G., and J. Monster. 1965. Sulfur-isotope geochemistry of petroleum, evaporites, and ancient seas. In A. Young and J. E. Galley (eds.), *Fluids in subsurface environments*. AAPG Memoir 4. Tulsa: American Association of Petroleum Geologists, pp. 367–377.

Thode, H. G., R. K. Wanless, and R. Wallouch. 1954. The origin of native sulfur deposits from isotope fractionation studies. *Geochim. Cosmochim. Acta*, 5, 286–298.

Thomas, B. M., P. Moller-Pedersen, M. F. Whitaker, and N. D. Shaw. 1985. Organic facies and hydrocarbon distributions in the Norwegian North Sea. In B. M. Thomas et al. (eds.), *Petroleum geochemistry in exploration of the Norwegian Shelf*. London: Graham & Trotman for the Norwegian Petroleum Society, pp. 3–26.

Thomas, B. M., et al. (eds.). 1985. *Petroleum geochemistry in exploration of the Norwegian Shelf*. London: Graham & Trotman for the Norwegian Petroleum Society.

Thomas, B. R. 1969. Kauri resins–Modern and fossil. In G. Eglinton and M. T. J. Murphy (eds.), *Organic geochemistry: Methods and results*. New York: Springer-Verlag, pp. 599–618.

Thomas, O. D. 1980. North Sea petroleum: Past and future. *Proceedings of the Tenth World Petroleum Congress*. London: Heyden & Son, vol. 2, pp. 177–182.

Thompson, C. L., and H. Dembicki Jr. 1986. Optical characteristics of amorphous kerogen and the hydrocarbon-generating potential of source rocks. *Internat'l J. of Coal Geology*, 6, 229–249.

Thompson, K. F. M. 1983. Classification and thermal history of petroleum based on light hydrocarbons. *Geochim. Cosmochim. Acta*, 47, 303–316.

Thompson, K. F. M. 1987. Fractionated aromatic petroleums and the generation of gas-condensates. *Org. Geochem.*, 11 (6), 573–590.

Thompson, K. F. M. 1988. Gas-condensate migration and oil fractionation in deltaic systems. *Marine and Petroleum Geol.*, 5, 237–246.

Thompson, R. G., F. D. Singleton Jr., and L. L. Raymer. 1986. Today's low oil prices spell trouble for consumers tomorrow. *World Oil*. May, 32–34.

Thompson, S., B. Cooper, R. J. Morley, and P. C. Barnard. 1984. Oil-generating coals. In B. M. Thomas et al. (eds.), *Petroleum geochemistry in exploration of the Norwegian Shelf*. London: Graham & Trotman for the Norwegian Petroleum Society, pp. 59–73.

Thompson-Rizer, C. L. 1987. Some optical characteristics of solid bitumen in visual kerogen preparations. *Org. Geochem.*, 11, 385–392.

Thompson-Rizer, C. L., and R. A. Woods. 1987. Microspectrofluorescence measurements of coals and petroleum source rocks. *Internat'l J. of Coal Geology*, 7, 85–104.

Ting, T. C. 1975. Reflectivity of disseminated vitrinites in the Gulf Coast region. Pétrographie de la matière organique des sédiments relations avec la paléotempérature et le potentiel pétrolier. Paris: Centre national de la recherche scientifique, September 15–17, 1973.

Tissot, B. 1969. Primières données sur le mécanismes et la cinétique de la formation du pétrole dans les sédiments: Simulation d'un schema réactionnel sur ordinateur. *Rev. l'Inst. français du pétrole,* 24 (4), 470–501.

Tissot, B., B. Durand, J. Espitalié, and A. Combaz. 1974. Influence of nature and diagenesis of organic matter in formation of petroleum. *AAPG Bulletin,* 58, 499–506.

Tissot, B., and J. Espitalié. 1975. L'évolution thermique de la matière organiques des sédiments: Application d'une simulation mathématique. *Rev. l'Inst. français pétrole,* 30, 743–777.

Tissot, B., R. Pelet, and Ph. Ungerer. 1987. Thermal history of sedimentary basins, maturation indices, and kinetics of oil and gas generation. *AAPG Bulletin*, 71, 1445–1466.

Tissot, B., and D. H. Welte. 1984. *Petroleum formation and occurrence.* 2nd ed. Heidelberg: Springer Verlag.

Toland, W. G. 1960. Oxidation of organic compounds with aqueous sulfate. *J. Amer. Chem. Soc.,* 82, 1911–1916.

Towler, B. F. 1993. Statistical, economic analyses given for surface exploration projects. *Oil & Gas Journal,* May 10, 69–72.

Trask, P. D. 1936. Proportion of organic matter converted into oil in Santa Fe Springs Field, California. *AAPG Bulletin,* 20, 245–257.

Trask, P. D., H. E. Hammar, and C. C. Wu. 1932. *Origin and environment of source sediments of petroleum.* Houston: Gulf Publishing.

True, W. R. 1991. Worldwide gas-processing activity levels out. *Oil & Gas Journal,* July 22, 41–44.

Tuttle, R. N. 1983. High-pour-point and asphaltic crude oils and condensates. *J. of Petrol. Tech.*, 35, 1192–1196.

Twenhofel, W. H. 1961. *Treatise on sedimentation*. New York: Dover. First pubiished 1932 by Williams & Wilkins.

Tyson, R. V., and T. H. Pearson (eds.). 1991. *Modern and ancient continental shelf anoxia*. Geological Society special publication 58. London: Geological Society, p. 113.

Unomah, G. I., and C. M. Ekweozor. 1993. Application of vitrinite reflectance in reconstruction of tectonic features in Anambra Basin, Nigeria: Implication of petroleum potential. *AAPG Bulletin,* 77, 436–451.

Vail, P. R., R. M. Mitchum Jr., and S. Thompson III. 1977. Global cycles of relative changes of sea level. In C. E. Payton (ed.), *Seismic stratigraphy—Applications to hydrocarbon exploration.* AAPG Memoir 26. Tulsa: American Association of Petroleum Geologists, pp. 83–97.

van den Bark, E., and O. D. Thomas. 1981. Ekofisk: First of the giant oil fields in Western Europe. *AAPG Bulletin,* 65, 2341–2363.

Vandenbroucke, M., B. Durand, and J. L. Oudin. 1983. Detecting migration phenomena in a geological series by means of C_1–C_{35} hydrocarbon amounts and distributions. In M. Bjorøy et al. (eds.), *Advances in organic geochemistry 1983*, pp. 147–155.

Vandenburg, L. E., and E. A. Wilder. 1970. The structural constituents of carnauba wax. *J. Amer. Oil Chem Soc.,* 47, 514–518.

van Gijzel, P. 1979. Manual of the techniques and some geological applications of fluorescence microscopy. Workshop sponsored by the American Association of Stratigraphic Palynologists, 12th Annual Meeting, Dallas, October 29–November 2, 1989. Dallas: Core Laboratories.

van Gijzel, P. 1980. Characterization and identification of kerogen and bitumen and determination of thermal maturation by means of qualitative and quantitative microscopical techniques. SEPM short course 7: How to assess maturation and paleotemperatures. Denver, June 7–8.

van Gijzel, P. 1990. Transmittance colour index (TCI) of amorphous organic matter. In W. J. J. Fermont and J. W. Weegink (eds.), *Proceedings of International Symposium on Organic Petrology, Zeist, The Netherlands, January 7–9, 1990*. Mededelingen rijks geologische dienst, vol. 45, pp. 50–63.

van Graas, G. W. 1990. Biomarker maturity parameters for high maturities: Calibration of working range up to the oil-condensate threshold. *Org. Geochem.*, 16, 1025–1032.

van Krevelen, D. W., 1961, *Coal: Typology–Chemistry–Physics–Constitution.* Amsterdam: Elsevier Science.

Vassoevich, N. B., Y. I. Korchagina, N. V. Lopatin, and V. V. Chernischev. 1969. *The main stage of petroleum formation.* Moscow University Vestnik 6, pp. 3–37. (in Russian); English trans. in *Int. Geol. Rev.*, 12 (11), 1970, 1276–1296.

Vassoevich, N. B., I. V. Visotskii, V. A. Sokolov, and Ye. I. Tatarenko. 1971. Oil–gas potential of late Precambrian deposits. *Internat Geol. Rev.*, 13 (3), 407–418.

Vernadskii, V. I. 1934. Outlines of geochemistry. *ONTI, Gornogeolog. Neft. Izd.*, 152–153.

Vernon, J. W., and R. A. Slater. 1963. Submarine tar mounds, Santa Barbara County, California, *AAPG Bulletin*, 47, 1624–1627.

Versluys, J. 1932. Factors involved in segregation of oil and gas from subterranean water. *AAPG Bulletin*, 16, 924–942.

Vidal, G. 1984. The oldest eukaryotic cells. *Scientific American*, 250 (2), 48–57.

Vincent, P. W., I. R. Mortimore, and D. M. McKirdy. 1985. Hydrocarbon generation, migration and entrapment in the Jackson–Naccowlah area, ATP 259P, southwestern Queensland, *APEA Journal*, 25 (1), 62–84.

Volkman, J. K., R. Alexander, R. I. Kagi, S. J. Rowland, and P. N. Sheppard. 1984. Biodegradation of aromatic hydrocarbons in crude oils from the Barrow Sub-basin of Western Australia. *Org. Geochem.*, 6, 619–632.

Volkman, J. K., D. I. Allen, P. L. Stevenson, and H. R. Burton. 1986. Bacterial and algal hydrocarbons in sediments from a saline Antarctic lake, Ace Lake. *Org. Geochem.*, 10, 671–681.

Volkman, J. K., J. W. Farrington, R. B. Gagosian, and S. G. Wakeham. 1983. Lipid composition of coastal marine sediments from the Peru upwelling region. In M. Bjorøy et al. (eds.), *Advances in organic geochemistry 1981.* Chichester: Wiley, pp. 228–245.

Volkman, J. K., R. B. Johns, and F. T. Gillian. 1980. Microbial lipids of an intertidal sediment: 1. Fatty acids and hydrocarbons. *Geochim. Cosmochim. Acta*, 44, 1133–1141.

von der Dick, H., J. D. Meloche, J. Dwyer, and P. Gunther. 1989. Source-rock geochemistry and hydrocarbon generation in the Jeanne D'Arc Basin, Grand Banks, offshore Eastern Canada. *J. Petrol. Geol.*, 12 (1), 52–68.

Vonk, C. G. 1976. On two methods for determination of particle size distribution functions by means of small-angle X-ray scattering. *J. Appl. Crystallogr.*, 9 (44), 433–440.

Vredenburgh, L. D., and E. S. Cheney. 1971. Sulfur and carbon isotopic investigation of petroleum, Wind River Basin, Wyoming. *AAPG Bulletin*, 55, 1945–1975.

Wade, W. J., J. S. Hanor, and R. Sassen. 1989. Controls on H_2S concentration and hydrocarbon destruction in the eastern Smackover trend. *Transactions–Gulf Coast Association of Geological Societies,* 34, 309–320.

Waldron, J. D., D. S. Gowers, A. C. Chibnal, and S. H. Piper. 1961. Further observations on the paraffins and primary alcohols of plant waxes. *Biochem. J.,* 78, 435–442.

Walker, P. L. 1981. Microporosity in coal: Its characterization and its implications for coal utilization. *Philos. Trans. R. Soc. London,* 300, 65–81.

Walter, M. R. 1983. Archean stromatolites: Evidence of the earth's earliest benthos. In J. W. Schopf (ed.), *Earth's earliest biosphere, its origin and evolution.* Princeton, NJ: Princeton University Press, pp. 187–213.

Waples, D. W. 1984. Thermal models for oil generation. In J. Brooks and D. H. Welte (eds.), *Advances in petroleum geochemistry.* London: Academic Press, vol. 1, pp. 7–67.

Waples, D. W. 1985. *Geochemistry in petroleum exploration.* Boston: International Human Resources Development Corporation.

Waples, D. W. 1994. Maturity modeling: Thermal indicators, hydrocarbon generation, and oil cracking. In L. B. Magoon and W. G. Dow (eds.), *The petroleum system–From source to trap.* AAPG Memoir 60. Tulsa: American Association of Petroleum Geologists, pp. 285–306.

Waples, D. W., and T. Machihara. 1991. *Biomarkers for geologists: A practical guide to the application of steranes and triterpanes in petroleum geology.* AAPG Methods in Exploration Series 9. Tulsa: American Association of Petroleum Geologists.

Ward, G. S. 1988. A hydrodynamic evaluation of the Venture gas field, Scotian Shelf, Canada. Unpublished manuscript of Ward Hydrodynamics, Kalispell, Montana.

Warner, M. A., and F. Royse. 1987. Thrust faulting and hydrocarbon generation: Discussion. *AAPG Bulletin,* 71, 882–889.

Webster, R. L. 1984. Petroleum source rocks and stratigraphy of the Bakken Formation in North Dakota. In J. Woodward, F. F. Meissner, and J. L. Clayton (eds.), *Hydrocarbon source rocks of the greater Rocky Mountain region.* Denver: Rocky Mountain Association of Geologists, pp. 57–81.

Wedlandt, W. W. 1985. *Thermal analysis.* 3rd ed. New York: Wiley, pp. 74–86.

Weeks, L. G. (ed.). 1958. *Habitat of oil: A symposium.* Tulsa: American Association of Petroleum Geologists.

Weeks, L. G. 1961. Origin, migration, and occurrence of petroleum. In G. B. Moody (ed.), *Petroleum exploration handbook.* New York: McGraw-Hill, chap. 5.

Weeks, L. G. 1978. *". . . A lifelong love affair": The memoirs of Lewis G. Weeks, geologist*. Westport, CT: Lewis G. Weeks Memoirs.

Wehner, J., and H. Hufnagel. 1986. Some characteristics of the inorganic and organic composition of oil shales from Jordan. In E. T. Degens et al. (eds.), *Biogeochemistry of black shales*. Hamburg: Geologisch–Pälaontologischen Institut der Universität Hamburg, 60, 381–389.

Wells, P. E. 1990. Porosities and seismic velocities of mudstones from Wairarapa oil wells of North Island, New Zealand, and their use in determing burial history. *New Zealand Journal of Geology and Geophysics*, 33, 29–39.

Welte, D. H. 1972. Petroleum exploration and organic geochemistry. *J. Geochem. Explor.*, 1, 117–136.

Welte, D. H., W. Stoessinger, R. G. Schaefer, and M. Radke. 1984. Gas generation and migration in the deep basin of western Canada. In J. A. Masters (ed.), *Elmworth: Case study of a deep basin gas field.* AAPG Memoir 38. Tulsa: American Association of Petroleum Geologists, pp. 35–47.

Wenger, L. M., and D. R. Baker. 1986. Variations in organic geochemistry of anoxic–oxic black shale–carbonate sequences in the Pennsylvanian of the Mid-Continent, U.S.A. *Org. Geochem.*, 10, 85–92.

Wetmore, D. E., C. K. Hancock, and R. N. Traxler. 1966. Fractionation and characterization of low molecular weight asphaltic hydrocarbons. *Anal. Chem.*, 38 (2), 225–230.

Whelan, J. K. 1979. C_1 to C_7 Hydrocarbons from IPOD Hole 397/397A. In W. B. F. Ryan and Ulrich von Rad (eds.), *Initial reports of the Deep Sea Drilling Project*. Washington, DC: U.S. Government Printing Office, vol. 47, pp. 531–539.

Whelan, J. K., R. Carangelo, P. R. Solomon, and W. G. Dow. 1990. TG/plus—A pyrolysis method for following maturation of oil and gas generation zones using T_{max} of methane. *Org. Geochem.*, 16 (4–6), 1187–1201.

Whelan, J., and J. W. Farrington (eds.). 1992. *Organic matter: Productivity, accumulation, and preservation in recent and ancient sediments*. New York: Columbia University Press.

Whelan, J. K., J. M. Hunt, and A. Y. Huc. 1980. Applications of thermal distillation–pyrolysis to petroleum source rock studies and marine pollution. *Journal of Analytical and Applied Pyrolysis*, 2, 79–96.

Whelan, J. K., M. C. Kennicutt II, J. M. Brooks, D. Schumacher, and L. B. Eglinton. 1994. Organic geochemical indicators of dynamic fluid flow processes in petroleum basins. *Org. Geochem.*, 22 (3–5), 587–615.

Whelan, J. K., R. Oremland, M. Tarafa, R. Smith, R. Howarth, and C. Lee. 1986. Evidence for sulfate-reducing and methane-producing microorganisms in sediments from Sites 618, 619, and 622. In *Initial reports of the Deep Sea*

Drilling Project. Washington, DC: U.S. Government Printing Office, vol. 96, pp. 767–775.

Whelan, J. K., and C. Thompson-Rizer. 1993 Chemical methods for assessing kerogen and protokerogen types and maturity. In M. H. Engel and S. A. Mako (eds.), *Organic geochemistry: Principles and applications*. New York: Plenum Press, pp. 289–353.

White, D. 1915. Geology: Some relations in origin between coal and petroleum. *J. Wash. Acad. Sci.*, 5 (6), 189–212.

White, S. M. 1975. Interstitial water studies, Leg 31. In D. E. Karig and J. C. Ingle Jr. (eds.), *Initial reports of the Deep Sea Drilling Project*. Washington, DC: U.S. Government Printing Office, vol. 31, pp. 639–653.

Whiticar, M. J. 1994. Correlation of natural gases with their sources. In L. B. Magoon and W. G. Dow (eds.), *The petroleum system—From source to trap*. AAPG Memoir 60. Tulsa: American Association of Petroleum Geologists, pp. 261–283.

Wilhelms, A., and S. R. Larter. 1994a. Origin of tar mats in petroleum reservoirs: Part I. Introduction and case studies. *Marine and Petroleum Geology*, 11 (4), 418–441.

Wilhelms, A., and S. R. Larter. 1994b. Origin of tar mats in petroleum reservoirs: Part II. Formation mechanisms for tar mats. *Marine and Petroleum Geology*, 11 (4), 442–456.

Williams, B. 1993. Seepage from the land of the odd. *Oil & Gas Journal*, May 31, 11.

Williams, J. A. 1974. Characterization of oil types in Williston Basin. *AAPG Bulletin*, 58, 1243–1252.

Williams, J. A., M. Bjorøy, D. L. Dolcater, and J. C. Winters. 1986. Biodegradation in south Texas Eocene oils–Effects on aromatics and biomarkers. *Org. Geochem.*, 10, 451–461.

Williams, P. F. V., and A. G. Douglas. 1980. A preliminary organic chemistry investigation of the Kimmeridgian oil shales. In A. G. Douglas and J. R. Maxwell (eds.), *Advances in organic geochemistry 1979*. Oxford: Pergamon Press, pp. 531–545.

Willis, C. 1991. Core tests speed coalbed methane gas development. *Oil & Gas Journal*, June 3, 93–96.

Wilson, R. D., P. H. Monaghan, A. Osanik, L. C. Price, and M. A. Rogers. 1973. Estimate of annual input of petroleum to the marine environment from natural marine seepage. *Transactions—Gulf Coast Association of Geological Societies*, 23, 182–193.

Wingert, W. S. 1992. G.C.-m.s. analysis of diamondoid hydrocarbons in Smackover petroleums. *Fuel* 71, 37–43.

Winniford, R. S., and M. Bersohn. 1962. The structure of petroleum asphaltenes as indicated by proton magnetic resonance. Paper presented at Symposium of Tars, Pitches, and Asphalts. In *Preprints, American Chemical Society, Division of Fuel Chemistry*. Washington, DC: American Chemical Society, pp. 21–32.

Winters, J. C., and J. A. Williams. 1969. Microbiological alteration of crude oil in the reservoir. Paper presented at Symposium on Petroleum Transformation in Geologic Environments, American Chemical Society, Division of Petroleum Chemistry, September 7–12, New York City, paper PETR 86, pp. E22-E31.

Wood, D. A. 1988. Relationships between thermal maturity indices calculated using Arrhenius equation and Lopatin method: Implications for petroleum exploration. *AAPG Bulletin*, 72, 115–134.

Woodward, J., F. F. Meissner, and J. L. Clayton (eds.). 1984. *Hydrocarbon source rocks of the greater Rocky Mountain region*. Denver: Rocky Mountain Association of Geologists.

Wroblewski, E. F. 1975. Developments in Oklahoma and panhandle of Texas in 1974. *AAPG Bulletin*, 59, 1401–1403.

Wyman, R. E. 1984. Gas resources in Elmworth coal seams. In J. A. Masters (ed.), *Elmworth case study of a deep basin gas field*. Tulsa: American Association of Petroleum Geologists, pp. 173–187.

Wyman, R. E. 1985. *The future of natural gas*. Calgary: Canadian Hunter Exploration.

Yang, H. S., and H. Y. Sohn. 1984. Kinetics of oil generation from oil shale from Liaoning province of China. *Fuel*, 63, 1511–1514.

Yang, W. 1985. Daqing oil field, People's Republic of China: A giant field with oil of nonmarine origin. *AAPG Bulletin*, 69, 1101–1111.

Yang, W., L. Yongkang, and G. Ruiqi. 1985. Formation and evolution of nonmarine petroleum on Songliao Basin, China. *AAPG Bulletin*, 69, 1112–1122.

Yarullin, K. S. 1961. Characteristics of the distribution of gas and oil pools in the Cis–Uralian Trough. *Dokl. Akad. Nauk SSSR*, 141 (1), 1142–1145.

Yasenev, B. P. 1962. New data on direct geochemical methods in the exploration of oil and gas fields. *Geol. Nefti Gaza*, 6 (12), 54–58.

Yen, T. F. 1974. Structure of petroleum asphaltene and its significance. *Energy Sources*, 1 (4), 447–463.

Yen, T. F., and J. M. Moldowan (eds.). 1988. *Geochemical biomarkers*. New York: Harwood Academic Press.

Youngblood, W. W., M. Blumer, R. L. Guillard, and J. Fiore. 1971. Calculation of ages of hydrocarbons in oils: Physical chemistry applied to petroleum geochemistry; part 1. *AAPG Bulletin,* 61, 573–600.

Youtcheff, J. S., P. H. Given, Z. Baset, and M. S. Sundaram. 1983. The mode of association of alkanes with coals. *Org. Geochem,*. 5 (3), 157–164.

Yukler, M. A., and W. G. Dow. 1990. Temperature, pressure and hydrocarbon generation histories in San Marcos Arch area, Dewitt County, Texas. In D. Schumacher and B. F. Perkings (eds.), *Gulf Coast oils and gases: Their characteristics, origin, distribution, and exploration and production significance. Proceedings of the Ninth Annual Research Conference, Society of Economic Paleontologists and Mineralogists Foundation*, pp. 99–104.

Zhuze, T. P., V. I. Sergeyevich, V. F. Burmistrova, and Y. A. Yesakov. 1971. Solubility of hydrocarbons in water under stratal conditions. *Dokl. Akad. Nauk SSSR,* 198 (1), 206–209.

Zhuze, T. P., G. S. Usakova, and G. N. Yushkevich. 1962. The influence of high pressures and temperatures on the content and properties of condensate in the gas phase of gas–oil deposits. *Geochemistry*, 8, 797–806.

Zieglar, D. I. 1992. Hydrocarbon columns, buoyancy pressures, and seal efficiency: Comparisons of oil and gas accumulations in California and the Rocky Mountain area *AAPG Bulletin*, 76, 501–508.

Zinger, A. S. 1962. Molecular hydrogen in gas dissolved in waters of oil–gas fields, lower Volga region. *Geokhim,* 10, 890–898.

Zor'kin, L. M., and E. V. Stadnik. 1975. Unique features of gas saturation of formation waters of oil and gas basins in relation to the genesis of hydrocarbons and formation of their accumulations. *Izv. Vyssh. Uchebn. Zaved., Geol. Razved*, 6, 85–99.

Zor'kin, L. M., E. V. Stadnik, V. K. Soshnikov, and G. A. Yurin. 1972. Geochemistry of gases in ground water in Devonian terrigenous rocks of the Russian Platform. *Dokl. Akad. Nauk SSSR,* 203, 200–202.

Zumberge, J. E. 1983. Tricyclic diterpane distributions in the correlation of Paleozoic crude oils from the Williston Basin. In M. Bjorøy et al. (eds.), *Advances in organic geochemistry 1981.* Chichester: Wiley, pp. 738–745.

Zumberge, J. E. 1984. Source rocks of the La Luna Formation (Upper Cretaceous) in the Middle Magdalena Valley, Colombia. In J. G. Palacas (ed.), *Petroleum geochemistry and source rocks potential of carbonate rocks.* AAPG Studies in Geology 18. Tulsa: American Association of Petroleum Geologists, pp. 127–133.

Name Index

Subject Index